Methods in Enzymology

Volume 103
HORMONE ACTION
Part H
Neuroendocrine Peptides

METHODS IN ENZYMOLOGY

EDITORS-IN-CHIEF

Sidney P. Colowick   Nathan O. Kaplan

Methods in Enzymology

Volume 103

Hormone Action

Part H
Neuroendocrine Peptides

EDITED BY

P. Michael Conn

DEPARTMENT OF PHARMACOLOGY
DUKE UNIVERSITY MEDICAL CENTER
DURHAM, NORTH CAROLINA

1983

ACADEMIC PRESS
A Subsidiary of Harcourt Brace Jovanovich, Publishers

New York   London
Paris San Diego San Francisco São Paulo Sydney Tokyo Toronto

COPYRIGHT © 1983, BY ACADEMIC PRESS, INC.
ALL RIGHTS RESERVED.
NO PART OF THIS PUBLICATION MAY BE REPRODUCED OR
TRANSMITTED IN ANY FORM OR BY ANY MEANS, ELECTRONIC
OR MECHANICAL, INCLUDING PHOTOCOPY, RECORDING, OR ANY
INFORMATION STORAGE AND RETRIEVAL SYSTEM, WITHOUT
PERMISSION IN WRITING FROM THE PUBLISHER.

ACADEMIC PRESS, INC.
111 Fifth Avenue, New York, New York 10003

United Kingdom Edition published by
ACADEMIC PRESS, INC. (LONDON) LTD.
24/28 Oval Road, London NW1 7DX

LIBRARY OF CONGRESS CATALOG CARD NUMBER: 54-9110
ISBN 0-12-182003-3

PRINTED IN THE UNITED STATES OF AMERICA

83 84 85 86   9 8 7 6 5 4 3 2 1

Table of Contents

CONTRIBUTORS TO VOLUME 103 ix

PREFACE xv

VOLUMES IN SERIES xvii

Section I. Preparation of Chemical Probes

1. Synthesis and Use of Neuropeptides	JOHN M. STEWART	3
2. Preparation of Radiolabeled Neuroendocrine Peptides	RICHARD N. CLAYTON	32
3. Ligand Dimerization: A Technique for Assessing Receptor–Receptor Interactions	P. MICHAEL CONN	49
4. Photoaffinity Labeling in Neuroendocrine Tissues	ELI HAZUM	58
5. Microisolation of Neuropeptides	FRED S. ESCH, NICHOLAS C. LING, AND PETER BÖHLEN	72

Section II. Equipment and Technology

6. Electrophysiological Assays of Mammalian Cells Involved in Neurohormonal Communication	BERNARD DUFY, LUCE DUFY-BARBE, AND JEFFERY BARKER	93
7. Electrophysiological Techniques in Dissociated Tissue Culture	GARY L. WESTBROOK AND PHILLIP G. NELSON	111
8. Neuroendocrine Cells *in Vitro*: Electrophysiology, Triple-Labeling with Dye Marking, Immunocytochemical and Ultrastructural Analysis, and Hormone Release	JAMES N. HAYWARD, TROY A. REAVES, JR., ROBERT S. GREENWOOD, AND RICK B. MEEKER	132
9. Single Channel Electrophysiology: Use of the Patch Clamp	FREDERICK SACHS AND ANTHONY AUERBACH	147
10. Push–Pull Perfusion Technique in the Median Eminence: A Model System for Evaluating Releasing Factor Dynamics	NORMAN W. KASTING AND JOSEPH B. MARTIN	176
11. Horseradish Peroxidase: A Tool for Study of the Neuroendocrine Cell and Other Peptide-Secreting Cells	RICHARD D. BROADWELL AND MILTON W. BRIGHTMAN	187

12. Visualization of Enkephalin Receptors by Image-Intensified Fluorescence Microscopy	STEVEN G. BLANCHARD, KWEN-JEN CHANG, AND PEDRO CUATRECASAS	219
13. Flow Cytometry as an Analytic and Preparative Tool for Studies of Neuroendocrine Function	JOHN C. CAMBIER AND JOHN G. MONROE	227

Section III. Preparation and Maintenance of Biological Materials

14. Optimization of Culture Conditions for Short-Term Pituitary Cell Culture	NIRA BEN-JONATHAN, EDNA PELEG, AND MICHAEL T. HOEFER	249
15. Separation of Cells from the Rat Anterior Pituitary Gland	WESLEY C. HYMER AND J. MICHAEL HATFIELD	257
16. Culture of Dispersed Anterior Pituitary Cells on Extracellular Matrix	RICHARD I. WEINER, CYNTHIA L. BETHEA, PHILIPPE JAQUET, JOHN S. RAMSDELL, AND DENIS J. GOSPODAROWICZ	287
17. Continuous Perifusion of Dispersed Anterior Pituitary Cells: Technical Aspects	WILLIAM S. EVANS, MICHAEL J. CRONIN, AND MICHAEL O. THORNER	294
18. Preparation and Maintenance of Adrenal Medullary Chromaffin Cell Cultures	STEVEN P. WILSON AND NORMAN KIRSHNER	305
19. Techniques for Culture of Hypothalamic Neurons	CATHERINE LOUDES, ANNIIE FAIVRE-BAUMAN, AND ANDRÉE TIXIER-VIDAL	313
20. Techniques in the Tissue Culture of Rat Sympathetic Neurons	MARY I. JOHNSON AND VINCENT ARGIRO	334
21. Methodological Considerations in Culturing Peptidergic Neurons	WILLIAM J. SHOEMAKER, ROBERT A. PETERFREUND, AND WYLIE VALE	347
22. Preparation and Properties of Dispersed Rat Retinal Cells	JAMES M. SCHAEFFER	362
23. Punch Sampling Biopsy Technique	MIKLÓS PALKOVITS	368

Section IV. Use of Chemical Probes

24. Excitotoxic Amino Acids as Neuroendocrine Research Tools	JOHN W. OLNEY AND MADELON T. PRICE	379
25. Use of Excitatory Amino Acids to Make Axon-Sparing Lesions of Hypothalamus	J. VICTOR NADLER AND DEBRA A. EVENSON	393

26. Use of Specific Ion Channel Activating and Inhibiting Drugs in Neuroendocrine Tissue	P. Michael Conn	401

Section V. Quantitation of Neuroendocrine Substances

27. Development and Use of Ultrasensitive Enzyme Immunoassays	Glennwood E. Trivers, Curtis C. Harris, Catherine Rougeot, and Fernand Dray	409
28. Strategies for the Preparation of Haptens for Conjugation and Substrates for Iodination for Use in Radioimmunoassay of Small Oligopeptides	William W. Youngblood and John Stephen Kizer	435
29. Preparation and Use of Specific Antibodies for Immunohistochemistry of Neuropeptides	Lothar Jennes and Walter E. Stumpf	448
30. Production of Monoclonal Antibodies Reacting with the Cytoplasm and Surface of Differentiated Cells	Richard M. Scearce and George S. Eisenbarth	459
31. Voltammetric and Radioisotopic Measurement of Catecholamines	Paul M. Plotsky	469
32. Microradioenzymic Assays for the Measurement of Catecholamines and Serotonin	Ralph L. Cooper and Richard F. Walker	483
33. Hypothalamic Catecholamine Biosynthesis and Neuropeptides	David K. Sundberg, Barbara A. Bennett, and Mariana Morris	493
34. Methods for the Study of the Biosynthesis of Neuroendocrine Peptides *in Vivo* and *in Vitro*	Jeffrey F. McKelvy, James E. Krause, and Jeffrey D. White	511
35. Methods for Investigating Peptide Precursors in the Hypothalamus	David K. Sundberg, Mariana Morris, and Kenneth A. Gruber	524
36. Measurement of the Degradation of Luteinizing Hormone Releasing Hormone by Hypothalamic Tissue	James E. Krause and Jeffrey F. McKelvy	539
37. Measurement of β-Endorphin and Enkephalins in Biological Tissues and Fluids	Jau-Shyong Hong, Kazuaki Yoshikawa, and R. Wayne Hendren	547
38. Assay of Corticotropin-Releasing Factor	Wylie Vale, Joan Vaughan, Gayle Yamamoto, Thomas Bruhn, Carolyn Douglas, David Dalton, Catherine Rivier, and Jean Rivier	565

39. Direct Radioligand Measurement of Dopamine Receptors in the Anterior Pituitary Gland — MARC C. CARON, BRIAN F. KILPATRICK, AND DAVID R. SIBLEY — 577

40. Ornithine Decarboxylase: Marker of Neuroendocrine and Neurotransmitter Actions — THEODORE A. SLOTKIN AND JORGE BARTOLOME — 590

Section VI. Localization of Neuroendocrine Substances

41. Secretion of Hypothalamic Dopamine into the Hypophysial Portal Vasculature: An Overview — JOHN C. PORTER, MARIANNE J. REYMOND, JUN ARITA, AND JANICE F. SISSOM — 607

42. Neurotransmitter Histochemistry: Comparison of Fluorescence and Immunohistochemical Methods — ROBERT Y. MOORE AND J. PATRICK CARD — 619

43. Simultaneous Localization of Steroid Hormones and Neuropeptides in the Brain by Combined Autoradiography and Immunocytochemistry — MADHABANANDA SAR AND WALTER E. STUMPF — 631

44. Immunocytochemistry of Steroid Hormone Receiving Cells in the Central Nervous System — JOAN I. MORRELL AND DONALD W. PFAFF — 639

45. Techniques for Tracing Peptide-Specific Pathways — LARRY W. SWANSON — 663

46. Immunocytochemistry of Endorphins and Enkephalins — FLOYD E. BLOOM AND ELENA L. F. BATTENBERG — 670

Section VII. Summary

47. Previously Published Articles from Methods in Enzymology Related to Neuroendocrine Peptides — 691

AUTHOR INDEX 693

SUBJECT INDEX 721

Contributors to Volume 103

Article numbers are in parentheses following the names of contributors.
Affiliations listed are current.

VINCENT ARGIRO (20), *Biological Sciences Group, University of Connecticut, Storrs, Connecticut 06268*

JUN ARITA (41), *Second Department of Physiology, Yokohama City University School of Medicine, Yokohama, Japan*

ANTHONY AUERBACH (9), *Department of Biophysical Sciences, State University of New York, Buffalo, New York 14214*

JEFFERY L. BARKER (6), *Laboratory of Neurophysiology, National Institute of Neurological and Communicative Disorders and Stroke, National Institutes of Health, Bethesda, Maryland 20205*

JORGE BARTOLOME (40), *Department of Pharmacology, Duke University Medical Center, Durham, North Carolina 27710*

ELENA L. F. BATTENBERG (46), *Division of Pre-Clinical Neuroscience and Endocrinology, Research Institute of the Scripps Clinic, La Jolla, California 92037*

NIRA BEN-JONATHAN (14), *Department of Physiology, Indiana University School of Medicine, Indianapolis, Indiana 46223*

BARBARA A. BENNETT (33), *Department of Medicine and Clinical Pharmacology, University of Colorado Health Science Center, Denver, Colorado 80262*

CYNTHIA BETHEA (16), *Department of Reproductive Physiology, Oregon Regional Primate Center, Beaverton, Oregon 97006*

STEVEN G. BLANCHARD (12), *Molecular Biology Department, The Wellcome Research Laboratories, Research Triangle Park, North Carolina 27709*

FLOYD E. BLOOM (46), *Division of Pre-Clinical Neuroscience and Endocrinology, Research Institute of the Scripps Clinic, La Jolla, California 92037*

PETER BÖHLEN (5), *Laboratories for Neuroendocrinology, The Salk Institute, La Jolla, California 92037*

MILTON W. BRIGHTMAN (11), *Section on Neurocytology, Laboratory of Neuropathology and Neuroanatomical Sciences, National Institute of Neurological and Communicative Disorders and Stroke, National Institutes of Health, Bethesda, Maryland 20205*

RICHARD D. BROADWELL (11), *Department of Pathology, Division of Neuropathology, University of Maryland School of Medicine, Baltimore, Maryland 21201*

THOMAS BRUHN (38), *Peptide Biology Laboratory, The Salk Institute, La Jolla, California 92037*

JOHN C. CAMBIER (13), *Division of Basic Immunology, Department of Medicine, National Jewish Hospital Research Center/National Asthma Center, Denver, Colorado 80206*

J. PATRICK CARD (42), *Department of Neurology, State University of New York at Stony Brook, Stony Brook, New York 11794*

MARC G. CARON (39), *Departments of Physiology and Medicine, Duke University Medical Center, Durham, North Carolina 27710*

KWEN-JEN CHANG (12), *Molecular Biology Department, The Wellcome Research Laboratory, Research Triangle Park, North Carolina 27709*

RICHARD N. CLAYTON (2), *Department of Medicine, University of Birmingham, Birmingham B15 2TH, England*

P. MICHAEL CONN (3, 26), *Department of Pharmacology, Duke University Medical Center, Durham, North Carolina 27710*

ix

RALPH L. COOPER (32), *Center for the Study of Aging and Human Development and Department of Psychiatry, Duke University Medical Center, Durham, North Carolina 27710*

MICHAEL J. CRONIN (17), *Department of Physiology, University of Virginia Medical Center, Charlottesville, Virginia 22908*

PEDRO CUATRECASAS (12), *Molecular Biology Department, The Wellcome Research Laboratories, Research Triangle Park, North Carolina 27709*

DAVID DALTON (38), *Peptide Biology Laboratory, The Salk Institute, La Jolla, California 92037*

CAROLYN DOUGLAS (38), *Peptide Biology Laboratory, The Salk Institute, La Jolla, California 92037*

FERNAND DRAY (27), *Radioimmunoassay Unit, INSERM U.207 and Institut Pasteur, 75015 Paris, France*

BERNARD DUFY (6), *Laboratoire de Neurophysiologie, Université de Bordeaux II, 33076 Bordeaux Cedex, France*

LUCE DUFY-BARBE (6), *Laboratoire de Neurophysiologie, Université de Bordeaux II, 33076 Bordeaux Cedex, France*

GEORGE S. EISENBARTH (30), *Joslin Diabetes Center/Brigham Hospital, Harvard Medical School, Boston, Massachusetts 02215*

FRED S. ESCH (5), *Laboratories for Neuroendocrinology, The Salk Institute, La Jolla, California 92037*

WILLIAM S. EVANS (17), *Department of Internal Medicine, University of Virginia Medical Center, Charlottesville, Virginia 22908*

DEBRA A. EVENSON (25), *Department of Pharmacology, Duke University Medical Center, Durham, North Carolina 27710*

ANNIE FAIVRE-BAUMAN (19), *Groupe de Neuro-endocrinologie Cellulaire, Collège de France, 75231 Paris Cedex 05, France*

DENIS J. GOSPODAROWICZ (16), *Cancer Research Center, University of California School of Medicine, San Francisco, California 94143*

ROBERT S. GREENWOOD (8), *Department of Neurology, University of North Carolina, Chapel Hill, North Carolina 27514*

KENNETH A. GRUBER (35), *Department of Medicine, Bowman Gray School of Medicine of Wake Forest University, Winston-Salem, North Carolina 27103*

CURTIS C. HARRIS (27), *Laboratory of Human Carcinogenesis, National Cancer Institute, National Institutes of Health, Bethesda, Maryland 20205*

J. MICHAEL HATFIELD (15), *Department of Biochemistry, Microbiology, Molecular and Cell Biology, Pennsylvania State University, University Park, Pennsylvania 16802*

JAMES N. HAYWARD (8), *Department of Neurology, University of North Carolina, Chapel Hill, North Carolina 27514*

ELI HAZUM (4), *Department of Hormone Research, The Weizmann Institute of Science, Rehovot 76100, Israel*

R. WAYNE HENDREN (37), *Chemistry and Life Sciences Group, Research Triangle Institute, Research Triangle Park, North Carolina 27709*

MICHAEL T. HOEFER (14), *Department of Physiology, University of Pittsburgh School of Medicine, Pittsburgh, Pennsylvania 15260*

JAU-SHYONG HONG (37), *Laboratory of Behavioral and Neurological Toxicology, National Institute of Environmental Heatlh Sciences, Research Triangle Park, North Carolina 27709*

WESLEY C. HYMER (15), *Department of Biochemistry, Microbiology Molecular and Cell Biology, Pennsylvania State University, University Park, Pennsylvania 16802*

PHILIPPE JAQUET (16), *Clinique Endocrinologique, Hopital la Conception, 13385 Marseille, France*

LOTHAR JENNES (29), *Department of Anatomy, University of North Carolina School* of Medicine, Chapel Hill, North Carolina 27514, and Department of Pharmacology, Duke University Medical Center, Durham, North Carolina 27710

MARY I. JOHNSON (20), Department of Anatomy and Neurobiology, Washington University School of Medicine, St. Louis, Missouri 63110

NORMAN W. KASTING (10), Department of Physiology, Faculty of Medicine, University of British Columbia, Vancouver, British Columbia V6T 1W5, Canada

BRIAN F. KILPATRICK (39), Departments of Physiology and Medicine, Duke University Medical Center, Durham, North Carolina 27710

NORMAN KIRSHNER (18), Department of Pharmacology, Duke University Medical Center, Durham, North Carolina 27710

JOHN STEPHEN KIZER (28), Biological Sciences Research Center, University of North Carolina School of Medicine, Chapel Hill, North Carolina 27514

JAMES E. KRAUSE (34, 36), Department of Neurobiology and Behavior, State University of New York at Stony Brook, Stony Brook, New York 11794

NICHOLAS C. LING (5), Laboratories for Neuroendocrinology, The Salk Institute, La Jolla, California 92037

CATHERINE LOUDES (19), Groupe de Neuro-endocrinologie Cellulaire, Collège de France, 75231 Paris Cedex 05, France

JOSEPH B. MARTIN (10), Harvard Medical School and Massachusetts General Hospital, Boston, Massachusetts 02114

JEFFREY F. MCKELVY (34, 36), Department of Neurobiology and Behavior, State University of New York at Stony Brook, Stony Brook, New York 11794

RICK B. MEEKER (8), Department of Neurology, University of North Carolina, Chapel Hill, North Carolina 27514

JOHN G. MONROE (13), Department of Pathology, Harvard Medical School, Boston, Massachusetts 02115

ROBERT Y. MOORE (42), Department of Neurology, State University of New York at Stony Brook, Stony Brook, New York 11794

JOHN I. MORRELL (44), The Rockefeller University, New York, New York 10021

MARIANA MORRIS (33, 35), Department of Physiology and Pharmacology, Bowman Gray School of Medicine of Wake Forest University, Winston-Salem, North Carolina 27103

J. VICTOR NADLER (25), Department of Pharmacology, Duke University Medical Center, Durham, North Carolina 27710

PHILLIP G. NELSON (7), Laboratory of Developmental Neurobiology, National Institute of Child Health and Human Development, National Institutes of Health, Bethesda, Maryland 20205

JOHN W. OLNEY (24), Department of Psychiatry, Washington University School of Medicine, St. Louis, Missouri 63110

MIKLÓS PALKOVITS (23), First Department of Anatomy, Semmelweis University Medical School, Budapest H-1450, Hungary

EDNA PELEG (14), Department of Physiology, Indiana University School of Medicine, Indianapolis, Indiana 46223

ROBERT A. PETERFREUND (21), Peptide Biology Laboratory, The Salk Institute, La Jolla, California 92037

DONALD W. PFAFF (44), The Rockefeller University, New York, New York 10021

PAUL M. PLOTSKY (31), Peptide Biology Laboratory, The Salk Institute, La Jolla, California 92037

JOHN C. PORTER (41), Cecil H. and Ida Green Center for Reproductive Biology Sciences, Departments of Obstetrics and Gynecology and Physiology, The University of Texas Health Science Center at

Dallas, Southwestern Medical School, Dallas, Texas 75235

MADELON T. PRICE (24), *Department of Psychiatry, Washington University School of Medicine, St. Louis, Missouri 63110*

JOHN S. RAMSDELL (16), *Laboratory of Toxicology, Harvard School of Public Health, Boston, Massachusetts 97005*

TROY A. REAVES, JR. (8), *Department of Neurology, University of North Carolina, Chapel Hill, North Carolina 27514*

MARIANNE J. REYMOND (41), *Division of Endocrinology, Department of Internal Medicine, C.H.U.V., Lausanne, Switzerland*

CATHERINE RIVIER (38), *Peptide Biology Laboratory, The Salk Institute, La Jolla, California 92037*

JEAN RIVIER (38), *Peptide Biology Laboratory, The Salk Institute, La Jolla, California 92037*

CATHERINE ROUGEOT (27), *Radioimmunoassay Unit, INSERM U.207 and Institut Pasteur, 75015 Paris, France*

FREDERICK SACHS (9), *Department of Biophysical Sciences, State University of New York, Buffalo, New York 14214*

MADHABANANDA SAR (43), *Department of Anatomy, University of North Carolina, Chapel Hill, North Carolina 27514*

RICHARD M. SCEARCE (30), *Department of Medicine, Duke University Medical Center, Durham, North Carolina 22710*

JAMES M. SCHAEFFER (22), *Department of Reproductive Medicine, University of California, San Diego, La Jolla, California 92093*

WILLIAM J. SHOEMAKER (21), *A. V. Davis Center, The Salk Institute, La Jolla, California 92037*

DAVID R. SIBLEY (39), *Departments of Biochemistry and Medicine, Duke University Medical Center, Durham, North Carolina 27710*

JANICE F. SISSOM (41), *Cecil H. and Ida Green Center for Reproductive Biology Sciences, Department of Obstetrics and Gynecology, The University of Texas Health Science Center at Dallas, Dallas, Texas 75235*

THEODORE A. SLOTKIN (40), *Department of Pharmacology, Duke University Medical Center, Durham, North Carolina 27710*

JOHN M. STEWART (1), *Department of Biochemistry, University of Colorado School of Medicine, Denver, Colorado 80262*

WALTER E. STUMPF (29, 43), *Department of Anatomy, University of North Carolina School of Medicine, Chapel Hill, North Carolina 27514*

DAVID K. SUNDBERG (33, 35), *Department of Physiology and Pharmacology, Bowman Gray School of Medicine, Wake Forest University, Winston-Salem, North Carolina 27103*

LARRY W. SWANSON (45), *The Salk Institute, La Jolla, California 92037*

MICHAEL O. THORNER (17), *Department of Internal Medicine, University of Virginia Medical Center, Charlottesville, Virginia 22908*

ANDRÉE TIXIER-VIDAL (19), *Groupe de Neuro-endocrinologie Cellulaire, Collège de France, 75231 Paris Cedex 05, France*

GLENNWOOD E. TRIVERS (27), *Laboratory of Human Carcinogenesis, National Cancer Institute, National Institutes of Health, Bethesda, Maryland 20205*

WYLIE VALE (21, 38), *Peptide Biology Laboratory, The Salk Institute, La Jolla, California 92037*

JOAN VAUGHAN (38), *Peptide Biology Laboratory, The Salk Institute, La Jolla, California 92037*

RICHARD F. WALKER (32), *Department of Anatomy and Sanders-Brown Research Center on Aging, University of Kentucky Medical Center, Lexington, Kentucky 40536*

RICHARD I. WEINER (16), *Department of Obstetrics, Gynecology and Reproductive Sciences, University of California School of Medicine, San Francisco, California 94143*

GARY WESTBROOK (7), *Laboratory of Developmental Neurobiology, National Institute of Child Health and Human Development, National Institutes of Health, Bethesda, Maryland 20205*

JEFFREY D. WHITE (34), *Department of Neurobiology and Behavior, State University of New York at Stony Brook, Stony Brook, New York 11794*

STEVEN P. WILSON (18), *Department of Pharmacology, Duke University Medical Center, Durham, North Carolina 27710*

GAYLE YAMAMOTO (38), *Peptide Biology Laboratory, The Salk Institute, La Jolla, California 92037*

KAZUAKI YOSHIKAWA (37), *Laboratory of Behavioral and Neurological Toxicology, National Institute of Environmental Health Sciences, Research Triangle Park, North Carolina 27709*

WILLIAM W. YOUNGBLOOD (28), *Biological Sciences Research Center, University of North Carolina School of Medicine, Chapel Hill, North Carolina 27514*

Preface

A survey of the current biological literature will convince even the most skeptical individual of the substantive and pervasive advances that have occurred in neuroendocrinology. It was the observation that this rapid growth resulted in large part from methodological improvements which first suggested the need for this volume.

Some of the advances can be easily recognized. High pressure liquid chromatography, once a tool of the solvent chemist, is now a routine analytical and preparative method in biological laboratories. It has enhanced the speed and resolution with which neuroendocrine substances can be separated. Likewise, peptide synthesis, modification, and analysis are no longer the ken of those whose work is dedicated solely to this task, rather, as one of the authors put it, "almost anyone can synthesize a peptide." Analytical advances have extended the sensitivity and specificity of assay techniques. Culture techniques have been devised which allow us to purify and maintain isolated cells and tissue. In turn, these enhancements have revealed the presence and biological roles of new neuroendocrine substances; in other cases, novel and surprising sites of action have been identified for substances already described.

Every effort has been made to provide a comprehensive volume. In some instances, techniques relevant to neuroendocrinology have been described in previous volumes. To facilitate locating these chapters, a cross-index is provided at the end of the volume. Omissions are certainly present. These result from prior commitments of some potential authors, editorial oversight, or timing of new developments relative to the publication deadlines.

I certainly appreciate the guidance of Drs. Colowick and Kaplan as well as the staff of Academic Press. Particular thanks are also extended to the individual authors for their effort, patience, and cooperation in meeting editorial deadlines.

P. Michael Conn

METHODS IN ENZYMOLOGY

EDITED BY

Sidney P. Colowick and Nathan O. Kaplan

VANDERBILT UNIVERSITY
SCHOOL OF MEDICINE
NASHVILLE, TENNESSEE

DEPARTMENT OF CHEMISTRY
UNIVERSITY OF CALIFORNIA
AT SAN DIEGO
LA JOLLA, CALIFORNIA

I. Preparation and Assay of Enzymes
II. Preparation and Assay of Enzymes
III. Preparation and Assay of Substrates
IV. Special Techniques for the Enzymologist
V. Preparation and Assay of Enzymes
VI. Preparation and Assay of Enzymes (*Continued*)
  Preparation and Assay of Substrates
  Special Techniques
VII. Cumulative Subject Index

METHODS IN ENZYMOLOGY

EDITORS-IN-CHIEF

Sidney P. Colowick   Nathan O. Kaplan

VOLUME VIII. Complex Carbohydrates
Edited by ELIZABETH F. NEUFELD AND VICTOR GINSBURG

VOLUME IX. Carbohydrate Metabolism
Edited by WILLIS A. WOOD

VOLUME X. Oxidation and Phosphorylation
Edited by RONALD W. ESTABROOK AND MAYNARD E. PULLMAN

VOLUME XI. Enzyme Structure
Edited by C. H. W. HIRS

VOLUME XII. Nucleic Acids (Parts A and B)
Edited by LAWRENCE GROSSMAN AND KIVIE MOLDAVE

VOLUME XIII. Citric Acid Cycle
Edited by J. M. LOWENSTEIN

VOLUME XIV. Lipids
Edited by J. M. LOWENSTEIN

VOLUME XV. Steroids and Terpenoids
Edited by RAYMOND B. CLAYTON

VOLUME XVI. Fast Reactions
Edited by KENNETH KUSTIN

VOLUME XVII. Metabolism of Amino Acids and Amines (Parts A and B)
Edited by HERBERT TABOR AND CELIA WHITE TABOR

VOLUME XVIII. Vitamins and Coenzymes (Parts A, B, and C)
Edited by DONALD B. MCCORMICK AND LEMUEL D. WRIGHT

VOLUME XIX. Proteolytic Enzymes
Edited by GERTRUDE E. PERLMANN AND LASZLO LORAND

VOLUME XX. Nucleic Acids and Protein Synthesis (Part C)
Edited by KIVIE MOLDAVE AND LAWRENCE GROSSMAN

VOLUME XXI. Nucleic Acids (Part D)
Edited by LAWRENCE GROSSMAN AND KIVIE MOLDAVE

VOLUME XXII. Enzyme Purification and Related Techniques
Edited by WILLIAM B. JAKOBY

VOLUME XXIII. Photosynthesis (Part A)
Edited by ANTHONY SAN PIETRO

VOLUME XXIV. Photosynthesis and Nitrogen Fixation (Part B)
Edited by ANTHONY SAN PIETRO

VOLUME XXV. Enzyme Structure (Part B)
Edited by C. H. W. HIRS AND SERGE N. TIMASHEFF

VOLUME XXVI. Enzyme Structure (Part C)
Edited by C. H. W. HIRS AND SERGE N. TIMASHEFF

VOLUME XXVII. Enzyme Structure (Part D)
Edited by C. H. W. HIRS AND SERGE N. TIMASHEFF

VOLUME XXVIII. Complex Carbohydrates (Part B)
Edited by VICTOR GINSBURG

VOLUME XXIX. Nucleic Acids and Protein Synthesis (Part E)
Edited by LAWRENCE GROSSMAN AND KIVIE MOLDAVE

VOLUME XXX. Nucleic Acids and Protein Synthesis (Part F)
Edited by KIVIE MOLDAVE AND LAWRENCE GROSSMAN

VOLUME XXXI. Biomembranes (Part A)
Edited by SIDNEY FLEISCHER AND LESTER PACKER

VOLUME XXXII. Biomembranes (Part B)
Edited by SIDNEY FLEISCHER AND LESTER PACKER

VOLUME XXXIII. Cumulative Subject Index Volumes I–XXX
Edited by MARTHA G. DENNIS AND EDWARD A. DENNIS

VOLUME XXXIV. Affinity Techniques (Enzyme Purification: Part B)
Edited by WILLIAM B. JAKOBY AND MEIR WILCHEK

VOLUME XXXV. Lipids (Part B)
Edited by JOHN M. LOWENSTEIN

VOLUME XXXVI. Hormone Action (Part A: Steroid Hormones)
Edited by BERT W. O'MALLEY AND JOEL G. HARDMAN

VOLUME XXXVII. Hormone Action (Part B: Peptide Hormones)
Edited by BERT W. O'MALLEY AND JOEL G. HARDMAN

VOLUME XXXVIII. Hormone Action (Part C: Cyclic Nucleotides)
Edited by JOEL G. HARDMAN AND BERT W. O'MALLEY

VOLUME XXXIX. Hormone Action (Part D: Isolated Cells, Tissues, and Organ Systems)
Edited by JOEL G. HARDMAN AND BERT W. O'MALLEY

VOLUME XL. Hormone Action (Part E: Nuclear Structure and Function)
Edited by BERT W. O'MALLEY AND JOEL G. HARDMAN

VOLUME XLI. Carbohydrate Metabolism (Part B)
Edited by W. A. WOOD

VOLUME XLII. Carbohydrate Metabolism (Part C)
Edited by W. A. WOOD

VOLUME XLIII. Antibiotics
Edited by JOHN H. HASH

VOLUME XLIV. Immobilized Enzymes
Edited by KLAUS MOSBACH

VOLUME XLV. Proteolytic Enzymes (Part B)
Edited by LASZLO LORAND

VOLUME XLVI. Affinity Labeling
Edited by WILLIAM B. JAKOBY AND MEIR WILCHEK

VOLUME XLVII. Enzyme Structure (Part E)
Edited by C. H. W. HIRS AND SERGE N. TIMASHEFF

VOLUME XLVIII. Enzyme Structure (Part F)
Edited by C. H. W. HIRS AND SERGE N. TIMASHEFF

VOLUME XLIX. Enzyme Structure (Part G)
Edited by C. H. W. HIRS AND SERGE N. TIMASHEFF

VOLUME L. Complex Carbohydrates (Part C)
Edited by VICTOR GINSBURG

VOLUME LI. Purine and Pyrimidine Nucleotide Metabolism
Edited by PATRICIA A. HOFFEE AND MARY ELLEN JONES

VOLUME LII. Biomembranes (Part C: Biological Oxidations)
Edited by SIDNEY FLEISCHER AND LESTER PACKER

VOLUME LIII. Biomembranes (Part D: Biological Oxidations)
Edited by SIDNEY FLEISCHER AND LESTER PACKER

VOLUME LIV. Biomembranes (Part E: Biological Oxidations)
Edited by SIDNEY FLEISCHER AND LESTER PACKER

VOLUME LV. Biomembranes (Part F: Bioenergetics)
Edited by SIDNEY FLEISCHER AND LESTER PACKER

VOLUME LVI. Biomembranes (Part G: Bioenergetics)
Edited by SIDNEY FLEISCHER AND LESTER PACKER

VOLUME LVII. Bioluminescence and Chemiluminescence
Edited by MARLENE A. DELUCA

VOLUME LVIII. Cell Culture
Edited by WILLIAM B. JAKOBY AND IRA PASTAN

VOLUME LIX. Nucleic Acids and Protein Synthesis (Part G)
Edited by KIVIE MOLDAVE AND LAWRENCE GROSSMAN

VOLUME LX. Nucleic Acids and Protein Synthesis (Part H)
Edited by KIVIE MOLDAVE AND LAWRENCE GROSSMAN

VOLUME 61. Enzyme Structure (Part H)
Edited by C. H. W. HIRS AND SERGE N. TIMASHEFF

VOLUME 62. Vitamins and Coenzymes (Part D)
Edited by DONALD B. MCCORMICK AND LEMUEL D. WRIGHT

VOLUME 63. Enzyme Kinetics and Mechanism (Part A: Initial Rate and Inhibitor Methods)
Edited by DANIEL L. PURICH

VOLUME 64. Enzyme Kinetics and Mechanism (Part B: Isotopic Probes and Complex Enzyme Systems)
Edited by DANIEL L. PURICH

VOLUME 65. Nucleic Acids (Part I)
Edited by LAWRENCE GROSSMAN AND KIVIE MOLDAVE

VOLUME 66. Vitamins and Coenzymes (Part E)
Edited by DONALD B. MCCORMICK AND LEMUEL D. WRIGHT

VOLUME 67. Vitamins and Coenzymes (Part F)
Edited by DONALD B. MCCORMICK AND LEMUEL D. WRIGHT

VOLUME 68. Recombinant DNA
Edited by RAY WU

VOLUME 69. Photosynthesis and Nitrogen Fixation (Part C)
Edited by ANTHONY SAN PIETRO

VOLUME 70. Immunochemical Techniques (Part A)
Edited by HELEN VAN VUNAKIS AND JOHN J. LANGONE

VOLUME 71. Lipids (Part C)
Edited by JOHN M. LOWENSTEIN

VOLUME 72. Lipids (Part D)
Edited by JOHN M. LOWENSTEIN

VOLUME 73. Immunochemical Techniques (Part B)
Edited by JOHN J. LANGONE AND HELEN VAN VUNAKIS

VOLUME 74. Immunochemical Techniques (Part C)
Edited by JOHN J. LANGONE AND HELEN VAN VUNAKIS

VOLUME 75. Cumulative Subject Index Volumes XXXI, XXXII, and XXXIV–LX
Edited by EDWARD A. DENNIS AND MARTHA G. DENNIS

VOLUME 76. Hemoglobins
Edited by ERALDO ANTONINI, LUIGI ROSSI-BERNARDI, AND EMILIA CHIANCONE

VOLUME 77. Detoxication and Drug Metabolism
Edited by WILLIAM B. JAKOBY

VOLUME 78. Interferons (Part A)
Edited by SIDNEY PESTKA

VOLUME 79. Interferons (Part B)
Edited by SIDNEY PESTKA

VOLUME 80. Proteolytic Enzymes (Part C)
Edited by LASZLO LORAND

VOLUME 81. Biomembranes (Part H: Visual Pigments and Purple Membranes, I)
Edited by LESTER PACKER

VOLUME 82. Structural and Contractile Proteins (Part A: Extracellular Matrix)
Edited by LEON W. CUNNINGHAM AND DIXIE W. FREDERIKSEN

VOLUME 83. Complex Carbohydrates (Part D)
Edited by VICTOR GINSBURG

VOLUME 84. Immunochemical Techniques (Part D: Selected Immunoassays)
Edited by JOHN J. LANGONE AND HELEN VAN VUNAKIS

VOLUME 85. Structural and Contractile Proteins (Part B: The Contractile Apparatus and the Cytoskeleton)
Edited by DIXIE W. FREDERIKSEN AND LEON W. CUNNINGHAM

VOLUME 86. Prostaglandins and Arachidonate Metabolites
Edited by WILLIAM E. M. LANDS AND WILLIAM L. SMITH

VOLUME 87. Enzyme Kinetics and Mechanism (Part C: Intermediates, Stereochemistry, and Rate Studies)
Edited by DANIEL L. PURICH

VOLUME 88. Biomembranes (Part I: Visual Pigments and Purple Membranes, II)
Edited by LESTER PACKER

VOLUME 89. Carbohydrate Metabolism (Part D)
Edited by WILLIS A. WOOD

VOLUME 90. Carbohydrate Metabolism (Part E)
Edited by Willis A. Wood

VOLUME 91. Enzyme Structure (Part I)
Edited by C. H. W. HIRS AND SERGE N. TIMASHEFF

VOLUME 92. Immunochemical Techniques (Part E: Monoclonal Antibodies and General Immunoassay Methods)
Edited by JOHN J. LANGONE AND HELEN VAN VUNAKIS

VOLUME 93. Immunochemical Techniques (Part F: Conventional Antibodies, Fc Receptors, and Cytotoxicity)
Edited by JOHN J. LANGONE AND HELEN VAN VUNAKIS

VOLUME 94. Polyamines
Edited by HERBERT TABOR AND CELIA WHITE TABOR

VOLUME 95. Cumulative Subject Index Volumes 61–74, 76–80
Edited by EDWARD A. DENNIS AND MARTHA G. DENNIS

VOLUME 96. Biomembranes [Part J: Membrane Biogenesis: Assembly and Targeting (General Methods, Eukaryotes)]
Edited by SIDNEY FLEISCHER AND BECCA FLEISCHER

VOLUME 97. Biomembranes [Part K: Membrane Biogenesis: Assembly and Targeting (Prokaryotes, Mitochondria, and Chloroplasts)]
Edited by SIDNEY FLEISCHER AND BECCA FLEISCHER

VOLUME 98. Biomembranes [Part L: Membrane Biogenesis (Processing and Recycling)]
Edited by SIDNEY FLEISCHER AND BECCA FLEISCHER

VOLUME 99. Hormone Action (Part F: Protein Kinases)
Edited by JACKIE D. CORBIN AND JOEL G. HARDMAN

VOLUME 100. Recombinant DNA (Part B)
Edited by RAY WU, LAWRENCE GROSSMAN, AND KIVIE MOLDAVE

VOLUME 101. Recombinant DNA (Part C)
Edited by RAY WU, LAWRENCE GROSSMAN, AND KIVIE MOLDAVE

VOLUME 102. Hormone Action (Part G: Calmodulin and Calcium-Binding Proteins)
Edited by ANTHONY R. MEANS AND BERT W. O'MALLEY

VOLUME 103. Hormone Action (Part H: Neuroendocrine Peptides)
Edited by P. MICHAEL CONN

VOLUME 104. Enzyme Purification and Related Techniques (Part C) (in preparation)
Edited by WILLIAM B. JAKOBY

VOLUME 105. Oxygen Radicals in Biological Systems (in preparation)
Edited by LESTER PACKER

Section I

Preparation of Chemical Probes

[1] Synthesis and Use of Neuropeptides

By JOHN M. STEWART

The entire range of naturally occurring neuropeptides (NPs)[1] as well as many analogs has been made available by aggressive commercial synthesis and marketing, and newly described NPs usually appear on the market soon after their structures are announced. Although the market prices of many of these are quite high, the person wishing to investigate biological properties of NPs is well advised to buy them from a reliable source if funds are available and the amounts needed are not large. If large quantities are needed, or if the investigator needs analogs not commercially available, synthesis may be both necessary and practical. Even in these cases, if the investigator is not experienced in peptide chemistry, custom synthesis by one of the established laboratories should be considered. This chapter is written for the investigator inexperienced in peptide synthesis who decides to synthesize his own NP.

Although the procedures of solid-phase peptide synthesis (SPPS) described in the following sections appear to be very simple, and indeed are so in the hands of persons experienced in this area, a biologist lacking experience in handling peptides and proteins can lose much time in attempting synthesis of new peptides. In spite of the great improvements and simplifications in peptide synthesis during the last few years, a considerable amount of skill and art is still involved in this work.

---

[1] Abbreviations: AA, amino acid, general; Acm, acetamidomethyl (—SH blocking group); Anh, anhydride; Boc, *tert*-butyloxycarbonyl; BrZ, 2-bromobenzyloxycarbonyl; BuOH, butanol; Bzl, benzyl; CCD, Countercurrent distribution; CDF, D-4-chlorophenylalanine; Chl, chloroform; ClZ, 2-chlorobenzyloxycarbonyl; Dcb, 2,6-dichlorobenzyl; DCC, dicyclohexylcarbodiimide; DCM, dichloromethane; DMAP, 4-dimethylaminopyridine; DMF, N,N-dimethylformamide; EtOAc, ethyl acetate; HBT, 1-hydroxybenzotriazole; HOAc, acetic acid; HPLC, high-performance liquid chromatography; k, partition coefficient; LHRH, luteinizing hormone releasing hormone; MBHA, 4-methylbenzhydrylamine resin (polystyrene); MeOH, methanol; NP, neuropeptide; ONp, 4-nitrophenyl ester (active ester); OHBT, hydroxybenzotriazole active ester; —Ⓟ, polymer (symbol for the peptide-carrying resin); PC, partition chromatography; SPPS, solid-phase peptide synthesis; TEA, triethylamine; TFA, trifluoroacetic acid; TLC, thin-layer chromatography; Tos, 4-toluenesulfonyl; Xan, 9-xanthenyl; Z, benzyloxycarbonyl. Representation of blocking groups on amino acids: A symbol to the left and hyphenated is on the α-amino group: Boc-Gly = N^α-Boc-glycine. A symbol to the right and hyphenated is an ester on the α-carboxyl: Gly-OHBT = hydroxybenzotriazole ester of glycine. A symbol after the amino acid symbol and in parentheses is a blocking group on the side chain: Tyr(Bzl) = O-benzyltryosine. Boc-Glu(OBzl)-ONp = N^α-Boc-glutamic acid-γ-benzyl ester-α-nitrophenyl ester.

Peptides can be synthesized either by solution chemistry or by SPPS. Solution methods will not be discussed in this chapter, since it is there that the greatest degree of skill and experience is needed. Even if a journal description of the exact synthesis needed is available, the person lacking experience may still encounter major problems. Only SPPS will be discussed, since the techniques of SPPS are for the most part sufficiently routine and easily described that the novice can often obtain very satisfactory results.

This chapter contains descriptions of general methods for synthesis, purification, and characterization of peptides and specific directions for a number of NPs. If a peptide not described is desired, these procedures can be readily modified to yield the particular analog needed. In addition, a section is included on the handling and use of NPs, since many of the problems involved may not be readily apparent to the biologist. These problems are of sufficient magnitude that much work in the literature is probably meaningless, and every necessary precaution should be taken not to perpetuate this unfortunate state of affairs.

Solid-Phase Peptide Synthesis

Description of the System

In the system of SPPS introduced by Merrifield[1a] (see Fig. 1) the amino acid that will become the carboxy-terminal (C-terminal) residue of the peptide is attached covalently to an insoluble polymer matrix —Ⓟ, and additional amino acids are attached to it, usually one at a time, until the desired peptide sequence is assembled on the polymer (**III**). A reagent is then applied to cleave the peptide–polymer bond and liberate the peptide free in solution. Purification and characterization of the synthetic peptide complete the process. The most popular polymer support is polystyrene cross-linked with 1% divinylbenzene. For synthesis of peptides having free C-terminal carboxyl groups, the polymer is functionalized with hydroxymethyl groups (**II**), and the C-terminal amino acid is esterified to these hydroxyl groups. The amino acid resin ester has most often been synthesized directly from a chloromethyl resin (**I**). The directions given herein are appropriate for the synthesis of peptides having free C-terminal carboxyl groups and containing perhaps as many as 20 amino acids. The peptide–resin link in the Merrifield resin is not adequately stable for synthesis of long peptides. A more stable resin is the phenylacetamidomethyl

[1a] R. B. Merrifield, *J. Am. Chem. Soc.* **86,** 1385 (1963).

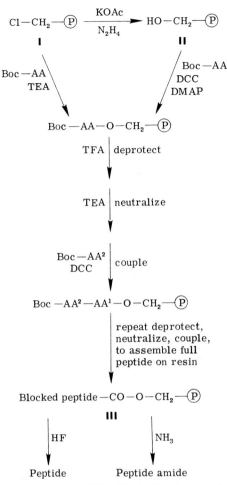

FIG. 1. The reaction scheme of Merrifield solid-phase peptide synthesis.

(Pam) polystyrene resin,[2-4] which will not be described here but should give better results for long peptides.

Many NPs have C-terminal amides. In some cases these can be satisfactorily synthesized on this same type of polymer; the peptide is re-

---

[2] G. Barany and R. B. Merrifield, in "The Peptides" (E. Gross and J. Meienhofer, eds.), Vol. 2, p. 1. Academic Press, New York, 1979.

[3] J. M. Stewart and J. D. Young, "Solid Phase Peptide Synthesis." Freeman, San Francisco, California, 1969.

[4] J. M. Stewart and J. D. Young, "Solid Phase Peptide Synthesis." Pierce Chemical Co., Rockford, Illinois, 1983.

full to assure thorough wetting) (types B and C vessels). One advantage of these vessels is that they can be used in systems that allow addition of liquids without opening the vessel and pouring noxious liquids or exposing the reaction to moisture. These reaction vessels are described and illustrated[4] and are available from Rocky Mountain Scientific Glass (5620 Kendall Ct., Arvada, CO 80002). A motorized rocker or "shaker" for these vessels can be purchased from Colorado Bioengineering. (Bioengineering Department, Univ. of Colorado HSC, Denver, CO 80262). The resin suspension can be mixed by stirring if no contact of the stirrer with walls or filter disk is permitted (to avoid grinding resin beads). A complete system of reaction vessel, valves, and associated tubing for SPPS is also available (the Peptider; Peninsula Laboratories, 611 Taylor Way, Belmont, CA 94002); this system uses nitrogen bubbling for mixing.

Apparatus for Cleavage of the Peptide from the Resin

The most rapid and convenient reagent for cleavage of the peptide from the resin is liquid anhydrous HF. This toxic and corrosive reagent (which dissolves glass rapidly) can easily be handled in a commercial all-plastic vacuum line (Peninsula; or Protein Research Foundation, 476 Ina Minoh-Shi, Osaka 562, Japan). Directions for its use are given below. HF is the only practical reagent for deblocking Arg(Tos). The HF cleavage of peptide-resins can be obtained commercially (Peninsula). Some chemists have carried out HF cleavage in an all-Teflon apparatus constructed from commercially available heavy-wall bottles, tubing, and valves. Polyethylene or polypropylene cannot be used; they become very brittle and shatter at low temperatures. This type of apparatus cannot be used with vacuum; HF is removed by a stream of nitrogen. There are many potential hazards in the use of such homemade equipment, and it cannot be endorsed.

Peptides can be cleaved from the Merrifield resin by HBr-TFA; side-chain deprotection is also complete if the peptide does not contain arginine. HBr-TFA cleavage is carried out in a "cleavage vessel,"[3,4] which is also available from Rocky Mountain Scientific Glass. A SPPS synthesis vessel of type B or C from Rocky Mountain Scientific Glass (180-degree rotation vessels) can be fitted with an adapter for introduction of HBr and used for cleavage.

Silanization of Glassware

Peptide–resins are quite sticky, and can be washed from glassware much more readily if the glassware is silanized before use. Peptides also adsorb tightly to glass; this adsorption can be minimized by silanization.

Reagents

1. Dichlorodimethylsilane, 10% by volume in dry toluene
2. Anhydrous MeOH
3. Dry toluene
4. Acetone

Procedure. Acid-wash the glassware and rinse and dry thoroughly (oven-dry). Fill the vessel with reagent 1 and allow to stand for 15 min. Be sure that the entire inner surface is wetted. *Caution:* The silanizing reagent is corrosive and volatile. Use in a good fume hood. Pour out the silanizing reagent and rinse thoroughly with dry toluene. Do not dry. Fill immediately with anhydrous MeOH and allow to stand for 15 min to "cap" all unreacted chlorosilyl groups. Rinse well with MeOH and acetone and dry.

Starting Materials for SPPS

All solvents are reagent grade, unless otherwise specified. *N,N*-dimethylformamide (DMF) for use in coupling reactions must be free from dimethylamine; a procedure for testing DMF using fluorodinitrobenzene is available.[3,4] Dimethylamine and formic acid can be removed from DMF by storage over molecular sieve 4A. Dichloromethane (DCM) must not contain HCl; to test, shake DCM with water and measure the pH of the aqueous phase. All solvent ratios in reagents are by volume.

The Boc amino acids are available from several sources (Bachem,[4a] Fluka, Peninsula, Protein Research Foundation, Vega, and others). They must be checked for homogeneity by TLC before use. Run TLC on silica gel in systems A (Chl:MeOH:HOAc, 85:15:5) and B (Chl:HOAc, 95:5); tables of R_f values are available.[3,4] Optical purity of Boc-D-amino acids has sometimes been a problem. They should be checked at least for optical rotation in a polarimeter, and the rotations should be compared with literature values.[4] While this method is not very precise, at least gross errors can be detected before a synthesis is completely ruined.

The most commonly used side-chain derivatives are: Ser and Thr: benzyl ethers; Asp and Glu: benzyl esters; Asn and Gln: xanthenyl (or none); Tyr: 2,6-dichlorobenzyl or 2-bromobenzyloxycarbonyl; Arg: toluenesulfonyl (Tos); Lys: 2-chlorobenzyloxycarbonyl; His: Tos; Cys: *p*-methylbenzyl or acetamidomethyl. Asn and Gln present special problems

---

[4a] Bachem Fine Chemicals, 3132 Kashiwa St., Torrance, CA 90505; Bachem Feinchemikalien AG, Hauptstrasse 144, CH-4416 Bubendorf, Switzerland; Fluka Chemical Corp., 255 Oser Ave., Hauppauge, NY 11787; Fluka AG, Buchs, Switzerland; Vega Biochemicals, PO Box 11648, Tucson, AZ 85734.

in coupling reactions; they can be dehydrated by DCC to the omega nitrile. To couple these residues with DCC, use Boc-Asn(Xan) and Boc-Gln(Xan). However, these bulky derivatives may not always couple well. Nitrophenyl esters have classically been used for Asn and Gln, although they react slowly and sometimes may not go to completion. Boc-Asn and Boc-Gln symmetrical anhydrides react rapidly. The DCC-HBT method can also be used. An additional problem with N-terminal Gln peptide–resins is cyclization to the pyroglutamyl peptide, which terminates the chain. This can be minimized by use of Boc-Gln(Xan) and by using the reverse order for steps 7 and 8 when coupling the following residue.

Chloromethylated poly(styrene–1% divinylbenzene) (chloromethyl resin) (**I**) is available from the above sources as Merrifield peptide synthesis resin. A relatively low degree of substitution is desirable (0.4–0.7 mmol/g). Hydroxymethyl resin (**II**) is not commercially available at present and must be synthesized from chloromethyl resin by the investigator.[4] If only a small amount of SPPS will be done, it may be advisable to purchase the resin, with the first amino acid already attached, from one of the above sources (Boc-amino acid resin esters for SPPS). A desirable level of substitution of amino acid on the resin is 0.2–0.4 mmol/g.

Attachment of Boc-Leu to Chloromethyl Resin

Mix 5 g of chloromethyl resin, 1.15 g of Boc-leucine (5.0 mmol; 1 mmol per gram of resin) and 25 ml of absolute EtOH in a 250-ml round-bottom flask containing a magnetic stirring bar. When the Boc-Leu is dissolved, add 0.70 ml (4.5 mmol) of triethylamine and fit the flask with a reflux condenser and drying tube. Heat under reflux, with stirring, in an oil bath (90°) for 48 hr. Cool the mixture, and filter the resin. Wash the resin three times each with EtOH, water, EtOH, and DCM, allowing the solvent to soak into the resin for about a minute each time. Before this resin is used for peptide synthesis, one must remove fine particles and analyze the resin for amino acid substitution (see below).

Do not use this procedure for Boc-Met, Boc-His(Tos), Boc-Glu(Bzl), Boc-Asp(Bzl), or Boc-Cys derivatives. Use the following procedure for these.

Attachment of Boc-Leu to Hydroxymethyl Resin

Place 2.0 of hydroxymethylpolystyrene resin (—OH substitution 0.7–0.9 mmol/g) in an SPPS synthesis vessel. Add 40 ml of DCM and rock for 5 min to swell the resin. Filter. Add a solution of 462 mg (2 mmol) of Boc-leucine in 20 ml of DCM and rock for 1 min. Add 2.0 ml (2.0 mmol) of 1.0 M DCC in Chl (206 mg/ml) and 25 mg (0.2 mmol) of dimethylaminopyridine (DMAP) and rock overnight. Filter; wash the resin three times each with DCM, DMF, EtOH, and DCM and dry. Analyze for amino acid substitution. If too low, repeat the above procedure. If the analysis is satisfactory (0.2–0.5 mmol/g), benzoylate the remaining free hydroxyl groups as follows: Swell the resin in 20 ml of DMF and add 2.8 g (2.3 ml, 20 mmol) of benzoyl chloride. Rock for 2 min and add 2.0 g (2.75 ml, 20 mmol) of triethylamine. Rock for 2 hr, then filter and wash three times each with DMF, EtOH, and DCM. Remove fine particles (see below) and dry the resin.

Attachment of Boc-Met to Methylbenzhydrylamine Resin

Place 2.0 g of MBHA resin in a SPPS synthesis vessel and swell in 40 ml of DCM. Attach Boc-Met by using steps 4–9 of the SPPS DCC Synthesis, Schedule A, below. Use 500 mg (2.0 mmol) of Boc-Met and 2.0 ml of 1.0 M DCC solution. After step 9, remove an aliquot (about 15 mg) of resin, deprotect the Boc group by allowing it to stand in TFA for 15 min, wash five times with DCM, and dry, finally in high vacuum. Weigh a 5-mg sample for amino acid analysis. If the substitution of Boc-Met is satisfactory (0.2–0.5 mmol/g), benzoylate the remaining amino groups on the resin as described above for the hydroxymethyl resin. Wash the resin three times each with DMF, EtOH, and DCM. Remove fine particles by sizing as described below.

Removal of Fine Resin Particles

Fine resin particles will block fritted disks and cause trouble during synthesis. Transfer the Boc-aminoacyl resin to a separatory funnel with DCM (20 ml/g). Swirl and allow to stand while the resin floats to the top of the solvent. If no fine resin particles are present, the bottom of the resin layer will show a sharp line of demarcation. If there are fine particles suspended below, rising more slowly, run them out with the solvent below the resin layer and repeat this flotation twice. When the resin floats cleanly, transfer it to a fritted-glass Büchner funnel and dry it well.

Stepwise Synthesis of Peptides on the Resin

Two detailed schedules are given for assembly of peptides on SPPS resins, one where the Boc-amino acid is activated by addition of a reagent such as DCC (DCC Synthesis, Schedule A) and the second for use with previously activated Boc-amino acid derivatives, such as active esters (AE) or symmetrical anhydrides (AE Synthesis, Schedule B). Schedules for other types of synthesis are given in Stewart and Young.[4]

Greatest simplification and least chance of errors come when the syntheses are standardized by the amount of amino acid on the resin. Two practical scales are 0.4 mmol and 1.0 mmol of aminoacyl resin. The 0.4 mmol scale is good for most syntheses where only a modest amount of peptide is needed, and the schedules are described for that scale of work. For DCC-mediated coupling reactions, it is suggested that 2.5-fold amounts of Boc-amino acid and DCC be used (1.0 mmol) for each coupling reaction (larger excesses are suggested for active ester and symmetrical anhydride reactions). Select a 25-ml beaker and scribe its accurate tare weight on it with a diamond; use this beaker for all Boc-amino acid weighings. Prepare a table of all the amino acid derivatives to be used in synthesis; this table should contain the molecular weight of the Boc-amino acid (do not forget the side-chain blocking group) and a column containing the gross weight of the weighing beaker plus 1.0 mmol of Boc-amino acid. Another column contains the gross weight of the beaker plus 2.5 mmol of Boc-amino acid, for use in 1.0-mmol syntheses. Since not all of the amino acid derivatives are adequately soluble in DCM, and DMF is used to dissolve these, another column in the table identifies these derivatives: Arg(Tos), Asn(Xan), Cys(Acm), Gln(Xan), His(Tos), Leu · H_2O, PyroGlu, Trp.

Solid-phase peptide synthesis involves many boring, repetitive operations. It is very difficult to remember just what has been done and when. To help keep track of the synthesis and provide a permanent record, prepare a work sheet on which every operation is to be recorded as it is done. Examples of such worksheets have been published.[3,4] At the head of the sheet, enter pertinent details of the Boc-aminoacyl resin used as starting material (identity of amino acid, substitution level, identifying lot number, total amount taken), nature of deprotection reagent, concentration of DCC solution and amount used, molar excess of each Boc-amino acid, and weight of finished peptide–resin. In the body of the table provide a horizontal line for each residue added and vertical columns for identity of Boc-amino acid, molecular weight of Boc-amino acid, weight of Boc-amino acid used, the manufacturer and lot number of the Boc-amino acid, and each of the steps of the DCC Schedule A. An additional column allows for recording comments such as: result of monitoring tests, use of DMF solvent. For each of the timed steps (deprotection, neutralization, and coupling) enter in the appropriate column the time of day when the operation began; a glance at the clock will then tell you when to stop that operation. For the nontimed operations (washes), enter a check mark for each wash as it is added to the reaction vessel.

These descriptions are written for a 0.4-mmol run. For a 1.0-mmol run, multiply amounts by 2.5. If you use synthesis vessels of type B or C, be sure to use enough solvent so that the resin suspension wets the entire inner walls of the vessel when it is inverted.

DCC Synthesis, Schedule A

Place an amount of Boc-aminoacyl resin containing 0.4 mmol of amino acid (usually about 1.0 g of resin) in an appropriate-sized synthesis vessel. The amino acid is the one occupying the C-terminal position in the peptide to be synthesized. Couple on the additional Boc-amino acids in sequence, working toward the amino end of the peptide, using the following procedure. This cycle of operations is for coupling one amino acid to the aminoacyl resin, using DCC as the coupling agent.

Reagents

1. Deprotection reagent: 25% TFA in DCM. Place 200 mg of indole in a 250-ml glass-stoppered Erlenmeyer flask. Add 150 ml of DCM and 50 ml of TFA. Stopper. Swirl to dissolve the indole. Allow to stand overnight before use. A dark burgundy color develops in the reagent, indicating that the indole has scavenged harmful substances from the reagent.
2. Neutralization reagent: Add 10 ml of triethylamine (TEA) to 90 ml of DCM in a glass-stoppered flask. Stopper and mix. Do not keep this reagent more than a few days.
3A. Boc-amino acid solution. For a 0.4-mmol run, weigh 1.0 mmol of Boc-amino acid into the standard weighing beaker. Just before time to add it to the synthesis mixture, dissolve it in 10 ml of DCM. For those Boc-amino acids not soluble in DCM alone (see list above), dissolve the Boc-amino acid first in DMF, then add DCM to make up the volume to 10 ml.
3B. Boc-amino acid–HBT mixture. Make up as for 3A and add an equivalent amount of HBT (153 mg, 1.0 mmol).
4. Coupling reagent: 1.0 M DCC in Chl. Dissolve DCC (206 mg/ml) in instrument-analyzed Chl (this grade of Chl is alcohol-free; alco-

SCHEDULE A

Step	Reagent	Volume (ml)	Time (min)
1	DCM wash (3 times)	15 each	3 × 1
2	#1 (TFA) prewash	15	1.5
3	#1 (TFA) deprotect	15	30
4	DCM wash (6 times)	15 each	6 × 1
5	#2 (TEA) neutralize	15 each	2 × 1.5
6	DCM wash (6 times)	15 each	6 × 1
7	#3A Boc-AA in DCM	10	0.5
8	#4 DCC in Chl	1.0	120
9	DCM wash (6 times)	15 each	6 × 1

hol reacts slowly with DCC) in a small glass-stoppered flask or test tube. DCM may be used, but it is difficult to pipette because of its low boiling point. Use 1.0 ml of this reagent (1.0 mmol of DCC) per coupling step in a 0.4-mmol run. *Warning:* DCC is a potent contact allergen. Avoid all contact with the skin.

The prewash (step 2) is essential to avoid excessive dilution of the deprotection reagent by solvent retained in the resin.

After the neutralization (step 5) proceed rapidly through the coupling reaction (steps 7 and 8). Excessive time here may cause problems.

In step 8, the DCC solution is added to the reaction mixture containing the peptide–resin and the Boc-amino acid solution.

When coupling on the third residue in a peptide (coupling a Boc-amino acid to a dipeptide–resin), invert the order of steps 7 and 8, as follows:

| 7 | #4 DCC in Chl | 1.0 | 0.5 |
| 8 | #3A Boc-AA in DCM | 10 | 120 |

This will avoid loss of dipeptide from the resin as diketopiperazine. This should also be done when coupling a Boc-amino acid to an N-terminal Gln peptide–resin to avoid chain termination by conversion of Gln to pyroglutamyl peptide.

When coupling Boc-Asn or Boc-Gln, use reagent 3B at step 7, or use Schedule B, below.

After the coupling reaction, monitor for completeness of coupling with the Kaiser test (see below). If the test is not negative, recouple the same Boc-amino acid by use of steps 4–9 of Schedule A. Again monitor for completeness of coupling. If the test is still positive, indicating the presence of unreacted free amino groups, you may wish to repeat the recoupling with addition of 1.0 mmol (153 mg) of HBT or 0.1 mmol (12 mg) of DMAP to the coupling reaction, or to try coupling with a different Boc-amino acid derivative, such as the symmetrical anhydride (see Schedule B). If it is still impossible to obtain complete coupling, terminate unreacted peptide chains by benzoylation, using the procedure given above in the section Attachment of Boc-Leu to Hydroxymethyl Resin. *Warning:* DMAP has been found to cause partial racemization of amino acids in some cases.

After completion of assembly of the blocked peptide sequence on the resin, do a final deprotection by use of steps 1–4 of Schedule A. Use EtOH to transfer the deprotected peptide–resin to a fritted-glass Büchner funnel. Wash three times with DCM and dry thoroughly.

Some idea of the effectiveness of a synthesis can be obtained by weighing the final peptide–resin and comparing the weight gain with that expected for 0.4 mmol of the blocked peptide synthesized.

Synthesis with Preactivated Boc-Amino Acids—Schedule B

Place the Boc-aminoacyl resin containing 0.4 mmol of the C-terminal amino acid in the synthesis vessel. Use Schedule B for attaching one residue of Boc-amino acid to the aminoacyl resin.

Reagents

1. Deprotection reagent: 25% TFA in DCM. Prepare as described above.
2. Neutralization reagent: 10% TEA in DCM. Prepare as described above.
3A. Boc-amino acid *p*-nitrophenyl ester (Boc-AA-ONp). For a 0.4-mmol run, weigh 2.0 mmol of Boc-amino acid *p*-nitrophenyl ester into the standard weighing beaker. Dissolve in 10 ml of purified DMF and add to the resin.
3B. Boc-amino acid HBT active ester (Boc-AA-OHBT). For a 0.4-mmol run, weigh 2.0 mmol of Boc-amino acid into the standard weighing beaker and dissolve in 5.0 ml of purified DMF. Transfer to a small Erlenmeyer flask or test tube that can be closed with a drying tube. Add 306 mg (2.0 mmol) of HBT and dissolve. Chill the solution to 0° in an ice bath and add 412 mg of DCC; dissolve. Warm to room temperature and allow to stand for 30 min. Add this HBT ester solution to the reaction vessel.
3C. Boc-amino acid symmetrical anhydride (Boc-AA-Anh). For a 0.4 mmol run, weigh 2.0 mmol of Boc-amino acid into the standard weighing beaker and dissolve in 8 ml of DCM, using DMF if necessary. Chill to −5° in ice–salt. Add 1.0 mmol of DCC (1.0 ml of the

SCHEDULE B

Step	Reagent	Volume (ml)	Time (min)
1	DCM wash (3 times)	15 each	3 × 1
2	#1 (TFA) prewash	15	1.5
3	#1 (TFA) deprotect	15	30
4	DCM wash (6 times)	15 each	6 × 1
5	#2 (TEA) neutralize	15 each	2 × 1.5
6	DCM wash (6 times)	15 each	6 × 1
7A	DMF wash (3 times)	15 each	3 × 1
7B	#3A Boc-AA-ONp in DMF	10	30
7C	Add 153 mg (1.0 mmol) of HBT	—	240
8A	DMF wash (3 times)	15 each	3 × 1
8B	DCM wash (3 times)	15 each	3 × 1

1.0 M DCC–DCM reagent 4. Mix. Keep at 0° for 30 min. Add to the reaction vessel containing peptide–resin.
  4. 1.0 M DCC in Chl. Prepare as in Schedule A above.

Use Schedule B also for Boc-amino acid HBT active esters (reagent 3B). For Boc-amino acid symmetrical anhydrides, use the following modification of steps 7 and 8:

7A	#3C Boc-AA-Anh in DCM	10	30
7B	Add 1.4 ml of reagent 2 (TEA)		120
8	DCM wash (6 times)	15 each	6 × 1

The HBT added to the active ester couplings in step 7C and the TEA added to the symmetrical anhydride couplings at step 7B are catalysts. Add them to the reaction mixture containing the peptide–resin and the activated Boc-amino acid derivative.

Monitoring Coupling Reactions. The Kaiser Test

One can make a qualitative estimate of the completeness of the coupling reaction by removing a few beads after coupling and applying a ninhydrin reagent. Many coupling reactions are complete within a few minutes, especially with symmetrical anhyrides. If free amino groups are present on the peptide–resin, the beads will turn blue. If the coupling is complete, the beads will remain a white or buff color.

Reagents

1. Cyanide: Dissolve 33 mg of KCN in 50 ml of water. Dilute 1 ml of this solution with 49 ml of pyridine. *Caution:* POISON
2. Ninhydrin: Dissolve 500 mg of ninhydrin in 10 ml of n-BuOH.
3. Phenol: Dissolve 40 g of phenol in 20 ml of n-BuOH
4. Wash mixture: EtOH : HOAc, 1 : 1

Equipment

Heating block for test tubes maintained at 100°
Test tubes, 12 × 75 mm
Disposable (Pasteur) pipette with a length of polyethylene tubing slipped on the tip

Procedure. Preheat the block to 100°. Withdraw a few beads of peptide–resin from the coupling mixture during coupling or after coupling is complete, and transfer them to a test tube. Wash three times by decantation (or aspiration with a suction tube) with reagent 4 and then three times with absolute EtOH. Add 2 drops each of reagents 1, 2, and 3. Place the test tube in the preheated block and heat for 3 min. Remove the tube from the block and observe the beads against a white background in good light.

If coupling is complete, the beads will be white and the solution yellow. A fully deprotected peptide–resin, on the other hand, gives intensely dark blue beads and a blue solution. N-terminal proline peptide–resins give brown beads. Thorough washing of the beads as described is important.

Cleavage of the Peptide from the Resin

Cleavage with HF. Anhydrous HF can be purchased from Matheson (PO Box 85, East Rutherford, NJ 07073) in small cylinders. It is liquid (bp 19°) in the cylinder and is distilled into the all-plastic reaction system by vacuum. Place a few grams of cobalt trifluoride (drying agent for HF) and a Teflon-covered magnetic stirring bar in the HF reservoir of the vacuum line. Chill the reservoir in alcohol–Dry Ice and evacuate the line with a water aspirator (use a drying tube in the line to prevent back-streaming of water vapor). Carefully distill HF from the cylinder into the reservoir until 20–30 ml is collected. Close the cylinder, remove the Dry Ice bath and stir the HF while it warms to room temperature.

Place 0.2 mmol of peptide–resin in the Teflon reaction vessel of the HF vacuum line. Add 1.0 ml of anisole and a magnetic stirring bar. Chill the reaction vessel in alcohol–Dry Ice, evacuate the line as above, and carefully distill HF from the reservoir into the reaction vessel. This is done by placing a beaker of room-temperature water around the HF reservoir, starting the magnetic stirrer in the reservoir, and carefully opening the stopcock to transfer HF to the reaction vessel. Take care that the cobalt trifluoride and HF do not "bump" up; only vapor is to be transferred to the reaction vessel. When 10 ml of HF has been distilled into the reaction vessel, close the stopcock, stop the reservoir stirrer, place a beaker of crushed ice and water around the reaction vessel, and start the magnetic stirrer under it. The suspension of resin in HF and anisole is a maroon color.

Stir the cleavage reaction mixture for 45 min, replenishing the ice as necessary. At the end of this time, chill the waste HF trap in alcohol–Dry Ice and carefully distill the HF from the reaction vessel into the waste trap, using aspirator vacuum. Continue ice-bath cooling and stirring of the reaction vessel during this time. When all liquid has been distilled from the reaction vessel, leaving a pasty residue of anisole and resin, attach a well-trapped high-vacuum pump and pump until the resin is visibly dry and light colored. Carefully allow dry air to enter the vacuum line and transfer the peptide and resin mixture to a fritted-glass Büchner funnel with EtOAc. Use enough EtOAc to wash away as much anisole as possible. Allow excess EtOAc to evaporate, and extract the peptide from the resin with a appropriate solvent, usually 1 N HOAc (or glacial HOAc for insoluble peptides). Lyophilize the peptide in a tared vial. The peptide can be examined by TLC or paper electrophoresis and then purified.

The entire HF cleavage operation must be carried out in a good fume hood. Respect the extremely corrosive and toxic properties of HF. Keep on hand a saturated solution of calcium gluconate in glycerol for treating HF burns. With care, there is no reason for anyone to be harmed by HF. To dispose of the waste HF collected in the trap, remove the trap from the vacuum line while the HF is still cold. Place it in the back of the hood and direct a stream of water into it from a wash bottle. Once the HF is dilute, it can be flushed down the drain.

Cleavage with HBr-TFA. The apparatus setup for HBr cleavage consists of a cylinder of anhydrous HBr gas (Matheson; Air Products and Chemicals, Inc., Emmaus, PA 18049), a safety flask, a scrubbing tube, the cleavage vessel in which HBr is bubbled through the suspension of peptide–resin in TFA, and a gas outlet tube protected by a calcium chloride drying tube. The safety flask is of a size sufficient to contain the entire contents of the scrubber and cleavage vessel to prevent suck-back of liquid into the HBr cylinder. In the scrubbing tube the HBr gas is bubbled through a solution of 2 g of anisole or resorcinol in 50 ml of TFA; this removes any bromine present in the HBr gas. The gas inlet tube in the cleavage vessel extends nearly to the fritted disk. This setup is illustrated in the literature.[3,4]

Suspend the deprotected peptide–resin in TFA (10 ml per gram of resin) in the cleavage vessel. Add 1.0 ml of anisole per 10 ml of TFA. Bubble HBr through the suspension for 30–90 min. In some cases it may be advisable to do two successive short cleavage operations. Much peptide is cleaved rapidly from the resin during the first few minutes, and this procedure minimizes exposure of sensitive peptides (for example, those containing tryptophan) to the strong acid. Use a rapid stream of HBr at first to flush out the system, then reduce the flow to the point just necessary to keep the resin evenly suspended in the TFA.

At the end of the selected time of cleavage, close the cylinder valve and clamp off the inlet tube just before the cleavage vessel. Open the stopcock at the bottom of the cleavage vessel and draw the TFA solution of peptide into a round-bottom flask, using suction. Wash the resin three times with 5–10 ml of TFA; add the washes to the main peptide solution. Evaporate the TFA under reduced pressure in a rotary evaporator without heating. Dissolve the peptide residue in a suitable solvent (water, HOAc, MeOH–water) and again evaporate, using minimal heating. Repeat this procedure again to remove excess HBr. Extract the crude peptide several times with ether to remove most of the anisole. Transfer the peptide to a suitable vial with water or HOAc and lyophilize. Carry out the entire cleavage process in a good hood.

Purification of Peptides

If all the operations of SPPS proceed perfectly, the peptide as cleaved from the resin should be essentially homogeneous. Although this goal has been achieved in a few cases, generally synthetic peptides must be rigorously purified before they can be used in biological systems. SPPS does provide crude NPs in many cases that need only relatively simple purification procedures. Suitable analytical procedures must be used (see next section) to follow the progress of purification and characterize the final product. Both purification and analysis of peptides are discussed in much greater detail in the laboratory books[3,4] on SPPS and in this series.[5]

Partition methods and molecular sizing methods are most widely used in purification of synthetic peptides. Countercurrent distribution (CCD) is very versatile and rapid and provides adequate purification in overnight automatic runs for useful quantities (up to several hundred milligrams) of many synthetic NPs. The classic Craig instrument (Spectrum Medical Industries, 430 Middle Village Sta., New York, NY 11379) for CCD is quite expensive, however. One limitation of CCD is that it does not handle very hydrophilic peptides, since it is difficult to find all-volatile systems that will give a sufficiently high partition coefficient (k) for these materials. Good results can also be obtained by column partition chromatography (PC), which requires much less investment in equipment and can handle peptides that may be too hydrophilic to be readily purified by CCD. The quantities of peptide that can be purified in a single run are generally less than with CCD, unless very large columns are set up. Peptide PC is generally done using hydrophilic gels (Sephadex, BioGel) as the stationary phase. This classic form of column PC operates in a fashion analogous to paper chromatography and TLC, in that the peptide is held in the aqueous phase associated with the stationary hydrophilic support. One particular advantage of partition purification of peptides synthesized by SPPS is that the anisole derivatives formed in the cleavage reaction are generally much more hydrophobic than the peptides and can be very readily removed. The most recent development in partition methods is HPLC, which is generally done in the reversed-phase mode. The stationary support is a hydrophobic hydrocarbon layer chemically bonded to silica particles. The peptide adsorbs to this hydrophobic support with a strength generally proportional to its hydrophobicity[6] and is eluted from the column by increasing the amount of organic solvent in the eluent.

Molecular sizing on Sephadex or BioGel columns is very useful, particularly for the larger NPs. Quite long gel columns are necessary to separate peptides that differ in size by one amino acid residue. The very

---

[5] This series, Vols. 11, 25, and 47.
[6] J. L. Meek, *Proc. Natl. Acad. Sci. U.S.A.* **77**, 1632 (1980).

recent development of molecular sizing materials for HPLC offers greatly increased speed and versatility, although these materials are still new and must be used with caution until they are completely understood.

Ion-exchange chromatography has been a major tool for purification of proteins for over two decades. The limited capacity and slow operation of conventional ion-exchange resins and the lack of suitable materials for ion-exchange HPLC make it generally less used for purification of small synthetic NPs.

Countercurrent Distribution

The first step is to select a suitable two-phase solvent system in which the peptide has a partition coefficient in the range between 0.2 and 4.0, which does not degrade the peptide, from which the peptide can be readily recovered (generally this means an all-volatile system) and that will separate readily into two phases. The following systems have been useful for the listed peptides, as well as others.

n-BuOH : HOAc : H_2O, 4 : 1 : 5; angiotensin II, k = 0.3; LHRH, k = 0.67; substance P, k = 0.15; TRH, k = 0.12; Met-enkephalin, k = 1.3

n-BuOH : EtOAc : HOAc : H_2O, 2 : 2 : 1 : 5; hydrophobic LHRH antagonists

n-BuOH : 1% aqueous TFA, 1 : 1; bradykinin, k = 1.2

n-BuOH : 2% aqueous TFA, 1 : 1; acid-stable hydrophilic peptides

s-BuOH : 2% aqueous TFA, 1 : 1; very hydrophilic peptides

n-BuOH : benzene : 2% aqueous TFA, 1 : 1 : 1

n-BuOH : 0.4 M ammonium acetate (pH 7.0), 1 : 1; angiotensin I, k = 0.8; angiotensin II, k = 0.13; for peptides that must be kept neutral

n-BuOH : pyridine : HOAc : H_2O, 8 : 2 : 1 : 9; very weakly acidic system; pyridine is noxious

n-BuOH : 2% aqueous formic acid, 1 : 1

Do not use strongly acidic systems for peptides that may not be stable, such as those containing tryptophan or those containing glutamine or asparagine residues adjacent to arginine or lysine. Ammonium acetate is volatile, but complete removal requires repeated lyophilization from water.

To find a suitable system for CCD, determine a preliminary k for the peptide in the system. Prepare about 10 ml of the system and pipette 2 ml of each phase into a glass-stoppered test tube. Add a small amount of the crude peptide. Mix by inverting the tube 15 times. After the phases separate, remove a sample from each phase and determine peptide quantitatively, preferably by using a color reaction specific for an amino acid in the desired peptide, such as the Sakaguchi test for arginine or the Pauly test for tyrosine and histidine. If absorbance at 280 nm is used, the k found will be falsely high owing to the presence of the anisole-related compounds, which absorb at 280 nm and partition exclusively into the upper phase.

$$k = \frac{\text{concentration in upper phase}}{\text{concentration in lower phase}}$$

The peak concentration of a substance having $k = 0.2$ will be in tube 16 after 100 transfers if equal volumes of upper and lower phase are used. The movement of the peptide in the train of tubes is faster if the volume of upper phase is greater than that of lower phase; in the example cited, if the upper phase volume is twice that of the lower phase, the peak will be centered in tube 32. Additional transfers will also move the peptide peak out from the beginning of the train. This is necessary in order to separate the peptide from salts that do not move in the train owing to their very low k. On the other hand, if the peptide k is too high, the peptide will not be separated from the anisole-related products, which move rapidly with the solvent front to the end of the train.

Fill the Craig CCD instrument with lower phase by adding lower phase along the train and then doing several "decant-transfer" steps. Add slightly more lower phase than will be needed to bring each tube up to the decantation level (slightly more than 10 ml per tube for a 10 ml per phase instrument). Do not add lower phase to the first two tubes. Fill the upper-phase reservoir with an amount of upper phase sufficient to do the planned run. For a 10 ml per phase instrument, dissolve the peptide (not more than 100 mg) in 10 ml of upper phase and 10 ml of lower phase. Add to the first tube. Rinse the peptide container with another 10 ml of each phase and add to the second tube. If you wish to purify more than 100 mg of peptide in a 10 ml per phase instrument, you can put the crude peptide in additional tubes (up to 5) without significantly broadening the peptide peak after 100 transfers. Run the distribution, being careful that no significant change in room temperature occurs during the run.

After completion of the distribution, sample appropriate tubes and analyze for peptide. For preparative purposes, most rapid results can be obtained by sampling only lower phase from the lower-numbered half of the tubes and upper phase from the higher-numbered tubes; these phases contain most of the peptide. If the peptide contains tyrosine or tryptophan, absorbance at 280 nm is useful. Add a little MeOH to each sample to dissolve tiny droplets before reading the samples in the spectrophotometer. Since this analysis is not destructive, return the withdrawn samples to the appropriate peptide-containing fractions. If UV-absorbing residues are not present, use a color reagent (as noted above) specific for an amino acid residue in the peptide. If none of these is present, use the quantitative ninhydrin reaction. If the peptide does not contain a free amino group (such as is the case with TRH), use rapid alkaline hydrolysis to hydrolyze the peptide partially so the ninhydrin reagent will yield a color. Specific directions for these analytical procedures are given in the laboratory manuals on SPPS[3,4] as well as in other sources. Qualitative monitoring of CCD runs can be done by spotting aliquots on TLC plates and running them to locate the peptides.

The theoretical shape of the distribution curve can be calculated using manual or computer methods,[7] and coincidence of the theoretical with the experimental curve will give some indication of purity of the peptide after CCD, unless an impurity having the same k is present. When the desired peptide peak has been located, withdraw the peptide solution from the appropriate tubes. Best purity is achieved if conservative cuts are taken. Even if the entire material in the peptide peak will be saved, divide it at least into a center and side cuts. Evaporate the solvent in a rotary evaporator with minimal heating and lyophilize the peptide from an appropriate solvent.

Partition Chromatography

Use a grade of Sephadex or BioGel P that will allow the peptide to penetrate the gel particles fully. Swell the gel in the lower (aqueous) phase of the desired system and pour the column, using additional lower phase; n-BuOH : HOAc : H_2O, 4 : 1 : 5 is the system most used. TLC on cellulose is a good indicator of the partition properties of a peptide in a given system. After the column is poured, run through about two column volumes of upper phase. Dissolve the crude peptide in a minimum volume of upper phase and apply to the column. Continue to run upper phase until the peptide emerges from the column. Monitor the effluent by an appropriate method.

If the peptide has not emerged after several column volumes of upper phase have been run, the peptide is evidently too hydrophilic to move in the organic phase. The peptide can still be eluted by running a gradient to a more polar solvent. Set up a two-chamber gradient former. In the chamber nearest the column, place two column volumes of the upper phase being run. In the supply chamber place the mixture upper phase: MeOH : H_2O, 1 : 1 : 1. If the peptide still is not eluted, use more methanol and water; if it emerged too rapidly and is impure, use a flatter gradient (less water).

[7] T. P. King and L. C. Craig, in "Methods of Biochemical Analysis" (D. Glick, ed.), p. 201. Wiley (Interscience), New York, 1962.

High-Performance Liquid Chromatography

The usual HPLC columns available commercially (such as standard reversed-phase C_{18} or C_8 silica) will be found to have a disappointingly low capacity for peptides. These materials are based on silica having 60-Å pores; the pore size is much reduced by the organic coating, leaving inadequate space for the peptide to penetrate the pores. For example, a "preparative" 9 × 250 mm column may hold only 2–4 mg of a decapeptide without serious degradation of resolving power. Very large instruments, such as the Waters Associates (Milford, MA 01757) Prep 500, are very expensive to purchase and operate. Preparative reversed-phase chromatography can be done by pouring bulk C_8 silica into glass columns. The researcher can achieve even greater savings by attaching the alkylsilyl phase to commercial silica. In either case it is not necessary to use very fine particles of the solid phase for this preparative work.

Several volatile buffer systems are available for preparative HPLC of peptides. Most commonly used are dilute ammonium acetate or ammonium formate buffers used with addition of sufficient methanol, isopropanol, or acetonitrile to elute the peptide from the column. Repeated lyophilization will remove these buffers from peptides. If the acetate concentration is not above 0.02 M, many HPLC spectrophotometric monitors can be used at around 220 nm to follow the run; higher acetate concentrations will necessitate monitoring at longer wavelengths. TFA (0.01 or 0.02 M) used with isopropanol provides high resolution for preparative work, and is volatile. The pH of TFA systems may be low enough to harm some peptides.

Other Purification Methods

The methodology of gel chromatography and ion-exchange chromatography are well known and adequately described in many manuals[3,4] and in earlier volumes of this series.[5] They will not be described here.

Analysis of Peptides

Purification of peptides must be followed by analytical methods that will give significant information on the purity of the peptide. An important rule is that analytical methods used to establish homogeneity of peptides must use criteria different from those used in the purification process itself. For example, if the peptide is purified by a partition method such as CCD, PC, or reversed-phase HPLC, an analytical technique based on partition, such as TLC or HPLC, will not necessarily detect impurities. In such cases paper or thin-layer electrophoresis or ion-exchange TLC will usually reveal more impurities. If the peptide was purified by molecular sizing or ion-exchange methods, then TLC and HPLC can be expected to be most useful. Paper or thin-layer electrophoresis is particularly useful for examination of crude peptides after cleavage from the resin, since direct information on the charge : mass ratio in buffers of different pH will usually tell whether the desired peptide is present or not, and will show the characteristics of impurities present. Use of a general peptide spray reagent, such as the chlorine reagent, which reacts with most peptide bonds, in conjunction with sprays for specific amino acid residues is most informative. Specific directions for these procedures are available[3,4,8,9]

Amino Acid Analysis of Peptides and Peptide–Resins

A pure synthetic peptide must show the proper ratios of the constituent amino acid residues after hydrolysis. Analysis for C, H, and N, the organic chemist's standby, is not very useful with peptides.

Hydrolysis of Peptide–Resins and Aminoacyl Resins

The degree of substitution of the Boc-amino acid on the resin must be determined before the chain elongation part of the synthesis can be begun. Direct peptide–resin hydrolysis is often valuable for following the progress of synthesis of a long peptide and for evaluating the effectiveness of cleavage methods.

Hydrolyze an accurately weighed aliquot of the dry aminoacyl resin or peptide–resin (10 mg or less) and run the hydrolyzate on the amino acid analyzer. Weigh the aminoacyl resin or peptide–resin into a 15 × 150 mm screw-capped culture tube; the cap must have a Teflon liner. Add 0.5 ml of propionic acid. Evacuate the tube on the water aspirator to remove air trapped within the resin beads; break the vacuum and allow air to enter the tube. Repeat this evacuation twice more. Add 0.5 ml of 12 N HCl. Briefly flush the tube with nitrogen (do NOT bubble through the liquid; you will lose HCl), cap and heat in a block at 130°. Heat aminoacyl hydroxymethyl resins for 2 hr; heat aminoacyl MBHA resins and peptide–resins for 16 hr. Remove the tube from the block (careful: pressure!) and cool. Dilute the hydrolyzate with water and filter into a 10-ml volumetric flask. Wash the tube three times, and add the washings to the flask. Make up to volume and mix. Apply an appropriate aliquot of this solution to the amino acid analyzer.

[8] G. Pataki, "Techniques of Thin Layer Chromatography." de Gruyter, Berlin, 1966.
[9] G. Zweig and J. R. Whitaker, "Paper Chromatography and Electrophoresis," Vol. 1. Academic Press, New York, 1967.

Hydrolysis by this method may give low values for sensitive amino acids such as Ser, Thr, Tyr, and Cys; Trp may be largely destroyed. Accurate analysis of Cys, either on resins or in free peptides, requires previous oxidation to cysteic acid with performic acid.[4,5] Values for the other amino acids listed may be improved by including phenol and 2-mercaptoethanol (1 mg of each per milliliter) in the acid hydrolysis mixture.

Hydrolysis of Peptides

The following method is simple and gives good results for most amino acids except Cys; it must be oxidized first. Tryptophan may be recovered in excellent yield if the hydrolyzate is worked up and applied to the amino acid analyzer without any delay.

Materials

Heating block to hold 15-mm tubes at 110°
Culture tubes, 15 × 150 mm, with Teflon-lined screw caps
Disposable pipettes
Nitrogen

Reagent

Constant-boiling HCl containing, per milliliter, 1 mg each of 2-mercaptoethanol and phenol. Add 10 µl each of liquefied 88% phenol and 2-mercaptoethanol to 10 ml of constant-boiling HCl. Prepare fresh each week. It is important to use distilled HCl; reagent concentrated HCl contains amino acids.

Procedure. Place an accurately weighed or measured aliquot of peptide into the culture tube. Add 1 ml of the HCl reagent. With a disposable pipette, bubble a moderate stream of nitrogen through the HCl for 2 min. Rapidly cap the tube tightly and heat at 110° for 22 hr. Remove the tube from the block and cool. Depending on the amino acid analyzer being used, the hydrolyzate may be diluted with sample buffer, and an aliquot applied to the analyzer or the acid may need to be evaporated first. In the latter case, use a good vacuum and work rapidly to avoid loss of tryptophan by oxidation after exposure to the air.

Sterically hindered sequences (e.g., Ile-Ile or Val-Ile) will not hydrolyze completely in 22 hr. Use 48- or 72-hr hydrolysis, which may further degrade sensitive amino acid residues (see above). Glutamine and asparagine are determined as glutamic and aspartic acids, respectively. Total enzymic hydrolysis is necessary to preserve the amides.[3,4]

Analysis for Methionine Sulfoxide

Several NPs contain methionine residues; these peptides are often totally inactive (or much reduced in potency) if the Met residue has been oxidized to the sulfoxide. Oxidation can occur readily, especially in dilute solution or when the peptide is adsorbed to a surface. Met-sulfoxide can be reduced to methionine by treatment with 2-mercaptoethanol or sodium borohydride. HCl hydrolysis converts Met(SO) largely to methionine. To preserve the sulfoxide during hydrolysis, hydrolyze the peptide with 3 N methanesulfonic acid.

Reagents

1. Methanesulfonic acid, 3 N. Dissolve 2.9 g (2.2 ml) of methanesulfonic acid in water and make up to 10 ml.
2. Sodium hydroxide, 3 N. Dissolve 12.0 g of reagent sodium hydroxide in water and make up to 100 ml.

Procedure. Weigh the peptide sample into the hydrolysis tube as described above. Add 0.5 ml of reagent 1. Bubble with nitrogen as above. Close tightly and heat at 110° for 22 hr. Cool and add 0.48 ml of reagent 2. The solution should then be still acidic. Note: This solution is 1.5 N in sodium and may need to be diluted further before application to the amino acid analyzer. Most analyzers operate with 0.2 N sodium buffers.

Synthesis of Specific Neuropeptides

Synthesis of Thyrotropin Releasing Hormone, ⟨Glu-His-Pro-amide

Attach Boc-Pro to MBHA resin by the procedure given above. Determine the substitution of Pro on the resin, and, if it is satisfactory, benzoylate the remaining amino groups. Remove fine resin particles by flotation. Place 0.4 mmol of Boc-Pro-resin in a synthesis vessel and attach Boc-His(Tos) and pyroglutamic acid using Schedule A above. Remember that both of these amino acid derivatives require some DMF for solubility. Remember to use the inverted order of steps 7 and 8 when attaching the pyroglutamyl residue. Transfer the tripeptide–resin to a fritted-glass Büchner funnel with EtOH, wash three times with DCM, and dry thoroughly. Weigh the product and calculate the weight gain.

Cleave the peptide from the resin with HF or HBr-TFA as described above. Weigh the lyophilized crude peptide and examine it by paper electrophoresis and silica gel TLC. Purify the peptide by CCD for 200 transfers in n-BuOH : HOAc : H$_2$O, 4 : 1 : 5, $k = 0.12$. Locate the peak by use of the quantitative Pauly reaction; the peak should be centered on tube 22. The product may also be purified by PC on Sephadex G-25 in the same solvent system. Remove the material from the peak tubes and evaporate it in a rotary evaporator. Lyophilize the product from glacial HOAc. Hydrolyze an aliquot for amino acid analysis.

Synthesis of Luteinizing Hormone Releasing Hormone (LHRH),
Glu-His-Trp-Ser-Tyr-Gly-Leu-Arg-Pro-Gly-amide

Synthesize, analyze, benzoylate, and float Boc-Gly-MBHA resin as described above. Place 0.4 mmol of this Boc-Gly-resin in the synthesis vessel and attach Boc-Pro, Boc-Arg(Tos), Boc-Leu, Boc-Gly, Boc-Tyr(BrZ), Boc-Ser(Bzl), Boc-Trp, Boc-His(Tos) and pyroglutamic acid, in that order, using Schedule A. Boc-Tyr(Dcb) may be used instead of Boc-Tyr(BrZ). Remember to use the inverted order of steps 7 and 8 when attaching Boc-Arg(Tos). Transfer the peptide–resin to a fritted-glass Büchner funnel with EtOH, wash with DCM, and dry. Weigh the product, and calculate the weight gain.

Cleave the peptide from the resin with HF-anisole. Evaporate the HF, wash the resin with EtOAc to remove anisole, and extract the peptide with glacial HOAc. Purify the LHRH by CCD for 100 transfers in n-BuOH : HOAc : H$_2$O, 4 : 1 : 5, $k = 0.67$ (peak in tube 40); locate by reading absorbance at 280 nm. You may also use PC in the same system. Concentrate the peptide-containing fractions in a rotary evaporator, being careful not to heat above 40°. Lyophilize from glacial HOAc.

Synthesis of an LHRH Superagonist,
Glu-His-Trp-Ser-Tyr-D Lys-Leu-Arg-Pro-ethylamide

Prepare Boc-Pro-hydroxymethyl resin, including analysis and flotation. Place 0.4 mmol of this resin in the synthesis vessel and couple in order Boc-Arg(Tos), Boc-Leu, Boc-DLys(ClZ), Boc-Tyr(BrZ), Boc-Ser(Bzl), Boc-Trp, Boc-His(Tos) and pyroglutamic acid, using Schedule A. Remember to use the inverted order of steps 7 and 8 for coupling Boc-Leu. Boc-Tyr(Dcb) may be used instead of Boc-Tyr(BrZ). Transfer the peptide–resin to a fritted-glass Büchner funnel with EtOH, wash with DCM, and dry thoroughly. Weigh, and calculate the weight gain.

Cleavage of the peptide from the resin with ethylamine. Place the peptide–resin and a magnetic stirring bar in a 250-ml glass-stoppered round-bottom flask and add 25 ml of peroxide-free dioxane.[3,4] Chill in an ice bath. Collect 25 ml of ethylamine (monoethylamine, bp 17°) in a 50-ml graduated cylinder in an ice bath by distilling it from a commercial cylinder (Matheson; Air Products and Chemicals). Keep the gas cylinder upright so that only gas is taken from the cylinder; lead it to the bottom of the graduated cylinder in the ice bath with a rubber tube and a disposable pipette. Add 25 ml of the liquid ethylamine to the resin suspension in the flask. Stopper the flask, tape the stopper in place securely, and stir the resin suspension for 5 days at room temperature. Evaporate the solvent on a rotary evaporator and extract the peptide with glacial HOAc. Lyophilize.

The blocking groups on the side chains must now be removed. Treat the blocked peptide–ethylamide with HF-anisole in the usual way. After the HF is evaporated, extract as much as possible of the anisole with ether; lyophilize the peptide from HOAc.

Purify the peptide by CCD for 100 transfers in the system n-BuOH : 1% TFA, 1 : 1; $k = 0.96$ (peak in tube 49). Locate the peptide peak by reading absorbance at 280 nm. Concentrate the peptide fraction under reduced pressure in a rotary evaporator and lyophilize from HOAc. Work rapidly so that the peptide, which contains tryptophan, is not exposed to the acid any longer than absolutely necessary. The peptide can also be purified by PC on Sephadex G-25 in n-BuOH : HOAc : H$_2$O, 4 : 1 : 5, although the k is quite low (0.04) in this system.

Synthesis of a LHRH Antagonist, Acetyl-CDF-CDF-D Trp-Ser-Tyr-DArg-Leu-Arg-Pro-DAla-amide

Load 0.4 mmol of Boc-DAla-MBHA resin into the synthesis vessel. Using Schedule A, couple in order Boc-Pro, Boc-Arg(Tos), Boc-Leu, Boc-DArg(Tos), Boc-Tyr(BrZ), Boc-Ser(Bzl), Boc-DTrp, Boc-p-chloro-DPhe, and Boc-p-chloro-DPhe. Boc-Tyr(Dcb) may be used instead of Boc-Tyr(BrZ). Remember to use the inverted order of steps 7 and 8 when coupling the third residue (Boc-Arg). Acetylate the peptide–resin using Schedule B, as follows: at step 7B, use reagent 3D: 10 ml of purified DMF containing 10 mmol each of acetic anhydride (1.02 g, 0.94 ml) and TEA (1.01 g, 1.40 ml); prepare this reagent just before use. Omit step 7C. The remainder of the synthesis and HF cleavage are carried out as described above for LHRH.

Purify the antagonist by CCD for 100 transfers in the system n-BuOH : EtOAc : HOAc : H$_2$O, 2 : 2 : 1 : 5; $k = 2.0$ (peak in tube 67). Locate the peptide by reading absorbance at 280 nm. Concentrate, lyophilize, and analyze the peptide as described above.

Synthesis of Substance P,
Arg-Pro-Lys-Pro-Gln-Gln-Phe-Phe-Gly-Leu-Met-amide

Load 0.4 mmol of Boc-Met-MBHA resin into the synthesis vessel. Using Schedule A, couple in order Boc-Leu, Boc-Gly, Boc-Phe, and Boc-

Phe. Remember to use the inverted order of steps 7 and 8 when coupling the Boc-Gly. Then, using Schedule B two times, couple Boc-Gln as the symmetrical anhydride (use reagent 3C and the modified steps 7 and 8). Then, using Schedule A, couple in order Boc-Pro, Boc-Lys(ClZ), Boc-Pro, and Boc-Arg(Tos). Finally, deprotect the peptide–resin using steps 1–4 of Schedule A. Transfer the peptide–resin to a fritted-glass Büchner funnel with EtOH, wash with DCM, and dry. Weigh and calculate the weight gain.

Cleave and deprotect the peptide with HF-anisole using the standard procedure. Purify the product by CCD for 200 transfers in n-BuOH-HOAc : H_2O, 4 : 1 : 5, $k = 0.15$ (peak in tube 26 after 200 transfers). Do not use BuOH-TFA to obtain a higher k; substance P is not fully stable under these conditions (hydrolysis of Gln). The product may be contaminated by substance P sulfoxide, which has a lower R_f on TLC and emerges from reverse-phase HPLC runs ahead of substance P. Hydrolysis of an aliquot with methanesulfonic acid will allow accurate analysis for Met(SO).

Synthesis of Methionine-enkephalin, Tyr-Gly-Gly-Phe-Met

Load 0.4 mmol of Boc-Met-hydroxymethyl resin into the synthesis vessel. Using Schedule A, couple in order Boc-Phe, Boc-Gly, Boc-Gly, and Boc-Tyr(BrZ). Remember to use the reversed order of steps 7 and 8 when coupling the first Boc-Gly residue. Finally, deprotect the Boc group, using steps 1–4 of Schedule A. Transfer the peptide–resin to a fritted-glass Büchner funnel with EtOH, wash with DCM, and dry. Cleave with HF-anisole in the usual way, and purify the product by CCD for 100 transfers in n-BuOH : HOAc : H_2O, 4 : 1 : 5, $k = 1.3$ (peak in tube 57). Locate the peptide peak by absorbance at 280 nm. Concentrate the peptide-containing fractions, and lyophilize the product from glacial HOAc. Characterize and analyze. This same procedure may be used to synthesize leucine-enkephalin, which has $k = 1.2$ in the same CCD system. Met-enkephalin oxidizes readily to the sulfoxide, which is more polar than the desired peptide. Methanesulfonic acid hydrolysis will allow quantitation of the Met(SO).

Use of Neuropeptides

Researchers not familiar with the use of peptides should realize that molecules of this class have many properties that give rise to problems so great as to invalidate experimental work if adequate precautions are not taken. Probably a significant amount of work in the literature is meaningless for these reasons. Problems with NPs are physical, chemical, and biological in nature.

Many peptides, especially those that are basic and hydrophobic, adsorb avidly to glass and some plastic surfaces. Since many NPs are extremely potent and are used in biological experiments in very dilute solutions, this problem is very significant. Basic peptides, being positively charged at physiological pH, adhere by ionic bonding to glass surfaces, which are anionic. This can be most dramatically demonstrated by placing a solution of a radioactive NP—such as substance P or bradykinin—in a glass tube, pouring out the solution, and following the radioactivity as the tube is rinsed. Only acid or detergents will completely remove the radioactivity. Persons familiar with radioimmunoassay techniques know that rinsing a glass tube with a dilute protein solution will prevent adsorption of peptides added subsequently. Many researchers, however, do not wish to add protein to their experiments. Much of the "stickiness" of glass can be neutralized by silanizing the glassware, using the procedure given above. Rinsing glassware with dilute acetic acid solutions before using it with peptides or including very dilute acetic acid in the peptide solution can produce a great improvement. Experiments on behavioral effects of substance P showed a 1000-fold increase in the potency of substance P when the solution contained 0.01 M HOAc.

Plastics such as polystyrene also adsorb peptides actively. Solid-phase immunoassays use this principle to advantage. The only plastics that should be used for peptide solutions when adsorption is not wanted are polyethylene and polypropylene; even these are not safe with some peptides. Pipettes are a special problem; many disposable pipettes are made of polystyrene and may adsorb a large part of the peptide from a dilute solution when it is pipetted.

Ease of solubilization of some hydrophobic peptides depends on the way they are lyophilized. If the peptide is dispersed in a fine, fluffy state, it will usually dissolve readily. On the other hand, if the peptide was evaporated or lyophilized from a very concentrated solution, the particles of peptide may be dense and difficult to dissolve in water. Moreover, peptide particles may have an index of refraction almost identical to that of water and be essentially invisible in suspension. Erratic results may be the first indication that the investigator did not really dissolve his peptide. Some hydrophobic peptides dissolve much less readily in physiological saline than in water.

The chemical problems have to do principally with oxidation and hydrolysis. One is not immune from these even while the peptide is stored "safely" in its vial. Most lyophilized peptides contain solvent adsorbed on the peptide; in some cases this solvent may amount to 40% of the total weight of the peptide. If such "dry" peptides are stored for extended periods at room temperature, sensitive bonds may be broken. Scission of the peptide chain at aspartic acid residues has been seen, for example, in angiotensin I. Amides may hydrolyze, either C-terminal amides or those of glutamine and asparagine. The latter are particularly labile when those residues occur in the peptide chain adjacent to a basic arginine or lysine residue. One amide is known to have a half-life of about 5 hr in solution at physiological pH and temperature. Such hydrolysis is much worse when the solution is not neutral. An additional problem with glass is that its surface is naturally alkaline, especially if the glass is old, and this promotes rapid hydrolysis.

A surprising chemical change is cyclization and cleavage of a dipeptide from the end of a peptide chain to form a diketopiperazine. This occurs mainly in peptides containing proline residues, or analogs containing other secondary amino acids. It can occur in peptide amides most readily.

Peptides containing methionine or cysteine residues are prone to oxidation by air. This occurs spontaneously and is worse in dilute solution, where the ratio of dissolved oxygen to peptide is greatest. Adsorption to a surface also promotes oxidation and can be seen dramatically in TLC of peptides. In most cases studied, oxidation of methionine residues to the sulfoxide causes either partial or total loss of biological activity. In the case of human calcitonin, biological activity is totally lost, but the peptide still reacts normally with the usual antibodies.

Disulfide interchange is a problem peculiar to cystine-containing peptides and occurs at neutral or higher pH. If a solution of oxytocin or vasopressin is allowed to stand at pH 7.4, the biological activity disappears rapidly as the peptide dimerizes and polymerizes.

Every researcher should remember that most peptides are ideal food for microorganisms—they usually grow faster on peptides than on amino acids. Therefore even those peptides that do not contain easily oxidizable or hydrolyzable groups may become contaminated and disappear in full view of the technician. Colonies of microorganisms have become established in Sephadex gel columns and have rapidly degraded peptides subsequently passed through the column.

The presence of significant and variable amounts of solvent in "dry" peptides requires that the actual peptide content of each sample be known if results of experiments are to be reported on a molar basis. Some commercial suppliers of peptides analyze their final product and sell on the basis of actual "bare-bones" peptide, without any salt or solvent. If a small sample of an expensive peptide is purchased from such a supplier and used directly without weighing, the results will obviously be very different from those obtained using the same peptide bought in bulk and weighed before dissolving—especially if the peptide is one of those that normally contains 40% solvent. The same peptide may vary from preparation to preparation and from supplier to supplier in this regard. For accurate work you must know the actual peptide content of the sample being used. Another way of expressing this is in a "working molecular weight" or a "molecular weight found by amino acid analysis."

In view of these problems, the following suggestions may be helpful.

1. Store NP in tightly sealed vials in the freezer, preferably in a desiccator containing drying agent. Remove the vial from the freezer and allow it to warm to room temperature at least 1 hr before opening the vial, or moisture from the air will condense on the remaining peptide and hydrolyze it on continued storage.

2. Dissolve peptides first in fairly concentrated solutions (for example, 0.1 mg/ml) in water or 0.01 M HOAc in a treated glass tube. Examine the "solution" carefully with a magnifier to be sure that all particles are dissolved. Divide the solution into small aliquots, flush with nitrogen, and freeze the aliquots.

3. Thaw each day only the amount needed for the day's work; keep diluted solutions cold during the day. Make dilutions in acidified saline in polyethylene or treated glass tubes. *Never* blow air through the solution to aid mixing upon dilution, or shake or vortex; this almost guarantees oxidation.

4. Surface-treat all glassware by silylation or at least rinse with dilute acetic acid.

5. Remove all polystyrene and acetate plastic ware from the laboratory.

6. Know the actual "working molecular weight" of your peptide sample. If you synthesized the peptide, this value can be obtained from the amino acid analysis. If you purchased the peptide, the manufacturer should provide this information.

[2] Preparation of Radiolabeled Neuroendocrine Peptides

By RICHARD N. CLAYTON

The innovative hypothesis of Harris (see review[1]) that factors from the hypothalamus control the release of pituitary hormones led to the birth of the modern discipline of neuroendocrinology, which has expanded and advanced rapidly in recent years. In the late 1960s and early 1970s came

---

[1] G. W. Harris, "Neural Control of the Pituitary Gland." Arnold, London, 1955.

the isolation, chemical characterization, and synthesis of the small hypothalamic peptides that release thyrotropin (TRH) and luteinizing hormone and follicle-stimulating hormone (LHRH) (reviewed by Schally et al.[2]). These were a tripeptide (TRH) and a decapeptide (LHRH or GnRH), respectively. Subsequently, a tetradecapeptide—somatostatin, or growth hormone release-inhibiting hormone (GHRIH)—was isolated from the median eminence.[3] The structure and synthesis of corticotropin-releasing factor (CRF)[4] and growth hormone-releasing factor (GRF)[5] were announced in 1981 and 1982, respectively. Although these latter two factors are peptide molecules, they are substantially larger than the early releasing hormones (41 and 40 amino acids, respectively). The reason for this is as yet not known. With the widespread availability of synthetic releasing hormones, enormous advances have been made in our understanding of pituitary physiology, biochemistry, and function at an ultrastructural level. Many advances have relied upon the development of specific radioimmunoassays and radioreceptor assays for the peptides. These have demonstrated the widespread and characteristic specific distribution of releasing factors throughout the central nervous system and, more recently, in other tissues—the gastrointestinal tract, pancreas, and gonads, to name a few. Apart from the above-mentioned peptides, many other small peptides, originally isolated from the gastrointestinal tract, have been found in the hypothalamus and brain, e.g., cholecystokinin (CCK) and neurotensin (NT), although their physiological function in these sites is currently not fully elucidated.

As previously indicated, these developments have relied largely upon reproducible and sensitive detection systems for analysis of many samples for specific peptide content. The major technique has been radioimmunoassay (RIA) or, more recently, radioreceptor assay (RRA), a more "biological" type of *in vitro* detection system. For both RIA and RRA the key to success lies in the preparation of a stable, biologically active radiolabel. The purpose of this chapter, therefore, is not to describe the detail of RIA or RRA systems for the multitude of these peptides, but rather to discuss the various methods for preparation and assessment of the radiolabeled materials. GnRH will be used as an example, but many of the principles and the methodology described have been applied, with minor modifications, to the preparation of other radiolabeled neuropeptides.

[2] A. V. Schally, A. Arimura, and A. J. Kastin, *Science* **179**, 341 (1973).
[3] P. Brazeau, W. Vale, R. Burgus, N. Ling, M. Butcher, J. Vivier, R. Guillemin, *Science* **179**, 77 (1973).
[4] W. Vale, J. Spiess, C. Rivier, and J. Rivier, *Science* **213**, 1394 (1981).
[5] J. Rivier, J. Spiess, M. Thorner, and W. Vale, *Nature (London)* **300**, 276 (1982).

Methods for Radiolabeling of Neuropeptides

Theoretical Considerations

The major consideration in choosing from the many available radiolabeling procedures is the use for which the label is required. For RIA the three-dimensional integrity of the label is not so critical as for RRA or bioassay, where preservation of the active conformation of the molecule is critical. This is particularly important with small peptide molecules, where the incorporation of a large iodine atom may easily perturb the tertiary structure of the molecule. Labeling by the addition of iodine-125 (^{125}I) requires the presence of reactive hydroxyl groups on an amino acid. Thus, the presence of a tyrosine or histidine residue in the peptide is essential. Oxidation reactions preferentially iodinate tyrosine residues prior to histidine, which is more difficult to label with ^{125}I. For some neuropeptides e.g., somatostatin (SST), the native molecule contains neither tyrosine nor histidine. In such circumstances it has been possible to synthesize analogs containing tyrosine, which retain full biological activity and can be easily iodinated, e.g. [Tyr^1]SST and [Tyr^1]CRF. The advantage of an iodinated radiolabel is the high specific radioactivity of commercially available carrier-free ^{125}I. Additionally, the γ emission of ^{125}I makes detection easy and avoids the time-consuming addition of scintillants required for β-emitters, such as ^{3}H. Some neuropeptides, notably TRH, which contains a histidine residue, can be iodinated, but in the process it loses bioactivity, although ^{125}I-labeled TRH is still satisfactory for radioimmunoassay. To produce a suitable biologically active TRH radioligand synthesis with ^{3}H-containing amino acids is essential, and all RRAs for TRH employ [^{3}H]TRH, or a tritiated higher-affinity analog, ^{3}H-labeled 3-Me-His-TRH.[6] Another example of a neuropeptide, which when iodinated is perfectly satisfactory for RIA but is devoid of bioactivity, is the posterior pituitary hormone arginine vasopressin (AVP).

In general terms, it is desirable to radiolabel a neuropeptide so that it may be used in an RRA or biological assay, since this will also be suitable for RIA, though as previously indicated the converse is not always true. Thus, the methods to be described will reflect the former requirements.

Radiolabeling of Gonadotropin-Releasing Hormone (GnRH)

Iodination of GnRH. Since GnRH contains both a tyrosine and a histidine residue, oxidative iodination presents no theoretical problems. The original method described by Greenwood *et al.*[7] uses large quantities

---
[6] R. L. Taylor and D. R. Burt, *Neuroendocrinology* **32**, 310 (1981).
[7] F. C. Greenwood, W. M. Hunter, and J. S. Glover, *Biochem. J.* **89**, 114 (1963).

TABLE I
BIOLOGICAL (LH-RELEASING) ACTIVITY OF IODINATED GnRH

Preparation	Potency[a] (%)	Assay system	Iodination method	References
[Iodo-Tyr^5]GnRH	100	Hemipituitary (HP) *in vitro*	Glu Ox/LPO[b]	Miyachi *et al.*[14]
Iodo-GnRH (His^2/Tyr^5 ND[c])	13	*In vivo*	H$_2$O$_2$/LPO and CT[d]	Arimura *et al.*[17]
Iodo-GnRH (His^2/Tyr^5 ND)	100	Pituitary cell (PC) *in vitro*	Glu Ox/LPO	Marshall and Odell[9]
[Iodo-Tyr^5]GnRH	25	PC *in vitro*	Glu Ox/LPO	Wagner *et al.*[18]
[Iodo-His^2]GnRH	13	HP *in vitro*	Condensation, synthesis	Terada *et al.*[19]
[Iodo-Tyr^5]GnRH	4	HP *in vitro*	Condensation, synthesis	Terada *et al.*[19]
[Diiodo-Tyr^5]GnRH	0	HP *in vitro*	Condensation, synthesis	Terada *et al.*[19]
[Iodo-Tyr^5]GnRH	4	HP *in vitro*	Direct iodination	Terada *et al.*[19]

[a] Unlabeled GnRH = 100%.
[b] Glucose oxidase (Glu Ox)/lactoperoxidase (LPO)/β-D-glucose.
[c] ND, site of ^{125}I atom not determined, i.e., on His^2 or Tyr^5.
[d] CT, chloramine-T.

of chloramine-T (CT; 100 μg) as the oxidizing agent and is unsuitable for iodination of small peptides since di- or triiodinated derivatives form a large proportion of the reaction products. Nevertheless, Lefkowitz *et al.*[8] indicated that stoichiometric concentrations of CT (250 ng) and ^{125}I produced ^{125}I-labeled ACTH suitable for receptor binding studies, and was therefore presumably bioactive. It has been shown that diiodinated GnRH is devoid of biological activity, while the monoiodotyrosyl product is variably described as having full activity, or much less bioactivity, than unlabeled GnRH (Table I).

Thus, a more "gentle" and carefully regulated oxidation reaction, in which excess oxidant is rapidly removed, is desirable when attempting to produce biologically active iodinated oligopeptides suitable for RRA. A well-developed system employs glucose oxidase to generate H$_2$O$_2$ from β-D-glucose, and lactoperoxidase to remove excess H$_2$O$_2$. This "enzymic" iodination technique produces bioactive monoiodo-GnRH suitable for receptor binding studies (Table I). It is evident from Table I that there is

[8] R. J. Lefkowitz, J. Roth, W. Price, and I. Pastan, *Proc. Natl. Acad. Sci. U.S.A.* **65**, 745 (1970).

considerable variation in the LH-releasing potencies of "enzyme"-iodinated GnRH preparations. These may, in part, reflect the degree of "contamination" of the labeled material by unlabeled peptide owing to inadequate separation of the products during the purification procedure (see below). In early receptor studies we routinely employed the "enzymic" method of Marshall and Odell[9] for preparation of ^{125}I-labeled GnRH. However, we have also iodinated GnRH with stoichiometric (250 ng) concentrations of CT, and there did not appear to be any substantial difference in receptor binding activity when compared with "enzyme"-iodinated GnRH (R. N. Clayton, MD Thesis, University of Birmingham, U.K.). As will be discussed later, further refinements to the GnRH iodination protocols were not pursued, since ^{125}I-labeled GnRH has been replaced as the ligand in receptor binding studies by iodinated nondegradable agonist analogs (see below).

For the "enzymic" iodination, individual enzyme preparations can be diluted immediately prior to the iodination. Alternatively, a fixed, stable, combination of glucose oxidase/lactoperoxidase, covalently bound to hydrophilic spheres (Enzymobeads, Bio-Rad Laboratories) is more convenent to handle.

For radioimmunoassay purposes the conformational integrity of GnRH is not so critical, so it can be iodinated using either CT, "enzyme" methods, or liquid H_2O_2 plus lactoperoxidase. Using the latter (protocol B in Table II), 85–90% of added ^{125}I-labeled GnRH is bound by an excess of anti-GnRH serum. Similar binding to excess antiserum is observed if CT iodination is performed.[10–13]

Separation of Products of Iodination

The products of GnRH iodination include mono- and diiodo derivatives, in addition to unreacted ^{125}I and unlabeled GnRH. It is important to separate these, especially the latter, since contamination by unlabeled GnRH decreases the specific radioactivity of the label.

Early methodology employed gel filtration on Sephadex G-25 columns.[10,11,14] However, the specific activity of ^{125}I-labeled GnRH was

[9] J. C. Marshall and W. D. Odell, *Proc. Soc. Exp. Biol. Med.* **149**, 351 (1975).
[10] T. M. Nett, A. M. Akbar, G. D. Niswender, M. T. Hedlund, and W. F. White, *J. Clin. Endocrinol. Metab.* **36**, 880 (1973).
[11] S. L. Jeffcoate, D. T. Holland, H. M. Fraser, and A. Gunn, *Immunochemistry* **11**, 75 (1974).
[12] M. L. Aubert, M. M. Grumbach, and S. L. Kaplan, *J. Clin. Endocrinol. Metab.* **44**, 1130 (1977).
[13] J. Sandow and W. Konig, *J. Endocrinol.* **81**, 175 (1979).
[14] Y. Miyachi, A. Chrambach, R. Mecklenburg, and M. B. Lipsett, *Endocrinology* **92**, 1725 (1973).

TABLE II
Protocols for Preparation and Purification of ^{125}I-Labeled GnRH for Radioimmunoassay

1. To 5 μg of GnRH in 10 μl of 0.1 M phosphate buffer (pH 7.5), add 10 μl of 0.1 M phosphate buffer and 1 mCi ^{125}I-Na (1–2 μl of IMS 300 Amersham)

Protocol A (Glu Ox/LPO)	Protocol B (H$_2$O$_2$/LPO)
+ Lactoperoxidase (Sigma Chemicals) 50 ng in 10 μl of 0.1 M sodium acetate, pH 5.6	+ Lactoperoxidase 10 μl of 100 μg/ml (Sigma liquid LPO diluted in 0.1 M sodium acetate, pH 5.6)
+ 5 μl of glucose oxidase (Miles Laboratories)	+ 10 μl of 0.003% H$_2$O$_2$ (diluted from 0.3% immediately prior to use)
+ 25 μl of 0.1% β-D-glucose	
Mix for 90 sec	Mix for 2 min

   Reagents are added, in order as above, to a 0.5-ml glass tube at room temperature (~22°)
2. Quench reactants by addition of 0.4 ml of 0.1 M phosphate buffer. Transfer to 200 mg of Dowex 1-X10 (200–400 mesh) (Sigma) in 0.3 ml of 0.1 M phosphate buffer.
3. Mix for 30 sec, centrifuge at 1000 rpm for 2 min.
4. Carefully aspirate supernatant; count supernatant, Dowex, and iodination vial to give indication of ^{125}I incorporation into peptide.
5. Apply supernatant to column of 20 × 0.9 cm CM-52 cellulose (Whatman) equilibrated with 0.002 M NH$_4$ acetate (pH 4.5) and pretreated with 2 ml of 2% BSA in 0.002 M NH$_4$ acetate.
6. Elute with 20 ml of 0.002 M NH$_4$ acetate, then change eluent to 0.06 M NH$_4$ acetate (pH 4.5). Collect 300 ml in 5-ml fractions.
7. Store peak fractions at 4°.

---

rather low (<500 μCi/μg), probably owing to incomplete separation of ^{125}I-labeled GnRH from unlabeled peptide. To effect complete separation of labeled from unlabeled peptide, Marshall and Odell[9] applied the iodination mixture to a cation-exchange column of CM-cellulose and eluted initially with 0.002 M NH$_4$ (pH 4.5) for 20 ml and then with 0.06 M NH$_4$ acetate (pH 4.5) for a further 250–300 ml. Iodinated GnRH elutes in a broad peak between 170 and 220 ml (Fig. 1) and has a specific activity of about 1350 μCi/μg, which closely approaches the theoretical specific activity of monoiodo-GnRH (1600 μCi/μg).

Anion-exchange chromatography of the iodination mixture on QAE-Sephadex G-25-120 eluted with 0.1 M ammonium carbonate buffer, pH 9.2, is also effective at separating unlabeled from mono- and diiodo-GnRH.[15]

As a preliminary to the chromatographic purification of ^{125}I-labeled GnRH, unreacted ^{125}I may be removed by adding the reaction mixture to 200 mg of Dowex 1-X10 (200–400 mesh) dissolved in 0.3 ml of 0.1 M

---

[15] T. M. Nett and T. E. Adams, *Endocrinology* **101**, 1135 (1977).

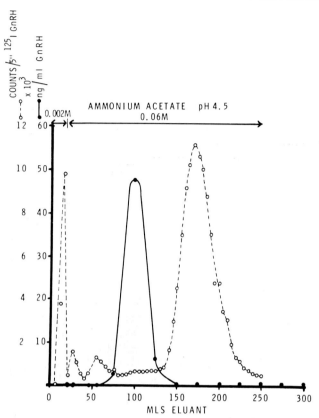

FIG. 1. Elution profiles of ^{125}I-labeled (GnRH) (○---○) and unlabeled GnRH (●——●) from a 20 × 0.9 cm column. Unlabeled GnRH was measured by double-antibody radioimmunoassay.[10] There is good separation of labeled from unlabeled peptide; also the specific activity of ^{125}I-labeled GnRH ≃ 1300 μCi/μg.

phosphate buffer. After mixing and centrifugation, to precipitate the Dowex, the supernatant is applied to the appropriate column. By prior γ-counting of the supernatant, an indication of the efficiency of incorporation of ^{125}I into GnRH is obtained.

Receptor Binding Activity of ^{125}I-Labeled GnRH. As indicated in Table I, LH releasing activity of ^{125}I-labeled GnRH preparations is variable, although that from "enzymic" iodinations is approximately equivalent to that of the unlabeled peptide. On this basis ^{125}I-labeled GnRH should have excellent receptor binding properties. Many initial studies (see review[16]) indicated that ^{125}I-labeled GnRH, however prepared, was an unsatisfac-

[16] R. N. Clayton and K. J. Catt, *Endocrine Rev.* **2**, 186 (1981).

tory ligand for GnRH receptor measurement. The problems were as follows: (1) The majority (75%) of specific ^{125}I-labeled GnRH binding to pituitary membrane preparations was to low-affinity sites ($K_d \simeq 10^{-6}$ M); only 25% was to sites of high affinity ($K_d \simeq 3 \times 10^{-9}$ M); (2) Significant tracer ligand degradation (30–40%) occurred even during 40-min incubations at 4°; (3) These problems precluded accurate measurement of changes in GnRH receptor numbers or affinity in different physiological conditions. While these data may account for some of the discrepancy between the apparent full LH-releasing potency of ^{125}I-labeled GnRH and its poor receptor-binding properties, other, as yet undetermined, factors may play a role.

The availability of "enzyme-resistant" superactive agonist analogs of the native decapeptide, substituted with D-amino acids in position 6 and with the Fujino modification at the carboxy terminus, heralded the advent of a respectable GnRH radioreceptor assay.

Radiolabeling of Superactive Agonist Analogs of GnRH (GnRH-A)

The first report of bioactive mono-[^{125}I]-GnRH agonist analogs, prepared by oxidation with H_2O_2 and lactoperoxidase, indicated that the radioactive GnRH analogs were specifically bound by anterior pituitary tissue and released LH *in vivo*.[20] This study further showed that the prolonged retention of ^{125}I-labeled GnRH-A in the pituitary, when compared to ^{125}I-labeled GnRH decapeptide, was associated with sustained increases in serum gonadotropin values. The iodinated analogs used by Reeves *et al.*[20] were purified by anion-exchange chromatography on QAE-Sephadex G-25 and were contaminated with 11% of unlabeled peptide, which probably accounted for the prolonged LH release. Although the specific radioactivity of the iodinated analogs was not measured, it was likely to be low.

Consequently, we applied the principles described above for the iodination of GnRH and the separation of products to the iodination and purification of GnRH-A, and have used the "enzymic" method for iodination.[21] The initial studies employed [DSer(tBu)^6]-des-Gly^{10}-GnRH-N-

---

[17] A. Arimura, H. Sato, T. Kumasaka, R. S. Worobec, L. Debeljuk, J. Dunn, and A. V. Schally, *Endocrinology* **93**, 1092 (1973).

[18] T. O. F. Wagner, T. E. Adams, and T. M. Nett, *Biol. Reprod.* **20**, 140 (1979).

[19] S. Terada, S. H. Nakagawa, D. C. Young, A. Lipkowski, and G. Flouret, *Biochemistry* **19**, 2572 (1980).

[20] J. J. Reeves, G. K. Tarnavsky, S. R. Becker, D. Coy, and A. V. Schally, *Endocrinology* **101**, 540 (1977).

[21] R. N. Clayton, R. A. Shakespear, J. A. Duncan, and J. C. Marshall with appendix by P. J. Munson and D. Rodbard, *Endocrinology* **105**, 1369 (1979).

TABLE III
PROTOCOL FOR PREPARATION AND PURIFICATION OF ^{125}I-LABELED GnRH-A FOR
RADIORECEPTOR ASSAY

1. To 5 μg of GnRH-A in 10 μl of 0.1 M phosphate buffer (pH 7.5), add the following reagents, in order, in a 0.5-ml glass tube at room temperature: 10 μl of 0.1 M phosphate buffer, 1.5 mCi ^{125}I-Na (2–3 μl of IMS 300, Amersham Radiochemicals), and 5 μg of chloramine-T in 10 μl 0.1 M phosphate buffer (diluted just prior to addition).
2. Mix reactants for 5 min.
3. Quench reactants by addition of 0.4 ml of 0.1 M phosphate buffer. Transfer to 200 mg of Dowex 1-X10 (200–400 mesh) in 0.3 ml of 0.1 M phosphate buffer.
4. Mix for 30 sec, centrifuge, and carefully aspirate supernatant. Count supernatant, Dowex, and iodination vial.
5. Apply supernatant to a 55 × 1 cm column of Sephadex G-25 (fine) (Pharmacia) and elute with 0.1 N acetic acid containing 0.25% bovine serum albumin. Collect 1.5-ml fractions for 125 ml at room temperature.

---

ethylamide[21]; [DAla^6]-des-Gly^{10}-GnRH-N-ethylamide[21]; or [DTrp^6]-des-Gly^{10}-GnRH-N-ethylamide analogs.[22] The original iodination protocol for [DSer(tBu)^6]-des-Gly^{10}-GnRH-N-ethylamide and [DAla^6]-des-Gly^{10}-GnRH-N-ethylamide was identical to that for GnRH, as shown in A of Table II. The only difference was in the chromatographic separation, where it was found that a shorter (0.9 × 10 cm) column of CMC-52 was preferable. By use of the shorter column a more compact peak of ^{125}I-labeled-GnRH-A was obtained. This protocol produces ^{125}I-labeled GnRH-A with specific activity of around 1000 μCi/μg,[21] and about 50% can be bound by excess pituitary receptors.

We have, however, found that iodination with CT and chromatographic separation on Sephadex G-25 (fine) produces ^{125}I-labeled GnRH-A of identical quality. Our currently preferred protocol for iodination and purification of ^{125}I-labeled GnRH-A is given in Table III.

By performing a rapid γ count of the supernatant from the Dowex, the Dowex itself, and the reaction vessel, an indication of the efficiency of the iodination is obtained. Thus, after a 5-min reaction with CT, the proportion of counts in the supernatant is 54% ± 3.8 (SD n = 13 iodinations). Thus, we would suspect a problem with the iodination if incorporation was <45%, and would reiodinate rather than apply this material to the Sephadex G-25 column. Experience has indicated that iodination failures are largely due to an outdated aliquot of GnRH-A. It is wasteful of material to weigh out a fresh aliquot of GnRH-A for every iodination. We make

---

[22] B. S. Conne, M. L. Aubert, and P. C. Sizonenki, *Biochem. Biophys. Res. Commun.* **90**, 1249 (1979).

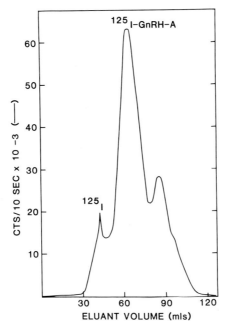

FIG. 2. Typical profile of ^{125}I-labeled [DSer(tBu)^6]-des-Gly^{10}-GnRH-N-ethylamide from a 55 × 1-cm Sephadex G-25 column eluted with 0.1 N acetic acid–0.25% bovine serum albumin at a flow rate of 0.5 ml/min. Between 40 and 50% of label from peak I binds to excess pituitary receptors.

a small quantity (about 0.5 ml) of a solution of 500 μg of GnRH-A per milliliter in 0.1 M phosphate buffer and store as 50-μl aliquots at −20°. A fresh aliquot is thawed for each iodination and then discarded. Such aliquots remain suitable for iodination for up to 6 months. Advantages of the CT iodination are listed below.

1. Avoidance of batch-to-batch variation in enzymes.
2. Simplicity and low cost.
3. Highly reproducible elution profile from Sephadex G-25 column, which can be reused many times.
4. A concentrated peak of ^{125}I-labeled GnRH-A, which leads to greater stability of stored tracer.
5. Constant specific activity and maximum bindability.

A typical elution profile from the Sephadex G-25 column is shown in Fig. 2. ^{125}I-labeled GnRH-A elutes in a rather broad peak, after ^{125}I. We routinely use the six fractions from this peak for RIA and RRA. Material

TABLE IV
METHODS USED FOR IODINATION OF GnRH AGONIST ANALOGS[a]

Authors[b]	Analog (concentration)	Iodination protocol	^{125}I (mCi)	Purification system	MXB (%)	SA (μCi/μg)
Conne et al.[22]	[dTrp[6]]-des-Gly[10]-GnRH-Eth (5 μg)	Glu Ox/LPO (Enzymobeads)	1	G-25	50–60	NG
Clayton et al.[21]	[dSer(tBu)[6]]-des-Gly[10]-GnRH-Eth (5 μg)	CT (250 ng)	1	CMC (6 × 0.9 cm)	50–60	850–1300
	[dAla[6]]-des-Gly[10]-GnRH-Eth (5 μg)	Glu Ox/LPO				
Sandow and Konig[13]	[dSer(tBu)[6]]-des-Gly[10]-GnRH-Eth (5 μg)	CT (5–10 μg)	1–2	G-25	ND	220
Reeves et al.[23]	[dSer(tBu)[6]]-des-Gly[10]-GnRH-Eth (2.5 μg)	CT (600 ng)	0.5	G-25	30–40	300–380
Savoy-Moore et al.[24]	[dAla[6]]-des-Gly[10]-GnRH-Eth (5 μg)	Glu Ox/LPO	1	CMC	50–60	900
Marian and Conn[25]	[dSer(tBu)[6]]-des-Gly[10]-GnRH-Eth (5 μg)	CT (250 ng)	1	CMC	30–50	850–1250
White-Smith and Ojeda[26]	[dAla[6]]-des-Gly[10]-GnRH-Eth (5 μg)	CT (10 μg)	1	CMC	NG	800
Adams and Spies[27]	[dAla[6]]-des-Gly[10]-GnRH-Eth (10 μg)	Glu Ox/LPO	1	QAE-Sephadex G-25 anion exchange	45–60	1330
Meidan and Koch[28]	[dSer(tBu)[6]]-des-Gly[10]-GnRH-Eth	LPO	1	G-25	NG	1000
Perrin et al.[29]	[dAla[6],Nα,MeLeu[7]]-des-Gly[10]-GnRH-Eth (1 μg)	CT (800 ng)	1	CMC → HPLC	NG	NG
Conn et al.[30]	[dLys[6]]GnRH (5 μg)	CT (250 ng)	1.0	CMC 200 mM NH$_4$Ac (pH 4.6)	NG	800
Clayton et al. (1982)	[dSer(tBu)[6]]-des-Gly[10]-GnRH-Eth (5 μg)	CT (5 μg)	1.5	G-25	40–50	900–1100

[a] MXB, maximum fraction of tracer bindable; SA, specific activity; Glu Ox/LPO, glucose oxidase/lactoperoxidase; CT, chloramine-T; ND, not determined; NG, not given; CMC, (carboxymethyl)cellulose; G-25, Sephadex G-25; HPLC, high-pressure liquid chromatography.
[b] Superscript numbers refer to references.

from the minor second peak also binds to pituitary receptors, but with 70% of the activity of the major peak material.

From Table IV it is evident that many protocols for GnRH agonist analog iodination and purification are currently employed.[13,21–30] There is little to choose in terms of specific activity and maximum bindability obtained with the various methods described.

It should be noted that when purifying GnRH analogs containing basic amino acids (e.g., [DLys⁶]GnRH) by CM-cellulose cation-exchange chromatography, the eluent should be 200 mM ammonium acetate (pH 4.6).[30] Failure to use the higher molarity ammonium acetate results in all the peptide being retained on the column. No such problem would be anticipated with purification of ^{125}I-labeled [DLys⁶]GnRH by gel filtration on Sephadex G-25.

For those laboratories with the facilities, purification of the iodination mixture by high-pressure liquid chromotography (HPLC), as described by Perrin et al.[31], produces an unequivocal monoiodinated GnRH-A peak, clearly separated from unlabeled GnRH-A. However, this refinement is not essential for routine preparation of RRA quality ^{125}I-labeled GnRH-A.

Assessment of ^{125}I-Labeled GnRH-A Quality

Specific Activity

The term specific activity refers to the amount of radioactivity incorporated into each molecule of iodinated product and can be expressed as curies per mole or microcuries per microgram. For receptor binding assays a high specific activity label is desirable so that many counts per minute can be added with a minute mass of the ligand. The high specific activity of such tracers can be, and often is, reduced by the addition of a known amount of unlabeled hormone, in a characterstic receptor competitive binding–inhibition curve as shown in Fig. 3A. The specific activity must be accurately known, for measurement of the total ligand concentra-

---

[23] J. J. Reeves, C. Seguin, F.-A. Lefebvre, P. A. Kelly, and F. Labrie, *Proc. Natl. Acad. Sci. U.S.A.* **77**, 5567 (1980).
[24] R. Savoy-Moore, N. B. Schwartz, J. A. Duncan, and J. C. Marshall, *Science* **209**, 942 (1980).
[25] J. Marian and P. M. Conn, *Life Sci.* **27**, 87 (1980).
[26] S. White Smith and S. R. Ojeda, *Endocrinology* **108**, 347 (1981).
[27] T. E. Adams and H. G. Spies, *Endocrinology* **108**, 2245 (1981).
[28] R. Meidan and Y. Koch, *Life Sci.* **28**, 1961 (1981).
[29] M. H. Perrin, Y. Haas, J. E. Rivier, and W. W. Vale, *Mol. Pharmacol.* **23** (in press).
[30] P. M. Conn, R. G. Smith, and D. C. Rogers, *J. Biol. Chem.* **256**, 1098 (1981).
[31] M. H. Perrin, J. E. Rivier, and W. W. Vale, *Endocrinology* **106**, 1289 (1980).

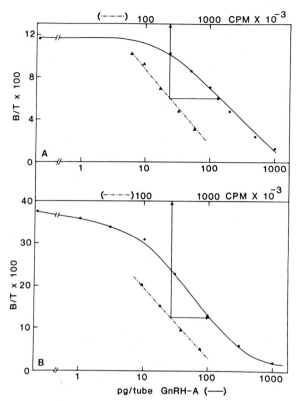

FIG. 3. Measurement of specific radioactivity of ^{125}I-labeled GnRH-A by radioreceptor (A) and radioimmunoassay (B). In either system a small amount (approximately 20,000 cpm of ^{125}I-labeled GnRH-A is incubated with the receptor or antibody, in the presence of increasing counts per minute of tracer (▲---▲), or unlabeled GnRH-A (●——●). B = specifically bound counts per minute (cpm); T = total cpm added. In both RRA (A) and RIA (B), displacement by tracer parallels that by unlabeled hormone. The counts per minute that produce the same B/T as a known amount of unlabeled hormone is indicated by the arrows (→). Different ^{125}I-labeled GnRH-A preparations were used for A and B. The RRA was performed with a 10,800 g rat pituitary membrane fraction as described by Clayton and Catt.[16] The RIA used antiserum R104 at an initial dilution of 1/20,000 and was performed according to Nett et al.[10]

tion, when calculating equilibrium association constants (K_a) by the method of Scatchard or when estimation of total receptor numbers by single-point saturation analysis.[16] Specific activity is determined by displacement of ^{125}I-labeled GnRH-A bound to either antiserum or receptor preparations by increasing amounts (cpm) of labeled peptide. The displacement of tracer binding produced by increasing counts per minute of added tracer is compared with that produced by known amounts of unlabeled hormone. From Fig. 3A a B/T of 6% is produced by 240,000 cpm and 145 pg of unlabeled GnRH-A. In this instance, the counting efficiency of the gamma spectrometer was 55%: ∴ 240,000 cpm ≡ 145 pg of GnRH-A, or 436,363 dpm ≡ 145 pg of GnRH-A. Now 1 μCi = 2.2 × 10^6 dpm; ∴ 436,363/(2.2 × 10^6) μCi ≡ 145 pg of GnRH-A. Thus, 0.198 μC ≡ 145 pg of GnRH-A; ∴ 1 μg of GnRH-A ≡ (1 × 10^6)/145 × 0.198 μCi ≡ 1366. Thus the specific activity of this preparation was 1366 μCi/μg.

Specific radioactivity can be equally well assessed by radioimmunoassay, provided an appropriate antiserum is available. The principle is identical to that of the RRA determination; RIA is a more convenient and less expensive means of assessment, and the one we routinely employ. An example of "self-displacement" in an RIA is shown in Fig. 3B. Specific activity of ^{125}I-labeled [DSer(tBu)[6]]-des-Gly[10]-GnRH-NEA measured by RIA, from 7 chloramine-T iodinations was 947 ± 106 (SD) μCi/μg (range 780–1087). The equilibrium association constant (K_a) for the interaction of ^{125}I-labeled GnRH-A preparations with pituitary GnRH receptors ranges from 3 to 6 × 10^9 M^{-1}.

Maximum Bindability

For calculation of receptor concentration, especially by single-point saturation analysis, it is essential to make a correction for the maximum fraction of the tracer preparation that is able to bind to receptors. For all radioiodinated ligands, not all the counts per minute are able to bind, presumably because some of the labeled hormone has been significantly "altered" or "damaged" in the process of iodination/separation. The method for assessment of the maximum bindable fraction of labeled GnRH-A is to incubate, to equilibrium, a low concentration of labeled ligand with a high concentration of the receptor preparation. As illustrated in Fig. 4 there is a curvilinear relationship between receptor concentration and proportion of hormone bound until a maximum value is reached. For most ^{125}I-labeled GnRH-A preparations, this is in the range of 40–60% (Table IV). Routinely, using CT iodination and Sephadex G-25 separation we achieve a bindable fraction of 40–50%.

The ^{125}I-labeled GnRH-A prepared as described here is not only eminently suitable for radioreceptor assays with tissue homogenates, but also excellent for binding studies with cultured pituitary cells.

Storage of ^{125}I-Labeled GnRH-A/^{125}I-Labeled GnRH

All iodinated GnRH peptides should be stored at acid pH (4.5 approximately) at 4°. An aliquot is removed and brought to pH 7.4 by addition of

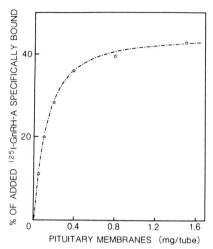

FIG. 4. Assessment of maximum bindable fraction of ^{125}I-labeled GnRH-A. A low concentration (15,000 cpm ≡ 1.5 × 10^{-11} M) of ^{125}I-labeled GnRH-A is incubated with increasing amount of pituitary plasma membrane protein (enriched bovine plasma membranes prepared according to Clayton et al.[21] A plateau of binding is reached when 42% of added tracer can be bound by excess receptor.

a small quantity (a few microliters) of 10 N NaOH immediately prior to use in the RRA or RIA. ^{125}I-labeled GnRH-A retains full receptor binding activity for up to 3–4 weeks when stored in this manner. It is not necessary to repurify the tracer on Sephadex G-25 prior to use, and we have not investigated the nature of the tracer after 4 weeks of storage at 4°. We would, therefore, recommend preparation of fresh tracer each month. Perrin et al.[31] have reported that HPLC-purified ^{125}I-labeled [DAla^6,N^α-MeLeu^7]-des-Gly^{10}-GnRH-N-ethylamide is stable for up to 8 weeks of storage. We do not routinely freeze our tracer, but freezing and thawing once does not appreciably reduce its ability to bind to pituitary receptors (unpublished observation).

Radiolabeling of GnRH Antagonist Analogs

For catecholamine (β- and α-adrenergic and dopamine), receptors considerable differences exist between the binding of radiolabeled agonists and antagonists. Hence, different subtypes of receptor have been defined, β_1 and β_2, according to their relative affinity for agonists or antagonists, respectively. A similar subclassification of dopamine receptors into D_1 and D_2 sites is also made on the basis of differential agonist–antagonist binding and activation. Such a heterogeneity of receptors has not been described for larger peptide hormones, e.g., insulin, gonadotropins, epidermal growth factor, lactogen receptors, although receptors may progress from a high affinity state to a low one (e.g., insulin). With the availability of a wide range of GnRH antagonist peptides, it is of considerable theoretical interest to ascertain whether receptors for a small peptide (intermediate in size between catecholamines, opiates, and larger peptide hormones) can also be subdivided. Hence comparison of labeled agonist and labeled antagonist binding to the pituitary is of importance. Meidan and Koch[28] reported that the affinity of ^{125}I-labeled [DpGlu^1,DPhe^2,DTrp^{3,6}]GnRH for receptors on cultured pituitary cells was one-third less than that of labeled agonist. Similarly, when calculated from competition curves for ^{125}I-labeled GnRH agonist binding, the affinity of unlabeled antagonist was the same as for labeled antagonist. The ^{125}I-labeled GnRH-antagonist analog labeled the same number of receptors as the ^{125}I-labeled GnRH agonist analog. Differential effects of cations and receptor pretreatment with sulfhydryl-blocking agents on the binding of ^{125}I-labeled GnRH antagonist and ^{125}I-labeled GnRH agonist analogs have been reported,[32] although the author concluded that the labeled peptides bound differently to the same receptor. Furthermore, using iodinated GnRH antagonist analog to analyze GnRH receptor numbers in pituitary cells obtained from female rats at different stages of the estrous cycle, Meidan and Koch[33] reported qualitatively identical changes as when ^{125}I-labeled GnRH agonist was employed as the radioligand. These data provide no evidence of separate receptors for GnRH antagonist analogs in the pituitary.

In the aforementioned studies [DpGlu^1,DPhe^2,DTrp^{3,6}]GnRH was iodinated using the glucose oxidase/lactoperoxidase method (as previously described for GnRH). Purification was effected by gel filtration on a 4 × 35 cm Sephadex G-25 column, eluted with 0.01 M acetic acid. Specific activities between 1000 and 1500 μCi/μg are quoted, though the maximum bindability is not given.

Perrin et al.[29] have radioiodinated the potent GnRH antagonist [acetyl-Δ^3-Pro^1,para-F-DPhe^2,DTrp^3,DLys^6]GnRH (1 μg) with 800 ng of chloramine-T, followed by HPLC purification. This tracer has the same receptor affinity as an iodinated agonist analog and labels the same number of high-affinity sites. Cations and guanyl nucleotides had the same effects on both antagonist and agonist analog binding. These data support that of earlier groups that there appears to be only one form of the GnRH receptor in the pituitary.

---

[32] E. Hazum, *Mol. Cell. Endocrinol.* **23**, 275 (1981).
[33] R. Meidan and Y. Koch, *FEBS Lett.* **132**, 114 (1981).

Tritiated GnRH

Grant et al.[34] first reported the use of [^{3}H-Pro^9]GnRH as ligand for receptor studies with cultured pituitary cells. However, as with ^{125}I-labeled GnRH, two classes of binding site were identified, and most of the [^{3}H-Pro^9]GnRH binding was to lower-affinity sites. [^{3}H-Pro^9]GnRH used for these studies was synthesized by solid-phase methodology. More recently, Perrin et al.[31] have reevaluated [^{3}H-Pro^9]GnRH and [^{3}H-pGlu^1]GnRH as a radioligand using pituitary membrane preparations. [Dehydro-DLPro^9]GnRH, synthesized in an automated synthesizer, was tritiated, then purified on CM-cellulose chromatography; this material was subjected to reverse-phase HPLC to obtain pure [^{3}H-Pro^9]GnRH. Specific activity was about 20 Ci/mmol and retained full biological potency. Such material was stable for 8–14 weeks. Even this sophisticated preparative procedure produces a ligand that binds to high-capacity, low-affinity sites and is therefore unsuitable for receptor quantitation. A range of agonist and antagonist GnRH analogs gave binding affinity constants in the same relative order as their biological potencies in this assay. Although much improved on the original RRA of Grant et al.,[34] the current preparations of tritiated GnRH ligands do not result in as precise or robust receptor assays as when ^{125}I-labeled GnRH agonist analogs are employed. Furthermore, the complicated and expensive methods for preparation of tritiated ligands are not available in most laboratories.

Conclusions

A typical oligoneuropeptide has been taken as an example for discussion of the methods available and problems encountered preparing a suitable radioligand for biological studies (i.e., radioreceptor assay using either tissue homogenates or intact cells). As with many other receptors, an analog of the native hormone has proved to be the most effective ligand for both in vivo and in vitro studies. Details for the iodination, purification, and assessment of tracer quality of GnRH-related peptides have been described. The principles for each step are the same for other neuropeptides, although the details, especially those of purification, will clearly differ according to the chemical characteristics of the neuropeptide concerned.

Acknowledgments

I am grateful to Miss L. C. Bailey for technical assistance and to Mr. L. S. Young and Mrs. S. Naik for permission to use their data. This work was supported, in part, by Medical Research Council grant. R. N. C. is a Medical Research Council Senior Clinical Research Fellow.

[34] G. Grant, W. Vale, and J. Rivier, Biochem. Biophys. Res. Commun. **50**, 771 (1973).

[3] Ligand Dimerization: A Technique for Assessing Receptor–Receptor Interactions

By P. Michael Conn

Evidence is available which suggests that many polypeptide hormone receptors activate their effector systems after receptor microaggregation.[1-3] This process is distinct from large-scale patching and capping.[4-6] Such microaggregation (i.e., the association of small numbers of receptors, perhaps dimerization alone) can be provoked by divalent, but not by monovalent, receptor antibodies (i.e., papain or pepsin fragments[7-9]). Because antibodies to receptors cannot always be conveniently obtained, it is advantageous to identify chemical means to construct divalent ligands or otherwise devise methods to prepare molecules that bind and cross-link receptors.

Synthesis of Divalent Ligands: General Considerations

The simplest synthesis of dimeric ligands can be effected with specific bifunctional cross-linking compounds that introduce a short "bridge" (8–15 Å) between the ligand molecules. The monomeric ligand should have exactly one reactive site at a location that is not requisite for biological activity. For the bifunctional cross-linking agent, a large number are available from Pierce Chemical Company (Box 117, Rockford, IL 61105) and elsewhere. These may be heterobifunctional (different reactive groups) or homobifunctional (same reactive groups). In addition, various bridge lengths are available. Some compounds have the additional advantage of being cleavable, although in some cases the conditions required

---

[1] A. C. King and P. Cuatrecasas, *N. Engl. J. Med.* **305**, 77 (1981).
[2] J. Foreman, *Trends Pharmacol. Sci.* December issue, p. 460 (1980).
[3] P. M. Conn, *in* "Biochemical Actions of the Hormones" (G. Litwack, ed.), Vol. 11. Academic Press, New York, 1983. In press.
[4] P. M. Conn, *in* "Cellular Regulation of Secretion" (P. M. Conn, ed.), p. 460. Academic Press, New York, 1982.
[5] P. M. Conn and E. Hazum, *Endocrinology* **109**, 2040 (1981).
[6] E. Hazum, P. Cuatrecasas, J. Marian, and P. M. Conn, *Proc. Natl. Acad. Sci. U.S.A.* **77**, 6695 (1980).
[7] S. Jacobs, K.-J. Chang, and P. Cuatrecasas, *Science* **200**, 1283 (1978).
[8] C. R. Kahn, K. L. Baird, D. B. Jarrett, and J. S. Flier, *Proc. Natl. Acad. Sci. U.S.A.* **75**, 4209 (1978).
[9] Y. Schechter, L. Hernaez, J. Schlessinger, and P. Cuatrecasas, *Nature (London)* **278**, 835 (1979).

BIFUNCTIONAL CROSS-LINKING AGENTS[a]

Drug	Short name	Molecular weight	Function ?	Cleavable ?	Photoreactive ?	Approximate span length (if known)
Bis[2-(succinimidooxy-carbonyloxy)ethyl] sulfone	BSOCOES	436.4	Homo	At basic pH, 37°	No	13 Å
Carbonylbis(L-methionine-p-nitrophenyl ester)	—	566.6	Homo	With CNBr	No	—
Dimethyl adipimidate · 2 HCl	DMA	245.1	Homo	No	No	8.6 Å
Dimethyl pimelimidate · 2 HCl	DMP	259.2	Homo	No	No	9.2 Å
Dimethyl suberimidate · 2 HCl	DMS	273.2	Homo	No	No	11 Å
Dimethyl-3,3'-dithio-bis(propionimidate) · 2HCl	DTBP	281.2	Homo	Yes	No	11.9 Å
Disuccinimidyl (N,N'-diacetylhomocystine)	—	546.6	Homo	With thiols	No	—
Disuccinimidyl suberate	DSS	368.5	Homo	No	No	—
Disuccinimidyl tartrate	DST	344.2	Homo	With periodate	No	—
Dithiobis(succinimidyl propionate)	Lomant's reagent I	404.4	Homo	With DTT	No	—
Erythritol biscarbonate	—	174.1	Homo	With periodate	No	—
Ethyl 4-azido-1,4-dithiobutyrImidate · HCl	EADB	332.9	Hetero	With thiols	Yes	—
Ethylene glycol bis(succinimidyl succinate)	EGS	456.4	Homo	With hydroxylamine at pH 8.5, 37°	No	12 Å

Name	Abbrev.	MW	Type	Cleavable	Iodinatable	
m-Maleimidobenzoyl N-hydroxysuccinimide ester	MBS	314.2	Hetero	No	No	—
Methyl 4-azidobenzimidate · HCl	—	212.6	Hetero	No	Yes	—
N-(4-Azidophenylthio)phthalimide	APTP	296.3	Hetero	With DTT	Yes	—
N-Hydroxysuccinimidyl 4-azidobenzoate	—	260.2	Hetero	No	Yes	—
N-Hydroxysuccinimidyl 4-azidosalicylic acid	NHS-ASA	272.2	Hetero	No	Yes (also can be iodinated)	—
N-Succinimidyl (4-azidophenyl)-1,3'-dithiopropionate	SADP	352.4	Hetero	With BME	Yes	—
N-Succinimidyl 3-(2-pyridyldithio)propionate	SPDP	312.4	Hetero	With BME	No	—
N-Succinimidyl 6-(4'-azido-2'-nitrophenylamino)hexanoate	Lomant's reagent II	390.4	Hetero	No	Yes	—
N-5-Azido-2-nitrobenzoyloxysuccinimide	ANB-NOS	305.2	Hetero	No	Yes	—
p-Azidophenyl glyoxal	—	193.2	Hetero	No	Yes	—
p-Azidophenylacyl bromide	—	240.1	Hetero	No	Yes	—
p-Nitrophenyl 2-diazo-3,3,3-trifluoropropionate	—	276.2	Hetero	No	Yes	—
Succinimidyl 4-(N-malemidomethyl)cyclohexane-1-carboxylate	SMCC	334.3	Hetero	No	No	—

(*continued*)

BIFUNCTIONAL CROSS-LINKING AGENTS (*continued*)

Drug	Short name	Molecular weight	Function ?	Cleavable ?	Photoreactive ?	Approximate span length (if known)
Succinimidyl 4-(*p*-maleimidophenyl)butyrate	SMPB	356.3	Hetero	No	No	—
1,5-Difluoro-2,4-dinitrobenzene	—	204.1	Homo	No	No	—
2-Iminothiolane · HCl	Traut's reagent	137.6	Homo	With DTT	No	—
4,4'-Difluoro-3,3'-dinitrodiphenyl sulfone	—	344.3	Homo	No	No	—
4,4'-Diisothiocyano-2,2'-disulfonic acid stilbene	DIDS	552.5	Homo	No	No	—
4,4'-Dithiobisphenylazide	—	300.4	Homo	With BME	Yes	—
4-Fluoro-3-nitrophenyl azide	FNPA	182.1	Hetero	No	Yes	—

[a] For each compound, the chemical and trivial names or abbreviations, as well as molecular weights, are given. Some compounds have identical reactive functional groups (homo), while others are nonidentical (hetero). Some compounds are cleavable under conditions that are not cytotoxic. Such cleavable compounds are useful in that the cross-linking can be reversed. On occasion, the reactive group may be photoreactive, and some compounds can be made radioactive with iodine-125. When known, the bridge length is given. The reaction specificity with peptides varies and may include phenolic or amino groups, O—H, N—H, C—H bonds; some are reported to be amino acid specific for lysine, serine, or tyrosine.

for cleavage are sufficiently harsh to result in damage to biological ligands. For others, cleavage can be effected by thiol reducing agents, and their use is therefore precluded in biological assays that are hampered by such agents. For convenience a summary of agents is provided in the table.

Model Cross-Linking: Production of ([DLys^6]GnRH)-EGS-([DLys^6]GnRH)

We have found that ethylene glycol bis(succinimidyl succinate) (EGS) is a particularly useful cross-linker, since it attacks amino groups, reacts at physiological pH values (although the rate of reaction can be greatly increased at slightly basic pH levels), and is stable under conditions that are consistent with conducting the biological assays. In addition, dry EGS is stable in excess of 6 months at $-20°$ in the dark.

Procedure[10]

[DLys^6]GnRH, 50 nmol in 50 μl of 10 mM sodium phosphate (pH 8.5) containing approximately 0.05–0.15 μCi of ^{125}I-labeled [iodoTyr^5-DLys^6]-GnRH in 200 mM sodium phosphate (pH 8.7) is combined with 25 nmol of ethylene glycol bis(succinimidyl succinate) (EGS, Pierce Chemical Company, Rockford, IL) in 15 μl of dimethyl sulfoxide (DMSO, redistilled from Fisher). The EGS is dissolved in DMSO immediately prior to use. It has 2 mol of reactive groups per mole; the reactive groups are present in equimolar concentration with the lysyl ε-amino in [DLys^6]-GnRH. After 6–8 hr at 18° the reaction mixture is applied to a Sephadex G-25 column (superfine, 10 × 260 mm) equilibrated in 10 mM phosphate, pH 8.5, 40% ethanol; 20-drop fractions were collected. We routinely identify positions of eluting substances by radioactivity and confirm them by radioimmunoassay (using antibody No. 9113 at a final titer of 1 : 7000, radioiodinated [D-Lys^6]GnRH, and authentic [DLys^6]GnRH standard). A typical elution pattern of the material from the Sephadex column is shown in Fig. 1. In addition, the dimer pool used in subsequent studies is shown in this figure.

It is sometimes convenient to increase the distance between ligands in excess of that obtainable by available cross-linkers (8–15 Å). One means for doing so is to use an antibody directed against the ligand in the presence of excess ligand dimer. Under these conditions, two ligand dimer molecules (if sufficiently short) will saturate both antibody binding sites and leave a free ligand unbound to the antibody. This ligand will now confer a new binding specificity on the antibody, enabling it to bind to the

[10] P. M. Conn, D. C. Rogers, and R. McNeil, *Endocrinology* **11**, 335 (1982).

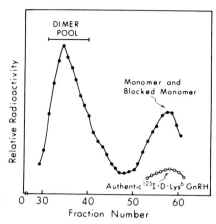

FIG. 1. Elution profile (Sephadex G-25, superfine) of ^{125}I-labeled [Tyr^5-dLys^6]GnRH following a 7-hr dimerization reaction (●) as described in the text. Shown for comparison is the elution of authentic ^{125}I-labeled [Tyr^5-dLys^6]GnRH (○). The pool of dimer is indicated. Reprinted by permission of the Endocrine Society from Conn et al.[10]

ligand binding site of the receptor, since one ligand extends from each binding site. The other ligand, bound by the antibody, is not accessible to the receptor. The accessible ligands are separated by 120–150 Å depending on the bridge length on the cross-linker selected. This procedure is summarized in Fig. 2 for a GnRH analog and detailed below.

Practical Use: Approaches Providing Evidence That a Receptor Microaggregate Is the Functional Unit of Biological Action

Potentiation of the Action of an Agonist[10]

In this approach[10] the ethylene glycol bis(succinimidyl succinate) (EGS) dimer of a GnRH agonist, [dLys^6]GnRH, was prepared. A dose-response curve for this compound was obtained (Fig. 3). The ED$_{10}$ dose [i.e., the amount of dimer sufficient to stimulate LH release equal to 10% of that released in response to 10^{-7} M GnRH (a saturating dose)] was then incubated with varying concentrations of antibody that binds to [dLys^6]GnRH. The shape of this antibody dose-response curve (Fig. 4) is distinctive. Optimum potentiation was seen at antibody titer 1:10^5–1:10^6. At higher and lower antibody concentrations, responsiveness drops. At low dilutions (high concentration) the antibody is in excess and probably cannot cross-link receptors. At high dilutions (low concentrations) there is insufficient antibody to bind and cross-link substantial num-

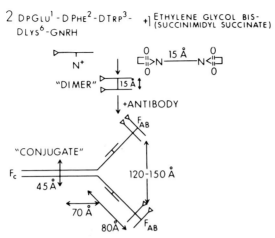

FIG. 2. Preparation of GnRH–antagonist [DpGlu^1-DPhe^2-DTrp^3-DLys^6]GnRH) conjugate. Two molecules of the antagonist are dimerized through the ε-amino (Lys^6). The resulting dimer is mixed with cross-reactive antibody. The "conjugate" molecule has an antagonist dimer at each Fab arm. One end of each dimer is bound by the antibody and therefore is unavailable to the receptor. The remaining antagonists can be bound by receptors and are held rigidly about 120–150 Å apart.

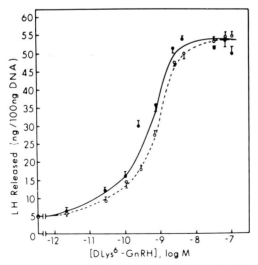

FIG. 3. Luteinizing hormone (LH) release in response to the indicated concentration of [DLys^6]GnRH (●) or ([DLys^6]GnRH)$_2$-EGS dimer (○). Cells were prepared and incubated as described in the text. Released LH was measured by radioimmunoassay. Reprinted by permission of the Endocrine Society from Conn et al.[10]

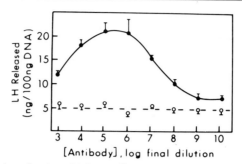

FIG. 4. Potentiation of action of 1.5×10^{-11} M dimer by indicated addition of antisera as described under Model Cross-linking (●). Also shown is lack of activity of antisera alone (○). Maximal LH release was 58 ± 6 ng of LH/100 ng of DNA in response to 10^{-7} M [DLys⁶]GnRH. Unstimulated release was 5 ± 0.5 ng/100 ng of DNA. Reprinted by permission of the Endocrine Society from Conn et al.[10]

bers of receptors. Schematic drawings of likely molecular species involved in the mechanism are shown in Figs. 5A and 5B.

For comparison, 10^{-7} M [DLys⁶]GnRH-stimulated LH release was 58 ± 6 ng per 100 ng of DNA in this study. At maximum the antibody cross-linked dimer thus stimulated response to about the ED_{40} level. Monovalent antibody (i.e., reduced pepsin fragments) was ineffective in stimulating release in the presence of 1.5×10^{-11} M dimer (mean release = 5.1 ± 0.5 ng per 100 ng of DNA at all titers). This control indicates that the bivalent nature of the compound is the essential feature needed in order to cross-link receptors. Addition of 5 μM Pimozide or 2 mM EGTA diminished LH release to near basal, at all titers, in the presence of 1.5×10^{-11} M dimer. Accordingly, the release mechanism involved is Ca^{2+}-dependent and involves calmodulin, much as in release in response to GnRH itself.[4] Addition of antibody alone (open symbol) did not stimulate LH release over basal levels (5.0 ± 0.5 ng/100 ng DNA).

Conversion of Antagonist to an Agonist[11]

Using the methods described above, it has been possible[11] to dimerize a pure GnRH antagonist, [DpGLu¹-DPhe²-DTrp³-DLys⁶]GnRH, then form the antibody conjugate. The product is an agonist. This observation suggests that the agonistic action can be provoked by bringing the receptors within a critical distance. An antagonist probably does not allow this interaction to occur. This behavior was modeled[12] on the assumption that

[11] P. M. Conn, D. C. Rogers, J. M. Stewart, J. Neidel, and T. Sheffield, *Nature (London)* **296,** 653 (1982).
[12] J. J. Blum and P. M. Conn, *Proc. Natl. Acad. Sci. U.S.A.* **79,** 7307 (1982).

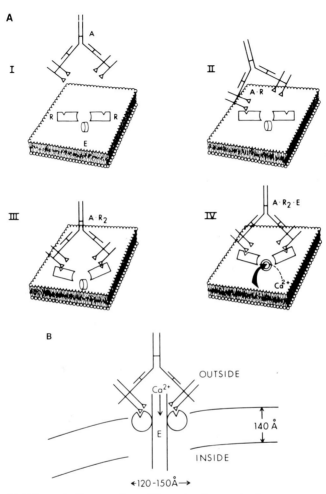

FIG. 5. (A) Schematic diagram of possible molecular species in a system containing a divalent antibody conjugate, A (AB[(GnRH-Ant)$_2$-EGS]$_2$), a receptor, R, and an effector, E, which is here represented as a channel for the passage of Ca^{2+} ions, but other cell systems could be a nucleotide cyclase or other generator of a second messenger. Reprinted, by permission, from Blum and Conn.[12] (B) Molecular species in Fig. 5A viewed from a position perpendicular to the plane of the plasma membrane.

the antibody conjugate, A, can react with a receptor, R, to form a complex, A · R, which in turn can react with another receptor to form A · R$_2$. This dimer then can react with a quiescent effector, E (e.g., a closed Ca^{2+} ion channel), to form A · R$_2$ · E, which contains activated effector and leads to cellular responses (gonadotropin release). The equilibrium equations governing the behavior of this model (Fig. 5) were derived, solved, and found to yield a good fit to the experimental data.[12]

Conclusions

Insofar as receptor–receptor interactions appear to be consequential in systems as diverse as gonadotropes, adipocytes, hepatocytes, thyroid cells, platelets, and mast cells, among others, it is likely that preparation of ligand dimers and antibodies that bind receptors will continue to provide useful tools for probing many systems.

[4] Photoaffinity Labeling in Neuroendocrine Tissues

By ELI HAZUM

Photoaffinity labeling of biological systems provide an important experimental tool for the detection and isolation of membrane components comprising binding sites for hormones and for the elucidation of interactions involved in biological processes at the molecular level (see reviews[1–4]).

The most widely used photolabel probes are the aryl azides. The advantage of photogenerated nitrene intermediates is that they are highly reactive and are capable of insertion into all protein amino acid side chains. In addition, they can also react with other components of the membrane (e.g., lipids and sugars). However, the price paid for high reactivity of these reagents is the multiplicity of potential products and low levels of labeling. Another attractive feature of the aryl azides stems from the ability to manipulate the reaction mixture in the dark with no reaction occurring at all. Thus, it is possible to measure the equilibrium binding constant and biological activity of an unphotolyzed photoaffinity-derivatized hormone. In addition, it is possible to load the binding sites, or let the photoreactive hormone enter inside components of a cell, prior to photolysis.

The following sections summarize the synthesis, stability, storage, and some properties of a few photoreactive hormones. Finally, the appli-

[1] K. Peters and F. M. Richards, *Annu. Rev. Biochem.* **46,** 523 (1977).
[2] H. Bayley and J. R. Knowles, this series, Vol. 46, p. 69.
[3] V. Chowdhry and F. H. Westheimer, *Annu. Rev. Biochem.* **48,** 293 (1979).
[4] J. V. Staros, *Trends Biochem. Sci.* **5,** 320 (1980).

cations of photoaffinity-derivatized hormones are described to highlight the potential uses of this method. Methodology related to gonadotropin releasing hormone (GnRH), insulin, and adrenocorticotropic hormone (ACTH) receptors will be presented.

Synthetic Approaches

General Considerations

Before design or synthesis of a photoreactive hormone, it is important to determine the effects of irradiation at 250, 350, and >400 nm on the binding capacity and the biological activity of a given system. This limits the molecules that might be used as reagents. Radiolabeled molecules should be synthesized so that the labeled atom is located as close to the photolabile group as possible, and it should not be separated from it by any linkages that might be cleaved during subsequent manipulations. It is essential to have the radiolabeled and nonradiolabeled photoreactive hormone of high purity, because small quantities of unlabeled hormone may prevent labeling, whereas other impurities may increase nonspecific labeling. Finally, it is important to design the synthesis of photoreactive hormone in such a way that it retains high binding affinity and biological activity. If the dissociation rate of the ligand is high and the lifetime of photogenerated species is long, the ligand may become attached to parts of the receptor remote from the ligand binding site. The situation then becomes analogous to normal affinity labeling and has been termed pseudo-photoaffinity labeling.[5]

Synthesis of Photoaffinity Derivatives of GnRH, Insulin, and ACTH

Synthesis of [DLys^6-N$^\varepsilon$-azidobenzoyl]GnRH.[6] pGlu-His-Trp-Ser-Tyr-DLys-(N$^\varepsilon$-azidobenzoyl)-Leu-Arg-Pro-Gly-NH$_2$ is prepared by reaction of [DLys^6]GnRH (0.6 mg: 0.45 mmol) (Peninsula) with 2 molar equivalents of (4-azidobenzoyl)-N-hydroxysuccinimide (Pierce) in methanol in the presence of 1.2 equivalents of triethylamine. After standing at 24° for 3 hr, protected from light, the product is precipitated by the addition of ether and washed three times with ethyl acetate in order to remove unreacted (4-azidobenzoyl)-N-hydroxysuccinimide. Thin-layer chromatography (silica gel) reveals an R_f value of 0.75 in BuOH : AcOH : H$_2$O : ethyl acetate (1 : 1 : 1 : 1, by volume).

[5] A. E. Ruoho, H. Kiefer, P. E. Roeder, and S. J. Singer, *Proc. Natl. Acad. Sci. U.S.A.* **70,** 2567 (1973).
[6] E. Hazum, *FEBS Lett.* **128,** 111 (1981).

Preparation of Photoaffinity Derivatives of Insulin

Synthesis of 4-Azido-2-nitrophenylinsulin.[7] Insulin hydrochloride (20 mg; 0.0033 mmol) is dissolved in dimethylformamide (3 ml). To the stirred solution is added triethylamine (12 mg; 0.12 mmol), followed by the addition of 4-fluoro-3-nitrophenylazide (182 mg; 1.0 mmol) in dimethylformamide (0.5 ml), and the reaction is stirred for 5 hr at room temperature, protected from light. The reaction is terminated by the addition of ether (50 ml), and the resultant precipitate is washed with ether (3 × 10 ml), isolated by centrifugation, and dried under vacuum over P_2O_5. The product can be used without further purification.[8] Alternatively, the several derivatives (glycine^{A-1}-azidonitrophenylinsulin, phenylalanine^{B-1}-azidonitrophenylinsulin, and lysine^{B-29}-azidonitrophenylinsulin) can be isolated in pure form by DEAE-Sephadex chromatography.[7]

Synthesis of Azidobenzoylinsulin.[9] Insulin (48 mg, 0.008 mmol) is suspended in 2 ml of dimethylformamide and 40 μl of triethylamine. To this suspension is added (4-azidobenzoyl)-N-hydroxysuccinimide (32 mg, 0.123 mmol), and the reaction mixture is stirred in the dark for 60 min. Azidobenzoylinsulin is precipitated by the sequential addition of 3 ml of ice-cold absolute ethanol and 5 ml of ice-cold diethyl ether. After 60 min in an ice bath, the precipitate is recovered by centrifugation and lyophilized. The products, lysine^{B-29}-azidobenzoylinsulin and glycine^{A-1}, lysine^{B-29}-diazidobenzoylinsulin are separated and purified on an SP-Sephadex column.

Synthesis of [(2-Nitro-5-azidophenylsulfenyl)Trp^9]adrenocorticotropic Hormone (2,5-NAPS-ACTH).[10] ACTH (8.45 mg, 1.86 mmol) is dissolved in 0.2 ml of distilled water. Methionine (excess) and glacial acetic acid (2.8 ml) are added. 2,5-NAPS-Cl (20 mg, 0.087 mmol) is added, and the mixture is left in the dark at room temperature for 4 hr. Distilled water (30 ml) is then added, and the excess reagent is extracted with ethyl acetate. The organic phase is extracted with 0.1% acetic acid, and the combined aqueous phase is lyophilized and subjected to gel filtration on a Sephadex G-25 column. The modified ACTH is further purified by partition chromatography on Sephadex G-50.

Characterization of Derivatized Hormones

Spectral Properties. The aryl azide-derivatized hormones have strong ultraviolet absorptions often with characteristic shoulders on the long

---

[7] D. Levy, *Biochim. Biophys. Acta* **322**, 329 (1973).
[8] S. Jacobs, E. Hazum, Y. Shechter, and P. Cuatrecasas, *Proc. Natl. Acad. Sci. U.S.A.* **76**, 4918 (1979).
[9] C. C. Yip, C. W. T. Yeung, and M. L. Moule, *Biochemistry* **19**, 70 (1980).
[10] K. Muramoto and J. Ramachandran, *Biochemistry* **19**, 3280 (1980).

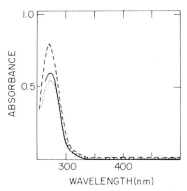

FIG. 1. The ultraviolet absorption spectra of 10^{-4} M GnRH (——) and of 5×10^{-5} M [DLys^6-N^ε-azidobenzoyl]GnRH before (---) and after (·····) 5 min of photolysis in 0.1 M acetic acid. Data from Hazum.[6]

wavelength side of the peak, e.g., [DLys^6-N^ε-azidobenzoyl]GnRH, λ_{max} 270 (ε 16,000); glycine^{A-1}-azidonitrophenylinsulin, λ_{max} 265 (ε 31,000), and λ_{max} 460 (ε 4800); 2,5-NAPS-ACTH, λ_{max} 283 (ε 17,600) and λ_{max} 325 (ε 9400). These bands usually disappear or are considerably diminished upon photolysis (Fig. 1). The absorption characteristics can be used to estimate the number of photoreactive groups that are attached to the hormone.

Stability, Reactivity, Iodination, and Storage of Photoreactive Hormones. Most aryl azide derivatives are stable in the solid state or in solution. Nevertheless, excessive heating and strong oxidizing or reducing conditions should be avoided during experimental procedures. The nitroaryl azides are generally less stable to heat and light and should be recrystallized with care to avoid thermal decomposition. Low-molecular weight aryl azides are explosive when heated and should be handled carefully, especially in the solid state.

Substitution of the aryl azide ring with electron-withdrawing groups, e.g., a nitro group, increases the reactivity and electrophilicity of the intermediate nitrene. The increased electrophilicity can be a disadvantage owing to the more facile reaction with the solvent water as compared to hormone–receptor complexes. The presence of a nitro group adjacent to the azido group must be avoided because of the formation of benzofuroxan derivatives.[11] Another reason why the nitro-aryl derivatives are usually preferred for photolabeling purposes is that they shift the λ_{max} to longer wavelengths, thus avoiding photochemical damage to the biological system.

[11] A. S. Bailey and J. R. Case, *Tetrahedron* **3**, 113 (1958).

Photoreactive derivative of hormones can be iodinated by the chloramine-T method[12] or by the lactoperoxidase method.[13] Alternatively, the hormone can be first iodinated, then allowed to react with the photoaffinity reagents.[14]

Aryl azide derivatives are best stored in the solid state under an atmosphere of nitrogen at $-20°$ in the dark. Radioactive azides stored as solids often darken, but at $-20°$ their radiochemical purity remains stable for at least a month. Nevertheless, they are best stored in solution to minimize radiolytic decomposition.

Photolysis

General Considerations

The duration of photolysis is very important. Prolonged photolysis can result in destruction or alteration of the binding sites, and it increases nonspecific labeling. The photolysis time can be estimated from the half-life of the photoaffinity derivative alone by monitoring the appropriate wavelength. In general, the duration of photolysis of hormone–receptor complexes is longer than for hormone alone (in the range of 1–10 min) because of the screening effects in biological preparations. In some systems, e.g., ACTH,[15] it is better to do repeated photolysis for short periods rather than for a single long period of irradiation. In such treatment it is important to remove the noncovalently bound products because they can block receptor sites. Other important factors are the concentration and affinity of the photoreactive hormone. The tighter the binding affinity of the photolabile ligand to the receptor, the more successful labeling will generally be. In those cases where the binding affinity of the photoreactive hormone is low, it is recommended to work near saturation condition. However, when the binding affinity is high, it is possible to saturate receptor sites and then remove excess of unbound hormone. Nonspecific labeling that particularly occurs with low-affinity binding ligands, can be minimized by performing the photolabeling experiments in the presence of scavengers such as *p*-aminobenzoic acid, soluble proteins, dithiothreitol, hydroquinone, *p*-aminophenylalanine, and Tris buffer.

Photolysis of hormone–receptor complexes is conducted after incubating (in the dark under equilibrium conditions) the photoreactive hormone with cells or membrane preparations in the appropriate buffer or

---

[12] W. M. Hunter and F. C. Greenwood, *Nature (London)* **194**, 495 (1962).
[13] J. J. Marchalonis, *Biochem. J.* **113**, 299 (1969).
[14] C. C. Yip, C. W. T. Yeung, and M. L. Moule, *J. Biol. Chem.* **253**, 1743 (1978).
[15] J. Ramachandran, J. Hagman, and K. Muramoto, *J. Biol. Chem.* **256**, 11424 (1981).

medium. Irradiation in the ultraviolet, near-ultraviolet, or visible can be achieved by simple means. Adrenocortical cells incubated with 2,5-NAPS-ACTH are irradiated by using a Blak-Ray UV lamp emitting principal radiation at 366 nm.[15] The cell suspensions are kept stirring throughout photolysis (at 0° or 24°) 10–15 cm from the lamp. Photolysis is generally conducted for 4 min, twice, with a 2-min pause between exposures to irradiation. Azidobenzoylinsulin derivatives incubated with liver or adipocyte membrane preparations are irradiated at 4° for 15 min by using a 250-W General Electric sun lamp at a distance of approximately 15 cm.[9,14] Photolysis of 4-azido-2-nitrophenylinsulin with liver membranes is carried out for 5 min, 20 cm from a Hanovia 450-W mercury arc lamp fitted with a Pyrex filter.[8] In the case of B^2-(4-azido-2-nitrophenylacetyl)-des-Phe^{B1}-insulin incubated with adipocytes, irradiation is carried out for 1 min with a Philips HPK 125-W/L UV lamp equipped with a black glass filter (UVW-55, Hanau Wertheim). The lamp is 8 cm from the surface of the stirred adipocyte suspension.[16] Photolysis of [DLys^6-N^ε-azidobenzoyl]GnRH with pituitary membrane preparations,[17] granulosa cells, and membranes[18] is conducted at 4° for 5 min, 10 cm from a Wild-Universal lamp equipped with a high-pressure 200-W mercury lamp.

Details about filters and further useful information of photolysis equipment may be found in the review by Bayley and Knowles[2] and references cited therein.

Applications of Nonradioactive Photoreactive Hormones

Two main approaches are used for determining whether covalent labeling has occurred. The first is to assess the number of binding sites irreversibly blocked. The second is the persistent activation of a given system. However, consideration must be given to the following points: removal of unreacted hormone, effect of irradiation on the receptor alone, and the possible photooxidation of the receptor sites by the bound ligand.

Photoaffinity Inactivation of GnRH Receptors[6]

Pituitary membrane preparations and iodination of Buserelin ([DSer(tBu)^6]-des-Gly^{10}-GnRH ethylamide) are conducted as previously described.[6]

---

[16] D. Brandenburg, C. Diaconescu, D. Saunders, and P. Thamm, *Nature (London)* **286**, 821 (1980).
[17] E. Hazum, *Endocrinology* **109**, 1281 (1981).
[18] E. Hazum and A. Nimrod, *Proc. Natl. Acad. Sci. U.S.A.* **79**, 1747 (1982).

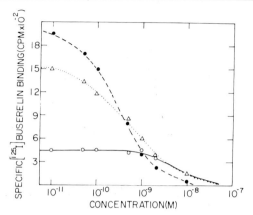

FIG. 2. Competition of binding of ^{125}I-labeled Buserelin by unlabeled Buserelin to control pituitary membranes (●---●), pituitary membranes photolyzed in the presence of 10^{-7} M [DLys^6-N^ε-azidobenzoyl]GnRH (○——○), or 10^{-7} M prephotolyzed [DLys^6-N^ε-azidobenzoyl]GnRH (△···△). Modified from Hazum.[6] See the text for the binding assay.

The membranes are incubated with 10^{-7} M [DLys^6-N^ε-azidobenzoyl]GnRH or 10^{-7} M prephotolyzed [DLys^6-N^ε-azidobenzoyl]GnRH for 90 min at 4° in 1 ml of 10 mM Tris · HCl, pH 7.4, containing 0.1% bovine serum albumin (BSA), protected from light. At the end of the incubation period the mixtures are photolyzed for 6 min at 4° with a mercury lamp at a distance of 15 cm. After irradiation, the membranes are washed extensively to remove noncovalently bound hormone. Competition in binding of ^{125}I-labeled Buserelin by unlabeled Buserelin to control pituitary membranes and membranes preincubated with the photoaffinity derivative is shown in Fig. 2. The binding is carried out by incubating the radioactive Buserelin (40,000 cpm) with 20–30 μg of protein of pituitary membranes and different concentrations of unlabeled Buserelin in a total volume of 0.5 ml of 10 mM Tris–0.1% BSA. After 90 min at 4°, the homogenate is filtered under vacuum through Whatman GF/C filters, presoaked in 0.1% BSA, and washed with 10 ml of ice-cold incubation buffer; the filters are counted in a gamma counter.

Covalent Linking of Photoreactive Insulin to Adipocytes Produces a Prolonged Signal[16]

A fat-cell assay[19] is used to determine the biological effects of B^2-(2-nitro-4-azidophenylacetyl)-des-Phe^{B1}-insulin. Adipocytes are first incubated with the photoreactive insulin derivative (80 ng/ml) to obtain a

---

[19] M. J. Rodbell, *J. Biol. Chem.* **239,** 375 (1964).

TABLE I
EFFECT OF COVALENT BINDING OF PHOTOLABELED
INSULIN ON LIPOGENESIS IN ADIPOCYTES[a]

Addition	Percentage of maximal lipogenesis[b]	
	No irradiation	Irradiation
Control	7	13
Insulin (14 pmol/ml)	10	22
Photoinsulin (14 pmol/ml)	12	80

[a] Preparation of adipocytes and lipogenesis assay are conducted as described by Rodbell,[19] Moody et al.[20] and J. Gliemann, *Horm. Metab. Res.* **6**, 12 (1974).
[b] Percentage values are the ratios of basal lipogenesis to maximal response (10 ng of insulin per milliliter). The incubation time for lipogenesis assay is 2 hr. Modified, by permission, from D. Brandenburg et al.[16]

receptor occupancy of more than 90%. After photolysis by short exposure (1 min) to UV light of wavelengths 300–400 nm, the cells are washed extensively to dissociate noncovalently bound insulin from the receptor. In control experiments, adipocytes are incubated in the dark or under UV light with or without unmodified insulin. As shown in Table I,[19,20] only the covalent binding of the photoreactive insulin generates a prolonged biological response.

Persistent Activation of Steroidogenesis in Adrenocortical Cells by Photoaffinity Labeling[15]

Adrenocortical cells are incubated with 2,5-NAPS-ACTH (50 nM) in medium 199 containing 0.5% BSA. After a 30-min incubation in the dark, the cell suspension is irradiated using a Blak-Ray UV lamp. The cell suspensions are kept stirring throughout the preincubation and photolysis at the same temperature (4° or 24°) 10–15 cm from the lamp. Photolysis is conducted for 4 min, twice, with a 2-min pause between exposures to irradiation. At the end of photolysis, the cells are washed to remove noncovalently bound peptide. The cells are reincubated in medium 199–0.5% BSA at 37° for 1 hr and assayed for corticosterone production. Steroidogenesis is maximal when photolysis is performed at 24° but remains the same as control cells when irradiation is conducted at 4°.

[20] A. J. Moody, M. A. Stan, M. Stan, and J. Gliemann, *Horm. Metab. Res.* **6**, 12 (1974).

Applications of Radiolabeled Photoreactive Ligands

Identification of Receptor Macromolecules

General. Photoaffinity labeling has been used to identify various receptors, e.g., insulin receptors,[8,9,14,21,22] GnRH receptors,[17,18,23] and corticotropin receptors.[24] Labeling can be achieved by incubating the radiolabeled photoreactive hormone with cells or membrane preparations at optimal binding conditions in the dark. After photolysis, the noncovalently bound ligand is removed and the sample is subject to gel electrophoresis. The gels can be sliced and their radioactivity determined or can be dried for autoradiography.

General Controls

1. The labeling site should show saturation kinetics for photolabeling.
2. Specificity of photolabeling. The ability of native hormone and related peptides to inhibit photolabeling of specific bands should correlate with their potency for receptor binding.
3. Prephotolyzed ligand or incubations with photoreactive hormone in the dark without irradiation should not produce specific labeling.
4. Antibodies directed against the binding region of the receptor should inhibit photolabeling.
5. Physiological alterations in receptor content should be accompanied by similar changes in the radioactivity incorporated into specific bands.

Identification of receptor macromolecules can be followed by elucidating their structural features. This can be achieved by digestion of the receptors with proteolytic and glycosidic enzymes, precipitation with antibodies against the receptors, and other manipulations. An excellent example is the studies with insulin receptors.[25–30]

---

[21] M. H. Wisher, M. D. Baron, R. H. Jones, P. H. Sonksen, D. J. Saunders, P. Thamm, and D. Brandenburg, *Biochem. Biophys. Res. Commun.* **92**, 492 (1980).
[22] C. W. T. Yeung, M. L. Moule, and C. C. Yip, *Biochemistry* **19**, 2196 (1980).
[23] E. Hazum and D. Keinan, *Biochem. Biophys. Res. Commun.* **107**, 695 (1982).
[24] J. Ramachandran, K. Muramoto, M. Kenez-Keri, G. Keri, and D. I. Buckley, *Proc. Natl. Acad. Sci. U.S.A.* **77**, 3967 (1980).
[25] S. Jacobs, E. Hazum, and P. Cuatrecasas, *Biochem. Biophys. Res. Commun.* **94**, 1066 (1980).
[26] C. C. Yip, M. L. Moule, and C. W. T. Yeung, *Biochem. Biophys. Res. Commun.* **96**, 1671 (1980).
[27] S. Jacobs, E. Hazum, and P. Cuatrecasas, *J. Biol. Chem.* **255**, 6937 (1980).
[28] S. Jacobs and P. Cuatrecasas, *Endocrine Rev.* **2**, 251 (1981).
[29] M. P. Czech, J. Massague, and P. F. Pilch, *Trends Biochem. Sci.* **6**, 222 (1981).
[30] C. C. Yip, M. L. Moule, and C. W. T. Yeung, *Biochemistry* **21**, 2940 (1982).

Validation of the Method: Binding of Photosensitive Hormone to Antibodies[14]

^{125}I-labeled (azidobenzoyl)insulin (0.8 µCi) is incubated in the absence or the presence of insulin with a 1:32 dilution of guinea pig anti-insulin (bovine) serum for 3 hr at 37° in the dark in 0.2 ml of phosphate-buffered saline, pH 7.0. Serum from nonimmunized guinea pig is used as a control. The reaction mixture is photolyzed for 15 min using a 250-W General Electric sun lamp at a distance of 30 cm. The antibody–antigen complex is precipitated (at 5° overnight in the dark) by the addition of a rabbit anti-guinea pig γ-globulin serum. Approximately 80–90% of the insulin derivative is precipitated. The immunoprecipitates are washed once with 0.2 ml of phosphate-buffered saline, boiled in 4% SDS containing 10 mM dithiothreitol, and analyzed by disc gel electrophoresis in SDS according to the method of Weber and Osborn.[31] The tube gel is cut into 1.5-mm sections for the determination of radioactivity in a gamma counter.

Photoaffinity Labeling of GnRH Receptor of Rat Pituitary Membranes[17]

Pituitary membranes (0.8 g) are incubated with ^{125}I-labeled [DLys^6-N^ε-azidobenzoyl]GnRH (10^6 cpm) in the absence or the presence of various concentrations of Buserelin ([DSer(tBu)^6]-des-Gly^{10}-GnRH ethylamide) in 1.5 ml of 10 mM Tris · HCl, pH 7.4, containing 0.1% BSA at 4° in the dark. After 90 min (equilibrium conditions) the membranes are photolyzed for 7 min at 4° with a mercury lamp at a distance of 10 cm. The membranes are then washed (twice) with the same buffer by centrifugation, and the pellet is boiled in 1% SDS containing 10 mM dithiothreitol. Gel electrophoresis is performed according to the method of Laemmli,[32] modified by using 7.5–15% linear gradient SDS–polyacrylamide slab gels. Autoradiograms are prepared from the dried gels using a DuPont Cronex Lightning-Plus intensifying screen. The specificity of photolabeling of the 60 kilodalton band by ^{125}I-labeled [DLys^6-N^ε-azidobenzoyl]GnRH is shown in Table II.

Morphological Studies: Localization and Processing of Receptors

Localization and processing of receptors in cells with ^{125}I-labeled photoaffinity-derivatized hormones have two major advantages: (1) After photolysis, the hormone is bound covalently to the receptors, thus circumventing the problem of rapid dissociation from its receptor. (2) Other methods, such as fixation of cells or cross-linking reagents, may induce

---

[31] K. Weber and M. Osborn, *J. Biol. Chem.* **244**, 4406 (1969).
[32] U. K. Laemmli, *Nature (London)* **227**, 680 (1970).

TABLE II
SPECIFICITY OF PHOTOLABELING OF THE
60 KILODALTON BAND[a]

Concentration of peptide (M)	Counts per minute incorporated
None	1210 ± 90
Buserelin	
10^{-10}	890 ± 70
10^{-9}	640 ± 60
10^{-8}	250 ± 10
10^{-7}	220 ± 20
TRH, 10^{-7}	1230 ± 100

[a] Pituitary membranes are incubated with ^{125}I-labeled [DLys^6-N^ε-azidobenzoyl]GnRH in the presence of various concentrations of Buserelin. After photolysis, the membranes are subjected to SDS–polyacrylamide gel electrophoresis and the amount of radioactivity incorporated into the 60-kilodalton band is determined ($n = 2$). Data are from Hazum.[17]

nonspecific cross-linking of cell surface proteins, thus disturbing the overall membrane architecture.

Localization of GnRH Receptors in Gonadotropes[33]

Pituitary cells (5×10^6) are incubated with ^{125}I-labeled [DLys^6-N^ε-azidobenzoyl]GnRH (10^6 cpm) in 1.0 ml of medium 199, in the presence or the absence of unlabeled [DLys^6-N^ε-azidobenzoyl]GnRH. After 90 min at 4° or 30 min at 37°, in the dark, the cells are washed and suspended in 1.0 ml of phosphate–buffered saline (PBS) containing 0.1% BSA. The cells are then photolyzed for 5 min at 4°, washed three times with PBS–0.1% BSA, and finally fixed with 3% glutaraldehyde, 1% formaldehyde in 0.1 M cacodylate buffer, pH 7.4. The cells are centrifuged in a 0.4-ml polyethylene tube at 10,000 g for 10 min. The pellets are postfixed for 2 hr at 4° with 1% osmium tetroxide in the cacodylate buffer, dehydrated in ethanol and propylene oxide, and embedded in Epon. For light microscopic autoradiography, the cells are sectioned (1 μm thickness), and processed using Ilford L4 emulsion (exposure time, 30 days). Autoradiographs are photographed through a phase-contrast microscope (Zeiss photomicroscope

[33] E. Hazum, R. Meidan, D. Keinan, E. Okon, Y. Koch, H. R. Lindner, and A. Amsterdam, *Endocrinology* **111**, 2135 (1982).

III). For electron microscopic autoradiography, thin sections of 100 nm are cut with a Porter Blum MT_2 ultramicrotome, and a monolayer of Ilford L4 emulsion is applied. After exposure for 60 days, the autoradiograms are developed in D19 Kodak developer. The sections are stained with uranyl acetate and lead citrate and examined under a Jeol 100B electron microscope. Light microscopic and electron microscopic autoradiograms are shown in Fig. 3. Only gonadotropes (about 4% of total pituitary cells) are labeled.

Localization and Processing of Insulin Receptors in Hepatocytes[34]

Suspensions of isolated hepatocytes (10^6 cells/ml) and ^{125}I-labeled B^2-(2-nitro-4-azidophenylacetyl)-des-Phe^{B1}-insulin are incubated for 2–4 hr at 15°, conditions that permit steady-state binding. At the end of this association step, cells are collected by centrifugation, resuspended in the same volume of insulin-free buffer, and exposed or not to UV light. Hepatocytes are then diluted 20-fold in insulin-free buffer and incubated for 2–4 hr at 37°. Nonspecific binding is determined in simultaneous experiments where unlabeled insulin (5 μM) is added to the incubation at the beginning of the association period. At the end of incubation period, hepatocytes are centrifuged and the cell pellets are then processed for electron microscopic studies or solubilized in SDS sample solution for slab-gel electrophoresis.

Regulation of GnRH Receptors during the Rat Estrous Cycle[23]

Wistar-derived female rats (150–200 g; 60 days old) that exhibited at least two consecutive 4-day estrous cycles, as assayed by daily vaginal smears, are used. Pituitary membranes derived from 4 female rats at various stages of the estrous cycle (1.6–2.0 mg of protein per milliliter) are incubated with ^{125}I-labeled [DLys^6-N^ε-azidobenzoyl]GnRH (10^6 cpm) in the presence or the absence of 10^{-6} M Buserelin ([DSer(tBu)^6]-des-Gly^{10}-GnRH ethylamide) in 1.5 ml of 10 mM Tris buffer containing 0.1% BSA at 4° in the dark. After 90 min the membranes are washed by centrifugation and photolyzed (7 min at 4°) with a mercury lamp at a distance of 10 cm. The membranes are then washed (twice) with the same buffer by centrifugation, and the pellet is boiled in 1% SDS containing 10 mM dithiothreitol. Aliquots are prepared and analyzed in a 7.5% slab gel. After staining with Coomassie blue and destaining, the gels are dried for autoradiography. The results are shown in Table III.

---

[34] M. Fehlmann, J. L. Carpenter, A. L. Cam, P. Thamm, D. Saunders, D. Brandenburg, L. Orci, and P. Freychet, *J. Cell Biol.* **93**, 82 (1982).

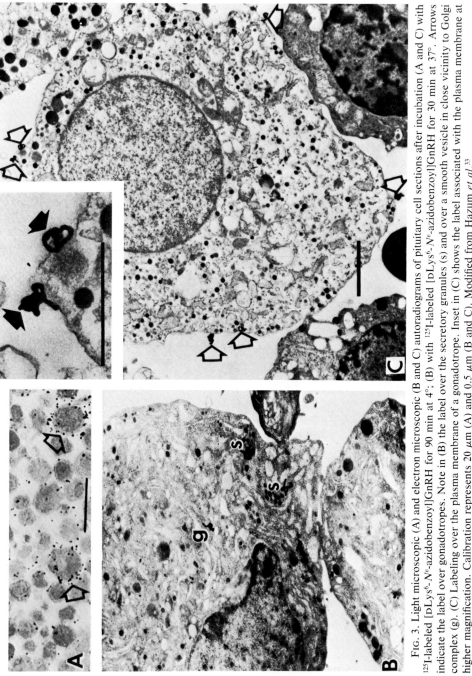

FIG. 3. Light microscopic (A) and electron microscopic (B and C) autoradiograms of pituitary cell sections after incubation (A and C) with ^{125}I-labeled [DLys^6-N^ϵ-azidobenzoyl]GnRH for 90 min at 4°; (B) with ^{125}I-labeled [DLys^6-N^ϵ-azidobenzoyl]GnRH for 30 min at 37°. Arrows indicate the label over gonadotropes. Note in (B) the label over the secretory granules (s) and over a smooth vesicle in close vicinity to Golgi complex (g). (C) Labeling over the plasma membrane of a gonadotrope. Inset in (C) shows the label associated with the plasma membrane at higher magnification. Calibration represents 20 μm (A) and 0.5 μm (B and C). Modified from Hazum et al.[33]

TABLE III
RADIOACTIVITY INCORPORATED INTO THE
60 KILODALTON BAND DURING THE
RAT ESTROUS CYCLE[a]

Animal status	Specific counts per minute incorporated[b]
Metestrus	1220 ± 130[c]
Diestrus	2930 ± 240
Proestrus	2780 ± 170
Estrus	1350 ± 150

[a] Data from Hazum and Keinan.[23]
[b] Pituitary membranes are photolabeled and then subjected to SDS–polyacrylamide gel electrophoresis. The amount of specific radioactivity (cpm in the absence of Buserelin minus cpm in the presence of 10^{-6} M Buserelin) incorporated into the 60 kilodalton band determined ($n = 3$).
[c] The values are corrected with respect to protein concentration.

Comments

Photoaffinity labeling has evolved into a widely used technique for studying complex biological systems. These include photoaffinity inactivation of receptors; persistent activation of biological functions; identification of receptor macromolecules; localization, processing, and regulation of receptors. One of the more exciting areas of future applications will be mapping of binding sites within receptor macromolecules and the organization of receptor–effector complexes within the membrane assembly. Synthesis of reversible photoaffinity probes and improvement in the yields of labeling (generally 3–10%) will lead to a more sophisticated applications of photogenerated reagents in biochemical studies.

Acknowledgments

This work was supported by the Ford Foundation and the Rockefeller Foundation, New York, and by the fund for basic research administered by the Israel Academy of Sciences and Humanities. I am grateful to Mrs. M. Kopelowitz for typing the manuscript.

[5] Microisolation of Neuropeptides

By FRED S. ESCH, NICHOLAS C. LING, and PETER BÖHLEN

The development of highly efficient procedures for the isolation of neuropeptides has provided a major impetus for rapid advances in neuroendocrinology. Improvements have included better extraction procedures that yield a relatively protein-free peptide fraction, the use of affinity chromatography employing antibodies to the peptide or a related peptide of interest, and high-performance liquid chromatography (HPLC) on preparative and analytical levels to achieve final peptide homogeneity. This chapter describes a general methodology that has been applied successfully in our laboratory to isolate neuroendocrine peptides; specific examples will be cited to illustrate the methods used. This approach involves: (a) the efficient extraction of peptides from tissues, (b) pre-HPLC concentration and peptide purification, and (c) preparative and analytical HPLC purification of the peptides. Throughout it is essential that a suitable assay be available to detect and measure the peptide being purified.

Extraction of Peptides from Tissues

The efficient extraction of peptides from fresh or frozen tissues requires the inactivation of endogenous proteases and the removal of large proteins and lipids from the peptide-containing fraction of the extract. Typically this may be accomplished with an acidic extraction medium containing enzyme inhibitors, e.g., a 2 N acetic acid solution containing, per liter, 10 mg of phenylmethylsulfonyl fluoride (a serine proteinase inhibitor), and 10 mg of pepstatin A (a carboxyl proteinase inhibitor).[1] Since most enzymes are inactive in solutions of low pH and high salt concentration, Bennett *et al.*[2] have devised another extraction medium consisting of 5% formic acid, 1% trifluoroacetic acid, and 1% (w/v) NaCl in 1 M hydrochloric acid. This medium has the advantage of precipitating most of the cellular proteins so that an enriched peptide extract is obtained. The tissue is usually homogenized in a volume of extraction medium equivalent to 5–10 times the tissue weight in a glass or Polytron homogenizer (Brinkmann Instruments). The homogenate is centrifuged, and the super-

---

[1] A. J. Barrett, *Fed. Proc., Fed. Am. Soc. Exp. Biol.* **39**, 9 (1980).
[2] H. P. J. Bennett, C. A. Browne, D. Goltzman, and S. Solomon, *in* "Peptides: Structure and Biological Function" (E. Gross and J. Meienhofer, eds.), p. 121. Pierce Chemical Co., Rockford, Illinois, 1979.

natant is decanted. The pellet is rehomogenized in 0.5 volume of the extraction medium and centrifuged. The supernatants are combined, and the lipids are removed by three extractions with equal volumes of petroleum ether or dichloromethane. The trace amount of organic solvent in the aqueous phase is removed under aspirator vacuum. Further treatment of the extract depends upon the nature of the pre-HPLC step to be used; e.g., the extract could be equilibrated in the appropriate buffer for affinity chromatography, lyophilized for gel filtration, or pumped directly onto a reverse-phase column.

Pre-HPLC Purification Steps

Affinity Chromatography. If a suitable antibody is available, immunoaffinity chromatography of the crude extract may greatly purify and concentrate the peptide of interest. The extract is passed through a column containing a covalently coupled antibody–Sepharose 4B conjugate at neutral pH to retain selectively the desired peptide. The nonretarded material is then washed off the column with 10 mM phosphate-buffered saline, pH 7.4, and the retained peptide is desorbed using a low pH medium, such as 25 mM HCl or 1 M acetic acid. The recovered peptide fraction may then be further purified by Sephadex gel filtration if desired. In preparing the immunoaffinity column, the immunoglobulins in the antiserum are usually concentrated by precipitation with a saturated solution of ammonium sulfate equal to one-half the volume of serum.[3] The precipitated immunoglobulin is redissolved in a volume of borate-buffered saline[3] equal to the volume of original serum sample, and the excess ammonium sulfate is removed by dialysis in the borate-buffered saline. The supporting gel matrix for the affinity column is usually cyanogen bromide activated Sepharose 4B (Pharmacia Fine Chemicals). The antibody solution is coupled to the activated Sepharose by gently mixing equal volumes of the enriched IgG solution and the Sepharose at 4° overnight. The unreactive sites in the Sepharose are blocked by further gentle mixing with an excess of ethanolamine at 4° overnight. After washing with phosphate-buffered saline, the column is ready for application of the sample.

Gel Filtration Chromatography. Gel filtration chromatography will selectively remove small molecules, e.g., small peptides, metal ions, and color pigments, as well as larger proteins to yield an enriched peptide fraction containing material in the molecular weight range of the desired

---

[3] J. S. Garvey, N. E. Cremer, and D. H. Suffdorf, *in* "Methods in Immunology," 3rd ed., p. 218. Benjamin, New York, 1977.

peptide. Typically this is done in a 5 × 150 cm glass column (3-liter gel bed volume) packed with Sephadex G-50 or G-75 (Pharmacia Fine Chemicals) employing 3–5 M acetic acid as the developing solvent. The high acetic acid concentration acts as a protein denaturant and suppresses hydrophobic interactions between the peptide and the Sephadex gel. Up to 3 g (corresponding to approximately 200 g of tissue) of crude extract can be applied onto this column in a single pass. Fractions containing the desired peptide may then be further purified by pumping them directly onto a preparative HPLC column. This is preferable to lyophilization and redissolution of the fractions, which sometimes results in large losses ($\geq 50\%$) of the peptide of interest (see below).

Octadecyl Silica (ODS) Chromatography. The hydrophobic alkyl chain of octadecyl-bonded silica (ODS) selectively retains peptides but not small polar molecules (amino acids, inorganic salts), thus making ODS useful for purifying and/or concentrating peptides from large volumes of solvent. The apparatus used in this procedure can be a preparative HPLC system or a simple device as illustrated in Fig. 1. The assem-

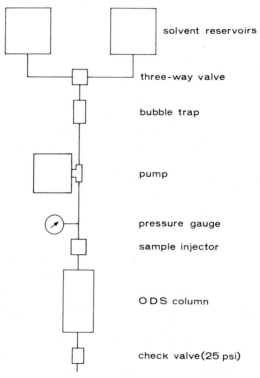

FIG. 1. A schematic diagram of the apparatus employed to concentrate and/or purify crude peptide extracts via octadecyl silica (ODS) chromatography.

bly of this apparatus has been described,[4] and the necessary parts can be obtained from Altex (Berkeley, CA) and various suppliers of Swagelok components. LRP-2 ODS, 37–53 μm particle size, is from Whatman Inc. (Clifton, NJ). The pH of the peptide extract is adjusted to pH 3, and the solution is pumped directly onto the ODS column equilibrated in 0.2 M acetic acid or 0.1% trifluoroacetic acid. The salt is washed off the column with the equilibration solvent, and the retained peptide material is desorbed with the equilibration solvent containing 60% n-propanol. Alternatively, it is possible to purify the peptide of interest from at least some of the contaminating peptide material by employing judicious n-propanol step gradients. This peptide solution may then be further purified by pumping it directly onto a preparative HPLC column after appropriate dilution to reduce the concentration of n-propanol.

Preparative and Analytical HPLC Purification

Reverse-phase high-performance liquid chromatography (RPLC) is the most powerful method available for peptide purification and is relied upon for the ultimate isolation of a peptide after the pre-HLPC purification step. Peptides with different polarity are eluted sequentially as a function of their hydrophobicity with mobile phase gradients of increasing organic solvent concentration. For the proper selection of a chromatographic system (i.e., type of column and mobile phase), it is necessary to take into account (1) the quantity of proteinaceous material to be chromatographed, and (2) the complexity of the peptide mixture or the relative abundance of the peptide of interest. The first RPLC step is usually done on a preparative (50 × 2.5 cm; capacity <1 g of peptide material) or a semipreparative (25 × 0.9 cm; 50–100 mg capacity) column. Subsequent RPLC steps are carried out on semipreparative and/or analytical columns (25 × 0.46 cm; 10 mg capacity) until the peptide is purified to homogeneity. With all but the simplest peptide mixtures it is advantageous to change the solute selectivity of the chromatographic system at each step. This optimizes the separation of peptides that have similar hydrophobic character and are therefore difficult to separate. Selectivity can be modified by changing either the type of column or the mobile phase. Changes in the latter are most effective, as peptide polarity and hence retention on reverse-phase columns is greatly modified by pH as well as the type and concentration of buffer salt present in the mobile phase. The pH controls protonation of peptide carboxyl groups and thus influences the hydrophobic character of a peptide. Buffer anions, on the other hand, may form ion pairs with positively charged peptide groups in a concentration-dependent

---

[4] P. Böhlen, F. Castillo, N. Ling, and R. Guillemin, *Int. J. Pep. Protein Res.* **16**, 306 (1980).

manner, thereby forming complexes whose hydrophobicities are largely governed by the nature of the buffer ion. For example, phosphate buffers will form hydrophilic ion pairs, whereas ion pairs with heptafluorobutyrate are strongly hydrophobic.

These factors have been incorporated into the general purification scheme outlined in Table I. The systems used employ acidic (pH <4) mobile phases (protonation of carboxyl groups increases hydrophobicity), which improve the chromatographic behavior and maximize separation for most peptides. Mobile phases of high buffering capacity (pyridine formate–n-propanol or TEAP–acetonitrile systems) are employed to chromatograph large peptide quantities in the early steps of isolations. This minimizes peptide-induced pH changes within the column and thus prevents peptide elution under uncontrolled conditions. These mobile phases facilitate the elution of hydrophobic peptides because formate and, particularly, phosphate anions form hydrophilic ion pairs with peptides at low pH. Furthermore, the pyridine content of the pyridine formate–n-propanol solvent system contributes greatly to rapid elution of peptides. For later isolation steps, TFA or HFBA in combination with acetonitrile are employed. In contrast to phosphate or formate, trifluoroacetate and heptafluorobutyrate form hydrophobic ion pairs with peptides, thus introducing strong changes of solute selectivity. HFBA, in particular, interacts with positively charged peptide groups in such a way that separation primarily occurs as a function of the number of positively charged moieties in a peptide and thus causes separation behavior approaching that seen in cation-exchange chromatography. Most often TFA or HFBA concentrations of 0.1% (v/v) have been used. At this concentration the mo-

TABLE I
SEPARATION SCHEME FOR THE ISOLATION OF PEPTIDES USING VARIOUS RPLC SYSTEMS

Step	Column	Mobile phase[a]	Detection
1	C_{18}; preparative or semipreparative	Pyridine formate/n-propanol	Fluorescence
2	C_{18}; semipreparative or analytical	Pyridine formate/n-propanol or TEAP/acetonitrile	Fluoresence or UV 210/280
3	C_{18}; analytical	TFA/acetonitrile	Fluorescence or UV 210/280
4	C_{18}; analytical	HFBA/acetonitrile	Fluorescence or UV 210/280

[a] TEAP, triethylammonium phosphate; TFA, trifluoroacetic acid: HFBA, heptafluorobutyric acid.

bile phases are still transparent at 210 nm for peptide detection, and the pH of the mobile phase is within the limits prescribed for use with silica-based reverse phases. It should be pointed out that significantly better resolution is obtained when these mobile phases contain higher concentrations (0.5% v/v) of TFA or HFBA. Their use may therefore be advantageous in difficult separation problems. However, the low pH of these solvent systems can accelerate degradation of the bonded phases produced by some manufacturers and precludes peptide detection at 210 nm. Finally, the mobile-phase systems in Table I are employed because, with the exception of TEAP–acetonitrile, they are volatile, and this permits their removal by lyophilization. Thus, they are completely compatible with immunoassays, bioassays, and all types of structural analyses, which is of considerable importance in isolation projects. The exact compositions and the preparation of the above mobile phases are listed in Table II.[2,5-8]

Additionally, these chromatographic systems employ the relatively hydrophobic C_{18} columns, which appear to be satisfactory for most separations; they are sufficiently hydrophobic to retain most small polar peptides and yet they usually afford good recovery of relatively large hydrophobic peptides. Other types of stationary phases, e.g., C_8 columns, may be used, but usually they offer only minor differences in selectivity. In certain situations, however, a change of stationary phase may be useful. For example, the diphenyl column interacts selectively with aromatic peptides and thus may be employed in the purification of such a peptide. Chromatographic theory predicts that resolution improves with smaller stationary-phase particles and low mobile-phase flow rates. Consequently, columns with 5-μm particles and flow rates of 0.5–1.0 ml/min are now employed most frequently.

Highly sensitive detection of all peptides is possible by determination of their UV light absorption at 200–220 nm with a HPLC UV detector. With the exception of pyridine formate or acetate–n-propanol, the mobile phases listed in Table I are compatible. However, the materials used in the preparation of mobile phases must be free of UV absorbing impurities; otherwise, problems with the chromatographic baseline (e.g., appearance of artifactual peaks, baseline drifts) may occur and seriously limit the usefulness of peptide detection in this wavelength range. For peptides

[5] S. Kimura, R. V. Lewis, L. Gerber, L. Brink, S. Rubenstein, S. Stein, and S. Udenfriend, *Proc. Natl. Acad. Sci. U.S.A.* **76**, 1756 (1979).

[6] M. Rubenstein, S. Stein, L. Gerber, and S. Udenfriend, *Proc. Natl. Acad. Sci. U.S.A.* **74**, 3052 (1977).

[7] J. Rivier, *J. Liq. Chromatogr.* **1**, 343 (1978).

[8] H. P. J. Bennett, C. A. Browne, and S. Soloman, *J. Liq. Chromatogr.* **3**, 1353 (1980).

TABLE II
Preparation and Composition of RPLC Mobile Phases[a]

1 M Pyridine formate–n-propanol, pH 3.0[5]
  Buffer A: Mix 29 ml of pyridine, 60 ml of formic acid, and 911 ml of water. Titration to the exact pH may be necessary.
  Buffer B: Mix 29 ml of pyridine, 60 ml of formic acid, 600 ml of n-propanol, and 311 ml of water.
1 M Pyridine acetate–n-propanol, pH 4.0[6]
  Buffer A: 29 ml of pyridine, 85 ml of acetic acid, and 886 ml of water
  Buffer B: 29 ml of pyridine, 85 ml of acetic acid, 600 ml of n-propanol, and 286 ml of water
0.25 M TEAP–acetonitrile, pH 3.0[7]
  Buffer A: 16 ml of triethylamine, 5.75 ml of phosphoric acid, 978.25 ml of water. Titration to the exact pH may be necessary.
  Buffer B: 16 ml of triethylamine, 5.75 ml of phosphoric acid, 800 ml of acetonitrile, 178.25 ml of water
0.1% (v/v) TFA–acetonitrile[2]
  Buffer A: 1 ml of TFA with water, 999 ml of water
  Buffer B: 1 ml of TFA, 800 ml of acetonitrile, 199 ml of water
0.1% (v/v) HFBA–acetonitrile[8]
  Buffer A: 1 ml of HFBA, 999 ml of water
  Buffer B: 1 ml of HFBA, 800 ml of acetonitrile, 199 ml of water

[a] The necessary reagents were prepared as follows: Reagent grade solvents were distilled over ninhydrin (1 g/liter) once (acetic acid, formic acid, n-propanol, triethylamine) or twice (pyridine). Reagent grade TFA was distilled. Phosphoric acid (reagent grade), UV grade acetonitrile (Burdick & Jackson), and HFBA (Sequenal grade) were used as commercially available. High-purity water was obtained from a Milli-Q water system (Millipore, Bedford, MA). If necessary TFA and HFBA can be further purified (to produce a very stable baseline for UV detection at 210 nm) by passing a 1% (v/v) solution through a bed (2 × 10 cm) of ODS (LRP-2, Whatman) as described in the text under Octadecyl Silica (ODS) Chromatography.

containing aromatic residues, less-sensitive UV detection is also possible at 280 nm with less stringent requirements for the purity of the solvent system. Although most peptides are not inherently fluorescent, their fluorometric detection is possible with the fluorescamine-stream sampling detection system.[9] In this system small aliquots of column effluent are removed automatically via a sampling valve at regular intervals and allowed to react on-line with fluorescamine for highly sensitive detection of peptide fluorophors. The advantages of this detection system are its compatibility with most mobile phases, including those not suitable for UV detection, and the highly sensitive and selective detection of polypeptides. The system is not commercially available but can be easily assembled using Milton-Roy minipumps (Laboratory Data Control, Riviera

[9] P. Böhlen, S. Stein, J. Stone, and S. Udenfriend, *Anal. Biochem.* **67**, 438 (1975).

Beach, FL) and Altex valves and plumbing material. Instructions for assembly of the system have been published[9,10] and are also available in detailed form from the authors.

In microisolation projects good peptide recovery is often essential, and therefore recoveries should be determined. Peptide recoveries from RPLC columns are with few exceptions satisfactory (>90%). If low yields are incurred, it may be necessary to replace the column or switch to another column brand containing the same or a less hydrophobic stationary phase. If recovery problems still persist, the mobile phase should also be changed. Losses are easily incurred if column fractions are lyophilized prior to further chromatography. Therefore, lyophilization should be avoided whenever possible. Fortunately, peptide-containing solutions need not be lyophilized (or concentrated) prior to a RPLC step. Aqueous samples can be applied to a reverse-phase column by simply pumping them onto the column, where peptides will be adsorbed until eluted with a gradient of aqueous-organic mobile phase. For this purpose the HPLC apparatus is modified to include a manually operated three-way valve (Altex) in the inlet line of pump A. In this way the pump can be fed with either the sample solution or HPLC buffer A. Even very large sample volumes, such as gel filtration fractions may be pumped through the column as long as the column capacity for peptides is not exceeded. Column fractions, originating from a reverse-phase column, that contain some organic solvent, can be pumped onto another RPLC column after sufficient dilution (usually 2- to 3-fold) to prevent unwanted migration of the peptide during loading of the column.

We describe below in detail the use of these procedures in three different neuropeptide isolation projects. A brief outline of the protocols employed in each of these projects is shown in Fig. 2. The four-step procedure of extraction, pre-HPLC purification, and RPLC employing 10×250 mm semipreparative and 4.6×250 mm analytical HPLC columns yields a homogeneous peptide preparation suitable for structural characterization by amino acid and sequence analysis.

Isolation of Rat Hypothalamic Somatostatin-28[11]

Extraction

Lyophilized rat hypothalami (96,800; 93 g) were homogenized in 2 liters of 2 M acetic acid containing the protease inhibitors pepstatin and

---
[10] S. Stein and J. Moschera, this series, Vol. 79, p. 7.
[11] P. Böhlen, P. Brazeau, F. Esch, N. Ling, and R. Guillemin, *Regul. Pept.* **2**, 359 (1981).

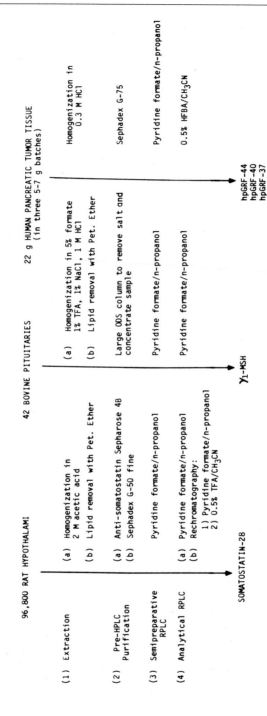

FIG. 2. Isolation protocols for rat hypothalamic somatostatin-28, bovine pituitary γ_1-MSH, and human pancreas growth hormone-releasing factors (hpGRF).

phenylmethylsulfonyl fluoride (10 mg/liter each) with a Polytron homogenizer. After centrifugation (17,700 g, 30 min) the pellet was reextracted twice with 500 ml of 2 M acetic acid and centrifuged. The supernatants of the three centrifugations were combined and extracted with 0.33 volume of petroleum ether (three times) for lipid removal. The aqueous phase was lyophilized and the dry residue (26.6 g) was redissolved in 4 liters of 10 mM sodium phosphate, and buffered to pH 7.4 with NaOH.

Pre-HPLC Purification: Affinity Chromatography and Gel Filtration

After centrifugation the supernatant was passed twice through a Sepharose 4B anti-somatostatin immunoaffinity column (25 × 7.5 cm), prepared as described above, at a flow rate of 250 ml/hr. The column was washed with 750 ml of 10 mM sodium phosphate, pH 7.4, and the bound material was eluted with 1 liter of 25 mM HCl. This eluate was lyophilized and subjected to gel filtration on Sephadex G-50 fine as described in the legend of Fig. 3. A radioimmunoassay (RIA) using sheep antiserum,

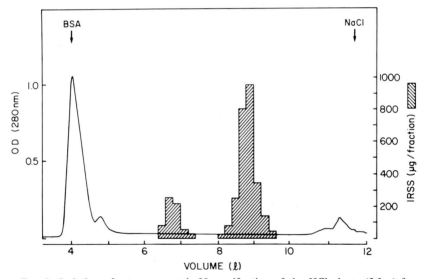

FIG. 3. Isolation of rat somatostatin-28: purification of the HCl eluate (5.3 g) from immunoaffinity chromatography by gel filtration on a Sephadex G-50 fine column (145 × 10 cm). The sample was dissolved in 80 ml of 5 M acetic acid for loading and eluted with 5 M acetic acid at a flow rate of 200 ml/hr. The column was calibrated with bovine serum albumin (BSA) and NaCl, as indicated. Fractions of 20 ml were collected, and aliquots from 10 fractions were pooled for radioimmunoassay. Fractions eluting between 6.4 and 7.2 l were pooled and lyophilized for further purification. IRSS, immunoreactive somatostatin.

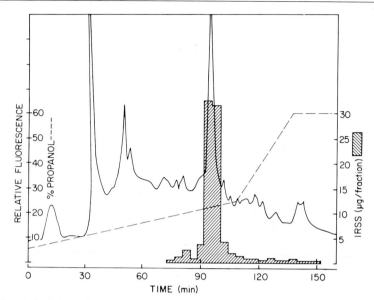

Fig. 4. Isolation of rat somatostatin-28: RPLC of the material contained in the early eluting zone of immunoreactive somatostatin from gel filtration (Fig. 3). A semipreparative RP-18 column (250 × 10 mm, 5 μm particle size) and the pyridine formate–n-propanol mobile phase were used. The sample (13 mg) was dissolved in 5 ml of 0.2 M acetic acid and loaded via a 5-ml sample injection loop. Fractions of 3.2 ml were collected at a flow rate of 0.8 ml/min.

BARBAR-78, was used to detect somatostatin-like immunoreactivity.[12] This antiserum, which was raised against synthetic somatostatin-14, cross-reacts with synthetic ovine somatostatin-28 in an equimolar ratio. Two major zones of somatostatin-like immunoreactive material, compatible with peptides possessing molecular weights (M_r) of 3000 and 1500 were obtained. The zone corresponding to M_r 1500 had previously been shown to contain somatostatin-14[13] and was not further investigated.

Semipreparative and Analytical RPLC Purification

Further purification of the M_r 3000 immunoreactive material was achieved by reverse-phase high-performance liquid chromatography (RPLC) employing an Altex Model 332 microprocessor-controlled liquid chromatograph system equipped with an automatic stream-sampling

[12] J. Epelbaum, P. Brazeau, D. Tsang, J. Bower, and J. B. Martin, *Brain Res.* **126,** 309 (1977).
[13] P. Böhlen, R. Benoit, N. Ling, R. Guillemin, and P. Brazeau, *Endocrinology* **108,** 2008 (1981).

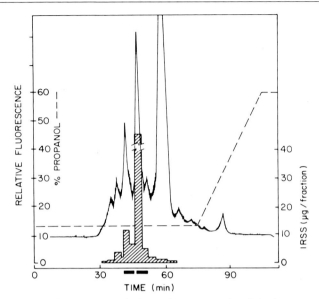

FIG. 5. Isolation of rat somatostatin-28: rechromatography of the immunoreactive peak from Fig. 4 using an analytical RP-18 column (4.6 × 250 mm, 5 μm particle size) and pyridine formate containing 13% (v/v) n-propanol as mobile phase. The sample, consisting of the pooled fractions containing immunoreactive material, was diluted to 30 ml with 0.2 M acetic acid and loaded by pumping the solution directly through the column. Fractions of 1.8 ml were collected at 0.6 ml/min, and aliquots were used for radioimmunoassay. Fractions used for further purification are designated by horizontal bars.

fluorescamine detection device. Initial chromatography on a semipreparative Ultrasphere RP-18 (Beckman) column (10 × 250 mm) using the pyridine formate–n-propanol solvent system yielded a single zone of immunoreactive material (Fig. 4).

Upon rechromatography under isocratic conditions on an analytical Ultrasphere RP-18 (4.6 × 250 mm) column with the same solvent system, this material resolved into two immunoreactive peaks (Fig. 5). The major immunoreactive fraction was further purified by rechromatography, first in the same system and then in the 0.5% trifluoroacetic–acetonitrile system, which exhibits different solute selectivities (Fig. 6). Subsequent characterization showed that this material possessed a structure identical to that of somatostatin-28 previously isolated from porcine[14,15] and ovine[16]

[14] L. Pradayrol, J. A. Chayvialle, M. Carlquist, and V. Mutt, *Biochem. Biophys. Res. Commun.* **85,** 701 (1978).
[15] L. Pradayrol, H. Jörnvall, V. Mutt, and A. Ribet, *FEBS Lett.* **109,** 55 (1980).
[16] F. Esch, P. Böhlen, N. Ling, P. Brazeau, and R. Guillemin, *Proc. Natl. Acad. Sci. U.S.A.* **77,** 6827 (1980).

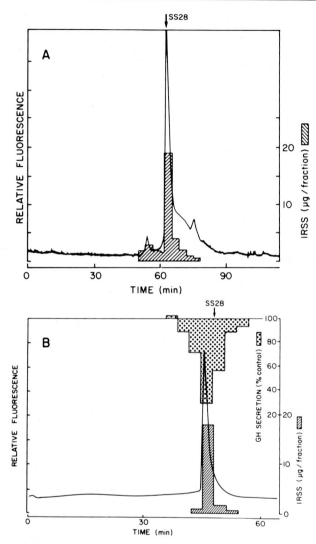

FIG. 6. Isolation of rat somatostatin-28: purification of the major immunoreactive peak from Fig. 5 under isocratic conditions in two different HPLC systems. (A) An analytical RP-18 column (250 × 4.6 mm, 5 μm particle size) and pyridine formate containing 12% n-propanol (v/v) were employed. The sample was diluted to 12 ml with 0.2 M acetic acid and loaded by pumping the solution through the column. (B) An analytical RP-18 column (250 × 4.6 mm, 5 μm particle size) and 0.5% trifluoroacetic acid containing 32.4% acetonitrile (v/v) were used. The fraction containing most of the immunoreactive material (Fig. 6A) was diluted to 10 ml with 0.5% trifluoroacetic acid and pumped onto the column. Other conditions were as in Fig. 5. The retention time of synthetic somatostatin-28 (SS28) is indicated by an arrow. GH, growth hormone.

sources. The minor immunoreactive material from Fig. 4, which corresponds to [Met(O)^8]somatostatin-28, was purified to homogeneity in the same manner (data not shown). This finding is consistent with the tendency of methionine-containing peptides to oxidize during isolation.[13]

Isolation of a γ_1-Melanotropin-like (γ_1-MSH) Peptide from Bovine Neurointermediate Pituitary[17]

Extraction

The neurointermediate lobes of 42 bovine pituitaries were collected into liquid nitrogen within 10–30 min of death. The tissue (11.5 g) was homogenized in 230 ml of 5% (v/v) formate, 1% (v/v) trifluoroacetic acid, 1% (w/v) NaCl in 1 M HCl[2] first in a small blender for 2 min and then in a Polytron homogenizer for 1 min. The homogenate was centrifuged (17,700 g, 30 min), and the pellet was reextracted with 50 ml of the same solvent and centrifuged again. The supernatants from the two centrifugations were combined and adjusted to pH 2.4 with 5 M NaOH. Lipids were removed by two extractions with equal volumes of petroleum ether. The aqueous phase was then degassed under a water aspirator vacuum to remove residual organic solvent.

Pre-HPLC Purification: Desalting and Concentration via ODS Chromatography

The defatted extract was pumped through a 2.5 × 20 cm glass column (Beckman) filled with octadecasilyl silica (ODS, LRP-2, 37–60 μm particle size; Whatman, Clifton, NJ) and equilibrated in 0.2 M acetic acid. The column was then washed free of salt with 0.2 M acetic acid. The peptide material retained on the column was eluted with 0.36 M pyridine formate, pH 3.0, in 60% (v/v) n-propanol and lyophilized to yield 137 mg.

Semipreparative and Analytical RPLC Purification

All RPLC purification steps were accomplished with Ultrasphere RP-18 columns, the pyridine formate–n-propanol solvent system, and the automatic stream-sampling fluorescamine detection system. Aliquots of chromatography fractions were monitored by radioimmunoassay for γ_1-MSH.[18] Semipreparative RPLC purification of the lyophilized material

---

[17] P. Böhlen, F. Esch, T. Shibasaki, A. Baird, N. Ling, and R. Guillemin, *FEBS Lett.* **128**, 67 (1981).

[18] T. Shibasaki, N. Ling, and R. Guillemin, *Biochem. Biophys. Res. Commun.* **96**, 1393 (1980).

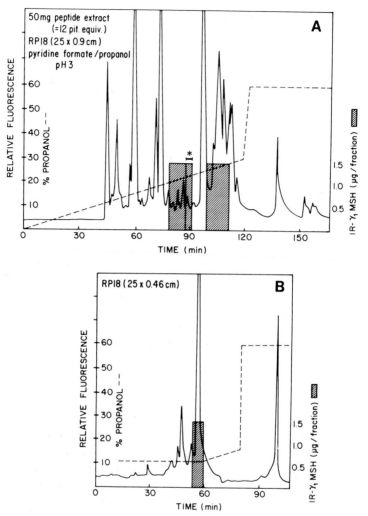

FIG. 7. Isolation of bovine γ_1-MSH. (A) RPLC of 50 mg of peptide extract on a semi-preparative Ultrasphere RP-18 column (10 × 250 mm, 5 μm particle size) with the pyridine formate–n-propanol solvent system. The sample was dissolved in 10 ml of 0.2 M acetic acid and loaded by pumping it directly onto the column. Fractions of 2.4 ml were collected at 0.8 ml/min. (B) Rechromatography of the active fraction from Fig. 7A (marked with an asterisk) on an analytical Ultrasphere RP-18 column (4.6 × 250 mm, 5 μm particle size). For loading, the fraction was diluted to a volume of 5 ml with 0.2 M acetic acid and injected via a 5-ml sample injection loop. Fractions of 1.8 ml were collected at 0.6 ml/min. All other conditions were as in Fig. 7A.

yielded two broad zones of γ_1-MSH immunoreactive material as shown in Fig. 7A. The first zone contained several relatively well defined peaks, aliquots of which were used for amino acid analysis. From the fraction identified with an asterisk (Fig. 7A), an amino acid analysis was obtained that closely matched that of the pro-opiomelanocortin fragment Lys^{-56}-Tyr-Val-Met-Gly-His-Phe-Arg-Trp-Asp-Arg-Phe^{-45}. Rechromatography of this fraction by analytical RPLC (Fig. 7B) yielded 40 nmol of pure peptide possessing the expected structure as deduced by amino acid analysis and automated Edman degradation.

Isolation of Human Pancreas Growth Hormone-Releasing Factors (hpGRF)[19]

Extraction

A tumor of the human pancreas suspected of ectopically producing the growth hormone-releasing factor was diced and collected into liquid nitrogen within 2–5 min of resection. The tissue (5–7 g) was extracted in 40 ml of 0.3 M HCl containing pepstatin A and phenylmethylsulfonyl fluoride (10 mg/liter) using a glass homogenizer. After centrifugation (20,000 g, 30 min), the pellet was reextracted once in 10 ml of the same solvent, and the mixture was centrifuged again.

Pre-HPLC Purification: Sephadex G-75 Chromatography

The combined supernatants were filtered on a Sephadex G-75 column as shown in Fig. 8A. Small aliquots (5–10 μl) of the fractions from column effluents were dried in 17 × 100-mm polypropylene tubes containing 100 μl of a 1 mg/ml human serum albumin solution in a Speed-Vac vacuum centrifuge (Savant). These were then assayed for their ability to stimulate the release of growth hormone from pituitary cells in culture.[20] Growth hormone released into the medium was quantitated by radioimmunoassay. Zones of growth hormone-releasing activity with an apparent molecular weight of 4000–5000 from three separate gel filtration columns (from three 6–8 g extractions totaling 22 g of tissue) were pooled for further purification.

[19] P. Böhlen, P. Brazeau, F. Esch, N. Ling, and W. B. Wehrenberg, *Regul. Peptides* (in press).
[20] P. Brazeau, N. Ling, F. Esch, P. Böhlen, R. Benoit, and R. Guillemin, *Regul. Pept.* **1**, 255 (1981).

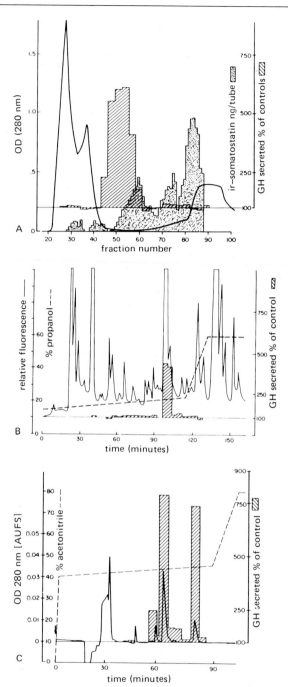

Semipreparative and Analytical RPLC Purification

Ultrasphere RP-18 semipreparative and analytical columns were employed in conjunction with the pyridine formate–n-propanol or 0.5% heptafluorobutyric acid–acetonitrile solvent systems, respectively, to obtain homogeneous peptide preparations. HPLC column effluents were monitored with either the automatic stream-sampling fluorescamine detection system or a variable-wavelength UV detector (Kratos/Schoeffel). Semipreparative RPLC purification of the pooled bioactive, gel-filtered material produced a single zone of bioactivity as shown in Fig. 8B. The active material was then subjected to analytical RPLC using a buffer system with different selectivity (Fig. 8C). This step yielded three bioactive and apparently homogeneous fractions, each corresponding to a distinct peak of UV-absorbing peptide material. Amino acid and sequence analyses[21] of the three peptides showed that they contained 44 (hpGRF-44), 40 (hpGRF-40), and 37 (hpGRF-37) amino acids in identical sequences from their amino termini. As shown in Fig. 8C, these peptides elute from the HPLC column in accordance with size and basicity, the smallest and least basic eluting first.

---

[21] F. S. Esch, P. Böhlen, P. Brazeau, N. Ling, P. E. Brazeau, W. B. Wehrenberg, and R. Guillemin, *J. Biol. Chem.* **258**, 1806 (1983).

FIG. 8. Isolation of the human pancreas growth hormone-releasing factors (GRF). (A) Gel filtration of the acidic tissue extract (50 ml) from 7 g of GRF containing tumor tissue on a Sephadex G-75 column (120 × 4.5 cm) equilibrated and run in 5 M acetic acid at a flow rate of 60 ml/hr with fraction volumes of 15 ml. The first and last absorbance peaks correspond to exclusion and salt volumes of the column, respectively. (B) RPLC of the pool of bioactive gel filtration fractions 43–58 on two semipreparative RP-18 columns (10 × 250 mm, 5 μm particle size) using a pyridine formate–n-propanol mobile phase. To avoid losses due to lyophilization, the pooled gel filtration fractions were pumped directly onto the reverse-phase column prior to starting the elution gradient. Fractions of 3.2 ml were collected at 0.8 ml/min. (C) RPLC of the bioactive fractions from semipreparative chromatography (Fig. 8B) using an analytical Ultrasphere RP-18 column (4.6 × 250 mm, 5 μm particle size) in conjunction with the 0.5% heptafluorobutyric acid–acetonitrile solvent system. The sample was loaded after a 3-fold dilution with 0.5% heptafluorobutyric acid. Fractions of 1.8 ml were collected at 0.6 ml/min.

Section II

Equipment and Technology

[6] Electrophysiological Assays of Mammalian Cells Involved in Neurohormonal Communication

By BERNARD DUFY, LUCE DUFY-BARBE, and JEFFERY L. BARKER

Many studies have shown that the release of hormones from cells that store their secretory products in granules can be evoked by chemically or electrically induced depolarization of the cell membrane. Douglas[1] observed that incubation of terminal endings in the isolated neurophypophysis in excess K^+ ions stimulated hormone release and that the presence of extracellular Ca^{2+} was required. He proposed that hormone release was initiated from nerve endings following influx of Ca^{2+} through membrane depolarized during regenerative action potential activity. Action potential discharge has now been associated with secretory activity in many, if not all, types of excitable cells, including anterior pituitary cells. Thus, study of action potential generation in excitable cells should provide insight regarding regulation of the secretory activity. By examining what membrane mechanisms regulate action potential discharge, we should eventually be able to achieve a more complete understanding of input–output relations in excitable cells.

Three, quite different, strategies for studying the excitability of cells, and for the purposes of this chapter, those involved in neurohormonal communication, have evolved over the past 10 years, each having its own advantages and limitations, each directed at a different level of understanding. One level involves extracellular recordings *in vivo* from central nervous system (CNS) neurons that are either unidentified or identified by a characteristic firing pattern and/or by antidromic activation. The basic data that emerge at this level involve (1) classification of cells according to their characteristic action potential discharge rate, which may or may not be correlated with the release of a specific transmitter or hormone; (2) description of the responses of individual cells to different inputs as a means of identifying those cells involved in a particular neuroendocrine reflex or behavior; and (3) characterization of the pharmacological responses of individual cells to specific endogenous or exogenous substances. However, extracellular recordings utilized in this manner reveal only the relative rate of action potential discharge rather than the mechanisms underlying the observed patterns of electrical activity or changes in discharge rate. Furthermore, it is difficult to determine whether the ef-

---

[1] W. W. Douglas, *Br. J. Pharmacol.* **34,** 451 (1968).

fects result directly from actions on neuronal membrane properties, or indirectly, from changes in levels of synaptic inputs. Another problem arising in the use of *in vivo* preparations involves the need for general anesthesia, which may obscure, in a selective fashion, the response of a particular set of neurons. Finally, the extracellular recording technique itself predisposes to recording larger-sized neurons, thereby selecting a population for study. Thus, the membrane mechanisms that underlie patterns of electrical activity endogenous to specific types of cells and the actions of transmitters, hormones, or drugs on these patterns cannot presently be analyzed *in vivo*.

In order to study the cellular basis of excitability in neuroendocrine circuits in mammals, many investigators have begun to use *in vitro* preparations of cells from the hypothalamus and anterior pituitary. Two types of *in vitro* preparations have been developed. In the "slice technique," thin tissue slices are carefully made from the hypothalamus (or other) regions of the CNS and incubated for the duration of the electrophysiological experiment in a well-defined saline. In this acute preparation of CNS tissue the cells exhibit electrophysiological viability for many hours. Electrical properties can be recorded with relative ease using either intracellular or extracellular recording techniques. The second strategy, developed over the past 15 years, involves dissection of cells from different regions of the embryonic CNS including the hypothalamus for maintenance in culture. Clonal or transformed cells expressing excitable membrane properties have also been successfully maintained in culture. Since, in culture, the cells grow as a monolayer, they are readily accessible to electrophysiological assays. However, the advantages gained in the accessibility of monolayers of cultured cells are compromised to some, as yet uncertain, degree either by the loss of the proper developmental signals in primary cultures of CNS tissue, or by the transformed character of the clonal lines, whose phenotypic expressions may also be transformed from those normally present *in vivo*.

In this chapter we briefly review some of the protocols involved in making electrophysiological assays of excitability from cells *in vivo* and *in vitro*, detailing some of the experimental requirements necessary for studying electrically detectable events of cells directly involved in neuroendocrine reflexes and neurohormonal forms of intercellular communication. We also summarize some recent observations on the excitable membrane properties of anterior pituitary cells. The results of these studies show that even cells not derived from the neural crest have complex electrophysiological properties that are undoubtedly related to the secretory activities of these cells.

Recording from Hypothalamic Cells *in Vivo*

Extracellular recordings have often been used to study spontaneous patterns of discharge in physiological or experimental situations related to hormone secretion from the hypothalamus. This part of our chapter deals with studies performed on the magnocellular and parvocellular neurosecretory circuits in the mammalian hypothalamus. We shall briefly outline three aspects of electrical recording of neuroendocrine cells *in vivo:* (1) maintenance of animals and insertion of recording electrodes in the hypothalamus; (2) technical problems of the recording; and (3) identification of the cells being recorded.

Preparation of the Animals for Acute and Chronic Recordings

Electrical activity can be recorded *in vivo* in either acute or chronic preparations. In acute experiments, the animal is used for recording immediately after the completion of surgical procedures and is usually sacrificed at the end of the experimental session. In chronic experiments, the animal bears either chronically implanted microelectrodes or a chronic device allowing the introduction of acute microelectrodes during multiple recording sessions. Such animals can be used during several days or weeks or even longer periods of time.

In acute experiments the animal's head is rigidly held by a stereotaxic instrument, which provides both a secure platform with which to control and locate the recording electrodes in specific regions of the CNS and a means of preventing movement of the brain. Anesthetics are used both to eliminate painful and other sensory stimuli arising during operative and recording procedures and to paralyze the preparation effectively. The most commonly used general anesthetics include the barbiturates, chloralose, and urethane. Phencyclidine and ketamine hydrochloride, which are more effective analgesics than the previously mentioned anesthetics, are used less frequently, owing in part to the fact that pharmacological effects on CNS excitability are even less well understood than those of the other anesthetic drug classes. Since anesthetics alter the excitability of central neurons, the results of electrophysiological studies carried out *in vivo* under anesthesia must be considered in this light. For example, we have found that, in the monkey, sodium pentobarbital (30 mg intravenously every hour) reduces the spontaneous electrical activity of neurons recorded in the hypothalamus by over 90%.[2] Moreover, barbiturates and other anesthetics are known to interfere with a variety of neurohormonal

---

[2] B. Dufy, L. Dufy-Barbe, J. D. Vincent, and E. Knobil, *J. Physiol. (Paris)* **75**, 105 (1979).

reflexes. Therefore, electrical activity recorded *in vivo* under anesthesia should be compared, if possible, to activity recorded in decerebrate preparations. This comparison should reveal similarities and differences in the spontaneous and transmitter- or drug-induced electrical activity.

Animals chronically prepared for electrophysiological recordings require a stable, relatively fixed relationship between the microelectrodes and the neurons under study. This may be achieved by using either stereotaxic or nonstereotaxic procedures. For recording from awake animals, nonstereotaxic devices of various kinds have been used by investigators. Some of them are described in detail by De Valois and Pease.[3] These devices are inserted during surgery under general anesthesia several days before the beginning of recording sessions. For electrophysiological recordings from neurons in the hypothalamus of the awake rabbit or monkey, we have found suitable the technique developed by Evarts[4] and modified by Hayward and Vincent.[5] Without excising the dura mater, the calvarium of the vertex is opened aseptically in order to place a stainless steel cylinder stereotaxically on the dura, immediately above the area in the hypothalamus of interest. This cylinder is cemented to an elevated Lucite platform that is then affixed to four additional stainless steel epidural screws. Between recording sessions, the cylinder is filled with bone wax and capped. For recordings, a calibrated hydraulic microdrive device is fixed to the cylinder and to its attached stainless steel guide tube. The guide tube penetrates the bone wax, the dura matter, and the brain. The microelectrode can then be lowered out of its protective position to the hypothalamic area. This type of device allows the exploration of a hypothalamic area 5 mm in diameter and 1 cm in vertical depth.

It should be pointed out that conditioning animals to the stress of chair restraint is important in providing a stable baseline of electrical activity and hormone release, since even simple chair restraint can have profound effects on neuroendocrine activities.[6]

Microelectrodes

The microelectrodes used for recording electrical activity *in vivo* are usually made of metal, since glass microelectrodes are too fragile for penetrating the necessary distance through the CNS. Three types of mi-

[3] R. L. De Valois and P. L. Pease, *in* "Bioelectric Recording Techniques" (R. F. Thompson and M. M. Patterson, eds.), Part A. "Cellular Processes and Brain Potentials," p. 97. Academic Press, New York, 1973.
[4] E. V. Evarts, *Electroenceph. Clin. Neurophysiol.* **24,** 83 (1968).
[5] J. N. Hayward and J. D. Vincent, *J. Physiol.* (*London*) **210,** 947 (1970).
[6] L. Dufy-Barbe and B. Dufy, unpublished observations.

croelectrodes are commonly used: glass-coated platinum, tungsten, and stainless steel. Both platinum and tungsten electrodes are made by electrolytically etching wire blanks to a suitable taper. Electrolytic etching, insulation and its removal from the electrode tip have been described in detail for both types of electrodes by Snodderly.[7] Platinum electrodes insulated with glass for adequate rigidity and plated with platinum black[8] are probably the best choice for most recording situations, since they usually give stable recordings with a high signal-to-noise ratio. They are usually made from 70% platinum–30% iridium wire either 0.005 or 0.010 in. in diameter. Even under optimal circumstances recordings with rigid microelectrodes in either acute or chronic preparations are usually limited to no more than 4–5 hr. The time course of most neuroendocrine reflexes, as for example, those associated with gonadotropin regulation during the estrous cycle (days to months), make such short-term recordings less than ideal.

However, successful recordings have been made from hypothalamic neurons using a technique initially developed by Strumwasser[9] and Olds[10] that allows long-term recording from the same cell or group of cells. This technique involves the use of electrodes that are so flexible that they allow the position of the electrode tip to be maintained in relation to the neuron(s) under study. These "floating" electrodes are usually made of fine (30–60 μm in diameter) metal wire (platinum or nickel–chrome) that has been preinsulated with enamel or varnish. For implantation, a small hole is drilled into the skull, and the dura is punctured with a sharp surgical needle. Some investigators lower each electrode after attaching it directly to a micromanipulator according to a procedure described in detail by Olds.[10] For recordings in the hypothalamus, we have found it more convenient to glue the electrodes with sucrose (solution of sucrose in water at its boiling point of 115°) to a rigid guide, according to the method of Burns et al.[11] A straight etched tungsten needle of 150 μm can be used for this purpose. We have found that 3 or 4 electrodes can be glued to the same needle, the tip of which may be positioned 1–2 mm in advance of the tip of the electrodes. Electrodes mounted in this way must

[7] D. M. Snodderly, Jr., in "Bioelectric Recording Techniques" (R. F. Thompson and M. M. Patterson, eds.), Part A. "Cellular Processes and Brain Potentials," p. 137. Academic Press, New York, 1973.

[8] M. L. Wolbarsht, E. F. MacNichol, and H. G. Wagner, Science **132**, 1309 (1960).

[9] F. Strumwasser, Science **127**, 469 (1958).

[10] J. Olds, in "Bioelectric Recording Techniques" (R. F. Thompson and M. M. Patterson, eds.), Part A. "Cellular Processes and Brain Potentials," p. 165. Academic Press, New York, 1973.

[11] B. D. Burns, J. P. Stean, and A. C. Webb, Electroenceph. Clin. Neurophysiol. **36**, 314 (1974).

be stored in a desiccator. For implantation, the guide tube is stereotaxically lowered into the appropriate hypothalamic area by stereotaxy. After 5–10 sec, the sucrose glue dissolves and the tungsten needle is withdrawn, leaving the electrodes in place. The free end of the electrode protruding from the skull can be either soldered to a miniature electrical socket bolted to the skull[12] or wrapped around a screw through which the plug contacts the electrode.[10]

Electrodes of this type can be used to record either single- or multiunit electrical activity. Using this technique in the monkey, we have been able to record from the same hypothalamic units for 35 days, as judged from the similarities in the size and shape of the extracellular recorded action potentials. Thus, this technique offers several major advantages: recordings can be made in conscious and calmly behaving animals, simultaneously from several electrodes, and the same cell or cells can be recorded for several consecutive days or weeks.

Identification of Neurons in Vivo

Topographical Identification. The use of a brief electrical stimulus to activate hypothalamic neurons either antidromically or orthodromically was first introduced by Kandel[13] for locating magnocellular neurosecretory neurons in the goldfish that project to the posterior pituitary. Magnocellular neurons in the vertebrate hypothalamus can also be activated by antidromic invasion of their cell body following an electrical shock applied to the posterior pituitary. For stimulation in monkeys we have used a concentric stainless steel electrode implanted in the posterior pituitary with a frontal transcranial approach under both frontal and lateral radiographic control.[14] Antidromic potentials should meet the following minimum criteria: constant latency to onset of the evoked potential, ability to follow with constant latency during the application of trains of pulses at high frequency, and cancellation of the evoked potential after collision with a spontaneously occurring orthodromic action potential. The same antidromic activation technique can be used to activate cells in the parvocellular system by implanting the tip of the stimulating electrode in the median eminence, where most parvocellular neurons terminate.[15] However, we should mention that, in rabbits, for example, this technique

---

[12] A. J. S. Summerlee, *J. Physiol. (London)* **321**, 1 (1981).
[13] E. R. Kandel, *J. Gen. Physiol.* **47**, 691 (1964).
[14] E. Arnauld, B. Dufy, and J. D. Vincent, *Brain Res.* **100**, 315 (1975).
[15] G. B. Makara, M. C. Harris, and K. M. Spyer, *Brain Res.* **40**, 283 (1972).

allows the activation of no more than 3% of hypothalamic neurons.[16] Marking the recording site (by passing a direct current via the electrode when the animal is still alive) and postmortem histological control should be performed whenever possible.

Functional Identification. Physiological studies performed on cells involved in neurohormonal communication *in vivo* frequently involve measurement of both electrical activity and hormone release. Blood samples are often collected via catheters implanted in the right atrium through the left internal jugular. In the chronic situation, the distal end of the catheter may be threaded under the skin of the neck to a cranial platform. For capping the Silastic catheter, Baker *et al.*[17] have designed an elegant device consisting of a rubber-tipped Luer-lok cannula attached to a needle fixed to the catheter. This device not only permanently secures the distal end of the catheter outside the animal, but allows intravenous injections and withdrawal of blood samples without disturbing the animal. However, more simple devices consisting of a rubber stopper can be used as well.

Some neurosecretory neurons are relatively easy to identify because they exhibit a fairly well-established pattern of action potential discharge related to a particular physiological stimulus. For example, some magnocellular neurons recorded in the lactating rat can be tentatively identified as those that probably synthesize and secrete oxytocin. Oxytocin release in systemic blood can be measured using variations in intramammary pressure as a monitor, since this method offers a sensitive and specific bioassay of the hormone. Some neurons in the hypothalamus exhibit a low level of ambient electrical activity for most of the suckling period, but 10–20 sec before each milk ejection they display a brief high-frequency discharge of action potentials lasting about 4 sec followed by a period of silence. This pattern of discharge is observed in about half the population of cells recorded in the supraoptic and paraventricular nuclei. However, the activation of parvocellular neurons is made more difficult since specific stimuli have not been established for the release of specific hypophysiotropic hormones. Thus, it is difficult, if not impossible, to assign a particular function to a given parvocellular neuron. Sometimes, the variation of electrical activity recorded from the neuron coincides temporally with changes in hormone concentration assayed in the plasma (Fig. 1). This suggests, but does not prove, that the activity of the neuron recorded

---

[16] B. Dufy, C. Partouche, D. Poulain, L. Dufy-Barbe, and J. D. Vincent, *Neuroendocrinology* **22**, 38 (1976).

[17] M. A. Baker, E. Burrell, J. Penkhus, and J. N. Hayward, *J. Appl. Physiol.* **24**, 577 (1968).

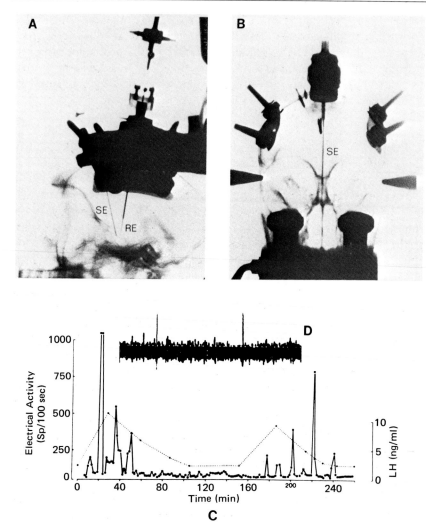

FIG. 1. Lateral (A) and frontal (B) X-rays of the head of a rhesus monkey prepared for chronic recording from hypothalamic neurons. The animal bears lateral screws for clamping the head. A stainless steel cylinder has been positioned on the dura matter through which a guide tube protecting the recording electrode (RE) is lowered during each recording session. The stimulating electrode (SE) for antidromic activation of cells in hypothalamus has been permanently implanted in the median eminence and has been positioned under radiographic control. (C) The variation in extracellularly recorded electrical activity (shown in D) of an antidromically activated neuron located in the arcuate nucleus is plotted over a 4-hr recording period. From the relationship between changes in electrical activity and alterations in the level of luteinizing hormone in the plasma (also shown in C), it can be tentatively concluded that this particular neuron is related to the secretion of luteinizing hormone releasing hormone.

is related in some way, to secretion of the hormone. Clearly, the precise relationship between the electrical activity of individual cells in the hypothalamus and the levels of hormone released from the anterior pituitary is far from clear. These difficulties have prompted the development of various *in vitro* preparations, consisting of different parts of the neuroendocrine circuit.

In Vitro Preparations

Slice Preparations of the Hypothalamus

Thin slices of different regions of the CNS including the hypothalamus can be prepared from guinea pigs or rats. After decapitation and brain removal, sagittally or coronally oriented hypothalamic slices of 300–500 μm thickness can be cut with a tissue chopper. The slices are then transferred to a petri dish containing oxygenated medium as described by Yamamoto.[18] The slices are positioned on a nylon net in a chamber with the upper surface of the slice exposed to an atmosphere of humidified and warmed 95% O_2, 5% CO_2.[19] The slice is bathed in a physiological saline. It should be noted that for recording spontaneous activity from cells of the paraventricular nucleus, the $[Ca^{2+}]$ of the medium bathing the hypothalamic slices is critically important. $[Ca^{2+}]$s greater than 1 mM *abolish* spontaneous activity. In 0.75 mM $[Ca^{2+}]$ phasic electrical discharges similar to those attributed to vasopressin neurons *in vivo* can be recorded in slices. Usually the recordings begin after a 1-hr recovery period following acute preparation and continue for 8–10 hr, although most observations are made between 2 and 6 hrs after preparation. Intracellular and extracellular recordings are made using glass fiber-containing microelectrodes. For intracellular recording of hypothalamic cells, microelectrodes having a direct current resistance of 50–100 megohms are required, and extracellular recordings can be carried out with microelectrodes having a resistance of 5–10 megohms. Activation of neurons projecting to the median eminence or to the posterior pituitary can be accomplished by electrical stimulation of axonal endings using stimulating electrodes.[20,21]

Slice preparations allow drugs to be applied either in droplets or by superfusion at known concentrations. When drugs are administered in droplet form the substances are dissolved in the bathing medium and

---

[18] C. Yamamoto, *Exp. Brain Res.* **14**, 423 (1972).
[19] H. J. Spencer, V. K. Bribkoff, C. W. Cotman, and G. S. Lynch, *Brain Res.* **105**, 471 (1976).
[20] F. E. Dudek, G. T. Hatton, and B. A. MacVicar, *J. Physiol. (London)* **301**, 101 (1980).
[21] M. J. Kelly, U. Kuhnt, and W. Wuttke, *Exp. Brain Res.* **40**, 440 (1980).

ejected by pressure from micropipettes positioned close to the neuron being recorded. To avoid unwanted diffusion of drugs, the pipette usually has to be held in the air above the slice and lowered to the surface of the slice only at the time of administration. This technique allows relatively rapid delivery. When drugs are dissolved in the superfusate, the fluid in the recording chamber may require many minutes of superfusion to be completely exchanged. Using slices of hypothalamic tissue, stable intracellular recordings have been maintained for many hours.

Since patterns of electrical activity recorded in these neurons appear similar to those reported *in vivo,* acutely prepared slices may be a useful system for studying the excitability of neurons involved in neurohormonal communication. The presence of anatomical landmarks and different functional inputs allow relatively precise identification of the cells under study. However, the viability of cellular elements in a slice is always suspect, and the level of electrophysiological resolution afforded by the preparation, albeit improved over blind experiments *in vivo,* is still limited.

Pituitary Cells in Culture

Cells dissociated from the hypothalamus can be maintained in culture as can cells derived from pituitary tumors. The monolayer character of the culture makes the cells readily accessible to a variety of electrophysiological recording techniques, including single- and two-electrode voltage clamp methods. Other types of newly developed electrical assays of excitable membrane events including patch clamp of micrometer-sized areas of membrane[22] and whole-cell patch clamp of intact cells[23] can also be applied to excitable cells. In fact, patch clamp analysis of Ca^{2+} channels in clonal pituitary cells has already been carried out, revealing the elementary properties of these channels.[24]

For electrophysiological experiments, the plastic petri dish containing the monolayer is placed on the stage of an inverted microscope and the cells are viewed with phase contrast optics at high (250–500×) magnification. Electrophysiological recording can be carried out either at room temperature (20–24°) or using various heating devices to raise the temperature closer to 37° ± 1°. The excitable membrane properties of anterior pituitary cells appear to be particularly sensitive to temperature. For example, the GH3/B6 pituitary clone does not display any spontaneous

---

[22] O. P. Hamill, A. Marty, E. Neher, B. Sakmann, and F. J. Sigworth, *Pfluegers Arch.* **391,** 85 (1981).
[23] K. S. Lee, N. Akaike, and A. M. Brown, *J. Gen. Physiol.* **71,** 489 (1978).
[24] S. Hagiwara and H. Ohmori, *J. Physiol. (London)* **331,** 231 (1982).

electrical activity if the ambient temperature falls below about 19°. It is likely that the physiochemical properties of the cell's membrane, including its microviscosity, is critical to the functioning of the ionic channels that underlie excitable processes.[25]

In a typical intracellular recording, microelectrodes with tips less than 1 μm in diameter are filled with either 3 M KCl, 2 M potassium citrate, or 1 M K$_2$SO$_4$. The composition of the conducting solution used to fill the microelectrodes is important, since different values of electrical properties of pituitary cells are obtained with different conducting solutions. For example, the cells are significantly more polarized when using K$_2$SO$_4$-filled electrodes than when KCl-filled electrodes are used. It is probable that leakage of citrate interacts with internal Ca^{2+} and this in turn affects the excitability of pituitary cells. This sensitivity of membrane properties to [Ca^{2+}]$_i$ is not peculiar to these cells, since effects of changes in [Ca^{2+}]$_i$ alter the excitability of several different types of nerve cells including insect motoneurons[26] and bullfrog sympathetic ganglion cells.[27]

Simultaneous intracellular recordings and current injections can be made through a conventional bridge circuit. The high-impedance preamplifiers necessary for making these recordings and a high-frequency switching system that allows control of membrane voltage while measuring membrane current response are all commercially available (e.g., from the Dagan Corporation). However, two-microelectrode voltage clamp systems need to be tailor-made to the particular cell type under study.[28] The data from such recordings are usually displayed on an oscilloscope (with storage capabilities) and recorded on a pen-recorder and/or acquired by computer, depending on the type of electrophysiological experiment. The ready accessibility of monolayered cells allows direct application of known concentrations of drugs and substances to the cell under study. This is usually accomplished by pressure applied to a micropipette (3–5 μm tip-diameter) containing recording medium plus drug positioned close to the cell surface. Thus, monolayers of excitable cells afford the opportunity to study the basis of cellular excitability. The remainder of our chapter is a brief summary of recent studies on the excitability of anterior pituitary cells. These studies may serve as a catechism for electrophysiological assays in all excitable cells whether or not they participate directly in neuroendocrine reflexes.

[25] E. Zyzek, J. P. Desmazes, D. Gourdji, B. Dufy, and G. Georgescauld, *Experientia* **39**, 56 (1983).
[26] R. M. Pitman, *J. Physiol. (London)* **291**, 327 (1979).
[27] K. Morita, K. Koketsu, and K. Kuba, *Nature (London)* **283**, 204 (1980).
[28] T. G. Smith, J. L. Barker, B. M. Smith, and T. R. Colburn, *J. Neurosci. Methods* **3**, 105 (1980).

Excitable Membrane Properties of Cultured Pituitary Cells

Current-Clamp Observations. There is now ample evidence that pituitary cells, like nerve cells, are electrically excitable[29,30] and that the electrically excitable membrane properties of the cells are sensitive to a variety of transmitter substances.[31-39] These properties of pituitary cells have been studied by a variety of laboratories using clonal prolactin-secreting cells derived from the GH cell line, since these cells exhibit electrical and chemical excitability and are relatively easy to maintain in culture. Simple single-electrode recordings under current clamp reveal that not all cells recorded in any one culture plate appear to be excitable. Inexcitable elements may represent a subpopulation of cells whose phenotypic expression is devoid of excitable membrane properties, or, alternatively, they may reflect a stage in the development of phenotypic properties following cell division. However, all laboratories report that a majority (ca. 80%) of GH3 cells are electrically excitable (Figs. 2 and 3), if not spontaneously active (Fig. 3A). Injection of depolarizing current stimuli triggers regenerative action potential responses that are graded in character, rather than clearly all-or-none (Fig. 2A). These potentials are blocked by Co^{2+} (Fig. 3D), and, in some reports, by tetrodotoxin. The action potentials are prolonged by tetraethylammonium (TEA) (Fig. 3F) and in some reports by 4-aminopyridine (4-AP) as well.

Injection of hyperpolarizing and depolarizing currents shows that the "steady-state" current-voltage (I-V) relations of many of these cells are quite nonlinear at the level of the resting membrane potential (-40 to -50 mV) (Fig. 2). This nonlinearity becomes even more noticeable when depolarizing currents are injected from the level of the resting membrane potential, with substantial rectification occurring during the brief (50 msec) depolarizing current step. Rectification of membrane voltage responses to depolarizing current stimuli has functional implications for

---

[29] Y. Kidokoro, *Nature (London)* **258**, 741 (1975).

[30] B. Biales, M. A. Dichter, and A. Tischler, *Nature (London)* **267**, 172 (1977).

[31] B. Dufy, J. D. Vincent, H. Fleury, P. du Pasquier, D. Gourdji, and A. Tixier-Vidal, *Science* **204**, 509 (1979).

[32] B. Dufy, J. D. Vincent, H. Fleury, P. du Pasquier, D. Gourdji, and A. Tixier-Vidal, *Nature (London)* **282**, 855 (1979).

[33] S. Ozawa and N. Kimura, *Proc. Natl. Acad. Sci. U.S.A.* **76**, 6017 (1979).

[34] S. Ozawa and S. Miyazaki, *Jpn. J. Physiol.* **29**, 411 (1979).

[35] O. Sand, E. Haug, and K. M. Gautvik, *Acta. Physiol. Scand.* **108**, 247 (1980).

[36] P. S. Taraskevich and W. W. Douglas, *Neuroscience* **5**, 421 (1980).

[37] G. J. Kaczorowski, R. L. Vandlen, G. M. Katz, and J. P. Reuben, *J. Membr. Biol.* **71**, 109 (1983).

[38] B. Dufy and J. L. Barker, *Life Sci.* **30**, 1933 (1982).

[39] P. R. Adams, D. A. Brown, and A. Constanti, *J. Physiol. (London)* **330**, 537 (1982).

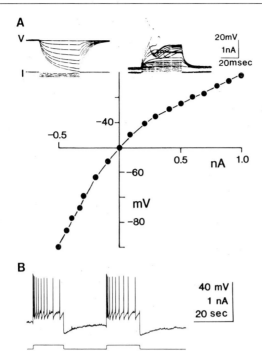

FIG. 2. GH3/6 cells are electrically excitable. (A) The cell was current-clamped at its resting potential (−50 mV), and a series of hyperpolarizing and depolarizing current stimuli were injected through a bridge circuit. The voltages occurring at the end of the 50-msec current stimuli are plotted as a function of the applied current. The resulting current–voltage (I-V) curve is clearly nonlinear about the level of the resting potential. The slope conductance at hyperpolarized levels is considerably greater than that at depolarized potentials, indicating a significant degree of rectification at depolarized levels. In addition, the inset shows graded regenerative action potentials triggered at the most depolarized levels. (B) A sustained current stimulus evokes repetitive action potential discharge in a GH3/6 cell. The initial high-frequency activity decreases during the stimulus as rectification of the membrane predominates. Termination of the current stimulus results in a prolonged period of hyperpolarization.

regulating the rate at which action potentials can be triggered (Fig. 2B). Sustained depolarizing current stimuli evoke an initial, though transient, burst of action potentials whose frequency of discharge declines rapidly over a 20-sec period. Thus the rectification evident during injection of both transient and sustained depolarizing current pulses acts to regulate the rate of action potential discharge. The termination of a sustained current injection is followed by a prolonged hyperpolarization of the membrane potential (Fig. 2B), which is associated with an increase in conductance. This potential response reflects the K^+ conductance(s) acti-

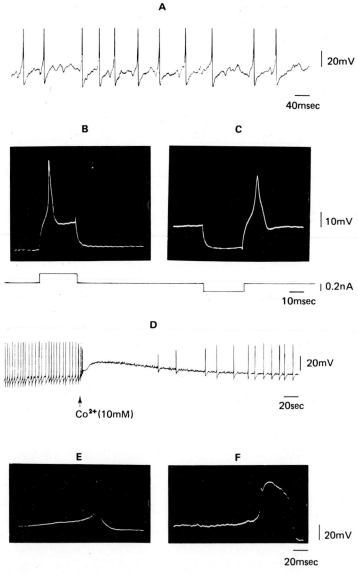

FIG. 3. Current-clamp observations on GH3/6 cells. (A) Some cells are spontaneously active; discharging action potentials at the level of the "resting" potential. (B and C) Other cells are not spontaneously active, but can be excited either by injecting depolarizing current (B) or by terminating the injection of hyperpolarizing current (C). (D) The action potentials are blocked, and the resting potential is depolarized by Co^{2+}. (E and F) The repolarization phase of the action potential is prolonged by 5 mM TEA (F) relative to control values (E). From B. Dufy and J. Barker,[42] with permission.

vated during the depolarization. Its sensitivity to agents that block Ca^{2+}-dependent conductance mechanisms suggests that the prolonged K^+ conductance is regulated in part by Ca^{2+}. In many GH3 cells the resting membrane potential is also sensitive to the same antagonists (Fig. 3D), indicating that it may be regulated in a complex way by Ca^{2+}.

Voltage-Clamp Observations. We have applied the single- and two-electrode voltage-clamp techniques to GH3/B6 cells to extend our electrophysiological analysis of excitability in these cells to a level of resolution where we can examine the membrane currents underlying the observed membrane potential behavior. Clamping the cells at hyperpolarized potentials and applying a simple voltage-jump protocol reveals a spectrum of inward and outward currents (Fig. 4). Over the most hyperpolarized region of the I–V curve, the relationship between membrane voltage and steady-state membrane current is constant. There are no detectable time- and/or voltage-dependent conductances. Over the region −60 to −40 mV, deviation from this simple I–V relationship occurs to a variable degree. There is little deviation in the I–V relations over this range of potential in the cells illustrated in Fig. 4, but significant time-dependent outward current is detectable over this range in the results illustrated in Fig. 5. This outward current is blocked by Co^{2+},[38] Mn^{2+} (unpublished observations) and Ba^{2+} (Fig. 5). It corresponds to the Ca^{2+}-dependent K^+ conductance whose activation sets the level of the resting membrane potential. The noninactivating character of this K^+ conductance is similar to the K^+ conductance described in sympathetic neurons, block of which accounts for some of the slow depolarizing responses activated by acetylcholine and peptides[39] in these cells. Ca^{2+}-dependent K^+ conductances at the level of the resting potential have also been recorded in cultured sensory neurons[40] and parasympathetic cells.[41] Thus, such an ion conductance regulated by voltage and by Ca^{2+} ions and activable at the level of the resting potential is not unique to GH3 cells.

Inspection of the membrane currents generated over the range of membrane potential about the resting level, subthreshold to the potential at which action potentials and their corresponding fast inward currents can be activated, shows that a rapidly arising, transient outward current (TOC) response occurs (Figs. 4 and 5). The TOCs evoked in these cells are variably sensitive to 4-aminopyridine, Co^{2+}, and Ba^{2+} (Fig. 5), but relatively insensitive to tetraethylammonium (TEA), which blocks the delayed, sustained outward current developed at depolarized potentials.[42]

[40] D. A. Mathers and J. L. Barker, *Brain Res.* **211**, 451 (1981).
[41] H. C. Hartzell, S. W. Kuffler, R. Stickgold, and D. Yoshikami, *J. Physiol.* (*London*) **271**, 817 (1977).
[42] B. Dufy and J. L. Barker, *C.R. Soc. Biol.* (*Paris*) **177**, 166 (1983).

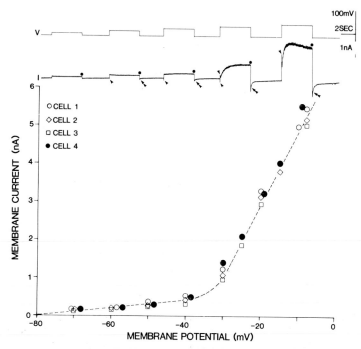

FIG. 4. Voltage-clamp observations in four GH3/6 cells. The cells were all clamped at −80 mV using two independent microelectrodes filled with 3 M KCl. A series of depolarizing voltage commands was applied to evoke membrane current responses about, and depolarized to the level of, the resting potential (−40 mV). A variety of inward and outward current responses are evident in the specimen records shown above the plot. Each current response has been identified with a different symbol: small filled circles represent the "steady-state" current response at the end of the 5-sec command (corresponding to TEA- and Co^{2+}-sensitive current); single, upward-tilted arrowheads reflect the fast Co^{2+}-sensitive inward current response (corresponding to the Ca^{2+}-dependent action potential); single, downward-tilted arrowheads are the transient outward current (TOC) (see text for details); and double, upward-tilted arrowheads point to an inward tail current. The "steady-state" membrane current has been plotted as a function of membrane potential for each of the four cells. The plots are highly nonlinear, yet remarkably similar for all four cells.

The TOCs can thus be distinguished from the later developing outward currents by their kinetics and pharmacological sensitivity. Similar TOCs have been reported in a variety of invertebrate[43,44] and vertebrate[45,46] nerve cells, so these too are not unique to GH3 cells. The TOCs are thought to

[43] J. A. Connor and C. F. Stevens, *J. Physiol.* (*London*) **213**, 21 (1971).
[44] S. H. Thompson, *J. Gen. Physiol.* **80**, 1 (1982).
[45] M. Segal, M. A. Rogawski, and J. L. Barker (submitted for publication).
[46] A. B. MacDermott and F. F. Weight, *Nature* (*London*) **300**, 185 (1982).

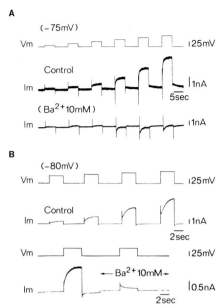

FIG. 5. Ba^{2+} blocks late outward currents in GH3/6 cells. (A) The cell was held at -75 mV using a Dagan single electrode voltage-clamp system and a series of depolarizing voltage commands (Vm) applied. Under control conditions time- and voltage-dependent inward and outward currents (Im) similar to those described in Fig. 4 are observed. The outward current generated is eliminated by 10 mM Ba^{2+}, and frank inward current is evoked. (B) Another GH3/B6 cell clamped with two microelectrodes at -80 mV and subjected to a paradigm similar to that used on the cell in (A). In this example Ba^{2+} selectively blocks the later developing outward current with little effect, if any, on the transient outward current (TOC).

affect membrane potential behavior in two ways: (1) as transient rectifiers of the membrane during rapid depolarizations from relatively hyperpolarized potentials; and (2) as a mechanism for repolarizing the action potential. Functionally then, the TOCs govern the rate of action potential discharge and the rate of action potential repolarization. The delayed outward current observed in Figs. 4 and 5 is regulated by both Ca^{2+} and voltage. The TEA sensitivity of this current suggests that it participates in the repolarization of the cell during an action potential (cf. Fig. 3F). Some fraction of this current likely underlies the prolonged hyperpolarizing response occurring at the end of a sustained depolarization (Fig. 2B).

Our initial observations of the electrophysiological properties of GH3/B6 cells under voltage clamp indicate that these cells are quite complex, expressing a wide spectrum of voltage-activated ion conductance mechanisms, some of which are regulated by Ca^{2+} ions. Furthermore, these excitable membrane properties do not appear to be unique to GH3 cells, since superficially similar ionic conductance mechanisms have been recorded in many types of excitable cells.

Peptides Alter the Excitability of GH3/B6 Cells. A variety of peptide and other transmitter substances alter the excitability as well as the secretory activity of anterior pituitary cells. For example, thyrotropin-releasing hormone (TRH) enhances the excitability of GH3/B6 cells and stimulates the secretion of prolactin, whereas somatostatin inhibits the excitability of these cells and depresses the secretion of prolactin.[31-37] We have begun to study the mechanisms underlying these peptide effects using the voltage-clamp technique. One type of excitable membrane property at which these two peptides have opposite actions is the TOC (Fig. 6). TRH depresses the amplitude of the TOC whereas somatostatin enhances the amplitude of the TOC. Although the mechanisms underlying these modulatory effects on TOC generation have not been elucidated, changes in the level of activation of this current may account in part for the changes in excitability observed under current clamp. Depression of the TOC would effectively increase the excitability of GH3 cells by attenuating the transient rectification occurring over the membrane potential region subthreshold to action potential generation, thus increasing the probability of action potential discharge. Enhancement of the amplitude of the TOC would have quite the opposite effect, dampening the excitability of the cell and reducing the probability of action-potential discharge. TRH increases the propensity of GH3/B6 cells to discharge action potentials, and somatostatin has the opposite effect. Hence, modulatory

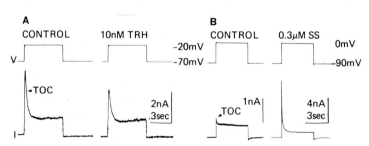

FIG. 6. Thyrotropin-releasing hormone (TRH) depresses, and somatostatin enhances, TOCs in GH3/B6 cells. (A) The cell was clamped with two KCl-filled electrodes at −70 mV and stepped to −20 mV for 0.5 sec. Under control conditions a TOC several nanoamperes in amplitude is triggered during such a command; 10 nM TRH blocks this TOC by about 30% (in a reversible manner). (B) A different cell was clamped at −90 mV and a 5-sec command to 0 mV was applied under control conditions and in the presence of 0.3 μM somatostatin (SS). The entire recording was carried out in the presence of 2 mM Ba^{2+} to eliminate delayed outward current. SS causes a 20-fold increase in the amplitude of the TOC. Note the 4X gain difference between the control and experimental current traces.

actions of these peptides on TOCs probably underlie their effects on unclamped membrane potential behavior. How these actions on the generation of TOCs are related to other effects on excitability and to the secretory activities of the cells remains to be studied.

Conclusion

In this chapter we have outlined several types of electrophysiological strategy for studying the excitability of cells involved in different levels of neuroendocrine integration and neurohormonal communication. We have concluded the chapter by reviewing some recent observations on the excitable membrane properties of anterior pituitary cells, since these are receiving increasingly more attention and provide a useful catechism of electrophysiological assays for determining not only the presence of different types of excitable membrane property, but also the functional roles these properties play in the input–output relations of the cells. The fairly homogeneous nature of the GH3 cells and the relative simplicity of their secretory output make them attractive for correlating the precise roles of certain conductance mechanisms in prolactin (and growth hormone) secretion. These studies may then serve as reference for examining the input–output relations of other cells at the same or at different levels in this and other neuroendocrine circuits.

[7] Electrophysiological Techniques in Dissociated Tissue Culture

By Gary L. Westbrook and Phillip G. Nelson

Dissociated cell cultures provide unique access for the electrophysiological analysis of central neurons. From a methodological point of view, intracellular recording in cell culture circumvents many of the classical problems faced by electrophysiologists recording *in vivo*. The ability to see the neurons with phase-contrast microscopy before impalement is the most important of these advantages. As a result, recording electrodes, iontophoretic pipettes, and "puffer" pipettes can be placed at any position over the neuronal surface. The absence of respiratory movements or cerebrospinal fluid pulsations allows stable impalements to be maintained for hours. The composition of the extracellular media is also easy to control. This additional access and stability create opportunities for more intricate and novel electrophysiological experiments that would be difficult or impossible in the intact nervous system. However, recording in cell culture can also place a number of limitations on recording techniques. In this chapter, we will focus on the practical aspects and limitations of intracellular recording in tissue culture. Primary dissociated cultures of fetal mouse spinal cord will be used as an example.[1] A number of excellent reviews are available for more extensive discussion of specific topics related to cell culture and electrophysiology.[2-10]

Intracellular recording and collection of physiological and pharmacological data can be straightforward in cell culture. Results can generally be obtained on a large number of neurons in a reasonable number of experiments. In addition to these obvious advantages, improvements in recording and culture techniques now allow electrophysiological approaches to a wide variety of interesting neurobiological questions. Some examples follow.

1. Synaptic interactions. Pairs of neurons can be impaled, monosynaptic connections identified, and the basic mechanisms and pharmacology of these interactions studied.[11,12] Although it is not possible to impale presynaptic terminals in CNS cultures, the use of the patch clamp[13] or

[1] B. R. Ransom, E. Neale, M. Henkart, P. N. Bullock, and P. G. Nelson, *J. Neurophysiol.* **40**, 1132 (1977).
[2] P. G. Nelson and M. Lieberman (eds.), "Excitable Cells in Tissue Culture." Plenum, New York, 1981.
[3] R. D. Purves, "Microelectrode Methods for Intracellular Recording and Ionophoresis." Academic Press, New York, 1981.
[4] W. L. Nastuk (ed.), "Physical Techniques in Biological Research," Vol. 5, "Electrophysiological Methods," Part II. Academic Press, New York, 1964.
[5] M. A. Dichter, *in* "Bioelectric Recording Techniques," Part A: "Cellular Processes and Brain Potentials" (R. F. Thompson and M. M. Patterson, eds.), p. 39. Academic Press, New York, 1973.
[6] J. L. Barker, D. L. Gruol, L. Y. M. Huang, J. F. MacDonald, and T. G. Smith, Jr., *in* "Role of Peptides in Neuronal Function" (J. L. Barker and T. G. Smith, Jr., eds.), p. 273. Dekker, New York, 1980.
[7] G. D. Fischbach and P. G. Nelson, *Handb. Physiol. Sect. 1; Nerv. Syst.* [*Rev. Ed.*], p. 714 (1977).
[8] D. Gottlieb (ed.), "Society for Neuroscience Short Course: New Approaches in Developmental Neurobiology." Society for Neuroscience, Bethesda, Maryland, 1981.
[9] W. B. Jacoby and I. B. Pastan (eds.), this series, Vol. 58.
[10] J. L. Barker (ed.), "Society for Neuroscience Short Course: Strategies for Studying the Roles of Peptides in Neuronal Function." Society for Neuroscience, Bethesda, Maryland, 1982.
[11] B. R. Ransom, C. N. Christian, P. N. Bullock, and P. G. Nelson, *J. Neurophysiol.* **40**, 1151 (1977).
[12] R. L. Macdonald and P. G. Nelson, *Science* **199**, 1449 (1978).
[13] F. Sachs and A. Auerbach, this volume [9].

voltage-sensitive dyes[14,15] may soon allow direct recordings from synaptic sites.

2. Chronic recordings. It is possible repeatedly to impale the same neuron or to record with an extracellular electrode matrix[16–18] and follow the electrical activity of an ensemble of neurons over several weeks.

3. Use of combined methodologies. The use of biochemical, immunohistochemical, or morphological methods in combination with electrophysiology can allow direct comparison of the properties of individual neurons.[19,20]

4. Identified cell types. It is not yet possible in our spinal cord cultures to identify neuronal subtypes, e.g., a motoneuron vs a Renshaw cell. However, techniques for maintaining cultures of known cell types as well as identification of subtypes within the culture are evolving[21–23] and should allow further correlation of physiology with neuronal type.

5. Developmental studies. Since the dissociated culture technique begins with disruption of the nervous system into single cells, the organization, development, and growth of neurons in culture provides ample opportunity for the study of morphological structure–function relationships, synapse formation and elimination,[24,25] development of chemosensitivity,[26] and formation of ionic channels.[27] Many of these developments may occur early during life in culture (i.e., hours–days) when the neurons are still small (diameter, 10 μm or less) and standard intracellular techniques

---

[14] A. Grinvald, W. N. Ross, and I. Farber, *Proc. Natl. Acad. Sci. U.S.A.* **78**, 3245 (1981).
[15] L. B. Cohen and B. M. Salzberg, *Rev. Physiol. Biochem. Pharmacol.* **83**, 35 (1978).
[16] J. Pine, *J. Neurosci. Methods* **2**, 19 (1980).
[17] C. A. Thomas, Jr., P. A. Springer, B. E. Loeb, Y. Berwald-Netter, and L. M. Okun, *Exp. Cell Res.* **74**, 61 (1972).
[18] G. W. Gross, *IEEE Trans. Biomed. Eng.* **BME-26**, 273 (1979).
[19] E. A. Neale, R. L. MacDonald, and P. G. Nelson, *Brain Res.* **152**, 265 (1978).
[20] E. A. Neale, P. G. Nelson, R. L. Macdonald, C. N. Christian, and L. M. Bowers, **49**, 1459 (1983).
[21] T. M. Jessell, *in* "Society for Neuroscience Short Course: Strategies for Studying the Roles of Peptides in Neuronal Function" (J. L. Barker, ed.), p. 190. Society for Neuroscience, Bethesda, Maryland, 1982.
[22] P. B. Guthrie and D. E. Brenneman, *Soc. Neurosci. Abstr.* **8**, 233 (1982).
[23] L. M. Okun, *in* "Society for Neuroscience Short Course: New Approaches in Developmental Neurobiology" (D. Gottlieb, ed.), p. 109. Society for Neuroscience, Bethesda, Maryland, 1981.
[24] R. F. Mark, *Physiol. Rev.* **60**, 355 (1980).
[25] M. C. Fishman and P. G. Nelson, *J. Neurosci.* **1**, 1043 (1981).
[26] M. B. Jackson, H. Lecar, D. E. Brenneman, S. Fitzgerald, and P. G. Nelson, *J. Neurosci.* **2**, 1052 (1982).
[27] N. Spitzer, *Annu. Rev. Neurosci.* **2**, 363 (1979).

difficult. In these cases, patch clamp may be the technique of choice.[26,28,29]

6. Receptor distribution. Focal iontophoresis of amino acids on spinal cord neurons in culture has revealed local "hot spots" suggesting nonuniform receptor distribution over the cell surface.[30–32] The development of this distribution, its interaction with electrotonic membrane properties and relationship to synaptic innervation are active areas of interest.

7. "Chronic" drug treatment. Cultured cells can be used to test the effects of agents over a period of weeks to mimic chronic treatments. For example, anticonvulsants have been screened in this manner to look for toxic effects.[33–35]

The remainder of this chapter will outline the electrophysiological methods that have been used in our laboratory for intracellular recording in dissociated tissue culture. We have directed our discussion toward investigators who are unfamiliar with intracellular recording. The equipment and methods suggested here are examples of what has worked in our laboratory, but are certainly not the only effective approaches.

Selection of Culture Conditions

Are the cell properties observed in culture the same as *in vivo*? The selection of culture conditions can certainly affect the answer to this question. Selective neuronal survival and changes in phenotypic expression can be produced by variations in growth conditions.[21,36] These effects are both interesting and potentially misleading. Electrophysiological experiments in culture may be altered by changes in growth media; e.g., nerve growth factor (NGF) is routinely added to dorsal root ganglion cells to enhance their growth. The use of conditioned media in addition to NGF can affect the peptide content of such cultures.[37] As another exam-

---

[28] O. P. Hamill, A. Marty, E. Neher, B. Sakmann, and F. J. Sigworth, *Pfluegers Arch.* **391**, 85 (1981).
[29] E. M. Fenwick, A. Marty, and E. Neher, *J. Physiol. (London)* **331**, 577 (1982).
[30] B. R. Ransom, P. N. Bullock, and P. G. Nelson, *J. Neurophysiol.* **40**, 1163 (1977).
[31] J. L. Barker and B. R. Ransom, *J. Physiol. (London)* **280**, 331 (1978).
[32] D. W. Choi and G. D. Fischbach, *J. Neurophysiol.* **45**, 605 (1981).
[33] K. F. Swaiman, B. K. Schrier, E. A. Neale, and P. G. Nelson, *Ann. Neurol.* **8**, 230 (1980).
[34] G. K. Bergey, K. W. Swaiman, B. K. Schrier, and P. G. Nelson, *Ann. Neurol.* **9**, 584 (1981).
[35] B. K. Schrier, in "Nervous System Toxicology" (C. L. Mitchell, ed.), p. 337. Raven, New York, 1982.
[36] P. H. Patterson, *Annu. Rev. Neurosci.* **1**, 1 (1978).
[37] A. W. Mudge, *Nature (London)* **292**, 764 (1981).

ple, Nelson et al.[38] found that glycine in the growth media resulted in a decrease in inhibitory activity between neurons as well as desensitization of the response to iontophoretically applied glycine. Growth media frequently contain horse serum or fetal bovine serum, which has many undefined components. The use of serum-free ("defined") media is an advantage in this regard,[39,40] but physiological results obtained in serum-free media may differ from those obtained in serum-containing media. Until more is known about the effect of components in the growth media on cell phenotype, any change in the growth media must be considered as a potential source of a change in physiological behavior.

Growth conditions can be tailored to meet experimental objectives. For example, cultures of the ventral half of the spinal cord are enriched with motoneurons compared to cultures of whole cord.[22] An ingenious method of obtaining pure cultures of motoneurons using retrograde transport of Lucifer Yellow conjugated with wheat germ agglutinin is another promising approach.[23,41] Cell density in culture may also be varied. A lower plating density (5×10^5 cells per 35-mm dish) reduces polysynaptic activity and simplifies studies of monosynaptic interactions in our spinal cord cultures. Microcultures[42] and special chambers[43] provide other means to study cell–cell interactions. If high-quality photomicroscopy is desired along with intracellular recording, cells can be grown on glass coverslips that provide superior optics to standard tissue culture plastic (see Landis[44] for details). In addition, the selection of the background surface may differentially affect the growth and survival of different cell types. For example, studies of peptide effects on dorsal horn neurons in culture may be critically dependent on the culture substratum, since it appears that the survival of dorsal horn neurons without reaggregation depends on an astrocyte background instead of a collagen background.[21,22] The presence of critical inputs may also be required in order to obtain certain classes of physiological responses, as has been found

---

[38] P. G. Nelson, B. R. Ransom, M. Henkart, and P. N. Bullock, *J. Neurophysiol.* **40,** 1178 (1977).

[39] J. E. Bottenstein and G. Sato, *Proc. Natl. Acad. Sci. U.S.A.* **76,** 514 (1979).

[40] J. E. Bottenstein, S. D. Skaper, S. S. Varon, and G. H. Sato, *Exp. Cell Res.* **125,** 183 (1980).

[41] R. J. O'Brien, L. W. Role, and G. D. Fischbach, *Soc. Neurosci. Abstr.* **8,** 129 (1982).

[42] E. J. Furshpan, P. R. MacLeish, P. H. O'Lague, and D. D. Potter, *Proc. Natl. Acad. Sci. U.S.A.* **73,** 4225 (1976).

[43] R. B. Campenot, this series, Vol. 58, p. 302.

[44] S. C. Landis, in "Society for Neuroscience Short Course: New Approaches in Developmental Neurobiology" (D. Gottlieb, ed.), p. 100. Society for Neuroscience, Bethesda, Maryland, 1981.

with the effect of locus coeruleus explants on norepinephrine responses of spinal cord cells in culture.[45]

Support Equipment and Recording Solutions

Standard electrophysiological experiments can be performed on the stage of an inverted phase-contrast microscope. The microscope should have adequate clearance above the recording dish for placement of up to four electrodes. We use a Zeiss Model D Invertoscope, which fits these criteria. However, the addition of a fluorescent lamp housing to this microscope not only limits access to the stage, but involves a mechanically unstable arm that can transmit vibration to the recording chamber. A microscope with fluorescent and 35-mm photographic capability built into the base (e.g., Leitz Diavert, Nikon Diaphot, or Zeiss IM 35) is preferable if these features are necessary. Neurons can be well visualized and impaled at 160–200 ×. Higher magnifications are useful for photography, but it is difficult to change the objective during an impalement. Nomarski optics can also be used, but the accompanying condenser may limit the working distance above the recording chamber. Hoffman optics are another alternative.

Chambers to maintain the preparation need not be complex. Our chamber consists of a Perspex block with a center well to hold a 35-mm petri dish and a number of side wells for ground baths and iontophoresis current return. A channel, surrounding the center well, contains a heating wire and perforated polyethylene tubing through which a 5% CO_2–95% air mixture flows over the culture dish in order to buffer the bicarbonate-containing recording solution. In conjunction with a substage heating element, the temperature can be maintained within 1–2° across the 35-mm dish. See O'Lague et al.[46] for another chamber design. A recording chamber for cultured cells is also commercially available (Medical Systems Corp.). Note that recordings can be made at room temperature. This minimizes the evaporation from a static bath, but can affect results that are dependent on the kinetic behavior of ionic channels.

Recordings can be made in normal growth media (90% Eagle's minimum essential medium, 10% horse serum in our spinal cord cultures) buffered with bicarbonate, or in a defined salt solution. HEPES (5–10 mM) can be substituted for bicarbonate as a buffer. We add phenol red (10 mg/liter) to all our media and recording solutions to provide constant pH monitoring. The recording solution should be warmed prior to exchanging

[45] K. C. Marshall, R. Y. K. Pun, W. J. Hendelman, and P. G. Nelson, *Science* **213**, 355 (1981).
[46] P. H. O'Lague, D. D. Potter, and E. J. Furshpan, *Dev. Biol.* **67**, 384 (1978).

with the growth media. The osmolarity of the two solutions should be identical to prevent osmotic damage to the cells. We routinely use a static bath containing 1.5 ml of solution in a 35-mm dish (fluid depth 1.5 mm). The fluid is then covered with about 1 ml of lightweight mineral oil (Carnation white mineral oil, Ruger Chemical Co.) to minimize evaporation. Smaller volumes of recording solution will reduce capacitative coupling between recording electrodes, but they increase the risk of osmotic changes due to evaporation and often cause mechanical "drag" between electrodes as they are moved in the dish. The clarity of the phase-contrast image will also be affected by the depth of the fluid in the dish.

In addition to the static bath approach, chambers can be adapted for constant perfusion. The main problems with constant perfusion are mechanical vibration due to fluid turbulence and electrical noise created by the antenna effect of the fluid-filled tubing leading into the chamber. This can be partly circumvented by stopping the perfusion during periods of data collection and/or by shielding the perfusion tubing. Fluid in the reservoir should be warmed to prevent the formation of gas bubbles in the chamber.

The ionic composition of our growth media is similar to that of cerebrospinal fluid except for the increased osmolarity (325–330 mOsm) due to added glucose. We use a defined salt solution for recording that contains 135 mM Na$^+$, 5 mM K$^+$, 1.8 mM Ca^{2+}, 0.8 mM Mg^{2+}, 10 mM glucose, and 5–10 mM HEPES. Sucrose is added to raise the osmolarity to 325 mOsm, and the pH is adjusted to 7.3. However, the solution can easily be modified depending on experimental objectives. For example, we routinely use an elevated divalent cation solution (Ca^{2+} 6 mM, Mg^{2+} 6 mM) when studying monosynaptic connections. This reduces unwanted spontaneous activity, and the elevated Ca^{2+} tends to "stabilize" intracellular penetrations. Synaptic activity can be greatly suppressed with a reversed Ca^{2+}:Mg^{2+} ratio (1:10) or with tetrodotoxin (0.1–1.0 μM). However, certain extreme changes in ionic composition, such as sodium-free or calcium-free solutions, are not well tolerated by cultured cells and should be used for only brief exposures.

Control of Vibration

Control of vibration begins with the selection of the room for recording. If possible it is desirable to place the rig in a basement away from extraneous vibration due to elevators, centrifuges, or hallway traffic. However, low-frequency oscillations intrinsic to the building may still cause a problem. The type and severity of these oscillations can be assessed with an accelerometer if several possible sites for the rig are available. The acceleration (i.e., vibration) transferred from the floor to the recording table can be damped by increasing the compliance between floor and table and by increasing the mass of the table. The best (and most expensive) solution to this problem is a pneumatic table (e.g., Micro G, Technical Manufacturing Corp.). However, we have been successful with the less elegant combination of a marble balance table (700 lb, A. H. Thomas Co.) "floated" on partially inflated motorcycle inner tubes. A metal base plate rests on top of the stone table. The base plate should be as rigid as possible (e.g., 3/4-in. hardened steel) to minimize resonance caused by flexing of the base plate under the weight of instruments.[8] We have placed an additional 200 lb of lead bricks on the baseplate to further increase the mass of the table.

A degree of empiricism is useful, since not all experiments or rooms will require rigorous antivibration measures. In fact, one of our rigs has functioned well on a steel laboratory table with the legs in coffee cans full of sand. Furthermore a well-designed antivibration table can be rendered useless by vibration sources on the table. Cables from equipment racks containing cooling fans, turbulent flow in a perfusion system, or a loose connection in the probe arm holding the microelectrode are all frequent vibration sources of this type.

Micromanipulators

Micromanipulators for intracellular use should be precise and easy to manipulate, free of drift or backlash, and durable. Ease of manipulation is critical for several uses in tissue culture: to impale small cells, to place two electrodes in the same cell, or to position drug pipettes at exact locations over the neuronal surface. The manipulator must then remain in position without drift for several hours. A faulty manipulator is immediately apparent with tissue culture as you watch the electrode drift away from the impaled cell. Mechanical micromanipulators with a heavy base such as the Leitz provide good stability and adequate vertical positioning, but we have encountered problems with these manipulators drifting, especially when using the joystick. In addition, since the controls of this manipulator rest on the rig table, a slight bump with a hand can transmit vibration to the electrode and dislodge an impalement. For these reasons, we prefer a hydraulic manipulator with a remote controller placed off the rig table (Narishige MO-103). Movements of 0.5–1.0 μm in all planes can be made with the manipulator, which greatly facilitates twin impalements and slight adjustments of the electrode within the cell. Backlash has yet to be a problem with the Narishige. Best of all, the cost of the Narishige manipulator is one-third that of the Leitz manipulator. Another alternative is the Burleigh Inchworm micropositioner, which uses piezoelectric elements controlled by a remote microprocessor. However, manipulators are not the sole source of electrode drift. Loose connections in the electrode holder assembly, vibration or drag on the hydraulic lines leading to the manipulator controller, or backlash of the microscope stage movement can all cause drift.

For most recordings, the electrode–manipulator assembly rests on the rig table or on the nonmovable microscope stage, thus impalements are limited to one optical field (ca. 500 μm diameter at 160 ×). For some experiments it is desirable to be able to move the viewing field without disturbing an electrode already in a cell, e.g., to locate the postsynaptic contacts of a presynaptic neuron with long processes. This can be done with lightweight manipulators by mounting them directly on the movable recording chamber, or by fixing large manipulators to a movable framework driven by the stage carriage (see Calvet and Calvet[47] for one such design).

Electronics for Intracellular Recording

The basic elements of the recording circuit include a preamplifier ("head stage") usually of unity gain, a second stage for further amplification, and an output device such as an oscilloscope, chart recorder, FM tape recorder, or microprocessor. The use of output devices for recording in culture differs little from other preparations and therefore will not be discussed here. Rather, we will focus on the preamplifier and the problems created by the high-resistance microelectrodes (50–100 megaohms) used for intracellular recording in culture.

The most important specifications of the preamplifier are a high input impedance (at least 10^{11} with 100-megaohm electrodes) to prevent loss of signal due to a "voltage divider" effect; and a low leakage current to prevent flow of current out of the preamplifier and across the cell membrane. In addition, the preamplifier must have a sufficiently wide bandwidth to record accurately rapid time-varying signals. However, the stray capacitance to ground due to a high-impedance glass microelectrode is the major limitation on the frequency response of the recording system. The microelectrode input circuit acts as a low-pass filter with a time constant, $R_e C_{tot}$, where R_e is the electrode resistance and C_{tot} is the total capacitance of the input to the preamplifier. For this reason, preamplifiers designed for intracellular recording incorporate a capacitance neutralization circuit to increase the frequency response of the input circuit. The "nega-

[47] J. Calvet and M. C. Calvet, *J. Neurosci. Methods* **4**, 105 (1981).

tive" capacitance can be adjusted by passing a square current pulse across the microelectrode and increasing the capacity neutralization until the voltage pulse is "squared off." Although this technique is straightforward, a number of limitations exist. The C_{tot} of the input circuit consists of several components including the distributed capacitance across the wall of the microelectrode to the bath (C_e), the capacitance to ground due to the cable from electrode to headstage (C_c), and the capacitance of the head stage (C_A). C_e is not adequately compensated by the neutralization circuit; and therefore must be reduced by other means, such as using microelectrodes with the lowest possible impedance, keeping the fluid level low in the dish, and when necessary shielding the microelectrode to as near the tip as possible. Microelectrodes can be shielded within 25–50 μm of the tip by coating them with silver paint (Silver Print, GC Electronics) and then covering the paint with an insulating layer of Loctite Super Bonder 420 (see Smith et al.[48]). A somewhat different method has been reported by Sachs and McGarrigle.[49]

C_c can be reduced by using a shielded cable and keeping the cable length to an absolute minimum. Both shields can then be "driven" to keep the shield and the input signal at the same potential, and thereby prevent current flow across the capacitance. It is important to realize that a grounded shield reduces electrical field interference and capacitative coupling between electrodes, but may actually *increase* the capacitance of the shielded electrode by effectively increasing the surface area of the capacitor. As with all recording methods, the need to minimize C_{tot} depends on the experiment. Voltage clamping of fast currents (e.g., spike and synaptic currents) with microelectrodes requires extensive efforts to reduce capacitance, whereas voltage recording of spikes and membrane potentials requires use only of the capacitance neutralization circuit of the preamplifier. However, optimal "tuning" of the capacitance compensation can cause oscillation of the amplifier and destruction of the cell if a sudden change occurs in C_{tot} (e.g., as can occur when the fluid level fluctuates during perfusion or the microelectrode impedance changes). Capacity compensation also increases the high-frequency electrical noise. For these reasons it is sometimes desirable to record with an undercompensated circuit that can improve the signal-to-noise ratio. Unwanted electrical noise can also be reduced with a simple low-pass filter (e.g., dc = 1 kHz). The appropriate cutoff frequency of the filter will depend on the frequency spectrum of the biological signal.

[48] T. G. Smith, Jr., J. L. Barker, B. M. Smith, and T. R. Colburn, *J. Neurosci. Methods* **3**, 105 (1980).
[49] F. Sachs and R. McGarrigle, *J. Neurosci. Methods* **3**, 151 (1980).

Another problem with single microelectrode impalements arises from the need to pass current and record voltage through the same microelectrode. Since current injection causes a voltage drop across the electrode resistance, most intracellular preamplifiers use a Wheatstone bridge circuit to "balance out" this voltage, so that only the transmembrane voltage is recorded. This method demands that the microelectrode resistance must not vary as a function of the current. Unfortunately, the resistance of microelectrodes becomes nonlinear (as well as "noisy") with currents larger than about 1 nA. This can make bridge balance difficult and the measurement of transmembrane potential subject to large errors. Furthermore, the bridge is usually balanced before impalement, but the balance frequently changes after impalement and continues to change during the experiment. Thus constant adjustment is necessary. For example, if the effect of a peptide on the cell membrane is monitored by trains of short current pulses to measure input resistance, then accurate bridge balance is crucial. There is no simple solution to this problem except constant vigilance and attempts to use lower-resistance microelectrodes. The best way to avoid this problem is to use two electrodes in the same cell, one for current passing and one for voltage recording. This is a far superior method if accurate measurements of membrane potential or passage of large currents are necessary to the experiment.

A discussion of the types of current sources is beyond the scope of this chapter. However, the source should produce a current independent of the load impedance (i.e., the microelectrode impedance). A so-called constant-current generator or current pump meets this requirement. These are usually driven by ±15-V power supplies and thus have a maximum current delivery of 150 nA through a 100-megaohm electrode. This is an optimistic maximum, since electrode resistance may increase markedly as large currents are passed and therefore further limit the current maximum. This may be inadequate for certain applications, such as voltage clamping of large currents, or iontophoresis. For such uses, a "high voltage" current source may be necessary.[48,50]

See Purves[3] and Fein[51] for further discussion of sources of current.

Grounding and Interference

All biological signals are measured relative to a reference or ground electrode that in intracellular recording is in contact with the chamber

---

[50] B. M. Smith and B. J. Hoffer, *Electroencephalogr. Clin. Neurophysiol.* **44,** 398 (1978).
[51] H. Fein, "An Introduction to Microelectrode Technique and Instrumentation." W-P Instruments, Inc., New Haven, Connecticut, 1977.

bath. We use an agar bridge (glass tubing filled with 1.5% agar in saline) to connect the bath to a side well of the recording chamber. The agar bridge serves to keep changes in the bath composition or temperature from affecting the reference electrode. The side well contains a chlorided silver wire immersed in 1 M KCl. The wire is then connected either directly to the oscilloscope frame that serves as system ground or through a current-to-voltage converter, "virtual ground." See Fein[51] for the uses of a "virtual ground" circuit. All pieces of equipment in the rig, including Faraday cage, amplifiers, iontophoresis units, camera motors, should connect directly to system ground to avoid "ground loops." Care should be taken to avoid any extra paths to ground from the bath as can occur owing to crusted salt on the edge of the 35-mm dish. Wiring should be kept as simple as possible to facilitate troubleshooting. All equipment should then be connected to the ground of the ac line voltage at a single point.

The high-impedance microelectrode input circuit acting as an "antenna" is a sensitive "receiver" of electrical interference transmitted from 60-Hz ac power lines via capacitative coupling. Approaches to interference reduction include shielding the sources (e.g., ac power cables), use of dc rather than ac power sources, especially for equipment within the Faraday cage (e.g., microscope lamp), physical separation of sources from the input circuit, and shielding the input circuit. Fluorescent room lights and the power cable to the microscope lamp have been major sources of interference in our setup.

It is best to try to prevent grounding and interference problems by assembling a rig within the general guidelines mentioned above. Inevitably, however, problems occur, and the frustrating process of troubleshooting begins. A volt-ohmmeter, and a "model cell-electrode" consisting of resistors and capacitors with realistic values for electrode impedance and cell membrane are essential to this process. For a number of practical troubleshooting tips and sample interference problems, see Purves.[3]

Electrodes

Micropipettes for intracellular recording can be reliably fabricated with a two-stage puller such as the Brown–Flaming design (Sutter Instrument Co.). An adjustable jet of nitrogen gas serves to reduce the length of the shank without affecting tip size.[52] The most critical component in day-to-day operation of this puller is the platinum heating coil. This coil can be easily damaged when inserting the glass pipette, and excessive heating can cause "sagging" of the coil. The choice of glass depends on the application. "Thin-walled" (e.g., 0.9 mm i.d., 1.2 mm o.d.; Frederick

[52] K. T. Brown and D. G. Flaming, *Neuroscience* **2**, 813 (1977).

Haer Co.) borosilicate micropipettes make good electrodes for intracellular dye injection, iontophoresis, or micropressure. Thin-walled glass is desirable also to use for recording electrodes owing to its lower impedance (20–50 megaohms) compared to thick-walled pipettes (0.6 mm i.d., 1.2 mm o.d.) of the same tip size (80–100 megaohms). However our cultured cells frequently deteriorate after impalements with thin-walled micropipettes. Use of fiber-filled glass tubing allows back-filling of the micropipette immediately before use. A 27-gauge needle is ideal for this purpose. A 0.22-μm filter should be used on the filling syringe to prevent clogging of the electrode tip with crystals or debris. Either 3 M KCl or 3 M K$^+$ acetate (pH adjusted to 7.0 with KOH) are our electrolytes for recording. K$^+$ acetate or K$^+$ methylsulfonate do not affect the chloride equilibrium potential of the cell and therefore are useful in studies involving IPSPs or the chloride channel. For special purposes other electrolytes can be used, such as 1–2 M CsCl to block outward potassium currents or Lucifer Yellow to record and inject dye simultaneously.

Other glass tubing such as theta glass (e.g., TST 150, W-P Instruments), star glass (Radnoti), or multibarreled pipettes have all been advocated for special purposes, but we have not routinely used these in our cultures. The same is true of the beveling technique, although beveling has a number of potential advantages.[53,54]

After filling, the micropipette is coupled with a silver–silver chloride "half-cell" (e.g., WPI electrode holder) and connected by shielded cable to the head stage. Several DC offset potentials occur in this system: (1) liquid junction potentials at the interface of different electrolytes, e.g., at the electrode tip where cytoplasm and 3 M KCl interface; (2) electrode potentials at the interface of the silver–silver chloride with an electrolyte; and (3) "tip" potentials[55] presumably due to charge separation near the tip of fine-glass micropipettes. The first two of these potentials are usually stable and thus easily offset; however, tip potentials (and liquid junction to a lesser degree) may change greatly with impalement and thus give an inaccurate membrane potential. If this is suspected, the tip of the electrode can be broken and the change in dc potential taken as a rough estimate of the tip potential. A number of detailed discussions of glass microelectrode behavior are available.[3,56,57]

---

[53] K. T. Brown and D. G. Flaming, *J. Neurosci. Methods* **1**, 25 (1979).
[54] T. E. Ogden, M. C. Citron, and R. Pierantoni, *Science* **201**, 469 (1978).
[55] D. P. Agin, in "Glass Microelectrodes" (M. Lavallee, O. F. Schanne, and N. C. Hebert, eds.), p. 62. Wiley, New York, 1969.
[56] L. A. Geddes, "Electrodes and the Measurement of Bioelectric Events." Wiley (Interscience), New York, 1972.
[57] M. Lavallee, O. F. Schanne, and N. C. Hebert (eds.), "Glass Microelectrodes." Wiley, New York, 1969.

Cell Impalement

Many intracellular preamplifiers now incorporate a "buzzer" circuit to facilitate entry into the cell. This circuit delivers a high-frequency current pulse to the cell membrane. A similar effect can be obtained by overcompensating the capacitance neutralization until the amplifier oscillates. We position the microelectrode with the tip just touching the neuronal membrane and then delivering a short "buzz" via a pushbutton on the panel of the preamplifier. The time-honored method of tapping a finger on the table is usually too traumatic for small spinal cord cells, but can be used for cells with "tougher" membranes, such as dorsal root ganglia neurons. After the sudden DC shift signaling entry into the cell, there is usually a period of a few minutes until the membrane "seals" around the microelectrode, as evidenced by an increase in membrane potential and input resistance. It can be helpful to inject DC and pulsed hyperpolarizing current during this period. Twin impalements by two independent microelectrodes are feasible if the cell soma is about 15 μm or larger. Waiting until the first electrode has sealed in the cell helps to prevent dislodging the first when the second electrode is buzzed into the cell. We position the two electrodes as far away from each other in the soma to lessen mechanical and capacitative coupling between electrodes.

After impalement, an assessment of cell "health" can be made on the basis of its physiological behavior. Reliance on any one criterion can be misleading. Criteria used for healthy cells, such as membrane potential, input resistance, spike amplitude, must be evaluated on the basis of ionic composition of the bath, cell type, age of the cells in culture, and the parameter under investigation. The reported values for healthy impalements have gradually improved as techniques have improved, so that criteria must be constantly reevaluated. We have also found that cells frequently "run down" after a period of impalement despite a well maintained membrane potential and spike. This "run down" may be manifest as an increased leak conductance, decrease in EPSP amplitudes, or a decrease in ionic currents (e.g., calcium current) under voltage clamp. The explanation for this behavior is unclear.

Iontophoresis

Techniques of iontophoresis have been well established *in vivo*.[58-62] Iontophoresis has been equally well applied in tissue culture to study chemosensitivity and receptor distribution over the neuronal surface,[30-32]

---

[58] D. R. Curtis, *in* "Physical Techniques in Biological Research," Vol. 5: "Electrophysiological Methods," Part I (W. L. Nastuk, ed.), p. 88. Academic Press, New York, 1964.

as well as for noise analysis.[63,64] Iontophoresis has the advantage of rapid discrete application, but the disadvantage that an unknown concentration of substance is delivered to the cell surface (but see Dreyer, et al.[65]). In addition, since ejection of substances with a low pK_a requires a low solution pH, artifacts due to iontophoresis of H^+ ions can occur. Hydrogen ion artifacts have been reported with pH values as high as 4–5.[66] The iontophoretic current may also cause a membrane response that can be confused with a drug response. The use of an automatic "current return" through another electrode helps to avoid this problem.[3,50]

Microperfusion

To circumvent some of the problems of iontophoresis, the delivery of substances to the neuronal surface by pressure ejection from a small-tipped (2–10 μm) micropipette has become a widely used technique.[6,12,32] This technique, variably called microperfusion, miniperfusion, or "puffer" application, is well suited to dissociated cultures. Some advantages of this approach include delivery of a known concentration of substance; the ability to test substances of low solubility or those available only in minute quantities; rapid onset and washout of application compared to bath perfusion; and use of multiple concentrations to obtain dose–response information.[67]

A very simple delivery system can be constructed using a nitrogen gas cylinder as a pressure source, pressure regulator, pressure valves, noncompliant tubing, and a micropipette holder with a port for gas delivery.[68] Delivery can be controlled manually or via a pulse generator and solenoid circuit. A commercial system is also available (Picospritzer, General Valve Corp.). Pipettes can be prepared from thin-walled glass and pulled

---

[59] K. Krnjevic, in "Methods of Neurochemistry" (R. Fried, ed.), Vol. I, p. 129, Dekker, New York, 1971.
[60] A. Globus, in "Bioelectric Recording Techniques" (R. F. Thompson and M. M. Patterson, eds.), Part A, p. 23. Academic Press, New York, 1973.
[61] F. E. Bloom, Life Sci. 14, 1819 (1974).
[62] J. S. Kelly, in "Handbook of Psychopharmacology" (L. L. Iversen, S. H. Iversen, and S. D. Snyder, eds.), Vol. 2, p. 29, Plenum, New York, 1975.
[63] R. N. McBurney and J. L. Barker, Nature (London) 274, 596 (1978).
[64] H. Lecar and F. Sachs, in "Excitable Cells in Tissue Culture" (P. G. Nelson and M. Lieberman, eds.), p. 137. Plenum, New York, 1981.
[65] F. Dreyer, K. Peper, and R. Sterz, J. Physiol. (London) 281, 395 (1978).
[66] D. L. Gruol, J. L. Barker, L. Y. M. Huang, J. F. MacDonald, and T. G. Smith, Jr., Brain Res. 183, 247 (1980).
[67] L. M. Nowak, A. B. Young, and R. L. Macdonald, Brain Res. 244, 155 (1982).
[68] R. E. McCaman, D. G. McKenna, and J. K. Ono, Brain Res. 136, 141 (1977).

on a standard microelectrode puller. The tip is broken back to the desired size by touching the tip against a glass rod under a low-power microscope. Alternatively, the tip can be brushed against tissue paper, although this provides less consistency in tip size. Tip size can be adjusted to fit the experiment. To change the extracellular ionic environment (e.g., increase [K^+]), the tip can be made 5–10 μm in diameter, positioned 50–100 μm away from the impaled cell, and gentle pressure (less than 1 psi) applied to the back of the pipette. With proper technique, the effective concentration reaching the neuron should be within 10–20% of that in the pipette (see Choi and Fischbach[32]). To deliver a discrete application of a potent substance, a smaller tip (2–5 μm) and a higher pressure can be used. In general, all tips larger than 5 μm will leak significantly and should be kept out of the recording media between each application.

Since flow varies from one pipette to another, a number of precautions should be taken. Remove air bubbles with a finger tap or with a large pressure pulse before placing the pipette in the bath. The direction and intensity of flow can then be tested by aiming the pipette at a piece of debris on the bottom. We then lower the pressure to the lowest value that still gives a continuous flow. Excessive pressure causes artifacts (see below) and can dislodge the recording electrode. Flow can also be monitored by the use of a microcolloid. Drugs should be diluted in the recording media and the pH and osmolarity carefully matched. To avoid adsorption of peptide solutions onto glass or polystyrene surfaces, stock solutions can be stored frozen in polypropylene vials and dilute solutions placed in the glass pipettes just before use. Spontaneous hydrolysis of drugs may be reduced if solutions are kept on ice during the experiment.

Despite the apparent simplicity of this method, a number of artifacts can and do occur. These are particularly a problem when the anticipated effect of the applied substance is a small or subtle change in neuronal properties, such as a decrease of a small potassium current under voltage clamp. The most frequently observed artifact is a small hyperpolarization that may be accompanied by a small increase in input resistance. This artifact increases with increased pressure and can also be seen when pressure is directed at the recording electrode *outside* the neuron. Whether this effect is due to ionic changes at the cell surface (e.g., washout of accumulated K^+) or is a direct physicochemical effect is unclear. Subtle or gross artifacts can also be caused by the perfusion of hypoosmotic solution onto the cell. A difference between solutions of 10 mOsm is more than enough to cause this effect. When severe, this results in membrane depolarization, but less overt osmotic effects can occur. To safeguard against these problems, the use of a control perfusion with recording media from a pipette with similar tip size, location, and applied pressure compared to the drug pipette is essential.

Pressure systems have also been designed for intracellular injection of substances through micropipettes with <1 μm tips,[69,70] but we have had no experience with this technique.

Intracellular Staining

A reliable and permanent record of the morphology of a physiologically studied neuron can be extremely valuable. An ideal intracellular dye should be easy to inject with a small amount of current and usable as a recording electrode, should diffuse rapidly and completely throughout the cell, be nontoxic to the cell, and be easy to fix for both light and electron microscopy. No dye at present meets all these criteria. However the complementary properties of the fluorescent dye, Lucifer Yellow (LY), and horseradish peroxidase (HRP) provide a reasonable pair of choices depending on experimental needs.[71-73]

We use a 4% solution of LY (Sigma) to fill standard recording micropipettes. The resistance of these electrodes is 3–4 times that of the same micropipettes filled with K^+ acetate. Neurons can be easily impaled, and spikes and synaptic potentials recorded. However, values of membrane potential may be unreliable owing to large DC offset potentials with LY electrodes. Cultured spinal cord neurons can be well filled with small hyperpolarizing currents (1 nA × 1 min). Fine processes of some cells, such as dorsal root ganglia neurons, are less well filled. Photographs should be taken immediately if possible, since prolonged viewing under fluorescent optics causes bleaching and is toxic to the dye-filled cell. LY can be fixed with paraformaldehyde for light microscopy, but special procedures are needed for electron microscopy.[74] We have also used another fluorescent dye, carboxytetramethylrhodamine (Molecular Probes), which can be diluted in 50 mM KCl and injected with hyperpolarizing current. Filling of fine processes is not as good as with LY, but cell toxicity does not appear to be a problem.

HRP (type VI, Sigma) provides excellent definition of fine dendrites and synaptic boutons. However, HRP is more difficult to inject than the fluorescent dyes, and must be fixed and allowed to react with a chromagen before it is visible. Injection and fixation techniques in our laboratory

---

[69] Y. Kallstrom and S. Lindstrom, *Brain Res.* **156**, 102 (1978).
[70] M. Sakai, B. E. Swartz, and C. D. Woody, *Neuropharmacology* **18**, 209 (1979).
[71] W. Stewart, *Cell* **14**, 741 (1978).
[72] W. W. Stewart, *Nature (London)* **292**, 17 (1981).
[73] G. A. Bishop and J. King, in "Tracing Neural Connections with Horseradish Peroxidase" (M.-Marsel Mesulam, ed.), IBRO Handbook Series: Methods in the Neurosciences, p. 185. Wiley, New York, 1982.
[74] A. R. Maranto, *Science* **217**, 953 (1982).

have been previously described.[19] The large depolarizing currents (3 nA × 10 min) required for adequate filling with HRP can cause deterioration of the cell. To avoid this problem, the neuron can be impaled with a second microelectrode to deliver a balancing hyperpolarizing current. Microelectrodes filled with 4% HRP in 0.5 M KCl-Tris buffer have resistances about twice that of K^+ acetate microelectrodes. Electrode blocking is a problem with HRP electrodes, but can be reduced by using thin-walled glass or beveled electrodes.

Voltage Clamp

The use of voltage clamp methods with microelectrodes has been reviewed,[48] therefore only a few points will be mentioned here. The power of the voltage clamp technique is the ability to measure membrane conductance directly and determine the time- and voltage-dependent characteristics of the ionic currents that contribute to the membrane conductance. The voltage clamp is particularly applicable to the study of substances that may act on time- and voltage-dependent currents, such as the Ca^{2+} current or the M current (e.g., see Dunlap and Fischbach[75] and Adams et al.[76]). Current clamp studies on such effects can be quite misleading, since an *apparent* resistance increase (i.e., conductance decrease) under current clamp may actually be due to a membrane chord conductance increase under voltage clamp.

Voltage clamp amplifiers have traditionally been designed for two electrodes, one for current passing and one for voltage recording. The two-electrode clamp is generally superior in performance if it is possible to penetrate the neuron with two independent microelectrodes. However, several one-electrode voltage clamps have also been designed using either a switch and hold circuit[77,78] or an iterative voltage sampling process.[79] The primary limitations of the two electrode voltage clamp with cultured spinal cord neurons are related to the use of high-impedance microelectrodes. The large stray capacitance and limited current-carrying capabilities of the microelectrodes must be overcome in order to clamp the membrane voltage rapidly and accurately. The methods we use to reduce the stray capacitance have already been discussed earlier in this chapter. A high-voltage (e.g., 100 V) source may be necessary to overcome the limited current-carrying ability of the microelectrode.

[75] K. Dunlap and G. D. Fischbach, *J. Physiol. (London)* **317**, 519 (1981).
[76] P. R. Adams, D. A. Brown, and A. Constanti, *J. Physiol. (London)* **332**, 223 (1982).
[77] W. A. Wilson and M. M. Goldner, *J. Neurobiol.* **6**, 411 (1975).
[78] R. E. Wachtel and W. A. Wilson, *J. Neurosci. Methods* **4**, 87 (1981).
[79] M. R. Park, W. Leber, and M. R. Klee, *J. Neurosci. Methods* **3**, 271 (1981).

The design of the voltage clamp and the severity of microelectrode limitations depend on the speed and amplitude of the membrane currents under study. We have utilized a "mixed" slow-fast clamp designed by T. G. Smith, Jr. and B. M. Smith (see Smith et al.[48] for details) to clamp "slow" membrane currents and iontophoretic responses effectively, using only a driven shield on the voltage electrode above the surface of the bath. The "settling time" of the membrane voltage with this arrangement is approximately 0.5 msec with 60-megaohm microelectrodes, but the complete decay of capacitative current takes 1–2 msec, presumably reflecting less than ideal space clamp. The error produced by an inadequate space clamp will vary with the shape and electrotonic properties of different cells, but it does not appear to cause major qualitative errors in our cultured spinal cord and dorsal root ganglia neurons where length constants are close to one.[80,81]

A method combining intracellular perfusion with voltage clamp analysis has been developed for use with both invertebrate and vertebrate cells.[82,83] This method has the obvious advantage of controlling the ionic composition on both sides of the membrane and allows the use of low-resistance electrodes greatly to increase the current-passing ability of the clamp. The patch clamp technique also provides a method to voltage clamp a patch of membrane and thus avoids the space clamp limitations of the traditional voltage clamp. Patch clamp techniques have been recently modified for use as a "whole cell" clamp.[28,29]

Patch Electrodes for Intracellular Recording

The patch technique was initially used to record single channels in the membrane under the electrode. However the technique has been modified for intracellular recording of either currents (whole cell clamp) or voltages (current clamp).[28,29] Patch electrodes for current clamp recording have a number of significant advantages over conventional glass microelectrodes including a much lower impedance (2 MΩ vs 50–100 MΩ) and a higher current-carrying capacity, and they are especially useful for cells as small as 5 μm. Rapid diffusion from the large-tipped (1–2 μm) patch electrode allows for dye injection or control of intracellular ionic composition.

Electrodes can be fabricated in the same manner as for single-channel recording (e.g., two-stage pull on a vertical pipette puller, Kopf Model

---

[80] B. R. Ransom, E. A. Neale, M. Henkart, P. N. Bullock, and P. G. Nelson, *J. Neurophysiol.* **40**, 1132 (1977).
[81] T. H. Brown, D. H. Perkel, J. C. Norris, and J. H. Peacock, *J. Neurophysiol.* **45**, 1 (1981).
[82] P. G. Kostyuk, O. A. Krishtal, and V. I. Pidoplichko, *J. Neurosci. Methods* **4**, 201 (1981).
[83] K. S. Lee, N. Akaike, and A. M. Brown, *J. Neurosci. Methods* **2**, 51 (1980).

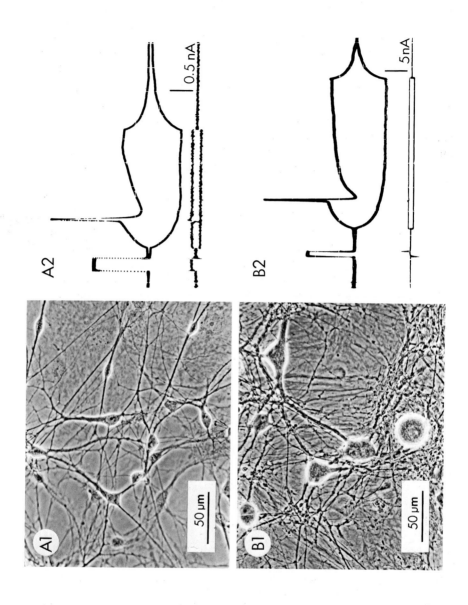

700). Fire polishing of the tip appears to improve seal formation, but may not be necessary for all applications. We use 1.5 mm o.d. microhematocrit capillary tubing (Fisher Scientific Co.) filled with a solution (K^+ 144 mM, Mg^{2+} 2 mM, EGTA 1.1 mM, pH 7.2, mOsm 310) that is slightly hypotonic to our media (320–325 mOsm). Filled electrodes can be dipped in melted wax (Fisher Pyseal 220) or coated with polystyrene Q-dope (GC Electronics) to reduce electrode capacitance and reduce the optic glare from the interface of the blunt electrode shaft with the bath. A current pulse is passed through the electrode to monitor contact with the cell surface. Mild suction by mouth or syringe is then adequate to create a seal. Bridge compensation usually increases from 2–4 MΩ to 15–20 MΩ after penetration as a result of the series resistance. The low Ca^{2+} concentration and hypoosmolarity of the patch solution seem to help break the cell membrane.

In our spinal cord cultures, this recording method seems to cause less cell injury than conventional glass microelectrodes based on the higher values of cell input resistance (R_N). It should be noted, however, that the measured R_N (either for standard microelectrodes or patch electrodes) is an equivalent resistance composed of the cell membrane in parallel with the seal resistance (R_s), and thus is accurate only if $R_s \gg R_N$. Intracellular recordings from small spinal cord neurons (5–10 μm) can be routinely obtained in this manner. In fact, the irregular surface of the large neurons after several weeks in culture makes the formation of seals more difficult. Figure 1 shows typical recordings from both small neurons (7-day culture), and large neurons (28-day culture). Many additional practical suggestions are included in the references cited at the beginning of this section.

Summary

The application of electrophysiological techniques to tissue culture is still evolving. We have attempted in this chapter to give a practical summary of intracellular recording techniques used in our laboratory, as well

---

FIG. 1. (A) Phase contrast photomicrograph (A1) and accompanying typical intracellular patch electrode recording (A2, upper trace: voltage; lower trace: current) of mouse spinal cord neurons after 7 days in culture. The small size of the neurons (5–10 μm) precludes use of conventional glass microelectrodes. (B) Phase contrast photomicrograph (B1) of 28 day coculture of spinal cord and dorsal root ganglion neurons, at which time routine use of conventional microelectrodes is possible. However, intracellular patch recordings can also be used (B2) on these "mature" cultures. Calibration pulse: A2, 50 mV, 10 msec; B2, 50 mV, 2 msec.

as give some examples of new experimental strategies and electrophysiological methods that should provide further information on a number of interesting neurobiological questions. The combination of an increasing knowledge of the cell biology of cultured neurons and advances in electrophysiology should continue to be a fruitful interaction.

Acknowledgments

We wish to thank Drs. Peter Guthrie and Douglas Brenneman for reading earlier versions of this chapter, Dr. Raymund Pun for help with the section on patch recording, and Linda Bowers for photographic assistance. G. L. W. was a PRAT Fellow of the Pharmacological Sciences Program, NIGMS.

[8] Neuroendocrine Cells *in Vitro:* Electrophysiology, Triple-Labeling with Dye Marking, Immunocytochemical and Ultrastructural Analysis, and Hormone Release

By JAMES N. HAYWARD, TROY A. REAVES, JR., ROBERT S. GREENWOOD, and RICK B. MEEKER

Recent physiological[1-3] and immunocytochemical[4] studies show that mammalian hypothalamic supraoptic (NSO) and paraventricular (NPV) nuclei contain heterogeneous populations of functional and peptidergic neuroendocrine cell types. As a result, rigorous new methods for analysis are necessary. Previously studies of vertebrate magnocellular neuroendocrine cells (MgC) *in vivo* and *in vitro* have used both extracellular and intracellular electrophysiological techniques in conjunction with dye-marking. Generally they lacked immunocytochemical identification, ultrastructural analysis, or measurement of secretory output.[1-3] Exceptions were our studies in the goldfish[5-7] and those of Kayser *et al.*[8] in the rat hypothalamic slice, where MgC were electrophysiologically analyzed,

---

[1] J. N. Hayward, *Physiol. Rev.* **57**, 574 (1977).
[2] J. N. Hayward and T. A. Reaves, Jr., in "Comprehensive Endocrinology, Endocrine Functions of the Brain" (M. Motta, ed.), p. 21, Raven, New York, 1980.
[3] D. A. Poulain and J. B. Wakerley, *Neuroscience* **7**, 773 (1982).
[4] L. W. Swanson and P. E. Sawchenko, *Neuroendocrinology* **31**, 410 (1980).
[5] T. A. Reaves, Jr. and J. N. Hayward, *Cell Tissue Res.* **202**, 17 (1979).
[6] T. A. Reaves, Jr. and J. N. Hayward, *Proc. Natl. Acad. Sci. U.S.A.* **76**, 6009 (1979).
[7] T. A. Reaves, Jr. and J. N. Hayward, *J. Comp. Neurol.* **193**, 777 (1980).
[8] B. E. J. Kayser, M. Muhlethaler, and J. J. Dreifuss, *Experientia* **38**, 391 (1982).

dye-marked, and immunocytochemically identified. Armstrong and Sladek[9] recorded extracellularly from phasic-firing antidromically identified rat supraoptic (NSO) neuroendocrine cells in the acute organ-perifused hypothalamoneurohypophysial complex (HNC). This HNC explant, for many years a chronic organ-cultured system for endocrinological study of vasopressin (AVP) and neurophysin (NP) secretion,[10] includes the supraoptic neurons (NSO) and the final neurosecretory pathway to the neural lobe. The following review discusses preliminary work on an *in vitro* system whereby the characteristics of mammalian neuroendocrine cells identified by antidromic stimulation, dye-marking, and immunocytochemical and ultrastructural analysis can be correlated with hormone release.[11,12]

Hypothalamoneurohypophysial Complex (HNC) *in Vitro*

The acute organ perifused HNC rat explant with tissue movement virtually eliminated and the anatomical sites under direct visual observation provides a significant advantage over *in vivo* preparations for intracellular electrophysiological analysis of magnocellular neuroendocrine cells (MgC).[1,2,11,12] Resting and stimulated firing patterns and membrane potentials recorded *in vitro* from "identified" MgC can be correlated with rat neurohypophysial hormone secretion under rigidly controlled conditions. In addition, complex extrahypothalamic influences, such as nociceptive input, state of arousal, body temperature, and cardiovascular state, are eliminated.[1-3]

Preparation of Explants. From decapitated adult male rats (Sprague–Dawley; 150–250 g) we excise a 15–30-mg diamond-shaped block of ventral hypothalamic tissue with the neurointermediate lobe (NIL) still connected, according to Sladek and Knigge[10] (Fig. 1). Using the caudal approach, we remove the brain quickly from the skull in serial steps: the skull cap is removed, the petrous temporal bones are crushed, the lower eight cranial nerves are cut, the dura is stripped from the posterior sella turcica, the dura, vessels, and nerves in the cavernous sinuses are cut, the pituitary is deflected toward the hypothalamus, the brain with pituitary resting on it is gently removed forward from the skull. All this is accomplished without traction on the pituitary stalk. After severance of the optic and olfactory nerves, the brain is removed from the skull. Under a

[9] W. E. Armstrong and C. D. Sladek, *Neuroendocrinology* **34**, 405 (1982).
[10] C. D. Sladek and K. M. Knigge, *Endocrinology* **101**, 411 (1977).
[11] T. A. Reaves, Jr., A. Hou-Yu, E. A. Zimmerman, and J. N. Hayward, *Neurosci. Lett.* **37**, 173 (1983).
[12] T. A. Reaves, Jr., M. T. Libber, and J. N. Hayward, *Soc. Neurosci. Abstr.* **8**, 531 (1982).

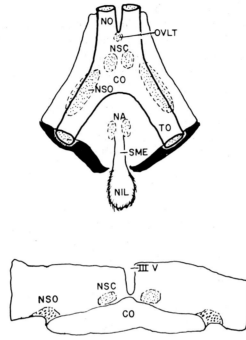

FIG. 1. Hypothalamoneurohypophysial complex (HNC) of the rat. *Upper:* Ventral surface of the HNC explant diagrammatically showing the optic nerves (NO), optic chiasm (CO), optic tracts (TO), stalk-median eminence (SME), and neurointermediate lobe (NIL). Projected deep structures (stippled) include the nuclei—suprachiasmatic (NSC), supraoptic (NSO), and arcuate (NA)—and a circumventricular organ, the organum vasculosum of the lamina terminalis (OVLT). *Lower:* Coronal section of the HNC through the rostral optic chiasm (CO) showing the locations of the NSO and NSC nuclei and the preoptic recess of the third ventricle (III V).

dissecting microscope (8 ×) with the ventral hypothalamus up and repeatedly bathed in cold (5°), oxygenated (95% O_2–5% CO_2) Yamamoto's[13] solution, we dissect the adenohypophysis free from the neurointermediate lobe (NIL). Care is taken to strip the carotid vessels, their branches, and the associated pia arachnoid from the hypothalamus without damage to the stalk. The NIL is then reflected over the median eminence, and the HNC explant is isolated by three horizontal and five vertical scissors cuts (Fig. 1). We remove the HNC explant on a moistened stainless steel spatula tip and place the ventral surface up on a nylon net in a Plexiglas perifusion chamber (Fig. 2). The time from decapitation to removal of the HNC explant is approximately 10–15 min.

[13] C. Yamamoto and H. McIlwain, *Exp. Brain Res.* **13**, 1333 (1966).

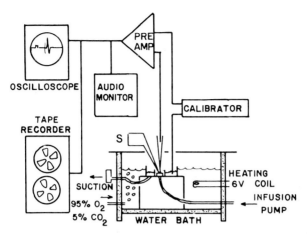

FIG. 2. Diagram of the Plexiglas chamber support system and instrumentation for electrophysiology and dye-marking of single cells and hormone release in the rate HNC explant *in vitro*. The Plexiglas chamber (lower right) consists of covered water bath, oxygenated by bubbled gas (95% O_2–5% CO_2), warmed by a resistance heating coil and 6-volt battery (6V). The HNC explant is fixed between two nylon nets in the isolated central compartment perifused by Yamamoto's solution (infusion pump, suction), impaled by a recording microelectrode and stimulating electrode (S). The microelectrode connects to the amplifier, calibrator, audiomonitor, oscilloscope, and tape recorder. Adapted from Sakai et al.,[14] Dingledine et al.,[15] and Hatton et al.[16]

Plexiglas Chamber Support System. As shown in Fig. 2 the Plexiglas chamber support system consists of an outer water bath maintained at 36–37° by a resistance heating coil (6 V battery) and oxygenated by bubbled gas (95% O_2–5% CO_2) after Sakai et al.,[14] Dingledine et al.,[15] and Hatton et al.[16] The central tissue chamber (30 mm in diameter × 14 mm deep), isolated from the water bath, consists of a central nylon net compartment (two concentric nylon net rings, i.d. 11 mm × 1 mm, 200-μl volume, containing the HNC explant on polyethylene mesh) into which flows the Yamamoto medium[13] (2.4 mM $CaCl_2$; osmolality 295–306 mOsm per kilogram of H_2O; oxygenated with 95% O_2–5% CO_2) at 0.38–0.76 ml/min at 36–37°. A pump removes the perifusion fluid from the central HNC explant compartment allowing the HNC explant to lie at a fluid–gas interface for optimal oxygenation.[15,16] The water bath cover allows for optimal humidity, temperature, and oxygenation of the HNC explant (Fig. 2).

[14] K. K. Sakai, B. H. Marks, J. M. George, and A. Koestner, *J. Pharmacol. Exp. Ther.* **190**, 482 (1974).
[15] R. Dingledine, J. Dodd, and J. S. Kelly, *J. Neurosci. Methods* **2**, 323 (1980).
[16] G. I. Hatton, A. D. Doran, A. K. Salm, and C. D. Tweedle, *Brain Res. Bull.* **5**, 405 (1980).

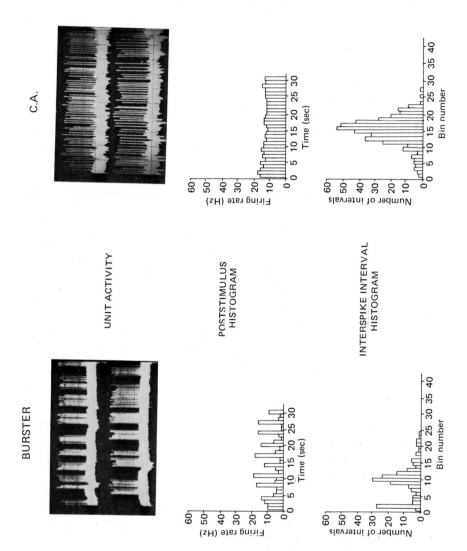

Recording, Stimulation, and Electrophysiological Data Analysis

Recording. Lucifer Yellow-CH (LY[17]; 2% in 0.1 M Tris buffer with 0.15 M lithium chloride, pH 7.4)-filled glass micropipettes beveled to an impedance of 5–40 megohms[18] sustain the intra- and extracellular recording. We connect the recording electrodes through a silver–silver chloride junction to the miniature probe of a unity gain "negative capacitance" broad-band "electrometer" preamplifier (W.P.I., M707), and to a stimulator that serves as the source of current for injecting the dye.[19] An electronically controlled stepping-microdrive (A.B. Transvertex) guides the microelectrode. After adequate amplification in a high-gain dc amplifier, the signals are displayed on the face of a dual-beam storage oscilloscope, photographed, and stored on an FM tape recorder as illustrated in Fig. 2. A reference electrode (silver–silver chloride junction) is placed in the outer compartment of the central tissue chamber (Fig. 2). We connect a calibrator (Bioelectric, CA-1) in series to the oscilloscope ground, trigger it by the oscilloscope time-base amplifier, and place the calibrator pulse at the beginning of each oscilloscope sweep (Fig. 2). The restoration of this square-wave signal, i.e., criteria for proper neutralization,[19] uses negative capacitance feedback to the microelectrode.

Stimulation. For antidromic activation of neurosecretory neurons, we place twisted bipolar Teflon-insulated platinum–iridium microwires (AM Systems, 0.18 mm in diameter, 0.5 mm tip separation) with exposed tips across the stalk median eminence (SME, Fig. 1). An electronic stimulator (Grass, S-88) with an isolated constant-current unit (WPI-601) delivers bipolar square-wave pulses (0.1–5.0 mA and 0.1–1.0 msec duration, 1/sec intervals) for antidromic activation of these neurosecretory neurons.

Electrophysiological Data Analysis. We analyze data on intracellular recording from antidromically identified NSO cells (action potentials 40 mV or greater) that remain stable for over 3 min, are labeled with fluores-

---

[17] W. W. Stewart, *Nature* (*London*) **292**, 17 (1981).
[18] K. T. Brown and D. G. Flaming, *Brain Res.* **86**, 172 (1975).
[19] J. N. Hayward, *J. Physiol.* (*London*) **239**, 103 (1974).

FIG. 3. Unit activity and histograms of two spontaneously active supraoptic neurons recorded extracellularly from the rat hypothalamic explant. *Left:* Burster. Cell firing with alternating periods of rapid firing bursts and periods of silence is shown in the photograph (top). Poststimulus histogram (center) demonstrates periodic firing pattern. Interspike interval histogram (lower) shows biphasic peaks, indicating short intraburst intervals and long intervals associated with periods of silence. *Right:* Continuously active (C.A.). Steady cell firing is shown in the photograph (top). Poststimulus histogram (center) is nonperiodic. Interspike interval histogram (lower) shows a monophasic peak indicating uniform distribution of intervals.

cent marker (LY), and prove to be peptide-containing by our double-labeling immunocytochemical criteria.[5-7] The characteristics of antidromic latencies, action potentials, and excitatory and inhibitory postsynaptic potentials are stored on magnetic tape as continuous dc recordings for later analysis. On-line and off-line photographs of the storage oscilloscope are shown in Fig. 3. We analyze these extracellular supraoptic neuronal spike trains on a PDP-11-03 digital computer (Digital Equipment). Single-unit responses, stored on the magnetic tape recorder, are monitored continuously with a pulse-detector window, displayed on polygraph paper (Offner Type R dynograph), stored in the digital computer. We analyze single supraoptic unit activity by means of a histogram program,[20] which displays interspike-interval, peri- (poststimulus, autocorrelation or joint) interval and provides for statistical descriptive and comparative analyses using parametric and nonparametric methods (histograms and firing-rate characteristics). Figure 3 shows the poststimulus and interspike interval histograms of "burster" and "continuously active" supraoptic neurons from the rat HNC explant *in vitro*.

Dye-Marking, Fixation, and Microtomy, Localization of Single-Label Cells

Dye-Marking. Electrophoretic injection intracellularly of the anionic fluorescent dye (LY[17]) according to the techniques of Hayward and co-workers[5-7,19] allows the localization of the physiologically studied supraoptic neurons (NSO) in the HNC explant. After intracellular recording, we inject dye into the cell by a constant hyperpolarizing current of 5–10 nA for 0.5–6.0 min. The fluorescent dye-marker (LY) labels one cell per supraoptic nucleus.[5-7,19]

Fixation and Microtomy. After each experiment, we immerse the HNC explant in freshly prepared 4% paraformaldehyde with 0.1–0.25% glutaraldehyde in 0.1 M phosphate buffer (PB; pH 7.4) for fixation. After postfixing the explant overnight at 4° in the same fixative, we embed it in 5% agar–saline, section it serially (30 μm, vibratome) in the frontal plane, collect sections in cold 0.1 M PB saline (PBS), and rinse them in cold PBS.

Localization of Single-Labeled Cells. In order to select serial sections containing the soma and processes of dye-filled supraoptic neurons, we examine wet vibratome sections in 0.1 M PBS on glass slides under a Leitz Orthoplan microscope with Ploem illumination from a HBO 200 mercury vapor lamp. A 4BG-38 red suppression filter (330–600 nm), a

---

[20] D. H. Perkel, G. L. Gerstein, and G. P. Moore, *Biophys. J.* **7**, 391 (1967).

3BG-12 excitation filter (330–500 nm), an effective dichroic combination of beam-splitting mirror (495 nm), and a built-in suppression filter (495 nm) are used to visualize the bright yellow (LY dye-filled) supraoptic (NSO) neuron bodies and processes against a dark background. In Fig. 4 the LY-filled NSO neuron soma and processes lie adjacent to the autofluorescent myelin of the optic chiasm (OC). These LY-filled cells are photographed with Kodak Tri-X Pan film. We usually see these injected cells in several sections, then reconstruct them with montage photographs or camera lucida tracing. In our estimates of cell size, we determine the two largest internal diameters at right angles and use the mean of these measurements.[5–7,19]

Immunocytochemistry and Identification of Double-Labeled Cells

Immunocytochemistry. We stain sections containing LY-filled somata by the indirect immunofluorescence technique,[21] incubate unmounted PBS-rinsed sections sequentially in (a) PBS-containing 1% normal sheep serum (NSS; 10 min); (b) mouse monoclonal antibody to arginine vasopressin 1 : 100 to 1 : 200 in PBS/NSS, overnight at room temperature (clone III D-7, compliments of Dr. E. A. Zimmerman[11,22] or rabbit anti-rat neurophysin antiserum 1 : 100 to 1 : 2000 in PBS/NSS; overnight at room temperature (Lot RN 4, compliments of Dr. A. G. Robinson[12,23]); (c) PBS/NSS, 10 min; (d) tetramethylrhodamine isothiocyanate (TRITC)-conjugated goat anti-mouse (or anti-rabbit) IgG antisera, 1 : 50 to 1 : 150 in PBS/NSS at room temperature for 1–4 hr (Cappel Laboratories); (e) PBS. We determine specificity of the antibody by substituting normal mouse (rabbit) serum for the primary antibody, omitting the primary antibody step, or absorbing (liquid or solid phase) the primary antibodies with excessive amounts of their respective antigens.[24] We scan sections for TRITC (rhodamine) immunofluorescence under incident UV light at 580 nm.

Identification of Double-Labeled Cells. As Fig. 5 shows, we first identify double-labeled supraoptic neurons of the HNC explant by examining the section under incident UV light at 495 nm for Lucifer Yellow (LY) fluorochrome fluorescence and then at 580 nm for TRITC-fluorochrome fluorescence.[5–7] By alternating the selection of exciter and barrier filters at

---

[21] A. H. Coons, in "General Cytochemical Methods" (J. F. Danielli, ed.), Vol. 1, p. 399. Academic Press, New York, 1958.
[22] A. Hou-Yu, P. H. Ehrlich, G. Valiquette, D. L. Engelhardt, W. H. Sawyer, G. Nilaver and E. A. Zimmerman, *J. Histochem. Cytochem.* **30,** 1249 (1982).
[23] H. W. Sokol, E. A. Zimmerman, W. H. Sawyer, and A. G. Robinson, *Endocrinology* **98,** 1176 (1976).
[24] L. A. Sternberger, "Immunocytochemistry." Wiley, New York, 1979.

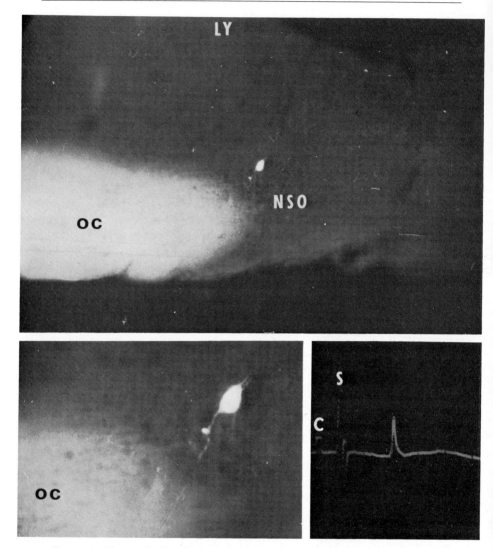

FIG. 4. Antidromically identified supraoptic neuron (NSO) *in vitro* in rat hypothalamo-neurohypophysial complex (HNC) injected intracellularly with Lucifer Yellow (LY). *Upper:* LY-filled cell in NSO of HNC explant adjacent to the optic chiasm (OC). Not immunocytochemically defined. *Lower left:* Enlarged view of LY-filled bipolar cell with processes emerging from upper and lower poles. *Lower right:* Antidromic action potential. C, calibrating square wave, 1 msec, 10 mV; S, stimulus artifact. Fixation of tissue with 5% glutaraldehyde enhances autofluorescence of optic chiasm at 495 nm and abolishes immunoreactivity.

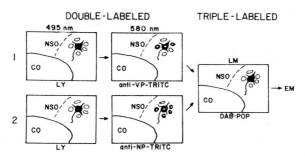

FIG. 5. Identification protocol for double- and triple-labeled neurons in rat supraoptic nucleus (NSO). *Left:* Double-labeled. Fluorescence microscopy (495 nm, left row) of vibratome sections (30 μm) shows the somata of two hypothetical neurons (black) that were injected intracellularly with Lucifer Yellow (LY) and found on sections in the NSO adjacent to the optic chiasm (CO). The right-hand row shows the same two sections (upper and lower) when viewed at 580 nm after applying different primary antisera (VP, arginine vasopressin; NP, neurophysin) to each section with the indirect immunofluorescence technique using tetramethylrhodamine isothiocyanate (TRITC)-tagged secondary antibody. The LY-injected neuron is identified as VP or NP, since its unique profiles are seen at both 495 and 580 nm. *Right:* Triple-labeled. Under light microscopy (LM) the LY-filled and immunocytochemically identified NSO cell develops a brown granular material in the nucleus and cytoplasm (DAB–POP) after photooxidation (POP) under UV light (495 nm) in a solution of diaminobenzidine (DAB). Osmication intensifies the electron density of this DAB–POP reaction product for studies at the ultrastructural (EM) level.

high power magnification (312–500 ×), we can see the unique correspondence between LY and rhodamine-fluorescent cell profiles. We photograph double-labeled supraoptic neurons at 495 nm and 580 nm using Kodak Tri-X Pan film for a permanent record. Camera lucida drawings of adjacent sections will permit reconstruction of the soma and processes of immunocytochemically identified neurons. Figure 6 shows the sequential analysis of a physiologically identified supraoptic neuron in the HNC explant. By comparing cell location, size, contour, and processes on the same section, a single LY-filled cell (Fig. 6A) can be identified as neurophysin (NP) immunoreactive.[5–7,12] These NP-identified NSO neurons show greater morphological detail at 495 nm (LY fluorescence) than at 580 nm (TRITC fluorescence) (Fig. 6 A,B).[5–7]

If the initial attempt to identify a LY-filled NSO neuron immunocytochemically is negative to vasopressin antibodies (Fig. 5, upper row), the section can be rinsed in PBS/NSS and then restained with another primary antibody, such as anti-neurophysin (Fig. 5, lower row) following methods outlined above in section Immunocytochemistry. This procedure can be repeated using different primary antisera until the supraoptic neuron is immunocytochemically identified.

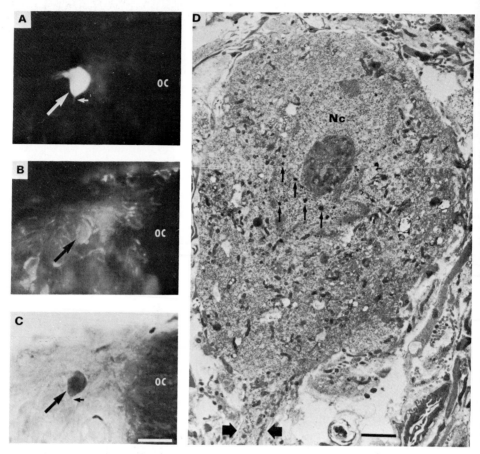

FIG. 6. Light microscopic and ultrastructural visualization of an electrophysiologically studied, neurophysin-containing neuron in the rat supraoptic nucleus (NSO) of HNC *in vitro*. (A) Lucifer Yellow (LY)-filled NSO magnocellular neuron near the optic chiasm of HNC explant. The large arrow denotes cell soma; the small arrow indicates a major process. Calibration in (C); OC, optic chiasm. (B) Immunocytochemical identification of the LY-filled neuron (arrow in A) using the indirect immunofluorescence method with an antiserum against neurophysin. The immunofluorescent profile (580 nm, TRITC) is nearly identical to the LY profile in (A). Calibration in (C). (C)Visualization of the DAB photooxidation product within the LY-injected neuron after irradiation (495 nm). Tissue was osmicated and embedded in Epon. The large arrow indicates cell soma; small arrow, the major process. Calibration, 30 μm; OC, optic chiasm. (D) Electron micrograph of contrasted ultrathin section cut from neuron in (C). Note intensely stained nucleolus (unlabeled) and electron-dense spots (small arrows) in the nucleus (Nc), vacuoles (unlabeled) in the cytoplasm, and the major process (large arrows) shown in panels (A) and (C). Calibration, 2 μm. Hayward, Reaves, Greenwood, Libber, and Meeker, unpublished study.

Osmophilic Polymer (DAB-POP), LM and EM Identification of Triple-Labeled Cells

Our double-labeling technique, combining immunocytochemistry with intracellular recording and the intracellular injection of Lucifer Yellow-CH (LY[17]), allows us to use the fluorescence light microscope (LM) exclusively for immunocytochemical identification.[5-7] Since these LY-filled cells are not electron dense,[17] the ultrastructural analysis (EM) of a physiologically characterized, dye-marked neuron was difficult or impossible until Maranto[25] observed that irradiation of a LY-filled neuron with intense blue light (430 nm) stimulates the photooxidation of diaminobenzidine (DAB) into an insoluble osmophilic polymer. This polymer becomes visible under white light microscopy as a reddish-brown reaction product and under the electron microscope as discrete granular deposits within the LY-injected cell.[25,26] Combining Maranto's[25] procedure with our double-labeling technique,[5-7] we find that an electrophysiologically studied single neuron can be LY-injected (single label), immunocytochemically identified (double label), UV-irradiated, and studied at the light (LM) and electron microscopic (EM) levels (triple label).[12,26]

Osmophilic Polymer (DAB–POP). We rinse unmounted sections containing LY-injected, double-labeled NSO neurons in Tris buffer (0.1 M, pH 7.6, 10 min), soak them in a solution of 3,3'-diaminobenzidine tetrahydrochloride (DAB; 1.5 mg per milliliter of Tris buffer; 15 min) and irradiate them at 495 nm in DAB until the LY fluorescence fades below visibility.[12,25,26] A final rinse in Tris buffer allows us to examine the irradiated sections with white light to confirm the presence of the reddish-brown reaction product within the injected cell. The designation "Triple-Labeled" in Fig. 5 indicates the presence of this reddish-brown reaction product within LY-injected and immunocytochemically identified (double-labeled) neurons.[12,25,26]

LM Identification of Triple-Labeled Cells. Unmounted sections containing Lucifer Yellow (LY)-injected triple-labeled NSO neurons are postfixed in phosphate-buffered 2% osmium tetroxide for 0.5–2 hr at room temperature. We then process the sections through graded alcohols and propylene oxide, infiltrate with Epon 812, and embed in Epon between Teflon-coated cover slips.[27] These triple-labeled cells we photograph under the light microscope (LM). The dense brown reaction product contrasts well with the surrounding neuropil (Fig. 5, Triple Labeled;

[25] A. R. Maranto, *Science* **217**, 953 (1982).
[26] T. A. Reaves, Jr., R. Cummings, M. T. Libber, and J. N. Hayward, *Neurosci. Lett.* **29**, 195 (1982).
[27] D. K. Ramanovicz and J. S. Hanker, *Histochem. J.* **9**, 317 (1977).

Fig. 6C).[12,25,26] Many details observed in LY-filled neurons, such as soma contour, processes, and location, are present in these diaminobenzidine–photooxidation product (DAB–POP) triple-labeled neurons[12,25,26] (Fig. 6C).

EM Identification of Triple-Labeled Cells. After trimming a small region containing these DAB–POP triple-labeled cells from the Epon wafers, we reembed them in Epon capsules. For single-cell localization, we trim with a Pyramitome (LKB) into a mesa as described by Berthold *et al.*[28] After first preparing 1-μm sections for orientation, we then cut ultrathin sections and mount them on copper-mesh grids for observation with a Zeiss EM109 electron microscope. Selected grids are contrasted with uranyl acetate and lead citrate.

At the ultrastructural (EM) level, each LY-neurophysin (LY-NP) soma has granular DAB–POP distributed throughout the cytoplasm and stands out clearly from the surrounding neuropil even in uncontrasted sections.[12,25,26] Uranyl acetate and lead citrate contrasting greatly increases the electron density of the LY-NP DAB–POP somata (Fig. 6D) and, to a lesser degree, of non-LY filled somata.[12,25,26] In general, the ultrastructural characteristics of injected and noninjected neurons are reasonably well preserved (Fig. 6D).[12,25,26] Some LY-NP DAB–POP neurons display ultrastructural alterations characteristic of mammalian Procion Yellow-injected neurons.[28] These alterations, presumably the result of injury from dye injection or from electrode penetration, include moderate swelling of intracellular organelles, the presence of intranuclear granules, vacuoles within the cytoplasm, and distribution of synaptic boutons.[12,25,26,28] Figure 6D shows electron-dense intranuclear granules, cytoplasmic vacuoles, and some disruption of the surrounding neuropil, perhaps reflecting these changes. Fibers stained with DAB–POP are usually observed within the initial segments of processes, perhaps related to the intensity of LY fluorescence (Fig. 6D).[12,25,26] The intense DAB–POP staining of the nucleolus in Fig. 6D may be the result of extensive accumulation of LY within this structure.[5]

Our triple-labeling identification technique of neurons studied electrophysiologically, cell-marking with a fluorescent dye, immunocytochemical and ultrastructural analysis, may allow researchers to describe functional subsets of magnocellular peptidergic neurons more precisely.

Radioimmunoassay of AVP, Data Analysis, AVP Release

The HNC explant in the rat provides an ideal model system for correlating the electrophysiologically studied, triple-labeled magnocellular

---

[28] C.-H. Berthold, J.-O. Kellerth, and S. Conradi, *J. Comp. Neurol.* **184**, 709 (1979).

neuron with hormone release under controlled conditions. While many endocrinological studies of the HNC explant utilize the 3–4-day organ-cultured HNC complex,[10] we use the acute (1–8 hr) perifused HNC explant. The release of arginine vasopressin (AVP) is measured by radioimmunoassay (RIA) from samples collected as we record and/or stimulate (local, general, electrical, osmotic, chemical, pharmacological) the magnocellular supraoptic neurosecretory neurons.

Radioimmunoassay of Arginine Vasopressin (AVP). We determine AVP levels in the HNC explant perifusate by a radioimmunoassay (RIA) system,[29] using synthetic AVP (Manning) and anti-AVP antibody (H-1; Hayward-1). We use carrier-free vasopressin, 8-arginine, ^{125}I-labeled monoiodinated with a specific activity of 2200 Ci/mmol (New England Nuclear, NEX-128) as a tracer. The H-1 antibody we developed earlier by immunizing rabbits with AVP–thyroglobulin–carbodiimide conjugate.[29] The H-1 antiserum detects a minimum of 0.28 pg (0.1 μU) of AVP standard. Synthetic lysine vasopressin (LVP), arginine vasotocin (AVT), and oxytocin (OT) were, respectively, about 1/3, 1/50, and 1/100 as reactive as AVP.[29] Since acetone extraction from Yamamoto's solution is not necessary, we have eliminated the extraction step and substitute 50 mM phosphate buffer, 100 mM NaCl, 0.1% BSA, pH 7.4, for the Veronal buffer. Recovery of AVP (5.6–14 pg/ml; 2–5 μU/ml) from Yamamoto's medium is 95–98%. Intraassay coefficient of variation is 7%, and interassay coefficient of variation is 17%.[29]

Data Analysis. We use a laboratory computing system (Hewlett–Packard, 9815A) with an RIA software program (HP, 09815-14251) for logit-log analysis to calculate AVP values. AVP values are expressed both as picograms per minute and as microunits per minute.

Vasopressin Release. For study of the central neural regulation of arginine vasopressin (AVP) or other neurohypophysial peptides, the HNC explant has several advantages over the intact rat or other slice techniques. (a) We can compare the responses of identified single cells (electrophysiologically studied triple-labeled AVP neurons) with cell-population responses (AVP release) either simultaneously or sequentially under rigidly controlled environmental conditions. (b) We can visualize directly various hypothalamic structures for positioning stimulating probes (electrical, osmotic, chemical, pharmacological). (c) We can compare stimulated AVP release from diffusely (perfusion media) vs locally (microinjection) applied osmotic, chemical, and pharmacological agents to the HNC explant. (d) We can work without the confounding influences of the extrahypothalamic responses (behavioral, thermoregulatory, cardio-

[29] J. N. Hayward, K. Pavasuthipaisit, F. R. Perez-Lopez, and M. V. Sofroniew, *Endocrinology* **98**, 975 (1976).

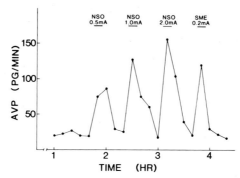

FIG. 7. Arginine vasopressin (AVP) release from the neurointermediate lobe (NIL) of rat hypothalamoneurohypophysial complex (HNC) *in vitro* following electrical stimulation. Bilateral supraoptic nucleus (NSO) stimulation with increasing current strengths of 0.5, 1.0, and 2.0 mA, resulted in incremental, pulsatile AVP releases of 87, 128, and 156 pg/min, respectively, above the baseline release of 12–25 pg/min. Stalk-median eminence stimulation at 0.2 mA resulted a comparable pulsatile AVP release of 121 pg/min. Note the stimulus-bound character of AVP release. Other NSO stimulation parameters include: 15 Hz, 0.5 msec bipolar pulse duration, 30-sec on–off for 10 min using deep, bilateral, bipolar insulated tungsten microelectrodes with tip separation of 0.8 mm. Other SME stimulation parameters include 15 Hz, 0.25-msec bipolar pulse duration; 5-min continuous stimulation using a single surface; bipolar, Teflon-insulated platinum iridium microwires with 0.5 mm tip separation. Hayward, Qasim, Greenwood, and Meeker, unpublished study).

vascular, others) to the applied stimuli. (e) We can collect the total and continuous output of neurohypophysial secretion. (f) We can study the secretory responses of the NSO in isolation from the paraventricular nuclei (NPV).[4]

As Fig. 7 shows, discrete electrical stimulation of the supraoptic nuclei (NSO), at increasing electrical current strengths results in a progressively larger pulsatile (stimulus-bound) release of AVP into the perfusion medium. In comparison, the electrical stimulation of the stalk median eminence (SME) at different parameters also has a similar pulsatile (stimulus bound) AVP release. We see similar pulsatile AVP release from osmotic and pharmacological stimuli.

Conclusions

We present a method for analyzing vasopressin (AVP) supraoptic (NSO) magnocellular neuroendocrine cells (MgC) in the rat hypothalamo-neurohypophysial complex (HNC) *in vitro*. Rapidly removed from ventral hypothalamus, the HNC complex consists of the complete neuroendocrine final common pathway—supraoptic neurons, supraopticoneurohypophysial tracts, and neural lobe. Under carefully controlled *in vitro* conditions of perifusion, we combine electrophysiological studies of identified single vasopressin-containing neurons and stimulated AVP release. Our single-cell identification techniques include cell-marking with a fluorescent dye, immunocytochemical double-labeling, and DAB–POP triple-labeling for ultrastructural studies. We expect that these methods will allow further detailed studies of other peptidergic MgC neurohypophysial neurons, such as oxytocin-, enkephalin-, dynorphin-, and cholecystokinin-containing cells.

Acknowledgments

The authors wish to thank D. Cronce, M. T. Libber, and K. Michels for skillful technical assistance, S. Curtis for helpful criticism of the manuscript, Dr. A. G. Robinson (Grant AM-16166) for providing the anti-neurophysin (RN 4) serum, Drs. R. Dingledine and G. King for advice on construction of the recording chamber, and Dr. C. D. Sladek for kindly demonstrating the HNC explant preparation. This work was supported, in part, by Grant NS-13411 from the National Institute of Health.

[9] Single-Channel Electrophysiology: Use of the Patch Clamp[1]

By FREDERICK SACHS and ANTHONY AUERBACH

Ion channels are integral membrane proteins that catalyze the diffusion of ions across cell membranes. A single ion channel may pass, e.g., 20 picoamperes (pA) of current, which is equivalent to a flow of some 10^8 ions/sec. Conformational changes in the channel can turn this flow on and off, thereby producing pulses of current. The patch clamp is a technique that can easily resolve these current pulses, making it the most sensitive assay known for the study of protein conformational changes. Using traditional macroscopic techniques, the kinetic and permeation properties of ion channels are inferred from measurements of the amplitude and time course of currents arising from ensembles of many thousand channels. With the patch clamp, the transition rates between various conformational states of a channel and the current amplitude of each state can be measured directly for individual channels.

[1] This work was supported by Grant NS-13194 from the National Institute of Neurological Diseases and Stroke, U.S. Public Health Service.

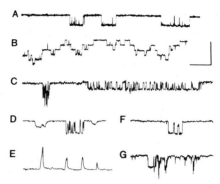

FIG. 1. Examples of single-channel currents from membrane patches attached to chick skeletal muscle cells. Unless otherwise noted, the pipette and bath solutions were physiological saline and the temperature was 22°. Inward current is down. (A) Currents from nicotinic acetylcholine receptor–channels activated by 200 nM acetylcholine in the pipette. The currents are rectangular pulses that are clustered into bursts. The baseline is defined as the current level between bursts. Many of the gaps within bursts are too brief to be completely resolved and thus appear as spikelike events that do not reach the baseline. The presence of both long and short populations of gaps indicates that the channel can exist in more than one closed state. The bandwidth (f_c) was 3.5 kHz, and the pipette potential (V_p) was +100 mV relative to the bath (calibration: 13 pA, 35 msec). (B) Nicotinic channels activated by 25 nM suberyldicholine in the pipette. The amplitude of the current at any given time is an integral multiple of the unitary current (2.4 pA). The maximum current level is 11.9 pA, so that at least five channels were active in the patch. Since nicotinic channels are known to carry inward current at the cell resting potential ($V_p = 0$), the baseline was arbitrarily defined as the level of least inward current. Unlike the record shown in (A), the time spent at any given current level cannot be unambiguously associated with the dwell time of a channel in an open state (f_c = 1.25 kHz, 10 pA, 105 msec). (C) Bursts of openings from two different types of channel in a single patch. The leftmost burst is from a Ca^{2+}-activated, K^+-selective channel. Both the open and closed dwell times in the burst are too brief to be resolved, hence the true open-channel current cannot easily be determined. The rightmost burst is from a channel of undetermined selectivity. These two types of channel can be clearly distinguished by their amplitude and kinetic properties (35°, V_p = +20 mV, pipette contained 140 mM KCl, 10 mM HEPES, and 1 mM EGTA, f_c = 8 kHz, 25 pA, 7 msec). (D) A record showing both rounded and square-edged channel currents. Currents of both types are from nicotinic channels activated by pentyltrimethylammonium, an agonist that can act also as a channel blocker. The rounded channels have smaller and more variable amplitudes than do the square ones. The difference in size and shape indicates that not all the channels in the patch have the same access resistance to the pipette interior. One explanation is that the rounded-channel currents arise from "rim" channels located in the sealing region of the patch (f_c = 1.25 kHz, V_p = +60 mV, 12 pA, 30 msec). (E) A record where all channel currents are rounded, probably owing to the formation of a vesicle in the tip of the pipette. The pipette contained the same solution as in trace C. The rising edge of the channel current is faster than the falling edge because during the rising edge the channel is open and the conductance is higher (V_p = −100 mV, f_c = 8 kHz, 18 pA, 40 msec). (F) A nicotinic channel burst showing more than one conducting state. The second gap within the burst appears to have an amplitude that is about one-fourth that of the full open-channel current (taken from the same record as trace F. The channel was activated by 50 nM acetylcholine (f_c = 2.5 kHz, V_p =

The essential feature of the patch clamp is the isolation of a small patch of cell membrane, perhaps a few square micrometers, within the tip of a glass micropipette. The currents flowing into and out of the pipette across the membrane patch are measured, and, under appropriate conditions, unitary channel currents can be detected. While there is wide variation in the characteristics of single-channel currents, prototypically they are rectangular pulses with an amplitude of a few picopamperes that persist for times ranging from milliseconds to seconds. Some typical single-channel currents are shown in Fig. 1.

In addition to its great sensitivity, the patch clamp has several advantages over macroscopic methods of measuring channel currents. The seal between the cell membrane and the electrode is mechanically stable so that a patch of membrane may be excised with either the cytoplasmic or extracellular face of the membrane facing the interior of the electrode. Thus, the composition of the solutions along either face of the membrane as well as the membrane potential can be controlled. In contrast to whole-cell recording techniques, patch clamp recording from isolated patches does not require a balance of ionic or osmotic strength between the solutions on either side of the membrane. Because the interior of the pipette is chemically, as well as electrically, isolated from the bath, when the patch is still attached to the cell, drugs may be applied either to the patch of membrane in the pipette or exclusively to the surrounding cell membrane. Since the currents are recorded from a small area of membrane, the patch clamp offers a high degree of spatial resolution so that channel characteristics in different regions of a single cell can be measured. The patch clamp can be used to record either voltage or current from small, whole, cells with time resolutions approaching that of axial-wire voltage clamps. Finally, owing to the high impedance of the patch relative to the pipette, unitary currents are virtually free from errors due to series resistance, incomplete space clamp, or ion accumulation effects.

We will briefly discuss some of the fundamental aspects of patch clamp electrophysiology, including cell preparation, electrode construction, seal formation, instrumentation, and the analysis of single-channel data. Detailed treatments of patch-clamp technology are contained in

---

+130 mV, 15 pA, 12 msec). (G) An example of "seal breakdown." The rectangular pulse to the left is a nicotinic channel current taken from the same record as trace F. The large, inward-going spikes are thought to be currents arising from the breakdown of the membrane and/or seal. These currents are most common when large voltages are applied to the pipette. When the pipette and bath solutions are symmetrical, these currents reverse polarity when the pipette potential is approximately that of the bath. Breakdown can occur in both cell-attached and excised patches.

Hamill et al.,[1a] Fenwick et al.,[2] and several excellent chapters of Sakmann and Neher.[3]

Cell Preparation, Electrode Construction, and Seal Formation

The key to successful patch clamping is the formation of a high-resistance seal between the cell membrane and the wall of a current-collecting pipette. In practice, seal resistances of greater than 10^{11} ohms (100 Gohms) have been obtained. These high values indicate that the membrane must be in close apposition to the glass, perhaps within molecular dimensions. The salient interaction in seal formation appears to be between the glass and the membrane lipids (rather than the membrane proteins or some other element), since high-resistance seals can be made to pure phospholipid membranes.

Cell Treatments

To obtain a tight seal, the cell surface must be free of connective tissue and, if possible, basement membrane. Intact tissues can be enzymatically cleaned of connective tissue by any one of a variety of methods. For example, snake neuromuscular junctions have been treated for 2 hr at room temperature in 2 mg of collagenase per milliliter (Advance Biofactures Corp., Lynwood, NY; Form TD) followed by 20–40 min in 0.02 mg of protease per milliliter (Sigma Chemical Co., St. Louis, MO: type XIV[4]). A more aggressive treatment for *Helix* neurons consists of treating the preparation with 0.5 mg of Pronase E per milliliter for 10 min at 20° followed by treatment in 1% trypsin for 1–2 hr at 37°.[5] Alternative treatments include 10–20 min in 0.2% trypsin for *Aplysia* neurons[6] or 1 hr in 5% papain for vertebrate cortical neurons.[7] Cells grown in tissue culture and "loose" cells, such as red blood cells and mast cells, can be sealed with no enzymic treatment.

With most cells it is possible to form seals with resistances greater than 10 Gohms. Some cells, such as locust striated muscle, resist forming seals above 1 Gohm. The reason for this difficulty in sealing is not known,

[1a] O. P. Hamill, A. Marty, E. Neher, B. Sakmann, and F. J. Sigworth, *Pfluegers Arch.* **39,** 85 (1981).
[2] E. Fenwick, A. Marty, and E. Neher, *J. Physiol.* (*London*) **331,** 557 (1982).
[3] B. Sakmann and E. Neher, eds. "Single Channel Recording in Biological Membranes" (B. Sakmann and E. Neher, eds.). Plenum, New York, 1983. In press.
[4] V. E. Dionne and M. D. Leibowitz, *Biophys. J.* **39,** 253 (1982).
[5] H. D. Lux and K. Nagy, *Pfluegers Arch.* **391,** 242 (1981).
[6] S. A. Siegelbaum, J. S. Camardo, and E. R. Kandel, *Nature* (*London*) **299,** 413 (1982).
[7] R. Numann, R. K. S. Wong, and R. B. Clark, *Soc. Neurosci. Abstr.* **8,** 413 (1982).

but in some preparations, such as the frog neuromuscular junction, seal formation may be difficult because of the convoluted microstructure of the cell surface. A general observation is that for a given cell type there is a wide variation from day to day and from preparation to preparation in the ability to form tight seals. This variability probably comes from the cell surface rather than the electrodes, but no controlling variables have yet been defined.

Pipette Construction

Patch-clamp electrodes are similar to standard microelectrodes and are only slightly more difficult to construct. A good patch electrode has a steep taper with a tip diameter of about 1 μm and has a resistance in the range of 1–10 Mohms when filled with physiological saline.

Because of its lower melting point, flint glass is easier to work with than either borosilicate or aluminosilicate glass and, with proper preparation, has equivalent noise and sealing properties. Hematocrit tubing or microcapillary glass with an outside diameter of about 1.5 mm is satisfactory. Either a two-stage horizontal or vertical electrode puller can be used for making electrodes. With the horizontal puller, the first stage should draw the glass to about 200 μm in diameter, and the second stage should be adjusted to give a weak pull so that the glass has a chance to cool before being broken. With the vertical puller, a calibrated mechanical stop is used to limit the first-stage travel to about 5 mm. The glass is then recentered in the heater, and the second-stage pull is performed at a lower temperature. For both stages of pull, the armature weight is sufficient for drawing the glass without using the solenoid. Electrodes may be made in batches and stored under cover for a day.

For noncritical work, the electrodes may be used directly from the puller, but, for more reliable use and for lower noise levels, the electrodes should be coated with a hydrophobic layer and then fire-polished. A major source of electrical noise is the saline meniscus, which creeps up the electrode exterior (see below). This meniscus can be broken by coating the electrode with Sylgard 184 (Dow Corning, Midland, MI), a clear two-part silicone rubber that has excellent insulating properties. Once the rubber and its catalyst have been combined, aliquots of the mixture can be stored in the freezer for several weeks without degradation. The electrode should be coated for a distance of several millimeters, starting about 200 μm from the tip. The Sylgard can be applied under a dissecting microscope using a glass wand or hook. It is important not to coat too close to the tip since there appear to be components of low molecular weight in the Sylgard that can migrate to the tip and interfere with electrode filling and sealing. Also, the effect of the coating on reducing the electrode-to-bath capacitance is negligible compared with other sources of input capacity so that there is little to be gained by attempting to coat very close to the tip. After coating, the Sylgard can be cured by advancing the electrode tip into a heated coil or the flow of air from a hot-air gun. Other materials for coating electrodes, such as waxes and lacquers, are inferior to Sylgard. Moisture seems to penetrate between the hydrophobic coating and the hydrophilic glass so that the noise reduction is transient, lasting only 10–15 min. Making the electrode surface hydrophobic through the use of silinizing agents has not been useful. These agents interfere with electrode filling and the ability to form seals.

Electrodes can be fire polished to burn off contaminating Sylgard and to smooth and blunt the tip in order to reduce accidental penetration of the cell. Drawn and coated electrode can be fire-polished using a standard compound microscope and a heater consisting of a hairpin of 0.005-in platinum or nichrome wire. A small bleb of soft glass can be melted onto the tip of the filament to prevent the deposition of evaporated metal on the electrode. A 40 × objective (preferably long working distance) is suitable, and standard lenses can be used without damage if minimal heat is used and if the lens itself is protected by a fragment of cover slip. Since the objectives are usually optically corrected for the cover glass, this addition also results in better optical resolution. The electrode is held in a lump of plasticine on a glass slide that is manipulated using the microscope's mechanical stage. The filament is held in a micromanipulator and is heated using a variable voltage source. A Variac voltage transformer connected to a 6 V/5 A filament transformer is satisfactory. To limit the fire polishing to the electrode tip, the temperature gradient can be sharpened by directing a mild (inaudible) flow of air against the filament through a hypodermic needle.

To fire-polish, the filament is heated to an orange-red color and the electrode tip is moved to within a distance of about 10 μm of the filament. Thermal convection reduces optical resolution so that the tip is not always resolved, and the melting appears as a slight darkening of the pipette walls at the tip. Note that the filament expands as it heats and should be positioned accordingly.

After polishing, the electrodes are filled by dipping the tip into the desired solution, which has been previously filtered through a 0.2-μm filter. Electrodes generally fill the first few hundred micrometers by capillarity in about 30 sec. Filling can be speeded up by applying suction to the back of the pipette. After the tip is filled, the shank is backfilled with filtered solution using a syringe needle. Air bubbles that remain in the pipette are removed by tapping the electrode or by inserting a glass fiber.

Filled electrodes do not store well, so they should be made only as needed.

Seal Formation and Patch Configuration

The electrodes should be clamped in a holder that allows for both electrical connection of the pipette interior to the amplifier and for the application of suction, usually by mouth. The holder is made of an insulating material, such as methacrylate, and the electrode should be held firmly enough so that the application of suction does not move the electrode tip. Electrical contact can be established by either a chlorided silver wire (which can be protected from scratching by covering it with a piece of small-diameter, perforated tubing) or by a silver–silver chloride pellet, which makes a fluid coupling to the filled electrode. Electrode holders are available from commercial sources (e.g., WPI, Inc., New Haven, CT). In either case, the fluid level should be kept to a minimum to reduce input capacity and to prevent solution from being drawn into the suction line and causing electrical interference.

As the pipette is lowered into the bath, a slight positive pressure is applied to the pipette in order to prevent the accumulation of debris at the tip. Once in the bath, the potential difference between the pipette and the bath is offset to zero. This adjusted potential serves as a reference for changes in patch potential.

The pipette resistance is measured by applying square waves to the amplifier reference input (see below). The amplitude of the resulting current is inversely proportional to the electrode resistance. Since the electrode and seal resistances may vary from 10^6 to 10^{11} ohms, calibration voltages in the range of 0.5–100 mV are necessary to provide adequate resolution.

To make a seal, the pipette is gently pressed against the cell surface while the pipette-to-bath resistance is monitored. As the pipette is gradually advanced, the resistance should increase (the pulse amplitude will fall) by two to five times its initial value and then level off. At this time, slight suction is applied to the pipette. In good experiments, the pipette resistance will suddenly increase to a value greater than 10 Gohms, and the baseline noise will be greatly reduced.

Sometimes the application of suction only increases the pipette resistance to several hundred megohms, a value too small for most single-channel recordings. In these cases it is occasionally possible to increase the seal resistance by the alternate application of suction and pressure, but more commonly, no tighter seal can be established. Occasionally, the application of suction causes the electrode tip to puncture the cell membrane. Penetration results in a large offset in current, an increase in the low-frequency current noise, and an increase in the capacity transient associated with the resistance testing pulse. The probability of penetration is higher with high resistance or non-fire-polished electrodes. There is also wide variation in the ability of the cell membrane to withstand suction without rupturing. If a high-resistance seal cannot be formed, it is advisable to change electrodes because used pipettes will rarely form tight seals.

Once the seal has been formed, the membrane patch can be excised with either the cytoplasmic surface toward the bath (inside-out, or "rip-off," patch), or with the extracellular surface toward the bath (outside-out patch). To form an inside-out patch, the pipette is simply withdrawn from the cell. Sometimes the membrane patch will form a small vesicle within the tip of the pipette when exposed to the bath solution (see below). The probability of vesicle formation may be reduced by using a low concentration of Ca^{2+} in the bath. A typical bath solution for making inside-out patches is 140 mM KCl, 10 mM HEPES (pH 7.4), and 0.1 mM EGTA (HEPES is N-2-hydroxyethylpiperazine-N'-2-acid; EGTA is ethylene glycolbis(β-aminoethyl ether)-N,N,N',N'-tetraacetic acid). If vesicle formation does occur, it is sometimes possible to rupture the bath side of the vesicle by bringing the tip of the pipette into the air for several seconds.

To make an outside-out patch, or to clamp whole cells, the electrode should be filled with a low Ca^{2+} solution. In most preparations, formation of high-resistance seals is not impaired by the removal of Ca^{2+} from the pipette solution. After the cell-attached patch has been formed, the membrane is ruptured by applying a sharp pulse of suction. Membrane rupture is accompanied by an offset current, a decrease in the pipette-to-bath resistance, and an increase in low-frequency noise. In this configuration, the current-collecting pipette has access to the interior of the cell and the whole cell may be voltage clamped.[2] To form the outside-out patch, the pipette is gradually (a few seconds) withdrawn from the cell until the patch is excised. Sometimes this method results in the formation of an inside-out patch even if Ca^{2+} is present in the bath. The patch configuration can be checked, if the gating properties of the channel are known, by applying the appropriate voltage or drugs to the patch. Differences in channel behavior have been noted between cell-attached, outside-out, and inside-out patches.

Characteristics of Patch-Clamp Data

The prototypical single-channel current is a rectangular pulse that rises and returns to the baseline at a rate limited by the system bandwidth

(Fig. 1A; also see Fig. 6, inset). Channel open times are commonly clustered into bursts as shown in Figs. 1A and 1C. In records where channel activity is low, the baseline can be reliably defined as the current level between bursts, but when more than one channel is open, as in Fig 1B, the baseline may be more difficult to define. With more than one channel of a given type active in the patch, the current level will be an integral multiple of the unitary current (Figs. 1B and 6). If the channel kinetics have rates comparable to the system bandwidth, the amplitudes and durations of individual events may be difficult to resolve (Fig. 1C). Multiple conducting states may be visible as variable-amplitude closed times (Fig. 1F) or open times (not illustrated). Fluctuations about the mean open-channel current are greater than those about the closed-channel current (background). This phenomenon is probably due to small conformational changes of the channel and can be seen in Fig. 1B as noise that increases with the number of open channels.

Commonly, more than one species of channel will be active in a given patch. In some cases, these different channel types can be selected on the basis of amplitude (Fig. 1C). The differences can be exaggerated by working near the reversal potential for the interfering channel type. More generally, the traditional techniques of ion substitution and pharmacological blocking are appropriate.

Whether the patch is attached to the cell or is excised, several types of distortion affect channel currents. If the membrane patch forms a vesicle within the pipette tip, channel currents may become rounded and exhibit distinct charging time constants (Fig. 1E). Sometimes the channel currents decrease and eventually disappear. If a vesicle is present, this loss of signal may be due to loss of the voltage gradient and/or depletion of ions in the vesicle. Additionally, channels may diffuse into the sealing region, or the sealing region may advance to cover the channels. A different pattern is occasionally seen in which some currents appear rounded while others rise squarely out of the baseline (Fig. 1D). This cannot be due to vesicle formation, and it suggests that some of the channels are in a region that has a high access resistance to the interior of the pipette. One possible explanation is that these rounded currents arise from "rim" channels located within the sealing region of the patch and thus share current with the pipette and bath. Rim channels are common when seal resistances are less than 1 Gohm. Alternatively, the presence of channels on a highly evaginated or invaginated portion of the cell surface, such as a T tubule, would also give rise to similarly distorted currents.

Bursts of irregular, monopolar noise are sometimes seen, particularly when large driving potentials are applied to the patch (Fig. 1G). While the cause of the large, spikelike currents is not clear, we believe that they arise from a breakdown of the membrane or seal.

Instrumentation

A high-gain current-to-voltage converter is used to record single-channel currents. Simple amplifiers can be built for a few hundred dollars, or more elaborate versions can be purchased commercially (Dagan Inc., 2855 Park Ave. Minneapolis, MN 55407; List Electronics, Pfungstaedter Strasse 18-20, D-6100 Darmstadt/Eberstadt, FRG, represented in the United States by Medical Systems Corp. 239 Great Neck Rd., Great Neck, NY 11021; Yale Instrument Shop, Physiology Dept., Yale Medical School, 333 Cedar St., New Haven, CT 06510.

The basic head stage is shown in Fig. 2. The pipette is connected to the inverting input of A1, an operational amplifier that has a field-effect transistor input stage. For single-channel recording, the value of the feedback resistor, R_f, is typically 10 Gohms. For whole-cell recordings, the value of the feedback resistor may be reduced to 0.1–1 Gohm in order to avoid amplifier saturation with large currents. The differential amplifier, A2, is a low voltage-noise amplifier with a gain of 10, which serves to subtract the command voltage, V_c, applied to the noninverting input of A1.

The following discussion describes the noise and speed limitations of the patch-clamp amplifier.

Noise Sources

The minimum noise background noise of the amplifier–patch combination is dominated by the thermal (Johnson) noise of all resistances (more precisely, the real part of all impedances) connected to the input of A1. The thermal component of noise is given by

$$I_{RMS} = (kTB/R)^{0.5} \tag{1}$$

where I_{RMS} is the equivalent input noise, k is Boltzman's constant, B is the noise bandwidth, T is the absolute temperature, and R is the parallel combination of R_f and the seal resistance. At room temperature,

$$I_{RMS} \text{ (pA)} = 0.127[B(\text{kHz})/R(\text{Gohm})]^{0.5} \tag{2}$$

Thus, low noise can be obtained only with high-value feedback and seal resistances and/or low bandwidths. For a 10-Gohm resistor at room temperature, the thermal noise density is $S(f) = 1.6 \times 10^{-30}$ A²/Hz corresponding to an RMS noise level of 0.04 pA (0.25 pA peak-to-peak) in a 1-kHz bandwidth.

Any conductive pathway to ground, coupled either directly or capacitatively (such as the solution meniscus), will increase the noise level. Cleanliness and amplifier layout are thus important during the assembly of the amplifier. Both the feedback resistor and A1 should be sonicated in

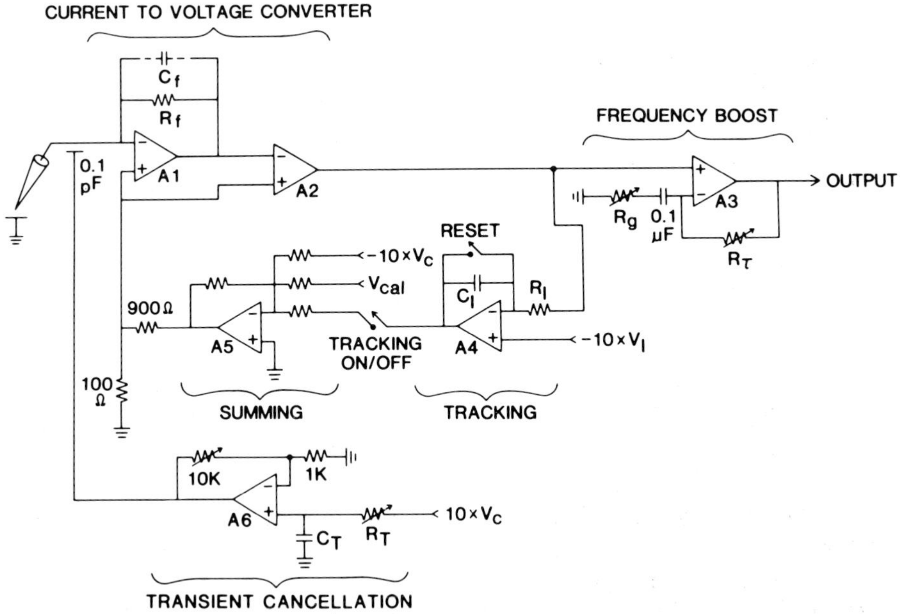

FIG. 2. Schematic diagram of a patch amplifier. Amplifier A1 determines the noise performance of the overall amplifier. Either discrete FET amplifiers or the Burr–Brown OPA-101 are satisfactory. An open-loop bandwidth of >5 MHz is necessary to avoid having the frequency response of the first stage depend upon input capacity. Amplifier A2 (run at a gain of 10) subtracts the command potential and should be a low-noise type (NE5534, Signetics). The overall frequency response is determined by the frequency boost section(s). A3 should have low crossover distortion for high-frequency compensation. R_g limits the maximum frequency of boosting and is typically 330 ohms for the highest frequency stage. R_T determines the frequency at which boosting begins and is in the range of 50 kohms. Amplifier A4 provides tracking (autozero) mode and current clamp. It should be a low-input current device such as an LF356 (National). For tracking, R_I is about 1 Gohm and C_I is 1 μF. The reset button is used for initial zeroing. For electrometer mode, C_I may be reduced to 100 pF and R_I to 10 kohms. For constant-current operation, command signals are fed into the noninverting input of A4 (V_I). The summing amplifier, A5, adds together the command and offset potentials, the tracking feedback signal, and the electrode-resistance measuring pulse, V_{cal}. The summed inputs are applied to A1 through a 10:1 voltage divider. The divider helps to reduce amplifier noise from earlier stages. A5 should be a low-noise amplifier (NE5534). Amplifier A6 serves to charge the stray capacity of the head stage during rapid changes in potential. The 0.1-pF capacitor, which couples the cancellation signal to the input, is simply an insulated wire wrapped once about the input connector. The actual value is not critical. R_T may be calibrated to read input capacity. To cancel the simple electrode capacity, C_T may be set to zero. To cancel the capacitance of a whole-cell clamp, an additional circuit needs to be added where $R_T C_T$ is about equal to the cell's time constant. A summing amplifier will be needed for adding the outputs of multiple cancellation circuits.

ethanol prior to installation on the circuit board. In humid environments, or where equipment might be subjected to temperature cycles producing condensation, it may be helpful to embed the head stage of Sylgard. To reduce leakage paths to ground, the negative input of A1 and the input end of R_f should be connected directly to a high-quality Teflon input connector, not wired to the circuit board. The input connectors and electrode holders should also be kept clean by washing them in ethanol or methanol.

The shot noise of the amplifier's input current produces additive white noise according to the relationship

$$I_{shot} = (2qB\langle I\rangle)^{0.5} \tag{3}$$

where $\langle I \rangle$ is the input current, q is the electronic charge (1.6 × 10^{-19} C), and B is the bandwidth. An input current of 5 pA produces a white noise contribution equal to that of a 10-Gohm resistor and is generally negligible in wideband recordings.

Additional noise comes from current driven through the input capacity by the voltage noise of the amplifier itself (typically 1–5 nV/$\sqrt{Hz}$ at 1 kHz), and by noise appearing on the command voltage, V_c. The passive voltage divider attached to the noninverting input of A1 (see Fig. 2) serves to reduce the effects of the noise coming through V_c. Since a 1-kohm resistor has a thermal noise of about 4 nV/$\sqrt{Hz}$, which is comparable to the amplifier's intrinsic voltage noise, low values of resistance are indicated in Fig. 2.

Voltage noise-induced current noise seems to be the dominant contribution to the total system noise at higher frequencies. As shown in Fig. 3, the high-frequency noise behavior is dominated by a spectral noise density that increases linearly with frequency. The linear, rather than f^2, dependence suggests that it arises in a distributed component. A semilogarithmic plot of RMS noise versus bandwidth (Fig. 4) shows how high-frequency noise dominants the overall noise level.

A third source of noise is the ever-present line-frequency interference. Like other noise sources, this is reduced as the seal resistance increases. To minimize line-frequency interference, it is desirable to use a single-point ground established at the positive input of A1. This point should be used as the only common ground for the amplifier circuits, the bath, and any Faraday shields around the setup.

Low-frequency noise is generated by voices and by building vibration. These vibrations shake the electrode, the bath, and components in the head stage, inducing extraneous currents. The lowest-frequency components may be reduced by mounting the components on a vibration isolation platform. Voice effects can be reduced by *not* speaking during high-resolution recordings.

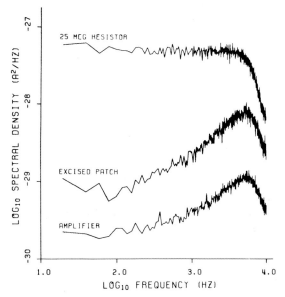

FIG. 3. Spectral density of the patch noise under different conditions. For all traces, the anti-aliasing filter (4-pole Butterworth) was set at 8 kHz and accounts for the roll-off at high frequencies. The lowest trace represents the patch amplifier with no load at the input (amplifier A1 was a Burr–Brown OPA-101). The noise up to about 500 Hz is white and is due to the noise of the 10-Gohm feedback resistor (1.6×10^{-30}) plus an equal amount from the shot noise of the amplifier input current (7 pA). The noise increase above 500 Hz is, in part, accounted for by the product of the voltage noise of the amplifier and the input capacitance, although such a mechanism predicts that the noise should increase as the square of frequency instead of linearly. The next highest trace shows the noise recorded from a membrane patch excised in symmetrical solutions with 0 mV applied across the patch. Clearly, additive amplifier noise is insignificant. The rough proportionality of the patch noise and the amplifier noise suggests that they are produced by a voltage noise source applied to the amplifier input. The uppermost trace shows the noise produced by a 25 Mohm resistor placed across the amplifier input. Since the resistor noise is white and much larger than the intrinsic amplifier noise, the resulting spectrum is proportional to the squared magnitude of the amplifier transfer function and can be used to correct other spectra for the amplifier and filter response.

In practice, the noise of the patch recording when the electrode is sealed to a cell membrane is higher than the noise when the electrode is sealed by other means, such as pressing it against the bottom of a plastic dish. The excess noise density is composed of a low-frequency $1/f$ component and a high-frequency component increasing as f^1-f^2 (Figs. 3 and 4). The $1/f$ component disappears when the pipette and bath solutions are symmetrical and the potential is zero. This suggests that the low-frequency component is due to unresolved channel openings and diffusion across the patch or seal. However, even with symmetrical solutions and

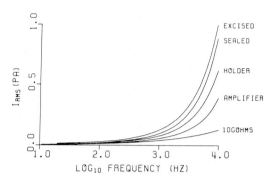

FIG. 4. The RMS current noise under different conditions. With the exception of the curve for the 10-Gohm resistor, which was calculated from theory, the curves were integrated spectra such as Fig. 3 and corrected for the frequency response of the amplifier–filter combination. The curve labeled "10-Gohms" is the lowest possible noise level for an amplifier using such a feedback resistor. The amplifier adds noise to that, the noise increasing much faster than expected for a resistive load. The curve labeled "holder" shows the noise after adding in an electrode holder with an electrode held just over the bath. Immersing the pipette in solution for several millimeters and sealing it against the bottom of a plastic tissue culture dish (>10^3-Gohm seal) gave the curve labeled "sealed." This additional noise is probably a combination of added capacitance from immersing the electrode in the bath and some dielectric loss in the glass of the tip. The noise of an excised patch in symmetrical solutions with zero transmembrane potential (curve "excised") is greater than the sealed pipette probably owing to lossy properties of the membrane-to-glass seal (F. Sachs, unpublished calculations).

zero transmembrane potential, there appears to be excess noise above 1 kHz. This noise is probably due to the distributed properties of the seal impedance, the real part of which is frequency dependent and contributes noise proportional to f. The noise at different stages of the recording system is shown in Fig. 4. The RMS noise level does not seem to be dominated by any one stage or component, so that dramatic improvements are not expected from improving the amplifier's noise performance. Note that the linear ordinate of Fig. 4 should not be taken to indicate that the noise from each successive stage is necessarily additive (in a sum-squared manner).

Frequency Response

The response time of the first stage is given by

$$\tau = R_f C_f \tag{4}$$

where C_f is the unwanted stray capacitance (~50 fF) in parallel with the feedback resistor, R_f.[8] A typical value for τ is 0.5 msec. While this response time may be adequate for some signals, it is too slow for study of rapid kinetic processes. The bandwidth of the system can be increased simply by adding to the output its derivative, as shown in Fig. 2 (frequency boost). This differentiation does not significantly increase the equivalent input noise, since the noise contributed by the differentiator is generally much smaller than the noise contributed by the first stage.

Unfortunately, Gohm resistors are not ideal and typically exhibit multiple time constants and thus require more than one differentiator to compensate properly the frequency response. Resistors will differ in the extent to which they deviate from ideal behavior, and it is thus important to sample resistors to find one that most closely approaches ideality. We have found that the best commercially available chip (2 × 2 mm) resistors are type 8-5 from IMS, Inc. (50 Schoolhouse Lane, Portsmouth, RI), and the best discrete resistors are type K01243 from K & M Electronics (123 Interstate Drive, West Springfield, MA 01089).

The frequency response of the amplifier can be tested by air-coupling a pure triangle wave to the open input of the amplifier. The triangle will be differentiated by the coupling capacitance, producing a square-wave output. The frequency response can then be adjusted to produce the squarest possible output. It is important to check the settling characteristics with both long (10 msec) and short (0.1 msec) pulses, since there may be more than one time constant to compensate. It is simple to obtain fast rise times, but the response to longer pulses may not be accurate. In practice, the rise time can be reduced to 15 μsec or less using two successive stages of differentiation. If power spectra can be calculated, the amplitude of the transfer function can be simply measured by using as a white noise source a relatively low-value resistor (25 Mohm) connected between at the input of A1 and ground (Fig. 3). Note that this method is not sensitive to phase distortions.

Finally, the linearity of the system should be checked by applying different amplitude signals to the input. We have observed nonlinear behavior in the form of a slow creep of the output and in the form of ringing. The ringing seems to be associated with crossover distortion in the differentiating amplifiers, and it is worst when attempting excessive frequency boosting.

Changing the Patch Potential

Voltage signals may be applied to V_c (Fig. 2) to control the patch potential. The accuracy of the voltage command is given by the ratio of

---

[8] F. Sachs and P. Specht, *Med. Biol. Eng. Comput.* **19**, 316 (1981).

the pipette resistance to the seal resistance. Since seal resistances are commonly 100 times larger than pipette resistances, the dc voltage error is negligible. When seals are not tight, or when whole cells are being clamped, there may be substantial errors. The most common limitation to applying voltage is that, with low resistances, the head stage amplifier saturates and noise becomes excessive.

For high-speed changes in potential, transient compensation is required. As shown in Fig. 2, the voltage clamp speed limitation comes from the current required to charge the input capacity. The input capacity may be 10 pF or more, so that, for a feedback resistance of 10 Gohms, the charging current cannot settle much faster than about 0.1 sec. If more rapid changes in potential are required, such as in the study of voltage-dependent channels, the input capacity must be charged from an external source.

Figure 2 shows a typical transient cancellation circuit. This circuit capacitatively couples to the input of A1 a signal with the proper amount of charge to raise the input capacity to the required voltage. Since this charge comes from a low-impedance source, A1 is not required to supply the current. In practice, two or three parallel stages may be necessary to compensate satisfactorily for the different time constants associated with the pipette capacity and the membrane time constant. With proper compensation, rise times of 20–30 μsec are possible when voltage-clamping small cells. Series-resistance compensation may be required if the IR drop through the pipette is significant with respect to the applied potential.

Tracking (Autozero) Operation

When the patch electrode is first placed in solution, the amplifier may have a voltage gain of 10^4 to 10^5 depending upon the electrode resistance. Small voltage offsets or drift (100 μV) can cause the amplifier to saturate or wander over the oscilloscope screen. To simplify the initial setup, amplifier A4 serves as an integrator that "bootstraps" the input so that there is no mean output. This auto-zeroing, or tracking, circuit effectively makes the system ac-coupled with a time constant of R_1C_1/G, where G is the net gain of A1 and A2 which depend upon the electrode resistance and R_f. The integrator is also useful for recording cell potentials (see below).

There are two types of errors in potential application and measurement, those due to diffusion potentials and those due to amplifier input current. When there is a difference in ion concentrations between the pipette and bath, some of the offset potential [applied to A5 through V_c (Fig. 2)] used to zero the amplifier initially was used to counter the diffusion potential. The diffusion potential may add or subtract from the applied potential, depending upon the ion gradients, and should be allowed for in precision measurements.[2] For a solution of 140 mM NaCl in contact with 140 mM KCl, the diffusion potential is about 5 mV, with the Na$^+$ side positive owing to the lower mobility of sodium compared to potassium. The diffusion potential may be strongly temperature dependent if strongly associating ions, such as fluoride, are present. Because of the unknown selectivity properties of the seal, dc measurements of patch currents are highly uncertain.

Amplifier input current causes a correctable error in the dc patch current. When the pipette is first placed in solution, the input current flows to ground through the pipette resistance, producing a voltage drop of only a few microvolts (5 pA × 1 Mohm). When a high resistance seal is formed, the current must be supplied through the feedback resistor. This error is significant only for dc measurements and can be compensated for if the input current and seal resistance are known. If the offset voltage trim of A2 is adjusted so that its output is zero with no input load on A1, and A1 is offset (using the command input bias pot) with the pipette in solution, the dc output will no longer be sensitive to changes in input resistance.

Recording Potentials

With whole-cell patch clamps, it may be useful to measure the cell potential. Feedback can be used to convert the patch clamp amplifier to a current clamp. If the output voltage is fed back through V_c (Fig. 2) in such a way as to reduce the input current to zero or some other constant value, the system behaves as an electrometer capable of constant current injection. In Fig. 2, the time constant of the integrator A3 can be reduced so that V_c tracks the input signal to A1. In voltage-recording mode, the amplifier can follow signals with a time resolution limited by the input capacity and the electrode resistance, just as in an ordinary electrometer.

Analysis of Single-Channel Data

The first stage of data analysis consists of acquiring, storing, filtering, and detecting events. The next stage is to measure the amplitude and duration of individual events. Finally, these data are collated and then fit to appropriate functions that characterize the population. An excellent discussion of these aspects has been given by Colquhoun and Sigworth.[9]

---

[9] D. Colquhoun and F. J. Sigworth, *in* "Single Channel Recording in Biological Membranes" (G. Sakmann and E. Neher, eds.). Plenum, New York, 1983. In press.

Theoretical Framework

The measurement of channel current amplitude is the simplest and least ambiguous of single-channel parameters. The magnitude of single-channel currents can be characterized by the ionic specificity of the reversal potential and the channel conductance. The reversal potential, V_r, is that transmembrane potential at which no current flows through the channel, and may be the same as the Nernst potential if only one ion species is permeable. The conductance, g, is then defined as

$$g = I/(V_m - V_r) \qquad (5)$$

where I is the observed channel current and V_m is the transmembrane potential. The conductance may vary with ion species, concentration, and membrane potential. Channel conductances typically vary from 1 to 300 pS.

For kinetic analysis, individual channels are usually assumed to undergo transitions between conformational states with time-independent probabilities. The time a channel resides (dwell time) in a given state is an exponentially distributed random variable. The mean duration of a given state is equal to the inverse of the sum of all possible exit rates from that state. For example, consider a channel that can be in either of two states, closed or open. The mean dwell time in the open state would be equal to the inverse of the channel closing rate, and the mean dwell time in the closed state would be equal to the inverse of the channel opening rate. To arrive at estimates of these mean durations, many events need to be measured because the standard deviation of an exponential distribution is equal to its mean.

If all states of a channel were distinguishable (i.e., by amplitude), then estimating transition rates from single-channel data would be a relatively simple matter of determining the mean dwell time in each state and measuring the ratios of the transition probabilities between different states. Most channels, however, can exist in more than one state, which cannot be unambiguously identified. The unconditional distribution of dwell times in these nonsingular states is described by a sum of exponentials, the number of which equals the number of states. For example, in the following model of a single channel

$$\text{CLOSED}_1 \underset{b_1}{\overset{f_1}{\rightleftarrows}} \text{CLOSED}_2 \underset{b_2}{\overset{f_2}{\rightleftarrows}} \text{OPEN}$$

the open times will be distributed as a single exponential with a mean equal to $1/b_2$, and the closed times will be distributed as two exponentials with a mean equal to $(f_1 + b_1)/(f_1 f_2)$. The mean closed time will be longer than the dwell time in either of the closed states, since transitions between closed states are undetected. As shown in Fig. 1A, channel events may consist of bursts of closely spaced openings separated by short gaps. The presence of bursts shows that the channel can exist in at least two closed states, since there are two kinetically distinguishable populations of gaps. In the above model, the short gaps within bursts would reflect residence of the channel in $CLOSED_2$. These gaps would have a mean duration equal to $1/(f_2 + b_1)$. Thorough treatments of the stochastic properties of single channels as Markov processes can be found in Colquhoun and Hawkes.[10,11] The analysis of multiple-channel activity may be possible using the techniques of Dionne and Liebowitz[4] and Horn and Lange.[12]

The analysis of kinetic data is further complicated if more than one channel is active in the record, because the times between observed transitions often cannot be attributed to a single channel. For example, in the record shown in Fig. 1B, virtually none of the dwell times in any of the six observable current levels can unambiguously be associated with the residence of an individual channel at that level. However, even when more than one channel is present, some dwell times at a given current level can be associated with a single channel. For example, the bursts of open periods shown in Fig. 1A–C, as well as the brief gaps within those bursts, can with confidence be attributed to a single channel because the likelihood of several independent openings separated by such short durations is extremely small.

The above considerations of channel kinetics have assumed that the data are stationary, that is, the statistics in question do not change with time in the record. In practice, the kinetics often show drift. Rates may change because of the nature of the channel kinetics, as in the case of sodium channels, or perhaps because of changes in the environment of a channel. When the kinetics of a single molecule are being observed, one must expect to see subtle changes that are averaged out in macroscopic records. The extreme nonstationarity in sodium channel kinetics, for example, may be treated using conditional probabilities that explicitly include time.[13] With more subtle forms of nonstationarity, such as drug desensitization and slow drifts in the kinetics, care must be used in the analysis and caution in the conclusions.

[10] D. Colquhoun and A. G. Hawkes, *Proc. R. Soc. London Ser. B* **300**, 1 (1982).
[11] D. Colquhoun and A. G. Hawkes, *Proc. R. Soc. London Ser. B* **211**, 205 (1981).
[12] R. Horn and K. Lange, *Biophys. J.* **43**, 207 (1983).
[13] R. W. Aldrich and G. Yellen, in "Single Channel Recording in Biological Membranes" (B. Sakmann and E. Neher, eds.). Plenum, New York, 1983. In press.

Data Acquisition

The raw data may be collected in different ways depending upon the type of experiment and the facilities available. For continuous random data, as observed with drug-activated channels, the analog tape recorder provides a very efficient means of storing data. Hours of data can be stored on a reel of tape, and several data channels are available to record voice comments as well as synchronization marks. Analog tape recorders are expensive, however, and do not provide a means of analysis.

For data that are not continuous, such as voltage step-induced sodium channel activity, the tape recorder is inefficient because it does not function efficiently in short start–stop operation. For these synchronized data, a computer or transient recorder provides a more efficient means of data acquisition. If only a single device is to be purchased, there is little question that a computer with disk or tape storage is the best choice.

In some cases the data may be so reproducible and of such high signal-to-noise ratio that simple analog detection circuitry may be used to measure kinetics. This eliminates the need to store the raw data, but makes it impossible to reexamine the data should some unexpected effect be observed.

The simplest recording device is the strip-chart recorder with a pair of calipers for measurement. If the data are sufficiently slow, the recording may be made directly. If the data are arriving too fast, a variable-speed tape recorder may be used to slow down the data. The strip chart may still provide the best global view of a long segment of data.

Computerized analysis methods vary from totally interactive systems where the user positions cursors to mark features in the data, to automated pattern recognition systems, the choice being made by available equipment and software.[9,14] Regardless of the system used, some features of the data collection and analysis are common to all.

Low-Pass Filtering and Event Detection

In the first phase of data analysis, the data are band-limited with a low-pass filter and digitized so that opening and closing transitions can be detected. The requirements for these different stages of analysis are interrelated, so that it is convenient to define the parameters to be used. Let f_c denote the cutoff frequency of the filter. For consistency, f_c refers to the filter's 3 db point, the frequency at which the signal power is reduced to half of its low-frequency value. The time constant of the filter, τ, is equal

---

[14] F. Sachs, in "Single Channel Recording in Biological Membranes" (B. Sakmann and E. Neher, eds.). Plenum, New York, 1983. In press.

to $1/(2\pi f_c)$. The rise time is taken as the inverse slope of the response to a step; for a four-pole Bessel filter, this is about $0.3/f_c$. The digitizing rate, f_s, is the inverse of the sampling interval, Δt.

The purpose of low-pass filtering is to limit the noise level so that false-positive detections are minimized and channel currents can be clearly visualized. After analog filtering, digital filters may be used further to limit the bandwidth. There is no clearly optimal filter for single-channel records because both the signal and the noise are broadband and the appropriate filter depends upon the type of measurement being made. The main requirement is that the filter order be higher than two so that the high-frequency noise, whose power increases linearly with frequency (Fig. 3), is adequately reduced.

Inevitably, low-pass filtering causes short-duration events to be attenuated. For the same f_c, sharp-cutoff filters, such as the Butterworth filter (flat amplitude), afford a lower noise and a slower rise time than do flat delay filters, such as the Bessel filter. The lower noise of the sharp-cutoff filters means that fewer false detections will occur, but the slower rise times means that more short-duration events will be missed. Perhaps the main reason for using slower roll-off filters is that their response to a step input does not overshoot significantly, so that the bias they produce in channel amplitude is in a consistent direction.

The filter's cutoff frequency is determined by the maximum allowable rate of false-positive detections. For Gaussian background noise, the number of threshold crossings per second (the false-positive rate) is closely approximated by

$$R = f_c/2 \times e^{-0.5(a/\sigma)^2} \tag{6}$$

where σ is the RMS variance of the noise and a is the detection threshold. The allowable value of R depends on the channel transition rate. Since false-positive defections cause errors in the observed distribution of event durations, the slowest transition rate of interest should be much greater than R. Assume that the detection threshold is set to half the channel amplitude (see below). The mean time between false-positive detections ($1/R$) is 100 msec/kHz of bandwidth when the peak-to-peak noise level (taken as 6 σ) is equal to the single-channel amplitude. This value increases to 36 hr/kHz when the channel amplitude is twice the peak-to-peak noise. With these extremely low false-positive rates, however, the assumption that the noise is Gaussian is no longer correct. The presence of baseline drift, unresolved channel currents, and power line interference become limiting. Also, the current fluctuations of the open channel may be much greater than those of the closed channel, so that false closures are more likely than false openings. As a rule of thumb, in order to obtain a tolerable false-positive rate, the peak-to-peak noise of the baseline should be less than two-thirds of the single channel amplitude.

If the data are to be digitized for computer analysis, they should be sampled at two to five times f_c. The lower limit is the Nyquist frequency and is the minimum sampling frequency that can recover a sine wave of frequency f_c. The higher the sampling rate the more accurately the digitized data represent the input data. However, above the Nyquist frequency, the extra information gained approaches zero because the signal amplitude between sample points can accurately be predicted from knowledge of the filter response. A sampling rate of $5f_c$ appears to be the highest useful rate for most applications.[9] For a fixed amount of storage space, higher sampling rates allow higher precision for individual events, but fewer events can be recorded, resulting in less precision in estimating the population parameters. The effect of the sampling rate on the observed distribution of event durations is considered below.

The optimal threshold for detection is one where the number of detected events is maximal and the false-positive rate is small. It can be shown that for a noise-free, bandlimited, exponential distribution of durations, the maximal number of opening and closing transitions are detected when the threshold is at one-half the unitary channel current.[9] In the presence of noise, the probability of detection is less sensitive to the precise position of the threshold. Because the half-amplitude threshold has no hysteresis, it may cause false triggering due to noise that is present on the rising and falling phases of the pulse. This kind of false triggering is a problem only when the data have been oversampled, because with the lower sampling rates the transitions are complete within a single sample period.

If an automated analysis computer program is used, two further considerations in event detection are necessary.[15] First, baselines are not perfectly flat, so an accurate and robust algorithm is needed to define the baseline. Second, additional criteria should be applied to detected events before a putative event is accepted for further analysis. This validation is necessary because records often contain currents from more than one type of channel (i.e., Fig. 1C) and "noisy" events such as the seal breakdown shown in Fig. 1G. Two useful criteria are that the event amplitude should lie within some window of current centered around the amplitude of the event of interest, and that the RMS deviation of the data from an idealized rectangular-pulse model of the event be less than some critical value. It is important to note that, when there is more than one type of ion channel in a patch, the current amplitude of the channels may be similar.

[15] F. Sachs, J. Neil, N. Barkakati, *Pfluegers Arch.* **395**, 331 (1982).

Thus, the channels must be distinguished by some other set of criteria besides amplitude, such as kinetic or selectivity properties.

Estimation of Event Durations

A simple and rapid method of determining event durations is to measure the time between half-amplitude crossings. Systematic errors in determining the parameters of the population arise because durations of brief events that *do* cross threshold are underestimated and events whose amplitude is less than half the unitary current are not measured. Figure 5 shows a plot of the observed duration distribution for an ideal exponential process that has passed through a single-pole band-limiting filter. The event durations are taken to be the time above threshold, and the calculation is done without noise to simplify the results. No events shorter than $0.1/f_c$ are recorded, and events up to about $0.3/f_c$ will have measured durations significantly less than the true event duration. For data that have been band-limited by an 8-pole Bessel filter, events shorter than $0.2/f_c$ (~30 μsec at a bandwidth of 5 kHz) do not reach half amplitude, and only events longer than $0.25/f_c$ (~50 μsec at 5 kHz) will be measured with greater than 90% accuracy. The addition of noise will broaden the observed distributions and will extend the minimum duration at which the data will match the true distribution.

Fitting exponentials to the observed distributions can either overestimate or underestimate (see curve labeled 0.1 in Fig. 5A) the population distribution parameters because the observed distributions are not really exponential. Note, however, that adequate exponential fits may be obtained to data that are monotonic, such as the curves labeled 1 and 3 in Fig. 5, but the extracted time constant will be too long by a factor of two or more. With multiple exponential processes, the errors in amplitude and duration due to band-limiting will change the proportion of fast and slow events. As a rule of thumb, the observed duration distributions should only be fit to exponential functions from times equal to $0.5/f_c$ or longer. When the sampling rate is equal to two times f_c, this allowance is equal to Δt.

As far as the sampling rate is concerned, if the observed distribution of durations is to be compared to theoretical models that do not include the effects of the recording system and noise, then sampling at the Nyquist limit ($f_s = 2f_c$) is satisfactory and, if storage space is limited, provides better precision for estimating the probability distributions than if the data were oversampled. Sampling at the Nyquist limit produces scatter in the measured durations of individual events, since the sampling clock is not synchronized with the channels. The duration of individual events, how-

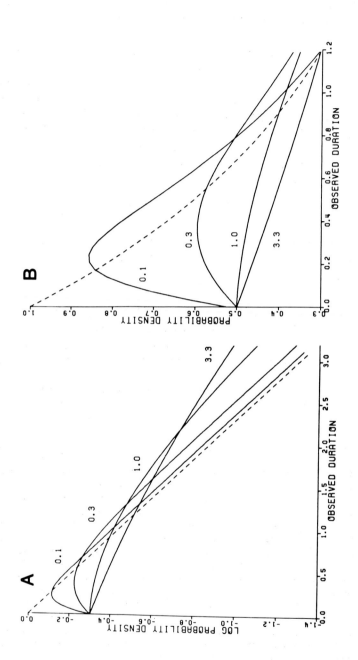

ever, is of little interest, since only population means have testable significance. For exponential distributions, it can be shown that the scatter does not bias or significantly alter the errors in the derived parameters.

An error in duration measurement occurs when short events are missed, because the period including the missed event appears longer than it actually is. For example, if a short closed period is missed, the apparent open time for the adjacent events is increased. If we assume that both the open and closed periods are exponentially distributed, that closed periods are much shorter than the open periods, and that events shorter than some dead time, T_d, are missed, the observed distribution is still exponential, but the observed rate constant is smaller than the true rate constant. The relationship is

$$\tau_t = \tau_o \times e^{-(\tau_c/T_d)} \tag{7}$$

where τ_o is the observed open time constant and τ_t is the true open time constant, and τ_c is the true closed time constant.

Fitting Histograms

There are a few points to keep in mind about fitting histograms. First, the population distortions discussed above limit the shortest events that can be included in the fit. Second, the residuals, (data minus theoretical),[2] must be weighted inversely with the data variance.[16] For histograms, this

[16] P. R. Bevington, "Data Reduction and Error Analysis for the Physical Sciences." McGraw–Hill, New York, 1969.

FIG. 5. (B) The distortion of an exponential distribution of pulse durations produced by band limiting, a noise-free system filtered by a single-pole filter with time constant τ. The population distribution has a time constant of 1. The dashed line is the distribution expected for an infinite bandwith recording ($\tau = 0$). The solid lines are the distributions that would be recorded for different values of τ. The number adjacent to each curve is the ratio of τ to the population time constant; i.e., 0.1 indicates that the distribution time constant is one-tenth that of the filter time constant. The time axis is in units of the population time constant; i.e., a duration of 1 corresponds to an event whose duration was equal to the mean duration. Note that significant deviations exist even for processes that are 10 times slower than the filter time constant. The presence of noise will broaden the curves and extend the time that significant deviations occur. For higher-order filters, using a time constant of $0.3/f_c$ gives reasonable correspondance to the curves shown. (A) A semilogarithmic presentation of the data in (B) showing the asymptotic approach of the band-limited distribution to the true distribution. For populations whose mean durations are comparable to the system time constant, a minimum delay time equal to three population time constants (not system time constants) is necessary before the observed slope is close to the unmodified exponential distribution.

variance is equal to the number of counts in each bin, since bins are filled in a Poisson manner. Third, histograms containing empty bins must be reformatted so that each bin to be fit contains at least 5 counts. The usual equations for nonlinear regression assume that the data variance is approximately Gaussian, and 5 counts per bin have been found to be adequate to give an unbiased estimate of the population mean. Bins should have variable spacing to preserve resolution at short times. To avoid errors due to finite bin widths, the theoretical function should be the integral of the appropriate probability density taken over each bin. Finally, to have confidence that the fitted parameters can be trusted, error estimates on the parameters of the fit should allow for the cross-correlation between parameters. This cross-correlation means that to estimate the time constant of a single exponential distribution to within 10% of its true value with 90% confidence, 500 or more counts are required. For multiple exponentials, no simple guidelines exist. To get some idea of the errors in fitting even a double exponential process, if the time constants are well separated (by a factor of 10) and the two components appear in equal proportions, more than 5000 events must be measured to extract time constants with less than 10% error with 90% confidence. Even more events are required for less well separated components. Such large numbers of events are usually unobtainable experimentally.

Estimating Event Amplitudes

If the amplitudes of the events of interest are well resolved, measuring current amplitudes presents no serious statistical problems. Amplitudes may be quantified by several means.[17] In one method, each data point is entered into an amplitude histogram that is referenced to the baseline. Total amplitude histograms can be constructed for the entire data set (Fig. 6) or for selected portions of the data, such as bursts. When event durations are long relative to the system bandwidth, the total amplitude histogram will appear as a series of roughly Gaussian peaks with means that differ by the single-channel amplitude. When multiple channels with different conductance are present, peaks will be present at the sum and difference amplitudes. Whether these peaks can be resolved or not depends upon the noise level, the difference in the current amplitudes, and the duty cycle. With seriously band-limited data (i.e., the leftmost channel shown in Fig. 1B), the total amplitude histogram peaks will be skewed to lower values and the histogram peak location will underestimate the single-channel amplitude. Inaccuracies in the estimation of the baseline will broaden the observed amplitude distribution.

[17] A. Auerbach and F. Sachs, *Biophys. J.* **42**, 1 (1983).

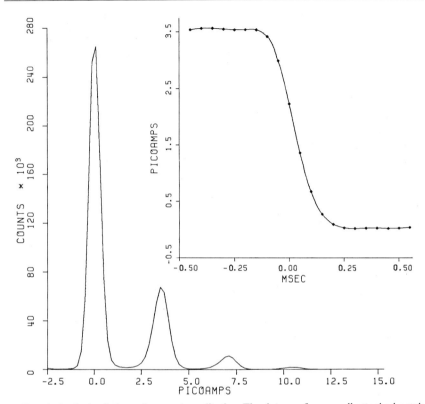

FIG. 6. Analysis of channel current amplitudes. The data are from a cell-attached patch were nicotinic channels were activated by 5 nM suberyldicholine. A baseline current level was estimated from portions of the record where no channels were open. The baseline estimate was updated approximately every 100 msec. Each data point was entered into the histogram with the baseline current level defined as zero. There are four roughly Gaussian peaks, corresponding to times when 0, 1, 2, and 3 channels were open. The means are at 0, 3.50, 7.00, and 10.46, pA, respectively. The distributions are skewed to lower values owing to the presence of band-limiting and subconducting gaps. The relative area under each peak approximates the fraction of the record spent at each current level (0.70, 0.25, and 0.04, and 0.01, respectively). The relative time spent at higher current levels may be underrepresented owing to bandwidth limitations. The variance of each distribution [86, 171, 208, and 264 (fA)^2, respectively] increases with the mean.

The inset shows the average burst trailing edge, from the same record (inward current is up). Bursts were defined as clusters of one or more open periods separated from each other by less than 5 msec. The burst edges were aligned at the point of threshold crossing before averaging. To allow for amplifier settling, only transitions in which the final open period remained above threshold for at least 250 μsec were included in the average. The mean transition amplitude was 3.54 pA. If continued over a longer time, the average current would eventually approach the mean of the entire record. The sampling interval was 50 μsec, and the system bandwidth was 2.5 kHz (4-pole Bessel low-pass filter). The transition rise time is that of the filter.

A second method of measuring event amplitudes is to measure transition amplitudes. Because of the short time scale involved, this method is less sensitive to inaccuracies in the baseline estimation and to the presence of independent overlapping channel currents, but more sensitive to the presence of subconducting states of the same channel.

Only channel currents that are steady both before and after the transition can be accurately measured. An allowance for settling time must be made preceding as well as following the transition, since multipole filters have a sigmoidal step response. The settling time allowance depends upon the system bandwidth, two rise times being a reasonable value. In the inset of Fig 6, all burst terminations that satisfied the settling time criteria were aligned at the moment of threshold crossing and averaged. From the averaged record, the mean transition amplitude is measured from one settling time before to one settling time after threshold crossing. The average current will fall on either side of the transition point owing to the exponential spread in durations. The average transition amplitude, however, does not give any information concerning the distribution of amplitudes. This can be obtained by estimating transition amplitudes on an event-by-event basis and then constructing histograms of the measured amplitudes.

A third method of characterizing amplitudes is simply to calculate the mean of the data points (after allowing for settling) within each open or closed period. Note that the number of points used to estimate the mean varies with the duration of the event, so that the amplitude of longer events is estimated with greater precision than the amplitude of short events. The amplitude distribution of these means is thus non-Gaussian. The presence of currents that do not cross threshold will bias the results. For example, open-channel currents will appear smaller owing to unresolved closures.

Estimating the Number of Active Channels in the Patch

If the channels of interest do not desensitize or inactivate, then estimating the number of channels active in the patch is a simple matter of giving a maximal stimulus (i.e., voltage step or agonist dose), measuring the total channel current, and dividing by the unitary current amplitude. With channels that cannot be driven to saturation, the number of channels active in the patch may be estimated by observing the fraction of time spent with 0, 1, 2, . . . , n channels open.[18] This discrete level histogram is expected to follow a binomial distribution. From a nonlinear regression

---

[18] J. Patlak and R. Horn, *J. Gen. Physiol.* **79**, 331 (1982).

of the data to a linearly interpolated binomial distribution, the total number of activatable channels, N, and the probability per channel of being open, p, can be estimated. Alternatively, a maximum likelihood estimate of N can be made as follows. The likelihood of n_i observations at level i is the product of the probabilities for each observation. Thus the likelihood for observing a set of n_i levels is given by,

$$L = \prod_{i=0}^{i_{max}} P^{n_i}(i,N,p) \tag{8}$$

where $P(i,N,p)$ is the binomial probability of i successes (open channels) out of N trials (channels), p is the probability of success (being open) per trial, and i_{max} is the maximum observed level. If L is maximized with respect to N and p, these numbers constitute the best guess for both. The maximization with respect to p can be done analytically and is simply the mean level divided by N, which has yet to be determined,

$$p = \sum_{i=0}^{i_{max}} in_i / N \sum_{i=0}^{i_{max}} n_i \tag{9}$$

The maximization with respect to N, using the above equation as a constraint, is most simply accomplished by calculating $\ln(L)$ for $N = i_{max}$ to perhaps 50 and searching for the maximum.

Two conditions limit the accuracy of binomial calculations. First, when the N_p product is less than about 0.1, the distribution appears Poisson and N and p are no longer separately distinguishable. The method does not work for sparse data. Second, the highest levels (for $p < 0.5$) will be underrepresented owing to bandwidth limitations and thus will bias the estimate to lower values of N. Estimating confidence limits on N requires some knowledge of the number of "experiments performed" or the degrees of freedom. This number is not the number of data points or the number of levels, but is approximately the number of independent opening events contributing to the data. The weighting and bin-width considerations discussed above for histograms of event durations must be followed when fitting discrete level histograms.

Summary

The patch clamp technique affords unparalleled resolution of the detailed properties of ion channel currents. Patch clamping is not difficult and has been used to record single-channel currents from many cell types. The number of artifacts associated with the method appears to be rather small. The main experimental difficulties arise from the need to process large amounts of data. With the development of inexpensive computers and mass storage devices, these problems should be alleviated. The study of membrane excitability is expanding rapidly owing to the introduction of high-resolution patch clamp techniques. The conceptual revolution that will inevitably follow is just beginning.

[10] Push–Pull Perfusion Technique in the Median Eminence: A Model System for Evaluating Releasing Factor Dynamics

By NORMAN W. KASTING and JOSEPH B. MARTIN

The hypothalamus regulates the release of hormones from the anterior pituitary by secreting hypophysiotropic factors into the hypophysial portal circulation. Most of these factors have now been characterized; they are predominantly small peptides, but also include the biogenic amine dopamine. Our laboratory has been interested in regulation of growth hormone (GH) secretion from the anterior pituitary and the interrelationship between the releasing and release-inhibiting factors that control GH secretion. These two factors are growth hormone-releasing factor (GRF), or somatocrinin, purified from a human pancreatic tumor and identified as a 44-amino acid peptide,[1,2] and somatostatin (SS), an inhibitory peptide that may occur either as 14- or 28-amino acid forms.[3,4] Secretion of GH in the male rat is characterized by intermittent pulsatile release with baseline values of <1 ng/ml and peak levels of >400 ng/ml.[5] Our work has been directed toward developing a suitable model for evaluating release of hypophysiotropic factors in the conscious animal. Specifically, we have investigated the SS–GH relationship in the male rat and hope to extend this work to somatocrinin when an appropriate radioimmunoassay (RIA) becomes available.

[1] R. Guillemin, P. Brazeau, P. Böhlen, F. Esch, N. Ling, and W. B. Wehrenberg, *Science* **218**, 585 (1982).
[2] J. Rivier, J. Spiess, M. Thorner, and W. Vale, *Nature (London)* **300**, 276 (1982).
[3] P. Brazeau, W. Vale, R. Burgus, N. Ling, M. Butcher, J. Rivier, and R. Guillemin, *Science* **179**, 77 (1973).
[4] A. V. Schally, A. Dupont, A. Arimura, T. W. Redding, N. Nishi, G. L. Linthicum, and D. H. Schlesinger, *Biochemistry* **15**, 509 (1976).
[5] G. S. Tannenbaum and J. B. Martin, *Endocrinology* **98**, 540 (1976).

Problems in Evaluating Dynamic Changes Occurring in the Hypophysial Portal System

The hypophysial portal system has unique properties that have caused considerable problems to investigators attempting to evaluate the precise nature of hypothalamic–anterior pituitary interactions. The portal system consists of two capillary beds, the first in the external layer of the median eminence (ME) and the second in the anterior pituitary and its connecting vessels.[6,7] The hypophysiotropic releasing and release-inhibiting factors, such as SS, are secreted by the nerve terminals of hypothalamic neurons into the extravascular space of the ME, where these factors subsequently pass into the portal blood of the first capillary bed. In this area, the tight junctions of the blood–brain barrier are absent. The factors reach very high concentrations in the portal blood as they travel to the second capillary bed and diffuse into the extravascular milieu of the cells of the anterior pituitary. These cells have specific receptors that allow specific interaction of the factors with their target cells. The adenohypophysial hormones and unbound hypophysiotropic factors then enter the vasculature, where the blood drains from the pituitary to enter the general venous drainage of the brain and body becoming greatly diluted. Because of the "closed" nature of this portal system, the small size of the vessels, and its location on the ventral aspect of the brain and within the skull, it has presented unique problems to the researcher.

Injection of one of these hypophysiotropic factors into the peripheral circulation (the general method of *in vivo* administration) will affect the anterior pituitary if it is given in a large enough dose to reach the concentration at which it is normally found in the portal circulation. However, this technique presents the whole animal, or at least the whole brain, to the same high concentration and consequently may also alter pituitary function nonspecifically, or indirectly by complex feedback loops. Conversely, measurement of these factors in the peripheral circulation has to take two major problems into account. First, there is the enormous dilution that occurs when the small blood volume of the pituitary venous drainage enters the general venous drainage of the whole brain making concentrations very low or unmeasurable and easily disguising fluctuations in secretion. Second, localization of many of these factors are not limited to the hypothalamic–pituitary portal system, and therefore measurements of blood levels may be obscured by release from neural or nonneural sources other than the nerve terminals in the ME. A good example of this situation is SS, which is found in high concentrations in

[6] B. Flerkó, *Neuroendocrinology* **30**, 56 (1980).
[7] R. B. Page, *Am. J. Physiol.* **243**, E427 (1982).

the ME but is also found throughout the central nervous system (CNS), as well as in several other organs of the body, such as pancreas and duodenum. As such, most of SS measured in blood is probably of extrahypothalamic origin.

Model systems currently in use are problematic and have obvious limitations. At present, *in vitro* systems have provided most of the information on hypothalamic–pituitary interactions using either cultures of pituitary cells[8] or medial basal hypothalamic explants, or both.[9] These studies provide potentially useful information, but interpretation is difficult because the cells are not exposed to their usual milieu, which contains numerous substances that potentially influence their function, synthesis, and releasing properties. *In vivo* research has been based primarily on a technique involving cannulation of the portal vessels of an anesthetized rat with the bottom of its skull surgically removed and small volumes of portal blood sampled at intervals.[10-12] Although this system potentially provides useful ideas regarding the hypothalamic factors and their response to various pharmacological agents, the effects of the surgical stress and anesthesia, both of which are known to alter profoundly pituitary hormone release, cannot allow accurate assessment of normal physiological *in vivo* changes. Furthermore, these systems allow evaluation of only one aspect of the system or the other, either release of hypophysiotropic factors or modulation of pituitary hormone release, but not both, and not under physiological conditions.

Because of these difficulties, we have sought to evaluate whether the push–pull perfusion technique combined with chronic intra-atrial catheterization could be used to determine temporal patterns of SS release in the ME and its relationship to the plasma levels of the anterior pituitary hormone that it regulates, namely GH. Because this technique can be used in the unanesthetized and freely behaving animal, physiological patterns of release may perhaps be determined.

Push–Pull Perfusion Technique

The idea of an *in vivo* perfusion system that could be used deep in the neuropil originated with Gaddum.[13] The push–pull perfusion cannula sys-

[8] W. Vale, G. Grant, M. Amoss, R. Blackwell, and R. Guillemin, *Endocrinology* **91**, 562 (1972).

[9] L. L. Iversen, S. D. Iversen, F. Bloom, C. Douglas, M. Brown, and W. Vale, *Nature (London)* **273**, 161 (1978).

[10] J. C. Porter and K. R. Smith, *Endocrinology* **81**, 1192 (1967).

[11] K. Chihara, A. Arimura, and A. V. Schally, *Endocrinology* **104**, 1434 (1979).

[12] P. Gillioz, P. Giraud, B. Conte-Devolx, P. Jaquet, J. L. Codaccioni, and C. Oliver, *Endocrinology* **104**, 1407 (1979).

[13] J. H. Gaddum, *J. Physiol. (London)* **155**, 1 (1961).

tem that we have used in these experiments was modified from that subsequently described by Myers.[14] This system was designed to wash a small volume of tissue (about 1 mm^3) with as little disruption as possible.

The push–pull perfusion system works on the following assumptions.

1. There is continuous and consistent exchange between the perfusate and the extracellular fluid at the tip of the cannula. A continuous and consistent exchange has been shown to occur if conditions are correctly controlled and blockage is prevented.[15]
2. The perfusate or the mechanical agitation, by itself, does not alter release of the substance being examined once the perfusion is established. This has been shown to hold for the perfusion system once the initial disruption of placing the cannulas has occurred, that is, the first several minutes.[16]
3. There is a sufficient number of undamaged neurons releasing the substance in response to a given stimulus in the area of the cannula tip. This assumption is proved by our work to be reported later in this chapter.

In practice, the usefulness of the technique will depend partially in the sensitivity of the assay used to measure the factor in question in the perfusate. That is, the more sensitive the assay, the less perfusate needed to determine concentrations and the more frequent the sampling interval possible. A more detailed description of the problems inherent in this technique has been previously published.[15] The push–pull cannulas consisted of an inner 30-gauge (o.d. 0.30 mm) infusion cannula and an outer 22-gauge (o.d. 0.71 mm) withdrawal cannula (Fig. 1).[17] These cannulas were attached by polyethylene tubing to matched glass syringes mounted on an infusion-withdrawal pump (Harvard Instruments, Millis, MA). The withdrawal line had parallel trap lines bathed in ice, both of which could be alternately removed with only brief interruption of flow. The pump was allowed to continue during trapline exchange. The flow rate was maintained at 40 μl/min, and 600-μl samples were collected every 15 min. The perfusion fluid was an artificial extracellular fluid[18] composed of the following concentrations of the chloride salts: 145 mM Na, 3.5 mM K, 1.3 mM Ca, and 1.0 mM Mg adjusted to a pH of 7.2.

Variations from this procedure have been adopted by other laboratories. Replacing the infusion-withdrawal pump with a roller pump allows greater flexibility with regard to duration of perfusion and ease with which

[14] R. D. Myers, *Physiol. Behav.* **5**, 243 (1970).
[15] T. L. Yaksh and H. I. Yamamura, *J. Appl. Physiol.* **37**, 428 (1974).
[16] M. P. Honchar, B. K. Hartman, and L. G. Sharpe, *Am. J. Physiol.* **236**, 1248 (1979).
[17] N. W. Kasting, J. B. Martin, and M. A. Arnold, *Endocrinology* **109**, 1739 (1981).
[18] R. D. Myers and W. L. Veale, *J. Physiol. (London)* **212**, 411 (1971).

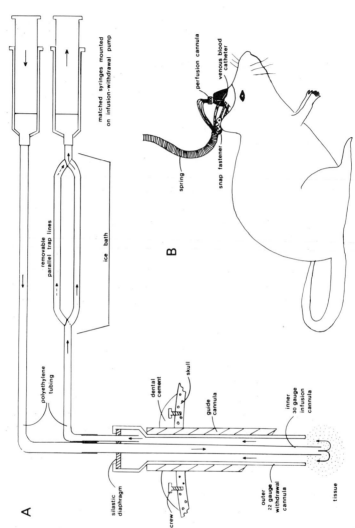

FIG. 1. (A) Schematic diagram of the push–pull perfusion system. (B) External appearance of an experimental rat. Reproduced, with permission, from Kasting et al.[17]

samples can be collected. Also, the perfusate may be buffered with phosphate or an artificial cerebrospinal fluid (CSF) may be used.

Animal Preparation and Experimental Protocol

Each rat was stereotaxically implanted with a guide cannula that was directed toward the ME and the tip of which rested 5 mm above the ME. The guide cannula was placed at a slight angle to prevent damage to the sagittal sinus. Each rat was also prepared with a chronic intra-atrial Silastic catheter to allow blood sampling.[19] The catheter was exteriorized and attached to the skull with dental cement and stainless steel screws. The animals were allowed approximately 7 days to recover to preoperative body weight before they were placed in the sampling cage. A spring-ball bearing assembly was attached to a snap fastener on the rat's skull to allow free movement while allowing passage of catheters for blood and push–pull apparatus to the outside of the cage. The sampling cage was placed in an isolation box. The rat was allowed 1 day to become accustomed to the cage before an experiment took place.

On the morning of the experiment, the push–pull cannula was allowed to run for several minutes and then inserted into the guide cannula of the conscious, unrestrained rat and removed several times to ensure there was no blood clot or tissue adhering to the cannula tip. After final placement of the cannula, the first 10 min of perfusate was discarded. At 10:00 hr the perfusion began and ran for 4 hr. Rats were typically undisturbed by this procedure as judged by their normal range of activities, including sleeping, eating, drinking, and exploratory behavior.

The rats had 0.6 ml of venous blood removed from intra-atrial catheter every 15 min. The blood was spun down immediately in a Eppendorf centrifuge, and the plasma was saved on ice. The red cells were resuspended in heparinized (40 U/ml) saline and reinjected when the subsequent sample was withdrawn. This method of plasma collection was well tolerated by rats for considerable periods of time. The perfusate was collected every 15 min, and, as represented in our data, each plasma GH value was paired with a value for the SS released into the perfusate for the 15-min period *preceding* that blood sample.

The RIA for GH permitted measurement of from 1 to 400 ng/ml in a 50-μl sample of plasma. The RIA materials were provided by NIAMMD, National Pituitary Agency. The RIA for SS was developed in our laboratory[20] and permitted measurement of from 1.25 to 320 pg per sample of

---

[19] M. R. Brown and G. A. Hedge. *Neuroendocrinology* **9,** 158 (1972).

[20] M. A. Arnold, S. M. Reppert, O. P. Rorstad, S. M. Sagar, H. T. Keutmann, M. J. Perlow, and J. B. Martin, *J. Neurosci.* **2,** 674 (1982).

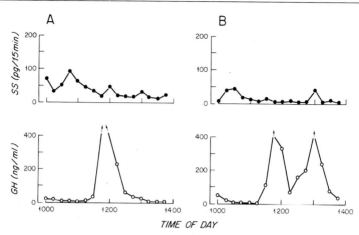

FIG. 2. Normal pulsatile growth hormone (GH) release and corresponding somatostatin (SS) secretion in two (A,B) unanesthetized, freely-behaving rats. Reproduced, with permission, from Kasting et al.[17]

200 μl of perfusate. Both perfusate and plasma samples were run in duplicate.

Evaluation of Technique Combining Push–Pull Perfusion with Chronic Intraatrial Catheterization in the Unanesthetized, Freely Behaving Rat

We undertook a series of experiments to evaluate the push–pull perfusion technique with regard to the modulation of growth hormone (GH) secretion (a) to determine whether SS is detectable in perfusates of the ME and to determine whether the release is related to GH secretion; (b) to determine whether the technique can be used to introduce exogenous substances into the ME and by such local application affect GH release.

Two typical experiments are illustrated in Fig. 2 in which normal episodic GH release into plasma was apparent, indicating a functionally intact hypothalamic–ME–pituitary pathway. That is, the normal release and transport of SS and GRF to the pituitary occurred despite the perfusion of the ME. The SS release as detected in the perfusate showed levels of from 10 to 90 pg of SS per 15-min interval. The SS secretion appeared intermittent, and often a peak in plasma GH was associated with a peak in SS secretion. However, the reverse is not true: SS peaks often occurred independently of GH peaks. This tentative relationship indicating a peak of SS secretion occurring with a peak of GH secretion may be used to

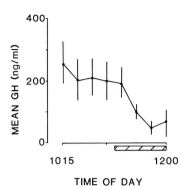

FIG. 3. Mean plasma growth hormone (GH) levels during control perfusion and perfusion with somatostatin (166 ng/ml) added to the perfusate (hatched bar) ($N = 5$).

support the short-loop negative-feedback theory[21,22] that increased GH levels stimulate SS secretion and consequently diminish further GH release. The meaning of the absolute values of SS in the perfusate are discussed later.

Thus, (a) SS is detectable in perfusates of the ME, without the perfusion disrupting normal regulation of pituitary hormone release; (b) the secretion is not constant, but tends to be of an intermittent nature; and (c) there appears to be a relationship between SS secretion and GH levels.

The second series of experiments sought to evaluate whether relevant exogenous substances could be introduced into the hypothalamic–pituitary system with this technique. In this experiment, control perfusion was carried out for 1 hr, and subsequently perfusion was done for 1 hr with SS (166 ng/ml) added to the perfusate. The results are illustrated in Fig. 3.

The mean GH levels for five rats are illustrated, and these data show that GH levels fall fairly quickly after SS perfusion begins. This indicates that exogenous SS introduced into the extracellular space of the ME can be transported to the pituitary and can be functionally effective at decreasing GH release. Another possibility is that SS introduced into the ME can presynaptically inhibit GRF release.

The next experiment was performed with a similar protocol, but perfusion during the second hour was performed with anti-somatostatin (ASS) serum diluted in the perfusate (1 : 12,000). The results are illustrated in Fig. 4, in which mean GH levels for 8 rats are presented. The data could be separated into two groups on the basis of concentrations of SS in the perfusate.

[21] G. S. Tannenbaum, *Endocrinology* **107**, 2117 (1980).
[22] M. E. Molitch and L. E. Hlivyak, *Horm. Metab. Res.* **12**, 559 (1980).

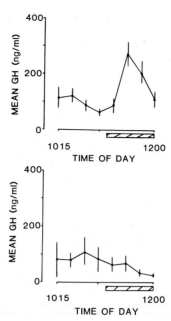

FIG. 4. *Upper graph:* Mean plasma growth hormone (GH) levels for rats ($N = 4$) with high somatostatin (SS) release from the median eminence (ME) as detected in perfusate, during control perfusion (no bar) and during perfusion with anti-somatostatin serum (ASS) added to perfusate (hatched bar). *Lower graph:* Mean plasma GH levels for rats ($N = 4$) with low SS release from the ME as detected in the perfusate, during control perfusion (no bar) and during perfusion with ASS (hatched bar).

Plasma levels of GH increased as expected in the four rats (upper graph) which had the highest SS levels in the perfusate prior to perfusion with ASS, whereas plasma levels of GH did not change in the four rats (lower graph) with the lowest, albeit detectable, SS levels in the perfusate. These data suggest that ASS introduced into the extravascular space of the ME can either alter SS output from the ME to the pituitary or affect SS influence on GRF neurons at the level of the ME, but this effect is dependent on the state of the system at the time of intervention, that is, on how much negative somatostatinergic influence is present. Thus, while GH is inhibited by relatively greater amounts of SS, negating the SS with ASS can increase GH release whereas, under conditions of low SS influence, tying up SS with ASS has correspondingly less or no influence on GH release.

Although we have determined that the ME can be perfused without functionally altering the hypothalamic–pituitary interaction, it was clear that the cannula does cause damage, as shown in Fig. 5. The inner cannula must penetrate the ME to perfuse the ME effectively, but the damage, although obvious, spares enough fibers and terminals to allow the normal release and subsequent transportation of SS and GRF in response to appropriate physiological signals.

Assuming a 5% exchange rate, the SS release of about 20–50 pg/15 min that is occurring under conditions of normal pulsatile GH secretion reflects about 1.3–1.6% of the total ME content of SS. The amount of SS entering the extravascular space of the ME during the perfusion with exogenous SS added to the perfusate is about 5 ng/15 min, which represents 16% of total ME content of SS. This indicates that a 10-fold increase in SS over baseline release values is sufficient to suppress GH secretion, and under physiological conditions perhaps less of an increase would be effective. We have observed[17] that in several rats where GH levels were suppressed in association with greatly elevated SS levels detectable in perfusates, the mean SS release associated with the suppressed GH levels in plasma was about 133 pg/15 min. Using the assumed 5% exchange rate, this represents a value of 8% of total ME SS content, which is consistent with the previous calculation of the amount of SS release necessary to

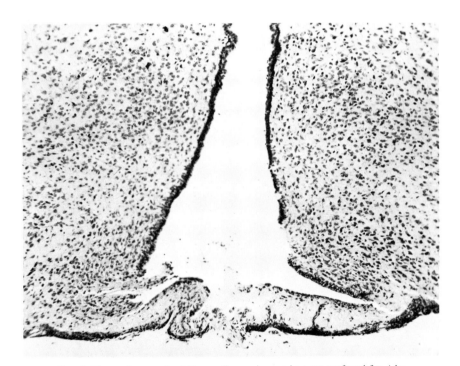

FIG. 5. Photomicrograph of the median eminence in a rat perfused for 4 hr.

INTERPRETATIONS OF DATA ON SOMATOSTATIN (SS) RELEASE
IN THE MEDIAN EMINENCE (ME) OF CONSCIOUS, FREELY
BEHAVING RAT ACQUIRED FROM PUSH–PULL
PERFUSION TECHNIQUE

1. Intermittent secretion of SS in ME
2. Short-loop negative feedback
3. Baseline-release equals 1.5% of ME content per minute
4. Fivefold increase of SS needed to suppress growth hormone
5. Secreted as two or more molecular weight forms[17]

suppress GH secretion derived from introducing exogenous SS. Thus, under normal conditions of pulsatile GH release, SS release as determined by the push–pull perfusion technique is about 1.5% of total ME content, whereas release on the order of 8% per 15-min period, or a 5-fold increase, is necessary to inhibit GH secretion such as might occur during stress.

This technique has also been adapted for evaluating the LHRH interaction with LH secretion[23] with encouraging results of a similar nature to those described above.

Problems

Despite the encouraging prospects for this technique in the evaluation of releasing-factor dynamics, there are several problems in the use of this technique. The problem of blockage of the cannulas with tissue is perhaps the worst ongoing problem, since blockage causes an artificial increase in substances in the perfusate and leads to a failure rate of about 50% or greater. Stereotaxic placement of the cannula into the ME has to be very precise, and histological examination shows that in about 10% of experiments the placement is unacceptable. Tissue damage must be a concern despite preservation of functional interaction between hypothalamus and pituitary and of normal behavior patterns. Several pilot experiments have been carried out to determine the effects of pharmacological stimulation on SS secretion as determined by this method, but results have been negative to date. Further study of this problem is underway in our laboratory.

Conclusions

These preliminary data suggest that push–pull perfusion of the ME combined with chronic atrial catheterization of the rat may be a valuable

[23] J. E. Levine and V. D. Ramirez, *Endocrinology* **111**, 1439 (1982).

tool for investigating dynamic changes in secretion of hypophysiotropic factors and subsequent release of pituitary hormones. This technique has allowed us to make some preliminary observations on secretion of SS in the unanesthetized rat (see the table). This technique has the potential of allowing much better understanding of the temporal patterns of hypophysiotropic factor secretion and consequently more precise knowledge about the interaction of these factors to regulate anterior pituitary hormone release into plasma. We look forward to future developments and refinements in this technique and the information it will provide.

Acknowledgments

This work was supported by U.S. Public Health Service Grant AM 26252. N. W. K. is a Postdoctoral Fellow, MRC of Canada.

[11] Horseradish Peroxidase: A Tool for Study of the Neuroendocrine Cell and Other Peptide-Secreting Cells

By RICHARD D. BROADWELL and MILTON W. BRIGHTMAN

Neurosecretory cells are defined as "nerve cells that specialize in the manufacture of chemical mediators to a degree greatly surpassing that of the more conventional neurons, and that have secretory activity comparable to that of gland cells."[1] All neurosecretory cells are similar insofar as they produce and release one or more different peptides as a chemical mediator. For this reason, such specialized neurons are collectively referred to as peptidergic; approximately two dozen varieties of peptidergic neurons have been identified within the mammalian central and peripheral nervous systems.[2] Many of the peptides are synthesized as part of larger molecules referred to as precursors. Posttranslational modification of the precursor results in the biologically active form of the peptide. Many peptidergic neurons in the brain establish synaptic contacts with neuronal cell bodies or processes; other centrally located peptidergic neurons send their axons to fenestrated blood vessels that do not have a blood–brain barrier. This latter class of peptidergic neurons is referred to as neuroendocrine, because their peptides, like hormones, are released into

[1] B. Scharrer, in "Peptides In Neurobiology" (H. Gainer, ed.), p. 1. Plenum, New York, 1977.
[2] F. E. Bloom, *Sci. Am.* **245**, 148 (1981).

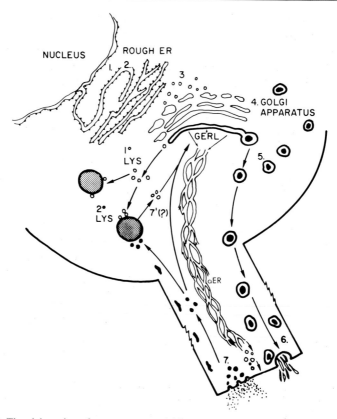

Fig. 1. The elaboration of secretory material in mammalian cells is believed to follow six successive steps collectively termed the secretory process. In peptidergic neurons and other peptide-secreting cells, such as those in the anterior pituitary, secretory material is (1) synthesized on ribosomes associated with the endoplasmic reticulum (ER); (2) segregated within the cisternal space of the rough ER; (3) transported, possibly by vesicles of rough ER origin, to the outer or forming face of the Golgi apparatus; (4) concentrated and further processed in saccules of the Golgi apparatus; (5) packaged and stored in secretory granules of GERL* origin; and (6) discharged from secretory granules to outside the cell. Step 7 involves the internalization of cell surface membrane, possibly that of the secretory granule membrane, and is a consequence of exocytosis. The intracellular pathways this internalized membrane may take can be traced with horseradish peroxidase delivered extracellularly. The membrane may be channeled directly back to GERL or indirectly to GERL by way of secondary lysosomes (2° LYS). In addition to producing secretory granules, GERL of secretory cells may also give rise to primary lysosomes (1° LYS) that convey acid hydrolases to secondary lysosomes. Our cytochemical studies of the ER, Golgi apparatus, and GERL indicate that GERL is associated structurally and functionally with the Golgi apparatus, not with the ER[7,8,71] [see also A. M. Cataldo and R. D. Broadwell, *J. Cell Biol.* **95**, 405a (1982)]. Secretory and nonsecretory material processed through the Golgi apparatus may be conveyed to the axonal endoplasmic reticulum (aER) in the neuron by vesicles in lieu of direct membrane continuities between the two organelles. Membranes appearing as

the blood for delivery to distant target tissues. Neuroendocrine cells with axons projecting into the posterior pituitary gland produce and secrete vasopressin and oxytocin. These cells are found principally in the supraoptic and paraventricular nuclei of the hypothalamus. Other neuroendocrine cells, distributed in discrete areas of the hypothalamus and associated limbic forebrain structures, send their axons to the median eminence, where their peptides are released into hypophysial portal blood for subsequent control of hormone secretion from the anterior pituitary gland.

The purpose of this chapter is to consider the usefulness of exogenous horseradish peroxidase (HRP; M_r 40,000) as a cytochemical tracer for morphological study of the neuroendocrine cell at the light and electron microscopic levels. Our attention shall focus on the value of this glycoprotein as an intracellular marker to visualize organelles participating in the secretory process within the supraopticoneurohypophysial system from normal mice and mice hyperosmotically stressed by imbibing 2% salt water.[3-7] Whenever possible, comparisons and contrasts will be made between this system and the peptide-secreting cells of the anterior and intermediate lobes of the pituitary gland.[8,9]

The secretory process refers to the elaboration of secretory protein destined for exocytosis from the cell and the events associated therewith.[10] This process is a basic function common to most eukaryotic cells. The sequence of events and subcellular organelles participating in the secretory process of peptidergic cells are diagrammed in Fig. 1 and are represented at the ultrastructural level in Fig. 2. The ensuing discussion will emphasize the transfer and/or recycling of membrane among intracellular compartments, including the plasmalemma. Inspection of membrane

---

[3] R. D. Broadwell and M. W. Brightman, *J. Comp. Neurol.* **166**, 257 (1976).
[4] R. D. Broadwell and M. W. Brightman, *J. Comp. Neurol.* **185**, 31 (1979).
[5] R. D. Broadwell, C. Oliver, and M. W. Brightman, *Proc. Natl. Acad. Sci. U.S.A.* **76**, 5999 (1979).
[6] R. D. Broadwell, C. Oliver, and M. W. Brightman, *J. Comp. Neurol.* **190**, 519 (1980).
[7] R. D. Broadwell and A. M. Cataldo, *J. Histochem. Cytochem.* (1983) (in press).
[8] R. D. Broadwell and C. Oliver, *J. Histochem. Cytochem.* **31**, 325 (1983).
[9] A. M. Cataldo and R. D. Broadwell (in preparation).
[10] G. Palade, *Science* **189**, 347 (1975).

---

thick lines or thin lines connote morphological similarities in membrane among the intracellular compartments. Note that both types of membrane exist in the Golgi apparatus/GERL where transition between the two member types may occur. *GERL is an acronym for acid phosphatase-positive, Golgi-associated, smooth Endoplasmic Reticulum from which Lysosomes (and secretory granules) arise [A. B. Novikoff, *Proc. Natl. Acad. Sci. U.S.A.* **73**, 2781 (1976)].

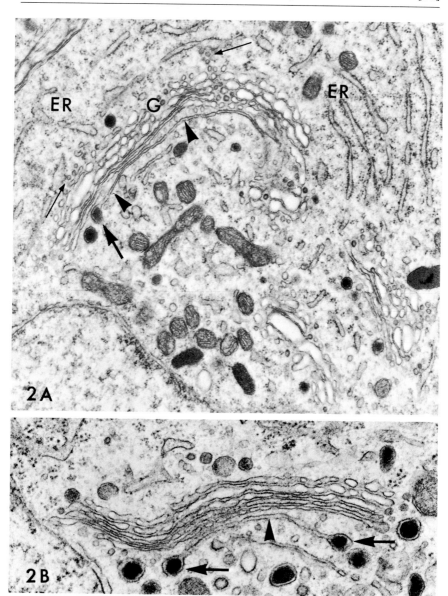

Fig. 2. Neuroendocrine cells of the supraopticoneurohypophysial system in the mouse contain parallel stacks of the rough endoplasmic reticulum (ER). Under normal conditions, secretory granules (arrows) are produced from GERL (arrowheads) and less so from Golgi (G) saccules. Transfer vesicles (small arrows) may provide communication between the rough ER and the outer face of the Golgi apparatus. The five types of peptide-secreting cells in the anterior pituitary gland exhibit similar organelle interrelationships.

dynamics with HRP as a tracer is accomplished by employing antibodies labeled with peroxidase for the intracellular localization of peptides serving as antigens and by fluid-phase endocytosis of native peroxidase delivered extracellularly.

Factors Affecting Peroxidase Cytochemistry

The most widely used cytochemical reaction to reveal the presence of HRP activity, and by inference the location of the enzyme, is with diaminobenzidine (DAB) and 1% hydrogen peroxide in Tris buffer (pH 7.2–7.4).[11] The reaction product is oxidized DAB, which is visible by light microscopy, and, after treatment with osmium tetroxide, by electron microscopy. The value of DAB as the chromagen lies at the ultrastructural level and is attributed to the unique properties of oxidized DAB: it is a polymer, homogeneous, osmiophilic (electron dense), and insoluble in organic solvents and embedding agents. These properties permit a high degree of resolution for the localization of reaction product. Chromagens such as o-dianisidine, tetramethylbenzidine, p-phenylenediamine dihydrochloride–pyrocatechol, and benzidine dihydrochloride are thought to be more sensitive than DAB in yielding reaction product,[12–15] particularly when small amounts of HRP are introduced to the tissue as in tracing neuronal connections or in immunocytochemistry. The problems with these chromagens are: the techniques in which they are utilized are more complex compared to the DAB procedure; the reaction products are crystalline and soluble in organic agents; and the reaction product observed ultrastructurally[16] is resolved poorly for localization in specific organelles. Although the DAB reaction is argued to be less sensitive than those with other chromagens, the DAB reaction can be intensified by adding imidazole to the incubation medium.[17]

Intensity of the staining reaction, regardless of the chromagen, is determined to a large extent by the concentration and activity of the HRP molecule. If HRP is injected into the tissue directly, the amount necessarily will be low (microliter quantities). Peroxidase possessing its maximum enzymic activity would be advantageous in detecting small, injected amounts of the protein. Peroxidase activity is somewhat dependent upon

[11] R. C. Graham and M. J. Karnovsky, *J. Histochem. Cytochem.* **14**, 291 (1966).
[12] M.-M. Mesulam and D. L. Resene, *J. Histochem. Cytochem.* **27**, 763 (1979).
[13] J. I. Morrell, L. M. Greenberger, and D. W. Pfaff, *J. Histochem. Cytochem.* **29**, 903 (1981).
[14] J. S. DeOlmos, *Exp. Brain Res.* **29**, 541 (1977).
[15] J. S. Hanker, P. E. Yates, C. B. Metz, and A. Rustioni, *Histochem. J.* **9**, B789 (1977).
[16] K. Carson and M.-M. Mesulam, *J. Histochem. Cytochem.* **30**, 425 (1982).
[17] W. Straus, *J. Histochem. Cytochem.* **30**, 491 (1982).

the purity of the molecule and the fixation process. Type VI HRP (Sigma Chemical Co., St. Louis, MO) and electrophoretically purified peroxidase (Worthington Corp., Freehold, NJ) are more enzymically active than the less pure and more toxic type II HRP (Sigma). The importance of purity assumes added significance when HRP is delivered locally or parenterally; HRP is known to initiate a degranulation of mast cells, resulting in an increased permeability of blood vessels and an anaphylactic response.[18] Mice and the Wistar/Furth strain of rats are resistant to the HRP stimulation of mast cells. In these species horseradish peroxidase is an excellent tracer to apply intravenously for, unlike other hemeproteins (i.e., microperoxidase), it will not bind to serum proteins.[19]

Several modifications of the original Graham and Karnovsky[11] protocol for the cytochemical reaction of peroxidase with DAB and hydrogen peroxide have been proposed. Three of these modifications[20-22] suggest that the reaction is more intense at pH 5 than at the neutral pH recommended by Graham and Karnovsky; however, Straus[23] has reported obtaining diffusion artifacts of the DAB reaction at this low pH. The modification suggested by Fahimi[24] is the one we follow routinely for our light and electron microscopic preparations. In lieu of Tris or phosphate buffer, the Fahimi incubation medium (pH 7.1–7.4) uses 0.1 M sodium cacodylate buffer, which is the same buffer as that in our fixative and wash solutions. The generated reaction product is judged not to diffuse from its sites of localization ultrastructurally.

Investigations tracing axonal connections by injection of HRP into the brain indicate that the reliability of the procedure, or the sustaining of the peroxidase activity, is sensitive to the type of fixative perfused through the brain.[22,25] The optimal situation would offer simultaneous preservation of the tissue and of HRP activity. Although many enzymes lose considerable activity upon exposure to certain fixatives, our experience indicates that the HRP molecule survives fixation for ultrastructural analysis. Cellular fine structure and HRP activity are preserved equally well in tissues exposed to peroxidase administered intravenously in mice prior to perfusion fixation with a 2% glutaraldehyde–2% formaldehyde mixture.[7-9,26]

[18] R. S. Cotran and M. J. Karnovsky, *Proc. Soc. Exp. Biol. Med.* **126**, 557 (1967).
[19] M. Bundgaard and M. Moller, *J. Histochem. Cytochem.* **29**, 331 (1981).
[20] E. E. Weir, T. G. Pretlow, A. Pitts, and E. E. Williams, *J. Histochem. Cytochem.* **22**, 1135 (1974).
[21] P. Streit and J. C. Reubi, *Brain Res.* **126**, 530 (1977).
[22] L. Malmgren and Y. Olsson, *Brain Res.* **148**, 279 (1978).
[23] W. Straus, *J. Histochem. Cytochem.* **28**, 645 (1980).
[24] H. D. Fahimi, *J. Cell Biol.* **47**, 247 (1970).
[25] D. L. Rosene and M.-M. Mesulam, *J. Histochem. Cytochem.* **26**, 28 (1978).
[26] R. D. Broadwell and M. Salcman, *Proc. Natl. Acad. Sci. U.S.A.* **78**, 7820 (1981).

TABLE I
Procedure for Horseradish Peroxidase Cytochemistry

1. Fix the tissue *in situ* by perfusion with a mixture of 2% glutaraldehyde–2% formaldehyde and 0.025% $CaCl_2$ in 0.1 M sodium cacodylate buffer, pH 7.2–7.4, for 10 min.
2. Transfer the tissue to cacodylate buffer.
3. Cut sections 25–75 μm thick on a vibratome or Smith–Farquhar tissue chopper.
4. Rinse the sections 3–5 times in cacodylate buffer. The glutaraldehyde must be washed out of the tissue prior to incubation in order to prevent the formation of nonspecific reaction product.
5. Incubate the sections for 15–30 min at room temperature in the Fahimi[24] medium: 10 mg of diaminobenzidine, 10 ml of 0.1 M sodium cacodylate buffer, and 0.2 ml of 1% H_2O_2, with a final pH of 7.1–7.4. Filter the medium through No. 1 Whatman filter paper prior to incubation of the tissue.
6. Rinse the tissue 3–5 times in cacodylate buffer.
7. For light microscopy (bright or dark field), mount the sections on slides freshly coated with albumin and air-dry the sections. Immerse the slide into xylene for 5 min, remove, and coverslip the sections with Permount.
8. For electron microscopy, trim down the sections to the area of interest. Postfix the sections for 2 hr in 2% osmium tetroxide (with 0.1 M cacodylate buffer) or in reduced osmium (2% osmium tetroxide, 1.5% potassium ferrocyanide, 0.1 M cacodylate buffer). The latter is particularly useful for enhancing membrane contrast. Rinse the sections briefly in distilled water, dehydrate, and embed in plastic. The tissue should not be stained *en bloc* nor poststained with heavy metals. Plastic sections 1 μm thick are placed on glass slides and viewed under dark-field are helpful for assessing the concentration of HRP-positive organelles within the cells.[6,60]

Adequate fixation of the tissue is important for minimizing staining artifact that may result from HRP diffusion. Our procedure for HRP cytochemistry is given in detail in Table I.

Immunocytochemistry

Methodologies

Immunoperoxidase procedures are more problematical than those associated with the incubation of tissues injected with peroxidase. The antigenic determinants of the substance to be localized must survive fixation to ensure that the antigen will complex directly with the HRP-labeled antibody or indirectly through linking antibodies. For ultrastructural immunocytology to be successful, a compromise must be reached between good tissue fixation and the preservation of antigenic determinants. The copreservation of tissue antigenicity and fine structure will be of little value if the antibodies employed fail to reach the antigen within the cells under investigation. The IgG molecule is about four times larger than

HRP, and its extracellular diffusion is more restricted; if this restriction is too severe, immunostaining may not occur. Similarly, oxidized DAB will be absent if the DAB molecule or the hydrogen peroxide cannot reach the peroxidase–antibody conjugate for interaction with the peroxidase molecule(s). Because maximum tissue penetration of DAB is approximately 20 μm,[27] the thickness of the incubated sections theoretically should not exceed 40 μm. The copreservation of antigenicity and morphology in ultrastructural immunocytochemistry is discussed in detail by us elsewhere.[5,28,29]

Covalent bonding of peroxidase to immunoglobulins, referred to as the labeled antibody–enzyme technique, was introduced by Nakane and Pierce[30] for immunocytochemical localization of tissue antigens. Covalent bonding necessarily decreases the efficiency of labeled antibody methods; it is destructive to the antibody and results in products that consist of mixtures of labeled and unlabeled antibodies. These problems are alleviated with the unlabeled antibody–enzyme or peroxidase–antiperoxidase (PAP) method of Sternberger.[31] The distinguishing feature of the PAP method is the high specificity of staining, which exceeds that of labeled antibodies by a factor of several orders of magnitude.

In the PAP method only immunological bonds are used to attach peroxidase to antigenic sites in the tissue. Tissue antigen is first allowed to react with a specific antibody (primary antiserum); this step is followed by exposure of the tissue to a secondary antiserum produced in another species and specific for the immunoglobulin of the primary antiserum. One of the two combining sites of the secondary antibody (anti-immunoglobulin) combines with the primary antibody, while the second site remains free. The PAP complex is added as a third step; it becomes bound to the secondary or linking antibody because the antibody of the PAP complex reacts as antigen with the free combining site of the secondary antibody. The HRP is allowed to react with its substrate, hydrogen peroxide, using DAB as the electron donor. This four-step incubation process is summarized in Table II.

Accurate interpretation of the results obtained with the PAP method is complicated by the presence of false-positive or false-negative staining.

[27] E. Essner, in "Electron Microscopy of Enzymes" (M. A. Hayat, ed.), p. 1. Van Nostrand-Reinhold, New York, 1974.
[28] R. D. Broadwell, in "Current Methods in Cellular Neurobiology" (J. L. Barker and J. F. McKelvy, eds.). Wiley, New York, 1983. In press.
[29] R. D. Broadwell, in "Strategies for Studying the Roles of Peptides in Neuronal Function." Soc. Neuroscience Short Course Syllabus, p. 27 (1982).
[30] P. K. Nakane and G. B. Pierce, Jr., *J. Cell Biol.* **33**, 307 (1967).
[31] L. Sternberger, "Immunocytochemistry," 2nd ed. Wiley, New York, 1979.

TABLE II
FOUR-STEP INCUBATION SEQUENCE FOR IMMUNOCYTOCHEMICAL STAINING USING THE
UNLABELED ANTIBODY–ENZYME METHOD[a]

Step 1. Primary antibody directed against a specific tissue antigen [rabbit anti-X (IgG)]
Step 2. Secondary antibody bridge [sheep anti-rabbit (IgG)]
Step 3. Tertiary antibody–enzyme complex [rabbit peroxidase–antiperoxidase complex]
Step 4. Detection of peroxidase activity[b,c] [cacodylate buffer (pH 7.1–7.4), DAB, 1% H_2O_2]

[a] Sternberger.[31]
[b] Fahimi.[24]
[c] For ultrastructural inspection, the oxidized DAB is converted to the electron-dense osmium black by postfixing the tissue is osmium tetroxide.

False-positive staining refers to immunoreactive cells or organelles that one would not expect to exhibit immunoreactivity; false-negative staining implies an absence of immunoreactivity in cells or organelles expected to be immunoreactive. Interpreting what is false-positive or false-negative staining can prove to be difficult in and of itself.

The PAP method is useful in neurobiology for localizing tissue antigens (i.e., peptides, hormones, neurotransmitters, enzymes, etc.) at light and electron microscopic levels by exposing tissue sections to the four-step incubation and staining process either before or after the sections are embedded in an appropriate medium. Postembedding staining has been highly successful on paraffin-embedded sections for tracing axonal connections of peptidergic, GABAergic (GABA, γ-aminobutyric acid), cholinergic, and monoaminergic neurons within the central and peripheral nervous systems. Similar antigens in neurons have been detected ultrastructurally using the postembedding staining regimen in aldehyde-fixed material. This approach entails incubating nonosmicated, plastic-embedded, ultrathin sections that have been placed on nickel or gold grids. The sections are etched with hydrogen peroxide prior to the incubation process in order to enhance antibody penetration through the plastic and into the sections. Osmication of the tissue follows the development of the reaction product in the DAB medium. In most instances the localization of antigen derived from postembedding staining is restricted predominantly to secretory granules and synaptic vesicles[32–36]; many antigens

[32] A. J. Silverman and E. A. Zimmerman, *Cell Tissue Res.* **159**, 291 (1975).
[33] A. J. Silverman and P. Desnoyers, *Cell Tissue Res.* **169**, 157 (1976).
[34] G. P. Kozlowski, S. Frenk, and M. S. Brownfield, *Cell Tissue Res.* **179**, 467 (1977).
[35] F. W. Van Leeuwen, D. F. Swaab, and C. de Ray, *Cell Tissue Res.* **193**, 1 (1978).
[36] G. Pelletier, L. Desy, J. Cote, G. Lefevres, H. Vawdry, and F. Labie, *Neuroendocrinology* **35**, 402 (1982).

have failed to be localized or have not been localized in expected intracellular sites with this staining approach. These false-negative results may be attributed to the destruction of antigenic determinants as a consequence of aldehyde fixation and/or the embedding procedure. Tissue processing by freeze substitution or freeze drying has not improved upon the false-negative results obtained with postembedding staining. The one asset postembedding staining offers is the possibility for localizing more than one antigen in any given cell by exposing thin sections through that cell to different primary antisera.

The alternative preembedding staining approach is by no means free of inherent difficulties in preparation and interpretation. As alluded to earlier, these problems are primarily the copreservation of cellular morphology and antigenicity along with achieving sufficient penetration of the IgG antibody (M_r 156,000) and the PAP complex (M_r 420,000) through cell membranes to reach the antigen to be identified and localized. Both false-positive and false-negative staining can be encountered, particularly in tissues prepared for ultrastructural analysis. Nevertheless, preembedding staining in immunocytology has revealed antigen in subcellular compartments where postembedding staining has failed.[5,37,38] This approach likewise has enabled specific peptidergic neurons to be identified ultrastructurally and their morphological characteristics to be defined.[39] Sections cut on a chopper or vibratome and prepared from aldehyde, perfused-fixed material are incubated free-floating in petri dishes through the four-step staining sequence (Table II). After osmication, the tissue is dehydrated and flat-embedded in plastic. Because the large molecular weight antibodies fail to penetrate cell membranes very far beneath the surfaces of the sections, extreme caution must be exercised to avoid discarding the immunostained thin sections cut from those surfaces. Our detailed procedure for the preembedding staining regimen is provided in Table III.

A peroxidase immunocytochemical alternative to the PAP method has been introduced that offers low background staining and greater sensitivity for localizing tissue antigens. This alternative is the avidin–biotinylated horseradish peroxidase complex (ABC).[40] Avidin (M_r 68,000) has a high affinity for biotin, a small molecular weight vitamin, and can bind four biotin molecules. Most proteins, including immunoglobulins and enzymes, can be conjugated with several molecules of biotin. Thus, macromolecular complexes can be established whereby avidin would serve as

---

[37] C. Tougard, R. Picart, and A. Tixier-Vidal, *Am. J. Anat.* **158**, 471 (1980).

[38] R. Y. Osamura, N. Komatsu, S. Izumi, S. Yoshimura, and K. Watanabe, *J. Histochem. Cytochem.* **30**, 919 (1982).

[39] G. P. Kozlowski, L. Chu, G. Hostetter, and B. Kerdelhue, *Peptides* **1**, 37 (1980).

[40] S. M. Hsu, L. Raine, and H. Fanger, *J. Histochem. Cytochem.* **29**, 577 (1981).

TABLE III
PREEMBEDDING STAINING APPROACH TO THE PEROXIDASE–ANTIPEROXIDASE METHOD
FOR ULTRASTRUCTURAL IMMUNOCYTOCHEMISTRY

1. Fix the tissue by perfusion for 5 min with a mixture of glutaraldehyde, paraformaldehyde, 0.025% $CaCl_2$, and 0.1 M cacodylate buffer at room temperature; pH 7.1–7.2. For our particular study[5] a 1.25% glutaraldehyde–1% paraformaldehyde mixture was successful. The fixative components and their concentrations eventually selected must depend on the properties of the antigen to be localized and must be compatible with maintaining cellular fine structure.
2. Remove the brain and place it into 0.1 M cacodylate buffer, pH 7.35.
3. Prepare 50 μm-thick sections by cutting the brain on a vibrating microtome or a tissue chopper. Rinse the sections in cacodylate buffer three times and store overnight at 4°.
4. Incubate the sections for 10–15 min in 0.1–0.2% Triton X-100 in cacodylate buffer at 25°. Detergents like Triton X-100 are believed to increase the penetration of the antibodies through cell and organelle membranes. Our experience suggests that detergents are of questionable value in this regard and serve to affect membrane morphology adversely.
5. Rinse the sections thoroughly in cacodylate buffer.
6. Incubate the sections for 1 hr at 25° in 10% normal serum (in cacodylate) of the same species that donates the secondary or linking IgG antiserum. Normal sheep serum was used in our experiments. The diluted normal serum eliminates false-positive staining that could result from nonspecific binding of PAP in the tissue.
7. Rinse the sections three times in cacodylate.
8. Incubate the sections for 24 hr at 4° in the primary antiserum mixed with 1% normal serum of the same species donating the secondary (linking) antibody. A dilution curve of the primary antiserum *must* be established to determine the proper concentration(s) of the primary antiserum that will yield specific staining. If the primary antibody is too concentrated, immunostaining will be nonspecific (false-positive).
9. Rinse the sections thoroughly in cacodylate.
10. Incubate the sections for 30 min at 25° in the secondary (linking) antiserum at 1:100 in 1% normal serum of the same species as the secondary antiserum.
11. Rinse the sections three times in cacodylate.
12. Incubate the sections for 30 min at 25° in PAP at 1:100 in 1% normal serum. The PAP should be of the same species that donates the primary antiserum.
13. Rinse the sections three times in cacodylate.
14. Incubate the sections in the diaminobenzidine (DAB)–H_2O_2 medium recommended by Fahimi[24] (see Table I). The incubation is conducted at room temperature and should be stopped as soon as background labeling becomes noticeable. Our best results were obtained with 7–10 min of incubation in the DAB–H_2O_2 medium.[a] At the light microscopic level the cytoplasm of the immunoreactive cell bodies did not appear to be homogeneously brown but contained reactive inclusions, which were seen ultrastructurally as immunostained organelles (Figs. 3 and 4). If the cells of interest and background begin to turn brown soon after the tissue is placed into the DAB–H_2O_2 medium, the primary antiserum may have been too concentrated. A primary antiserum that is too concentrated will generate false-positive immunostaining.
15. Rinse the sections three times in cacodylate.
16. Sections are now ready to be osmicated, followed by dehydration and plastic embedding. *En bloc* staining of the tissue and poststaining the ultrathin sections are not recommended.

[a] This short incubation time contrasts with the 15–30 min of incubation time required for tissue exposed to native peroxidase delivered extracellularly as a probe molecule.

an intermediary in linking a biotinylated antibody with a biotinylated enzyme such as HRP. The first step in the ABC procedure is identical to that in the PAP technique (Table I). In the second step a biotin-labeled secondary antibody (i.e., sheep anti-rabbit IgG) is introduced and complexes through immunological bonds with the primary antibody. The ABC is added as the third step; the avidin of the ABC binds to the biotin of the secondary antibody. The tissue antigen is localized by incubation in the $DAB-H_2O_2$ medium. The high sensitivity reported for the ABC method resides in the potentially large number of peroxidase molecules associated (indirectly through avidin molecules) with the secondary antibody. The high sensitivity permits lower concentrations of the primary antiserum to be used, thereby resulting in low background staining. A second important advantage of the ABC method, particularly with regard to membrane penetration of the complex for preembedding staining, is the lower molecular weight of the ABC (approximately 188,000) compared to that of the PAP molecules (420,000).

Light Microscopy

Application of the PAP technique for the study of neuroendocrine cells at the light microscopic level has been instrumental in demonstrating the locations and efferent projections of these neurons and the coexistence of two, and perhaps more, peptides within the same neuroendocrine cell. In this regard, the hypothalamic supraoptic and paraventricular nuclei have been the most thoroughly investigated. Immunostaining for oxytocin and vasopressin in the same section from the rat hypothalamus has shown that both hormones are present in the supraoptic and paraventricular nuclei, but in separate neurons.[41] Serial sections alternately stained for neurophysins I and II, the carrier proteins for oxytocin and vasopressin, respectively, have confirmed and extended this observation to the bovine hypothalamus.[42] These results initially support the notion of one cell–one hormone; however, subsequent immunocytochemical studies suggest that in some cells of the supraoptic and paraventricular nuclei, vasopressin may coexist with substance P[43] and/or dynorphin,[44] while oxytocin may coexist with cholecystokinin.[45] Enkephalin,[46] β-lipotropin, and ACTH[47]

[41] F. Vandesande and K. Dierickx, *Cell Tissue Res.* **164**, 153 (1975).
[42] F. Vandesande, K. Dierickx, and J. DeMay, *Cell Tissue Res.* **158**, 509 (1975).
[43] M. E. Stoeckel, A. Porte, M. J. Klein, and A. C. Cuello, *Cell Tissue Res.* **223**, 533 (1982).
[44] S. J. Watson, H. Akil, W. Fischli, A. Goldstein, E. Zimmerman, G. Nilaver, and T. B. van Wimersma Greidanus, *Science* **216**, 85 (1982).
[45] J. J. Vanderhaeghen, F. Lostra, F. Vandesande, and K. Dierickx, *Cell Tissue Res.* **221**, 227 (1981).
[46] J. Rossier, E. Battenberg, A. Bayon, R. J. Miller, R. Guillemin, and F. Bloom, *Nature (London)* **277**, 653 (1979).
[47] S. A. Joesph and L. A. Sternberger, *J. Histochem. Cytochem.* **27**, 1430 (1979).

may be present as well. Other neuroendocrine cells likewise may harbor more than one peptide. For example, some neurons in the rat arcuate nucleus are immunoreactive for β-endorphin and ACTH, both of which may be derived from a common precursor.[48]

In addition to sending axons into the neurohypophysis,[49] PAP immunoreactive oxytocin and vasopressin neurons of the paraventricular nucleus project to specific sites in the limbic forebrain, medulla, and spinal cord.[50] Whether or not supraoptic neurons contribute projections other than to the neurohypophysis has not been determined conclusively. A third population of vasopressin–neurophysin II neurons is located in the suprachiasmatic nucleus; these neurons do not direct their axons into the neurohypophysial system but project predominantly to forebrain structures such as the lateral septum and lateral habenula.[51] Oxytocin and vasopressin, therefore, may serve as neurotransmitters or neuromodulators within the central nervous system in addition to acting as neurohormones.

Neurons containing luteinizing hormone releasing hormone[52] (LHRH) or corticotropin releasing factor hormone[53] appear to be distributed within the limbic forebrain and hypothalamus. These neurons establish contacts with the median eminence as well as with widespread areas in the brain.

Primary antisera against either horseradish peroxidase or wheat germ agglutinin further permit application of the PAP method to identify neuroendocrine and peptidergic neurons that have accumulated these exogenous glycoprotein antigens by retrograde axoplasmic transport from axon terminals. By adopting this approach, the afferent projections to the median eminence[54] and efferent projections from the paraventricular nucleus to the medulla[55] have been described.

Electron Microscopy

To date, only three published immunocytochemical reports have achieved copreservation of ultrastructural detail and antigenicity of peptides or their precursors in organelles associated with the secretory process. Two of these studies deal with the prolactin cells of the rat anterior pituitary gland[37,38]; the third is our investigation of the supraopticoneuro-

---

[48] M. V. Sofroniew, *Am. J. Anat.* **154,** 283 (1979).
[49] M. W. Sofroniew, A. Weindl, I. Schinko, and R. Wetzstein, *Cell Tissue Res.* **196,** 367 (1979).
[50] R. M. Buijs, *Cell Tissue Res.* **192,** 423 (1978).
[51] M. V. Sofroniew and A. Weindl, *Am. J. Anat.* **153,** 391 (1978).
[52] A. J. Silverman and L. C. Krey, *Brain Res.* **157,** 233 (1978).
[53] S. A. Joesph and K. M. Knigge, *Neurosci. Lett.* **35,** 131 (1983).
[54] R. M. Lechan, J. L. Nestler, and S. Jacobson, *Brain Res.* **245,** 1 (1982).
[55] M. V. Sofroniew and U. Schrell, *Neurosci. Lett.* **22,** 211 (1981).

hypophysial system in hydrated mice and mice hyperosmotically stressed by imbibing 2% salt water.[5] The preembedding staining approach to the PAP method was employed in each of the studies.

Success in our study was obtained using a primary antiserum directed against neurophysins I and II, which permitted immunostaining of both vasopressin- and oxytocin-producing cells in the supraoptic nucleus. Neurophysin, as opposed to vasopressin or oxytocin, was chosen as the antigen to localize immunocytochemically because this M_r 10,000 protein possesses a greater number of antigenic determinants compared to either of the M_r 1000 nonapeptides. We speculated that the larger the number of antigenic determinants, the greater the likelihood that enough antigenic determinants would survive the fixation process to ensure their identification by the primary antiserum. With anti-neurophysin serum diluted 1 : 4000–7500, immunostaining in supraoptic cell bodies was restricted to membrane-delimited compartments. Immunoreactive organelles included the nuclear envelope, rough endoplasmic reticulum, Golgi apparatus, and secretory granules (Figs. 3 and 4). Vacuoles measuring 0.4–0.7 μm in diameter were also immunoreactive and very likely represent secondary lysosomes. Neurophysin immunoreactivity in the rough endoplasmic reticulum and Golgi apparatus may not be indicative of neurophysin per se but of neurophysin as a portion of a precursor molecule.[56] The major disappointment in our study was the repeated inability to demonstrate immunoreactivity in GERL (Fig. 4A), the structure from which the neurosecretory granules are formed under normal conditions (Fig. 2). Numerous secretory granules within the supraoptic cell bodies and axon terminals of the neurophypophysis also were not immunoreactive. These false-negative results may be the result of a major shortcoming of the preembedding staining approach to the PAP method—inadequate penetration of antibody and/or DAB through cellular membranes.

More recently, we[7] have focused our attention on the neurosecretory axons and terminals in the neurophypophysis. Many profiles of neurosecretory terminals contain populations of synaptic-type vesicles 40–70 nm wide and secretory granules of 100–300 nm wide. One group of investigators[57] has implicated the axonal endoplasmic reticulum and 40–70 nm vesicles of the neurophypophysial system in the transport, storage, and release of neurosecretory material. Our anti-neurophysin, immunocytochemical preparations do not support this belief.[7] Neurosecretory granules are immunoreactive in the axons and terminals in the neurophypophysis, but the axonal endoplasmic reticulum and synaptic-type

[56] H. Gainer, Y. Sarne, and M. Brownstein, *J. Cell Biol.* **73**, 366 (1979).
[57] G. Alonso and I. Assenmacher, *Cell Tissue Res.* **199**, 415 (1979).

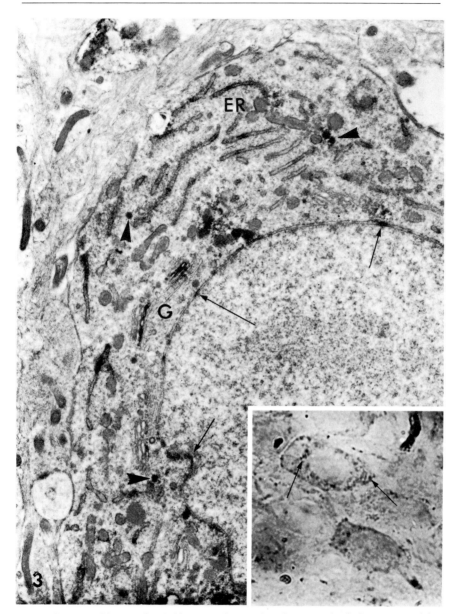

Fig. 3. Peroxidase–antiperoxidase reaction product for neurophysin immunoreactivity in supraoptic neurons of the mouse is localized within the nuclear envelope (arrows), rough endoplasmic reticulum (ER), the Golgi apparatus (G), and secretory granules (arrowheads). The inset shows that at the light microscopic level the immunoreactive organelles appear as granules (arrows) in the supraoptic cell body.

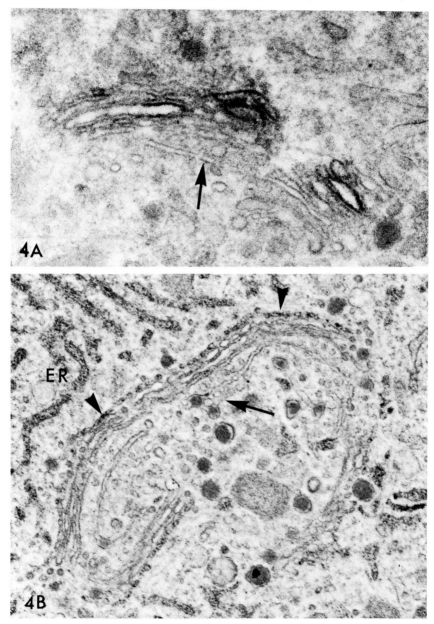

FIG. 4. (A) Peroxidase–antiperoxidase reaction product for neurophysin immunoreactivity is present within most saccules of the Golgi apparatus but is absent in GERL (arrow) of a supraoptic cell body from the mouse. (B) A similarity in contents between the rough endoplasmic reticulum (ER) and the Golgi apparatus in the supraoptic neuron is realized not

vesicles are not (Fig. 5). The absence of immunoreaction product in the axonal endoplasmic reticulum and vesicles is an unlikely example of false-negative staining by virtue of the fact that nearly all secretory granules are immunoreactive in the same axons and terminals harboring nonreactive synaptic-type vesicles, profiles of the endoplasmic reticulum, and mitochondria. These observations suggest that the transported and stored neurosecretory material in neurohypophysial axons and terminals is restricted to the secretory granules under normal conditions.

When mice are hyperosmotically stressed, we find that many neurosecretory terminals in the neurohypophysis take on a degenerating appearance (see Fig. 12B), which presumably is related to events associated with the stimulated release of hormone: internalization of axon terminal membrane following exocytosis, autophagy, and crinophagy.[7,58,59] Immunoreaction product in these "degenerating" terminals is present in intact secretory granules and in the surrounding cytoplasm. We believe that the extragranular reaction product may have escaped from secretory granules undergoing degradation or digestion (crinophagy) by acid hydrolase-containing lysosomes.[29]

The identification of anti-neurophysin reactivity in different organelles implies that the neurophysins and their nonapeptide hormones undergo a physical transfer from one subcellular compartment to another. Membrane necessarily accompanies the forming peptide throughout its course within the cell. The physical transfer of membrane or membrane molecules contributes to the formation of one subcellular compartment from another; thus, membrane may indeed arise from preexisting membrane. For example, neurophysin and associated membrane are delivered to the outer Golgi saccules from the rough endoplasmic reticulum. Vesicles blebbing off the endoplasmic reticulum may represent the morphological correlate for the transfer of membrane macromolecules and secretory product to the Golgi apparatus (see Figs. 1, 2, and 4). The secretory material and associated Golgi membrane eventually become incorporated into secretory granules, which are destined to fuse with the axon terminal

---

[58] R. D. Broadwell and A. M. Cataldo, *Soc. Neurosci. Abstr.* **8,** 789 (1982).
[59] R. D. Broadwell and A. M. Cataldo, in preparation.

---

only with immunocytochemistry but with cytochemical localization of glucose-6-phosphatase activity, a marker for the ER. Reaction product of glucose-6-phosphatase activity in the ER and outer Golgi saccule (arrowheads) of this supraoptic cell body suggests that the ER conveys the enzyme, and perhaps membrane or membrane molecules as well, to the Golgi apparatus [R. D. Broadwell and A. M. Cataldo, *J. Histochem. Cytochem.* **31,** 1077 (1983)]. Note the absence of reaction product in GERL (arrow) and its two forming secretory granules, further suggesting that GERL is not related to the ER.

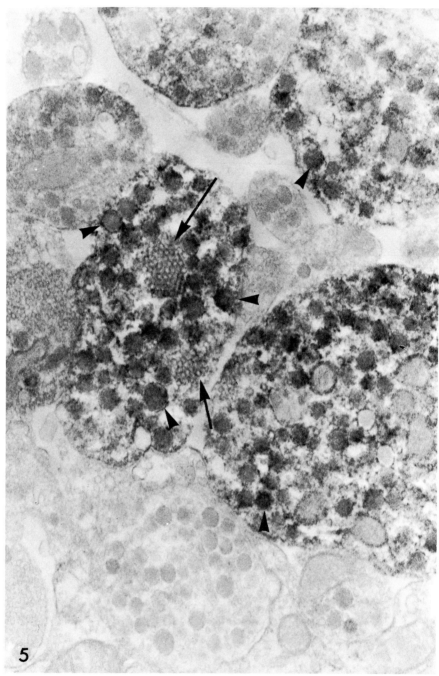

FIG. 5. Peroxidase–antiperoxidase reaction product for neurophysin immunoreactivity in the mouse neurohypophysis is present within secretory granules (arrowheads) but not in clusters of synaptic-type vesicles (arrows), which may represent internalized cell surface membrane (see Fig. 7).

plasmalemma for exocytosis of the secretory granule contents. The movement of membrane does not cease at the plasmalemma. Simultaneously with exocytosis and the concomitant addition of secretory granule membrane to the plasmalemma, compensatory mechanisms operating within the cell make the increase in surface area resulting from exocytosis a transient event. Axon terminal membrane is withdrawn into the cell. This retrieval of cell membrane, the intracellular pathway, and the possible fate of the membrane can be investigated with horseradish peroxidase delivered extracellularly. This topic is the focal point of the discussion in the following section.

Intracellular Transport of Horseradish Peroxidase

A second application for peroxidase is as a probe molecule for the identification of neuronal connections, based on the phenomenon of axoplasmic transport, as well as for the delineation of subcellular compartments associated with the transport and sequestration of the enzyme. When HRP is delivered extracellularly to *undamaged* neurons, it is incorporated by axon terminals and less so by cell bodies, dendrites, and myelinated axons.[4] The glycoprotein readily undergoes retrograde axoplasmic transport from the axon terminals to the parent cell bodies, thus permitting the identification of multiple sources of afferent input to a specific nuclear group or area within the central nervous system.

Peroxidase reaction product at the light microscopic level is not visible in undamaged axons and terminals. Organelles involved in transporting the enzyme in either the anterograde or the retrograde direction are beyond resolution with the light microscope[4]; however, once HRP enters the cell body and dendrites, its reaction product is localized in 0.2–0.8 μm wide granules visible throughout these portions of the neuron. The granules are equivalent to secondary lysosomes as determined ultrastructurally by acid phosphatase cytochemistry.[4,6,60] Available evidence suggests that the lysosomal system likewise is implicated in the anterograde transport of peroxidase out of the cell body.[4,6,58–60]

Light Microscopy

The retrograde axoplasmic transport of HRP was first applied to the neuroendocrine cell by Sherlock *et al.*,[61] who injected peroxidase directly into the posterior pituitary gland of the rat and observed retrogradely labeled cell bodies in the supraoptic and paraventricular nuclei and their

[60] R. D. Broadwell, *J. Histochem. Cytochem.* **28,** 87 (1980).
[61] D. A. Sherlock, P. M. Field, and G. Raisman, *Brain Res.* **88,** 403 (1975).

accessory nuclei. In the same year, we[62] first reported that presumptive neuroendocrine perikarya, the axons of which are directed to circumventricular organs possessing fenestrated capillaries (i.e., median eminence, neurohypophysis, organum vasculosum of the lamina terminalis), become labeled with peroxidase administered intravenously in mice. Blood-borne HRP leaked out of the fenestrated vessels and bathed the axon terminals innervating the circumventricular organs. The extracellular peroxidase was taken into these undamaged axon terminals and transported back to the cell bodies. HRP-positive perikarya were identified in the supraoptic, paraventricular, and accessory magnocellular nuclei, the arcuate nucleus, vertical limb of the nucleus of the diagonal band (Fig. 6), the periventricular stratum of cells, the area immediately surrounding the organum vasculosum and within the parenchyma of the median eminence.[3] Labeled cell bodies also were found in the brainstem noradrenergic A_1 and A_5 groups; we speculated that these latter two nuclei *may* represent the sources for the noradrenergic innervation of the median eminence.[3]

The major disadvantage in utilizing the vascular system to deliver peroxidase to neurosecretory axon terminals, as opposed to a focal injection into individual circumventricular organs, is that all the circumventricular organs are infiltrated simultaneously by the blood-borne tracer. For this very reason, we were cautious in our interpretation of the precise origins of afferent input to each of the circumventricular organs.[3] Direct injections of HRP or wheat germ agglutinin into the median eminence[54,63] or neurohypophysis[64] of the rat by other investigators have since confirmed our initial observations and further indicated that neurons in the diagonal band nucleus, medial division of the paraventricular nucleus, and in the vicinity of the ambiguous nucleus in the medulla (which corresponds in location to the A_1 nuclear group) are indeed afferent to the median eminence. Neurons in the diagonal band nucleus with a morphology similar to those that are retrogradely labeled with HRP in the same nucleus have been reported to be immunoreactive for LHRH.[39]

The most significant advantage to using the vascular route to deliver peroxidase to neurosecretory neurons is that the structural integrity of the axons and terminals is not compromised, an expectation that is not possible if a focal injection were made into the tissue. Our findings demonstrated conclusively that axon terminal arborizations are not required to be damaged for peroxidase uptake to occur. This fact has served us well for our cell biological investigations of peroxidase transport and related events in the undamaged neurosecretory cell.[4,6,7,58–60]

[62] R. D. Broadwell and M. W. Brightman, *Proc. Soc. Neurosci. Abstr.* **5**, 675 (1975).
[63] S. J. Wiegland and J. L. Price, *J. Comp. Neurol.* **192**, 1 (1980).
[64] J. Kelly and L. W. Swanson, *Brain Res.* **197**, 1 (1980).

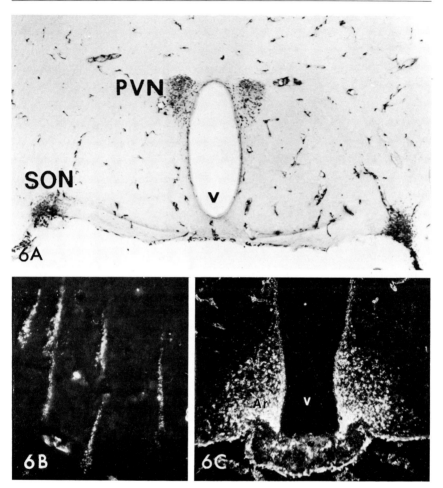

FIG. 6. Retrograde axoplasmic transport of native peroxidase administered intravenously in the mouse labels neuroendocrine cell bodies in the supraoptic and paraventricular nuclei (SON and PVN), vertical limb of the diagonal band nucleus (panel B), and in the arcuate nucleus (panel C, Ar) above the median eminence. v, the third ventricle.

Electron Microscopy

Peroxidase, either in its native form or conjugated to a lectin (i.e., wheat germ agglutinin, ricin), has proved to be an invaluable tracer ultrastructurally for investigating the process of endocytosis and the intracellular fate of internalized cell surface membrane in a host of cell types. Native or free HRP is considered not to bind to the plasmalemmal surface; it gains entry to the cell by fluid-phase endocytosis. This term implies that any substance in the extracellular fluid bathing a cell will be taken into the cell indiscriminately as a result of the normally occurring invagination of cell surface membrane in the form of vesicles and vacuoles. Unless the plasmalemma has been ruptured, HRP should not appear free in the cytoplasm. Conversely, HRP conjugated to a lectin enters the cell by adsorptive phase or receptor-mediated endocytosis.[65] In the case of wheat germ agglutinin–HRP, the wheat germ agglutinin will bind to surface membrane sites containing sialic acid and N-acetylglucosamine moieties; therefore, the cellular uptake of the lectin-conjugated HRP is highly membrane specific. Available evidence suggests that the fate of internalized cell surface membrane associated with fluid-phase endocytosis and with adsorptive phase endocytosis may be mutually exclusive.[65]

Neurosecretory terminals within the median eminence endocytose peroxidase predominantly in 40–70-nm-wide vesicles, which coalesce to form vacuoles, cup-shaped structures, and tubular profiles.[4,66] The internalization of axon terminal membrane with concomitant uptake of HRP in the neurohypophysis is more controversial than that in the median eminence. Douglas[67] and his co-workers were the first to propose that exocytosis of secretory granule contents from neurohypophysial terminals in the rat resulted in the retrieval of empty secretory granule membrane, with uptake of extracellular peroxidase, in the form of 40–70 nm-wide vesicles. Because the surface area of the neurohypophysial terminal failed to expand with successive exocytotic events, they calculated that as many as 10–25 of the endocytic vesicles were derived from a single 100–300-nm secretory granule. Other investigators were of the belief that the secretory granule membrane was retrieved as vacuoles similar in size to or larger than the secretory granule.[68–70] The number of HRP-positive vacuoles observed in these preparations appeared significantly greater than the concentration of similarly labeled vesicles. The speculation was offered that following exocytosis the secretory granule membrane is withdrawn intact from the plasmalemma. Our most recent studies[7] of the mouse neurohypophysis tend to confirm what Douglas and co-workers have reported. Within 5 min of an intravenous injection of HRP in mice, many endocytic vesicles exhibit peroxidase activity in neurohypophysial terminals. The larger peroxidase-reactive vacuoles, many of which are

[65] N. K. Gonatas, *J. Neuropathol. Exp. Neurol.* **41**, 6 (1982).
[66] R. Stoeckart, H. G. Jansen, and A. J. Kreike, *Cell Tissue Res.* **155**, 1 (1974).
[67] W. W. Douglas, *Prog. Brain Res.* **39**, 21 (1973).
[68] D. T. Theodosis, J. J. Dreifuss, M. C. Harris, and L. Orei, *J. Cell Biol.* **70**, 294 (1976).
[69] J. J. Nordmann and J. F. Morris, *Nature (London)* **261**, 723 (1976).
[70] J. F. Morris and J. J. Nordmann, *Neuroscience* **5**, 639 (1980).

similar in size to secretory granules, increase in number with time. Many of the large vacuoles appear to be formed by merger of the 40–70-nm vesicles (Fig. 7). Tubular profiles containing peroxidase probably are derived in a similar fashion. The exocytosis of secretory granule contents and the formation of endocytic, peroxidase-laden vesicles is increased demonstrably with hyperosmotic stress.[7] These observations serve to demonstrate that exocytosis and endocytosis are complementary events.

Endocytic vesicles also represent the major organelle incorporating extracellular HRP in the five peptide-secreting cell types of the anterior pituitary gland[8] and in cells of the pars intermedia.[9] Each of these cell types is characterized morphologically by the size and shape of its secretory granules. The labeled vesicles, but not vacuoles similar in size to the secretory granules, predominate in these different cells as quickly as 5 min after intravenous administration of peroxidase.

Native peroxidase, along with the cell surface membrane associated with it, is retrieved and transported by retrograde axoplasmic flow from neurosecretory terminals in the median eminence and neurohypophysis to the parent somata. Within these cell bodies the transported, HRP-positive vesicles, tubules, and vacuoles merge to form dense bodies or fuse with preexisting dense bodies (Fig. 8).[4,5] Supraoptic neurons from hyperosmotically stressed mice exhibit an increased concentration of labeled endocytic structures and dense bodies compared to hydrated controls.[6] Double incubation of neurosecretory perikarya for demonstration of peroxidase activity first, and for acid hydrolase activity second, revealed that at least some of the HRP-positive dense bodies are secondary lysosomes.[6,60] Many of the peroxidase-reactive dense bodies, although similar in size to lysosomes, may represent phagosomes or presecondary lysosomes that do not harbor acid hydrolase activity. A secondary lysosome is defined as containing acid hydrolytic enzymes and substrates upon which the enzymes can act. Phagosomes are derived from endocytic organelles and become secondary lysosomes upon receiving acid hydrolases, perhaps from primary lysosomes of GERL origin.[6,60] Secondary lysosomes would play a role in digesting the endocytic vesicles, vacuoles, tubules, and their contents. The lysosomal membrane is permeable to amino acids, small sugars, and other molecules released through the hydrolysis of macromolecules. These products of digestion can diffuse out of the lysosome and into the cytoplasm, where they become available for reutilization within the cell. An alternative possibility for the fate of endocytic membrane is that some endocytic organelles may deposit only their contents into the phagosome/lysosome and subsequently re-form to move on with their membrane intact to fuse with other depots, such as GERL or the Golgi apparatus. In these latter organelles, the endocytic membrane could be

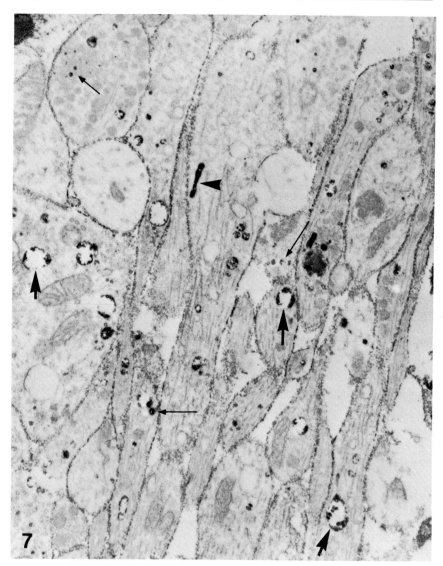

FIG. 7. Three hours after an intravenous injection of native HRP in the mouse, axons and terminals in the neurohypophysis exhibit peroxidase reaction product in endocytic vesicles (small arrows), vacuoles (large arrows), and tubules (arrowhead).

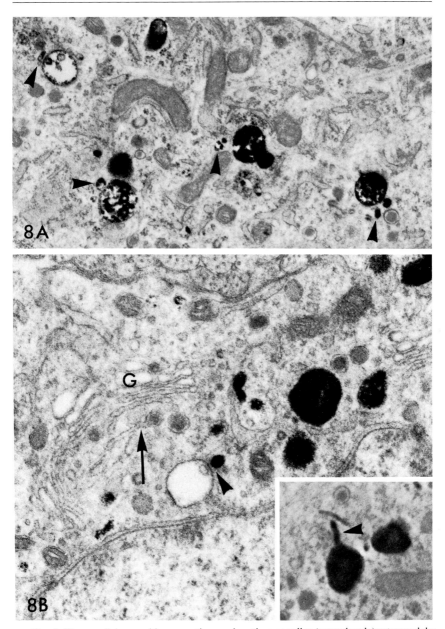

FIG. 8. Horseradish peroxidase-reactive, endocytic organelles (arrowheads) retrogradely transported to supraoptic perikarya in the mouse are clustered around or confluent with similarly reactive dense bodies. Note in (B) that the activity for native peroxidase is absent in the Golgi apparatus (G) and GERL (arrow).

reutilized in the formation of new membrane for other constituents of the endomembrane system (i.e., secretory granules, axonal endoplasmic reticulum, plasmalemma, lysosomes, etc.).

We have yet to observe a sequestration of native HRP within GERL or Golgi saccules in the neuron (Fig. 8B), including hyperosomotically stressed supraoptic neurons in which secretory granule production is increased demonstrably off GERL and all Golgi saccules.[71] These observations and reports that lectin-conjugated HRP labels GERL in nonneurosecretory cells[65,72] suggest that differences may exist between the fate of the contents and the fate of the membrane of endocytic structures. The absence of activity for native peroxidase in GERL and Golgi saccules of hyperosmotically stressed supraoptic neurons is difficult to rationalize if internalized cell surface membrane is directed to these organelles without first communicating with phagosomes/secondary lysosomes. If a direct intracellular pathway to GERL or the Golgi saccules exists for native HRP as it may for lectin-conjugated HRP,[72] a possible explanation for its apparent absence in the neurosecretory cell exposed to native peroxidase is that the enzyme may be too diluted within GERL and the Golgi saccules to visualize the enzyme activity cytochemically. This explanation seems implausible a priori, because we have localized native HRP activity not only within lysosomes but in GERL of somatotrophs in the anterior pituitary (Fig. 9).[8] The remaining cell types of the anterior pituitary and cells of the pars intermedia exhibit peroxidase activity only in endocytic structures and lysosomal dense bodies.[8,9] We considered that in the case of the somatotroph native HRP acts as a membrane-bound marker as opposed to a fluid-phase marker for most other cell types. The carbohydrate moieties of the HRP molecule may bind perferentially to cell surface sites containing mannose 6-phosphate.[73] At present, little information is available concerning the nature of the interaction between membrane and tracer molecules. Quite possibly, the binding of native HRP or lectin-conjugated HRP to the membrane could perturb the membrane to such a degree that the normal intracellular pathway and fate of the membrane are altered.

The preceding discussion has indicated that HRP undergoing retrograde axonal transport in the neuroendocrine cell, and in the neuron in general, is deposited eventually in perikaryal secondary lysosomes, where it is digested. For the most part, lysosomal acid hydrolase activity is minor in axons and terminals under normal conditions.[4,6,60,74] This fact

---

[71] R. D. Broadwell and C. Oliver, *J. Cell Biol.* **90**, 474 (1981).
[72] J. Q. Trojanowski and N. K. Gonatas, *Brain Res.* (in press).
[73] W. Straus, *Histochemistry* **73**, 39 (1981).
[74] E. Holtzman, in "Lysosomes, A Survey," Cell Biology Monographs, Vol. 3. Springer-Verlag, Berlin and New York, 1976.

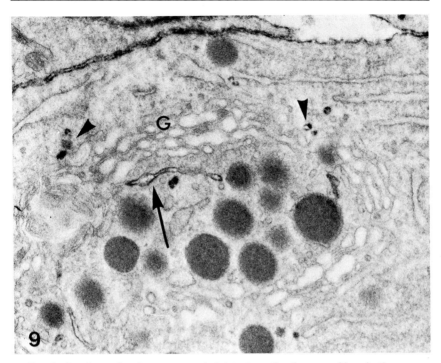

FIG. 9. One hour after intravenous administration of native peroxidase in the mouse, horseradish peroxidase reaction product is localized within GERL (arrow) and endocytic vesicles (arrowheads) of a somatotroph in the anterior pituitary gland.

provides additional functional significance for retrograde axonal transport in the neuron. Peroxidase taken up directly by cell bodies and dendrites also is carried to lysosomes (Fig. 10). Labeling of cell bodies by this route is not as great quantitatively as that derived by retrograde transport,[4] presumably because the internalization of cell surface membrane is more prominent in the axon terminal as a secondary consequence of exocytosis.

Endocytosed HRP, in addition to being useful for tracing cell surface membrane to perikaryal lysosomes, may be equally applicable for tracing membrane away from secondary lysosomes. Membrane-delimited tubules and vacuoles budding from perikaryal lysosomes may provide the basis for the joint anterograde axonal transport of acid hydrolases and peroxidase.[4,6] Employing peroxidase in such a fashion has enabled us to characterize the lysosomal system of organelles in the neuron. In this regard, the salt-stimulated neuroendocrine cell of the supraopticoneurohypophysial system has been ideal for demonstrating cytochemically and morphologically the dynamic properties of the lysosomal compartment.

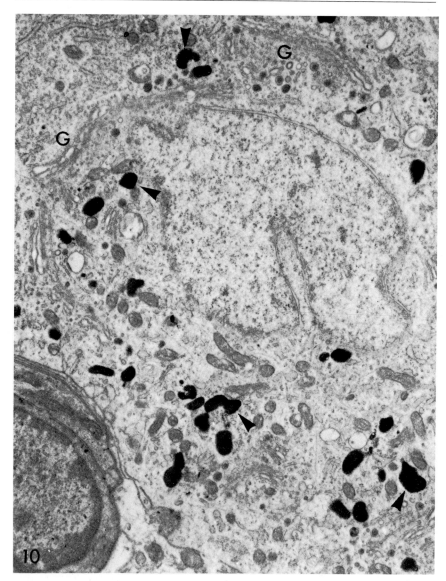

FIG. 10. Native peroxidase delivered into the cerebral ventricle of a salt-stressed mouse is endocytosed by supraoptic perikarya and subsequently directed to lysosomal dense bodies (arrowheads). Note the absence of peroxidase reaction product in the Golgi apparatus (G).

Supraoptic cell bodies and dendrites, but not axons or their terminals within the neurohypophysis, can be exposed to peroxidase circulating through the extracellular clefts of the neuropil subsequent to injection of the tracer into the lateral cerebral ventricle. Peroxidase is taken into the cells by endocytic vesicles. Dense body lysosomes in the perikarya become HRP-reactive (Fig. 10). Salt-treated mice so injected demonstrate numerous 50–100-nm-wide tubules and vacuoles that are peroxidase-positive in the pituitary stalk axons (Fig. 11). Some of the tubules appear varicose, while others have a smooth contour. Occasionally, a tubule and vacuole are confluent, as though perhaps the vacuole was in the process of separating from, or merging with, the tubule (Fig. 11). Not all axon terminals in the neurohypophysis contain peroxidase-labeled organelles; many terminals that do contain HRP have a degenerating appearance. Peroxidase-positive vacuoles predominate in these terminals and are similar morphologically and in size to autophagic and/or crinophagic vacuoles (Fig. 12). Acid phosphatase cytochemistry of the neurohypophysial system from salt-stressed mice indicates that organelles with a morphology similar to those that are HRP-positive are in fact lysosomes.[4,6,58,59] The neurohypophyses from hydrated mice, injected or not injected with peroxidase, served as controls and exhibited little or no evidence for anterograde transport of HRP and acid hydrolase enzymes.

The evidence presented suggests that the anterograde transport of peroxidase, when it occurs in the neuron, is in tandem with that of acid hydrolases; therefore, the lysosomal system of organelles appears to be intimately involved in this transport. The organelle that represents the common denominator as a depot for both HRP and acid hydrolases is the secondary lysosome, particularly the 0.2–0.8 μm wide secondary lysosomes in the cell body. These large lysosomes may provide the membrane, hydrolytic enzymes, and peroxidase (when present) to the smaller forms of lysosomes destined to be transported to the axon terminal, where they contribute the enzymes to the autophagic/crinophagic vacuoles. In support of this concept, we have observed HRP and acid phosphatase-reactive tubules, similar to those in the stalk axons, confluent with the large secondary lysosomes in supraoptic perikarya from hyperosmotically stimulated mice.[6]

The reactive tubules in the axon are not elements of the endoplasmic reticulum. Based on enzyme cytochemical analysis (i.e., glucose-6-phosphatase vs acid hydrolase) and morphological comparison, the endoplasmic reticulum and HRP–acid hydrolase-positive tubules represent two separate and distinct entities within the axon.[4,6,58,59] Nevertheless, the axonal endoplasmic reticulum does interact with the lysosomal system; it provides the membrane for the formation of the autophagic/crinophagic

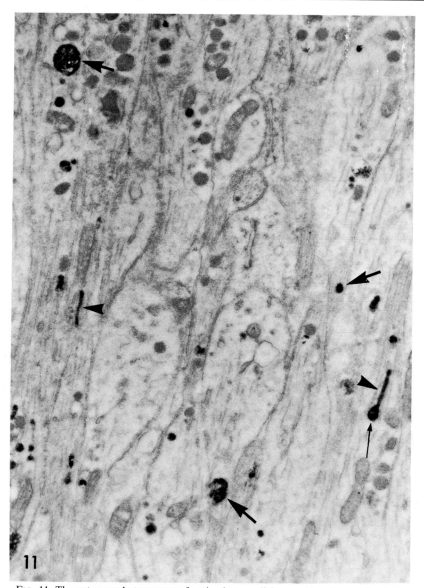

Fig. 11. The anterograde transport of native horseradish peroxidase is seen in pituitary stalk axons of salt-stressed mice, but not of control mice, 12 hr after a cerebral intraventricular injection of the protein. Labeled organelles include vacuoles (large arrows) of various sizes and tubules or cisterns (arrowheads), from which some of the smaller vacuoles may bud (small arrow). Similar organelles exhibit acid phosphatase activity and, therefore, are part of the lysosomal system.

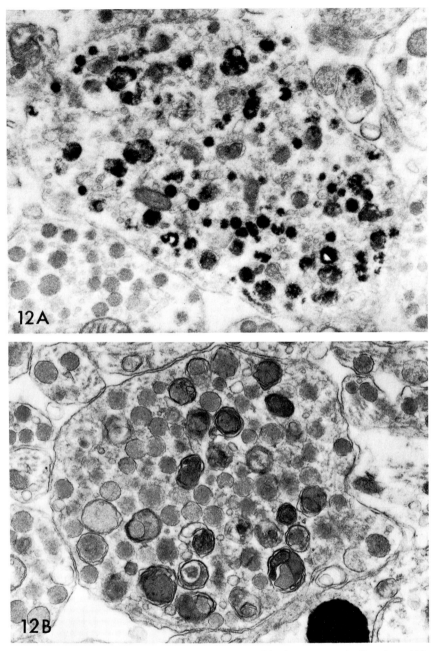

FIG. 12. Native horseradish peroxidase transported in the anterograde direction within axons of the neurohypophysial system of salt-stressed mice eventually accumulates in terminals (A) that have a degenerating appearance and contain numerous autophagic and/or crinophagic vacuoles (B). Acid phosphatase cytochemistry reveals that these vacuoles are lysosomes.

vacuoles.[58,59] Thus, we are dealing with what appears to be two functionally different compartments within the axon. The first, the endoplasmic reticulum, is an anabolic compartment involved in replenishing needed membrane-delimited organelles; the second compartment, the lysosomal system, is catabolic and functions for degradative purposes. The lysosomal system, though normally restricted to the cell body and proximal portion of dendrites, can extend itself into the axon when required to do so, as in the hyperosmotically stressed neuroendocrine cell or in the injured and degenerating neuron.[60]

Summary and Future Studies

The versatility of horseradish peroxidase is its usefulness both as an antigenic marker and as a probe molecule. We have demonstrated in the neuroendocrine cell that an HRP-bound antibody offers a high order of resolution for determining in which cellular compartment an antigen is located and where it is not. When native peroxidase is applied as an intracellular probe, it labels organelles associated with endocytosis in retrograde axonal transport and with the lysosomal system in both retrograde and orthograde axonal transport. The investigation that remains is the application of lectin-bound HRP to determine the pathways of membrane flow at the time when the neuroendocrine cell is stimulated to synthesize, transport, and secrete its peptide. For example, we are interested to know (1) whether internalized axon terminal membrane tagged with wheat germ agglutinin–HRP is channeled to all Golgi saccules engaged in the production of secretory granules in salt stimulated supraoptic neurons[71]; and (2) if internalized cell membrane of the supraoptic cell body is tagged with wheat germ agglutinin–HRP and channeled to GERL, will this membrane be transferred from GERL to secretory granules, lysosomes in the cell body and axon, the axonal endoplasmic reticulum, and to autophagic/crinophagic vacuoles in axon terminals of salt-stressed supraoptic neurons? These additional studies should provide a more comprehensive, morphological picture of membrane flow in a neuroendocrine cell that is responding to the metabolic demands placed upon it.

Acknowledgments

The typing of Ms. Elizabeth Tinnell is greatly appreciated. This work was supported, in part, by National Institute of Health NINCDS Grant NS18030-02.

[12] Visualization of Enkephalin Receptors by Image-Intensified Fluorescence Microscopy

By STEVEN G. BLANCHARD, KWEN-JEN CHANG, and PEDRO CUATRECASAS

Image-intensified fluorescence microscopy has proved to be a useful tool for visualization of the cellular distribution of ligand–receptor complexes in a number of systems.[1,2] The detection of fluorescence signals with an image-intensified television camera offers a number of advantages over traditional fluorescence microscopy. In particular, low levels of excitation light may be used so that photobleaching of fluorophore and photo damage to living cells are minimized, thus allowing study of the time course of ligand-induced redistribution of receptors. Furthermore, image intensification is often the only method sensitive enough to detect low light level signals that result from (a) a small number of receptors per cell, or (b) the limited number of reactive groups available for coupling of fluorophore to small molecular weight ligands, such as enkephalin. As an example of the application of this technique to neuroendocrine receptor systems, we describe here the preparation of two rhodamine derivatives of enkephalin and their use to study opiate receptors in a cultured mouse neuroblastoma system.

Choice of Fluorophore

The fluorophores suitable for image-intensified fluorescence microscopy are generally limited to fluorescein and rhodamine derivatives owing to their high quantum yields and emission at visible wavelengths. Rhodamines are more resistant to photobleaching than are the fluoresceins and are more often used for this reason. The use of n-propyl gallate to prevent photobleaching in fixed cell preparations has been described[3]; however, the suitability of this method for use with living cells has not been reported. The choice of a particular fluorophore also depends upon the type (and number) of reactive groups available for labeling (but not necessary for activity) in the ligand molecule used. A listing of commonly used fluorophores and their reactivities may be found in Table I of the

---

[1] J. Schlessinger, Y. Shechter, M. C. Willingham, and I. Pastan, *Proc. Natl. Acad. Sci. U.S.A.* **75**, 2659 (1978).
[2] E. Hazum, K.-J. Chang, and P. Cuatrecasas, *Neuropeptides*, **1**, 217 (1981).
[3] H. Giloh and J. W. Sedat, *Science* **217**, 1252 (1982).

chapter by Wang *et al.* "Fluorescent Localization of Contractile Proteins."[4]

Tyr-DAla-Gly-Phe-Leu-Lys
(I)

We have used the enkephalin analog [DAla^2,Lys^6]Leu-enkephalin **(I)** labeled at the lysine ε-amino group in our studies. The coupling of this peptide with tetramethylrhodamine isothiocyanate has been described[5] and will not be discussed further here. An advantage of compound **(I)** is that the lysine amino group can be coupled with other reactive groups to give ligands with useful properties (e.g., photoaffinity labels[5]). We have also coupled [DAla^2,Lys^6]Leu-enkephalin with the rhodamine sulfonyl chloride Texas Red (TxR, Molecular Probes Inc., Junction City, OR). This recently introduced fluorophore has a number of advantages over the more commonly used isothiocyanates: TxR has a higher quantum yield, the hydrolyzed TxR is highly water soluble and easy to remove from the product, and the sulfonamides formed are chemically very stable. A possible disadvantage of TxR is that its high excitation and emission maxima (~595 and 620 nm, respectively) do not correspond to filter packages commonly found on fluorescence microscopes. We have used a standard rhodamine filter set (Zeiss) to excite TxR (note shoulder at 550 nm; see Fig. 2), with good results. For cases where the resulting decrease in fluorescence intensity is undesirable, filters and a dichroic mirror that are optimal for the spectral properties of TxR may be used.[4]

Preparation of Texas Red Enkephalin

Boc-Tyr(OBut-)-DAla-Gly-Phe-Leu-Lys was the gift of Dr. S. Wilkinson, The Wellcome Research Laboratories (Beckenham, Kent, England). Peptide (9.2 mg; 10.8 μmol) and 6 μl of redistilled triethylamine were dissolved in 1 ml of dry dimethylformamide, and the resulting solution was added to 10 mg of TxR. The reaction mixture was protected from light. After 24 hr at room temperature, the solution was diluted with water and lyophilized to dryness. Deblocking was accomplished with anhydrous trifluoroacetic acid.[5] The acid was removed with a stream of dry nitrogen; the residue was dissolved in chloroform and exhaustively extracted with water. The chloroform layer, which contained the TxR-labeled [DAla^2,-Lys^6]Leu-enkephalin, was then brought to dryness. Final purification was by reverse-phase high-performance liquid chromatography (HPLC)

---

[4] K. Wang, J. R. Feramisco, and J. F. Ash, this series, Vol. 85, p. 514.
[5] E. Hazum, K.-J. Chang, Y. Shechter, S. Wilkinson, and P. Cuatrecasas, *Biochem. Biophys. Res. Commun.* **88**, 841 (1979).

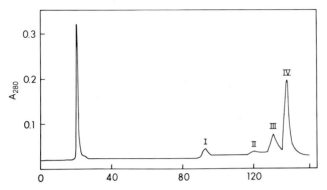

FIG. 1. Purification of Texas Red (TxR) enkephalin by reverse-phase HPLC. The absorbance profile at 280 nm as a function of time (minutes) after injection is shown. As evidenced by their pink color, peaks I, III, and IV contained TxR. The peak at ~21 min is an injection artifact.

(Whatman M9 10/50 ODS) using a gradient of 50 to 90% MeOH in 10 mM ammonium acetate, pH 4.2. The resulting elution profile is shown in Fig. 1. Peaks I, III, and IV contained TxR, as evidenced by their pink color. At equivalent concentrations (as estimated by absorption), material from all three peaks inhibited the binding of ^{125}I-labeled [DAla^2,D-Leu^5]enkephalin (DADLE) to rat brain membranes to the same degree. For ease of analysis only the largest peak (IV) was further characterized. The multiple peaks may be due to the presence of isomers of TxR and/or to labeling of a small amount of (unprotected) peptide at the amino terminal or the tyrosine hydroxyl. At any rate, this separation illustrates the importance of careful purification and characterization of fluorescently labeled ligand.

Properties of TxRE

The absorption spectrum of TxRE is shown in Fig. 2. Note the maximum at 590 nm and the distinct shoulder at 545 nm ($A_{590}/A_{545} \cong 3.3$). Since the wavelength of the shoulder is similar to the wavelength maximum for tetramethylrhodamine-labeled derivatives, a standard filter package can be used for visualization of TxRE fluorescence. The affinity of TxRE for opiate receptors was estimated from its potency in displacing receptor-bound ^{125}I-labeled DADLE.[6] Our tissue source was the N4TG1 neuroblastoma cell line (mouse), which contains a homogeneous population of δ (enkephalin) subtype opiate receptors.[7] As shown in Fig. 3, TxRE was

[6] R. J. Miller, K.-J. Chang, J. Leighton, and P. Cuatrecasas, *Life Sci*, **22**, 379 (1978).
[7] K.-J. Chang and P. Cuatrecasas, *J. Biol. Chem.* **254**, 2610 (1979).

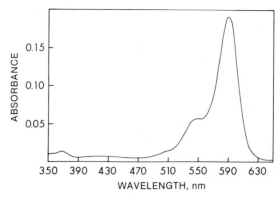

FIG. 2. Visible absorption spectrum of Texas Red enkephalin in water.

equally effective at displacing ^{125}I-labeled [DAla^2,DLeu^5]enkephalin binding to N4TG1 cells and membranes; the half-maximal inhibitory concentration (IC$_{50}$) was 1.9×10^{-8} M under these conditions. The affinity of TxRE for μ (morphine) subtype receptors was only slightly lower than for δ (enkephalin) subtypes. In rat brain membranes the IC$_{50}$ against ^{125}I-labeled [DAla^2,DLeu^5]enkephalin was 7.6 nM, and the corresponding value against ^{125}I-labeled FK-33,824, a μ subtype selective ligand, was 32 nM (data not shown). Thus, it should be possible to label both receptor subtypes and to distinguish between their labeling patterns by selective suppression of the binding to μ receptors using the highly selective μ-receptor ligand morphiceptin.[8]

In contrast to the tetramethylrhodamine-labeled enkephalin, which had to be repurified every few months, the purified TxRE still showed a single fluorescent peak on HPLC even after 16 months of storage at $-20°$. Thus, the fluorescently labeled peptide was a stable, high-affinity ligand for opioid receptors.

Instrumentation

Because research-quality fluorescence microscopes are commercially available and a discussion of the design of fluorescence microscopes is beyond the scope of this chapter, we present here only general requirements for any image-intensified microscope system. The microscope should have an auxiliary camera port available for attachment of the silicon-intensified target camera and an epifluorescence condenser

---

[8] K.-J. Chang, A. Killian, E. Hazum, P. Cuatrecasas, and J.-K. Chang, *Science* **212**, 75 (1981).

equipped with filter packages appropriate to the fluorophore(s) used. A high-pressure mercury arc lamp provides an intense, concentrated light source. We have used a Zeiss Photomicroscope III equipped with a III RS epifluorescence condenser and HBO 50W/2 arc lamp to fulfill the above requirements. Sample observation is generally by a Zeiss Planapochromat 63X oil immersion objective. This objective is an excellent choice, since it gives high resolution and its high numerical aperture (1.4) gives efficient transmission of low-level fluorescence signals to the detector (camera). The major requirement for the image-intensified video camera is that it be equipped with a *manual* gain control so that direct comparisons of control and labeled samples may be made at identical instrumental settings. We have used both RCA 1030H and Hamamatsu C1000 cameras. An additional consideration should be the compatibility of the camera with other video-processing equipment. For instance, the Hamamatsu Systems C1000 camera can be connected directly to a laboratory digital computer using optional analog-to-digital converter and interface boards; however, the Hamamatsu camera's scan rate is incompatible with other image-processing systems, such as those offered by Quantex Corp. Additional equipment required includes a videotape recorder and a video monitor to record and visualize the output of the video camera.

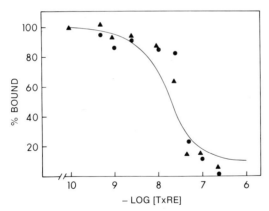

FIG. 3. Inhibition of ^{125}I-labeled [DAla^2,DLeu^5]enkephalin (DADLE) binding by TxRE. Aliquots (0.2 ml) of N4TG1 cells (▲) or membranes (●) ($\sim$3 × 10^6 cells) were incubated with 0.1 nM ^{125}I-labeled DADLE and the indicated concentration of TxRE for 1 hr at room temperature and the radioactivity associated with the samples were determined by a filtration assay (Whatman GF/C). Nonspecific binding was determined in the presence of 10^{-5} M unlabeled DADLE. The data are expressed as percentage of the control binding seen in the absence of TxRE. The line was drawn using an IC$_{50}$ of 1.9 × 10^{-8} M. This value was obtained from the x intercept of a linear least-squares fit of the data to the Hill equation (n_H = 1.06; correlation coefficient = 0.94).

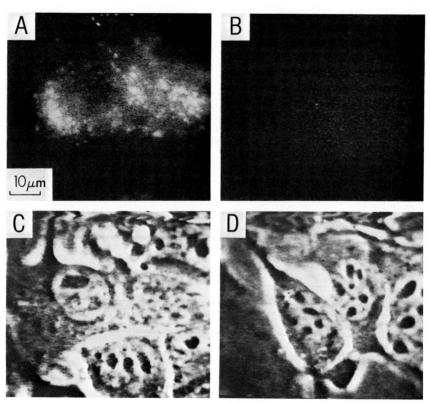

FIG. 4. Visualization of TxRE-labeled N4TG1 cells by image-intensified fluorescence microscopy. (A) Cells incubated with 50 nM TxRE in phenol red-free Dulbecco's minimum essential medium (DMEM) for 1 hr at 37°. (B) As panel A, except that 10^{-5} M [DAla^3,DLeu^5]enkaphalin was also included in the incubation medium. (C) and (D) show phase-contrast micrographs of the fields in (A) and (B), respectively.

Visualization of Opiate Receptors

Incubation of N4TG1 neuroblastoma cells (or NG108-15 neuroblastoma–glioma hybrids) with tetramethylrhodamine-labeled enkephalin for 40 min at 25° results in the appearance of fluorescent clusters on the cells.[2,9-11] Clusters were not observed when excess nonfluorescent opiate ligands (DADLE, morphine, naloxone) were present in the incubation

[9] E. Hazum, K.-J. Chang, and P. Cuatrecasas, *Science* **206**, 1077 (1979).
[10] E. Hazum, K.-J. Chang, and P. Cuatrecasas, *Nature* (*London*) **282**, 626 (1979).
[11] E. Hazum, K.-J. Chang, and P. Cuatrecasas, *Proc. Natl. Acadl. Sci. U.S.A.* **77**, 3038 (1980).

medium. Incubation of cells with $5 \times 10^{-8} \, M$ TxRE for 1 hr at 37° results in the appearance of patches (Fig. 4) that are visually indistinguishable from those observed for the tetramethylrhodamine isothiocyanate-labeled compound; however, as discussed below, interpretation of the observed fluorescence patterns requires careful consideration of the experimental conditions used and of supporting biochemical data.

Interpretation of Results

The patterns of cellular fluorescence observed by microscopy are limited to three general types: (a) no fluorescence, (b) a uniform distribution, and (c) "patches" or "clusters." Patches are often not associated with any visible subcellular organelle. Furthermore, focusing is often an unreliable method of determining whether a particular fluorescent ligand is internalized, since fluorophores out of the plane of focus also contribute to the observed signal. Although this may not be a problem for large, rounded cells in suspension, the effect is considerable for flat cells grown in monolayer culture; therefore, decisions regarding the distribution of a particular receptor need to take into account both fluorescence data and biochemical studies performed under identical conditions.

Examination of the fluorescence patterns of tetramethylrhodamine enkephalin suggested that there was no ligand-induced internalization of enkephalin under the conditions tested (i.e., incubation of suspension cells in Tris–sucrose buffers at 25°).[9] These results are consistent with the finding that more than 95% of all bound ^{125}I-labeled DADLE was dissociable from cells, even after prolonged incubation.[12] Furthermore, numerous studies have shown that prolonged exposure to morphine, both *in vivo* and *in vitro*, does not result in the alteration of the number or affinity of opiate receptors. It has been reported[13] that incubation of N4TG1 cells with enkephalin (but not morphine) for one to several hours results in a decrease in enkephalin receptor numbers. Study of the association of ^{3}H-labeled DADLE with cells in the presence of various inhibitors showed that the receptor loss is probably due to internalization of receptor–ligand complexes and that internalized ligand accumulates in excess of the initial receptor numbers.[14,15] This process of receptor down regulation is temperature dependent[13]; both its rate and extent are much greater at 37° than at 25°. Furthermore, internalization of both enkephalin[15] and epidermal

---

[12] K.-J. Chang, R. J. Miller, and P. Cuatrecasas, *Mol. Pharmacol.* **14**, 961 (1978).
[13] K.-J. Chang, R. W. Eckel, and S. G. Blanchard, *Nature (London)* **296**, 446 (1982).
[14] S. G. Blanchard, K.-J. Chang, and P. Cuatrecasas, *Life Sci.* **31**, 1311 (1982).
[15] S. G. Blanchard, K.-J. Chang, and P. Cuatrecasas, *J. Biol. Chem.* **258**, 1092 (1983).

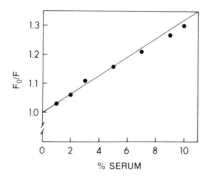

FIG. 5. Effect of serum concentration on TxRE fluorescence. Fluorescence was measured in a Perkin–Elmer Model 512 fluorometer. The excitation and emission wavelengths were 590 ± 3 nm and 610 ± 10 nm, respectively, and the concentration of TxRE was 5×10^{-7} M. The data are plotted according to the linearized form of the Stern–Volmer equation: $F_o/F = 1 + K[Q]$. The meaning of the symbols is given in the text. Linear least squares of the data gave $K = 0.032$ per percentage of serum and a y intercept of 1.005 ($\rho = 0.998$).

growth factor[16] are reduced for cells in suspension as compared to monolayers. Under conditions optimal for receptor down regulation (37° and monolayer cultures), pretreatment with TxRE does cause loss of enkephalin binding (unpublished data); therefore, it seems likely that the fluorescent clusters observed under these conditions correspond to internalized ligand. Consistent with this notion, incubation of cells with TxRE in the presence of the lysosomotrophic amine chloroquine (50 μM) causes an increase in the fluorescent clusters observed as compared to a control. Under the same conditions, chloroquine causes an increase in the amount of cell-associated ^{3}H-labeled DADLE due to a decrease in the rate of release of internalized ligand.[15] Thus, the above discussion illustrates that visually indistinguishable fluorescence patterns for the tetramethylrhodamine-labeled and TxR-labeled enkephalins may result even though the distributions of the receptor–ligand complexes were different (surface versus internalized, respectively). It should be clear then that patterns of fluorescence labeling need to be interpreted in terms of other known properties of the receptor system under study.

Controls

Because of the great sensitivity of image-intensifier systems, it is especially important to include proper controls. The background fluores-

---

[16] R. Zidovetzki, Y. Yarden, J. Schlessinger, and T. M. Jovin, *Proc. Natl. Acad. Sci. U.S.A.* **78**, 6981 (1981).

cence due to cells alone and the nonspecific labeling pattern on fluorophore in the presence of excess nonfluorescent ligand must be determined. In addition, one must be aware of a number of possible artifacts that can result in a loss of sensitivity of the fluorescent signal. The intensity of fluorescein fluorescence decreases with decreasing pH; therefore, the fluorescence of fluoresceinated ligand in an internal compartment of acidic pH may be quenched. Quenching of fluorescence may be caused by absorption of the exciting or emitting light, by a collisional process, or by oxidation–reduction of the fluorophore. Therefore, the buffer, as well as any test reagents that are to be added to the medium, must be checked in a fluorometer for any effects they may have on the fluorophore(s) used. For example, dithiothreitol had no effect on the fluorescence of TxRE. Newborn calf serum, however, quenches the fluorescence of TxRE (Fig. 5). This quenching can be described by the Stern–Volmer equation $F_o/F = 1 + K[Q]$, where F_o and F are the fluorescence intensities in the absence and in the presence of quencher, respectively, K is a constant for a particular quencher, and $[Q]$ is the concentration of quencher. For newborn calf serum $K = 0.032$ percentage of serum corresponding to a 23% decrease in fluorescence intensity at 10% serum concentration. Thus, for maximum sensitivity, experiments should be carried out in serum-free medium.

[13] Flow Cytometry as an Analytic and Preparative Tool for Studies of Neuroendocrine Function

By JOHN C. CAMBIER and JOHN G. MONROE

Investigation of the biology of virtually all cell populations that can be dissociated into single-cell suspensions has been revolutionized by the advent and development of flow cytometry. Modern flow cytometry is the product of the evolution of early rapid cell spectrophotometers developed solely for analytic purposes and of fluorescence-activated cell sorters developed to sort lymphocytes based upon fluorescence and light scatter. Modern technology allows measurement of RNA and DNA content, immunofluorescence, cell diameter and volume, membrane potential, membrane-bound calcium, membrane integrity, membrane fluidity, and intracellular pH and organellar content among other parameters. These have been the subject of excellent reviews.[1-3] It would appear that virtually any

[1] H. M. Shapiro, *Cytometry* **3**, 227 (1983).
[2] M. R. Loken and A. M. Stall, *J. Immunol. Methods.* **50**, R85 (1982).
[3] D. R. Parks, R. R. Hardy, and L. A. Herzenberg *Immunol. Today* **4**, 145 (1983).

cellular parameter that can be monitored based upon optical properties (e.g., light scattering or extinction), or by using fluorescent probes, is within the analytic capability of flow cytometry. Modern flow cytometers allow simultaneous analysis of multiple parameters on single cells. As a result, flow cytometry shows great promise for dissection and analysis of biological responses of subpopulations within complex cell mixtures. Many of the measurements referred to above, while originally described for blood cells, are applicable to studies of neurosecretory cells. The methodology for measurement of immunofluorescence, cell diameter, light scatter, and membrane potential is described below. This discussion will be preceded by a general overview of basic principles of flow cytometry.

Basic Principles of Flow Cytometry

In principle, flow cytometry involves analysis of cells passing orthogonally through a focused laser beam. Commercially available flow cytometers may be categorized based on the point at which the laser beam intersects the sample stream. Specifically, in the Cytofluorograf (Ortho Instruments, Westwood, MA) the laser beam intersects the stream as it transverses a quartz flow channel. In the FACS (Becton-Dickinson, FACS Division, Mountain View, CA) the stream is in air when intersected by the laser beam. Finally the EPICS V (Coulter Electronics, Hialeah, FL) utilizes exchangeable fluidic systems that allow "in air" or "in channel" analysis. For the applications described here, we have utilized a Cytofluorograf system 50H (Ortho Instruments). It is described in greater detail below.

An optics diagram of the Cytofluorograf system 50H is presented in Fig. 1. Monochromatic illumination from argon (laser 2) and helium–neon (laser 1) lasers are superimposed using a dichroic mirror and focused on the sample stream using crossed cylindrical lenses. For flow cytometric applications, crossed cylindrical lenses are advantageous over circular lasers because they focus light into an ellipse rather than a circle. The long axis of the ellipse is oriented perpendicular to the stream, and the short axis is parallel to the stream. This orientation allows the laser light to be focused to a width that approximates the diameter of a cell (5–10 μm) on the short axis while the long axis of the focused beam still extends the width of the stream. Thus, potential anomalous effects resulting from cells not being centered in the stream are minimized. As shown in the diagram (Fig. 1) laser light scattered at 90° and fluorescence are collected

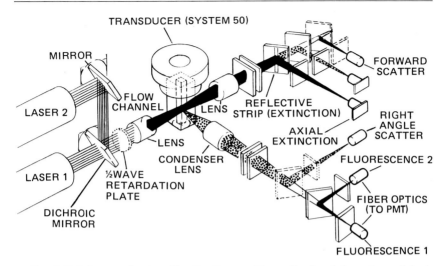

FIG. 1. Cytofluorograf system 50 optics diagram. Diagonally placed elements are dichroic filters. Orthogonally placed elements are filters based on wavelength, polarization, etc.

by an assembly that includes a condensing lens, appropriate filters and dichroic mirrors, and fiber optics positioned perpendicular to the axis formed by the intersection of the laser beam and the stream. Scattered (90%) laser light is reflected to a fiber optic by the dichroic mirror most proximal to the condensing lens. Red and green fluorescence are separated by a second dichroic mirror and collected by fiber optics. Signals are transmitted via these fiber optics to photomultipliers. Analog signals from the photomultipliers and the photodiode are processed and expressed in one of three modes: amplitude or peak height, total integrated signal or area, and duration of signal or pulse width (Fig. 2).

Incident laser light and forward light scatter are collected by a condensing lens located in line with the laser beam. After passing the lens, the laser beam is reflected by a metallic strip to a photodiode. The decrement or extinction of this signal, which occurs as a particle in the stream transverses (and partially blocks) the laser beam, is referred to as axial light extinction. The helium–neon and argon laser beams and light scattered by cells at <2° is not blocked by the metallic obscuration strip. Red (helium–neon) and blue (argon) forward scatter 2–19° are reflected by appropriate, more distally placed dichroic mirrors to fiber optics.

The flow of sample through the Cytofluorograf is demonstrated in the fluidic diagram in Fig. 3. A vessel containing sample is placed in the sample chamber, which is then pressurized to force the flow of sample

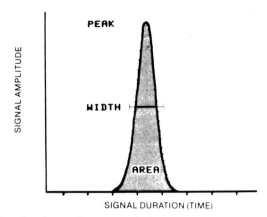

FIG. 2. Signal produced as a cell traverses the laser beam. This signal can be processed as the maximum signal obtained (PEAK), the signal integrated over time (AREA), or the duration of the signal (WIDTH). Changes in the signal width are directly proportional to the cell diameter because the signal is initiated as the leading edge of the cell enters the laser beam and terminated by the trailing edge of the cell leaving the laser beam.

(cells) through tubing and into the flow channel, where it is surrounded by a flow of sheath fluid (saline or distilled water). Integrity of the laminar flow of sheath and sample is maintained until they pass through a 75-μm orifice at the bottom of the flow channel. The sheath enters the flow channel under somewhat higher pressure than the sample. As a result, the sample stream is further constricted, causing alignment of particles on their long axes.[4] The combined effects of a piezoelectric transducer and passage of the sample and sheath through the orifice at the bottom of the flow channel causes atomization of the stream into 32,000 droplets per second. During sorting a charge is placed on the flow channel, and thus the stream, as the droplet containing the particle to be sorted is formed. The droplet retains this charge, as it falls between charged plates that cause it to be deflected toward an appropriately positioned collection vessel. Droplets containing unwanted cells are left uncharged and therefore are not deflected.

Forward and Right-Angle Scatter Analysis of Cell Populations

Differences in light-scattering properties, detected using light microscopy, have long been valuable in distinguishing cell types. The same

[4] T. Sharpless, F. Traganos, Z. Darzynkiewicz, and M. R. Melamed, *Acta Cytol.* **19**, 577 (1975).

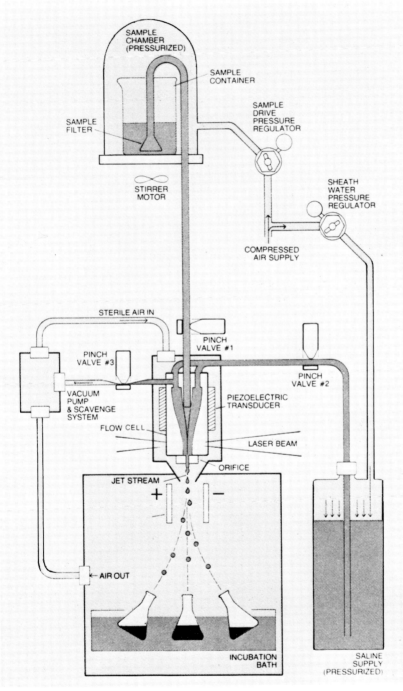

FIG. 3. Cytofluorograf system 50 fluidic diagram.

principles have been taken advantage of in flow cytometry to distinguish cell types,[5] to determine cell size[6] and viability,[7] and to identify cell asymmetry and orientation.[8]

Although light scattering has been useful in these assays, it is well established that cellular scattering of light at any angle is an extremely complex function of size and refractive and reflective properties of the particle (see review[2]). Thus, for example, light scatter does not represent an absolute indication of cell size. Light scattered at lowest angles is most proportional to cell size.[6,9] Mullaney and Dean[9] have found 0.5–2° to be optimal. Unfortunately, most available flow cytometric scatter detection systems collect scattered light at somewhat larger angles, e.g. 2–19° in the Cytofluorograph. Nonetheless, this parameter is useful in *approximating* cell size.

Discrimination of live and dead cells by light scatter is also most effective using only very low-angle scatter. Loken and Herzenberg[10] have found empirically that live–dead discrimination of lymphocytes improves as the lower angle of detection approaches 0°. Unfortunately this method of live–dead discrimination is effective only if the cells being analyzed are very restricted in size. For example, the approach is effective for lymphocytes, but not for more heterogeneous lymphoblast populations.

Light scattered at 90° from the laser path is contributed to significantly by light reflected by the nucleus and cytoplasmic organelles. This parameter has been used successfully in combination with narrow-angle light scatter to distinguish blood cell subclasses in peripheral blood.[11] This combination of parameters may have utility in defining pituitary cell subpopulations and therefore is the subject of the methodological discussion that follows.

Procedure

Sample Preparation and Analysis. Light scatter analysis is simplified by the fact that no special sample preparation or staining is required. Cell

---

[5] G. C. Salzman, J. M. Crowell, and C. A. Good, *Clin. Chem.* **21**, 1297 (1975).

[6] P. F. Mullaney and P. N. Dean, *Appl. Optom.* **8**, 2361 (1969).

[7] M. H. Julius, R. G. Sweet, C. G. Fathman, and L. A. Herzenberg, in "Mammalian Cells: Probes and Problems" (C. R. Richman, D. G. Peterson, P. R. Mullaney, and E. C. Anderson, eds.), p. 107. ERDA Tech. Information, Oak Ridge, Tennessee, 1975.

[8] M. R. Loken, D. R. Parks, and L. A. Herzenberg, *J. Histochem. Cytochem.* **25**, 790 (1979).

[9] P. F. Mullaney and P. N. Dean, *Biophys. J.* **10**, 764 (1970).

[10] M. R. Loken and L. A. Herzenberg, *Ann. N. Y. Acad. Sci.* **254**, 163 (1975).

[11] R. A. Hoffman, P. C. Kung, P. Hansen, and G. Goldstein, *Proc. Natl. Acad. Sci. U.S.A.* **77**, 4914 (1980).

populations are simply suspended in an isotonic solution (such as phosphate-buffered saline) containing 2% fetal calf serum at a concentration of 10^6 to 5×10^6/ml and subjected to analysis. It is important for all flow cytometric analyses that cell populations contain the fewest dead cells and cell aggregates possible. Routinely, cells are analyzed using integrated forward scatter vs integrated 90° scatter. In cytometers equipped for multiparametric analysis, these two parameters may be used in conjunction with fluorescence measurements.

Instrument Calibration. While light-scatter analysis is easily conducted, optimal optical alignment of the Cytofluorograf is of critical importance in obtaining satisfactory results. We routinely align the forward-scatter parameter first using 2-μm beads (Ortho, Westwood, MA) to achieve the best possible coefficient of variation. We then align forward scatter using glutaraldehyde-fixed chicken erythrocytes to achieve a bimodal distribution, as shown by Herzenberg and Herzenberg.[12] We then maximize the 90° scatter signal by adjusting the position of the condensing lens and the appropriate dichroic mirror in the 90° path. Aligned in this manner, the instrument is effective in live–dead discrimination of lymphocytes[7,10] and in three-part differential analysis.[11]

Determination of Cell Diameter by the Measurement of Axial Light-Loss Pulse Width

Subpopulations within complex cell mixtures can sometimes be distinguished based upon cell size. Direct measurement of cell size is now possible using commercial flow cytometers. Cell volume (Coulter volume) analysis capability is soon to be made available on Coulter, and Becton-Dickinson instruments. Cell diameter measurements by axial light-loss pulse-width analysis is possible only using standardly equipped Ortho (Cytofluorograf) flow cytometer. The resolution attainable with these two approaches is comparable.[13] Unlike Coulter volume, which is usable only for analysis, and axial light-loss pulse width may be used for analysis and sorting. Cell diameter determined by measurement of axial light-loss pulse width using the Cytofluorograf will be the subject of the succeeding discussion.

As discussed earlier, signals generated from the analysis of cells as they traverse the elliptically focused laser beam in a Cytofluorograf may be processed and presented in three different ways (Fig. 2): the pulse height (maximum signal), pulse area (total integrated signal), and, the

[12] L. A. Herzenberg and L. A. Herzenberg, *in* "Handbook of Experimental Immunology" (D. M. Weir, ed.), Chapter 12. Blackwell, Oxford, 1978.
[13] J. C. Cambier. Submitted for publication.

pulse width or time of flight (duration of the signal). When the half-intensity width of the laser beam is comparable to or less than the diameter of the cell (e.g., 5–10 μm), there is a direct relationship between the pulse width and the cell diameter. This analytical capability has been utilized by several investigators. Steinkamp and Crissman[14] reported, using human cervical specimens that had differentially stained nuclei and cytoplasm, that ratios of the pulse widths of different wavelengths of fluorescence can be used to distinguish among epithelial cells and other cell types. Sharpless et al.[4] have utilized fluorescence pulse width to differentiate single cells from cell aggregates in acridine orange-stained samples. Cambier et al.[15] have reported the use of the pulse width of axial light loss to determine cell size differences among bone marrow cell populations. The succeeding discussion describes the adjustment of the Cytofluorograf so that the relationship between 4.25 and 10 μm latex microspheres is linear. Since time, not light extinction, is the measurement utilized, it is possible to use the calibration to determine the diameter of any cell, isolated nuclei, or particles of similar size. This measurement can be used by itself or in conjunction with light scatter or fluorescence. Use of axial light-loss pulse width to determine the diameter of thymocyte nuclei is described in this protocol.

Procedure

Instrument

Configuration for	Calibration	Analysis
Time constant,	Slow	Slow
Y axis (parameter/mode)	Axial/width	Red fluorescence/area
X axis (parameter/mode)	Axial/width	Axial/width

Calibration of Cytofluorograf

1. Analyze a mixture containing equal numbers of 4.25- and 10-μm particles. Note: Any particles of known diameters in the range from 4 to 15 μm may be used.
2. Adjust the sample flow to ≃100 particles per second.
3. Initially switch the X and Y modes to PEAK and the X and Y parameters to AXIAL. Adjust the gain of the axial light loss detector

[14] J. A. Steinkamp and H. A. Crissman, *J. Histochem. Cytochem.* **22,** 616 (1974).
[15] J. C. Cambier, W. L. Havran, T. Fernandez, and R. B. Corley, *J. Immunol.* **127,** 1685 (1981).

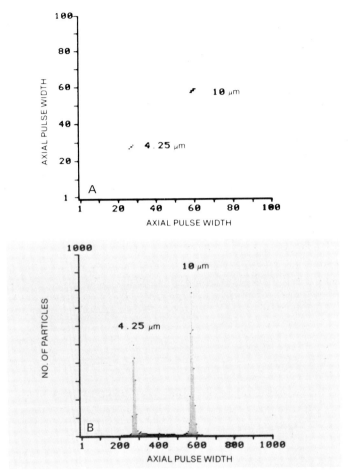

FIG. 4. (A) Cytogram of 4.25- and 10-μm particles (Axial/pulse width on both X and Y axis) with pulse width scale and offset adjusted. (B) Histograms of 4.25- and 10-μm particles. Peak channels were 253 and 586, indicating an error of less than 1%. The calibration factor (step 9) was calculated to be 0.017.

(PMT 4 Control) until both groups of particles are included in the cytogram. Optimize channel position, and switch the modes to WIDTH.

4. Adjust the pulse-width scale control so that the 10-μm particles are located just above the center of the cytogram. Then adjust the pulse-width offset control until the 4.25-μm particles are halfway between the origin of the cytogram and the 10-μm particles. If necessary, readjust the pulse-width scale control until both populations are distributed diagonally across the cytogram (Fig. 4A).

5. Check the offset by generating a histogram of the X axis and determining the peaks of the two populations of particles (Fig. 4B). The peak channel number of the 10-μm particles should be 2.3 times that of the 4.25-μm particles. If the two peaks are too close together, reduce the offset by turning the pulse-width offset control clockwise. If too far apart, increase the offset by turning the pulse-width offset counter clockwise. The amount of offset required is determined in step 6.

6. Calculate the offset required using Eq. (1):

$$N = L + \frac{RS - L}{1 - R} \qquad (1)$$

N is the new peak channel number of the large particles; L is the initial peak channel number of the large particles; S is the initial peak channel number of the small particles; and R is the ratio of the diameters of the small and large particles.

7. Make slight adjustments to the pulse-width offset control to place the peak channel of the 10-μm particles at the new channel number.

8. Recheck linearity several times. The results should be within 1% of the expected value. Readjust it if necessary.

9. Calculate the calibration factor using formula (2)

$$C = D/P \qquad (2)$$

where C is the calibration factor; D is the diameter of particles (μm); P is peak channel number of particles. Note: This calibration will no longer be valid if either the pulse-width offset or laser beam focus is changed.

Simultaneous Analysis of Cell Diameter and Red Fluorescence

1. Stain fixed calf thymocyte nuclei (Ortho) thymocytes by adding approximately 8 drops to 1 ml of phosphate buffer containing 50 μg of propidium iodide per milliliter.

2. Change the Y-axis parameter to red fluorescence and mode to AREA.

3. Generate a cytogram and select the $G_0 + G_1$, S phase, and G_2 subpopulations (Fig. 5A).

4. Generate the histograms of axial light-loss width (X axis) for subpopulations (Fig. 5B).

5. Determine the average nuclear diameter for each subpopulation (see the table) by multiplying its mean by the calibration factor.

Results and Comments

The histogram of the 4.25- and 10-μm particles (Fig. 4B) demonstrates the relationship between particle diameter and the pulse width of their

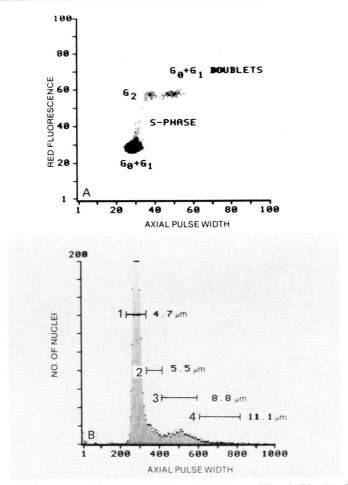

FIG. 5. (A) Cytogram of thymocyte nuclei stained with propidium iodide (Y axis, red fluorescence/area; X axis, axial light-loss/width) showing $G_0 + G_1$, S phase, and G_2 subpopulations and $G_0 + G_1$ doublets. (B) Histograms of axial pulse width of thymocyte nuclei showing the average (in micrometers) diameter of $G_0 + G_1$, S-phase, G_2 nuclei, and doublets of $G_0 + G_1$ nuclei.

respective axial light loss. Using this calibration, the pulse width of axial light loss and red fluorescence of propidium iodide-stained thymocyte nuclei were measured, and the $G_0 + G_1$, S phase, and G_2 subpopulations were identified (Fig. 5A). Histograms of the pulse width of axial light loss were then generated and used to determine the cell diameter of these subpopulations (Fig. 5B). The results (see the table) illustrate the ability of this measurement to determine cell diameter. This measurement is accu-

Phase of cell cycle	Mean channel no.	Average diameter (μm)
$G_0 + G_1$	277	4.7
S phase	324	5.5
G_2	515	8.8
Doublets ($G_0 + G_1$)	653	11.1

rate, regardless of the refractive or scatter characteristics of the particles measured, because time, not light loss, is being measured. This greatly simplifies both the calibration of the Cytofluorograf and the determination of cell sizes in populations containing many cell types or cells that have undergone treatment that could alter their refractive and scatter characteristics, e.g., fixation.

Immunofluorescence Analysis

Unquestionably the most widely utilized application of flow cytometry has been immunofluorescence analysis of cellular antigen expression. Immunofluorescence methodology has been reviewed in detail elsewhere.[1-3,16] Immunofluorescence may be undertaken using a variety of strategies, but the ultimate success depends primarily upon the specificity of the primary antibody. "Direct" immunofluorescence involves use of primary antibody, e.g., rabbit anti-cell surface antigen, to which fluorochrome has been directly conjugated. "Indirect" fluorescence invovles the use of a fluorochrome-labeled second antibody, e.g., goat anti-rabbit immunoglobulin, or other agent, e.g., staphlococcal protein A, which specifically bind the primary antibody. Other approaches utilize fluorescent avidin (Vector Laboratories, Burlingame, CA) in combination with biotin-conjugated primary antibodies. These indirect techniques are advantageous over direct fluorescence in providing more amplification of the primary antibody binding event but may suffer from a slight loss of specificity. Finally, primary antibody may be conjugated to fluorescent spheres before being used for localization. This approach is particularly appropriate when relatively few antigen molecules occur per cell and maximum amplification is essential. It has been used successfully for

[16] R. C. Nairn, in "Fluorescent Protein Tracing" (R. C. Nairn, ed.), Chapter 6. Churchill-Livingstone, Edinburgh, 1976.

definition and sorting of luteinizing hormone (LH)-bearing cells from the anterior pituitary.[17]

Also of critical importance to the success of immunofluorescence localization is selection of the appropriate fluorochrome. It must be excitable by a wavelength attainable from the available illumination source, it must have good quantum yield, and it must be readily coupled to antibody without affecting its specificity. Fluorescein and tetramethylrhodamine have long been the dyes of choice for fluorescence microscopy and flow cytometry. They excite efficiently at 488 and 514 nm, respectively, which are major emission lines for the argon lasers that are standard in most commercial flow cytometers. Other labels described to date include derivatives of rhodamine 101, which are excited by the 568-nm line of a krypton laser. They are XRITC[18] (Research Organics, Cleveland, OH), which is conjugated via an isothiocyanate linkage, and Texas Red[19] (Molecular Probes, Johnson City, OR), which is conjugated via a sulfonylchloride group. Finally, Oi et al.[20] have described the use of a series of highly fluorescent phycobiliprotein photosynthetic pigments isolated from red and blue-green algae. These pigments excite at wavelengths ranging from 488 nm to 650 nm. Phycobiliprotein-labeled antibodies are currently available through at least one commercial source (Becton-Dickinson, Mountain View, CA). Phycobiliproteins are being marketed by Molecular Probes (Johnson City, OR). The later fluorochromes, including XRITC, Texas Red, and phycobiliproteins, will probably be most useful in combination with fluorescein for simultaneous multi-color immunofluorescence analysis discussed below.

There is ever-increasing interest in the simultaneous analysis of distribution of two or more antigens on cells within one population. This was first described using a single argon laser at 514 nm to excite fluorescein- and rhodamine-labeled antibodies of different specificities.[21] Although this approach is feasible when cells are brightly stained, it has several inherent problems. First, fluorescein is not excited very efficiently by 514 nm illumination (use of this wavelength is necessary to excite rhodamine). Second, to eliminate 514 nm laser light scattered at 90° from the green

---

[17] M. O. Thorner, J. L. C. Borges, M. J. Cronin, D. A. Keefer, P. Hellman, D. Lewis, L. G. Dabney, and P. J. Quesenberry, *Endocrinology* **110**, 1831 (1982).

[18] T. M. Chused, S. O. Sharrow, J. Weinstein, W. J. Ferguson, and M. J. Sternfeld, *J. Histochem. Cytochem.* (in press).

[19] J. H. Titus, R. Haugland, S. O. Sharrow, and D. M. Segal, *J. Immunol. Methods* **50**, 193 (1982).

[20] V. T. Oi, A. N. Glazer, and L. Stryer, *J. Cell Biol.* **93**, 981 (1982).

[21] M. R. Loken, D. R. Parks, and L. A. Herzenberg, *J. Histochem. Cytochem.* **25**, 899 (1977).

fluorescence signal, the window of green fluorescence detection must be narrowed from 40 nm (500 to 540 nm) to about 10 nm (530 to 540 nm). Third, the emission spectra of fluorescein and tetramethylrhodamine overlap significantly, necessitating electronic or computer software cross correction of signals. This situation exemplifies the types of problems that must be overcome for effective multicolor immunofluorescence analysis. Technical advances have overcome some of these problems. For example, the use of multiple lasers intersecting the sample stream at different positions in conjunction with appropriate offset detectors and timing logic to track single cells through multiple stations has lessened illumination and optical problems inherent in multicolor analysis (see review[1]). Availability of fluorochromes, such as fluorescein and phycoerythrins, that excite at the same wavelength (488 nm) but emit at different wavelengths (530 and 575 nm, respectively) have improved the feasibility of conducting two-color fluorescence analysis with a single laser.

Procedure

Staining of Cells

1. Approximately 1×10^6 cells are suspended in the primary antibody. The antibody is diluted appropriately in phosphate-buffered saline (PBS) containing 2% fetal calf serum (FCS) and 0.2% sodium azide. The volume is generally 100 μl, and this suspension is allowed to incubate on ice for 30 min. Incubation at 0–4° in the presence of sodium azide is required to prevent capping of the marker; FCS is added as a source of nonspecific protein to minimize nonspecific staining.

2. After incubation with the primary antibody, the cells are washed four times with 2 ml of cold PBS + azide + FCS.

3. After the fourth wash the wells are suspended in the fluorochrome-labeled secondary antibody or avidin (100 μl appropriately diluted as before) for 30 min on ice.

4. The cells, after incubation, are washed 4 times as per step 2. After the last wash, they are suspended in 1 ml of the PBS–azide–FCS buffer and stored on ice in the dark until analyzed.

Instrument Calibration. Before fluorescence analysis, the instrument is aligned and calibrated with glutaraldehyde-fixed chicken red blood cells, which, when illuminated at 488 nm, fluoresce detectably in the green (520–540 nm) and red (580–600 nm) ranges. After the forward scatter parameter is aligned using fixed chicken erythrocytes as described for light-scatter measurements (see above), green fluorescence is optimized by adjustment of the condensing lens in the 90° light path. Red fluorescence is then optimized by adjustment of the dichroic mirror, which reflects the red fluorescence signal to its detector. The instrument may then be calibrated by adjusting photomultiplier power gain settings to place chicken erythrocytes in the desired channel.

Results and Comments

Lymphocytes stained for marker expression are generally analyzed using as parameters narrow-angle light scatter to facilitate gating and exclusion of dead cells and integrated fluorescence. The gain settings are adjusted so that cells stained with an antibody of irrelevant specificity are <10% positive. The appropriate controls are antibodies of the same subclass in the case of monoclonal reagents or normal immunoglobulin of the same species in the case of heteroantisera.

Analysis can be performed using log or linear integrated (area) fluorescence. Figure 6a depicts the histogram of murine B lymphocytes stained using biotinylated anti-Ia antibody in conjunction with fluoresceinated avidin and analyzed using integrated linear green fluorescence. The beaded line in Fig. 6a represents the frequency histogram for murine B cells stimulated to increase in surface Ia antigen expression by anti-immunoglobulin. As can be seen, Ia antigen-specific immunofluorescence is elevated on stimulated cells relative to the unstimulated cells. However, this difference is much more pronounced in Fig. 6b, which depicts the same cell populations analyzed by log of integrated fluorescence. In the case of the logarithmic analysis, negatives are restricted to the first channel on the cytofluorograph distribution analyzer.

It should be noted that immunofluorescence is not restricted to the analysis of surface marker expression. The same techniques can be applied to the quantitative assessment of intracellular antigens. The only necessary modifications involve fixation of the cells in order to make them permeable to the antibody reagents.

Cytofluorimetric Measurements of Relative Membrane Potential

The electrical potential across the plasma membrane has been postulated to influence cell activity in a variety of somatic cell systems. Maintenance of a high transmembrane potential has been postulated to be a control mechanism serving to arrest cells in the resting phase of the cell cycle.[22,23] Decreases in plasma membrane potential have been demonstrated to accompany an insulin secretion from pancreatic β cells.[24] Fur-

[22] C. D. Cone. *Oncogenesis* **24**, 438 (1970).
[23] C. D. Cone, *J. Theor. Biol.* **30**, 151 (1971).
[24] H. P. Meissner and H. Schmelz, *Pfluegers Arch.* **351**, 193 (1974).

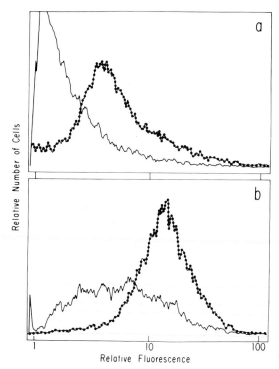

FIG. 6. Linear (a) and log (b) fluorescence histograms demonstrating expression of cell surface Ia antigen on mitogen-stimulated (·—·—·—·) and unstimulated (——) murine B lymphocytes.

ther, plasma membrane depolarization may follow interaction of gonadotropin-releasing hormone with receptors on rat pituitary cells.[25] Because of increasing evidence that changes in membrane potential accompany receptor-ligand induced activation and that these changes are very early events in activation, a rapid means of detecting these changes would greatly facilitate the assessment of roles for various ligands (i.e., neurotransmitters and hormones) in the activation of various neurosecretory cells. Further, such an analysis would allow an assessment of the activation state of a particular cell more quickly than conventional functional studies.

Traditionally, studies of potential changes across cell membranes have relied on the use of microprobes. These studies, however, are limited by physical damage inflicted by the probe itself and because only large cells

[25] P. M. Conn, J. Marian, M. McMillian, J. Stern, D. Rogers, M. Hamby, A. Penna, and E. Grant, *Endocrinol. Rev.* **2**, 171 (1981).

can be easily analyzed and only limited numbers of cells could be studied. More recently, description of charge-sensitive fluorochromatic dyes[26] coupled with the development of flow cytometry, has provided an alternative approach for determination of relative membrane potential.

The two most common classes of potential sensitive fluorochromatic dyes are the cyanines and oxonals. Each of the dyes functions by partitioning between the medium and the interior of the cell in a membrane potential-sensitive fashion. Cyanine dyes are permanent cations, whereas the oxonal dyes are anionic. Two excellent reviews discuss the various difference between these dyes.[27,28] In our work, we have commonly used the cyanine dye 3,3'-dipentyloxacarbocyanine iodide ($DiOC_5[3]$) for measurements of relative membrane potential of lymphocytes. As previously mentioned, this dye senses membrane potential by partitioning across the plasma membrane in a potential-sensitive fashion such that upon depolarization, when the interior of the cell become less electronegative with respect to the exterior, the positively charged dye leaves the cell. The observed result is a decrease in fluorescence intensity, which can be detected cytofluorimetrically. Conversely, the opposite effect is observed upon cell membrane hyperpolarization. It should be noted that extreme care must be taken in analysis of cells containing many mitochondria. In this situation, much of the fluorescence observed may be mitochondrial and a function of mitochondrial membrane potential.

In the succeeding paragraphs is presented a protocol developed for determination of the relative membrane potential following stimulation of lymphocytes via various ligand–receptor interactions. This method is a modification of the protocol described by Shapiro *et al.*[28] Although this protocol was developed for analysis of murine B lymphocytes, it should be applicable to other cell types.

If cells are to be analyzed within 10 min of stimulation, they are preequilibrated with dye as described below. If not, they are cultured in complete medium for an appropriate period of time with the stimulus before staining.

Procedure for Staining of Cells

1. Cells are centrifuged and resuspended in 1 ml of medium without FCS (generally with stimuli, i.e., antigens in the case of lymphocytes) at 2 to 5×10^5 cells/ml at room temperature. It is critical to maintain the cell concentration within this concentration range.

[26] A. S. Waggoner, *Annu. Rev. Biophys. Bioeng.* **8**, 4768 (1979).
[27] H. Shapiro, *Cytometry* **1**, 301 (1981).
[28] H. Shapiro, P. J. Natale, and J. A. Kamentsky, *Proc. Natl. Acad. Sci. U.S.A.* **76**, 5728 (1979).

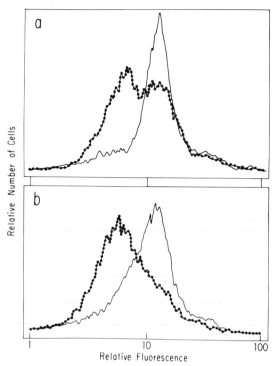

FIG. 7. Fluorescence histograms demonstrating relative membrane potential of control (———) murine B cells or B cells depolarized by exposure to mitogen (·—·—·—·—·) (a) or 50 mM K^+ (b).

2. To each sample (1 ml), add 10 μl of a 5 μM solution of DiOC$_5$[3] in dimethyl sulfoxide.
3. The cells are agitated gently and allowed to stain for 10 min.

Analysis

After staining, the cells are analyzed cytofluorometrically using the following parameters: forward narrow-angle light scatter and log of integrated green fluorescence. The Cytofluorograf should be calibrated and aligned for analysis of these parameters as described previously for light scatter and fluorescence measurements (see above). The dye is excited using the 488-nm argon laser line, and fluorescence is measured in the 500–540-nm range.

Forward narrow-angle light scatter is used as a second parameter to facilitate gating of dead cells and cell aggregates. Having selected the cells to be analyzed on the basis of upon forward light scatter, the control (unstimulated) cells are placed at the midpoint of the screen by adjusting photomultiplier power supply gain settings so as to achieve maximum sensitivity for both increases and decreases in fluorescence intensity. At this point, frequency histograms are constructed and compared for each sample. A typical result for mitogen-stimulated B lymphocytes is depicted in Fig. 7a. Figure 7b are B cells depolarized by supplementation of the medium with 50 mM K$^+$. As can be seen, cells depolarized (i.e., by K$^+$ or ligand) exhibit detectably less fluorescence intensity than control cells.

Comments

1. Maintenance of a constant number of cells in samples to be compared is essential. Variations in cell number will result in variations in the concentration of free dye and, therefore, in fluorescence intensity.

2. The time of staining is critical. Usually 7–10 min at room temperature is required to yield stable staining. However, these cells must be analyzed prior to 15 min. Past this point, the dye appears to be compartmentalized and analysis may not be representative of the plasma membrane potential. In any event, the staining is not stable past 15 min.

3. Appropriate gating of cells to assure analysis of those of homogeneous size is essential. Large cells take up more dye than do small cells and, therefore, appear brighter irrespective of their relative membrane potential.

4. Care must be taken to eliminate dead cells from the analysis. Dead cells exhibit significantly less fluorescence intensity than control live cells. If they are not excluded, they will confuse the analysis by appearing to be live cells that simply have undergone membrane depolarization.

Section III

Preparation and Maintenance of Biological Materials

[14] Optimization of Culture Conditions for Short-Term Pituitary Cell Culture

By Nira Ben-Jonathan, Edna Peleg, and Michael T. Hoefer

The use of pituitary cultures for studying the regulation of pituitary hormone secretion provides a number of advantages over *in vivo* experimentation. Such cultures allow the complete isolation of the pituitary from neural and humoral influences and the addition of test substances under a controlled environment. In addition, by using this experimental model, a correlation can be made between ligand binding, postreceptor events, and stimulation or inhibition of hormone secretion.

In the past, *in vitro* systems utilizing hemipituitaries were extensively used for the isolation and identification of the hypothalamic releasing–inhibiting factors.[1] However, such systems have many inherent limitations, such as the requirement for many donor animals for relatively few experiments, poor reproducibility, and short culture life. The recognition of these limitations has led researchers to the development of *in vitro* models consisting of dispersed pituitary cells,[2,3] which provide sensitivity, multiple treatment capability, and adaptability to various experimental designs. However, dispersion and culture conditions have distinct effects on cell functioning, and therefore the optimization of these parameters can greatly increase the usefulness of these cultures in endocrine research. A few such parameters are discussed below and can serve as a guideline for other investigators.

Cell Dispersion and Culturing

Dispersion of pituitary cells is accomplished by using a modification of the method of Nakano *et al.*[4] Rats are killed by decapitation, and the anterior pituitary is rapidly removed and placed in freshly prepared Krebs–Ringer bicarbonate buffer without Ca^{2+} and Mg^{2+}. The buffer

[1] R. Guillemin, W. R. Hearn, W. P. Cheek, and D. E. Housholder, *Endocrinology (Baltimore)* **60**, 488 (1957).
[2] T. Kobayashi, T. Kobayashi, T. Kagawa, M. Mizuno, and Y. Amenomari, *Endocrinol. Jpn.* **10**, 16 (1963).
[3] W. Vale, G. Grant, M. Amoss, R. Blackwell, and R. Guillemin, *Endocrinology (Baltimore)* **91**, 562 (1972).
[4] H. Nakano, C. P. Fawcett, and S. M. McCann, *Endocrinology (Baltimore)* **98**, 278 (1976).

(KRBGA) also contains 14 mM glucose, 1% bovine serum albumin (BSA), 2% MEM amino acids (Gibco), and 0.025% phenol red, with the pH adjusted to 7.35–7.4 with 1 N NaOH. This, and all other media used for culturing, are filtered through a 0.45-μm membrane filter (Falcon).

Pituitaries are rinsed with the buffer and cut into 1-mm pieces. The fragments are then incubated in KRBGA buffer containing 0.2% trypsin (Worthington; × 3 crystallized) for 35 min in a metabolic shaker at 37° under CO_2. After addition of DNase (1 mg/ml; Sigma) for 2 min, the fragments are washed several times with KRBGA buffer containing 0.2% lima bean trypsin inhibitor (Worthington). The fragments are then dispersed into individual cells by gentle trituration through a siliconized Pasteur pipette. All glassware is siliconized with Surfacil (Pierce Chemical Co.) and steam sterilized. The cell suspension is then filtered through a fine nylon mesh (160 μm; Tetco Inc.) and harvested by centrifugation at 70 g for 10 min.

The cell pellet is resuspended in a medium consisted of Dulbecco's modified Eagle medium, supplemented with 10% fresh rat serum, 3% horse serum, 2.5% fetal calf serum, 2% MEM nonessential amino acids, 10,000 U of mycostatin (all of the above from Gibco) and 10 mg of gentamicin (Shering Corp.) per milliliter. Cells are counted by means of a hemacytometer. The yield for one anterior pituitary is 3.5 to 4 × 10^6 cells, and cell viability as determined by trypan blue exclusion is greater than 90%.

Cells are plated in sterile tissue culture plates (Costar; Cluster 24) and incubated with 1.5 ml of Dulbecco medium in a metabolic incubator at 37° with 5% CO_2 and 95% air. After long-term incubation, cells are washed twice with medium 199 (1.25 g of $NaHCO_3$ per liter; Gibco) to remove all traces of serum. Short-term incubation is performed in 1 ml of medium 199 containing 1% BSA alone, (control) or with the various treatments. By the end of incubation, the medium is removed, centrifuged at 500 g for 10 min, and stored at −20° until analyzed by radioimmunoassay after appropriate dilution with 0.01 M phosphate-buffered saline with 1% egg white.

Several aspects of cell dispersion and culture conditions greatly influence cell responsiveness to various experimental manipulations. It should be emphasized, however, that the selection of the optimal conditions depends primarily on the purpose of the investigation (i.e., studying peptide and/or neurotransmitter uptake, membrane receptors, intracellular events, autoradiography, hormone secretion), and the type of the pituitary cell in question. The conditions discussed below have been optimized for studying the regulation of luteinizing hormone (LH) and prolactin

(PRL) secretion by releasing–inhibiting hormones and may not be ideal for other purposes.

Choice of Enzymes for Dispersion

Both collagenase and hyaluronidase,[5] and trypsin,[6,7] as well as a combination of the three[8] have been used successfully for dispersion of anterior pituitary cells. However, even the highly purified collagenase, contains several impurities, such as clostripain, trypsin-like activity, and other lytic enzymes, that might contribute to the enzymic dispersion, but can also be harmful to cell receptors and impair cell function. Therefore, if collagenase is used for dispersion, attention should be paid to the possible variability of contaminants from one batch of enzyme to another, and a sufficient quantity of a proven batch should be maintained to ensure reproducibility. Although the crude trypsin preparation also contains various contaminants, a highly purified preparation can be obtained at a relatively low cost, and it retains activity under refrigeration for at least than 2–3 years.

Choice of Medium for Long-Term Incubation

Several tissue culture media are available from commercial sources that are compatible for maintaining pituitary cells in culture for various periods of time. However, the maintenance of primary cultures of nonmalignant pituitary cells in a chemically defined, serum-free medium has been met with variable success. Although the procedure described above calls for the use of 10% fresh rat serum, this may introduce undesirable variability because of the different endocrine states of the pituitary donors from one experiment to another. Therefore, after careful evaluation, the fresh rat serum can be substituted with either horse serum or fetal calf serum. It is noteworthy, however, that such sera contain large concentrations of gonadal steroids, which might interfere with cell responsiveness to regulatory factors.[9] Therefore, in experiments wherein gonadal steroids may affect the results, the serum should be treated with dextran-coated charcoal,[10] and its steroid content should be checked prior to its addition to the incubation medium.

[5] W. A. Smith and P. M. Conn, *Endocrinology (Baltimore)* **112,** 408 (1983).
[6] R. Portanova and G. Sayers, *Neuroendocrinology* **12,** 236 (1973).
[7] W. C. Hymer, W. H. Evans, J. Kraicer, A. Mastro, J. Davis, and E. Griswold, *Endocrinology (Baltimore)* **92,** 275 (1973).
[8] J. E. Martin, D. W. McKeel, and C. Sattler, *Endocrinology (Baltimore)* **110,** 1079 (1982).
[9] U. Raymond, M. Beaulieu, and F. Labrie, *Science* **200,** 1175 (1978).
[10] L. Lagace, J. Massicotte, and F. Labrie, *Endocrinology (Baltimore)* **106,** 684 (1980).

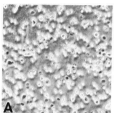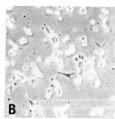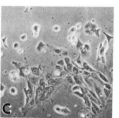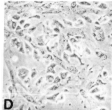

Fig. 1. Photographs of cultured anterior pituitary cells during the long-term incubation period. (A) Freshly dispersed cells. (B) Cells cultured for 24 hr. (C) Cells cultured for 48 hr. (D) Cells cultured for 96 hr. ×100.

Cell Adhesion and Aggregation

Maintenance in monolayer cultures involves attachment of the cells to a substrate. This depends on the nature of the available surface and its area. Enhancements of cell attachment can be achieved by initial plating of the cells in very small volumes and allowing them to settle for a few hours[11] or by coating the culture surface with positively charged polymers, such as poly(D-lysine).[12] Increase in the surface area can be accomplished by utilizing microcarrier beads made of plastic or organic material.[13] The advantage of the latter technique is that such beads with their attached cells can be used in a perifusion system for studying the dynamics of hormone secretion.[14]

The functional significance of the cellular organization of the pituitary gland and the role of intercellular communications between homologous as well as heterologous cells are not well understood. A method was introduced to facilitate the formation of three-dimensional pituitary cell aggregates by the use of constant gyratory shaking for several days.[15] However, dispersed pituitary cells in stationary monolayer cultures also undergo significant morphological reorganization and aggregation. As shown in Fig. 1A, freshly dispersed cells have round or spherical shapes, but after 24 hr in culture, they begin to flatten and extend cellular appendages (Fig. 1B). After 48 hr (Fig. 1C), the cells appear more planar and the

[11] L. Swennen and C. Denef, *Endocrinology (Baltimore)* **111**, 298 (1982).
[12] C. Denef, E. Hautekeete, R. Dewals, and A. DeWolf, *Endocrinology (Baltimore)* **106**, 724 (1980).
[13] A. L. Van Wezel, *Nature (London)* **216**, 64 (1967).
[14] M. O. Thorner, J. T. Hackett, F. Murad, and R. M. MacLeod, *Neuroendocrinology* **31**, 390 (1980).
[15] B. Vander Schueren, C. Denef, and J J. Cassiman, *Endocrinology (Baltimore)* **110**, 513 (1982).

TABLE I
INFLUENCE OF CELL DENSITY ON BASAL AND
DOPAMINE-INHIBITED PROLACTIN RELEASE

Cell number (× 10^3)	Prolactin in medium		
	Control (ng/ml)	Dopamine, 10^{-7} M	
		Ng/ml	Percent of control
25	290 ± 24[a]	214 ± 39	73.8
50	650 ± 46	362 ± 34	55.0
100	1290 ± 46	544 ± 67	42.9
200	3366 ± 101	986 ± 114	29.3
400	5612 ± 128	1818 ± 217	32.3
800	7188 ± 334	3156 ± 417	44.0

[a] Mean ± SE.

formation of aggregates is evident. By 96 hr (Fig. 1D), the cells are more spread, and cell-to-cell contact appears abundant.

Influence of Cell Density

In order to determine whether cell density affects cell function, dispersed cells were plated at densities ranging from 2.5 to 80 × 10^4 per each incubation well that has a surface area of 200 mm². After 4 days of long-term incubation, basal and dopamine (DA)-inhibited PRL secretion were determined after 3 hr of short-term incubation. As indicated in Table I, basal PRL release rose in a saturable manner, whereas PRL release in the presence of 10^{-7} M DA rose linearly with increased cell number. As evident, however, relative inhibition of PRL secretion by DA (expressed as percentage of control values) was greatest at 200,000 cells per well and decreased at both extremes of cell densities.

It has been suggested that high cell density may alter culture conditions because of accumulation of metabolic wastes, depletion of nutrients, change in the pH of the medium, or decreased oxygen tension,[16] but it may also directly alter the function of surface receptors.[17] Hence, the diminished pituitary cell response to DA might be due to changes in the DA receptor structure or number. The decreased response to DA at low

[16] M. P. G. Stocker and H. Rubin, *Nature* (*London*) **215**, 171 (1967).
[17] K. D. Brown, Y. C. Yeh, and R. W. Holley, *J. Cell. Physiol.* **100**, 227 (1979).

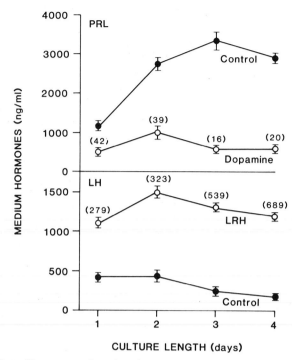

Fig. 2. Effect of long-term culture length on basal and DA (10^{-7} M) inhibited prolactin secretion (upper panel), and basal and luteinizing hormone releasing hormone (10^{-8} M) stimulated luteinizing hormone release (lower panel) by cultured anterior pituitary cells. Cell density was 20×10^4 cells per well, and short-term incubation was 3 hr. Means ± SE of 4–8 determinations are shown. Numbers in parentheses designate percentage of control values.

cell density may indicate that cell-to-cell contact is important. Pituitary cells have been shown to be united by gap junctions providing electrotonic coupling between cells,[18] although their role in facilitating pituitary hormone secretion is not completely understood.

It can be concluded from our results that cell-to-cell contact is required for optimal cell function in culture, but that overcrowding is inhibitory. With the culture system described above, 20×10^4 cells per incubation well provide a subconfluent density and excellent responsiveness to DA. Similar results were obtained after scaling down the culture system and utilizing 96-well plates (each with a surface area of 28 mm^2). Under these conditions, as little as 25 to 50×10^3 cells can be used with optimal responsiveness. The advantages, in terms of reduced number of donor

[18] W. H. Fletcher, N. C. Anderson, and J. W. Everett, *J. Cell Biol.* **67**, 469 (1975).

animals, increased number of treatments, and ease of handling, are obvious.

Effect of Long-Term Culture Length

Enzymic dispersion of anterior pituitary cells reduces their response to releasing–inhibiting factors for a variable length of time. Nakano *et al.*,[4] demonstrated that cell response to LH-releasing hormone (LHRH) and thyrotropin-releasing hormone (TRH) improved after overnight culture. On the other hand, Hopkins[19] suggested that at least 48 hr of culture were required for cellular responsiveness to LHRH to be regained.

In order to assess the effect of culture duration of pituitary cell response to DA and LHRH as well as on the basal secretion of PRL and LH, cells (20×10^4 per well) were incubated for 1–4 days, and their PRL and LH secretion following a 3-hr short-term incubation period was determined. As illustrated in Fig. 2, upper panel, basal PRL release rose significantly between 1 and 3 days in culture and reached a plateau after 3–4 days. DA inhibition of PRL secretion was significantly less effective (42% of control) after 1 day, and maximal (16–20%) of (control) after 3–4 days in culture. In contrast, basal LH release reduced with time (Fig. 2, lower panel), but a significant stimulation by LHRH was already evident after 1–2 days in culture. Still, relative LHRH stimulation (percentage above control), was also optimal after 3–4 days in culture. This experiment suggests that some damage to cell function (possibly to membrane receptors) occurs during the enzymic or mechanical dispersion procedure, but it can be repaired during culture. Long-term incubation for 4 days appears to be optimal for the secretion of both LH and PRL.

Short-Term Incubation Length

Although superfusion of pituitary cells provides a more appropriate system for studying the dynamics of hormone secretion,[20] it is of interest to know the time course of hormone release in static monolayer cultures. Table II shows the results of varying the short-term incubation length from 15 to 180 min on basal and DA-inhibited PRL secretion by cells (20×10^4 per well) incubated for 4 days. Basal PRL secretion (control) rose linearly throughout the incubation period. Relative dopaminergic inhibition (percentage of control) was greatest after 120 min and remained relatively constant for the next 60 min. These results suggest that for studies of DA inhibition of PRL secretion, an incubation length of between 120 and 180 min should be selected.

[19] C. R. Hopkins, *J. Cell Biol.* **73**, 685 (1977).
[20] G. Schettini, A. M. Judd, and R. M. MacLeod, *Endocrinology* (*Baltimore*) **112**, 64 (1983).

TABLE II
EFFECT OF SHORT-TERM INCUBATION LENGTH ON
DOPAMINE INHIBITION OF PROLACTIN RELEASE

Time (min)	Prolactin in medium		
	Control (ng/ml)	Dopamine (10^{-7} M)	
		ng/ml	Percent of control
15	473 ± 24[a]	349 ± 60	74.0
30	483 ± 50	365 ± 22	76.0
60	749 ± 59	481 ± 54	64.0
120	1244 ± 89	468 ± 70	38.0
180	1865 ± 102	765 ± 101	41.0

[a] Mean ± SE.

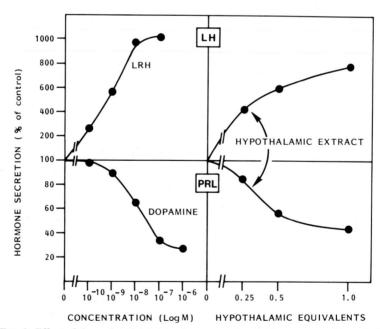

FIG. 3. Effect of dopamine, luteinizing hormone releasing hormone, and hypothalamic extract on prolactin (lower panels) and luteinizing hormone (upper panels) release by dispersed anterior pituitary cells. Cells (20 × 10^4 per well) were cultured for 4 days and then incubated for 3 hr. Each point represents a mean of four determinations in a typical experiment.

Utilization of Cultured Cells for Simultaneous Hormone Determination

Once the optimal conditions have been determined for the secretion of more than one pituitary hormone, the system can be effectively used for variable experimental designs. Figure 3 illustrates the simultaneous determination of the dose-response characteristics of DA inhibition of PRL secretion (lower panel, left side), LHRH stimulation of LH secretion (upper panel, left side), and the profile of the release of these hormones following incubation with hypothalamic extracts (right side, lower and upper panels).

In summary, the use of cultured pituitary cells under optimal conditions provide the investigator with a powerful tool for differentiating between hypothalamic and pituitary sites of hormone action, for probing the interactions between hypothalamic and peripheral hormones at the level of the pituitary, for studying receptor parameters, and for elucidating the mechanism of hormone action at the cellular level.

[15] Separation of Cells from the Rat Anterior Pituitary Gland

By WESLEY C. HYMER and J. MICHAEL HATFIELD

The rat anterior pituitary gland contains six hormone-producing cell types (Table I). Cell separation techniques can be used to obtain enriched populations of these different cell types. In turn, these separated cells are useful reagents for mechanism of action studies in which one wishes to investigate intracellular events leading up to hormone release. The topic of pituitary cell separation has been reviewed.[1]

General Considerations

The purity, viability, and responsiveness of the separated cells will depend upon (a) the physiological state of the pituitary donor, (b) the tissue dissociation procedure, (c) the staining technique used for identification of cell type, and (d) the cell separation technique employed.

[1] W. C. Hymer and J. M. Hatfield, *in* "Cell Separation: Methods and Selected Applications" (T. G. Pretlow and T. P. Pretlow, eds.), Vol 3. Academic Press, New York, in press.

TABLE I
DISTRIBUTION OF HORMONE-PRODUCING CELL TYPES IN THE RAT ANTERIOR PITUITARY

Cell type	Pituitary donor	Approximate percentage of cell type	References
GH	Up to 300 g	32–36	a–d
	>300 g	45–50	e,f
PRL	200–300 ♀	35–38	g,h
	200 g ♂	25	i
	15 day ♀	8	j
LH	200–300 g	7–10	e,h,k–m
	15 day	11–15	d,j,m
FSH	200–300 g ♂	8	k
	200–250 g ♀	12.0	h
	15 day	12–16	d
TSH	200–300 g	5–9	h,n,o
	15 day ♀	5	j
	Thyroidectomized	34–43	p,n
ACTH	200–300 g	3–4	q–s

[a] W. C. Hymer, W. H. Evans, J. Kraicer, A. Mastro, J. Davis, and E. Griswold, *Endocrinology* (Baltimore) **92**, 275 (1973).

[b] J. Kraicer and W. C. Hymer, *Endocrinology* (Baltimore) **94**, 1525 (1974).

[c] G. Snyder and W. C. Hymer, *Endocrinology* (Baltimore) **96**, 792 (1975).

[d] C. Denef, E. Hautekeete, A. deWolf, and B. Vanderschueren, *Endocrinology* (Baltimore) **103**, 724 (1978).

[e] R. V. Lloyd and W. H. McShan, *Endocrinology* (Baltimore) **92**, 1639 (1973).

[f] J. M. Hatfield and W. C. Hymer, unpublished data, 1981.

[g] W. C. Hymer, J. Snyder, W. Wilfinger, N. Swanson, and J. Davis, *Endocrinology* (Baltimore) **95**, 107 (1974).

[h] C. L. Hyde, G. Childs, L. M. Wahl, Z. Naor, and K. Catt, *Endocrinology* (Baltimore) **111**, 1421 (1982).

[i] J. M. Hatfield and W. C. Hymer, unpublished data, 1982.

[j] J. E. Martin, D. W. McKeel, Jr., and C. Sattler, *Endocrinology* (Baltimore) **110**, 1079 (1982).

[k] G. V. Childs, D. G. Ellison, J. R. Lorenzen, T. S. Collins, and N. B. Schwartz, *Endocrinology* (Baltimore) **111**, 1318 (1982).

[l] M. O. Thorner, J. L. Borges, M. J. Cronin, D. A. Keefer, P. Hellmann, D. Lewis, and L. G. Dabney, *Endocrinology* (Baltimore) **110**, 1831 (1982).

[m] Z. Naor, G. V. Childs, A. M. Leifer, R. N. Clayton, A. Amsterdam, and K. Catt, *Mol. Cell. Endocrinol.* **25**, 85 (1982).

[n] M. I. Surks and C. R. DeFesi, *Endocrinology* (Baltimore) **101**, 946 (1977).

[o] M. P. Leuschen, R. B. Tobin, and C. M. Moriarty, *Endocrinology* (Baltimore) **102**, 509 (1978).

[p] W. C. Hymer, J. Kraicer, S. A. Bencosme, and J. S. Haskill, *Proc. Soc. Exp. Biol. Med.* **141**, 966 (1972).

[q] J. Kraicer, J. L. Gosbee, and S. A. Bencosme, *Neuroendocrinology* **11**, 156 (1973).

[r] B. L. Baker, *Handb. Physiol., Sect. 7: Endocrinol.* **4** (pt. 1), 45 (1974).

[s] J. M. Hatfield and W. C. Hymer, *Endocrinology* (Baltimore) **108** (Suppl.), 223a (1981).

Physiological State. A majority of the hormone contained within a pituitary cell is stored in secretion granules. These granules are of varying size and density; as such, they affect the sedimentation properties of the parent cell. For example, (1) GH cells from 2-week-old rats are only sparsely granulated whereas after 8 weeks GH cells are the most dense of the total cell population. (2) Estrogen supplementation *in vivo* will increase PRL synthesis in the mammotroph population. In turn these cells will have increased sedimentation rates in gradients. (3) Removal of target organs will lead to increased division in that population of pituitary cells

TABLE II
METHODS COMMONLY USED FOR DISSOCIATION OF RAT ANTERIOR PITUITARY TISSUE

Base medium[a]	Method[b]	Reference
HEPES saline	Mince, rinse tissue → 0.1% hyaluronidase + 0.35% collagenase; impeller, 45 min, 31°, Pasteur pipette every 10 min → 0.25% viokase, 30 min, impeller → wash 6 X in DMEM + 2.5% FCS + 10% HS + nonessential amino acids.	c
sMEM	Mince, rinse tissue → 0.1% trypsin (c); impeller, 2 hr, 37°, Pasteur pipette every 30 min → wash 2 X in sMEM + 0.1% BSA.	d
KRBG	Mince, rinse tissue → 0.1% trypsin (P); shaker bath, 15 min, 37° → 0.002% DNase → 0.1% trypsin inhibitor, 37°, 5 min → Ca^{2+},Mg^{2+}-free (CMF) 2 mM EDTA, 37°, 5 min → CMF, 1 mM EDTA + 0.0008% neuraminidase, 37°, 15 min → wash 3 X in CMF → Pasteur pipette, CMF → low Ca^{2+} medium + 0.5% BSA → sediment through 4% BSA.	e

[a] Hepes saline: NaCl, 137 mM; KCl, 5 mM; Na_2HPO_4, 0.7 mM; HEPES, 25 mM, pH 7.2; glucose, 10 mM; $CaCl_2$, 360 μM; 3% BSA. sMEM: Ca^{2+}-free Spinner's minimum essential medium + 0.1% BSA. KRBG: Krebs–Ringer bicarbonate buffer + glucose + 0.5% BSA.

[b] Tissue is minced into ~1 mm^3 blocks with a razor blade. Impeller is the stirring device used in a conventional spinner flask. "Pasteur pipette" refers to mechanical disruption of tissue caused by extrusion through pipette tip. Arrows indicate transfer steps. Trypsin (C) = crude (1:250) preparation. Trypsin (P) = purified preparation. Wash is accomplished by low speed (200–250 g) centrifugation. DMEM = Dulbecco modified Eagle's medium.

[c] W. Vale, M. Grant, M. Amoss, R. Blackwell, and R. Guillemin, *Endocrinology* **91,** 562 (1972).

[d] W. C. Hymer, W. H. Evans, J. Kraicer, A. Mastro, J. Davis, and E. Griswold, *Endocrinology* **92,** 275 (1973).

[e] C. R. Hopkins and M. G. Farquhar, *J. Cell Biol.* **59,** 276 (1973).

regulating function of the organ removed. Castration will lead to increased numbers of gonadotrophs; thyroidectomy to thyrotrophs, etc. Thus, the pituitary donor will have a significant influence on the purity and function of the cell type being separated.

Tissue-Dissociation Procedure. A single pituitary gland contains between 3 and 4.5×10^6 cells depending upon the sex and age of the animal. Several methods have been developed to obtain suspensions of single, viable, hormone-containing cells (Table II). In general, cell yields range from 1 to 2.5×10^6 cells/gland with viabilities >95%. There is little indication that certain cell types are preferentially destroyed during dissociation. In terms of cell separation, the most critical requirement is that the starting suspension contain single cells, not aggregates. In some instances, agents that inhibit basal release of hormone can be added to the dissociation medium to prevent hormone loss (e.g., inclusion of dopamine will yield cells with significantly higher contents of PRL).

Identification of Cell Type. The most sensitive method for differentiating the different cell types is the immunoperoxidase technique.[2] In general, histological stains are less sensitive and therefore less useful. A possible exception is the Herlant's tetrachrome technique for somatotrophs, which stains GH cells yellow.

Pituitary Cell Separation Procedures. Virtually all the studies reported to date have used methods that separate cells on the basis of differences in size and/or density. The technique of velocity sedimentation at unit gravity (1G), which separates cells primarily on the basis of size differences, is used most often. Centrifugation of pituitary cells in gradients of BSA or Percoll to isodensity is the next most commonly used procedure. Finally, techniques that separate cells on the basis of affinity and surface charge are beginning to be applied to the pituitary.

Unit Gravity Separation

Theory

When pituitary cells are enzymically dissociated from intact tissue, they are spherical particles. According to the Stokes equation, the sedimentation rate of a spherical particle is

$$S = 2(P_c - P_m)r^2g/9\eta$$

where P_c and P_m are, respectively, densities of the cell and surrounding medium, η is the viscosity of the fluid, r is the cell radius, and g is the earth's gravitational acceleration. A number of statements and assump-

---

[2] P. K. Nakane, *J. Histochem. Cytochem.* **18**, 9 (1970).

tions can be made concerning the various factors involved in separating cells by this simple procedure. Diffusion, degree of hydration, and electrical charge have minimal effects in cell sedimentation with this technique. Sedimentation rates are therefore dependent on the size and density of the cells being separated. Since separations are a linear function of cell density and a squared function of cell diameter, differences in cell size constitute the primary factor in sedimentation rate. Rat pituitary cells range in size between 7 and 18 μm depending in part upon the physiological state of the donor. The density of different rat pituitary cell types covers a range of 1.05–1.08 g/cm^3. Since 1% and 3% BSA solutions have densities of 1.010 and 1.016 g/cm^3, respectively, the difference between pituitary cell density and surrounding medium can be as much as 0.06 g/cm^3 depending upon the specific cell type and its sedimentation rate in the chamber. Thus, the density of each cell subpopulation minus the density of the gradient at the location of this subpopulation can vary by more than 100%. Since velocities of particular cells are functions of their effective densities, velocities of cells of equal diameter may vary by more than 100% on the basis of differences in effective density. The sedimentation behavior of pituitary cells in shallow BSA gradients maintained at unit gravity is therefore *at least partially* dependent upon cell density differences.

Certain procedures used in velocity sedimentation can have significant and limiting effects on the quality of the results obtained. These include convection currents, cell interaction, and streaming. Convection currents in shallow (usually 0.3–3.0%) BSA gradients can be generated by heating caused by incandescent backup lighting, or more typically, by room vibrations or people clustering around the separation chamber. This problem can be eliminated by using fluorescent lighting behind the chamber to visualize separation (see Fig. 1). Gradient instability caused by people opening drawers close to the chamber is hopefully controllable. In our experience, cell–cell interactions occurring *during* the sedimentation run is not a real problem, provided that the concentration of cells in the initial band is below a certain limit (see below). However, resolution is seriously compromised when the starting suspension contains cell "doublets" or "triplets." It is not uncommon to find a doublet consisting of two small cells sedimenting to the same gradient region as a single large cell with comparable total volume.

The issue of "streaming" or "band capacity" is certainly one of the most critical variables in unit gravity sedimentation. It has been pointed out that probably the most commonly observed artifact in the published literature on cell separation by gradient centrifugation results from overloading the density gradient. When too many cells are layered over a gradient, ideal sedimentation no longer occurs and predictable cell separation is not possible. The number of particles that will exhibit ideal sedimentation in a stable gradient increases as a function of the gradient slope (g ml^{-1} cm^{-1}). When the band capacity is exceeded, gradients can become locally unstable, and well defined peaks become broader, frequently with the development of leading shoulders. According to Pretlow et al.[3] a lack of awareness of the concept of band capacity appears to be the rule among many who use gradients to separate cells and has resulted in some "amazing conclusions." Density inversions brought about by excessive cell loads cause streaming. This phenomenon has been characterized as that situation when the cell band assumes the appearance of an "upside-down grass lawn." Methods for minimizing streaming include lowering the concentration of cells applied and using a nonlinear gradient at the top (the so-called "buffered step gradient"—see below). When streaming occurs, it is probably wise to terminate the experiment.

The sedimentation distance, and therefore time taken to achieve a given degree of resolution, depends upon the thickness of the sample layer. This, in turn, is a function of sample volume and chamber diameter. Doubling the chamber diameter will reduce thickness of the sample layer by a factor of 4.

Unit Gravity Sedimentation Method ("Staput")

Chamber. The Lucite chamber we use for pituitary cell separations can be purchased from Bowers Instrument Co. (Davidsville, MD). It has a 500-ml capacity, an outside diameter of 11 cm, and is 11 cm high (Fig. 1). If the experiment calls for culture of the separated cells, the chamber can be sterilized either with gas or by rinsing in a 0.03% Clorox solution before use. Alcohol will destroy the chamber. For pituitary cells, the streaming limit for this chamber is 10 ml at a concentration of 1.4×10^6 cells/ml. We routinely use a concentration of 1.0×10^6 cells/ml to stay well below this streaming limit. Since $\sim 2 \times 10^6$ cells can be obtained from a single pituitary gland, 5–7 animals are required to obtain a sufficient number of cells for the usual experiment. Larger chambers (1- and 2-liter capacity) can be purchased from Bowers Instrument Co. They are useful for separating larger cell numbers.

Chamber-Gradient Setup and Operation. The temperature of the room should be kept relatively constant to maintain reproducibility between experiments. A flat, level surface for the chamber is essential. The gradient generating device is placed ~80 cm above the chamber (Fig. 1),

[3] T. G. Pretlow II, E. E. Weir, and J. G. Zettergren, *Int. Rev. Exp. Pathol.* **14**, 19 (1975).

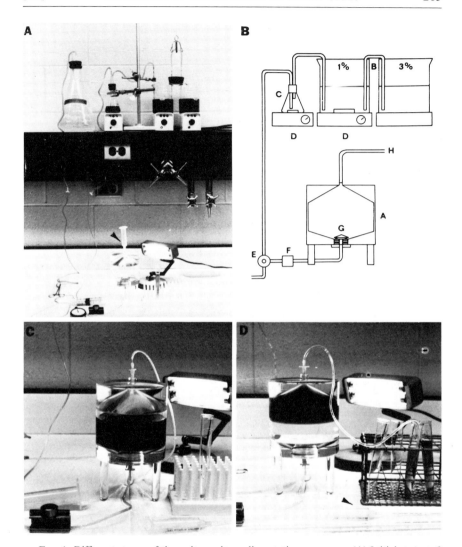

Fig. 1. Different stages of the unit gravity sedimentation process. (A) Initial status of apparatus; note sample loading device (arrowhead). (B) Line drawing of apparatus (see text for details of operation). (C) Chamber with three-fourths of gradient loaded. (D) Chamber at pump-out stage; note 7% sucrose and BSA solution interface, also note sample loading device (arrow). Modified from W. C. Hymer, *in* "The Anterior Pituitary" (A. Tixier-Vidal and M. G. Farquhar, eds.), p. 137. Academic Press, New York, 1975.

and the rates at which gradient fluids enter the chamber are carefully controlled by a sensitive clamp (Fig. 1C).

The gradient is generated using two beakers (the following letter designations refer to Fig. 1B) containing 300 ml of either 1% or 3% BSA prepared in medium 199 + 25 mM HEPES buffer, pH 7.3. Tubing connecting the two beakers is filled with 3% BSA and clamped. Bubbles in this line should be avoided. Vessel C contains 60 ml of 0.3% BSA in medium 199 + 25 mM HEPES. The line from vessel C to the chamber is filled with 0.3% BSA using a syringe attached to the three-way valve (E). This line must be bubble-free. Bubbles can be avoided by filling this line *slowly* and "bleeding-off" small volumes at E if necessary. The gradient solutions are mixed by magnetic stirrers (D).

The O ring at the base of the threaded baffle (G) should be lightly coated with grease to prevent leakage. Prior to charging the chamber with cells, the area under the baffle is filled with 0.3% BSA (by controlling valve F). The function of the baffle (G) is to minimize gradient mixing during filling of the chamber. Ten milliliters of cell suspension containing 10^7 cells is most conveniently applied at G using the sample loading device (arrow Figs. 1A and D) inserted at the top of the chamber. The loading time is ~15 sec. At this point several steps occurring in rapid succession are required. We find it useful to have two people doing these. The clamp at F is opened by one person to permit the gradient solution to enter the chamber while the cells are being loaded. It is critical that this rate be fast enough to displace the cells from the bottom cone surfaces within 5 min. The cell-loading device is removed from the top of the chamber by the other person and replaced with tubing (H) eventually used to harvest the cells. The clamp at B is removed. After 5 min the cells should therefore be located in the cylindrical body of the chamber. In our original paper[4] we suggested that this step take 15 min. However, we subsequently discovered that cell sedimentation occurs during this time and a significant percentage of cells can stick to the bottom chamber surface. This modification increases cell recovery from the 60–70% range to 85–95% range. Once the cells are in the body of the chamber, the flow rate should be adjusted to ~7.5 ml/min. At this rate, the time required to fill the chamber to the beginning of the upper cone is ~60 min.

Cell sedimentation occurs during the entire chamber-filling procedure. At the end of 1 hr, 2–4 cell bands are seen, those at the top being more discrete. The total sedimentation time will depend upon the nature of the cell sample and the cell type being sought. In most of our experiments

---

[4] W. C. Hymer, W. H. Evans, J. Kraicer, A. Mastro, J. Davis, and E. Griswold, *Endocrinology* (Baltimore) **92**, 275 (1973).

(with GH and PRL cell separations) an additional settling time of 20 min is used.

Cell removal from the chamber is accomplished by pumping the gradient upward and out through the top of the chamber (H) using a solution of 7% sucrose and 0.9% NaCl. This "pump-out" solution is introduced into the bottom of the chamber through a three-way valve (E). Pump-out flow rate is ~10 ml/min. The size of the fractions will vary depending upon the particular experiment. The smallest volume fraction we usually collect is 10 ml; the largest, 30 ml. Often some combination of these volumes is used after initial trials establish the sedimentation characteristics of the desired cell type.

Gradient Materials and Characteristics. The gradient device shown in Fig. 1 will generate a nonlinear gradient below the initial cell suspension. After the first 60 ml, the gradient becomes linear. The function of the nonlinear portion (the so-called "buffered-step" gradient) is to increase the streaming limit and therefore the concentration of cells that can be applied to the chamber.

Most of our experiments have utilized fraction V BSA as the gradient material. Serum can also be used, but it is more costly. Ficoll gradients should be avoided, since their intrinsic viscosity will hinder sedimentation. It is probable that Percoll, a nonviscous solution consisting of polyvinylpyrolidione-coated colloidal silica, may also prove to be suitable for 1 g separations. Any complete cell culture medium such as medium 199 can be used.

Analysis of Separated Cells. Unless stated otherwise, cells are kept at room temperature in protein-containing solutions during washing, counting, and other handling steps. We routinely use MEM + 0.1% BSA. *Counting:* We do not use an electronic counter since many fractions contain few cells and it is critical that debris not be counted. Furthermore, it is easy to see the cells increase in size as one monitors the gradient during counting. *Morphology:* A cytocentrifuge (Shandon Instrument Co., Sewickly, PA) is useful for centrifuging the cells onto a microscope slide prior to fixation. With this instrument cells are flattened so that satisfactory staining of the cytoplasm can be achieved. The BSA concentration should be no greater than 0.1% because higher protein concentrations will interfere with staining. We routinely centrifuge 1 to 2×10^5 cells/0.2 ml onto a slide. Although this centrifugation step may not be quantitative, there is no indication that cell losses are preferential to one cell type. However, the various cell types are usually not randomly distributed on the slide. As such, systematic counting procedures must be followed (e.g., center to outside rim). See Hymer et al.[4], for details of fixation and staining of such preparations using the Herlants technique.

Immunocytochemical staining of cell fractions can be done on cells centrifuged onto a slide or on cells bound to polylysine-coated slides. In the latter approach, 25,000 cells in 100 μl are allowed to settle onto a slide previously coated with 0.01% poly(l-lysine). Settling is done at 37° in a humidified atmosphere. See Denef et al.[5] for details of the immunocytochemical staining procedure. To achieve a quantitative assessment of the degree of separation achieved, it is essential to process the unfractionated cells as well. *Hormone content:* The degree of separation achieved can also be estimated on the basis of hormone contained in the cell fraction. This is usually accomplished by standard RIA procedures. To do this, 10^5 cells are pelleted by centrifugation and the hormone is extracted from the cells with 1.0 ml of 0.01 N NaOH. After 24 hr at 4°, 0.5 ml of the extract is diluted with 3.5 ml of RIA buffer (phosphosaline–1% BSA) and stored at −20° until analysis. As a rough approximation, there are 50–100 ng of GH per 1000 cells; 1–10 ng of PRL per 1000 cells, 100–1000 pg of ACTH per 1000 cells, 1–10 ng of FSH per 1000 cells, 5–30 ng of LH per 1000 cells, and 50–100 ng of TSH per 1000 cells. Purifications are estimated by relating the intracellular hormone content contained in various fractions to the content of the starting cell suspension. *Combination analysis:* There are numerous reports in the literature showing that functional heterogeneity exists within a population of cells producing any given hormone. For example, some GH cells contain more hormone per cell than others, and some GH cells release more hormone in culture than others. This means that assessment of cell purity on the basis of hormone content alone can be misleading if the fraction contains a minority population with disproportionately high hormone content. Most meaningful data are generated when both morphological and hormonal analyses are done. An example of separation of GH from PRL cells using the 1 G technique is given in Fig. 2. The appearance of separated PRL cells is shown in Fig. 3. Finally, the effect of the "estrogenic status" of the pituitary donor on PRL cell sedimentation is documented in Fig. 4.

Other Velocity Sedimentation Techniques: Celsep

A new unit gravity cell separation device, Celsep (Wescore, Inc., Logan, UT), appears to provide a useful alternative to the Staput chamber. The Celsep is a 1-liter chamber with a diameter of 20 cm and a height of 3.2 cm. The general strategy involved is given in Fig. 5. The reorienting chamber permits rapid loading and unloading of shallow-gradient materials. It also permits maximum utilization of gradient volume for sedimenta-

---

[5] C. Denef, E. Hautekeete, A. deWolf, and B. Vanderschueren, *Endocrinology* (*Baltimore*) **103**, 724 (1978).

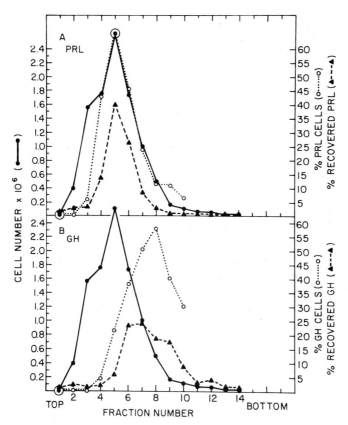

FIG. 2. Distribution of PRL, PRL cells, GH, and GH cells in consecutive 30-ml gradient fractions. Tissue source: random cycle ♀ rats. Results represent the average of three experiments. Average cell recoveries from the gradients were 90%. The percentage of PRL cells (panel A) and GH cells (panel B) was determined by differential cell counts on 500 cells per experiment using immunocytochemistry and Herlants tetrachrome stain, respectively (○····○). On the average, 33% of the cells in the initial suspension were PRL cells, and 32% were GH cells. The percentage of recovered radioimmunoassayable hormone (PRL panel A; GH, panel B) is given as (▲——▲). On the average, 103 μg of PRL and 403 μg of GH were applied to the gradient. Recoveries averaged 85% and 79%, respectively. Modified from W. C. Hymer, J. Snyder, W. Wilfinger, N. Swanson, and J. Davis, *Endocrinology* **95**, 107 (1974).

tion. The actual separation is based on principles *identical* to those already discussed for the Staput. Results from preliminary experiments indicate that the Celsep works as well as the Staput device.

The gradient is generated using two graduated cylinders containing 600 ml of either 1% or 3% BSA prepared in medium 199 + 25 mM HEPES

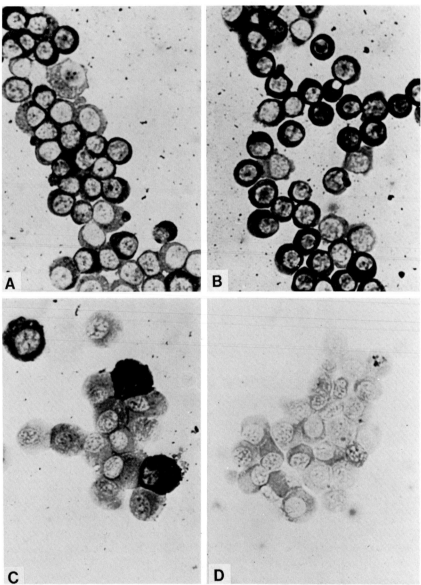

FIG. 3. Photomicrographs of dispersed and separated pituitary cells stained by the immunoperoxidase technique. Cells with the dark-staining cytoplasm are mammotrophs. Cells with the dark-staining cytoplasm are mammotrophs. Cells were prepared from diestrous rats (×320). (A) Freshly dispersed cells prior to separation. Approximately 35% of the cells are mammotrophs. (B) Cells recovered from fraction 5 (Fig. 3). (C) Cells recovered from fraction 7 (Fig. 3). Note that the mammotrophs in this fraction are larger than those in (B). (D) Cells that were not treated with anti-rat prolactin, but were treated with conjugate and substrate. The level of staining in these cells represents background. From W. C. Hymer, J. Snyder, W. Wilfinger, N. Swanson, and J. Davis, *Endocrinology* **95**, 107 (1974).

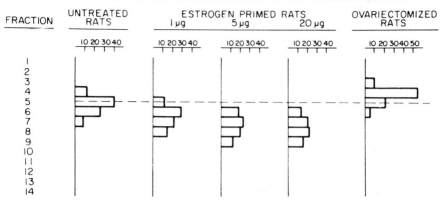

FIG. 4. Separation of PRL cells from Sprague–Dawley ♀ rats by velocity sedimentation at unit gravity. Recovery of cell-associated PRL is represented by open bars; only the four major fractions are shown. Cells in upper fractions are small; those in lower fractions are large. From W. C. Hymer and A. Signorella, *Adv. Exp. Med. Biol.* **138**, 251 (1982).

buffer, pH 7.3. The 1% BSA solution is mixed using a magnetic stirrer. (The following letter designations refer to Fig. 6.) A linear gradient of 1.0–2.5% BSA, in 930 ml, is generated by allowing the BSA to enter the chamber in raised position (A). This filling time takes ~1 hr using gravity flow. Faster filling rates can be achieved using a peristaltic pump. Then 70 ml of cushion solution (7% sucrose + 0.9% NaCl) (B) flows in under the BSA gradient. To load cells into the chamber, 20 ml of cell suspension at a concentration of no more than 1×10^6 cells/ml (in 0.5% BSA), is layered on the top of the gradient using a syringe (C). Cell loading is accomplished by withdrawing the cushion (20 ml) from the chamber through tubing connected to the "bleed off" valve (E). The rate of cell loading should be on the order of 3 min. Finally, 25 ml of isotonic phosphate-buffered saline (PBS) overlay (D) is drawn into the chamber, displacing an equal volume of cushion (E). The chamber is then reoriented to its horizontal position by the motor. This takes 5 min. After a settling time of about 80 min, the chamber is again reoriented to the raised position (5 min). Sampling of the gradient is accomplished either by displacement out the top of the chamber with cushion (B) or out the bottom of the chamber with PBS (D). The choice of top or bottom removal will depend upon the sedimentation rate of the cells of interest. For cells that sediment rapidly, bottom removal is the preferred method since their recovery is enhanced. Total cell recovery from the Celsep chamber routinely runs 85–95%. We find it useful to use hemostats as well as pinch clamps on all tubing entering the chamber. This safety feature ensures that the layering solutions are applied with a

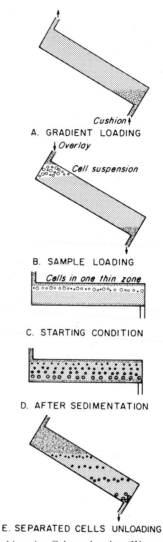

FIG. 5. Procedure followed in using Celsep chamber (Wescor Instrument Co.). Gradient 1–3% BSA; cushion 0.9% NaCl, 7% sucrose; cell suspension in ~0.5% BSA; overlay isotonic phosphate-buffered saline. For additional details see instrument manual. From Wescor Instrument Brochure.

minimum of mixing at solution interfaces. We have found that elevated head pressure can occasionally cause the O rings at the chamber faces to bulge and leak. Use of a peristaltic pump as described in the instruction manual may prevent this problem.

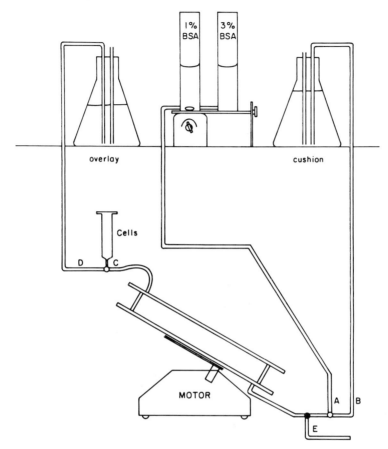

FIG. 6. Celsep line drawing (see text for details).

The Celsep chamber appears to offer a number of advantages over the Staput device. These include (a) generation of a thinner, more uniform, initial cell band; (b) versatility in terms of top vs bottom cell removal depending upon cell sedimentation velocities; and (c) generation of a preformed gradient that permits the option of a shorter sedimentation time.

Other Velocity Sedimentation Techniques: Elutriator

The technique of counterstreaming centrifugation (centrifugal elutriation) has been applied for pituitary cell separation. This procedure also separates cells on the basis of size differences. In this application, cells are centrifuged in a rotor (Elutriator, Beckman Instruments) and prevented from pelleting by continuous centripetal flow of media into the chamber. The rate of cell movement is a function of the sum of the centrifugal force and counterstreaming medium. For the pituitary, this procedure appears to be well suited for separations of minority populations (e.g., gonadotrophs), since cell load is not a problem. The advantages and disadvantages of this technique have been reviewed by Pretlow et al.[3]

Density Gradient Centrifugation

In isopycnic centrifugation, cells are centrifuged in linear density gradients for sufficient duration and centrifugal force to bring all the cells in the gradient to their isodensity location. Use of linear gradients of 14 to 28% BSA has established that three cell peaks at densities 1.0601 ± 0.0007 g/cm^3, 1.0695 ± 0.0007 g/cm^3, and 1.0761 ± 0.0007 g/cm^3 are contained in the rat pituitary. Between 80 and 92% of the cells banding between 1.0705 and 1.0850 g/cm^3 are somatotrophs. Approximately 50–65% of the somatotrophs contained in the gland fall in this density range. Basophils (gonadotrophs, thyrotropes) and PRL cells band at densities 1.0575–1.068 g/cm^3 with little evidence of separation.

The use of discontinuous density gradients, while technically more simple, may be less satisfactory for certain applications, since all the cells in the preparation cannot be brought to their buoyant density. Other disadvantages to the use of discontinuous gradients include "turnover" effects, loss of purity, and the illusion of clear-cut separation.[3] Nevertheless, discontinuous gradients of BSA have been used to considerable advantage for separating GH cells.[6] In these instances density layers were carefully chosen on the basis of the results first obtained with linear gradients. Using discontinuous density gradients of BSA, it is possible to prepare purified (~85%) GH cells in ~1.5 hr.

Methods

GRADIENT MATERIALS

BSA. The procedure of Shortman is used to prepare the BSA.[7] Basically this involves exhaustive dialysis against distilled water, centrifugation to remove debris, freeze-drying of the supernatant fraction, and reconstitution in an unbalanced salt solution (UBSS) to a stock of 40%

---

[6] G. Snyder and W. C. Hymer, *Endocrinology* (*Baltimore*) **96**, 792 (1975).
[7] K. Shortman, *Aust. J. Exp. Biol. Med. Sci.* **46**, 375 (1968).

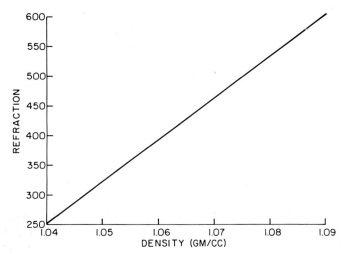

Fig. 7. Transformed standard curve of density vs percentage of sucrose for BSA solutions at 4°. Percentage of sucrose was measured on a Bausch & Lomb hand refractometer, Catalog No. 33-45-30 (0–60% or 298CD) at 4°. Refractive index readings of the same solutions were measured with a Bausch & Lomb Abbe refractometer (Catalog No. 33.45.71) at 20°. Refraction = (refractive index of sample − refractive index of water at 20°) × 10^4. The refractive index of water at 20° is 1.3330. Refraction = 7047.2 × density − 7076.8. Percentage of sucrose = 400 × density − 399. Solutions measured at 20° and then cooled to 4° should have the indicated densities to an average of 2.5%.

(w/w). UBSS is made by mixing isotonic solutions in deionized distilled H_2O as follows: 0.147 M NaCl, 121 volumes; 0.147 M KCl, 4 volumes; 0.098 M $CaCl_2$, 3 volumes; 0.147 M $MgSO_4$, 1 volume; 0.147 M KH_2PO_4, 1 volume. Further dilutions to suitable densities are also done with UBSS. The osmolarity of the BSA is about 0.294, and the final pH is 5.1. A plot relating refraction to BSA density is given in Fig. 7.

Percoll. Percoll is a polyvinylpyrrolidone-coated colloidal silica distributed by Pharmacia Fine Chemicals (Uppsala, Sweden). For cell separation its advantages over BSA include (a) use at room temperature and physiological pH without cell aggregation; (b) low osmolality (<20 mOs/kg, H_2O) and viscosity (1.2–1.8 cP for Percoll between 1.05 and 1.12 g/ml at 22°); (c) ease of preparation (exhaustive dialysis, lyophilization, and tedious reconstitution are not required with Percoll); and finally, (d) BSA is about five times more expensive than Percoll.

A 90% Percoll solution is used for generating the gradient. This solution is made by mixing 9 parts of Percoll stock with 1 part of 10 × Hanks' balanced salt solution (HBSS) without Ca^{2+} or Mg^{2+} containing 250 mM HEPES buffer, pH 7.3, 1% (w/v) BSA and 10% (v/v) penicillin-streptomy-

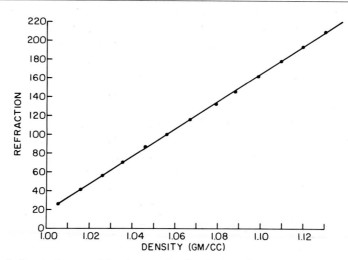

FIG. 8. Standard curve of density vs refraction for Percoll solutions at 22°. The refractions of Percoll solutions were measured with an American Optical 10400A TS meter (see Fig. 7 for definition of refraction). The densities of the same solutions were measured using a 5-ml bulb pycnometer calibrated with water at 22°. See text for details of Percoll solutions. Refraction = 1461.52 × density − 1443.38.

cin solution (GIBCO Laboratories, Grand Island, NY). Further dilution of the 90% Percoll solution to achieve desired densities for a particular experiment is done using a 1 × HBSS prepared from the same 10 × HBSS stock above. Densities of the Percoll solutions are measured by refraction (Fig. 8). Densities estimated on the basis of percentage can vary by more than 5% and are not recommended.

OPERATION

BSA Continuous Gradient. For pituitary cells, details of the gradient generating device and its operation, centrifugation, and subsequent sampling are all exactly as described in Shortman's report.[7] An advantage is that selected cell fractions containing 95% somatotrophs can consistently be obtained using this procedure. Disadvantages are as follows: (a) The technique is involved and quite cumbersome. (b) Centrifugation at 4° is required, some of the pituitary cell types (e.g., PRL cells) show altered function after exposure to cold. (c) As described by Shortman, a pH 5.1 buffer is required to prevent cell clumping. Trial runs using higher temperatures at neutral pH have been unsuccessful. (d) Since ~50% of the somatotrophs in the pituitary have a density <1.070, the high-purity populations are not representative of the total GH population.

BSA Discontinuous Gradients for GH Cells. Two Step Procedure.[6] A discontinuous gradient is prepared in a plastic centrifuge tube. The lower BSA layer is 1.087 g/ml (~3 ml) and the upper layer is 1.068 g/ml (~4 ml). The large density difference between these two solutions permits easy layering. A total of 7 to 10 × 10^6 cells in medium 199 + 1.0% BSA in 3–5 ml is layered over the upper BSA solution. Centrifugation is done in a refrigerated centrifuge (4°) at 2700 g for 30 min using a IEC 253 swinging-bucket rotor. After this time, ~65% of the cells are at the layering interface, ~27% are at the 1.068–1.087 interface, and the remainder are in a small pellet. The middle layer is recovered with a Pasteur pipette, diluted with a large volume (~10 ml) of phosphosaline buffer, pH 7.4, and washed. The resulting cell pellet is resuspended in medium 199 + 1.0% BSA and layered on a second discontinuous gradient consisting of two BSA layers: 1.087 and 1.071 gm/ml. Centrifugation conditions for the second gradient are identical to the first. After centrifugation, 62% of the cells applied sediment at the 1.071–1.087 interface. These cells constitute the purified somatotrophs. Their purity is 84 ± 6%. Basophils are the main contaminating cell type. Advantages are that (a) the procedure is relatively rapid, requiring about 2 hr; (b) somatotroph purity is comparable to that achieved with continuous BSA gradients; (c) the numerous manipulations required for the preparation and sampling of the continuous gradients are completely eliminated. Disadvantages are that (a) centrifugation must be carried out under Shortman's conditions (4° at pH 5.1); and (b) as with continuous gradients, the method selects 50% of the total somatotroph population.

One-Step Procedure. A discontinuous gradient consisting of a 1.087 g/ml layer and a 1.071 g/ml layer is prepared as above. Cell loads, centrifugation, and sampling are all as described for the two-step procedure. Cells collecting at the 1.071–1.087 g/ml interface are 80–85% somatotrophs. These somatotrophs are heavily granulated. In culture, they release GH, which has high biological–immunological activity. GH from the less dense somatotrophs has a ratio close to unity.[8]

Percoll Continuous Gradient. Cells can be applied to the gradient in three different ways: on the top, at the bottom, or mixed within the gradient at the time of its formation. While cell loading on top of preformed gradients is convenient, wall effects caused by radial sedimentation does occur. With bottom application, interface barriers between cells in the bottom cushion and the gradient can sometimes hinder cells from achieving their isodensity point. Adding the cells to the dense Percoll solution during gradient formation is the preferred method.

[8] R. E. Grindeland, W. C. Hymer, P. Lundgren, and C. Edwards, *Physiologist* **25,** 256 (1982).

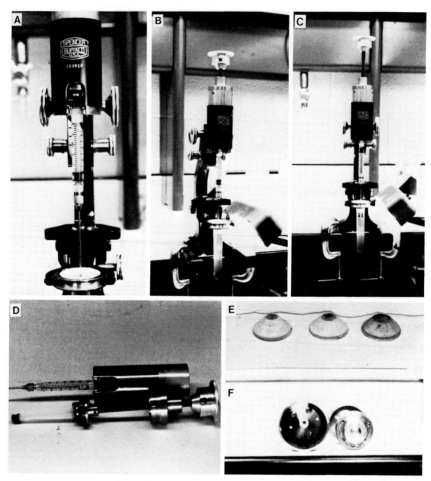

FIG. 9. Monocular microscope-gradient fractionator. (A) The microscope tube is shown with the fractionator inserted. The window is used to insert a 1- or 3-ml syringe. The syringe flanges must be modified to fit into the plate (see plate in F), as do the veins on the syringe plunger. (B) Tip of needle is positioned to withdraw the upper band using the microscope coarse adjustment knob. The fractionator knob attached to the threaded rod is used to withdraw the syringe plunger. (C) The band is shown almost completely withdrawn. (D) The fractionator is shown partially disassembled. The plunger socket is attached to the threaded rod by a bearings assembly to prevent rotation of the plunger and movement of the needle in the gradient. (E) These inexpensive cones (ISCO part No. 68-0643-266) allow the use of centrifuge tubes ranging in size from 4 to 38.5 ml. The cones are inverted and inserted into the microscope condenser assembly (see A) which has been machined to a 45° bevel. (F) Note the four screws in the horseshoe-shaped top plate (down the barrel left). The top two screws must be removed to insert the syringe barrel. The screws are then tightened to align and firmly hold the syringe barrel.

We separate 10 to 13 × 10⁶ cells in 16-ml linear Percoll gradients (1.040 to 1.120 gm/ml) contained in 16 × 102 mm cellulose nitrate tubes. Any standard two-chamber linear gradient maker will suffice. The densities of the Percoll solutions used to generate the gradient are determined by refraction (Fig. 8). Gradients are centrifuged for 30 min at 22° and 2300 g (av). We use the HS-4 swing-out rotor in a Sorvall RC5B centrifuge with a rate control setting of 60 (0–4000 rpm in 10 min and decelerate 4000–0 rpm in 15 min). Acceleration and deceleration control is essential to help prevent swirling within the gradient.

The Percoll manual states: "When Percoll is used in salt solutions under certain conditions, artifactual bands may form during centrifugation at high speed." We are presently evaluating pituitary cell separations achieved at reduced g force [575 g (av)].

After centrifugation, gradients can be sampled in several different ways. For example, a solution such as Fluoroinert (ISCO) can be pumped into the bottom of the gradient to displace fractions out the top. This material has proven to be superior to dense sucrose or bromobenzene. Alternatively, we have designed a gradient-sampling device that removes cell bands quantitatively without use of a pump. This fractionater, constructed from a monocular microscope (Fig. 9), enables one to control precisely the position of the sampling needle as well as rate of band removal.

Somatotrophs separate to high purity on Percoll gradients. For example, Hall et al.[9] report 90% purity of GH cells at densities >1.075 g/ml. Corticotrophs are recovered at density regions between 1.055 and 1.070 g/ml, but are heavily contaminated by PRL cells.[10]

Other Methods for Separating Rat Pituitary Cells

Electrophoresis. Using the Ficoll density gradient electrophoresis technique of Boltz et al.,[11] we have achieved additional separation of somatotrophs when the 1.071 to 1.087 g/ml cell band is applied to the electrophoresis column.[12] The apparent electrophoretic mobility of these heavily granulated cells is less than for the other pituitary cell types. However, it is likely that the relatively high density of the heavily granu-

---

[9] M. Hall, S. L. Howell, D. Schulster, and M. Wallis, *J. Endocrinol.* **94**, 257 (1982).
[10] J. M. Hatfield and W. C. Hymer, *Endocrinology (Baltimore)* **108**, Suppl. 223a (1981).
[11] R. C. Boltz, Jr., P. Todd, R. A. Gaines, R. P. Milito, J. J. Docherty, C. J. Thompson, M. F. D. Notter, L. S. Richardson, and R. Mortel, *J. Histochem. Cytochem.* **24**, 16 (1976).
[12] P. Todd, W. C. Hymer, L. D. Plank, G. M. Marks, M. E. Kunze, V. Giranda, and J. N. Mehrishi, in "Electrophoresis '81" (R. C. Allen and P. Arnaud, eds.), p. 871. de Gruyter, Berlin, 1981.

TABLE III
STUDIES UTILIZING PITUITARY CELL SEPARATION METHODS: MAMMOTROPHS

Reference	Pituitary donor[a]	Dissociation method[b]	Technique	Cell separation parameters[c]					Separated cells			Cell location relative to main peak[g]
				Gradient (% BSA)	Temperature (°C)	Sedimentation time (hr)	Diameter (cm)	Volume (ml)	Cell removal	% Cells stained[d]	Cell enrichment[e]	Hormone enrichment[f]	
h–j	SD ♀	Hymer + dopamine	1G	0.3–2.4 + dopamine	RT	1.25 (2)	10	500	Top				
	Random cycle									66₁	1.9	2–3	At main peak
	Estrus									68₁	1.9	—	At main peak
	Diestrus									66₁	1.9	—	At main peak
	E₂ primed									30–40₁	[0.8–1.1]	3–4	Lower half main peak
	Lactating									63₁	[1.8]	—	At main peak
	Lactating (pup removed)									40₁	[1.1]	—	Lower half main peak
	Ovariectomized									75₁	[2.1]	—	Upper half main peak
k,l	W, ♀ 14 d Adult, random cycle	Denef	1G	0.3–2.4	RT	3.25 (4)	14	1000	Top	—	—	2.7₅	—
m	SD ♂	Hopkins and Farquhar	1G	1.5–2.4	4	2.5 ( )	16 (reorienting)	1650	Bottom	—	—	—	—
n										—	—	—	At main peak
o,p	W ♂, 38–44 d	Hymer	1G	0.3–2.4	RT	1.5 ( )	10	600	Top	—	—	2.5–4.3	At main peak

q	SD ♀, 15 d	Denef	1G	1.0–3.0	RT	2.25 ( )	10	500	Top	32₁	4	—	At main peak
r,s	SD ♀, 15 d	Hymer	1G	0.5–2.0	RT	5 (7)	—	3000	Top	—	—	10	Lower half main peak
t	W ♀, 14 d	Hymer	Elutriation, 1300 rpm, 2.5–25 ml/min, RT, 70 min							—	—	—	Below main peak
u,v	SD ♀, 200–250 g	Denef	Elutriation, 1960 rpm, 11.8–39.5 ml/min, 22°, 80 min							50–88	2	3	At main peak

[a] SD, Sprague-Dawley; W, Wistar; d, days old.
[b] See Table I or text references.
[c] RT, room temperature; ( ) = sedimentation time including "pump out."
[d] I = immunocytochemistry; T = Herlant's tetrachrome stain; E = electron microscopy.
[e] Values in brackets were estimated from assumptions indicated in text (last page).
[f] Enrichment is based on cellular hormone content; s = enrichment based on secretion from purified cells.
[g] Cell location is expressed relative to sedimentation rate.
[h] W. C. Hymer, W. H. Evans, J. Kraicer, A. Mastro, J. Davis, and E. Griswold. *Endocrinology* **92**, 275 (1973).
[i] W. C. Hymer, J. Snyder, W. Wilfinger, N. Swanson, and J. Davis, *Endocrinology (Baltimore)* **95**, 107 (1974).
[j] J. Snyder, W. Wilfinger, and W. C. Hymer, *Endocrinology (Baltimore)* **98**, 25 (1976).
[k] B. Vanderscheuren, C. Denef, and J. Cassiman, *Endocrinology (Baltimore)* **110**, 513 (1982).
[l] L. Swennen and C. Denef, *Endocrinology (Baltimore)* **111**, 398 (1982).
[m] T. G. Gard, D. Atkinson, B. L. Brown, J. F. Tait, and G. D. Barnes, *J. Endocrin.* **67**, 40P (1975).
[n] G. D. Barnes, B. L. Brown, T. G. Gard, D. Atkinson, and R. P. Ekins, *Mol. Cell. Endocrinol.* **12**, 273 (1978).
[o] W. H. Rotsztejn, L. Benoist, J. Besson, M. T. Beraud, M. T. Bluet-Pajot, C. Kordon, G. Rosselin, and J. Duval, *Neuroendocrinology* **31**, 282 (1980).
[p] L. Benoist, M. Le Dafniet, W. H. Rotsztejn, J. Besson, and J. Duval, *Acta Endocrinol. (Copenhagen)* **97**, 329 (1981).
[q] J. E. Martin, D. W. McKeel, Jr., and C. Sattler, *Endocrinology (Baltimore)* **110**, 1079 (1982).
[r] G. Snyder, Z. Naor, C. P. Fawcett, and S. M. McCann, *Endocrinology (Baltimore)* **107**, 1627 (1980).
[s] G. Snyder, Z. Naor, C. P. Fawcett, and S. M. McCann, *Am. J. Physiol.* **241**, E298 (1981).
[t] M. de Monti, A. Enjalbert, and J. Duval, *C. R. Hebd. Seances Acad. Sci., Ser. D* **292**, 697 (1981).
[u] C. L. Hyde, G. Childs, L. M. Wahl, Z. Naor, and K. Catt, *Endocrinology (Baltimore)* **111**, 1421 (1982).
[v] G. Aquilera, C. L. Hyde, and K. J. Catt, *Endocrinology (Baltimore)* **111**, 1045 (1982).

TABLE IV
STUDIES UTILIZING PITUITARY CELL SEPARATION METHODS: SOMATOTROPHS[a]

Reference	Pituitary donor	Dissociation method	Cell separation parameters							Separated cells			
			Technique	Gradient (% BSA)	Temperature (°C)	Sedimentation time (hr)	Diameter (cm)	Volume (ml)	Cell removal	% Cells stained	Cell enrichment	Hormone enrichment	Cell location relative to main peak
b–e	SD ♂, 36–45 d Thyroidectomized	Hymer	IG	0.3–2.4	RT	1.25–3.25 (2–4)	10	500	Top	60–70$_T$	2.2–2.6	3.0–5.0	Bottom of gradient
	SD ♀, random cycle									0.9$_T$	None	—	Bottom of gradient
	Lactating (pup removed)									58$_T$	2.0	4.0–5.0	Bottom of gradient
										50$_T$	2.0	—	Bottom of gradient
f	SD ♂, 300–400 g	Lloyd	IG	0.5–2.0	4°	4.5 (5.8)	9.7	780	Bottom	87.4$_E$	1.7	—	Lower half main peak
g	W ♂, 38–44 d	Hymer	IG	0.3–2.4	RT	1.5 ( )	10	600	Top	—	—	10	Bottom of gradient
h,i	W ♂ adult	Denef	IG	0.3–2.4	RT	3.25 (4)	14	1000	Top	82$_T$	2.3	—	Bottom of gradient
	♂ 14 d									43$_T$	1.2	—	Lower half main peak
	♀ 14 d									49$_T$	1.4	—	Lower half main peak
j	SD ♀, 15 d	Hymer	IG	0.5–2.0	RT	5 ( )	—	3000	Top	—	—	5.8	Lower half main peak
k	SD ♀, 15 d	Denef	IG	1.0–3.0	RT	2.25 ( )	10	500	Top	—	—	—	Lower half main peak
c,e	SD ♂, 230–300 g	Hymer	IG and density: 1.058–1.086 g/cm³, 3500 rpm, 30 min, 4°	0.3–2.4	RT	1.25 ( )	10	500	Top	87–92$_T$	3	4–5	At densities >1.070 g/cm³

l,m	SD ♂, 65 d	Hymer	Density: 2 2-step BSA gradients, 3500 rpm, 30 min, 4°	85–95$_T$	2.2–2.5	—	Interface between 1.071–1.087 g/cm^3
m–q	SD ♂, 250 g	Hymer	Snyder and Hymer^j	>90$_T$	2.5	—	Interface between 1.071–1.087 g/cm^3
r	SD ♂, 150–200 g	Hymer	Density: 25–80% hyperbalic Percoll density gradient, 800 g, 25 min, 19–21°	90$_T$	3	—	At densities 1.075–1.082 g/cm^3
s	SD ♂	Hymer	Density: 1 2-step BSA gradient, 2000 rpm, 45 min, 4°	80–85	2.5	—	Interface between 1.071–1.087 g/cm^3
t	SD ♀, 200–250 g	Denef	Elutriation: 1960 rpm, 11.8–39.5 ml/min, 22°, 80 min	—	—	3.8	Below main peak
u	Mouse	Vale	Cell Sorter FACS II, forward angle light scatter, fluorescent conjugated GH antibody	—	—	—	

^{a} See Table III for table legend.
^{b} W. C. Hymer, J. Kraicer, S. A. Bencosme, and J. S. Haskill, *Proc. Soc. Exp. Biol. Med.* **141**, 966 (1972).
^{c} W. C. Hymer, W. H. Evans, J. Kraicer, A. Mastro, J. Davis, and E. Griswold, *Endocrinology (Baltimore)* **92**, 275 (1973).
^{d} W. C. Hymer, J. Snyder, W. Wilfinger, N. Swanson, and J. Davis, *Endocrinology (Baltimore)* **95**, 107 (1974).
^{e} J. Kraicer and W. C. Hymer, *Endocrinology (Baltimore)* **94**, 1525 (1974).
^{f} R. V. Lloyd and W. H. McShan, *Endocrinology (Baltimore)* **92**, 1639 (1973).
^{g} W. H. Rotsztejn, L. Besson, J. Besser, M. T. Beraud, M. T. Bluet-Pajot, C. Kordon, G. Rosselin, and J. Duval, *Neuroendocrinology* **31**, 282 (1980).
^{h} C. Denef, E. Hautekeete, A. deWolf, and B. Vanderschueren, *Endocrinology* **103**, 724 (1978).
^{i} C. Denef, in "Synthesis and Release of Adenohypophyseal Hormones" (M. Justisz and K. W. McKerns, eds.), p. 659. Plenum, New York, 1980.
^{j} G. Snyder, Z. Naor, C. P. Fawcett, and S. M. McCann, *Endocrinology (Baltimore)* **107**, 1627 (1980).
^{k} J. E. Martin, D. W. McKeel, Jr., and C. Sattler, *Endocrinology (Baltimore)* **110**, 1079 (1982).
^{l} G. Snyder and W. C. Hymer, *Endocrinology (Baltimore)* **96**, 792 (1975).
^{m} G. Snyder, W. C. Hymer, and J. Snyder, *Endocrinology (Baltimore)* **101**, 788 (1977).
^{n} S. Sheppard, J. W. Spence, and J. Kraicer, *Endocrinology (Baltimore)* **105**, 261 (1979).
^{o} J. W. Spence, M. S. Sheppard, and J. Kraicer, *Endocrinology (Baltimore)* **106**, 764 (1980).
^{p} J. Kraicer and J. W. Spence, *Endocrinology (Baltimore)* **108**, 651 (1981).
^{q} J. Kraicer and A. Chow, *Endocrinology (Baltimore)* **111**, 1173 (1982).
^{r} M. Hall, S. L. Howell, D. Schulster, and M. Wallis, *J. Endocrinol.* **94**, 257 (1982).
^{s} R. E. Grindeland, W. C. Hymer, P. Lundgren, and C. Edwards, *Physiologist* **25**, 256 (1982).
^{t} G. Aquilera, C. L. Hyde, and K. J. Catt, *Endocrinology (Baltimore)* **111**, 1045 (1982).
^{u} G. M. Gollapudi, J. E. T. Kelly, F. Ramjattansingh, J. Depasquale, P. Chandra, Y. N. Sinha, and E. P. Cronkite. *Endocrinology (Baltimore)* **108**, Suppl. 223a (1981).

TABLE V
STUDIES UTILIZING PITUITARY CELL SEPARATION METHODS: GONADOTROPHS[a]

Reference	Pituitary donor	Dissociation method	Technique	Gradient (% BSA)	Temperature (°C)	Sedimentation time (hr)	Diameter (cm)	Volume (ml)	Cell removal	% Cells stained	Cell enrichment	Hormone enrichment	Cell location relative to main peak
c–e	SD ♂, 300–400 g Castrate	Lloyd	IG	0.5–2.0	4	4.5 (5.8)	9.7	780	Bottom	L[b] 42.8$_E$	4.9	—	Bottom of gradient
										F 25.2$_E$	6.1	—	Bottom of gradient
										B 65$_T$	3.8	—	Bottom of gradient
f, g	W ♂, 38–44 d	Hymer	IG	0.3–2.4	RT	1.5 ( )	10	600	Top	L —	—	12	Below main peak
h–m	W ♂ adult	Denef	IG	0.3–2.4	RT	3.25 (4)	14	1000	Top	L 15$_I$	[2]	—	Bottom of gradient
										F 15$_I$	[2]	—	
										C 15$_I$	[2]	—	
										L 52$_I$	4.7	—	Bottom of gradient
		♂ 14 d								F 62$_I$	5.2	—	
										C 74$_I$	5.7	—	
		♀ 14 d								L 72$_I$	5.5	—	Bottom of gradient
										F 79$_I$	5.0	—	
										C 81$_I$	4.5	—	
n	SD, ♀ 15 d	Hymer	IG	0.5–2.0	RT	5 ( )	—	3000	Top	B 70–80$_T$	7–10	6	Bottom of gradient
o	♀ 12 d	Vale	IG	0.5–2.0	RT	4 (5.5)	—	3000	Top	L 65$_I$	8	3.1	Bottom of gradient

p	SD ♀, 15 d	Denef		IG	1.0–3.0	RT	2.25 ( )	10	500	Top	L 58₁	5	4–10	Bottom of gradient
q	W ♀, 40 d	Hymer	Elutriation: 1300 rpm, 2.5–25 ml/min, RT, 70 min								—	—	—	Below main peak
r,s	SD ♀, 200–250 g	Denef	Elutriation: 1960 rpm, 11.8–39.5 ml/min, 22°C, 80 min								L 40₁, F 47₁	5, 3.4	6, —	Bottom of gradient
t	SD, 150–400 g	—	Cell Sorter FACSIV: narrow angle light scatter, fluorescent probes LHRH and LH antibody								L 52₁	7.1	—	

[a] See Table III for table legend.
[b] L = LH, F = FSH. C = combined (FSH and LH), B = basophil.
[c] R. V. Lloyd and W. H. McShan, *Endocrinology (Baltimore)* **92**, 1639 (1973).
[d] R. V. Lloyd and W. H. McShan, *Proc. Soc. Exp. Biol. Med.* **151**, 160 (1976).
[e] R. V. Lloyd and H. J. Karavolas, *Endocrinology (Baltimore)* **97**, 517 (1975).
[f] W. H. Rotsztejn, L. Benoist, J. Besson, M. T. Beraud, M. T. Bluet-Pajot, C. Kordon, G. Rosselin, and J. Duval, *Neuroendocrinology* **31**, 282 (1980).
[g] L. Benoist, M. Le Dafniet, W. H. Rotsztejn, J. Besson, and J. Duval, *Acta Endocinol. (Copenhagen)* **97**, 329 (1981).
[h] C. Denef, E. Hautekeete, and L. Rubin, *Science* **194**, 848 (1976).
[i] C. Denef, E. Hautekeete, A. de Wolf, and B. Vanderschueren, *Endocrinology (Baltimore)* **103**, 724 (1978).
[j] C. Denef, E. Hautekeete, and R. Dewals, *Endocrinology (Baltimore)* **103**, 736 (1978).
[k] C. Denef, E. Hautekeete, R. Dewals, and A. de Wolf, *Endocrinology (Baltimore)* **106**, 724 (1980).
[l] C. Denef, *Neuroendocrinology* **29**, 132 (1979).
[m] C. Denef, in "Synthesis and Release of Adenohypophyseal Hormones" (M. Justisz and K. W. McKerns, eds.), p. 659. Plenum, New York, 1980.
[n] G. Snyder, Z. Naor, C. P. Fawcett, and S. M. McCann, *Endocrinology (Baltimore)* **107**, 1627 (1980).
[o] Z. Naor, G. V. Childs, A. M. Leifer, R. N. Clayton, A. Amsterdam, and K. Catt, *Mol. Cell. Endocrinol.* **25**, 85 (1982).
[p] J. E. Martin, D. W. McKeel, Jr., and C. Sattler, *Endocrinology (Baltimore)* **110**, 1079 (1982).
[q] M. de Monti, A. Enjalbert, and J. Duval, *C. R. Hebd. Seances Acad. Sci. Ser. D* **292**, 697 (1981).
[r] C. L. Hyde, G. Childs, L. M. Wahl, Z. Naor, and K. Catt, *Endocrinology (Baltimore)* **111**, 1421 (1982).
[s] G. Aquilera, C. L. Hyde, and K. J. Catt, *Endocrinology (Baltimore)* **111**, 1045 (1982).
[t] M. O. Thorner, J. L. Borges, M. J. Cronin, D. A. Keefer, P. Hellmann, D. Lewis, and L. G. Dabney, *Endocrinology (Baltimore)* **110**, 1831 (1982).

TABLE VI
STUDIES UTILIZING PITUITARY CELL SEPARATION METHODS: THYROTROPHS[a]

Reference	Pituitary donor	Dissociation method	Cell separation parameters							Separated cells			Cell location relative to main peak
			Technique	Gradient (% BSA)	Temperature (°C)	Sedimentation time (hr)	Diameter (cm)	Volume (ml)	Cell removal	% Cells stained	Cell enrichment	Hormone enrichment	
b,c	SD ♂, 200–250 g	Hymer	1G	0.5–2.0	4	4 (4.5)	9.7	780	Bottom	60–80₁ (after culture)	8–16	—	Below main peak (before culture) Upper half main peak (after culture)
d,e	SD ♂	Hopkins and Farquhar	1G	1.5–2.4	4	2.5 ( )	16 (reorienting)	1650	Bottom	—	—	—	Below main peak
f,g	SD ♂ adult	Denef	1G	0.3–2.4	RT	3.25 (4)	14	1000	Top	10₁	[2]	—	Below main peak
	♂ 14 d									30₁	[6]	—	Below main peak
	♀ 14 d									20₁	[4]	—	Below main peak

h	SD ♀, 15 d	Denef	1G	1.0–3.0	RT	2.25 ( )	10	500	Top	19_I	4	—	Below main peak
i	SD ♀	Hymer	1G	0.5–2.0	RT	5 (7)	—	3000	Top	—	—	12	Bottom of gradient
j	SD ♂, thyroid-ectomized	Hymer	1G	0.3–2.4	RT	1.25 (2)	10	500	Top	60_T	1.4	—	Bottom of gradient
k	SD ♀, 200–250 g	Denef	Elutriation; 1960 rpm, 11.8–39.5 ml/min, 22°, 80 min							—	—	4	Bottom of gradient
l,m	Sabra ♂, 30–45 d	Hymer	Affinity, TRH derivativized nylon fibers							80–85	[16–17]	11.0	Bound to fiber

[a] See Table III for table legend.
[b] C. M. Moriarty and M. P. Leuschen, *J. Cell Biol.* **67,** 295a (1975).
[c] M. P. Leuschen, R. B. Tobin, and C. M. Moriarty, *Endocrinology (Baltimore)* **102,** 509 (1978).
[d] T. G. Gard, D. Atkinson, B. L. Brown, J. F. Tait, and G. D. Barnes, *J. Endocrinol.* **67,** 40P (1975).
[e] G. D. Barnes, B. L. Brown, T. G. Gard, D. Atkinson, and R. P. Ekins, *Mol. Cell. Endocrinol.* **12,** 273 (1978).
[f] C. Denef, E. Hautekeete, A. deWolf, and B. Vanderschueren, *Endocrinology (Baltimore)* **103,** 724 (1978).
[g] C. Denef, *in* "Synthesis and Release of Adenohypophyseal Hormones" (M. Justisz and K. W. McKerns, eds.), p. 659. Plenum, New York, 1980.
[h] J. E. Martin, D. W. McKeel, Jr., and C. Sattler, *Endocrinology (Baltimore)* **110,** 1079 (1982).
[i] G. Snyder, Z. Naor, C. P. Fawcett, and S. M. McCann, *Am. J. Physiol.* **241,** E298 (1981).
[j] W. C. Hymer, W. H. Evans, J. Kraicer, A. Mastro, J. Davis, and E. Griswold, *Endocrinology (Baltimore)* **92,** 275 (1973).
[k] G. Aquilera, C. L. Hyde, and K. J. Catt, *Endocrinology (Baltimore)* **111,** 1045 (1982).
[l] E. Tal, S. Savion, N. Hanna, and M. Abraham, *J. Endocrinol.* **78,** 141 (1978).
[m] E. Tal and S. Friedman, *Experientia* **34,** 1286 (1978).

TABLE VII
STUDIES UTILIZING PITUITARY CELL SEPARATION METHODS: CORTICOTROPHS[a]

Reference	Pituitary donor	Dissociation method	Cell separation parameters							Separated cells				Cell location relative to main peak
			Technique	Gradient (% BSA)	Temperature (°C)	Sedimentation time (hr)	Diameter (cm)	Volume (ml)	Cell removal	% Cells stained	Cell enrichment	Hormone enrichment	
[b]	SD ♂, 400 g	Hymer	1G	0.3–2.4	RT	1.25 (2)	10	500	top	11.3₁	3.5	3.5	Lower half main peak
[c]	W, 60 g, adrenalectomized	Hymer	1G	0.3–2.0	—	2 ( )	—	—	top	—	—	2–3	At main peak
[d]	W ♂, 38–44 d	Hymer	1G	0.3–2.4	RT	1.5 ( )	10	500	top	—	—	10	Below main peak
[b]	SD ♂, 400 g	Hymer								—	—	16	Lower half main peak
[b]			Density: continuous Percoll density gradient, 1.04–1.12 g/cm³, 2100 g, 35 min							12.1₁	3.8	3.6	At main peak, 1.041–1.061 g/cm³

[a] See Table III for table legend.
[b] J. M. Hatfield and W. C. Hymer, *Endocrinology (Baltimore)* **108**, suppl., 223a (1981).
[c] H. Kling and R. Martin, *Acta Endocrinol. (Copenhagen)* **87** (S215), 19 (1978).
[d] W. H. Rotsztejn, L. Benoist, J. Besson, M. T. Beraud, M. T. Bluet-Pajot, C. Kordon, G. Rosselin, and J. Duval, *Neuroendocrinology* **31**, 282 (1980).

lated somatotroph promotes droplet sedimentation in the gradient. What percentage of somatotroph separation actually achieved can be attributed to cell surface charge is not known.

Affinity. This procedure depends upon specific cell surface receptors interacting with ligand coupled to a support phase. In the case of TSH cells, TRH bound to nylon fibers is capable of selecting out thyrotrophs to a purity of 80–85%.[13]

Cell Sorting. A report[14] shows that LH-containing cells can be enriched 7× using a fluorescence-activated cell sorter. In this case, the fluorescent probes were directed to LHRH receptors and LH on the cell surface, and sorting was accomplished on this basis.

Pituitary Cell Separations Quality of Results to Date

Typical results for the separation of each of the six pituitary hormone-producing cell types are listed in Tables III–VII.

Acknowledgments

Some of the work in this report was supported in part by Grant CA 23248 from the National Cancer Institute and NASA grants NAS 9-15566 and 2-OR-589-101. The superb secretarial assistance of Mrs. Eileen McConnell is acknowledged.

[13] E. Tal, S. Savion, N. Hanna, and M. Abraham, *J. Endocrinol.* **78,** 141 (1978).
[14] M. O. Thorner, J. L. Borges, M. J. Cronin, D. A. Keefer, P. Hellmann, D. Lewis, and L. G. Dabney, *Endocrinology (Baltimore)* **110,** 1831 (1982).

[16] Culture of Dispersed Anterior Pituitary Cells on Extracellular Matrix

By RICHARD I. WEINER, CYNTHIA L. BETHEA, PHILIPPE JAQUET, JOHN S. RAMSDELL, and DENIS J. GOSPODAROWICZ

Numerous studies have utilized dispersed anterior pituitary cells cultured on plastic dishes to study the hypothalamic regulation of the secretion of anterior pituitary hormones.[1] This approach is particularly appropriate for testing whether anterior pituitary hormone regulation by hypothalamic hormones was modified in surgically removed anterior pitu-

[1] W. Vale and G. Grant, this series, p. 82.

itary tumors. We hypothesized that, in the case of prolactin (PRL)-secreting adenomas, tumor development and uncontrolled secretory activity could be due to a resistance to the inhibitory action of the hypothalamic hormone dopamine. To our dismay, in initial studies we had little success in culturing PRL-secreting adenomas, since the cells tended to reaggregate following dispersion and attached poorly to plastic dishes. We reasoned that, since anterior pituitary cells are normally organized around an extensive capillary plexus, cell attachment could be facilitated by culturing cells on extracellular matrix (ECM). Extracellular matrix produced by cultured bovine corneal endothelial cells has been extensively used for the culture of various normal and tumor cells derived from mesenchyme or epithelium.[2] After the successful application of this technique to PRL-secreting adenomas,[3] we also utilized this approach with dispersed anterior pituitaries from the rat, bovine, and monkey. This chapter describes methods used in the production of ECM, dispersion of anterior pituitary tissue, separation of dispersed cells, and development of a serum-free medium for maintenance of cultured cells.

Production of Extracellular Matrix[4]

All procedures are carried out in a laminar-flow hood, and all media are filtered through a 45-μm Millipore filter immediately before use to ensure sterility. Fresh eyes from adult cattle are obtained from a local slaughterhouse and transported to the laboratory at room temperature. The eye is placed on an absorbent paper and rinsed with 70% ethanol. The cornea is pierced with a large needle near the sclera, excised using iridectomy scissors, and placed in a clean 10-cm culture dish with the inner surface facing up. The surface is then rinsed with sterile phosphate-buffered saline (PBS, pH 7.3), which is immediately removed by aspiration. The corneal endothelial surface is gently scraped with a sterile spatula to remove endothelial cells. A 60-mm culture dish is inoculated with endothelial cells by rinsing the spatula in 5 ml of Dulbecco's modified Eagle's medium (DME) supplemented with 1 g of glucose and 110 mg of pyruvate per milliliter, 10% fetal calf serum, 5% calf serum, 2 mM glutamine, 50 μg of gentamicin per milliliter (Schering), and 2.5 μg of Fungizone (E. R. Squibb) per milliliter (growth medium). The scraping and inoculation are repeated several times for each cornea.

The endothelial cells attach to the dish after 3–4 days of incubation at 37° in a humidified 10% CO_2 atmosphere. The medium is changed 4 or 5

[2] D. Gospodarowicz, D. Cohen, and D. K. Fujii, *Cold Spring Harbor Conf. Cell Proliferation* **9**, 95 (1982).
[3] C. L. Bethea and R. I. Weiner, *Endocrinology* **108**, 357 (1981).
[4] D. Gospodarowicz, H. L. Mesher, and C. R. Birdwell, *Exp. Eye Res.* **25**, 75 (1977).

days after the inoculation, and media containing fibroblast growth factor (FGF, 100 ng/ml) are added. The FGF is added every 2 days for 1 week; it is purified from bovine brains according to the procedure of Gospodarowicz et al.[5] or can be purchased from several commercial sources. Aliquots of a stock solution of FGF (10–50 μg/ml in DME + 0.5% bovine serum albumin (Schwarz–Mann) can be stored at −20°. After approximately 1 week of exposure to FGF, colonies of endothelial cells can be detected.

The primary colonies of endothelial cells are detached from the plate with 0.1% trypsin–Versene in saline (STV) and resuspended in 5 ml of growth medium. To maintain the cell line, a stock plate is made by adding an aliquot of cell suspension containing 300,000 cells to a 10-cm tissue culture dish containing 15 ml of growth medium. If the stock plates are preincubated with 0.2% PBS–gelatin for 2 hr at 4°, then removal of the cells is facilitated in the next passage. The stock plates reach confluency by 1 week when cultured in the presence of FGF and are again passaged. The cells can generally be passaged 7–13 times, but a culture is discarded when the cells become large and the rate of cell division decreases.

During passage the remainder of the cell suspension not used for the stock plates is diluted to 25,000 cells/ml with growth medium containing 5% dextran (Sigma, M_r 40,000). The cell suspension is dispensed into appropriate-sized tissue culture dishes for production of ECM. These plates are referred to as experimental plates. Experimental plates are cultured in the growth media containing FGF and dextran for approximately 2 weeks. The cultures reach confluency in 1 week, and during the additional week they elaborate the ECM on the surface of the culture dish.

The final step involves removal of cells, leaving the ECM on the dish. The medium is removed and a dilute solution of ammonium hydroxide is added (20 mM; 138 μl of 30% reagent grade NH_4OH plus 100 ml of distilled water). The cells are rapidly lysed, the ammonium hydroxide solution is removed (less than 5 min), and the dishes are washed twice with PBS. Then PBS containing gentamicin, Fungizone, and penicillin–streptomycin is added to the plates, which are stored at 4°. The storage solution of PBS and antibiotics is aspirated immediately prior to seeding cells onto the ECM.

Dispersion of Anterior Pituitary Tissue

We have developed a series of cell dispersion techniques for the anterior pituitary in a variety of species. We focused on obtaining prepara-

---

[5] D. Gospodarowicz, H. Bialecki, and G. Greenburg, *J. Biol. Chem.* **2523**, 3736 (1978).

tions that were acutely functional rather than on obtaining high yields of cells. For this reason we avoided using proteolytic enzymes, since we have shown that the use of Viokase destroys cell surface receptors for dopamine in bovine anterior pituitary cells.[6] It should be remembered that collagenase contains a small amount of trypsin-like activity, which can be inactivated using soybean trypsin inhibitor.

Rat Anterior Pituitary

Anterior pituitaries (10–15) are removed under sterile conditions and placed in a 100-mm petri dish containing Hanks' calcium, magnesium-free (CMF) medium with 25 mM HEPES (pH 7.2). The tissue is finely minced with a razor blade and transferred to a 15-ml plastic centrifuge tube. The fragments are allowed to settle, and the medium is aspirated off; then the fragments are resuspended in fresh medium. The washing procedure is repeated twice. Tissue fragments are then transferred to a 25-ml plastic Erlenmeyer flask containing 5 ml of 0.3% collagenase (Worthington enzymes type I) in Hanks' CMF and incubated for 10 min at 37° in a shaking incubator. Fifty microliters of DNase I (1 mg/ml; Sigma Chemicals) in Hanks' CMF is then added, and the fragments are mechanically dispersed by trituration using a 5-ml serological pipette. Cells are separated from undispersed fragments and connective tissue by filtration through cheesecloth. Two layers of cheesecloth are cut to size and placed in a Swinnex filter holder (Millipore) and autoclaved. The filtrate is then centrifuged at 150 g for 10 min, the medium is aspirated, and the cell pellet is resuspended in the medium to be used for culturing. This washing is repeated once. The cell yield is approximately 10^6 cells per pituitary.

Bovine Anterior Pituitary

Bovine pituitaries (10–20) are obtained from the slaughterhouse in Hanks' CMF medium containing 25 mM HEPES at room temperature. The anterior pituitaries are washed several times with sterile medium, minced as described above, and placed in a 50-ml plastic centrifuge tube containing medium. The fragments are washed three times in medium and transferred to a 50-ml spinner flask (Bellco) containing 50 ml of 0.1% collagenase, 0.1% hyaluronidase (Sigma, type I-S), 0.1% bovine serum albumin, 10 μM bacitracin, 0.01% soybean–trypsin inhibitor (Sigma, type I-S), and 1 mg of DNase I per milliliter. The suspension is gently stirred for 30 min at 37° on a magnetic stirrer at slow speed. The fragments are

---

[6] J. S. Ramsdell, F. Monnet, and R. I. Weiner, *Endocrinology* (*Baltimore*) (sumitted for publication).

transferred to a 50-ml plastic centrifuge tube, washed three times in Hanks' CMF, and mechanically dispersed by approximately 20 passages through a 10-ml serological pipette with the tip broadened to approximately 3 mm in diameter. The tube is centrifuged at 35 g for 1 min, and the supernatant containing the cells is transferred to a 100-ml bottle; the remaining pieces are again mechanically dispersed. This procedure is repeated four or five times until remaining fragments consist primarily of connective tissue. The pooled supernatant containing the cells is filtered through cheesecloth and centrifuged (150 g for 10 min). The cell pellet is resuspended in medium, and the washing procedure is repeated twice more. Cell yield is approximately 10^7 cells per pituitary.

One problem with bovine cells is that fibroblasts proliferate very rapidly, and pituitary cells are overgrown within 1 week. Replacement of L- with D-valine inhibits fibroblast growth; however, preliminary data suggested that the presence of D-valine for more than 3 days decreased the responsiveness of lactotrophs to the inhibitory action of dopamine.

Primate Anterior Pituitary

The anterior pituitary from a rhesus or cynomolgus macaque is dense and fibrous, weighing between 40 and 100 mg. The anterior pituitary is placed in Hanks' CMF at room temperature. The tissue is minced in fresh Hanks' CMF, and the fragments are transferred to a 15-ml conical tube and allowed to settle. The medium is aspirated, and 10 ml of Hanks' CMF containing 0.3% collagenase and 1 mg of DNase I is added. The tube is incubated at 37° for 1 hr with occasional agitation. The fragments are gently pelleted and washed with 10 ml of Hanks' CMF. The Hanks' CMF is removed and replaced with 10 ml of 0.25% Viokase (GIBCO) containing 1 mg of DNase in Hanks' CMF. The fragments are incubated at 37° for 15–30 min with occasional agitation. The fragments are pelleted, and 5 ml of Hanks' CMF is added. The tissue is mechanically dispersed by trituration, and the suspension is filtered through cheesecloth to remove small fragments. The filtrate is then centrifuged at 150 g for 5 min, and the cell pellet is resuspended in the desired medium. Optimally, 10^6 cells can be recovered per 10 mg of tissue, but yields of less then 0.5×10^6 cells/10 mg are not uncommon, depending on the age and condition of the donor animal.

Human Adenomas

Surgically removed tumor fragments are placed in Hanks' CMF and transported to the laboratory at room temperature. Markedly hemorrhagic tissue is dissected from the fragments and discarded. The fragments are then minced and triturated through a 5-ml disposable serological pipette. In most instances tumors can be dispersed without the addition of enzymes. However, if necessary the initial mechanical dispersion can be followed by incubation in 10 ml of Hanks' CMF containing 0.35% collagenase and 50 μl of DNase (1 mg/ml) for up to 30 min at 37°. The suspension is again triturated, and the remaining fibrous tissue fragments are discarded. The cells are pelleted by centrifugation at 150 g for 10 min and resuspended in Hanks' CMF. The cells are again pelleted and resuspended in the culture medium. With various tumors the final pellet contained 1 to 30 × 10^6 cells.

Anterior Pituitary Cell Isolation

Dispersed anterior pituitary cells are separated from debris and red blood cells by centrifugation through Ficoll–Hypaque as previously described for the isolation of mouse lymphocytes.[7] The Ficoll–Hypaque is the method of choice based on ease of preparation, cell recovery, and loading capacity. Stock solutions of Ficoll and Hypaque are prepared as follows: Ficoll 400 (Pharmacia Fine Chemicals) is prepared as a 14% solution (w/v) in double-distilled water. Hypaque (Winthrop Laboratories) is purchased as a 50% solution and diluted to 32.8% with double-distilled water. Twelve parts of the Ficoll are mixed with five parts of the Hypaque solution; the mixture is then filtered through a 45-μm Millipore filter and stored in the dark at 4°. Volumetric glassware should be used to ensure accuracy in preparing the Ficoll–Hypaque to a density of 1.09 g/cm^3 and osmolarity of 280–308 mOsm.

The centrifugation of cells through Ficoll–Hypaque should be performed at 20°. Four milliliters of the Ficoll–Hypaque is added to the bottom of 17 × 100 mm polypropylene tubes, and 2–6 ml of the cell suspension (0.5–1.0 × 10^7 cells/ml) is carefully layered on top of the Ficoll–Hypaque. The cells are centrifuged through the Ficoll–Hypaque at 2000 g for 20 min.

The anterior pituitary cells form a band just below the interface of medium and Ficoll–Hypaque with red blood cells collected in the pellet. The supernatant is removed from the red blood cell pellet and diluted in 10 ml of Hanks' CMF containing 25 mM HEPES. The pituitary cells are pelleted by centrifugation at 350 g for 15 min followed by two washes at 150 g for 10 min.

[7] B. B. Mishell and S. M. Siigi, "Selected Methods in Cellular Immunology." Freeman, San Francisco, California, 1980.

Serum-Free Medium

Numerous laboratories have successfully used dispersed anterior pituitary cell cultures to study hormone interactions *in vitro*. This approach has been limited by the stability of hormone secretion with time in the culture and the necessity of adding poorly defined supplements to the medium, i.e., serum. Relying heavily on previously published studies that described totally defined medium for dispersed anterior pituitary cells and GH cells,[8] we have developed a totally defined culture system for human prolactin-secreting adenomas.[9]

Prolactin-secreting adenomas are dispersed as described and plated on ECM-coated plates in DME containing 4.5 g of glucose per liter, 10% FCS, 0.2 mM glutamine (Sigma, Grade III, highest purity), 25 mM HEPES, 100 units of penicillin–streptomycin and 2.5 μg of Fungizone per milliliter. It is essential to culture cells from 3 days in serum containing medium to obtain high plating efficiencies. The FCS is then removed from the medium and replaced with 5 μg of insulin per milliliter (Sigma, bovine pancreas), 20 μg of transferrin per milliliter (Sigma, siderophilin 98%), 10^{-8} M selenium (Sigma, Analyzed Reagent, selenous acid neutralized with NaOH), 10^{-5} M putrescine (Sigma, free acid), and EGF (the concentration of EGF, purified from mouse submaxillary glands, varies from 10 to 50 ng/ml depending on the potency of the preparation and must be titered). In this medium prolactin secretion was stable for up to 21 days and represented from 60 to 80% of the hormone released from cells cultured in DME containing 10% fetal calf serum.

---

[8] J. P. Mather and G. H. Sato, *Exp. Cell Res.* **124**, 215 (1979).
[9] P. Jacquet, C. B. Wilson, R. B Jaffe, and R. I. Weiner, *J. Clin. Endocrinol. Metab.* (submitted for publication).
[10] G. R. Savage and S. Cohen, *J. Biol. Chem.* **23**, 7609 (1972).

[17] Continuous Perifusion of Dispersed Anterior Pituitary Cells: Technical Aspects[1]

By WILLIAM S. EVANS, MICHAEL J. CRONIN, and MICHAEL O. THORNER

Although certain questions concerning hormone secretion by the anterior pituitary gland lend themselves to *in vivo* protocols, others are better addressed by isolating the pituitary from the multiple extrapituitary trophic substances (e.g., neural and gonadal). Consequently, several *in vitro* methodologies have been developed to introduce trophic factors and to monitor concurrently the effects of these agents on hormone secretion. Such techniques have utilized large pieces of pituitary tissue (e.g., intact glands, hemi- and quartered pituitaries) and enzymically dissociated pituitary cells. The culture systems have included static incubation and, more recently, continuous perifusion. Clearly, each of these methods has distinct advantages and disadvantages that must be considered. The topic of this chapter is limited to *in vitro* techniques, with primary emphasis on the system of continuous perifusion.

Use of Pituitary Fragments versus Dispersed Cells

Having removed the anterior pituitary gland from its normal vascular perfusion, severe experimental limitations are associated with a reliance on simple diffusion to deliver substrate and remove metabolites as well as to maintain buffering capacity and ionic homeostasis. In this regard the morphological and functional characteristics of pituitary fragments and dispersed cells have been comprehensively reviewed by Farquhar *et al.*[1a] and thus will be mentioned only briefly. Farquhar confirmed the studies of Samli *et al.*,[2] which demonstrated that pituitary fragments were capable of linear incorporation of amino acids into protein for up to 4 hr. Furthermore, fine morphological structure remained intact for the first hour of incubation. After 1 hr, however, a progressive deterioration took place that included nuclear pycnosis, autophagy, and death. Such changes, while minimal on the surface of the fragments, increased markedly toward

---

[1] Supported in part by grants HD00439 (W. S. E.), RCDA 1K04NS500601, NS18409, AM22125 (M. J. C.), and HD13197, AM22125 (M. O. T.).
[1a] M. G. Farquhar, E. H. Skutelsky, and C. R. Hopkins, *in* "The Anterior Pituitary" (A. Tixier-Vidal and M. G. Farquhar, eds.), p. 83. Academic Press, New York, 1975.
[2] M. H. Samli, M. F. Lai, and C. A. Barnett, *Endocrinology* (*Baltimore*) **88**, 540 (1971).

the center. Autoradiographic studies involving tritiated leucine revealed a similar pattern: a gradient of incorporation was documented in which the labeled amino acid was found in a high concentration peripherally, but decreased rapidly as a function of distance from the surface of the fragment. These findings prompted the conclusion that the penetration of even small molecules through the fragments is poor, accounting for the morphological and functional changes noted between surface and center; thus, Farquhar cautioned that such fragments could not be considered as representative of normally perfused tissue. These ultrastructural observations were confirmed (I. S. Login and M. J. Cronin, unpublished observations) in the hemipituitary preparation routinely used in our laboratories. Beyond a depth of about 4–5 cells, clear signs of necrosis were visible after a 5-hr incubation.

When similar studies were performed on pituitary tissue that had been enzymically dissociated into single cells,[3] fewer morphological and functional problems were noted. The incorporation of amino acids into protein over 4 hr was linear both in freshly dispersed cells and in cells 15 hr after dispersion and was 140% greater than the incorporation by hemipituitaries. Furthermore, the fine cellular structure was well preserved. Such morphological and functional integrity was considered to be due both to better access of the dispersed cells to nutrients and to the buffering capacity of the medium than was the case with fragments. The same general conclusions were reached using reaggregated pituitary cells that were studied for up to 3 weeks in culture.[4]

The disadvantages of using dispersed cells compared to fragments are esoteric at this point. It is becoming increasingly clear that the character of the basement membrane can determine differentiated properties of the cells in contact with this organic matrix.[5] The attachment of cells to inorganic substrates (e.g., polyacrylamide gels, Sephadex microcarriers, cell culture plastic) or to "foreign" organic matrices, such as rat tail collagen or fibronectin, may alter strategic properties, especially with longer-term cultures. Other relevant observations implying that dispersed cells may show altered phenotypes include the fact that there are specialized structural contacts between pituitary cells *in vivo*.[6–8] In addition, the

---

[3] C. R. Hopkins and M. G. Farquhar, *J. Cell. Biol.* **59,** 276 (1973).
[4] B. van der Schueren, C. Denef, and J. Cassiman, *Endocrinology (Baltimore)* **110,** 513 (1982).
[5] G. Gospodarowicz, I. Vlodavsky, G. Greenburg, and L. K. Johnson, *Cold Spring Harbor Symp. Quant Biol.* **42,** 561 (1978).
[6] C. Girod, *in* "Handbuch der Histochemie" (Suppl.) (U. Graumann and K. Neumann, eds.), Part 4, p. 99. Fischer, Stuttgart, 1976.
[7] W. H. Fletcher, N. C. Anderson, Jr., and J. W. Everett, *J. Cell, Biol.* **67,** 469 (1976).
[8] E. Horvath, K. Kovacs, and C. Ezrin, *IRCS Med. Sci.* **5,** 511 (1977).

distribution of the pituitary cell types is not random *in situ*.[8–10] Both of these demonstrations of cell-to-cell associations suggest functional correlates that would presumably be disrupted in dispersed and cultured pituitary cells.

Another proposed difficulty with the use of dispersed cells described by several laboratories[1,11,12] relates to the apparent damage to the cell surface in general and receptors in particular occurring during the process of dispersion: acutely dispersed cells were found to respond less well to secretagogues than did cells cultured for 15–120 hr prior to challenge. Our group[13] and others[14] have examined this question and, using the techniques of dispersion and perifusion described below, failed to confirm these results. Rather, we have shown that dispersed pituitary cells respond to stimulatory and inhibitory hormones within 4 hr of the dispersion procedure.

Use of Static Incubation versus Continuous Perifusion of Dispersed Anterior Pituitary Cells

Advantages of Monolayer Cultures

Monolayer primary cultures of anterior pituitary cells have certain advantages when compared to continuously perifused cells. Dose-response curves can be easily established using multiple wells. The naive cells of a given well may be challenged with a single dose of test substance, thus generating a statistically useful number. In contrast, using continuous perifusion, the possibility exists that a previous exposure to the test substance may affect the cellular response to a later exposure.

Often, the ability to monitor intracellular events in conjunction with hormone secretion is desirable. Thus, monolayers can be extracted for calcium, cyclic AMP, phospholipid, messenger RNA, protein, etc., at selected times during the course of an experiment. Although cells may be retrieved after perifusion and extracted, the number of perifusion columns (i.e., the individual perifusion systems) that may be run simultaneously is limited; thus, for practical reasons, retrieval of cells at various

[9] P. K. Nakane, *J. Histochem. Cytochem.* **18**, 9 (1970).
[10] E. Siperstein, C. W. Nichols, W. E. Griesbach, I. L. Chaikoff, *Anat. Rec.* **118**, 593 (1954).
[11] W. Vale, G. Grant, M. Amoss, R. Blackwell, and R. Guillemin, *Endocrinology* **91**, 562 (1972).
[12] A. Tixier-Vidal, B. Kerdelhue, and M. Jutisz, *Life Sci.* **12**, 499 (1973).
[13] K. A. Daniel, M. O. Thorner, and F. Murad, *Annu. Meet. Endocrine Soc., 61st*, Abstr. 938 (1979).
[14] G. Gillies and P. J. Lowry, *Endocrinology (Baltimore)* **103**, 521 (1978).

times during an experiment is not routinely undertaken. Similarly, cells in primary culture may be monitored microscopically during an experiment, an option not available using continuously perifused cells.

The monolayer culture system also lends itself to the use of organic matrices for the seeding of cells and the ability to work in a relatively sterile environment, neither of which is easily achieved with the cell column. Moreover, the plasticware used in primary culture is disposable, which is of advantage when using toxic, highly potent, or radioactive compounds, and the required amounts of media and test substances are significantly less than are needed for continuous perifusion.

When cells are maintained for prolonged periods of time, such as when the long-term impact or reversibility of an agent is of concern, both the time commitment required for monitoring the perifusion system and the overall expense is significantly greater than with the monolayer system. Furthermore, the time involved in achieving a suitable number of observations for proper statistical analysis is markedly less for monolayer static incubations compared to perifusion techniques. Thus, for an issue that can be resolved using either method, the monolayer preparation is more economical in terms both of personnel time and supplies.

Advantages of Continuous Perifusion

For studies focusing on the dynamics of hormone secretion, however, the technique of continuous perifusion offers significant advantages compared to monolayer culture. Perhaps foremost of these is the ability easily to challenge the cells with pulsatile (i.e., seconds or minutes) and/or prolonged (i.e., hours or days) exposures to test substances. Thus, the temporal aspects of a given secretory response may be defined by using this more dynamic system. Second, as medium is continually perifused around the cells, concerns are diminished regarding the accumulation both of proteolytic enzymes and of hormones or autocoids, which could, by means of short-loop feedback mechanisms, affect the relevant cellular processes. In addition, artifactual changes in cellular function, associated with the necessity of changing medium and washing the cells in static cultures, are negated with the perifusion system. Finally, when utilizing a standard monolayer preparation, time is required for cell attachment to the culture dish before testing. Using continuous perifusion, such a time lapse is not required and, as is the case with centrifuge assays in which recently dispersed cells or reaggregates are in suspension, experiments may be quickly performed, perhaps better representing the status of the donor animal.

Continuous Perifusion of Dispersed Anterior Pituitary Cells

The technique of continuous perifusion involves dissociation of the cells, addition of the single-cell suspension to the inert matrix, introduction of the cells and inert matrix mixture into the perifusion system, the perifusion process itself as the setting for the experiment, and data analysis following measurement of the cellular product of interest. In this section a detailed description of each procedure is provided, including the characteristics and source of the materials used. For the reader's convenience, however, a compilation of these materials is presented in the table.

Preparation of Dispersed Anterior Pituitary Cells

A number of techniques for dissociating anterior pituitary cells have been described.[3,11,15,16] Many of these procedures involve an initial incubation with enzyme (e.g., hyaluronidase, collagenase, neuramindase, trypsin) followed by mechanical dispersion either by a rotating device such as a Teflon-coated paddle[16] or by shearing the tissue with a pipette.[3] Utilizing the method of chemical dispersion with trypsin followed by mechanical dispersion with a pipette, we have found that (1) the duration of the process from removal of the gland from the animal to adding the cells to the inert matrix can be accomplished in less than 1.5 hr; (2) between 2.5 and 3.0 million cells per pituitary are routinely recovered; (3) cell viability, as estimated by the trypan blue exclusion technique, is always in excess of 95%; and (4) damage to the cell surface receptor appears minimal in that receptor-mediated function is present at least as early as 4 hr after dispersion.

For most protocols 3 or 4 rats provide pituitaries for each perifusion column. After decapitation, the neurointermediate pituitary is separated from the anterior pituitary, and the latter is placed in a petri dish containing Earle's balanced salt solution (EBSS; GIBCO, Grand Island, NY) which has a pH of 7.2 and is at room temperature. Using a No. 10 scalpel blade, the glands are diced into approximately 30 pieces each. A useful rule is that if a fragment is large enough to be easily bisected once again, then it should be. The EBSS is then removed with a Pasteur pipette and replaced with 10 ml of a trypsin-containing EBSS (0.2% w/v; trypsin: 206 U/mg; Worthington Biochemicals, Freehold, NJ). The elapsed time between decapitation and placement of the cells into trypsin averages 15 min. The tissue-containing trypsin solution is then transferred to a Teflon cup that is placed in a 37° water bath with ambient room gas. The frag-

[15] W. C. Hymer, W. H. Evans, J. Kraicer, A. Mastro, J. Davis, and E. Griswold, *Endocrinology (Baltimore)* **92**, 275 (1973).
[16] P. J. Lowry, C. McMartin, and J. Peters, *J. Endocrinol.* **59**, 43 (1973).

Chemicals, Media, and Equipment Utilized in the Perifusion System[a]

Supplies	Component	Description	Source
Chemicals and media	Earle's balanced salt solution	Normal	GIBCO Laboratories, Grand Island, NY
	Earle's balanced salt solution	Without calcium or magnesium	GIBCO Laboratories
	Trypsin	Lyophilized; 210 U/mg	Worthington Biochemicals, Freehold, NJ
	Trypan blue stain	0.4%	GIBCO Laboratories
	BioGel P-2 polyacrylamide gel	200–400 mesh	Bio-Rad, Richmond, CA
Equipment	Automatic pipette	1 ml	Pipetman, Woburn, MA
	Plastic syringe	2 ml, hollow barrel	Gilette Surgical, Enfield, Middlesex, England
	Nylon gauze	20 μm	Henry Simon Ltd., Stockton, England
	Tygon tubing	1/32 in., 3/32 in. i.d.	Norton Plastics, Pittsburg, PA
	Manifold pump tubing	0.015 in., 0.02 in., 0.45 in. i.d.	Fisher Scientific, Pittsburg, PA
	T-tube connector	Model AL-116B034-01	Acculab, Norwood, NJ
	3-Port fitting connector	Model AL-116-0202-03	Acculab
	Intravenous line extension kit	Twin size	Abbott Hospital Products, North Chicago, IL
	Fraction collector	Fractomette Alpha 400	Buchler Instruments, Fort Lee, NJ
	Peristaltic pump	Ismatek SA Micropump	Brinkmann Instruments, Westbury, NY

[a] Although these supplies and equipment have proved satisfactory in our laboratory, other brands and suppliers may be as suitable.

ments rest undisturbed on the bottom of the cup during the 20-min incubation. The fragments and solution are then poured into a 12-ml conical tube that is centrifuged at 400 g for 2 min. The supernatant is decanted, and the fragments are suspended in 3 ml of calcium- and magnesium-free EBSS (containing no trypsin). After mixing by inverting the tube several times, the solution is centrifuged again as described above. The pellet is again resuspended in 3 ml of calcium- and magnesium-free EBSS and is ready for mechanical dispersion using a pipette (Pipetman, Woburn, MA) with a 1-ml plastic tip. For 10–15 min the fragments are slowly taken into and then discharged from the pipette tip until the solution has become cloudy and few, if any, fragments remain. Fragments are removed with a Pasteur pipette, as are any strands of DNA that have formed. Although the addition of DNase can also reduce DNA strands, we have found it to be an unnecessary step. An aliquot (20 μl) is removed from the mixed cell suspension for cell counting with a standard hemacytometer and determination of viability using trypan blue (0.4%; GIBCO). Care must be exercised to exclude the red blood cells in the count. We have found that the numbers of cells obtained from a given number of rats (of the same strain, sex, and age) do not differ statistically among dispersions.

The Inert Matrix

After dissociation, pituitary cells tend to reaggregate unless preventive measures are taken. Lowry et al.[16] pioneered a methodology in which the dissociated cells are mixed with a polyacrylamide gel (BioGel) prior to being loaded into the perifusion chamber. More recently, Smith and Vale[17] described a technique in which Sephadex microcarrier spheres (Cytodex beads) may be used for the same purpose. Although we have used BioGel as the inert matrix for the past 6 years and have found it to be entirely satisfactory, there are no data to indicate that the use of Cytodex beads would result in any less acceptable results.

The process of preparing the BioGel and mixing it with the cells is simple. BioGel P-2 (200–400 mesh; Bio-Rad; Richmond, CA) is preswollen overnight prior to setting up a perifusion column (0.5 g of BioGel in 10 ml of 0.9% saline per perifusion column). When the cells have been dispersed (see above) the saline is decanted from the BioGel, and the latter is poured into a petri dish prior to adding the 3 ml of cell suspension (dispersed cells in Ca- and Mg-free EBSS). The covered dish is then gently moved on a flat surface in a circular manner for 30–45 sec or until the cellular solution and BioGel are mixed. The cell–BioGel mixture is now ready for introduction into the perifusion chamber.

The Continuous Perifusion Apparatus

The perifusion system used in our laboratory is detailed in Fig. 1. Central to the system is the cell and BioGel-containing chamber for which we use a 2-ml plastic syringe with a hollow barrel (Gilette Surgical, En-

[17] M. A. Smith and W. W. Vale, *Endocrinology* (*Baltimore*) **107**, 1425 (1980).

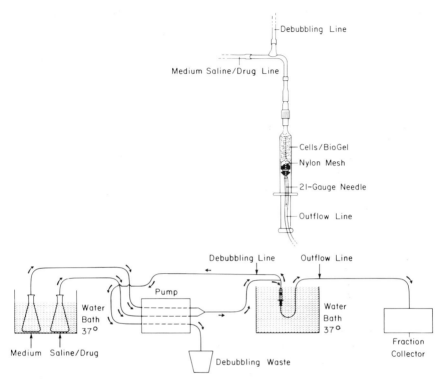

FIG. 1. The dispersed-cell perifusion system. The details of the construction of the perifusion chamber using a hollow-barrel 2-ml plastic syringe are shown at the upper right; the overall system is demonstrated below. Reproduced, with permission, from W. S. Evans et al.; Endocrinology (Baltimore) **112**, 535 (1983).

field, Middlesex, England). Construction of the chamber is detailed in the top part of Fig. 1. A 21-gauge needle punctures the septum of the plunger and is positioned flush with the inner surface. The plunger surface is then covered with nylon gauze (20 μm; Henry Simon Ltd., Stockton, England). Tygon tubing (1/32 in. i.d., Norton Plastics, Pittsburg, PA) is connected to the needle, thereby forming the outflow line. The inflow line, through which medium and test solution reach the chamber, is connected to the nozzle of the syringe. Medium and saline or test solution are in separate flasks situated in a 37° water bath. Medium is conveyed from the medium flask to the peristaltic pump via 3/32 in. i.d. Tygon tubing. Similarly, the vehicle or test solution is drawn through 1/32 in. i.d. Tygon tubing to the pump. Proximal to the pump (Ismatek SA, Brinkmann Instruments, Westbury, NY) the medium and test solution lines are joined to 0.045 in. and 0.015 in. i.d. manifold pump tubing (Fisher Scientific,

Pittsburg, PA), respectively. Distal to the pump the medium and test solution pump tubings connect to a T-tube (AL-116B034-01, Acculab, Norwood, NJ) from which a single line (3/32 in. i.d. Tygon tubing) carries the mixture of medium and test solution. The medium–test solution line enters the 37° water bath, where approximately 40 cm of tubing are wrapped around a glass rod to act as a warming coil. The tubing then joins the inflow–debubbling unit, which attaches to the plastic syringe nozzle as described above (depicted in Fig. 1). A 3-port fitting connector (AL-116-0202-03, Acculab) serves as the inflow–debubbling unit. The bottom port attaches to the syringe nozzle by means of a connecting tube cut from an intravenous line extension set (Abbott Hospital Products, North Chicago, IL). The medium–test solution line attaches directly to the lateral port. From the top port, a debubbling line (3/32 in. i.d. Tygon tubing) runs to the pump and attaches to manifold pump tubing (0.02 in. i.d., Fisher Scientific), beyond which its contents are discharged and discarded. This debubbling line removes approximately 10% of the medium–test solution mixture in addition to any bubbles that may have formed in the system.

Introduction of the Cell–BioGel Mixture into the Perifusion Chamber

For several hours prior to adding the cells, the perifusion system is washed with 70% ethanol followed for at least 1 hr with distilled water. Approximately 30 min prior to loading the cell–BioGel mixture, the medium and test solution lines are placed into their respective flasks to prime the system. To load the perifusion chamber, the outflow line is first clamped with a hemostat and all liquid is extruded from the plastic syringe. The cell–BioGel mixture is then drawn into the syringe, and the syringe tip is connected to the inflow–debubbling unit. The hemostat is removed and, after the syringe is vertically positioned, the cell–BioGel mixture is allowed to settle. When the settling process is complete (2–3 min), the hemostat is once again clamped on the outflow line and the syringe is disengaged from the inflow–debubbling unit. The medium above the cell–BioGel mixture is then carefully pushed out allowing only a 1–2-mm column of medium above the BioGel, and the syringe is reconnected. This procedure increases the rate of cellular response by reducing the time it takes for an agent to reach the cells from the top of the chamber. The outflow line is then connected to a fraction collector with which timed eluate samples are obtained.

Mixture Ratios and Flow Rates

By manipulating the internal diameter of the tubing and the pump speed, any dilution of test solution and desired flow rate may be obtained.

The design described here results in an approximate 1 : 10 dilution of test substance prior to reaching the cells. This dilution factor must be validated for any particular system. The tubing length in our system is such that the total transit time from test substance flask to fraction collector is approximately 10 min. As we frequently utilize 2.5- and 5.0-minute eluate fractions, the 10-min transit time is convenient for estimation of when the effluent from cells challenged with test solution will reach the fraction collector (Fractomette Alpha 400, Buchler Instruments, Fort Lee, NJ). Our pump speed (Ismatek setting No. 3) and tubing size results in a flow rate of approximately 0.43 ml/min.

Analysis of Data Obtained from Continuously Perifused Anterior Pituitary Cells

After an experiment, eluate fractions are properly stored until the substance of interest can be measured. Using a Digital Equipment Corporation PDP11/44 (Maynard, MA), the measured concentration (e.g., for LH: ng/ml) is corrected for cell number (expressed as 10^7 cells) and perifusate flow rate (ml/min) and plotted against time. From these data (ng/min per 10^7 cells vs time), mean secretion rates of the cells during any time interval can be calculated.

As a rule we express our data as mean hormone secretory rate ($\pm$ SEM) in which the same experiment has been performed a minimum of three times on different days. Although interexperiment error can be minimized by examining a given group using parallel perifusion columns on a single day (unpublished observations), the potential of biasing experimental results from secular trends seems high. By performing experiments on different days, a block design is created in which such secular variability is distributed evenly across the experimental groups. Secretory rates are then examined with analysis of variance to identify statistical differences among groups, and such differences are then identified with a secondary procedure such as Duncan's Multiple Range test.

Potential Problems and Concerns Associated with Continuous Perifusion

As with any *in vitro* system, certain difficulties may arise during continuous perifusion of dissociated cells. In our experience, the most frequently encountered problem is clogging of the perifusion chamber outlet by strands of DNA that have accumulated on the nylon mesh. When this occurs, the outlet line must be clamped and the syringe nozzle disengaged from the inflow–debubbling unit. Having manually occluded the nozzle, the syringe is inverted and the outflow line is unclamped. The syringe plunger is then carefully removed, and the nylon mesh is cleaned. The plunger is then reinserted, the outflow line is clamped, the syringe is reconnected to the inflow–debubbling unit, and the outflow line is unclamped. When performed in this manner, no obvious deleterious effects on cell viability have been noted. However, if the DNA is properly extracted from the dissociated cell preparation prior to loading the cell–BioGel mixture, problems with clogging rarely occur.

The possibility that test substance may bind to the inflow–outflow tubing, BioGel, or plasticware must also be considered. For all test substances for which an assay is available, concentrations in the storage flask and eluate should be measured and compared in a test column containing no cells. Alternatively, a similar procedure may be followed using radioactively labeled substance. Without these data, the effective concentration of test substance reaching the cells cannot be known with certainty. In addition, the recognition of binding of bioactive substance to the system can provide an awareness of potential artifact associated with the dissociation of the substance from the plastic.

Infection is an important concern which, fortunately, is rarely encountered during perifusion of anterior pituitary cells when a modest number of precautions are taken. For experimental protocols of all durations, the system is washed with 70% ethanol for several hours prior to use and antibiotics are added to the medium (i.e., penicillin, 10 U/ml; streptomycin, 2.5 μg/ml; gentamicin, 5 μg/ml; and amphotericin B, 187.5 ng/ml). For perifusions of less than 48 hr, it is unnecessary to employ sterile technique in preparing the cells. We recommend, however, that for perifusions longer than 48 hr, the entire dispersion process be performed with sterile technique (e.g., dispersion under a laminar-flow hood; rat heads washed with ethanol prior to opening the cranium; use of sterilized instruments, pipette tips, etc.).

The medium–buffering system requires comment. We have noted that for incubations of less than 12 hr, the addition of amino acids to a buffered salt solution appears not to alter a variety of basal or dynamic hormone release characteristics (unpublished observations). In contrast, cell responsiveness (e.g., GnRH-stimulated LH release) wanes markedly if cells are perifused overnight in the absence of amino acids. We therefore recommend the routine supplimentation of buffered salt solution with amino acids for perifusions of any duration. On the other hand, we did not find fetal calf serum (FCS) to be necessary in terms of maintaining basal and secretagogue-stimulated LH release for incubations lasting up to 43 hr.[18] Thus, as FCS is expensive and contains a number of poorly defined fac-

---

[18] J. L. C. Borges, D. L. Kaiser, W. S. Evans, and M. O. Thorner, *Proc. Soc. Exp. Biol. Med.* **170**, 82 (1982).

tors, we do not routinely employ it in our system for short-term studies. Whether or not FCS is required for long-term culture remains to be proved.

Summary and Conclusions

Continuous perifusion of dispersed anterior pituitary cells is a powerful dynamic *in vitro* technique that complements the static incubation technique of primary culture. The major advantages of perifusion include the ability to challenge cells with test substance in a pulsatile manner and to monitor the immediate response. Furthermore, the accumulation of hormonal product and proteolytic enzymes, with their potential effects on cell function, is essentially eliminated as a concern. Finally, although methods using static techniques are available, experiments using continuously perifused dispersed cells can be initiated soon after removal of the gland from the donor animal.

We and others[14,19] have found this system to be useful in the study of hormone secretion by the anterior pituitary. We expect that the system will lend itself equally well to investigations of other cellular processes involving the export of organic and inorganic substances.

[19] J. S. Loughlin, T. M. Badger, and W. F. Crowley, Jr., *Am. J. Physiol.* **240**, E591 (1981).

[18] Preparation and Maintenance of Adrenal Medullary Chromaffin Cell Cultures

By STEVEN P. WILSON and NORMAN KIRSHNER

Historically, the adrenal medulla has been useful for studying the mechanisms of catecholamine synthesis, storage, and secretion. These mechanisms, first elucidated in the adrenal, were shown subsequently to operate also in both peripheral and central catecholaminergic neurons. The development of methods for the dissociation and purification of adrenal medullary chromaffin cells in large yield[1-3] and for the maintenance of

[1] A. S. Schneider, R. Herz, and K. Rosenheck, *Proc. Natl. Acad. Sci. U.S.A.* **74**, 5036 (1977).
[2] J. C. Waymire, K. G. Waymire, R. Boehme, D. Noritake, and J. Wardell, *in* "Structure and Function of Monoamine Enzymes" (E. Usdin, N. Weiner, and M. B. H. Youdim, eds.), p. 327. Dekker, New York, 1977.
[3] E. M. Fenwick, P. B. Fajdiga, N. B. S. Howe, and B. G. Livett, *J. Cell Biol.* **76**, 12 (1978).

these cells in primary culture[2,4-8] has greatly simplified study of the functions of the gland and has resulted in an increased interest in the biology of these cells. We describe here a method for isolation of chromaffin cells from the bovine adrenal medulla,[5,8] based largely on the procedure of Livett and co-workers.[3,4] Maintenance of the chromaffin cells in serum-free medium is also described.[8]

Chromaffin Cell Isolation

Perfusion

Bovine adrenals are obtained at a slaughterhouse as quickly as possible after the death of the animal. It is important that the gland be intact, with the adrenal vein and part of the wall of the vena cava attached. Most of the fat is trimmed from the gland, and the glands are transported to the laboratory in cold, calcium- and magnesium-free, Locke's solution (154 mM NaCl–5.6 mM KCl–5.6 mM glucose–5.0 mM HEPES–200 units of penicillin G per milliliter–40 µg of gentamicin per milliliter–7 units of heparin per milliliter; pH 7.4). Although the Locke's solutions used in these procedures may be sterilized by filtration, experience has shown that this is not necessary. All the glassware is sterilized by autoclaving; pump tubing and cannulas are soaked in Zephiran chloride (1 : 100) and rinsed before use. Usually four glands are processed at one time. The adrenal vein is cannulated with the flared end of a 20-cm length of 3.25-mm o.d. polyethylene tubing attached to a syringe filled with Locke's solution. The cannula is secured by a ligature around the adrenal vein above the flared end of the cannula. Light pressure is applied to the gland by means of the attached syringe, and the glands are observed for points of leakage. Leaks are stopped with ligatures or the gland is discarded. Once the gland is leak-free, multiple slits, approximately 1 mm deep, are made in the capsule–cortex of the gland. Application of moderate pressure to the syringe should result in a slow flow of solution from these slits. The cannula and a second length of polyethylene tubing are then passed through a 6-mm hole in the center of a No. 9 rubber stopper and suspended in a 300-ml Fleaker beaker. After all four glands are prepared, the cannulas are attached to a multichannel peristaltic pump, and the glands

[4] B. G. Livett, D. M. Dean, and G. M. Bray, *Soc. Neurosci. Abstr.* **4**, 592 (1978).
[5] D. L. Kilpatrick, F. H. Ledbetter, K. A. Carson, A. G. Kirshner, R. Slepetis, and N. Kirshner, *J. Neurochem.* **35**, 679 (1980).
[6] J. M. Trifaro and R. W. H. Lee, *Neuroscience* **5**, 1533 (1980).
[7] K. Unsicker, G.-H. Griesser, R. Lindmar, K. Loffelholz, and U. Wolf, *Neuroscience* **5**, 1445 (1980).
[8] S. P. Wilson and O. H. Viveros, *Exp. Cell Res.* **133**, 159 (1981).

are perfused with Locke's solution at a flow rate of 5–10 ml per minute for 30 min at room temperature. The extra piece of tubing in each container is used to pump the perfusate to a waste receptacle. At the end of this perfusion, 12 ml of Locke's solution containing 0.05–0.1% collagenase and 12–15 μg of deoxyribonuclease per milliliter are perfused through each gland from a common reservoir (50 ml). This perfusate is discarded. The four glands are then perfused for 30 min with fresh collagenase solution using the extra piece of tubing in each container to return the perfused solution to a 100-ml reservoir. The glands should remain turgid throughout the perfusion period.

Further Collagenase Digestion

The glands are then removed from the perfusion apparatus, and the medullas are dissected from the cortices. The medulla should be flaccid and readily stripped from the cortex. All subsequent procedures are carried out at room temperature and in a sterile hood except as noted. The medullas are placed in a petri dish and minced with scalpels. Care should be taken to remove any remaining cortical tissue. The mince is washed with Locke's solution and transfered to a covered 250-ml beaker with a suspended stirring bar. The digestion is continued by slowly stirring the mince for two 30-min periods at 37° using 100–150 ml of fresh collagenase each time. The Locke's solution containing collagenase, debris, and dissociated cells is decanted at the end of each digestion period, and the cells are collected by centrifugation at 100–250 g for 10 min. The cells collected from both digestions are resuspended in Locke's solution and filtered through 250-μm and 105-μm nylon mesh (50-μm mesh can also be used). The cells are again collected by centrifugation. The crude cell preparation obtained at this point is heavily contaminated with cellular debris, erythrocytes, and other small cells.

Density Gradient Centrifugation

Percoll. The crude cell preparation is resuspended in 48 ml of Locke's solution, and the cell suspension is mixed with 60 ml of a balanced salt–bovine serum albumin–Percoll mixture prepared by mixing 6 ml of 1.54 M NaCl, 56 mM KCl, 56 mM glucose, 50 mM HEPES, 2% bovine serum albumin, and 0.01% Phenol Red, pH 7.4, with 54 ml of Percoll and adjusting to pH 7.2–7.4 with 1.0 M HCl. Addition of 3 mg of deoxyribonuclease to this cell suspension diminishes clumping of cells during centrifugation. This cell suspension is placed in 15-ml polycarbonate centrifuge tubes with caps and centrifuged at 20,000 rpm (49,000 g) for 20 min in a Sorvall SM-24 or SS-34 rotor. Centrifugation can be performed at 4–15°. At least four fractions can be isolated from the gradients: (a) cellular debris near the top, usually clumped; (b) a broad band of chromaffin cells extending from just under the debris to the middle of the gradient, sometimes in small clumps; (c) a diffuse band of small cells below the chromaffin cells; and (d) erythrocytes near the bottom of the gradient. The chromaffin cell fraction is collected from the gradient by aspiration and washed twice by centrifugation in Locke's solution. The cells are then resuspended in culture medium containing serum (see below) for counting and plating.

Renografin. An alternative cell purification procedure using a discontinuous gradient of Renografin can also be used. All steps are performed at 1–4°. The crude chromaffin cell pellet described above is washed in Locke's solution and resuspended in 60 ml of 14.4% Renografin (335 mOsm, $\rho = 1.07$, diluted from Renografin-60 with H_2O). This cell suspension is pipeted into centrifuge tubes (10 ml each) and overlayered with 2.5 ml of 7.2% Renografin (14.4% Renografin–Locke's solution, 1:1). The gradients are centrifuged at 7800 g in a swinging-bucket or fixed-angle rotor for 20 min. Because 14.4% Renografin has a density greater, and 7.2% Renografin a density lower, than the equilibrium density of the chromaffin cells, most of the chromaffin cells in the gradient accumulate at the interface between the two Renografin layers. Other cell types sediment through the more dense Renografin solution, while debris enters the lighter overlay. The chromaffin cell fraction is collected and washed as described for cells prepared using Percoll.

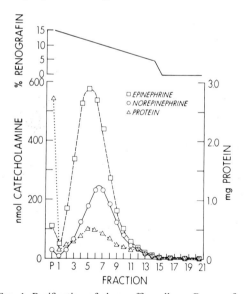

FIG. 1. Purification of chromaffin cells on Renografin.

CATECHOLAMINE CONTENT AND RECOVERY IN CHROMAFFIN CELLS PURIFIED BY
DENSITY GRADIENT CENTRIFUGATION[a]

Cell fraction	Catecholamine content		Percent epinephrine	Recovery of catecholamines
	nmol/mg protein	nmol/10^6 cells		
Crude	701 ± 41 (7)	ND	64 ± 2 (7)	—
Percoll	741 ± 72 (4)	108 ± 9 (6)	68 ± 2 (7)	65 ± 4 (4)
Renografin	998 ± 56 (6)	148 ± 16 (4)	61 ± 2 (6)	67 ± 4 (4)

[a] ND, not determined. The epinephrine, norepinephrine, and dopamine contents of acid extracts of chromaffin cells were determined by high-performance liquid chromatography on a reverse-phase column with electrochemical detection. Mean ± SEM. (n). Dopamine was approximately 1% of the total catecholamine content.

Chromaffin cells can also be purified on continuous gradients of Renografin (Fig. 1). The data in Fig. 1 suggest that norepinephrine cells have a lower bouyant density than epinephrine cells, a finding that may be useful in better separating the two cell types.

Purity and Yield of Chromaffin Cells

Purity. More than 90% of the cells isolated by these methods exclude trypan blue.[5,8] The cells recovered by density gradient centrifugation in Percoll show no purification over the unfractionated cells when expressed as nanomoles of catecholamine per milligram of protein (see the table). When the values are expressed as catecholamines per cell, a 1.6-fold purification is apparent.[8] Purification of the cells with Renografin as the density medium produced a 1.4-fold increase in the catecholamine content of the cells when normalized to protein. A higher catecholamine content per cell is also observed in Renografin-purified cells when compared to cells isolated from Percoll gradients; hence, the protein content per cell is similar in the two preparations (150 μg of protein per 10^6 cells). The catecholamine content of chromaffin cells purified by centrifugation in Renografin (148 nmol per 10^6 cells) approaches the theoretical value of 170 nmol per 10^6 cells calculated by Phillips.[9] Indeed, values of up to 176 nmol per 10^6 cells were obtained in certain preparations. The relative epinephrine content of the gradient-purified cells does not differ significantly from that of the starting cell preparation (see the table) or adrenal medullary homogenates,[10] although a slightly higher proportion of epi-

[9] J. H. Phillips, *Neuroscience* **7**, 1595 (1982).
[10] S. P. Wilson and N. Kirshner, *J. Neurosci.* **3**, in press (1983).

nephrine-containing cells is obtained with Percoll gradients than with Renografin gradients. The recovery of catecholamines by either gradient technique is similar. Assessment of cell purity by election microscopy has shown that cells purified by centrifugation in Percoll gradients are 80–85% pure[5,8]; no such determination has been performed on Renografin-purified cells. Higher levels of purity (up to 98% chromaffin cells) have been reported following differential plating techniques,[2,11] but these procedures have not been effective in our hands.

Yield. The above method using Percoll routinely yields 50 to 100 × 10^6 cells per adrenal gland. With one preparation of collagenase we have obtained yields as high as 200 × 10^6 cells per gland. As may be calculated from the similar recoveries of catecholamines in cells isolated by the two gradient techniques and the higher amount of catecholamines per cell (see the table), cell yields from Renografin gradients are about 75% of those stated here for Percoll gradients. The bovine adrenal medulla contains approximately 500 × 10^6 cells[9]; hence, even out best yields represent recovery of only 40% of the cells present.

Critical Factors in the Cell Preparation. Two factors seem to be important in obtaining the yields of chromaffin cells described above. First is the collagenase preparation. Commercially available collagenase is generally a mixture of collagenases and other proteases, polysaccharidases, and lipases. Because of the ill-defined nature of these preparations, there is considerable variation from lot to lot. At present, the only method of determining the usefulness of a particular lot of collagenase is strictly empirical. Second, perfusion with collagenase is important. Where this step is omitted, cell yields are dramatically lower. Lack of good perfusion pressure during perfusion with collagenase also reduces yields. Apparently, perfusion forces the enzymes into the gland in a way not duplicated by incubating small pieces of minced gland with the enzyme mixture.

Chromaffin Cell Cultures

Plating and Maintenance

The basic medium used for culture of the chromaffin cells consists of 50% Dulbecco's modified Eagle's medium (Grand Island Biological Co., catalog No. 430-1600) and 50% nutrient mixture F12 (catalog No. 430-1700) with 5 mM HEPES and 28.6 mM NaHCO$_3$ (pH 7.4 at 37° and 5% CO$_2$). This medium is filter-sterilized and may be stored at 4° up to 2 months before use. Chromaffin cells prepared by either of the methods described above are resuspended in approximately 100 ml of plating me-

[11] K. Unsicker and T. Muller, *J. Neurosci. Methods* **4**, 227 (1981).

dium (the basic medium supplemented with 5–10% newborn calf serum, 20 μM 5-fluorodeoxyuridine, 100 units of penicillin G per milliliter, 40 μg of gentamicin per milliliter, 50 units of Nystatin per milliliter, and 250 μM ascorbic acid) for counting. Cells are then diluted to an appropriate density with plating medium, and 0.25–0.5 ml of this cell suspension is plated per square centimeter of culture dish surface area. We routinely plate 0.2 to 0.5 × 10^6 cells per square centimeter in untreated plastic tissue culture ware; collagen or polylysine coating of the plastic is not required for cell attachment. Chromaffin cells also attach firmly to untreated glass surfaces. Chromaffin cells attach to fibronectin-treated culture dishes in the absence of serum, allowing for culture in a completely defined medium.

After 1–3 days in culture, the plating medium is removed from the cells and replaced with 0.25 ml of maintenance medium per square centimeter of surface area. This medium is identical to the plating medium, except that serum is omitted. The maintenance medium is replaced twice weekly, although chromaffin cells are well maintained for at least 8 days without replacement of medium.[8] The cultures are maintained at 37° in a humidified atmosphere containing 5% CO_2. An alternative method employs Dulbecco's modified Eagle's medium supplemented with 10% fetal bovine or newborn calf serum.[4,5]

Stability of the Cultures

Morphology. The chromaffin cells attach to the substratum within a few hours after plating and become flattened over the first 2–3 days in culture.[4–7] These cells are capable of extending long, varicose processes, a behavior observed primarily in low-density cultures. Transmission electron microscopy has shown that the cells initially retain the appearance of adrenal medullary chromaffin cells, including an abundance of electron-dense chromaffin vesicles.[5,7,8] After 7–9 days in culture, however, some swelling of chromaffin vesicles was observed and by 14–16 days occasional cells undergoing autolysis and disintegration of vesicles was seen. By 21 days in culture these effects were more pronounced. These changes in the morphology of the cells are similar whether the cultures are maintained with or without serum in the medium. In the serum-free medium, the proportion of chromaffin cells in the cultures remains constant for 3 weeks.[8]

Biochemistry. Chromaffin cell cultures maintained in serum-free medium retain high levels of catecholamines and opioid peptides.[8,10] After an initial loss of approximately 40% of the catecholamines and 25% of the opioid peptides between plating and 2 days in culture, the cellular stores of these substances remain constant over 2–3 weeks. The retention of catecholamines depends upon the initial seeding density and on the frequency and extent of replacement of medium, suggesting that medium "conditioning" is important in maintenance of the cells. The inclusion of ascorbic acid in the culture medium enhances retention of chromaffin cell catecholamines and the ability of the cells to resynthesize their amine stores following loss via evoked secretion, although stoichiometric quantities of this vitamin are not required for catecholamine synthesis. Addition of the synthetic glucocorticoid dexamethasone to the culture medium is without effect on the cellular content or proportion of epinephine and norepinephine. Addition of insulin (1 μM) to the cell cultures produces a doubling of the protein and opioid peptide contents within 6 days and steadily increasing catecholamine levels between day 6 and day 12 of culture.[10] Addition of transferrin, progesterone, putrescine, and selenium to the cultures[12] is without effect on chromaffin cell catecholamine and protein contents or light microscopic morphology.[8] Chromaffin cells retain the ability to secrete large amounts of catecholamines and other substances stored in the chromaffin vesicle for at least 3 weeks in culture.[5,8] The cells also retain high levels of the catecholamine biosynthetic enzymes tyrosine hydroxylase, dopamine β-hydroxylase, and phenylethanolamine N-methyltransferase.[5] The biosynthesis of Met- and Leu-enkephalin continue in cultured chromaffin cells, and the rate of this process can be regulated.[13] Chromaffin cells stimulated to secrete large portions of their catecholamine and opioid peptide stores resynthesize their lost norepinephrine content within 3 days, but recover only 20% and 50%, respectively, of their secreted epinephrine and opioid peptides.[10] The results suggest a difference in the responsiveness of the two types of chromaffin cells present in the cultures.

Conclusion. The cell cultures described here retain most of the functional properties of chromaffin cells *in vivo* and are suitable for a wide variety of studies on the biochemistry and neurobiology of the cells comprising the adrenal medulla. There remain, however, opportunities to improve the conditions for culture of the chromaffin cells to enhance retention of their *in vivo* characteristics.

[12] J. E. Bottenstein and G. H. Sato, *Proc. Natl. Acad. Sci. U.S.A.* **76,** 514 (1979).
[13] S. P. Wilson, K.-J. Chang, and O. H. Viveros, *Proc. Natl. Acad. Sci. U.S.A.* **77,** 4364 (1980).

[19] Techniques for Culture of Hypothalamic Neurons

By CATHERINE LOUDES, ANNIE FAIVRE-BAUMAN, and ANDRÉE TIXIER-VIDAL

The complexity of the structural organization of the hypothalamus and of its connections with other brain areas limits the analysis of the cellular mechanisms that regulate the synthesis and release of hypothalamic neuropeptides and neurotransmitters. There is, therefore, a need for *in vitro* systems that permit experimentation under simplified conditions. It is necessary, however, that the integrity of the neuronal structure from perikarya to axon terminals and synapses be maintained. Such a requirement can be met with culture methods only. Other *in vitro* systems, such as synaptosomal preparations or tissue slices, have been used for short-term incubations. Although they have brought about many new findings concerning the release of several neuropeptides, their major drawback resides in the possibility that an undetermined part of the released material represents leakage from damaged cells (see Robbins and Reichlin[1]). They will not be considered here.

The pioneering work of Hild,[2] who showed that explants of paraventricular nuclei (PVN) and of supraoptic nuclei (SON) remain viable for weeks in culture, where they form axonal processes, demonstrated the feasibility of this *in vitro* approach for the central nervous system. After about 10 years, this approach again attracted attention because of the discovery of several neuropeptides and the progress and refinement of methods to characterize and to measure small amounts of neuropeptides and to visualize peptidergic neurons by immunocytochemistry at both light and electron microscopic levels.

At present, there is a rapid increase in the number of reports on cultures of hypothalamic neurons, in parallel with those on cultures of other brain areas (see references in this volume). Several culture methods have now been applied to the hypothalamus. They are outlined, in the following sections with three objectives: (1) to describe technical conditions; (2) to survey neuron functions expressed in culture; and (3) to show the respective advantages or disadvantages of each.

Many culture methods have been applied to the hypothalamus: organ culture, where the whole hypothalamus or even the hypothalamohy-

---

[1] R. Robbins and S. Reichlin, *in* "Neuroendocrine Perspectives" (E. E. Muller and R. M. Macleod, eds.), Vol. 1, p. 111, Elsevier Biomedical Press, New York, 1982.
[2] W. Hild, *Z. Zellforsch. Mikrosk. Anat.* **40**, 257 (1954).

pophysial complex is maintained in culture without cell outgrowth (see Table I); tissue culture, in which small pieces of tissue are explanted into various conditions where they can grow (see Table II); and cell culture, where the cells are dissociated before being plated in liquid medium (see Table III). In the first case, the integrity of the hypothalamic organization is respected. In the second case, the original organization of the explant may be lost depending on the intensity of cell outgrowth. In the third case, the dissociated cells may develop in culture a new histiotypic organization.

Most of these studies have been performed in medium supplemented with biological fluids that have ill-defined supplies of hormones and trophic factors. In order to work in more defined conditions, the technology of serum-free hormonally defined medium, pionered by G. Sato and colleagues (see Bottenstein[3]) has also been applied to dissociated hypothalamic cell culture (see Table IV).

Last, the possibility of obtaining permanent cell lines derived from the hypothalamus has also been investigated, using viral transformation. This method as well as the potentialities of the cell lines isolated are described in the last section.

Organ Cultures

Organ cultures have been used essentially to maintain the adult hypothalamus (Table I). The explant is generally placed on a nylon mesh or metal grid in the central well of a plastic dish. In view of its relatively large size, a widespread cell degeneration occurs with time in culture.[4] This system was used for the first studies on the *in vitro* biosynthesis of the neuropeptide, vasopressin, which yielded the first evidences of a common precursor form of vasopressin and neurophysin. However, owing to the limit in cell viability, organ cultures were mostly used as short-term cultures (1–5 days) to study neuropeptide release (Table I). The release experiments are always performed in a medium different from the culture medium, that is, a salt buffer devoid of serum.

As compared to short-term incubation of fragments or slices, these organ cultures had disadvantages. Because of the large size of the explants, the penetration of reagents into the cells is certainly irregular and slower than in slices or small pieces. Moreover, although the survival time *in vitro* is longer, neuron degeneration progressively occurs and it is impossible to relate functional activities to a definite number of neurons.

[3] J. E. Bottenstein, *Adv. Cell. Neurobiol.* **4,** 333 (1983).
[4] H. Sachs, R. Goodman, J. Osinchak, and J. McKelvy, *Proc. Natl. Acad. Sci. U.S.A.* **68,** 2782 (1971).

Long-Term Explant Cultures in Serum Supplemented Medium

Long term (up to 3 months) explant cultures are initiated with small fragments of fetal or newborn rodent hypothalamus (Table II), except for one report where small pieces of adult bovine hypothalamus were used (Table II). Either specific areas of the hypothalamus were dissected out (coronal sections, punched nuclei) or the whole hypothalamus was randomly minced in small pieces (0.5–1.0 mm^3). The explants are generally attached to a collagen-coated coverslip or to a plasma clot. They are then kept in hanging drop in Maximow chambers or in roller tubes. In a few cases the coverslips are immersed in petri dishes or in multiwell trays. The basal nutrient medium generally consists of MEM solution (sometimes DMEM or Ham's F10 or medium 199) supplemented with glucose, glutamine, HEPES, insulin, NEAA, or Hank's salt solution, depending on the authors. It is always supplemented with a high concentration of serum (up to 40%) (horse serum, human placental serum, fetal calf serum) or of chick embryo extract (30%).

As compared to organ cultures, long-term explant cultures differ because of the progressive migration out of the explant of fibroblast-like cells, neuronal somas, and extending neurites. Numerous synapses develop in this cell outgrowth. The initial organization of the explant, which is maintained at the beginning of the culture, is progressively lost owing to cell necrosis. This process is not limited to fetal or young tissue, since it has also been reported, at least in light microscopy studies, in adult bovine explants.[5]

In addition to evidence from electron microscopy for neurosecretory neurons and synaptogenesis in culture, extracellular and intracellular recordings showing synchronous bursting discharges demonstrate the functional activity of cultured neurons. Several specific neuron products have been identified: catecholamines by endogenous fluorescence as well as several neuropeptides by immunocytochemistry or radioimmunoassay (LHRH, AVP, β-endorphin). Very few studies so far deal with neuropeptide secretion into the culture medium (Table II).

Explant cultures of fetal or newborn hypothalamic areas have been found particularly suitable for the study of factors involved in neuron development. The possibility of isolating anatomically defined hypothalamic areas has permitted the study of regional differences in the development of catecholamine neurons,[6] in the response of neurite growth and LHRH-containing neuron number to estrogen, which was correlated with

---

[5] D. M. Nicholson and W. T. Mason, *Brain. Res.* **249**, 123 (1982).
[6] S. C. Feldman, L. L. Brown, and M. B. Bornstein, *Cell. Mol. Neurobiol.* **1**, 279 (1981).

TABLE I
SHORT-TERM ORGAN CULTURES OF ADULT HYPOTHALAMUS

Structure	Species	Culture medium[a]	Time in culture	Functional markers[b]			References
				Neuro-transmitter	Neuro-peptide	Other markers	
Hypothalamoneuro-hypophysial complex	Guinea pig	DMEM + 10% FCS + glutamine + NEAA + Ab	Up to 6 days for functional studies	—	Vasopressin: RIA, biosynthesis	Proteins, RNA, DNA, electron microscopy	c, d
Hypothalamoneuro-hypophysial complex	Rat (150–250 g)	F12 + 10% or 20% FCS + glucose (1 mg/ml)	Up to 9 days	—	AVP: immunocyto-chemistry, RIA, release stimulated by acetylcholine and angiotensin	Electrical recording	e–h
Hypothalamoneuro-hypophysial complex	Guinea pig	MEM + 10% FCS + Ab	1–5 days	—	AVP: RIA, release stimulated by angiotensin and PGE$_2$	—	i, j
Hypothalamoneuro-hypophysial complex	Toad (Bufo marinus)	Amphibian culture medium (GIBCO) + Ab	1–3 days	—	Vasotocin: bioassay, release stimulated by osmotic shock	—	k
Median eminence	Guinea pig	F12 + FCS + Ab	Up to 13 days	—	TRH: bioassay, biosynthesis	Electron microscopy	l

Total hypothalamus	Rat (250 g)	F12 + 10% FCS + HEPES + glucose (6 mg/ml) + bacitracin	24 hr	—	TRH: RIA, release induced by NE, inhibited by SRIF [m] SRIF: RIA, release inhibited by acetylcholine and serotonin, stimulated by melatonin [n] LHRH: RIA, release Ca^{2+}-dependent, stimulated by acetylcholine and melatonin [o]

[a] DMEM, Dulbecco's modified Eagle's medium; FCS, fetal calf serum; NEAA, nonessential amino acids; Ab, antibiotics; MEM, minimal Eagle's medium; HEPES, N-2-hydroxyethylpiperazine-N'-2-ethanesulfonic acid.
[b] RIA, radioimmunoassay; AVP, arginine vasopressin; PGE_2, prostaglandin E_2; TRH, thyroliberin; NE, norepinephrine; SRIF, somatostatin; LHRH, luteinizing hormone releasing hormone.
[c] D. Pearson, A. Shainberg, S. Malamed, and H. Sachs, *Endocrinology* **96**, 982 (1975).
[d] D. Pearson, A. Shainberg, S. Malamed, and H. Sachs, *Endocrinology* **96**, 994 (1975).
[e] C. D. Sladek and K. M. Knigge, *Endocrinology* **101**, 411 (1977).
[f] C. D. Sladek and R. J. Joynt, *Endocrinology* **104**, 148 (1979).
[g] C. D. Sladek and R. J. Joynt, *Endocrinology* **104**, 659 (1979).
[h] W. E. Armstrong and C. D. Sladek, *Neuroendocrinology* **34**, 405 (1982).
[i] S. E. Ishikawa, T. Saito, and S. Yoshida, *Endocrinology* **106**, 1571 (1980).
[j] S. E. Ishikawa, T. Saito, and S. Yoshida, *Endocrinology* **108**, 193 (1981).
[k] P. Eggena and A. X. Polson, *Endocrinology* **94**, 35 (1974).
[l] J. F. McKelvy, M. Sheridan, S. Joseph, C. H. Phelps, and S. Perrie, *Endocrinology* **97**, 908 (1975).
[m] Y. Hirooka, C. S. Hollander, S. Suzuki, P. Ferdinand, and S. I. Juan, *Proc. Natl. Acad. Sci. U.S.A.* **75**, 4509 (1978).
[n] S. B. Richardson, C. S. Hollander, R. D'Eletto, P. W. Greenleaf, and C. Thaw, *Endocrinology* **107**, 122 (1980).
[o] S. B. Richardson, C. S. Hollander, J. A. Prasad, and Y. Hirooka, *Endocrinology* **109**, 602 (1981).

TABLE II. LONG-TERM EXPLANT

Hypothalamus	Species and age	Culture technique	Culture medium
Divided into 4 coronal sections	Mouse, 15-day fetus and newborn	Collagen-coated coverslip in Maximow chamber	MEM + 30% chick embryo extract + insulin + glucose (6 mg/ml) + glutamine
Mamillary region, 3 coronal sections	Mouse, newborn	Collagen-coated coverslip in Maximow chamber or roller tubes	MEM + 40% horse or human serum + glutamine + HEPES
Tuberal	Rat, newborn	Coverslip	
Divided into 6 coronal sections	Mouse, sexed newborn	Collagen-coated coverslip in Maximow chamber	MEM + 24% horse serum, HEPES + NEAA + glucose (9 mg/ml)
SO area fragments (<0.5 mm³), coculture with neurohypophysis	Rat, 1–7 p.n.d.[p]	Plasma clot on glass coverslip in Falcon roller tubes	MEM + 25% Hanks' + 25% horse serum + glucose (6 mg/ml)
Punched PVN, SON, basal hypothalamus, coculture with anterior pituitary	Rat, 6–7 p.n.d.	Plasma clot on coverslip in Falcon roller tube	MEM + 25% Hanks' + 25% horse serum + glucose (6 mg/ml)
6 coronal sections, median eminence, arcuate nucleus	Rat fetus, 21 day	Collagen-coated coverslip in Maximow chamber	MEM + 33% human placental serum + glucose (6 mg/ml)

CULTURES—SERUM SUPPLEMENTED MEDIUM

Time in culture	Functional markers			References
	Neurotransmitter	Neuropeptide	Other markers	
Up to 12 weeks	—	—	Electron microscopy, synapses	a
Up to 7 weeks	—	—	Cresyl violet or silver stain, architectonics	b
3 weeks	—	—	Electrical activity, effects of neurotransmitter	c, d
Up to 11 weeks	—	LHRH: immunocytochemistry, $\oplus$ neurons,[q] effect of E_2 on neuron number	Neurite outgrowth stimulated by E_2[q] autoradiography of E_2 binding	e–g
Up to 11 weeks	—		Electrical activity of magnocellular neurons Electron microscopy, neurosecretory granules	h
Up to 4 weeks	—	AVP[r]: RIA (tissue and medium), AVP in PVN only; effect on ACTH secretion by anterior pituitary; effect of anterior pituitary coculture on AVP content	—	i
Up to 8 weeks	Catecholamine: fluorescence of fibers excepted in PO area	LHRH: immunocytochemistry, appearance in culture of $\oplus$ neurons in arcuate nucleus	—	j, k

(continued)

TABLE II

Hypothalamus	Species and age	Culture technique	Culture medium
6 coronal sections, mostly SON, PVN	Guinea pig fetus, 40 day	Hanging drop in Maximow chamber	MEM + 30% human placental serum + 30% Hanks' + glucose 6 mg/ml
Anterior basal diencephalon, posterior basal diencephalon, Rathke pouch	Rat fetus, 12.5 day	Collagen-coated cellulose acetate membrane in center well of Falcon dish	Medium 199 + 10% horse serum + glucose (6 mg/ml) + Ab
Ventral diencephalon	Rat fetus, 14–17 day	Collagen-coated glass slide immersed in liquid medium	DMEM + 20% f.c.s. + glucose (6 mg/ml) + Ab
Mediobasal hypothalamus and arcuate nucleus, minced in 0.5–1 mm³ fragments	Bovine ♂ or ♀ adult	Polylysine-coated coverslip, plasma clot in Linbro mutiwell trays	Ham F10 + 15% horse serum + 2.5% FCS + a.b.

[a] E. B. Masurovsky, H. H. Benitez, and M. R. Murray, *J. Comp. Neurol.* **143,** 263 (1971).
[b] H. M. Sobkowicz, R. Bleier, and R. Monzain, *J. Comp. Neurol.* **155,** 355 (1974).
[c] H. M. Geller, *Brain Res.* **93,** 511 (1975).
[d] H. M. Geller, *Brain Res.* **105,** 423 (1976).
[e] C. D. Toran-Allerand, *Brain Res.* **106,** 407 (1976).
[f] C. D. Toran-Allerand, *Brain Res.* **149,** 257 (1978).
[g] C. D. Toran-Allerand, J. L. Gerlach, and B. S. McEwen, *Brain Res.* **184,** 517 (1980).
[h] B. H. Gähwiler, P. Sandoz, and J. J. Dreifuss, *Brain Res.* **151,** 245 (1978).
[i] A. J. Baertschi, J. L. Beny, and B. Gähwiler, *Nature (London)* **295,** 145 (1982).

(*continued*)

Time in culture	Functional markers			References
	Neuro-transmitter	Neuro-peptide	Other markers	
Up to 10 weeks	—	—	Electron microscopy, synapses after 10 days, neurosecretory granules after 15–20 days, magnocellular neurons	*l*
Up to 10 days	—	—	Growth of Rathke-pouch stimulated	*m*
Up to 4 weeks	—	β-Endorphin: immunocytochemistry ⊕ neurons at 3 weeks	—	*n*
Up to 6 weeks	Catecholamine: fluorescence, HPLC, disappearance with time in culture	LHRH, ACTH: immunocytochemistry, RIA, bioassay LHRH: numerous ⊕ neurons, high tissue content, no release ACTH : ⊖	Tetanus toxin binding to neurons	*o*

[j] S. C. Feldman, A. B. Johnson, M. B. Bornstein, and G. T. Campbell, *Neuroendocrinology* **28**, 131 (1979).
[k] S. C. Feldman, L. L. Brown, and M. B. Bornstein, *Cell. Mol. Neurobiol.* **1**, 279 (1981).
[l] A. M. Marson, and A. Privat, *Cell Tissue Res.* **203**, 393 (1979).
[m] S. Daikoku, T. Chikamori, T. Adachi, and Y. Maki, *Dev. Biol.* **90**, 198 (1982).
[n] L. W. Haynes, D. G. Smith, and S. Zakarian, *Brain Res.* **232**, 115 (1982).
[o] D. M. Nicholson and W. T. Mason, *Brain Res.* **249**, 123 (1982).
[p] p.n.d., postnatal days.
[q] E_2, 17β-estradiol; ⊕, positive.
[r] AVP, arginine vasopressin.

TABLE III. LONG-TERM DISSOCIATED CELL

Hypothalamus	Species, age	Dissociation method	Substratum	Culture medium
Total	Rat, 1–5 p.n.d. 4–5 weeks	Trypsin	Collagen-coated coverslip immersed in petri dishes	MEM + 5 to 25% FCS + 5–10% chick embryo extract + glutamine + glucose (6 mg/ml), 5% CO_2 atm
Total	Mouse fetus, 14–20 days	Mechanical dissociation	Plastic (Falcon)	DMEM or Ham F12 + 5 to 20% FCS or + 15% horse serum + glucose (6 mg/ml) + Ab 8% CO_2 atm
Total	Mouse fetus, 13–16 days	Mechanical dissociation	Plastic (Lux)	Ham F10 + 10% FCS + glutamine + glucose (6 mg/ml) + Ab 7% CO_2 atm
Total	Rat, 10–12 p.n.d.	Trypsin	Plastic (Falcon)	L15 + 5% FCS + glutamine + glucose (6 mg/ml) + BSA + vitamin cofactors, amino acids + Ab, Air atm
Total	Rat fetus, 18 days	Papain, DNase	Glass or plastic	MEM + 10% FCS for 3 days, then 10% heat-inactivated horse serum + glucose (6 mg/ml) ± Ara-C (2×10^{-6} M), 5% CO_2 atm

CULTURES SERUM SUPPLEMENTED MEDIUM

Time in culture	Functional markers			References
	Neurotransmitter	Neuropeptide	Others	
Up to 5 weeks	—	—	Silver stain	a
Up to 8 weeks	—	—	Electron microscopy, magnocellular neurons, neurosecretion granules, synapses	b
Up to 2 weeks	—	TRH: immunocytochemistry, RIA; ⊕ neurons (28% of total) tissue content	Electron microscopy, immunocytochemistry, ⊕ soma, ⊕ axons, ⊕ terminals	c
Up to 12 weeks	Catecholamine: fluorescence large ⊕ neurons	LHRH: immunocytochemistry ⊕ neurons (25% of total)		d
Up to 2 weeks	—	—	Androgen aromatase activity	e

(*continued*)

TABLE III

Hypothalamus	Species, age	Dissociation method	Substratum	Culture medium
Septal chiasmatic area, mediobasal area	Rat fetus, 16.5 days, 18.5 days	Trypsin + collagenase	Plastic Falcon dishes	Medium 199 + 20% FCS + Ab, 5% CO_2 atm
Total	Rat, 15 p.n.d.	Trypsin	Polylysine-coated dishes (Falcon)	MEM + 10% FCS
Total	Rat fetus, 18–19 days	Hyaluronidase, collagenase, viokase	Plastic (Corning)	DMEM + 10% FCS charcoal treated + glutamine + 1% NEAA 10% CO_2 atm
Septal chiasmatic area, retrochiasmatic area	Rat fetus, 18 days	Trypsin	Plastic (Falcon)	Ham's F10 + 15% FCS + glucose (6 mg/ml) + glutamine + Ab 5% CO_2 atm
Total	Rat fetus, 18 day	Trypsin	Plastic (Falcon)	Same as above
Anterior	Rat adult castrated females	Collagenase, DNase	Glass coverslip	DMEM/Ham's F12 + 10% FCS, 5% CO_2 atm

(*continued*)

Time in culture	Functional markers			References
	Neurotransmitter	Neuropeptide	Others	
Up to 4 weeks	—	SRIF: RIA, S.14, Sephadex, HPLC, tissue content increases in culture, release Ca^{2+} dependent, inhibited by GABA	—	*f*
Up to 4 weeks	Uptake of GABA	Substance P, neurotensin: RIA	Electrical activity	*g*
Up to 12 days	—	LHRH: immunocytochemistry, appearance in culture of LHRH neurons	—	*h*
Up to 4 days	—	Opiates (β-LPH, β-endorphin ACTH): gel filtration, immunological evidence for biosynthesis	—	*i*
Up to 7 days	—	Neurophysin: immunocytochemistry, RIA of tissue content Other peptides (MSH, ACTH, substance P, Leu enkephalin, oxytocin, LHRH, vasopressin, somatostatin): immunocytochemistry, negative; RIA, low levels at 4 days	Scanning electron microscopy	*j*
Up to 21 days	—	LHRH, MSH, AVP, neurophysin: immunocytochemistry, 1–2% of ⊕ neurons	—	*k*
Up to 29 days	—	VIP: immunocytochemistry, 0.01–0.03% of ⊕ neurons	—	*l*
Up to 8 weeks	—	Neurophysin, AVP; immunocytochemistry	—	*m*

(*continued*)

TABLE III

Hypothalamus	Species, age	Dissociation method	Substratum	Culture medium
Total	Rat fetus, 16–18 day	Trypsin, DNase	Polylysine-coated plastic dishes or preexisting monolayer	DMEM + 10% heat-inactivated FCS + FUdR (8×10^{-5} M) + NEAA + Ab
Total	Mouse fetus, 13–14 day	Mechanical	Plastic (Lux)	Ham's F12 + 10% FCS + glucose (1 mg/ml) + Ab (for the first 4 days), 5% CO_2 atm
Total	Rat fetus, 18 day	Collagenase, DNase	Polylysine-coated dishes (Falcon)	DMEM + 10% horse serum + 10% FCS + HEPES + Ab 15%, CO_2 atm
Total	Rat, 2–3 p.n.d.	Mechanical	Plastic (Falcon)	DMEM + 10% FCS (during the 1st 7 days) + 10% horse serum + glucose (6 mg/ml) + insulin + Ab 5% CO_2 atm
Total	Rat fetus, 17–19 days	Trypsin, collagenase	Reaggregated cultures (constant shaking)	MEM + 25% heat-inactivated FCS + insulin + glucose (6 mg/ml) 5% CO_2 atm

[a] M. Wilkinson, C. J. Gibson, B. H. Bressler, and D. M. Inman, *Brain Res.* **82**, 129 (1974).
[b] P. Benda, F. de Vitry, R. Picart, and A. Tixier-Vidal, *Exp. Brain Res.* **23**, 29 (1975).
[c] A. Faivre-Bauman, A. Nemeskeri, C. Tougard, and A. Tixier-Vidal, *Brain Res.* **185**, 289 (1980).
[d] K. M. Knigge, G. Hoffman, D. E. Scott, and J. R. Sladek Jr., *Brain Res.* **120**, 393 (1977).
[e] J. A. Canick, D. E. Vaccaro, K. J. Ryan, and S. E. Leeman, *Endocrinology* **100**, 250 (1977).
[f] R. Gamse, D. E. Vaccaro, G. Gamse, M. Di Pace, T. O. Fox, and S. E. Leeman, *Proc. Natl. Acad. Sci. U.S.A.* **77**, 5552 (1980).
[g] D. E. Vaccaro, A. Messer, M. A. Dichter, and S. E. Leeman, *J. Neurobiol.* **11**, 417 (1980).
[h] S. Daikoku, H. Kawano, and H. Matsumura, *Cell Tissue Res.* **194**, 433 (1978).
[i] A. S. Liotta, C. Loudes, J. F. McKelvy, and D. Krieger, *Proc. Natl. Acad. Sci. U.S.A.* **77**, 1880 (1980).
[j] F. Denizeau, D. Dube, T. Antakly, A. Lemay, A. Parent, G. Pelletier, and F. Labrie, *Neuroendocrinology* **32**, 96 (1981).

(*continued*)

Time in culture	Functional markers			References
	Neurotransmitter	Neuropeptide	Others	
Up to 2 weeks			Silver stain, thionine; neuron survival improved on feeder layer of homologous non-neuronal cells	n
Up to 8 weeks	—	−AVP, neurophysin II: immunocytochemistry, ⊕ neurons Oxytocin: immunocytochemistry, ⊖	Electrical activity, intracellular recording	o, p
Up to 2 weeks	—	SRIF (SS-14, SS-28): RIA, gel filtration, HPLC, culture medium content, release calcium dependent, effects of neurotransmitter	—	q, r
Up to 6 weeks	—	SRIF (SS-14, SS-28, 15 k): RIA, gel filtration, HPLC, cell content, release calcium dependent, gel filtration and immunological evidence for biosynthesis	—	s
Up to 4 weeks	—	—	Electron microscopy synaptogenesis, granular vesicles, release of PGE_2 (RIA)	t

---

[k] G. Jirikowski, I. Reisert, and C. Pilgrim, *Neuroscience* **6**, 1953 (1981).
[l] G. Jirikowski, I. Reisert, and C. Pilgrim, *Neurosci. Lett.* **31**, 75 (1982).
[m] H. S. U., G. F. Erickson, and W. B. Watkins, *Endocrinology* **108**, 1810 (1981).
[n] S. A. Whatley, C. Hall, and L. Lim, *J. Neurochem.* **36**, 2052 (1981).
[o] P. Legendre, D. Theodosis, I. Cooke, and J. D. Vincent, *in* "Neuroendocrinology of Vasopressin, Corticoliberin, and Opiomelanocortins" (A. J. Baertschi and J. J. Dreifuss, eds.), p. 137. Academic Press, New York, 1982.
[p] P. Legendre, I. M. Cooke, and J. D. Vincent, *J. Neurophysiol.* **48**, 1121 (1982).
[q] R. A. Peterfreund and W. Vale, *Brain Res.* **239**, 463 (1982).
[r] R. A. Peterfreund and W. Vale, *Endocrinology* **112**, 526 (1983).
[s] H. H. Zingg and Y. C. Patel, *J. Clin. Invest.* **70**, 1101 (1982).
[t] A. Gyevai, P. J. Chapple, and W. H. J. Douglas, *J. Cell Sci.* **34**, 159 (1978).

the distribution of estradiol binding sites.[7] This also permitted elegant studies on interactions between hypothalamic areas and anterior pituitary during development.[8,9]

One of the disadvantages of these culture systems lies in the small quantity of tissue and medium. This barely permits biochemical analysis, which indeed has not been reported to date in explant cultures of fetal or newborn hypothalamus. A recent report, however, suggests that cultures of adult bovine hypothalamic fragments may obviate that difficulty.[5] However their validity for biochemical studies remains to be proved. Finally such a system does not fundamentally differ from dissociated cell cultures: indeed after 3–4 weeks *in vitro* the explants disappear and a new cell organization develops.

Primary Cell Cultures in Serum-Supplemented Medium

Cells are mostly taken from fetal hypothalamus of the rat or mouse. They are enzymatically dissociated, using a large variety of enzymes either alone or in combination (Table III). However, we found it easy to use mechanical dissociation for cells taken from young fetuses (16-day fetuses or less). The cells are seeded directly on plastic dishes, sometimes coated with D-polylysine. The medium composition greatly varies with regard to the basal nutrient medium (Ham's F10 or F12, MEM, DMEM, 199, L15) as well as the proportion of serum. Fetal calf serum is always used (10–25%), sometimes heat-inactivated. Horse serum is sometimes added to fetal calf serum (10%). Glucose is often added (6 mg/ml) and sometimes glutamine and nonessential amino acids (NEAA), depending on the basal medium used. The CO_2 pressure in the atmosphere of the incubator varies from 5 to 15%.

As expected, the cells can be grown in culture for weeks or even months. Their behavior in culture is rather uniform, leading to an histiotypic pattern with a basal, multilayered, carpet of glial cells overlaid by neuron-like cells. Electron microscope observations confirmed this organization and, in addition, showed that synaptogenesis occurs with time in culture (Table III). Neurons do not divide in culture. The active proliferation of nonneuronal cells is sometimes prevented by addition of cytosine arabinoside (Ara C) or fluorodeoxyuridine (FUdR).

Several specific neuron products have been identified using various approaches: immunocytochemistry, radioimmunoassay (RIA), chromatographic characterization. Most of the known hypothalamic peptides have

[7] D. Toran Allerand, J. L. Gerlach, and B. S. McEwen, *Brain. Res.* **184,** 517 (1980).
[8] A. J. Baertschi, J. L. Beny, and B. Gähwiler, *Nature (London)* **295,** 145 (1982).
[9] S. Daikoku, M. Chikamori, T. Adachi, and Y. Maki, *Dev. Biol.* **90,** 198 (1982).

been detected in culture, however differences are found depending on the authors. This may be related to the culture technique or the peptide identification procedure. The percentage of specific neuron product, when determined, greatly varies with the authors. From a functional point of view, such cultures have been used to study opiate biosynthesis in short-term cultures (4 days),[10] somatostatin spontaneous and acute release in 10- to 12-day cultures,[11,12] and intracellular electrical recording after a long time in culture (3-8 weeks).[13]

Surprisingly, very few data on neurotransmitter neurons are available except for catecholamine fluorescence.[14]

Compared to explant or tissue cultures, dissociated cell cultures present several advantages. They are more appropriate to quantitative studies. The number of cells plated in culture is measurable; the exact number of neurons is measured with time in culture. They can be counted following immunostaining either with a general neuronal marker (tetanus toxin binding) or with a specific marker. The accessibility of the cells to reagents is more rapid than in explants. They therefore offer better conditions for functional or biochemical studies. From a developmental point of view, spontaneous cell reassociation *in vitro* can be followed. This is particularly favored in reaggregated cell cultures where hypothalamic neurons are functionally mature.[15] However, if one is interested in the factors involved *in vivo* in the establishment of neuronal connections, explant cultures of anatomically defined hypothalamic areas remain certainly the most suitable system.

Serum-Free Cultures of Dissociated Hypothalamic Cells

We have established the conditions for culturing fetal mouse hypothalamic cells in a serum-free medium. Hypothalamic cells taken from 16-day mouse fetuses are introduced into the culture. Cells are mechanically dissociated as described for serum-supplemented cultures. One major problem resulting from the absence of serum is related to cell attachment. By preincubating the petri dishes successively with gelatin and 10% fetal calf serum in PBS this has been resolved. After withdrawing the last

[10] A. S. Liotta, C. Loudes, J. F. McKelvy, and D. Krieger, *Proc. Natl. Acad. Sci. U.S.A.* **77**, 1880 (1980).
[11] R. Gamse, D. E. Vaccaro, G. Gamse, M. Di Pace, T. O. Fox, and S. E. Leeman, *Proc. Natl. Acad. Sci. U.S.A.* **77**, 5552 (1980).
[12] R. A. Peterfreund and W. Vale, *Endocrinology* **112**, 526 (1983).
[13] P. Legendre, I. M. Cooke, and J. D. Vincent, *J. Neurophysiol.* **48**, 1121 (1982).
[14] K. M. Knigge, G. Hoffman, D. E. Scott, and J. R. Sladek, Jr., *Brain Res.* **120**, 393 (1977).
[15] A. Gyevaï, P. J. Chapple, and W. H. J. Douglas, *J. Cell Sci.* **34**, 159 (1978).

TABLE IV
OPTIMUM COMPOSITION OF THE SERUM-FREE CHEMICALLY DEFINED MEDIUM

Components	Concentration	Effects	
Insulin	5 µg/ml		
Human transferrin	100 µg/ml		
Progesterone	2×10^{-8} M	Regular serum-free medium[a]	
Selenium	3×10^{-8} M		
Putrescine	10^{-4} M		
17β-Estradiol	10^{-12} M		
Arachidonic acid	1 µg/ml	Restore normal fatty acid composition of membranes[b]	Increase TRH release[d]
Docosohexaenoic acid	0.5 µg/ml		
Triiodothyronine	10^{-8} M	Neurite elongation[c]	
Corticosterone	10^{-7} M		

[a] A. Faivre-Bauman, E. Rosenbaum, J. Puymirat, D. Grouselle, and A. Tixier-Vidal, *Dev. Neurosci.* **4**, 118 (1981).

[b] J. M. Bourre, A. Faivre-Bauman, O. Dumont, A. Nouvelot, C. Loudes, J. Puymirat, and A. Tixier-Vidal, *J. Neurochem.* (in press).

[c] J. Puymirat, A. Barret, R. Picart, A. Vigny, C. Loudes, A. Faivre-Bauman and A. Tixier-Vidal, *Neuroscience* (in press).

[d] C. Loudes, A. Faivre-Bauman, A. Barret, D. Grouselle, J. Puymirat, and A. Tixier-Vidal, *Dev. Brain Res.* (in press).

coating solution, serum-free medium is added to the culture dish before plating the cells. The serum-free medium has the following composition: the basal nutrient medium is a mixture of equal volumes of Ham's F12 and DMEM, and 15 mM HEPES and 0.5 mM glutamine. As described by Bottenstein and Sato,[16] for the growth of a neuroblastoma cell line, various supplements that make up the serum are added: insulin, progesterone, putrescine, human transferrin, and selenium. To this medium we also add 10^{-12} M 17β-estradiol, which greatly improves the quality of the culture (Table IV).

Under these conditions, it is possible to culture fetal hypothalamic neurons for up to 1 month. The pattern of cell outgrowth is similar to that previously described for cultures in serum-supplemented medium (see preceding section). However, the basal layer of glial cells is noticeably reduced in serum-free medium and there is no need to use Ara C or FUdR to stop the proliferation of nonneuronal cells.

The neuronal nature of the small overlaying cells has been confirmed by cytochemical visualization of tetanus toxin binding.[17] Using an antise-

[16] J. E. Bottenstein, and G. H. Sato, *Proc. Natl. Acad. Sci. U.S.A.* **76**, 514 (1979).

[17] J. Puymirat, C. Loudes, A. Faivre-Bauman, A. Tixier-Vidal, and J. M. Bourre, *Cold Spring Harbor Conf. Cell Proliferation* **9**, 1033 (1982).

rum to GFA, the presence of numerous astrocytes in the basal layer has been demonstrated. This made it possible to count the cells and estimate a ratio of neurons to astrocytes based on the time in culture and on the culture conditions.[18] Electron microscope observations of serum-free cultures after 8 days *in vitro* have revealed the presence of synapses that are smaller and fewer than in serum-supplemented cultures of the same age. Several neuropeptides have been identified by immunocytochemistry. With the fixative we used (4% paraformaldehyde in Sorensen's buffer), numerous TRH-positive neurons and a few α-Melanotropin (αMSH)-positive cells were found. The TRH content increases in culture as revealed by radioimmunoassay of cell and culture medium extracts. Dopaminergic neurons have been visualized by autoradiography after uptake of exogenous [^{3}H]dopamine or by immunocytochemical detection of tyrosine 3-monooxygenase.

The main advantage of culturing in a serum-free medium is the possibility of studying the effects of environmental factors, for example, the effect of hormones on the development of hypothalamic neurons. In our system, they can be tested against several parameters: neuron survival, neurite elongation, synaptogenesis, expression of neuron specific functions such as TRH and dopamine. In the case of triiodothyronine (T_3), we have shown that it affects neuron survival and neurite elongation, but it does not seem to play a role in synaptogenesis and does not modulate TRH activity. In contrast it specifically increases the size of perikarya as well as the arborization of neurites and the [^{3}H]dopamine uptake of dopaminergic neurons.[19] Addition of polyunsaturated fatty acids was found necessary to restore the normal fatty acid composition of the membranes of the cultured cells.[18] Serum-free cultures can also be used as a model to study TRH release. However, to obtain a substantial release capacity of TRH-containing neurons, we found it necessary to supplement the regular serum-free medium with T_3 and corticosterone in addition to fatty acids (Table IV).

Clonal Hypothalamic Cell Lines

Clonal cell lines of hypothalamic origin have been obtained by *in vitro* viral transformation.[20] Briefly, primary cultures of dissociated hypotha-

[18] A. Faivre-Bauman, J. Puymirat, C. Loudes, and A. Tixier-Vidal, *in* "Methods for Preparation of Serum-Free Culture Media and Methods for Growth of Cells in Serum-Free Culture" (D. Barnes, ed.). Liss, New York. In press.

[19] J. Puymirat, A. Barret, R. Picart, A. Vigny, C. Loudes, A. Faivre-Bauman, and A. Tixier-Vidal, *Neuroscience* (in press).

[20] F. de Vitry, M. Camier, P. Czernichow, P. Benda, P. Cohen, and A. Tixier-Vidal, *Proc. Natl. Acad. Sci. U.S.A.* **71**, 3575 (1974).

lamic cells taken from 14-day-old mouse fetuses were grown for 6 days in serum supplemented medium (Ham's F10, 15% heat-inactivated horse serum, 2.5% fetal calf serum, 200 mM L-glutamine, antibiotics). After washing with serum-free medium, the cell layer was covered with 1 ml of SV40 suspension (7×10^7 plaque-forming units/ml). After adsorption for 30 min at 37°, the cells were covered with regular culture medium. One month later foci of dividing cells appeared. These cells were collected and serially transferred (three times) for $2\frac{1}{2}$ months. After the third passage, 10 clones were isolated by single-cell plating in Falcon microwell trays. Selection of colonies was done according to morphological criteria. These clones are grown in the same serum-supplemented medium as primary cultures. The medium is half renewed at each medium change. The cells are transferred by scraping without any enzyme or chelating agent.

Functional characterization was first performed for one of these clones (C7), which displayed neuron-like features based on phase-contrast observation, Gomori fuchsin-positive staining, and ultrastructural features suggestive of an intense secretory activity. These cells were found to synthesize ^{35}S-labeled components that possess biochemical and immunological properties of neurophysin and vasopressin, respectively.[20] However, the C7 cells cannot be considered as fully differentiated neurons. Indeed, they do not form axons and axon terminals. Several attempts, using high potassium concentrations, cooling, and neurotransmitters, failed to induce the release of radioimmunoassayable vasopressin into the medium.[21] Intracellular recording revealed that they do not display the phasic discharge pattern of fully differentiated magnocellular neurons.[22] As a whole these cells appear as immature neurosecretory cells that possess the capacity to synthesize a neuropeptide, but do not resume the complete pattern of neuronal differentiation.

However, they offer an interesting system for study of precursor cells of the hypothalamus. Indeed, one of the subclones that was isolated from the C7 after several passages does not display secretory features and, in contrast to C7, contains 5–6-nm fibrils[23] and possesses GFA immunoreactivity (F. de Vitry, personal communication). After *in vivo* passage it was able to regain some of the differentiated properties of the neurophysin-containing neuronal clone.[24]

Moreover, while grown *in vitro*, in regular medium, the primitive F7 cells have been found to synthesize materials possessing the immunoreac-

[21] A. Tixier-Vidal and F. de Vitry, *Int. Rev. Cytol.* **58**, 291 (1979).
[22] B. Bioulac, B. Dufy, F. de Vitry, H. Fleury, A. Tixier-Vidal, and J. D. Vincent, *Neurosci. Lett.* **4**, 257 (1977).
[23] A. Tixier-Vidal and F. de Vitry, *Cell Tissue Res.* **171**, 39 (1976).
[24] F. de Vitry, *Nature (London)* **267**, 48 (1977).

tivity of somatostatin[25] and of Met-enkephalin (up to 62 pg/mg protein).[26] This is of interest since such a coexistence in a same cell of somatostatin and met-enkephalin immunoreactivities has been also observed in adult hypothalamic neurons *in vivo*. However, in F7 cells this is associated with a very primitive organization of the cytoplasm which by no mean possesses any neuronal feature. This indicates that the expression of neuropeptide genes is independent of the expression of the neuronal morphological phenotype, a situation that has been reported *in vivo*.

Other cell lines that do not originate from the hypothalamus have been shown also to synthesize several neuropeptides. They have been isolated from tumors according to procedures previously reviewed.[27] The presence of immunoreactive TRH has been detected in homogenates of two clonal cell lines (BN 1010-1, BN 1010-3) derived from an ethylnitrosourea-induced central nervous system tumor.[28] The intracellular content is relatively low (up to 208.6 pg per milligram of alkali-soluble proteins) and it was not possible to detect TRH in the medium. These cells do not possess any of the ultrastructural features of secretory neurons. Like the SV40-transformed F7 clone, they contain numerous filaments (5–8 nm).[21] The presence of immunoreactive vasoactive intestinal peptide (VIP) in rather high concentrations has been reported in the glial cell line C6 derived from an ethylnitrosourea-induced central nervous system tumor[29] as well as in three neuroblastoma cell lines derived from the C1300 mouse neuroblastoma. The intracellular content reaches 3.6 ng per milligram of protein in the richest neuroblastoma line (C46), but only a very low amount could be detected in the medium.[30] Enkephalin-like peptides have been detected in neuroblastoma × glioma hybrid cells.[31] Neurotensin is contained in, and released into the medium by, the PC 12 rat pheochromocytoma cell line.[32]

Several neuropeptides are produced in the culture medium by cell lines that have been isolated from human small-cell carcinoma of the lung (SCCL)[33,34]; these include vasopressin, oxytocin, somatostatin, parathy-

[25] F. de Vitry, M. Dubois, and A. Tixier-Vidal, *J. Physiol. (Paris)* **75**, 11 (1979).
[26] F. Cesselin, M. Hamon, S. Bourgoin, N. Buisson, and F. de Vitry, *Neuropeptides* **2**, 351 (1982).
[27] D. Schubert and W. Carlisle, this series, Vol. 58, p. 584.
[28] Y. Grimm-Jorgensen, S. E. Pfeiffer, and J. F. McKelvy, *Biochem. Biophys. Res. Commun.* **70**, 167 (1976).
[29] P. Benda, J. Lightbody, G. Sato, L. Levine, and W. Sweet, *Science* **161**, 370 (1968).
[30] S. I. Said and R. N. Rosenberg, *Science* **192**, 907 (1976).
[31] T. Glaser, K. Hübner, and B. Hamprecht, FEBS *Lett.* **131**, 63 (1981).
[32] A. S. Tischler, Y. C. Lee, V. W. Slayton, and S. R. Bloom, *Regul. Peptides* **3**, 415 (1982).
[33] O. S. Penttengill, L. H. Maurer, and C. S. Faulkner, *Proc. Am. Assoc. Cancer Res.* **18**, 121 (1977).
[34] G. D. Sorenson, O. S. Pettengill, T. Brinck-Johnsen, C. C. Cate, and L. H. Maurer, *Cancer* **47**, 1289 (1981).

roid hormone, glucagon, ACTH, and calcitonin. One of these lines (DMS 79) produces immunoreactive materials with either large molecular weights or the same molecular weight as ACTH, β-LPH and β-endorphin.[35] It also produces large molecular weight materials with human calcitonin immunoreactivity.[36] Although cloning is difficult, it seems that each of these cell lines can produce several hormones at the same time. The fact that they secrete their products into the medium can be related to the presence in all of them of secretory granules or dense-core vesicles[37] and to their ability to generate calcium spikes.[38] The presence of neuropeptides in SCCL-derived cell lines may be explained by the hypothesis that they belong to the APUD series and are derived from the neural crest.[37]

Clonal cell lines would represent an ideal way to obtain a homogeneous population of neuropeptide-synthesizing nerve cells. However, the presently available strains, either SV40 transformed or tumor derived, seem to be arrested at different steps of the expression of a complete differentiation program of secretory neurons. In that respect, they offer interesting model systems.

Acknowledgments

We acknowledge the skillful assistance of A. Barret in the culture procedures and of Miss A. Bayon and Mrs. C. Scalbert in preparing the manuscript.

[35] X. Y. Bertagna, W. E. Nicholson, G. D. Sorenson, O. S. Pettengill, C. D. Mount and D. N. Orth, *Proc. Natl. Acad. Sci. U.S.A.* **75**, 5160 (1978).
[36] X. Y. Bertagna, W. E. Nicholson, O. S. Pettengill, G. D. Sorenson, C. D. Mount, and D. N. Orth, *J. Clin. Endocrinol. Metab.* **47**, 1390 (1978).
[37] G. D. Sorenson and O. S. Pettengill, *Biol. Cell.* **39**, 277 (1980).
[38] F. V. McCann, O. S. Pettengill, J. J. Cole, J. A. G. Russell, and G. D. Sorenson, *Science* **212**, 1155 (1981).

[20] Techniques in the Tissue Culture of Rat Sympathetic Neurons

By MARY I. JOHNSON and VINCENT ARGIRO

The tissue culture of sympathetic neurons has been advantageous in the study of a variety of questions concerning neuronal differentiation. The cells have been placed in culture either in an organotypic form as small chunks (explants) of a ganglion or as dissociated cells, depending upon which best suited a particular investigation. The age dependency of neurite outgrowth was thus most easily analyzed in explant cultures.[1] Dissociated neurons were used for intracellular recording of electrical activity,[2,3] biochemical and immunocytochemical assays of neurotransmitter enzyme activity,[4,5] and cinegraphic and electron micrographic analysis of growth cone motility and structure.[6,7] Coculture of the sympathetic neuron with other neuronal or nonneuronal tissue has allowed study of the specificity and the morphology of synapse formation between the spinal cord and sympathetic neurons[8,9] as well as the influence of ganglionic nonneuronal cells on neurotransmitter function.[10]

Tissue culture also permits the study of other environmental influences, such as soluble medium components and substratum composition. For example, studies have shown that corticosteroids as well as elevated K^+ concentrations can alter the development of neurotransmitter function in the sympathetic neuron when grown in serum containing media.[11,12] More recently, a chemically defined medium has been found to effect neurotransmitter choice as well as the appearance of electrical coupling in cultures of sympathetic neurons.[13,14] Substratum effects have been shown to be important not only in neuronal attachment and survival, but in neuronal growth and neurotransmitter development.[15]

Although the above discussion summarizes but a few studies that have utilized the sympathetic neuron in culture and serves to point out the potential advantages, it should also make clear the potential problems. An investigator undertaking a study of this neuron in culture must be aware of a number of factors that might influence his results.

This chapter describes in some detail the techniques for the tissue culture of rat sympathetic neurons. The superior cervical ganglion (SCG),

---

[1] V. Argiro and M. I. Johnson, *J. Neurosci.* **2**, 503 (1982).
[2] C.-P. Ko, H. Burton, M. I. Johnson, and R. P. Bunge, *Brain Res.* **117**, 461 (1976).
[3] D. Higgins, L. Iacovitti, T. H. Joh and H. Burton, *J. Neurosci.* **1**, 126 (1981).
[4] M. Johnson, D. Ross, M. Meyers, R. Rees, R. Bunge, E. Wakshull, and H. Burton, *Nature (London)* **262**, 308 (1976).
[5] L. Iacovitti, T. H. Joh, D. H. Park, and R. P. Bunge, *J. Neurosci.* **1**, 685 (1981).
[6] V. Argiro, M. B. Bunge, and M. Johnson, *J. Cell Biol.* **91**, 92a (1981).
[7] M. B. Bunge, *J. Cell Biol.* **56**, 713 (1973).
[8] M. I. Olson and R. P. Bunge, *Brain Res.* **59**, 19 (1973).
[9] R. P. Rees, M. B. Bunge, and R. P. Bunge, *J. Cell Biol.* **68**, 240 (1976).
[10] P. H. Patterson and L. L. Y. Chun, *Proc. Natl. Acad. Sci. U.S.A.* **71**, 3607 (1974).
[11] L. S. McLennan, C. E. Hill, and I. A. Hendry, *Nature (London)* **283**, 206 (1980).
[12] P. A. Walicke, R. B. Campenot and P. H. Patterson, *Proc. Natl. Acad. Sci. U.S.A.* **74**, 5767 (1977).
[13] D. Higgins and H. Burton, *Neuroscience* **7**, 2241 (1982).
[14] L. Iacovitti, M. I. Johnson, T. H. Joh, and R. P. Bunge, *Neuroscience* **7**, 2225 (1982).
[15] E. Hawrot, *Dev. Biol.* **74**, 136 (1980).

rather than the sympathetic chain ganglia, has been used as the neuronal source. The SCG has been well studied *in vivo* and, because of its size and accessibility, it is easily dissected.

Preparation of Explants

Embryonic Rats

Explants of the rat SCG can be prepared from as early as embryonic day 14 or 15 (E14 or E15) through full adulthood (200–400 g or 2–6 months postnatal). Pregnant female rats with known sperm-positive dates (sperm-positive day equals embryonic day zero) are etherized, and the uterine horns are removed (under sterile conditions) into a petri dish. The embryos are dissected free of the uterus and fetal membranes and stored in Hanks' Ca^{2+}, Mg^{2+}-free balanced salt solution or Leibovitz's L-15 medium [HBSS, No. 310-4170; L-15, No. 320-1415, Grand Island Biological Co. (GIBCO), Grand Island, NY]. The gestational age can be confirmed using the charts of Long and Burlingame.[16]

Cutting into the thorax and through the heart at this time will assure exsanguination and reduce troublesome bleeding during the dissection. Embryos are pinned supine into a petri dish half filled with wax using 25-gauge needles. The head is gently hyperextended to provide better exposure. Skin, glands, and muscle are removed, exposing the carotid bifurcation behind which lies the SCG. Care should be taken not to confuse the SCG with the more lateral nodose ganglion. After birth the nodose ganglion becomes progressively separated from the SCG, making such confusion less likely. Ganglia are collected in L-15 medium. A well defined capsule is present on the SCG from E19 through E21, and its removal is facilitated by using extra fine forceps (Biologie No. 5, Dumont et Fils, Switzerland). For SCG from E15 through E17 only thin wisps of probable capsule can be distinguished. Removing the capsule is desirable if non-neuronal cell-free cultures are needed.

The nerve trunks are removed from the decapsulated ganglia, which are then cut into approximately 1-mm chunks using two No. 11 blades (Sterisharps, Seamless Hospital Products Co., Wallingford, CT) crossed in a scissoring motion. Ganglia from E15 can be cut in half or more often are placed in culture intact. The explants are rinsed in L-15 to remove tissue debris and transferred in a small volume to the appropriate culture dish with a fire-polished pipette. To promote explant attachment, the culture dishes should initially contain a small volume of medium (3–4

[16] J. A. Long and P. L. Burlingame *Univ. Calif. Berkeley Publ. Zool.* **43**, 143 (1938).

drops in a dish 2.5-cm in diameter; see below). After 2–3 hr or preferably 12–18 hr, feed to cover the explants should be added gently to the side of the dish.

Postnatal Rats

The dissection of the SCG from postnatal rats differs only in some details from that described above for the embryonic ganglion. For all ages ether anesthesia can be used and the dissection done after appropriate sterile preparation of the neck. Three sets of sterile instruments for the dissection of the skin, glands, muscles, and SCG helps in the sterile removal of the ganglion. A faster method avoiding the possible membrane effects of ether anesthesia uses decapitation. For the younger postnatal animals appropriately sized sterilized scissors can be used, and a guillotine (No. 51330, Stoelting Co., Chicago, IL) is used for the older rats. The cut edge of the head is rinsed with a stream of sterile saline solution, and the carotid artery along with the vagus nerve is located between muscle layers. The SCG is found medial and posterior to the carotid bifurcation, again taking care to avoid the nodose ganglion as in the younger postnatal rats.

The ganglia are collected in L-15 containing penicillin (100 units/ml) and streptomycin (100 mg/ml), and any loose connective tissue and blood vessels are removed. In a manner similar to that used for the late embryonic rats, the ganglionic capsule can be removed from postnatal rats up to 5–7 days. By approximately 10 postnatal days it is virtually impossible to remove the true capsule without considerable damage to the ganglion. The ganglia are then freed of nerve trunks and cut into explants with a maximum diameter of 1 mm. The explants are rinsed several times in the antibiotic containing L-15 before transfer to culture dishes. Inclusion of antibiotics in the medium for several days may help prevent infection.

Preparation of Dissociated Neurons

Preparation of dissociated neurons from rats of all ages begins with the preparation of explants as described above. Dissociation procedures then differ according to the age of rat under study.

Embryonic Neurons

Rinsed explants can be treated by two methods to achieve dissociation. Mechanical dissociation as described by Bray[17] is preferred when

[17] D. Bray, *Proc. Natl. Acad. Sci. U.S.A.* **65**, 905 (1970).

exposure of the neurons to enzymes is for some reason undesirable. The SCG explants are gently pulled apart using two fine forceps, transferred to a test tube, and agitated with a Vortex mixer. This generates a cell suspension with some remaining fragments, which is then filtered as described below. This procedure has the disadvantage of a lower yield (10–15%) compared to enzyme dissociation (as high as 80%). Mechanical dissociation also gives less satisfactory results for the younger embryonic SCG (E15 to E17), as these neurons seem more adherent to one another and without enzyme treatment remain in clumps.

Enzyme dissociation utilizes 0.25% trypsin (TRL-3, No. 3707, Worthington Diagnostics, Freehold, NJ) in L-15 or HBSS. The explants are incubated with gentle rotation in the trypsin solution at 35° for 30–45 min. The tissue is then rinsed 3 or 4 times with L-15 and finally in the medium to be used for plating the cells. Using approximately a 1-ml volume, the chunks are triturated against the side of the tissue culture tube with a pipette fire-polished to reduce the bore at the tip to 0.5–1 mm. After 4 or 5 squirts remaining fragments are allowed to settle, the cell suspension is removed, and more medium is added to the chunks. Trituration is completed, and the two suspensions with remaining fragments are combined. This two-step procedure reduces trauma to the cells released in the first trituration.

The cell suspension, whether generated by mechanical or enzyme dissociation, may be filtered through a nylon mesh (pore size 15 μm, Nitex HD3-15, TETKO, Inc., Des Plaines, IL) to remove cell aggregates and other tissue fragments. The filtered suspension is then plated onto the prepared culture dishes. To promote uniform plating the suspension should be agitated frequently and only 4 or 5 dishes seeded between agitations. For optimum attachment and growth, the final layer of collagen should be applied to the dishes and air-dried just shortly before plating the cells. Under these conditions the cell suspension can be applied as several drops to the center of the dish. The neurons will be confined by surface tension to the area of the initial drop, facilitating the counting of all cells on the dish at a later date.

Postnatal Rat SCG Neurons

The dissociation procedure using trypsin for embryonic rats can be used for the SCG from young rats up to 3–4 days postnatal with similar results. Dissociated neurons can still be obtained from rats up to 10–12 days of life, but with reduced yields. With trypsin alone it is virtually impossible to obtain single neurons from rats over 2–3 weeks of age. Initial attempts to obtain cells from older rats involved the use of a series of enzymes including trypsin, Pronase, collagenase, and hyaluronidase.[18] The best dissociations and subsequent survival (5–10%) were obtained using the following procedure.

Ganglia are removed as described above for explants, stripped of connective tissue, and divided into 10–12 pieces. Incubation in 0.25% collagenase (No. 4194 CLS, Worthington Diagnostics, Freehold, NJ) is carried out with gentle agitation in a humidified atmosphere (35°, pH 7.4). After 45–60 min the explants, which are now quite sticky and clumped, are teased apart and each is gently stretched out using fine forceps. Removing the old collagenase, fresh collagenase is added and the chunks are incubated for another 45–60 min. If the tissue is incubated for 90–120 min without being pulled apart the yield is considerably reduced. The chunks are rinsed briefly in L-15 before incubation (45–60 min) in 0.25% Pronase (No. 53702 B Grade, Calbiochem-Behring Corp, La Jolla, CA) in L-15 (pH 7.4). Enzyme solutions should be prepared and filter-sterilized shortly before use. Empirically it has been found that not all lot numbers of either enzyme give the same results. For Pronase in particular some lots give few if any viable neurons. Therefore check several lots and order a supply of a given lot that gives satisfactory results.

After incubation in Pronase the chunks are transferred to a test tube and rinsed four times (3 ml each) in L-15 and once in standard medium. Trituration is carried out as for embryonic neurons. A two- or even three-step procedure with subsequent combining of supernatants is particularly important in avoiding unnecessary trauma to the neurons released during the initial 3 or 4 squirts through the pipette tip. Usually very few tissue fragments remain, but they can be allowed to settle and the supernatant is removed. Plating is done as described above for embryonic neurons. Because no inhibitors of collagenase or Pronase are available as for trypsin, extensive rinsing is particularly important. In addition, a thicker collagen substratum, i.e., 3–4 drops for the ammoniated layer and 2–3 drops for the air-dried layer, promotes initial attachment of the neurons and avoids breakdown of the substrate after 7–10 days of culture.

Plating Density

The surface density and total number of neurons per dish can be adjusted by dilution of the cell suspension. An 8-mm circle can contain from 10^2 to 10^4 cells without excessive clumping. Empirically a "conditioning effect" has been observed even in cultures containing only neurons; that is, with greater densities the neurons, particularly those from adult rats, seem to survive and grow better. For neurons from perinatal

---

[18] B. S. Scott *J. Neurobiol.* **8**, 417 (1977).

rats the activity of the enzyme choline acetyltransferase increases with increasing density to a plateau at about 3000 neurons per dish.[19] Thus the plating density may be a variable, depending on the type of study being done.

Culture Conditions

We have employed a system segregating small groups of cultures in sealed containers to help prevent spread of mold and fungus infections and obviate the need for CO_2-filled incubators. The cultures are housed in desiccator jars (No. 3118, 160-mm Pyrex, Corning Glass Works, Corning, NY) thoroughly cleaned with alcohol and autoclaved. Sterile water is added to the bottom to provide a humidified atmosphere. High-vacuum grease (Dow Corning Corp., Midland, MI) is used to seal ground-glass surfaces (use sparingly to prevent contamination of culture glassware). Depending upon the type of medium in use (see below), CO_2 is added to maintain pH 7.3. This measured amount is approximately 100 ml for our standard medium and approximately 180 ml for the defined medium. The desiccator jars are then placed in the incubator at 35°.

Culture Dishes

Our standard dish is heat-molded from a fluorocarbon plastic, Aclar (33 C, 5-mil, Applied Chemical, Morristown, NJ) and measures 12 or 24 mm in diameter and 44 mm in height. The dish is easily handled, and preparation for light and electron microscopy can be done directly in the dish. Production and preparation of this dish has been detailed by Wood and Bunge.[20] For photography, an adapted plastic petri dish with a glass coverslip bottom can be used to house the Aclar mini-dish.[20]

Cultures have been successfully grown using a variety of other types of dishes. Optically favorable 35-mm glass dishes (No. 1934-12030, Bellco Glass Inc., Vineland, NJ) hold up to five explants and are useful for the long-term study of neurite growth.[1] Plastic tissue culture dishes can also be used, but it is known that toxic effects are possible, probably from volatile components of the plastic.[21]

For long-term observation and micrography with maximum optical

---

[19] D. Ross and R. P. Bunge *Soc. Neurosci. Abstr.* **2,** 769 (1976).
[20] P. M. Wood *Brain Res.* **115,** 361 (1976).
[21] F. A. Mithen, M. Cochran, M. I. Johnson, and R. P. Bunge, *Neurosci. Lett.* **17,** 107 (1980).

clarity, the chamber design of Sykes and Moore[22] may be used. The coverslips with explants or dissociated cells are sealed into the commercially available 25-mm chambers (No. 1943-1111; Bellco) with approximately 0.3 ml of medium (see below). The balance of the chamber volume remains filled with air. Alternatively, for applications such as interference reflectance microscopy, for which a liquid–air meniscus is intolerable, or for use on an upright microscope, the chamber is filled with medium. The chambers are sufficiently well sealed to permit observation on the microscope stage for 5–7 days without risk of dehydration or microbial contamination.

Substrata

Collagen Preparation

Collagen is prepared by a modification of the method of Bornstein.[23] Because standardization of this procedure has proved to be critical for obtaining consistent culture results, it is given here in detail. Tails from 6-month-old to 1-year-old male rats are used because those from younger males or female rats give unsatisfactory preparations. Each tail is thoroughly scrubbed with antiseptic soap, rinsed once each in 80% alcohol and distilled tap water, placed in a filter paper-lined petri dish, and frozen for 24 hr or longer at −20°.

The tail is sterilized in 95% alcohol for 20 min and dried in a large petri dish with filter paper. With the tail held at the small end with a hemostat, the skin is cut completely around the tail using a bone-cutting forceps and starting 1–1.5 cm from the end. The bone-cutting forceps are rotated clockwise to break bones and pull out the tendons. Too much pressure on the tail with the bone-cutting forceps will cut the tendons, giving a low yield. The tendons are cut free and placed in double-distilled water. Moving 1.5 cm toward the larger end of the tail, the procedure is repeated until the last of the tail is used. The tendons are teased apart with small forceps to loosen up clumps and remove blood vessels and any other adherent nontendon connective tissue.

The tendons are extracted using 150 ml of sterile 0.1% acetic acid for each gram of tendons for 5 days at 4° with daily agitation. The dissolved tendons are centrifuged in tissue culture tubes (10,000 rpm for 1 hr), and the supernatant is harvested. The acid extract is dialyzed (dialysis tubing No. 8-667D, Fisher Scientific) against 50 volumes of sterile double-distilled water for 18 hr at 4°. The collagen is removed from the dialysis bags,

[22] J. A. Sykes and E. Bailey Moore, *Tex. Rep. Biol. Med.* **18,** 288 (1960).
[23] M. B. Bornstein, *Lab. Invest.* **7,** 134 (1958).

aliquoted into sterile containers, tested for sterility in soy broth, and stored at 4°.

Application of Collagen to Dishes

The detailed procedure for collagen-coating Aclar dishes is given below; appropriate modifications can be made for other culture dishes. Aclar dishes are sterilized in 80% alcohol, dried, and transferred to a 200-mm petri dish. Two or three drops of dialyzed collagen are spread evenly over the bottom, using a sterile disposable Pasteur pipette flamed to close the tip and bent to the shape of a hockey stick. Avoiding delay, the freshly spread collagen is exposed to ammonia vapor for 2 min. The dishes are rinsed twice with sterile double-distilled water, drained, and allowed to dry. Another 1 or 2 drops of collagen are then added, gently spread, and allowed to dry. Suspensions of dissociated cells are added directly to the air-dried layer without prior wetting (see above). For explants, the dishes are first wetted using L-15, which is then removed and replaced by 3 or 4 drops of the desired medium.

This double-layered collagen substratum has been very useful in the culture of either dissociated or explanted SCG neurons. Dissociated SCG neurons will not attach or grow well on the ammoniated collagen alone. Although explants will grow on this substratum, their rate of growth is slower, and after several weeks they tend to detach from the substrate. Furthermore, we have evidence that neurite growth and nonneuronal cell migration from explants is not only age dependent but also substrate dependent.[24] Thus E15 SCG explants on ammoniated collagen have neurites with accompanying nonneuronal cells; on air-dried collagen (double layered as described above) the neurites have but a few nonneuronal cells and grow more slowly. These results plus that of others (for review see Roufa *et al.*[24]) serve to alert any investigator of the possible effects of the substratum on neuronal growth and differentiation.

In order to study single neurons or a small number of neurons both electrophysiologically and morphologically, a method described by Furshpan *et al.*[25] has been adapted. Small drops of collagen are applied in arrays of 9–12 on the Aclar dish using a 25-gauge needle. Ammoniation, rinsing, and drying are carried out as above. Smaller drops overlapping the first are added and allowed to dry. The suspension of dissociated neurons is serially diluted, and neuronal attachment is observed visually before rinsing. By varying dilutions and time for attachment, collagen islands with single or a few neurons can be obtained.

[24] D. G. Roufa, M. I. Johnson, and M. B. Bunge *Dev. Biol.* (in press).
[25] E. J. Furshpan, P. R. MacLeish, P. H. O'Lague, and D. D. Potter, *Proc. Natl. Acad. Sci. U.S.A.* **76**, 4225 (1976).

Culture Media

Several medium formulations have been developed in our laboratory, each tailored for different types of studies. All are typically prepared days or weeks before use and stored frozen at $-80°$. Generally, the cultures are fed with 0.2–0.5 ml of fresh medium every 2–4 days; however, longer intervals can be used for some types of studies.

Standard Medium

One hundred milliliters of the standard medium includes 61 ml of Eagle's minimum essential medium (No. 320-1090 GIBCO, Grand Island, NY), 25 ml of human placental serum, 10 ml of 9-day chick embryo extract (50% in BSS), 3 ml of 20% glucose, 0.7 ml of 200 mM glutamine (No. 320-5030, GIBCO) and nerve growth factor (NGF). The NGF used is the 7 S form prepared according to the method of Bocchini and Angeletti[26] and assayed either on dissociated SCG neurons to determine the optimal concentration or on chick dorsal root ganglia by the standard bioassay.[27] The usual concentration is approximately 25 biological units per milliliter of feed.

Medium for Long-Term Observation

When cultures must be maintained outside the CO_2 environment for hours or days, such as for time-lapse cinematography of cultures in the Sykes–Moore chambers, a variant of the standard medium is used that maintains a stable pH of 7.3 indefinitely under an ambient air atmosphere. The medium is based on a modified formula of Eagle's medium in which the bicarbonate buffer is replaced by a phosphate buffer (Hanks' salts) and 25 mM N'-2-hydroxyethylpiperazine-N'-ethanesulfonic acid (HEPES) (No. 380-2370; GIBCO). This base is supplemented with 10% human placental serum, 0.6% glucose, 1% L-glutamine (200 mM), and 25 biological units of nerve growth factor (NGF) per milliliter.

Chemically Defined Medium

The formula for the defined medium we have used is modified from that designated N2 by Bottenstein and Sato.[28,29] To a basal mixture composed of equal parts of Dulbecco's modified Eagle's medium (No. 320-

---

[26] V. Bocchini and P. Angeletti *Proc. Natl. Acad. Sci. U.S.A.* **64**, 787 (1969).

[27] S. Varon, J. Nomura, J. Perez-Polo, and E. Shooter, in "Methods of Neurochemistry" (R. Fried, ed.), Vol. 3, p. 203. Dekker, New York, 1972.

[28] J. E. Bottenstein and G. H. Sato, *Proc. Natl. Acad. Sci. U.S.A.* **76**, 514 (1979).

[29] R. P. Bunge, M. B. Bunge, D. J. Carey, C. J. Cornbrooks, D. H. Higgins, M. I. Johnson, L. Iacovitti, D. C. Kleinschmidt, F. Moya, and P. Wood, *Cold Spring Harbor Conf. Cell Proliferation* **9**, 1017 (1982).

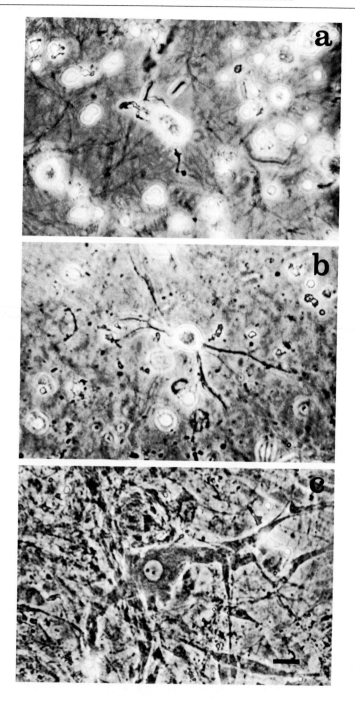

1965, GIBCO) and nutrient mixture F12 (Ham) (No. 320-1765, GIBCO) is added 5 µg of insulin (bovine crystalline, I-5500 Sigma, St. Louis, MO) and 100 µg of transferrin (human, iron free, T2252, Sigma) per milliliter, 20 nM progesterone (P0130 Sigma), 100 µM putrescine dihydrochloride (P7505, Sigma, St. Louis, MO), and 30 mM sodium selenite (SO7150, Pfaltz and Bauer, Stamford, CT). Unlike the initial N2, no HEPES, antibiotics, or bicarbonate are added. In addition, 1.4 mM L-glutamine and 100 ng of purified 2.5 S NGF per milliliter are added. Finally, the cultures are grown in a 7% CO_2 atmosphere (versus 5% for standard medium to maintain the pH at approximately 7.4). Since the pH of this medium drifts upward rapidly under ambient air, any handling and feeding of the cultures should be done expeditiously.

Antimitotic Agents

When nonneuronal cell-free cultures are desired, the antimitotic agent 5-fluorodeoxyuridine (FUdR; F-0503, Sigma) at 10 µM is added to the medium along with 10 µM uridine (U-3750, Sigma). Serum and embryo extract content are reduced to 10% and 2%, respectively, as this modification alone reduces proliferation of nonneuronal cells. Usually one feeding (for 24–48 hr) early in the first week of culture is sufficient to eliminate nonneuronal cells from dissociated SCG neurons even though the cultures may be maintained for months. For SCG explants, medium containing an antimitotic agent is given for 24–48 hr every alternate feed period for 10–14 days.

In the modified defined medium, elimination of nonneuronal cells from dissociated neuronal cultures taken from embryonic rats has proved to be more difficult. In the absence of antimitotic agents the growth of Schwann cells appears to be enhanced while fibroblast growth is inhibited.[29] Thus, the FUdR-containing defined medium is given as 2–4 pulses (24–48 hr) starting on the second or third day of culture and alternating with defined medium alone. Delay in use of the antimitotic treatment to the fourth or fifth day of culture will result in incomplete elimination of the nonneuronal cells.

FIG. 1. Dissociated superior cervical ganglion neurons taken from an adult rat (12 weeks old), grown in culture, and photographed in the living state. Twenty minutes after plating (a), the neuronal somata are often granular and have one or more processes with terminal swellings. Considerable debris is present. By 5 days in culture (b), the neurons are actively extending neurites and beginning to form a network. By 12 days in culture (c), the neurons show considerable maturation, having four to six processes and a central nucleus. In this culture nonneuronal cells were allowed to proliferate. Phase contrast; ×275. Bar, 30 µm.

Similarly, dissociated neuronal cultures taken from postnatal rats 3–4 weeks old or older and grown in either standard or defined medium have been difficult to free of nonneuronal cells. With some caution a schedule similar to that described above for defined media is used starting the second day of culture.

Comments concerning Media Components

Antimitotic agents are generally regarded to have little effect on the nondividing neuronal population. However, studies have shown a cumulative decrease in the rate of neurite extension from explants when FUdR is continuously present in the medium.[30] This effect is most dramatic on neurons derived from early embryonic and older postnatal rats.

Studies with a variety of media formulations have shown the importance of the concentration of certain components. Thus the concentration of human placental serum or embryo extract can affect the development of choline acetyltransferase activity.[31] In contrast, 10% serum and 2% embryo extract suffice to support neurite growth. Elevated K^+ levels are known to increase survival for some neuronal types,[32] but in SCG cultures it will also effect neurotransmitter synthesis of the neurons.[12] Because of this, hemolyzed serum must be used with caution, as the K^+ concentration may be increased.

We have used human placental serum almost exclusively, as it yields consistent results and is available to us from an affiliated hospital. Other sera (horse serum, human adult serum, fetal calf serum) may be used, but from our experience we recommend that the effects of any serum be well characterized before routine use. In particular, commercially available sera may vary considerably from lot to lot.

Use of media having L-15 as a base and containing 5% rat serum as described previously[33] results in dissociated cultures similar to those described for standard media above. Over the long term, the neurons remain somewhat smaller and the levels of tyrosine hydroxylase and choline acetyltransferase activity are lower.[34]

Description of Cultures

When grown in standard medium, embryonic neurons, whether plated as explants or as dissociated cells, begin to grow soon after plating. Disso-

[30] V. Argiro and M. Johnson, *Soc. Neurosci. Abstr.* **7,** 346 (1981).
[31] M. I. Johnson and R. P. Bunge, *in* "Spinal Cord Reconstruction" (C. C. Kao, R. P. Bunge, and P. J. Reier, eds.), p. 329. Raven, New York, 1983.
[32] B. S. Scott, *Exp. Neurol.* **30,** 297 (1971).
[33] E. Hawrot and P. H. Patterson, this series, Vol. 58, p. 574.
[34] L. Iacovitti, D. Higgins, and M. I. Johnson, unpublished data.

ciated neurons thus have been observed to extend neurites within hours of attachment and form an interlocking network within several days.[35] In cultures from perinatal rats, the neuronal somata increase in size from 20 μm at 1 week to 50 μm at 8 weeks, and the nuclei shift from an initially eccentric position to a more cental one. In the defined medium the neurons survive in similar numbers and show the same progression during initiation of neurites and early formation of the network. The neurons, however, remain smaller with eccentric nuclei, and the neurites tend not to fasciculate as well as when standard medium is used.[14]

Neurons taken from young postnatal rats behave as described above. With increasing age, the onset of neurite growth is delayed and explants from fully adult rats will show substantial growth only after several days in culture.[1] Dissociated neurons from adult rats are likewise distressing to view in the first few days of culture. They are often very granular with only a few irregular processes bearing bulbous swellings (Fig. 1a). By a week many more neurons are actively extending neurites, surviving in numbers greater than initially thought possible, and forming networks (Fig. 1b). By several weeks the neurons appear quite healthy and have central nuclei and 4 to 6 processes (Fig. 1c).[36]

[35] M. I. Johnson, D. C. Ross, M. Meyers, and R. P. Bunge, *J. Cell Biol.* **84,** 680 (1980).
[36] E. Wakshull, M. I. Johnson, and H. Burton, *J. Neurophysiol.* **42,** 1410 (1979).

[21] Methodological Considerations in Culturing Peptidergic Neurons

By William J. Shoemaker, Robert A. Peterfreund, and Wylie Vale

Introduction

The culturing of specific cell types from mammalian tissue has been a powerful tool in obtaining information about the metabolism, pharmacological responsiveness, and secretory role of many cell types. Culture techniques provide an important approach that often yields data that cannot be obtained by other means. The contributions of culture methods are no less important in the study of neuropeptides, where the controlled conditions of the culture situation can reduce the number of variables present in the intact nervous system, thereby allowing investigations to focus on the small proportion of nervous system cells that contain neuropeptides. This chapter will cover the methods that the authors have found useful in their studies of neuropeptide content and secretion for a variety of neuropeptides from a number of cultured brain regions. The methods described here are compatible with long-term (weeks to months) culturing of brain regions where maintenance of differentiated characteristics has been documented and survival of neuronal elements (as opposed to glial, endothelial, or other cell types) is optimal.

The two cultures methods described here, primary dispersed and primary explant, have both been utilized for neuropeptide studies. Dispersed cell cultures will be discussed initially; the description contains considerable detail, since this is a newer preparation that has only recently been described.[1] The explant culture technique is an adaptation from our earlier published reports on catecholamine neuron-containing explant cultures; the description of these techniques will omit many of the details common to both preparations. The advantages and disadvantages of each method will be discussed briefly.

Primary Dispersed Cultures

Primary dispersed cell cultures are preparations in which the individual cells comprising a tissue are dissociated from one another to yield, initially, a uniform suspension of cells. The suspended cells may then be used acutely or attached to various substrates for long-term manipulations. The preparation has the advantage that intact, healthy cells can be recovered for use in acute or long-term experiments. Access to cells for pharmacologic, electrophysiologic, or anatomic experiments is unimpeded, particularly in monolayer cultures. Homogeneity of cell behavior within (i.e., dish to dish) or between preparations is routinely observed. Primary culture, which utilizes normal cells taken directly from the animal, reduces the problems of interpreting and extrapolating results obtained from inherently abnormal experimental models, such as tumor cell lines. Disadvantages of primary dispersed cell culture preparations include disruption of normal anatomic relationships, which may be of particular importance for neural tissue, irreversible damage to cells during dissociation procedures, and artifacts associated with nonphysiologic nutritional and substratum conditions. With the possible exception of spinal cord cells, the problem of dispersing and maintaining in long-term culture adult CNS tissue has not been completely solved, thus, the most successful procedures employ fetal or neonatal tissues. Although this section is

---

[1] R. A. Peterfreund and W. W. Vale, *Brain Res.* **239**, 463 (1982).

devoted to cell culture methods for peptidergic neurons from both cerebral cortex and hypothalamus, in particular for those secreting the peptide somatostatin, these techniques should be applicable to all neurotransmitter cell types.

There appear to be four general technical considerations for preparing dispersed cell cultures of brain tissue.

1. *Source of tissue.* Fetal or early neonatal rodent tissues are amenable to dispersed cell techniques. In principle, any region which can be reliably dissected from the brain may be dissociated and cultured. However, rapid dissection with a minimum of anoxic and mechanical damage would appear to be preferable—this may render unfeasible dispersed cell preparations of deep structures with complex anatomy. The procedure described below has been employed for successful monolayer culturing of day 18 fetal rat superficial cerebral cortex and hypothalamus cells.[1]

2. *Dispersal of tissue.* Mechanical dispersal involves slicing or shearing (as by repeated passage though a small bore pipette or needle). This procedure has the advantage of speed and use of enzymes can be avoided. However, damage to cells can be a significant problem and it may be difficult to produce uniform, monodisperse suspensions of intact cells. Enzymatic dispersal depends on digestion of matrix material which bonds cells together. Dilute mixtures of trypsin, papain, and hyaluronidase have been employed. Addition of deoxyribonuclease (DNase) to the digestion mixture reduces adhesion attributed to DNA released from lysed cells. We describe below a procedure utilizing bacterial collagenase prepared from *Clostridium histolyticum* and porcine spleen DNase. Enzymatic dispersal may be less harsh than mechanical dissociation, thus preserving cell viability. However, the procedure is slow and could result in digestion of cell surface proteins such as receptors or recognition molecules.

3. *Attachment substratum.* For long-term maintenance and ease of manipulation, attachment of cells to some type of substrate may be desirable. Possible substrates include glass cover slips, tissue culture plates, dialysis fibers (i.e., Amicon Vitafibers), or beads (i.e., Cytodex II, Pharmacia). It may be necessary to prepare the attachment surface to permit adhesion. Useful surface coating materials include collagen, polyornithine, serum proteins, fibronectin, and extracellular matrix materials.[2] We routinely employ poly-D-lysine coated tissue culture treated plastic dishes (Falcon). The poly-D-lysine is believed to impart a positive charge to the dish surface which facilitates sticking of cells. The "D" configuration of the polymer protects against breakdown by enzymes from the cultured cells.

[2] D. Gospodarowicz and J. P. Tauber, *Endocrine Rev.* **3**, 201 (1980).

4. *Culture media.* In general, standard base medium may be used to maintain cells. Presence of serum for at least some portion of the time in culture appears to be necessary, although we have found that brain cells will retain neurosecretory properties after culturing for several days in serum-free media. Serum-free, hormone-supplemented media which were originally developed for cell lines (i.e., N1, or N2[3,4]) provide alternatives although serum-free conditions may be insufficient for initial attachment of cells. Cytotoxic agents, such as cytosine arabinoside, have been used to minimize the proliferation of dividing cells which coculture with neurosecretory cells. Addition of antibacterial and antimycotic agents to the nutrient media prevents contamination by microbial cells.

A useful general overview of *in vitro* techniques appears in *Methods in Enzymology,* Volume 58. Successful procedures for preparation of cultured brain cells have been described by Vaccaro and Messer[5] and by Tixier-Vidal and co-workers[6] (see also [19], this volume). Early work on hypothalamic neurons in culture has been reviewed.[7] Vernadakis and Culver surveyed several types of brain cell preparations used in biochemical and enzyme studies.[8]

Solutions, Buffers, and Media

Poly-D-*lysine stock solution,* used within 4–6 weeks, is 1 mg/ml poly-D-lysine (Sigma, St. Louis, MO, molecular weight > 200,000) in water, sterile filtered through a 0.45-μm membrane and stored in a glass vessel at 4°. Dish-coating solution, used within 2–3 weeks, consists of a 20 μg/ml solution in sterile water. (This solution can be refiltered through a 0.2-μm membrane.)

Hepes dissociation buffer (HDB) consists of 137 mM NaCl, 5 mM KCl, 0.7 mM Na$_2$HPO$_4$, and 25 mM Hepes (free acid), pH 7.3 to 7.4. This solution is autoclaved and stored at 4° in glass bottles.

Collagenase-DNase solution consists of 0.4% (w/v) clostridial collagenase type II (Worthington, Freehold, NJ) plus 0.4% fraction V bovine serum albumin (Calbiochem, La Jolla, CA) plus 0.4% D-glucose in HDB. The solution is first filtered through a 0.8-μm filter (Millipore type AA) followed by sterile filtration through a 0.45-μm filter. The mixture is ali-

[3] J. E. Bottenstein and G. H. Sato, *Proc. Natl. Acad. Sci. U.S.A.* **76,** 514 (1979).
[4] J. E. Bottenstein, S. D. Skaper, S. Varon, and G. H. Sato, *Exp. Cell Res.* **125,** 183 (1980).
[5] D. Vaccaro and A. Messer, *Tissue Culture Assoc. Manual* **3,** 561 (1977).
[6] A. Faivre-Bauman, E. Rosenbaum, J. Puymirat, D. Greuselle, and A. Tixier-Vidal, *Dev. Neurosci.* **4,** 118 (1981).
[7] J. J. Dreifuss and B. H. Gahwiler, *J. Physiol. (Paris)* **75,** 15 (1979).
[8] A. Vernadakis, B. Culver, and S. Kumar, eds., "Biochemistry of Brain." Pergamon, New York, 1980.

quoted and stored frozen. Deoxyribonuclease II (Sigma, type IV from porcine spleen) may be directly added to the collagenase solution or added during a dissociation procedure. Addition of DNase (3–5 × 10^3 kunitz units per 40 ml collagenase solution) reduces the tendency of cells to aggregate in a viscous mass during dissociation.

Culture medium for maintenance of cells consists of a serum-supplemented standard nutrient medium. Frozen bottles of fetal calf serum (FCS) and horse serum (HS) are thawed at 37° in a water bath and agitated to ensure mixing. Sera are then incubated at 56° for 45–60 min to inactivate complement proteins, aliquoted into sterile tubes, and stored frozen. The base medium is Hepes buffered Dulbecco's modified Eagle's medium (HDME), purchased from various suppliers (i.e., Gibco, Long Island, NY) or prepared from powered stock materials, supplemented with penicillin (50 units/ml) and streptomycin (100 µg/ml). On the day of dissociation, we prepare a mixture of 10% FCS (v/v) + 10% HS (v/v) in HDME and add antimycotics [Fungizone (Gibco), 0.25 µg/ml and/or mycostatin (Calbiochem) 20 µg/ml], glucose (additional 1–2 g/liter) and glutamine (additional 200–300 mg/liter). For feedings the original medium is diluted to 5% HS + 5% FCS with HDME.

Hepes buffered Krebs Ringer bicarbonate solution (HKRBG) for secretion experiments consists of 111 mM NaCl, 4.7 mM KCl, 2.5 mM $CaCl_2$, 1.2 mM $MgSO_4 \cdot 7H_2O$, 1.2 mM KH_2PO_4, 24.8 mM Na_2HCO_3, and 11.1 mM D-glucose buffered with 15 mM Hepes (free acid), at pH 7.35. The solution is gassed with carbogen and supplemented with 0.1% (w/v) crystalline BSA (Miles, Pentex) and bacitracin (30 µg/ml). For experiments requiring sterile conditions, this solution can be filtered through a 0.45-µm membrane. High potassium medium (i.e., 10 ×) is prepared by isotonically substituting KCl for NaCl. Choline chloride may be substituted for NaCl and calcium chloride may be omitted, depending on the particular experiment.

Siliconizing solution consists of a 1:100 dilution of Prosil-28 (PCR Research Chemicals, Gainesville, FL) in water. Clean glassware is immersed in the solution for 30–60 sec, extensively rinsed with water, and autoclaved.

Equipment

Essential equipment for cell culture procedures includes an incubator with precise carbon dioxide and temperature control, a sterile hood, an autoclave, a water bath, a table top centrifuge capable of low-speed operation with a capacity for 50 ml tubes, a microscope with inverted phase contrast optics, and a vacuum line. Useful items include an electronic cell counter (i.e., Coulter Counter), pipette aids or sterile Cornwall syringes, a burner for the hood, and standard laboratory equipment such as freezers, vortexers, hot plate stirrers, etc. For dissection, an ether bell, an ether nose cone made from a conical tube stuffed with cotton, a surgery board for the anesthetized mother, and syringes and needles for the euthanizing agent are useful. Surgical instruments, soaked in 70% ethanol before use, include heavy scissors and forceps to incise the maternal abdomen and iridectomy scissors and fine forceps for the fetal surgery. A squeeze bottle filled with 70% ethanol is used when sanitizing all work surfaces and the abdomen of the mother.

Items used in dissociation include siliconized, sterile spinner flasks (Bellco Glass, Vineland, NJ) and a submersible stirrer (i.e., TRI R model MS-7, TRI R Instruments, Rockville Center, NY). We also routinely prepare sterile, siliconized glass Pasteur pipettes which are stored in capped 18-mm glass test tubes.

Disposable tissue culture-treated plastic dishes come in a variety of sizes. We find that 60 mm dishes (Falcon #3002), 35 mm dishes (Falcon #3001), and multiwell plates (Falcon #3046 and #3047) are useful. Devices for sterile filtration are also available in several styles.

Collection of Tissues and Dissociation

In general, sterile surgical procedures are not required although attention should be given to cleanliness. Timed pregnant female Sprague–Dawley rats (Zivic Miller, Allison Park, PA) are anesthetized with ether on the eighteenth day of gestational age. The abdomen is soaked with 70% ethanol and the fetuses removed a few at a time through a long incision. The placental membranes are stripped away and the fetal spinal cord is immediately severed with an iridectomy scissors. After all fetuses are removed, the mother is killed with an intracardiac injection of pentobarbital. The fetal brain is exposed by first snipping through the skin and membranous skull bilaterally from the dorsal surface of the neck toward the eyes, with the carotid arteries as rostral landmarks. The skull is peeled forward with a fine forceps. We collect thin slices of rostral lateral cerebral cortex from each hemisphere. Hypothalamus tissue is removed by peeling back the brain from the base of the skull. Shallow cuts are made rostral to the anterior communicating artery, and bilaterally along the hypothalamic sulci. The hypothalamus fragment is then lifted out with a pair of curved fine forceps. When perfusion is maintained, the arterial Circle of Willis serves as a useful landmark for cuts.

Cortex (CC) and hypothalamus (H) tissue are separately collected in room temperature HDB solution containing 0.1% BSA. After tissue collection, all procedures are carried out under sterile conditions.

Tissue fragments are washed three times with 37° HDB–0.1% BSA and then placed into a siliconized sterile spinner flask. Collagenase solution is added and the tissues are digested at 37° in a water bath, gently agitated by the stirrer which is driven by a submersible stir plate. We typically employ 0.75 to 1 ml collagenase solution per hypothalamus fragment and 60 ml of collagenase solution for cortex fragments (about 3 cm^3 packed tissue). Digestion is allowed to proceed for 2–2.5 hr. Occasional trituration by gentle passage through a siliconized sterile Pasteur pipette helps to break up clumps. We add additional DNase at the time of the first trituration, generally after 45 min of digestion. Cerebral cortex tissue usually requires less digestion time than hypothalamus tissue and CC tissue is more sensitive to overly long digestion. It is not possible to entirely disperse all of the tissue. Some fragments remain undissociated, often aggregated in a gelatinous mass which settles to the bottom of the flasks or adheres to the stirrer.

After digestion, the cell suspension is decanted into conical centrifuge tubes (Corning, 50 ml) and cells are pelleted by gentle centrifugation in an I.E.C. clinical centrifuge. The dissociated cells are resuspended in a small volume (<1 ml) of 0.1% BSA–HDB and layered over a solution of 4% BSA in HDB in a centrifuge tube. This material is pelleted. The supernatant, containing cell fragments and debris, is decanted and discarded. The cell pellet is washed three times by vortexing and centrifugation with culture medium, resuspended, and cells are counted. We typically obtain 1.5×10^6 cells per hypothalamus fragment as determined with a Coulter counter. H and CC cells are plated at a density of 5×10^6 cells per 35-mm tissue culture dish (Falcon #3001) which has been coated for 4–6 hr at room temperature with 1 ml poly-D-lysine coating solution and washed once with HDME just prior to plating cells. Cells are maintained at 37°, 15% CO_2, 100% humidity in an incubator. We typically feed cells at least twice before experiments which are carried out on the tenth, eleventh, or twelfth day in culture.

Observations

Viewed in phase contrast optics, cells typically appear round or ovoid immediately after dispersion and float individually or in small clusters in the medium. A few multicell clumps may be found (see Fig. 1). Cells rapidly adhere to the dish (within 2 hr) and begin to send out branching processes, often to neighboring cells. Several cell types are readily distinguished. Phase bright, ovoid cells with a few branching processes have the appearance of neurons. Flat, pale cells with multiple processes may be glia. Lightly refractile, large flat cells possibly represent connective tissue elements. These latter cells proliferate to form a confluent "carpet"

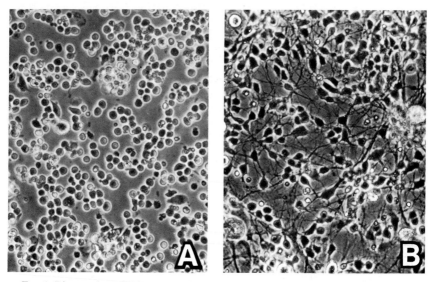

Fig. 1. Dispersed cerebral cortex cells as seen with phase contrast optics. (A) Two hours after plating; (B) 5 days after plating. Cell adhesion to the dish is rapid and extensive process formation is observed. Cell distribution is often nonuniform on the dishes, with greatest cell density at the periphery. Cells often form groups or clusters separated by relatively acellular patches. Cell body diameter typically ranges between 5 and 10 μm. Proliferation of dividing cells usually obscures the morphology of "neural" cells after 5 to 7 days *in vitro*.

upon which other cell types appear to rest. We have maintained cultured cells for up to several weeks, however, there appears to be progressive sloughing of cells with "neural" appearance after 12 to 14 days.

Plated at low density (less than 1.5×10^6 cells per 35-mm-diameter dish) the highly differentiated morphology of neurons is well seen in cultures of brain cells. At higher densities, close packing of cell bodies and multiplication of dividing cell types prevents detailed morphologic examination. However, for studies on secretion of the peptide somatostatin, high-density plating appears to be necessary because cell content and secretion of this peptide do not directly vary with cell number.

We typically perform secretion experiments on day 10, 11, or 12 of culture which is a period prior to the initiation of cell sloughing but well after damaged cells would be expected to die and wash off the plates. Other workers[9] have reported that 10- to 12-day-old cultures continue to increase their cell content of somatostatin.

[9] J. Delfs, R. Robbins, J. L. Connally, M. Dichter, and S. Reichlin, *Nature (London)* **283**, 676 (1980).

TABLE I
MODULATION OF SOMATOSTATIN SECRETION
FROM DISPERSED CEREBRAL CORTEX CELLS[a]

Treatment	Mean secretion ± SEM of SSLI (fmol/dish)	Difference from control[b]
A. Control	64.2 ± 3 (4)	—
CRF (100 nM)	106.6 ± 11 (4)	**
GABA (50 μM) + CRF (100 nM)	34.5 ± 4 (4)	**
B. Control	35.2 ± 3 (4)	—
Carbachol (100 μM)	56.7 ± 2 (4)	**
Acetylcholine (100 μM)	54.3 ± 3 (4)	**
Acetylcholine (1 μM)	46.4 ± 5 (4)	*
Carbachol (100 μM) + atropine (5 μM)	32.8 ± 1 (4)	NS
Acetylcholine (100 μM) + atropine (5 μM)	36.8 ± 3 (4)	NS
C. Control	49.0 ± 6 (5)	—
Picrotoxinin (5 μM)	134.6 ± 9 (5)	**
Bicuculline (2 μM)	111.5 ± 4 (5)	**

[a] Experiments were performed as described in legend to Fig. 2. For the experiment in A, incubation buffer contained ascorbic acid (10^{-4} M) as an antioxidant. Responses to CRF, carbachol, picrotoxinin, and bicuculline are blocked by pretreatment with cobalt (1.0 mM) in low calcium incubation medium (not shown). Tetrodotoxin (10^{-7} M) also suppresses secretion. Numbers in parentheses indicate dishes per treatment group.

[b] NS, Not significantly different from control; *, different from control, $p < 0.05$; **, different from control, $p < 0.01$.

A variety of substances modulate somatostatin secretion from H or CC cells including monoamines,[10,10a] peptides,[10b] and phorbol diesters.[10c] For example, carbachol, a cholinergic agonist, stimulates somatostatin-like immunoreactivity (SSLI) secretion in a dose-dependent manner (Fig. 2). The peptide ovine corticotropin-releasing factor (CRF) also stimulates secretion of SSLI, an effect prevented by the inhibitory neurotransmitter GABA (Table IA). Inhibitory actions of GABA can be blocked by bicuculline or by picrotoxinin (the active form of picrotoxin). Both GABA block-

[10] R. J. Robbins, R. E. Sutton, and S. Reichlin, *Brain Res.* **234**, 377 (1982).
[10a] R. A. Peterfreund and W. W. Vale, *Endocrinology* **112**, 526 (1983).
[10b] R. A. Peterfreund and W. W. Vale, *Endocrinology* **112**, 1275 (1983).
[10c] R. A. Peterfreund and W. W. Vale, *Endocrinology* **113**, 200 (1983).

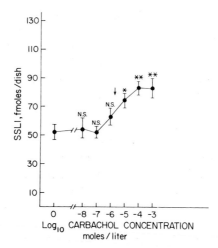

FIG. 2. Secretion of somatostatin by dispersed cerebral cortex cells in response to carbachol. On the tenth day in culture, cells were washed three times with HKRBG and allowed to equilibrate in the same medium for 1 hr. Medium was then changed and cells received either no treatment (control) or one of several doses of carbachol administered in a small volume. After 1 hr, medium was collected, acidified with an equal volume of 2 N HOAc on ice, heated for 5 min at greater than 85° in a water bath, and samples were dried in a Speed Vac (Savant Instruments). Somatostatin like immunoactivity (SSLI) was measured by radioimmunoassay. Each point represents mean secretion ± SE of four or five dishes. Results were analyzed with a multiple range test. NS = not significantly different from control. *, different from control, $p < 0.05$; **, different from control, $p < 0.01$. Arrow indicates dose of carbachol for approximate half-maximal response.

ers are also associated with direct stimulation of SSLI secretion (Table IC) which suggests that GABAergic cells may coculture with peptidergic neurons and tonically modulate peptide secretion.

The dispersed cell preparation provides a quantitative model system for examining the brain region at which various modulators exert their effects. However, in view of the possibility of cell interaction within the cultures, as suggested by our data and that of Gamse et al.,[11] the actions of some effectors may be mediated by one or more relays between cells of the heterogeneous culture population.

Explant Cultures

Our general procedure follows, although variations have been tried and will result in acceptable cultures. Brain tissue from fetuses less than 20 days of gestational age can be successfully maintained in explant cul-

---

[11] R. Gamse, D. Vaccaro, G. Gamse, M. D. Pace, T. Fox, and S. Leeman, *Proc. Natl. Acad. Sci. U.S.A.* **78**, 2315 (1981).

ture, but the dissection of specific brain regions becomes difficult with the smaller brain. The rate of brain growth is very rapid during late fetal stages[12] so that the brain of a fetus 24 hr earlier may be only half as large as the later fetus. We have found that taking explants from postnatal material is not very successful; the reason for this is not clear but the older the fetus or newborn, the more difficult it is for neurons to "take" to the culture conditions.

The following procedures are based on our previously published work on explant cultures for catecholamine-containing neurons[13] and are adapted for neuropeptide cultures. Fetuses are removed from Sprague–Dawley albino rats of 20 days gestation and kept on ice. Individual fetal brains are placed on sterile gelatin blocks and stabilized with Methocel-enriched medium. Under microscopic visual control, we dissect the brains at different anterior–posterior levels, into standard slices from which are dissected the regions of interest (see Fig. 3). Fragments of interest (approximately 3–4 mm^3) are excised from the appropriate slice and placed in 35-mm Falcon dishes freshly coated with collagen or poly-D-lysine and containing approximately 0.5 ml of Methocel medium. The tissue fragment is then teased into 20–30 smaller pieces whose final positions in the dish are established by a minimal amount of Methocel medium. The cultures are incubated at 35° in a 95% air/5% CO_2 atmosphere.

Media with Serum

The optimal medium as judged by morphologic and biochemical criteria is Dulbecco's modified Eagle's medium with N-2-hydroxyethyl piperazine-N'-2-ethanesulfonic acid (HEPES) buffer [Grand Island Biological Co. (Gibco), Grand Island, NY] enriched with serum. Ten percent heat-inactivated fetal calf serum (Gibco or Reheis Chemical Co., Chicago, IL), 10% heat-inactivated horse serum (Gibco), and 0.4% Methocel (Methocel 60 HZ, Dow Chemical, Midland, MI) are added for the first 5 days of culture (311 mOsm/kg). After 5 days, the cultures are continued in Dulbecco's modified Eagle's medium supplemented with 10% horse serum (298 mOsm/kg).

Serum-Free Media

The serum-free media formula is taken from Bottenstein and Sato[3] as previously described by us for primary explant cultures.[14] A 50:50 mix-

---

[12] M. Schlumpf, W. J. Shoemaker, and F. E. Bloom, *J. Comp. Neurol.* **192**, 361 (1980).

[13] M. Schlumpf, W. J. Shoemaker, and F. E. Bloom, *Proc. Natl. Acad. Sci. U.S.A.* **74**, 4471 (1977).

[14] W. J. Shoemaker, J. E. Bottenstein, R. J. Milner, B. R. Clark, and F. E. Bloom, *Soc. Neurosci. Abstr.* **5**, 758 (1979).

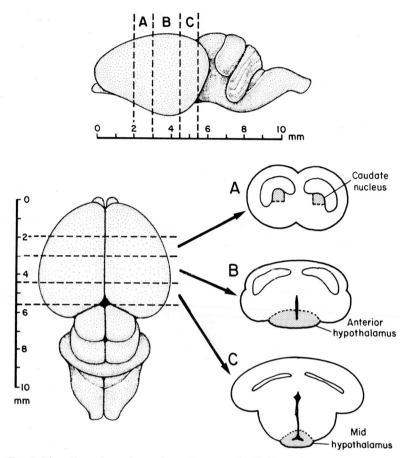

FIG. 3. Dissection scheme for explant cultures used in Tables II and III. Drawing of a 20-day rat fetal brain from top and side view to show placement cuts for further dissection. The brain is kept moist with sterile viscous media while dissections are performed under a dissecting microscope. Illumination is best using a fiber optic system that reduces heating of tissue. The final dissections of the regions (lower right hand) are done with microdissecting lances and chisels.

ture of Dulbecco's modified Eagle's media (Gibco) and F-14 (Gibco) is added to insulin (Sigma) 5 µg/ml, transferrin (Sigma) 5 µg/ml, progesterone (20 nM), putrescine 100 µM, and selenium (Apache Chem. Inc.) 30 nM. Explant cultures are fed twice weekly.

Neuropeptide Assays. Because neuropeptides are very susceptible to breakdown by peptidases during sample processing for radioimmunoassays, a procedure was devised based on the success of brain tissue assays. This procedure attempts to inactivate the peptidases by high heat before access can be gained to the peptide during tissue disruption procedures. After removal of the media, 1 ml hot acid (1 N acetic/0.1 N HCl) is added to the culture dish, followed by scraping of the surface for removal of the tissue. The tissue–acid mixture is placed in a small plastic tube in a heated bath. One additional milliliter of hot acid is used to rinse the culture dish, this is added to the plastic tube. The tissue–acid mix is then heated for an additional 20 min, cooled, sonicated for further disruption, and spun. The supernatant is then dried down for processing in a radioimmunoassay. The media is treated with a small amount (6 μl) of concentrated acid and placed in a boiling water bath for 20 min. After cooling, the media is spun and the supernatant dried for RIA processing similar to the cellular component.

We have used explant cultures to study several neuropeptide-containing neurons. As a first step, the explant from several regions is assayed for the neuropeptide levels in the culture media and in the explant tissue itself. One limitation of this approach is whether the peptide will be stable in the media nourishing the cells. The stability of the peptides in the media cannot be assumed, since peptidases may be released from the cells or attached to the outer cell surface. Serum is known to contain peptidases of various substrate specificity; we routinely culture explants in both media containing serum and serum-free media for that reason. The presence of peptide levels in media most likely represents a steady-state level where peptide breakdown is balanced by production from the cells. A similar situation has been described for catecholamine neurons in culture.[13]

In the examples given here, β-endorphin could be measured in all of the cultures from hypothalamic regions (Table II) that are known to contain β-endorphin neurons. Explant cultures from brain regions where positive staining β-endorphin cells have not been seen (e.g., cerebral cortex, ventral tegmentum) fail to show any assayable β-endorphin in media or cell extracts. Thus, for long-projecting neurons from a limited number of large cell bodies, such as β-endorphin neurons, cultures have been used to verify that projection regions do not, in fact, contain cell bodies, since regions rich in β-endorphin terminal fibers would be expected to lose their β-endorphin content within a few days of being placed in culture.

Some neuropeptides are contained in smaller neurons that have limited axonal and dendritic processes. Such is the case for enkephalin-containing neurons found in many brain regions. (We will use the term "enkephalins" rather than specifying leucine-enkephalin or methionine-enkephalin since these two forms are usually seen in a constant ratio in all brain regions and, since there are few good antisera against met-enkepha-

TABLE II
HYPOTHALAMUS CULTURES[a]

	With serum	Serum-free media
Anterior hypothalamus cultures		
Cellular β-endorphin content (fmol/culture)	60.8 ± 29.4	19.0 ± 9.65
Cellular β-endorphin concentration (fmol/g)	6080 ± 294	1900 ± 965
Media β-endorphin concentration (fmol/ml)	186.3 ± 36.7	72.5 ± 21.0
Mid hypothalamus (including arcuate nucleus) cultures		
Cellular β-endorphin content (fmol/culture)	51.4 ± 47.1	23.8 ± 6.5
Cellular β-endorphin concentration (fmol/g)	5140 ± 471	2380 ± 650
Media β-endorphin concentration (fmol/ml)	91.5 ± 20.0	52.4 ± 3.52

[a] Procedure for peptide assays of cultures and media as described in text for explant cultures. Values are ± standard error of the mean. Values represent mean of 7–8 for anterior hypothalamus and 5 for mid hypothalamus. The length of time in culture for these explants varies from 4 months to 1 month. There is no correlation of cellular or media peptide levels with age in culture. All cultures tested had measurable levels. The cellular concentration of β-endorphin was calculated assuming the tissue mass of the culture to be 10 mg normalized to 1 g.

lin, all of our data were acquired with a leu-enkephalin antiserum, which cross-reacts 3% with met-enkephalin.) Table III contains data from three brain regions, all giving positive results for enkephalin-containing neurons. A third opioid peptide was assayed in these cultures, dynorphin, a family of neuropeptides separate from the endorphins or enkephalins,[15] but with a distribution similar to enkephalin. Dynorphin 1-17 could only be detected in the media, thus no cellular levels are given in Table III. In general, we find higher concentrations of peptide in the media than in the cellular components of the cultures, but this may represent an artifact of the assay system. Because both media and cellular extracts are dried down and brought up to a constant volume in buffer before assay, the media samples represent the neuropeptide present in approximately 1 ml of media, whereas the cellular component is from 5 to 10 mg of tissue. It is not surprising that we can often find a peptide present in the media of a culture but none in the cells themselves. This is due to the limit of sensitivity of our assays (usually about 1.0 pg). We have attempted to correct for this by assuming a 10 mg mass for the cells in Table II and normalizing to 1 g tissue so that tissue *concentration* could be compared to media

[15] A. Goldstein, W. Fischli, L. Lowney, M. Hunkapiller, and L. Hood, *Proc. Natl. Acad. Sci. U.S.A.* **78**, 7219 (1981).

TABLE III
PEPTIDE ASSAYS OF CULTURES AND MEDIA[a]

	Media with serum	Serum-free media
Anterior hypothalamus		
L-Enkephalin media (fmol/ml)	1195 ± 259	342.8 ± 140.1
L-Enkephalin cells (fmol/culture)	16.8 ± 11.7	5.7 ± 1.6
Dynorphin media (fmol/ml)	56 ± 12.5	23.2 ± 10.6
Mid hypothalamus (including arcuate nucleus)		
L-Enkephalin media (fmol/ml)	962 ± 282	188.3 ± 15.7
L-Enkephalin cells (fmol/culture)	7.78 ± 3.33	9.81 ± 4.19
Dynorphin media (fmol/ml)	52.7 ± 12.5	11.5 ± .98
Caudate putamen		
L-Enkephalin media (fmol/ml)	795.0 ± 182.5	257.1 ± 112.1
L-Enkephalin cells (fmol/culture)	6.43 ± 2.04	4.12 ± 0.8
Dynorphin media (fmol/ml)	37.3 ± 10.3	38.0

[a] Procedures for peptide assays of cultures and media as described in text for explant cultures. Values represent the mean of 7–8 cultures for anterior hypothalamus, 5 for mid hypothalamus, and 7 for caudate putamen. Values are ± standard error of the mean. Enkephalin could be detected in nearly every culture measured, but dynorphin was detected in 26 out of 39 media samples and in none of the cellular extracts. The amount of dynorphin measured could not be due to cross-reactivity of the antiserum with enkephalin.

concentration. Under these circumstances the higher concentration of peptide in the small number of neurons is evident.

Summary

Both explant and dispersed cell culture preparations of brain tissue provide a means to directly assess the functions of neuropeptide-containing brain cells isolated from the complex influences present *in vivo*. The validity of the approach depends on reproducibility of observations. In dispersed cell cultures, we find that cell responses, as determined by secretion of the peptide somatostatin, have remained relatively constant both quantitatively and qualitatively over numerous preparations. In addition, for pharmacologic studies on somatostatin secretion, data from several laboratories are in good agreement. The validity of the dispersed cell approach also depends on whether the pharmacologic and physiologic behavior of cells parallels that expected of excitable tissue.

The variability of the explant cultures from culture dish to culture dish makes quantitative experiments, such as demonstrated with the dispersed cultures, difficult. On the other hand, explant cultures better maintain the integrity of the tissue components,[16] so that interactions between neurons and glial cells could occur as *in vivo*.

The long-term health and viability of neuropeptidergic cells in explant and dispersed culture make both preparations potentially useful models to examine central nervous system physiology. Future work with such preparations must eventually address the problems of culturing adult brain tissue, the precise nutrient and hormonal requirements of brain cells, so that undefined medium components, such as serum, can be eliminated from the culture environment, and the general question of whether observations made *in vitro* facilitate our understanding of intact brain physiology.

[16] R. J. Milner, Q. J. Pittman, W. J. Shoemaker, and F. E. Bloom, *Dev. Brain. Res.*, submitted.

[22] Preparation and Properties of Dispersed Rat Retinal Cells

By James M. Schaeffer

Our understanding of the biochemical response of photoreceptor cells to light stimulation have been greatly enhanced by the development of techniques for isolation of large quantities of purified outer rod segments.[1] However, biochemical events within other retinal cells have been more difficult to study. Many putative neurotransmitters have been identified within the mammalian retina, and, more recently, several of the biologically active peptides originally thought to be present only in the hypothalamus are now known to be present in retinal tissue.[2,3] The cellular heterogeneity of the retina precludes obtaining meaningful biochemical data using whole-retina homogenates. Our laboratory has developed a rapid method to disperse rat retinas enzymatically and separate large quantities of

[1] D. S. Papermaster, this series, Vol. 81, p. 48.
[2] J. M. Schaeffer, M. J. Brownstein, and J. Axelrod, *Proc. Natl. Acad. Sci. U.S.A.* **74**, 3579 (1977).
[3] O. P. Rorstad, M. J. Brownstein, and J. B. Martin, *Proc. Natl. Acad. Sci. U.S.A.* **76**, 3019 (1979).

various cell types by centrifugation through a Metrizamide density gradient. This method has been useful for biochemical and pharmacological characterization of several types of retinal cells.[4,5]

Preparation of Dispersed Cells

Sprague-Dawley rats (175–200 g) were decapitated between 8:00 and 9:00 AM; the retinas were quickly removed and placed into ice-cold calcium-free Krebs' bicarbonate buffer (124 mM NaCl, 4 mM KCl, 26 mM NaHCO$_3$, 1.2 mM KH$_2$PO$_4$, 1.3 mM MgSO$_4$) containing 0.1% bovine serum albumin (buffer A). Use of calcium-free media is critical in order to avoid cell clumping. The retinas (6 retinas/ml) were transferred into buffer A containing 0.3 mg of collagenase (Boehringer Mannheim, West Germany) per milliliter, minced with scissors, incubated at 37° for 12 min, and then mechanically dispersed by aspiration in a Pasteur pipette (inner diameter, 0.4 cm) for 15 sec before filtration through a single layer of nylon mesh. The dispersed cells are more than 95% viable as judged by trypan blue dye exclusion. The cells are layered onto a Metrizamide (Nyegaard) continuous density gradient (30 ml, 0–33% Metrizamide prepared in buffer A), centrifuged for 12 min at 1430 g at 4° and fractionated into 0.7-ml aliquots with a peristaltic pump. Each fraction is diluted 7-fold with buffer A, washed by centrifugation for 5 min at 1000 g, and resuspended in 250 μl of buffer A. The cell number of each fraction is estimated by measuring the absorption at 550 nm or determination of DNA content. Similar results were obtained with both techniques.

Identification of Isolated Cells

Using phase-contrast microscopy, we observed that the neurons are primarily rounded and could not be identified without further biochemical characterization (Fig. 1A). After centrifugation through the Metrizamide density gradient, several distinct bands of cells were observed (Fig. 1B).

Photoreceptor Cells. Layer V of the gradient contains only photoreceptor cells (>95%). These cells were morphologically identified at both the light and electron microscope level (see Fig. 2). The isolated photoreceptor cells are composed of outer rod segments and perikaryon (approximately 30% of the cells retain axonal processes). The electron micrographs (Fig. 2C) reveal the typical stacked membrane characteristic of the photoreceptor cells. Cyclic GMP is known to be concentrated within

---

[4] J. M. Schaeffer, D. E. Schmeckel, P. M. Conn, and M. J. Brownstein, *Neuropeptides* **1**, 39 (1980).

[5] J. M. Schaeffer, *Exp. Eye Res.* **30**, 431 (1980).

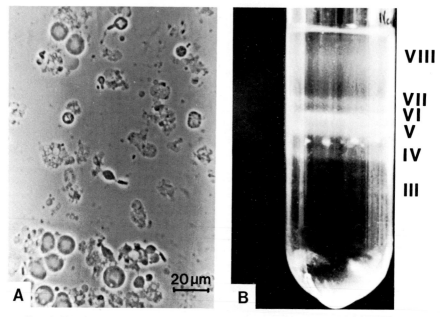

Fig. 1. (A) Phase-contrast photomicrograph of dispersed retinal cells. (B) Photograph of dispersed retinal cells separated by centrifugation through a continuous Metrizamide gradient. The gradient was fractionated into 0.7-ml aliquots, diluted with 7 volumes of buffer, and washed by centrifugation. The number of cells in each fraction was estimated by measuring the absorption at 550 nm.

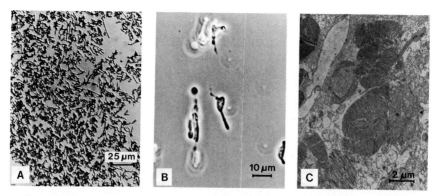

Fig. 2. Isolated photoreceptor cells. (A) The cells from layer V were washed by centrifugation for 5 min at 1000 g and then stained with cresyl violet. (B) Immediately after centrifugation through the Metrizamide gradient, the cells were observed with phase-contrast microscopy. (C) Cells from layer V were pelleted and viewed with an electron microscope (note the typical stacked membranes).

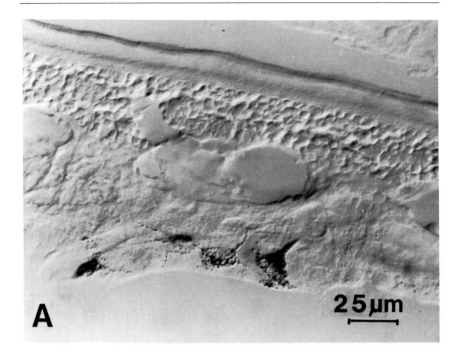
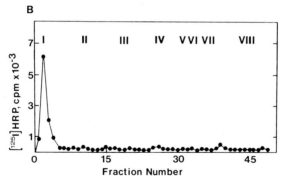

FIG. 3. (A) Cross-sectional view of a retina from a horseradish peroxidase-treated rat stained with diaminobenzidine and viewed with Nomarsky optics. (B) Profile of ^{125}I-labeled horseradish peroxidase in each fraction of the Metrizamide gradient after the retinas were dispersed and separated as described in the text.

photoreceptor cell outer rod segments and may be used as a biochemical marker for these cells. We measured cyclic GMP in each fraction of the Metrizamide gradient using a commercially available radioimmunoassay kit (New England Nuclear Corp., Boston, MA). All of the cyclic GMP was associated with cells present in layer V, demonstrating the exclusive localization of photoreceptor cells in layer V.

Ganglion Cells. We specifically labeled retinal ganglion cells *in vivo* by injecting horseradish peroxidase (HRP) into the lateral geniculate bodies and allowing the HRP to travel retrogradely to the retina. As seen in Fig. 3, HRP-stained cells are present exclusively in the ganglion cell layer. Six rats were then bilaterally injected with [^{125}I]HRP, and after 48 hr the retinas were removed, pooled with 18 unlabeled retinas, chemically dispersed, and separated by centrifugation through a Metrizamide gradient as previously described. All the radioactivity was present in fractions 2–4, coinciding with layer I (the layer that migrates to the most dense region of the Metrizamide gradient; see Fig. 3B).

Müller Cells. Glutamine synthetase activity has been observed in the retinas of several vertebrate species, and autoradiographic studies have demonstrated that glutamine synthetase is almost exclusively localized in the Müller cells of both adult rat[6] and chick[7] retina. In addition, the levels of glutamine synthetase activity during development parallel the appearance of Müller cells.[8] Consequently, glutamine synthetase may be used as an enzymic marker for Müller cells. Retinal cells were prepared and separated on a Metrizamide gradient as previously described. Glutamine synthetase activity was measured (using the assay described by Levintow[9]) in each fraction of the Metrizamide gradient and identified in only one of the cell layers (Fig. 4). These results demonstrate the localization of Müller cells in layer VII.

Amacrine Cells. Dopamine is known to be concentrated within a small percentage of amacrine cells in the rat retina.[10] Using high-performance liquid chromatography coupled to an electrochemical detector, dopamine-containing cells were found exclusively in layer VI of the Metrizamide gradient. Somatostatin, a neuropeptide originally identified in the hypothalamus, has been immunohistochemically identified in amacrine cells.[11] Studies in our laboratory have confirmed the presence of immunoreactive somatostatin in rat retinal tissue. Cells from each fraction were tested for somatostatin content. Although 75% of the somatostatin was lost during the dispersion and subsequent separation procedure, the re-

[6] R. E. Riepe and M. D. Norenberg, *Nature (London)* **268**, 654 (1977).
[7] P. Linser and A. A. Moscona, *Proc. Natl. Acad. Sci. U.S.A.* **76**, 6476 (1979).
[8] R. E. Riepe and M. D. Norenberg, *Exp. Eye Res.* **21**, 435 (1978).
[9] L. Levintow, *J. Natl. Cancer Inst.* **15**, 347 (1954).
[10] B. Ehinger, *in* "Transmitters in the Visual Process" (S. L. Bonting, ed.), p. 144, Pergamon, Oxford, 1976.
[11] W. Stell, D. Marchak, T. Yamadu, N. Brecha, and H. Karten, *Trends Neurosci.* **3**, 292 (1980).

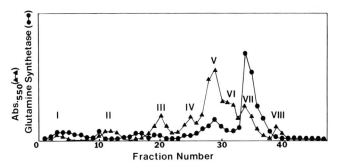

FIG. 4. Retinas were chemically dispersed, and cells were separated by centrifugation through a continuous Metrizamide density gradient. The gradient was fractionated, and the relative number of cells in each aliquot was determined by absorbance at 550 nm (▲). Glutamine synthetase activity (●) in each fraction was determined as described in the text.

covered somatostatin is associated with cells in layer VI. The absence of either of these amacrine cell markers in any of the other cell layers demonstrates the efficiency of our separation procedure. However, we cannot preclude the possibility that non-dopamine-containing amacrine cells are present in other cell layers. In addition, other cell types may be present in layer VI. Consequently, layer VI is enriched for specific cell types but may not represent a truly homogeneous population of amacrine cells.

Advantages and Disadvantages of This Procedure

Data describing the biochemical characterization of the whole retina are difficult to interpret because of the heterogeneity of cell types. Consequently, the isolation of enriched populations of purified retinal cells represents a breakthrough in the studies of retinal cell function. Preparations of chemically dispersed retinal cells were first described by Drujan and Svaetichin,[12] Anctil et al.,[13] and Lam.[14] Since that time several papers from Lam and his colleagues have described biochemical characteristics of isolated dispersed retinal cells from several different species.[15–17] Sarthy and Lam[18] have reported the isolation of several different identifiable cell types from adult male rat retina. These cells were chemically

[12] B. D. Drujan and G. Svaetichin, *Vision Res.* **12,** 1777 (1972).
[13] M. Anctil, M. A. Ali, and P. Coullard, *Rev. Can. Biol.* **32,** 107 (1973).
[14] D. M. K. Lam, *Proc. Natl. Acad. Sci. U.S.A.* **69,** 1987 (1972).
[15] D. M. K. Lam, *Nature (London)* **254,** 345 (1975).
[16] A. Kaneko, D. M. K. Lam, and T. N. Weisel, *Brain Res.* **105,** 567 (1976).
[17] P. V. Sarthy and D. M. K. Lam, *J. Neurochem.* **32,** 455 (1979).
[18] P. V. Sarthy and D. M. K. Lam, *Brain Res.* **176,** 208 (1979).

dispersed with collagenase and hyaluonidase and individually identified. Several biochemical parameters of the total dispersate were described.

The technique for isolation of retinal cells developed in my laboratory involves chemical dispersion of whole retina using collagenase followed by nonequilibrium centrifugation through a Metrizamide continuous density gradient. This technique has several advantages over previously described techniques. (1) Large quantities of cells are recovered, enabling thorough biochemical characterization of the cells. (2) The dispersion is rapid (the cells are exposed to collagenase for 12 min). (3) Metrizamide is isoosmotic and nontoxic and the cells remain viable (they exclude trypan blue and may be maintained in culture for several days).

The efficiency of cell separation using our technique is an important question. Several lines of evidence suggest that we have isolated enriched populations of different cells. (1) The cells of layer V have been morphologically identified as being >95% photoreceptor cells; (2) Biochemical markers are heterogeneously distributed in the different cell populations. The localization of specific cell markers, such as cyclic GMP, glutamine synthetase, and dopamine, in unique populations of cells demonstrates the efficiency of this separation procedure. (3) The DNA content of each fraction isolated on the Metrizamide gradient parallels the absorbance profile. Disadvantages of our technique include the loss of axonal processes, damage to cells during the dispersion procedure, and contamination from other cell layers due to the large number of cells in each fraction.

[23] Punch Sampling Biopsy Technique

By MIKLÓS PALKOVITS

"Micropunch technique," that is, microdissection of brain sections with a hollow needle,[1] has become a routine procedure in many neurobiological laboratories. Several macro- and microdissection methods for sampling brain regions have been reported, but only the micropunch technique suits all of the following requirements.

1. The technique should be fine enough to allow sampling individual brain nuclei (the lowest limit is about 10 μg wet weight = tissue pellet of 100 μm radius) separately.

[1] M. Palkovits, *Brain Res.* **59**, 449 (1973).

INSTRUMENTATION FOR MICROPUNCH SAMPLING

Procedure	Frozen brains	Fresh brains
Sectioning	Cryostat	Stainless steel razor blade, templates, or tissue chopper, or elastic (rubber) stage
Micropunch	Stereo or dissecting microscope	Stereo or dissecting microscope
	Cold stage	Elastic (rubber) stage
	Microdissection needles	Microdissection needles + stylet

2. The procedure should be rapid, since hundreds of brain samples are often needed within a limited period of time.
3. Postmortem changes should be minimized.
4. The technique should be employed without any complicated instrumentation.
5. During microdissection, the brain tissue should be preserved in the perfect condition required by postsampling procedures.
6. The technique should be reproducible: the standard error due to the microdissection should be kept at a level commensurate with general neurobiological standards (SEM $\approx \pm 10\%$ of the mean).
7. The technique should be easily validated by microscopic control of the dissection.

Using the micropunch technique, over 220 brain nuclei can be separately removed from rat brain.[2] The technique has been worked out for rat brain, but it can easily be adapted to other species. In this chapter only main principles and basic technical information and advice are summarized; for further methodological details, the reader is referred to a review article.[3]

Instrumentation

Microdissection can be performed on fresh or frozen brains. The technique itself is composed of two procedures: brain sectioning and micropunch. The instruments needed are related to these two procedures (see the table).

[2] M. Palkovits, "Guide and Map for the Isolated Removal of the Rat Brain Areas" (Hungarian text). Akadémiai Kiadó, Budapest, 1980.
[3] M. Palkovits and M. J. Brownstein, in "Brain Microdissection Techniques" (A. C. Cuello, ed.), p. 1 (IBRO Handbook Series). Wiley, New York, 1983.

Fig. 1. Micropunch needle.

The Cryostat. Most commercially available cryostats are suitable; the only requirement is that 300 μm-thick serial sections can be routinely cut and the specimen holder is fixed, or can be fixed, to a ball joint. The Cryo-Cut (American Optic) and Minotome (IEC) have proved to be the simplest and most convenient for this purpose.

Templates or Tissue Choppers. Templates from plexiglass can be fashioned that correspond to the shape of the brain, and cuts are made at 1-mm intervals across the block that razor blades can slide into.[3] Various types of tissue choppers are commercially available.[3]

Stereo- or Dissecting Microscopes. Any kind of microscope can be used. The only requirements are: (1) 6- to 20-fold magnification; (2) upper illumination with variable focused cool light (fiber optic system); (3) free space under the microscope tube high enough (minimum 5 in.) for the cold or rubber stages and for manipulations with the dissecting needles.

Cold or Elastic Stages. (1) Cold stage: a box or dish filled with Dry Ice powder or circulating cooling solution is suitable. The top of the stage should be metal (or a metal sheet should be placed on it), which provides good heat conductivity. (2) Microdissection from fresh brain is performed on a rubber or other elastic stage. A large black stopper is the most convenient and provides an excellent contrasting background.

Microdissecting Needles. These needles are constructed of hard stainless steel tubing mounted in a thicker handle (Fig. 1). For comfortable handling, needles should be 4–5 cm (2 in.) long with a thinner end, at least 0.5 cm long, so that the tip of the needle is visible under the microscope. The end of the needle should be sharpened. The inner diameters of the needles vary from 0.2 to 2.0 mm depending on the size of the brain nuclei to be microdissected. (A complete set comprises 0.2-, 0.3-, 0.5-, 1.0-, 1.5-, and 2.0-mm needles.) It is essential that the needle used be smaller than the smaller diameter of the nuclei to be removed. Needles can be equipped with a stylet. It should be fitted well into the lumen of the needle and reach 2–3 mm beyond the tip of the needle.

Brain Sectioning

The microdissection technique is suitable for fresh and frozen brain sections.

Fresh Brain. Slices should be as thin as possible. Manual brain slicing may be carried out with a wet razor blade on a black elastic (rubber) plate under a magnifier or a stereomicroscope. The brain, the blade, the rubber plate, and the operator's fingertips should all be kept moist while the brain is being sliced. Slices about 0.5 to 1.0 mm thick can be cut. Sections about 1 mm thick can be cut using a template or tissue chappers.[3,4] Fresh cut brain slices cannot be stored; microdissection must be performed within 5 min.

Frozen Brain. Serial sections can be cut from frozen brain. After removal, the brain is frozen on a microtome specimen holder with Dry Ice. Coronal serial sections, 300 μm thick, are cut in a cryostat at $-10°$. The 300-μm section thickness has been chosen empirically for sectioning rat brains; the sections may be 200 μm thick for mouse brains or 400–500 μm thick for dog or cat brains. It is more difficult to pick up the tissue pellet from a section that is considerably thinner than 200 μm; the upper limit of the section thickness is determined by the size of the brain nuclei in the rostrocaudal coordinates.

Sections are placed on histological slides (3–6 sections per slide), then removed from the cryostat and kept at room temperature for a few seconds. When the sections begin to thaw, the glass slide is placed on Dry Ice until micropunch sampling. Sections can be stored in a container (histological slide box) filled with Dry Ice. The box should be closed and insulated to avoid desiccation of the brain tissue.

Freehand Sections of Frozen Brains. Microdissection of brain nuclei of species with relatively large brains (sheep, cattle, dog, primates) is more reproducible from serial sections than their *in situ* preparation. Slightly frozen brains (about 0° to $-5°$) can easily be sliced with a knife on an elastic plate. Serial sections about 1.5–2.0 mm thick can be obtained. These are stored on Dry Ice until microdissection with a fine knife or large micropunch needles.

Micropunch Sampling

Microdissection from fresh brain slices is performed on a rubber plate, and microdissection from frozen brain sections is performed on histological slices kept in a freezing container, under a stereomicroscope (Fig. 2). In fresh brain sections (not later than 5 min after sectioning) the white matter (myelinated fiber systems) and the ventricles can be recognized; they provide excellent landmarks for micropunching the nuclei. Vital staining of the sections with methylene blue[4] before microdissection improves the orientation. Micropunching on frozen sections should be

[4] R. E. Zigmond and Y. Ben-Ari, *J. Neurochem.* **26**, 1285 (1976).

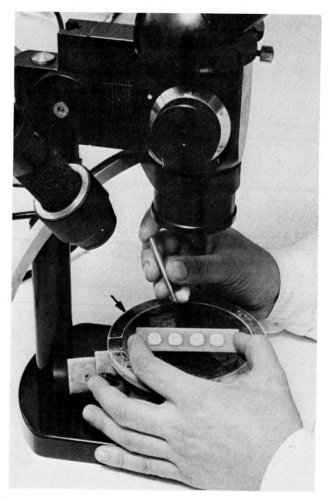

Fig. 2. Micropunch procedure under a stereomicroscope. Four brain sections are placed on a slide and kept frozen on a cold stage (arrow) during dissection. The lamp box of the stereomicroscope filled up with Dry Ice powder is used for that purpose.

quickly accomplished, since the visibility will gradually diminish owing to the accumulation of frost on the section surface; this is especially common in laboratories with high temperature and high humidity.

At the beginning of the micropunch procedure, the needle is moved in an oblique position while the localization of the microdissecting brain region is not fixed (Fig. 3). Then, the tip of the needle is fixed on the section exactly over the nucleus to be punched out, and the needle is

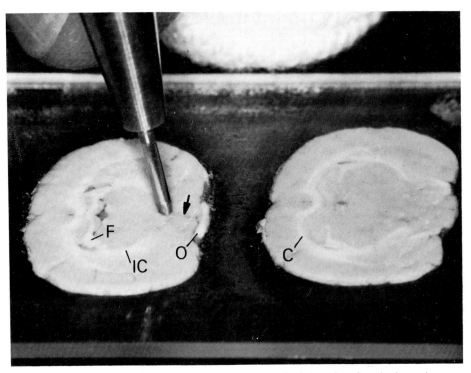

FIG. 3. Microdissection of a hypothalamic (paraventricular) nucleus from brain sections of rat. Landmarks: such major fiber bundles (C, corpus callosum; IC, internal capsule; F, fimbria hippocampi; O, optic chiasm) or brain ventricles (arrow points to third ventricle) can be recognized on the section.

brought into a vertical position and pressed into the tissue. When microdissecting fresh tissue, the needle should be pressed strongly into the rubber plate. The elasticity of the plate pushes the tissue pellet back into the needle. Manipulating on frozen sections, the needle should be pressed, slightly rotated, and quickly withdrawn.

The tissue pellet can be blown out of the needle directly into a tube or dish or onto the tip of a microhomogenizer. If a needle with stylet is used (mainly for micropunching from fresh brain slices) the tissue is just pushed out with the stylet.[3]

After the nuclei are removed, sharp-edged holes remain in the section (Fig. 4). Several nuclei can be removed from one section and over 200 from the whole brain. Rat brains can be cut with 72 coronal sections, each containing a number of individual nuclei. It is a general rule that the tissue pellet should be smaller than the nucleus to be removed; not the entire

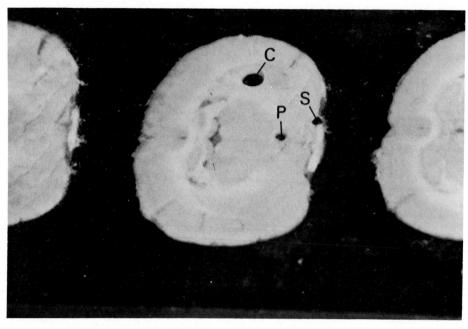

FIG. 4. Sharp-edged holes on the section (which is identical with section shown in Fig. 3) represent the nuclei after removal by micropunch. C, caudate-putamen; P, paraventricular nucleus; S, supraoptic nucleus.

nucleus, but a specimen from the nucleus should be taken out. When dissecting large brain nuclei, the micropunching should be done always on the same part of the nucleus or, if certain cell groups (subdivisions) can be recognized within the nucleus, they should be removed separately like an independent nucleus.

There are several ways to validate the microdissection. The exact topography of holes on the sections (Fig. 4) can be checked under a microscope either on unstained or stained sections. Immediately after the micropunching, slices are allowed to thaw and can be transilluminated. The microdissected area, fiber bundles, and ventricles can be recognized in detail even with low-power magnification. These sections, however, desiccate rapidly and therefore cannot be stored. For that reason, sections can be fixed with 4% formalin, embedded (or refrozen), and sliced for thin sections and stained.[3]

Microdissected tissue pellets are homogenized—except when brain nuclei have been removed for tissue culturing or histological (electron microscopic) studies—for further biochemical analysis with microhomogenizers or by sonication. Pellets are usually too small for direct weighing, therefore their protein contents are measured. The estimated weights of microdissected nuclei, if the punch was complete and no fragments were left on the section, taken from sections 300 μm thick are as follows: with a needle of 0.2 mm inside diameter, 10 μg; with 0.3 mm, 20 μg; with 0.5 mm, 60 μg; with 1.0 mm, 0.25 mg; with 1.5 mm, 0.5 mg; and with 2.0 mm, 1.0 mg.

The micropunch technique is simple; one person can process it alone, but it is more effective when performed by a team. In that case, one person slices the brain, one microdissects the nuclei, and the third one homogenizes the samples, removes aliquots for protein determination, and centrifuges the samples.

Microdissection of brain nuclei may be performed in any species. Since the technique has been developed for dissecting individual rat brain nuclei—and the rat is the most popular laboratory species—detailed microdissection guides and maps are available only for rat brains. A complete guide and map for dissecting over 200 nuclei from 72 coronal serial sections has been published.[2] Detailed description of removal of individual nuclei from larger brain areas has been reported separately: areas from the cerebral cortex,[5] from the hypothalamus,[6] from the diencephalon,[7] from the limbic system,[8] and from the brainstem[9] can be removed by the aid of these guides.

Special microdissecting maps for other species have not yet been published, but excellent topographical atlases are available as aids to recognizing brain nuclei. Recommended stereotaxic maps exist for pigeon,[10] chicken,[11] mouse,[12] rat,[2,13,14] opossum,[15] guinea pig,[16] rabbit,[17] goat,[11]

---

[5] M. Palkovits, L. Záborsky, M. J. Brownstein, M. I. K. Fekete, J. P. Herman, and B. Kanyicska, *Brain Res. Bull.* **4,** 593 (1979).

[6] M. Palkovits, in "Topographical Neuroendocrinology" (W. E. Stumpf and L. D. Grant, eds.), p. 72. Karger, Basel, 1975.

[7] M. J. Brownstein, R. M. Kobayashi, M. Palkovits, and J. M. Saavedra, *J. Neurochem.* **24,** 35 (1975).

[8] M. Palkovits, J. M. Saavedra, R. M. Kobayashi, and M. J. Brownstein, *Brain Res.* **79,** 443 (1974).

[9] M. Palkovits, M. J. Brownstein, and J. M. Saavedra, *Brain Res.* **80,** 237 (1974).

[10] H. J. Karten and W. Hodos, "A Stereotaxic Atlas of the Brain of the Pigeon (*Columba livia*)." Johns Hopkins Press, Baltimore, Maryland, 1967.

[11] T. Joshikawa, "Atlas of the Brains of Domestic Animals." Univ. Tokyo Press, Tokyo; and Pennsylvania State, Univ. Park Press, Baltimore, Maryland, 1968.

[12] R. L. Sidman, J. B. Angevine, Jr., and E. Taber Pierce, "Atlas of the Mouse Brain and Spinal Cord." Harvard Univ. Press, Cambridge, Massachusetts, 1971.

[13] J. F. König and R. A. Klippel, "The Rat Brain: A Stereotaxic Atlas of the Forebrain and Lower Parts of the Brain Stem." Williams & Wilkins, Baltimore, Maryland, 1963.

[14] G. Paxinos and C. Watson, "The Rat Brain in Stereotaxic Coordinates." Academic Press, New York, 1982.

sheep,[11] pig,[11] cattle,[11] horse,[11] cat,[18,19] dog,[20,21] and various primates.[22-26] All these maps are useful for micropunch sampling of brain nuclei.

In spite of the fact that the micropunch technique is rather simple and detailed guides and maps are available, a certain amount of practice is needed to administer this technique correctly. It is easy to punch samples out of brain sections without regard to validation of the micropunch. Incorrect topographical placement and identification of the punching sites lead to uninterpretable data.

---

[15] E. Oswaldo-Cruz and C. E. Rocha-Miranda, "The Brain of the Opossum (*Didelphis marsupialis*). A Cytoarchtectonic Atlas in Stereotaxic Coordinates." Inst. Biofisica Univ. Fed. do Rio de Janeiro, Rio de Janeiro, 1968.
[16] T. J. Luparello, "Stereotaxic Atlas of the Forebrain of the Guinea Pig." Williams & Wilkins, Baltimore, Maryland, 1967.
[17] M. Monnier and H. Gangloff, "Atlas for Stereotaxic Brain Research on the Conscious Rabbit." Elsevier, Amsterdam, 1961.
[18] R. S. Snider and W. T. Niemer, "A Stereotaxic Atlas of the Cat Brain." Univ. of Chicago Press, Chicago, Illinois, 1961.
[19] W. J. C. Verhaart, "A Stereotaxic Atlas of the Brain of the Cat." Van Gorcum, Assen, 1964.
[20] R. K. S. Lim, C. Liu, and R. Moffitt, "A Stereotaxic Atlas of the Dog's Brain." Thomas, Springfield, Illinois, 1960.
[21] S. Dua-Sharma, S. Sharma, and H. L. Jacobs, "The Canine Brain in Stereotaxic Coordinates." MIT Press, Cambridge, Massachusetts, 1970.
[22] R. Emmers and K. Akert, "A Stereotaxic Atlas of the Brain of the Squirrel Monkey (*Saimiri sciureus*)." Univ. of Wisconsin Press, Madison, 1963.
[23] S. L. Manocha, R. T. Santha, and G. H. Bourne, "A Stereotaxic Atlas of the Brain of the Cebus (*Cebus apella*) Monkey." Oxford Univ. Press, London and New York, 1968.
[24] R. S. Snider and J. C. Lee, "A Stereotaxic Atlas of the Monkey Brain (*Macaca mulatta*)." Univ. of Chicago Press, Chicago, Illinois, 1961.
[25] M. R. de Lucchi, B. J. Dennis, and W. R. Adey, "A Stereotaxic Atlas of the Chimpansee Brain (*Pan satyrus*)." Univ. of California Press, Berkeley, 1965.
[26] R. Davis and R. D. Huffman, "A Stereotaxic Atlas of the Brain of the Baboon (*Papio*)." Univ. of Texas Press, Austin, 1968.

Section IV

Use of Chemical Probes

[24] Excitotoxic Amino Acids as Neuroendocrine Research Tools

By JOHN W. OLNEY and MADELON T. PRICE

The putative excitatory transmitters L-glutamate (Glu) and L-aspartate (Asp), also known as excitotoxins (ETs), have the property, when systemically administered, of selectively penetrating circumventricular organ (CVO) regions of brain and, depending on dose, either exciting or destroying CVO neurons. The mechanism by which ETs destroy neurons is such that axons of passage are spared. Therefore, ETs are useful as either provocative or ablative systemic tools for studying the wide range of neuroendocrine functions that CVO neurons are believed to mediate. In addition, the neuroendocrine regulatory roles of other neurons not accessible to blood-borne ETs can be studied by injecting an appropriate ET directly into the locale of such neurons, using a low dose to provoke an acute reversible change in endocrine status or a high dose to delete neurons and thereby permanently eliminate their influence over the endocrine parameter concerned.

The Excitotoxic Concept

It is widely believed that Glu and Asp are the transmitters released at the majority of excitatory synapses in the mammalian central nervous system. The relevant pharmacological and electrophysiological evidence, developed primarily in the laboratories of Curtis and Watkins, has been previously reviewed.[1-3] The Glu neurotoxicity literature, also previously reviewed,[4,5] includes evidence that either oral or subcutaneous administration of Glu to either infant or adult animals of various species results in acute necrosis of neurons in several brain regions, most notably the arcuate nucleus of the hypothalamus (AH), and that the neonatally induced AH lesion is associated in rodents with obesity and multiple neuroendo-

---

[1] D. R. Curtis and J. C. Watkins, *J. Neurochem.* **6,** 117 (1960).
[2] D. R. Curtis and G. A. R. Johnston, *Rev. Physiol.* **69,** 97 (1974).
[3] J. C. Watkins, *in* "Kainic Acid as a Tool in Neurobiology" (E. G. McGeer, J. Olney, and P. McGeer, eds.), p. 37. Raven, New York, 1978.
[4] J. W. Olney, *in* "Kainic Acid as a Tool in Neurobiology" (E. McGeer, J. W. Olney, and P. McGeer, eds.), p. 95. Raven, New York, 1978.
[5] J. W. Olney, *in* "Glutamic Acid: Advances in Biochemistry and Physiology" (S. Garratini, L. J. Filer, M. R. Kare, W. A. Reynolds, and R. Wurtman, eds.), p. 287. Raven, New York, 1979.

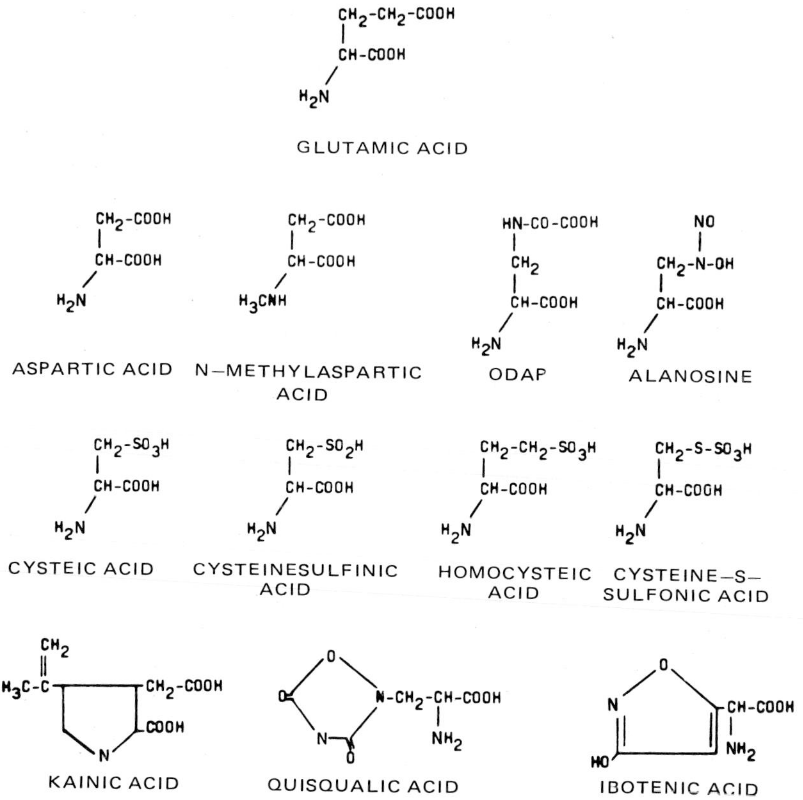

GLUTAMIC ACID

ASPARTIC ACID   N–METHYLASPARTIC ACID   ODAP   ALANOSINE

CYSTEIC ACID   CYSTEINESULFINIC ACID   HOMOCYSTEIC ACID   CYSTEINE–S–SULFONIC ACID

KAINIC ACID   QUISQUALIC ACID   IBOTENIC ACID

FIG. 1. Structural analogs of glutamic acid found to have neuroexcitatory activity in microelectrophoretic studies and neurotoxic activity when administered systemically and/or by direct intracranial injection. The order of toxic potencies is kainic > quisqualic > ibotenic > N-methyl-DL-aspartic > ODAP > DL-homocysteic > alanosine = cysteine-S-sulfonic > cysteine sulfinic = cysteic = aspartic = glutamic. Without significant exception, the same order of potencies has been described for their excitatory activities. Moreover, when administered systemically, those tested have the same order of potencies for releasing luteinizing hormone, an action thought to reflect excitation of AH neurons. ODAP = β-N-oxalyl-L-α,β-diaminopropionic acid.

crine disturbances in adult life. From molecular specificity studies,[1,6] it is known that certain structural analogs of Glu (Fig. 1) mimic both the excitatory and toxic activities of Glu and display a parallel order of potencies for the two phenomena. Ultrastructural studies[7,8] have localized the

[6] J. W. Olney, O. L. Ho, and V. Rhee, *Exp. Brain Res.* **14**, 61 (1970).
[7] J. W. Olney, *Neuropathol. Exp. Neurol.* **30**, 75 (1971).
[8] J. W. Olney, L. G. Sharpe, and R. D. Feigin, *J. Neuropathol. Exp. Neurol.* **31**, 464 (1972).

toxic action of Glu to dendritic and somal portions of the neuron, which contain the excitatory receptors through which the depolarizing effects of Glu putatively are mediated. These initial observations led to the "excitotoxic" hypothesis[6,9] that a depolarization mechanism underlies Glu neurotoxicity and that the toxic action may be mediated through dendrosomal excitatory synaptic receptors specialized for glutamergic (or aspartergic) transmission. Subsequent studies[10] have suggested that Glu/Asp synaptic receptors may be of at least three types: those differentially sensitive to N-methyl aspartate (NMA), kainic acid (KA), or quisqualic acid (QA). The most important advance has been the discovery that certain nonexcitatory Glu analogs, such as D-α-aminoadipate (D-αAA) and D-aminophosphonovalerate (D-APV), antagonize both the excitatory[10] and the toxic[11] activities of ETs at NMA receptors. Specific antagonists for the KA and QA receptor subtypes remain to be found.

Axon-Sparing Lesions

Use of ETs as axon-sparing lesioning agents is a concept that originated in early studies of immature mice[7] and monkeys,[8] which showed by electron microscopy that subcutaneous Glu destroys AH neurons without any pathological effects on axonal processes terminating in AH or coursing through AH to reach the median eminence. Thus, endocrine deficits resulting from the Glu lesion could more confidently be attributed to the loss of AH neurons than would be the case if, as occurs with other ablative techniques, all tissue components including numerous fiber tracts coursing through AH had been destroyed.

More recently, several ETs have been shown to induce Glu-like dendrosomatotoxic/axon-sparing lesions when introduced directly into various brain regions.[9,12] For neuroendocrine research purposes, therefore, axon-sparing lesions can be produced either by subcutaneous administration or by direct intrahypothalamic injection of an appropriate ET. Some advantages of the subcutaneous approach are (1) noninvasiveness, (2) ease of administration, (3) bilateral symmetry of lesions, (4) high yield of animals bearing lesions of uniform characteristics. A potential disadvantage is that the lesion is not specific for a single brain region, but rather for several brain regions (CVO) which lack blood–brain barriers (Fig. 2). This

[9] J. W. Olney, L. G. Sharpe, and T. deGubareff, *Neurosci. Abstr.* **1**, 371 (1975).
[10] J. C. Watkins, J. Davies, R. H. Evans, A. A. Francis, and A. W. Jones, in "Glutamate as a Neurotransmitter" (G. Di Chiara and G. L. Gessa, eds.), p. 263. Raven, New York, 1981.
[11] J. W. Olney, J. Labruyere, J. F. Collins, and K. Curry, *Brain Res.* **221**, 207 (1981).
[12] E. McGeer, J. W. Olney, and P. McGeer (eds.), "Kainic Acid as a Tool in Neurobiology." Raven, New York, 1978.

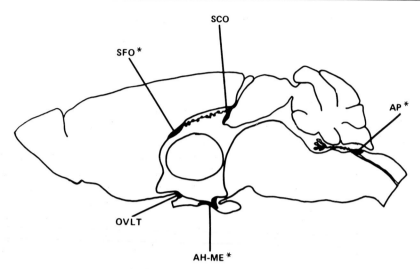

FIG. 2. Diagram of a midsagittal section of rat brain depicting the location of circumventricular organs (CVO), which are special midline periventricular zones that differ from other brain regions in having fenestrated capillaries that are permeable to various blood-borne substances. Neurons in or near CVO are subject to damage by systemically administered excitotoxins, whereas other brain regions, even in immature animals, are relatively well protected. The CVO include the area postrema (AP), subcommissural organ (SCO), subfornical organ (SFO), organum vasculosum of the lamina terminalis (OVLT), and the arcuate hypothalamic–median eminence (AH-ME) region. Asterisks indicate CVO most vulnerable to damage by circulating excitotoxins.

disadvantage may be more apparent than real, since AH is the CVO region that sustains the most severe damage and the other two CVO that are significantly damaged [area postrema (AP) and subfornical organ (SFO)] are not believed to share functions with AH neurons.[13] On the other hand, a very high ET dose sometimes destroys medial preoptic (MPO) neurons that possibly do share common functions with AH neurons. However, high ET doses that destroy 80–90% of the neurons in AH delete only 5–10% of those in MPO.[14]

An important question relative to either the systemic or direct injection of ET concerns cytospecificity. The bulk of evidence suggests that most central neurons are sensitive to both the depolarizing and neurotoxic actions of ET. However, some neurons appear to be relatively more sensitive to some ETs than to others,[10,15] and possibly there are central

[13] Also worth mentioning is the fact that vulnerability of AP and SFO neurons to circulating ET provides a means, thus far unexploited, of studying these neuronal groups.
[14] J. W. Olney, unpublished observation, 1980.
[15] J. W. Olney, in "Glutamate as a Neurotransmitter" (G. Di Chiara and G. L. Gessa, eds.), p. 375. Raven, New York, 1981.

neurons that are invulnerable to the toxic action of any ET.[16,17] Unfortunately, experience in the direct intrahypothalamic application of ET is so limited at present that very few tested guidelines for lesioning specific neuronal groups can be given.

Basis for Provocative Approach

Since transmitter-mediated excitation of hypothalamic neurons is a plausible mechanism by which release of hypophysiotrophic hormones from these or related neurons might be influenced, it should be possible to trigger an acute change in neuroendocrine parameters (e.g., circulating hormones) by administration of ET in nontoxic but excitatory doses either systemically or by direct injection into various hypothalamic nuclei (assuming that various neuroendocrine axes have neurons that possess ET-sensitive receptors and are programmed to give a neuroendocrine-relevant response to an ET stimulus). Experience with the provocative approach is limited almost entirely to systemic administration, but, in principle, considerable information might be obtained by direct application of ET in excitatory (nontoxic) doses to specific neuronal groups within the endocrine hypothalamus or elsewhere, provided the neuronal group is part of a neuroendocrine regulatory pathway.

Role of Blood–Brain Barriers and Age

The CVO distribution of lesions induced by systemic ETs is explained by the fact that CVO brain regions, unlike the brain proper, have permeable capillaries that permit relatively free access of circulating ETs to CVO neurons.[18] Although subcutaneous Glu destroys AH neurons in either infant or adult animals, the minimal effective dose for adults is about four times that for infants and a nearly lethal dose destroys <20% of AH in adults compared to >80% in infants.[5] The presumptive explanation for this is that both Glu metabolic systems in peripheral organs and Glu uptake–inactivation mechanisms in CVO brain regions become progressively more competent as a function of maturation.

Selecting an Excitotoxin

The basic properties and idiosyncrasies of various ETs will be discussed with an aim to identifying the agent(s) of choice for each neuroendocrine investigational purpose.

[16] C. deMontigny and J. P. Lund, *Neuroscience* **5**, 1621 (1980).
[17] J. W. Olney and T. deGubareff, *in* "Kainic Acid as a Tool in Neurobiology" (E. G. McGeer, J. W. Olney, and P. McGeer, eds.), pp. 201–217. Raven, New York, 1978.
[18] M. T. Price, J. W. Olney, O. H. Lowry, and S. Buchsbaum, *J. Neurochem.* **36**, 1774 (1981).

Kainic Acid. Although the powerful neurotoxic action of kainic acid (KA) was first demonstrated by subcutaneous administration to mice,[19] its usefulness as a systemic neuroendocrine investigational tool was immediately questioned because doses sufficient to damage CVO neurons were usually rapidly lethal. Moreover, in subsequent experiments with rats it was shown that KA, when injected either subcutaneously[20] or directly into various brain regions[17,21] causes sustained limbic seizures that are associated with a relatively widespread pattern of seizure-mediated brain damage. Furthermore, direct injection of KA into the hypothalamus has revealed that some hypothalamic neurons, including those in AH, are less sensitive than neurons elsewhere in brain to KA neurotoxicity.[17] Therefore KA can probably not be considered of much value for systemic neuroendocrine research, and its value as a direct neuroendocrine probe or lesioning agent is contingent upon the demonstration in future research that specific neuroendocrine systems are differentially sensitive to KA and that its action can be confined to specifically targeted systems (see the section on direct intracranial approach below).

Quisqualic Acia. Quisqualic acid (QA) is nearly as potent in excitatory and toxic activity as KA but is thought to act at a different subset of receptors[10] and, therefore, on theoretical grounds is worth considering as a neuroendocrine investigational tool. However, we have administered this agent subcutaneously to both infant and adult rodents and found that it exerts very little toxic action against AH or other CVO neurons.[14] It appears, therefore, that AH neurons may not possess receptors that are differentially sensitive to QA and that QA will probably not prove to be useful as a systemic investigational tool. Since QA has not been injected directly into any hypothalamic region, it remains to be clarified whether a given hypothalamic nucleus might be selectively sensitive to its excitatory or toxic effects.

Ibotenic Acid. Ibotenic acid (Ibo) has been proposed as a promising alternative axon-sparing lesioning agent to KA because it is relatively powerful in destroying local neurons when injected into various brain regions and does not cause the seizures or seizure-related pattern of remote brain damage associated with KA.[22,23] Some disadvantages of Ibo are that it is scarce, exorbitantly expensive, and unstable, a variable amount being converted in biological systems to the inhibitory compound

---

[19] J. W. Olney, V. Rhee, and O. L. Ho, *Brain Res.* **77**, 507 (1974).
[20] J. E. Schwob, T. A. Fuller, J. L. Price, and J. W. Olney, *Neuroscience* **5**, 991 (1980).
[21] Y. Ben-Ari, J. Lagowska, E. Tremblay, and G. LeGal LaSalle, *Brain Res.* **163**, 176 (1979).
[22] R. Schwarcz, T. Hökfelt, K. Fuxe, G. Jonsson, M. Goldstein, and L. Terenius, *Brain Res.* **37**, 199 (1979).
[23] S. Hansen, C. Kohler, M. Goldstein, and H. V. M. Steinbush, *Brain Res.* **239**, 213 (1982).

muscimol. These factors tend to disqualify Ibo, at least for systemic investigations. Moreover, all available evidence[22,24,25] suggests that NMA receptors mediate Ibo activity. Since NMA is a stable molecule, has the same spectrum of excitotoxic effects and roughly the same potency as Ibo,[26] is readily available at low cost, is suitable for both systemic and direct intracranial injection and, like Ibo, lacks the KA property of inducing seizure-mediated brain damage, we recommend trying NMA before Ibo for either provocative or ablative purposes.

N-Methyl Aspartate. Many of the properties of NMA were described in the preceding paragraph. Although NMA[26] is too toxic for administration to neonatal animals, it is well tolerated by 20-day-old mice or rats and is more effective than Glu in destroying AH neurons in rodents from this age to adulthood. Doses in the range of 75–100 mg/kg sc delete 15–20% of the neurons in AH, and increasing the dose to 300 mg/kg increases the neuronal loss to ~30%, which is a ceiling effect; higher doses increase mortality without destroying more AH neurons.[27] While this is not as large a lesion as might be desired, it is larger than we have achieved by treatment of *adult* rodents with Glu or other ETs; therefore, we consider NMA the preferred ET for systemic lesioning of *adult* rodent brain.

The toxic action of NMA on AH neurons is effectively blocked by D-αAA and D-APV (see the table and Fig. 3),[11] which specifically antagonize the excitatory action of NMA and of natural transmitter at NMA-sensitive synapses.[10] From this and related evidence,[4–6] it seems very likely that the locus of NMA action in AH is at NMA-type synaptic receptors on the dendrosomal surfaces of AH neurons. The power of NMA as an investigational tool is substantially enhanced by the availability of antagonists that simultaneously exert receptor-specific blocking action against both its excitatory and toxic activities; in any given experimental paradigm, such antagonists can be used to confirm the mechanism by which the tool is working.

N-Methyl aspartate has proved to be quite useful as a provocative systemic probe of the gonadotropin axis, and recent findings suggest that NMA can be similarly applied for studying other neuroendocrine axes (see below under Systemic Provocative Approach).

Although the direct injection of NMA into neuroendocrine-related

[24] R. Schwarcz, *Neuroscience* **7**, Suppl. S188 (1982).
[25] P. Krogsgaard-Larsen, T. Honore, and J. J. Hansen, *Nature (London)* **284**, 64 (1980).
[26] The D isomer of NMA is much more powerful than the L isomer, but the DL racemate, which is commercially available (Sigma) at low cost, is a suitable substitute for the D isomer. Doses given herein are for NM(DL)A. The potency of Ibo is about 1.5 times that of NM(D)A or 3 times that of NM(DL)A.
[27] J. W. Olney, A. C. Scallet, and T. A. Fuller, unpublished observation, 1981.

RELATIVE ANTAGONIST POTENCIES OF D-αAA AND D-APV[a]

n	Agents	Dose[b] (mg/kg)	Lesion severity[c]
24	NMA	50	23.8 ± 1.2
10	NMA-D-αAA	50–250	13.3 ± 2.1[d]
5	NMA-D-αAA	50–500	1.5 ± 0.7[d]
3	NMA-D-αAA	50–750	0.7 ± 0.6[d]
6	NMA-D-APV	50–2.5	17.0 ± 3.0[e]
6	NMA-D-APV	50–5.0	5.5 ± 2.1[d]
6	NMA-D-APV	50–7.5	0.5 ± 0.2[d]
9	NMA-D-APV	50–>7.5	0.0 ± 0.0[d]

[a] From Olney et al.[11]
[b] Agents were administered subcutaneously to 25-day-old mice, and necrotic neuronal counts were made 4 hr later. NMA, N-methyl aspartate; D-αAA, D-α-aminoadipate; D-AVP, D-aminophosphonovalerate.
[c] AH is a bilateral nucleus. Counts given here are mean (±SEM) number of necrotic AH neurons (per heminucleus) per transverse section at point of maximal damage to AH. A typical lesion is depicted in Fig. 3.
[d] $p < 0.001$ compared to NMA control.
[e] $p < 0.05$ compared to NMA control.

brain regions in either subtoxic or toxic doses should provide useful information, experience in this type of application is extremely limited at present. For investigators interested in exploring NMA for this type of use, tentative guidelines are suggested below in the section on direct intracranial approach.

In summary, although experience with some applications are limited, we believe that NMA may prove to be the ET of first choice for many if not all neuroendocrine research purposes other than systemic lesioning of CVO regions of neonatal brain, which is best achieved with Glu.

Homocysteic Acid. Homocysteic acid (HCA) is not as powerful as NMA in toxic activity but is more powerful than Glu or Asp. We explored both D- and L-HCA as systemic tools and found that the D form has more toxic side effects than Glu, but neither isomer at tolerated doses is more effective than Glu for neonatal lesioning, nor than NMA for adult lesioning.[28]

[28] A. C. Scallet and J. W. Olney, unpublished observation, 1981.

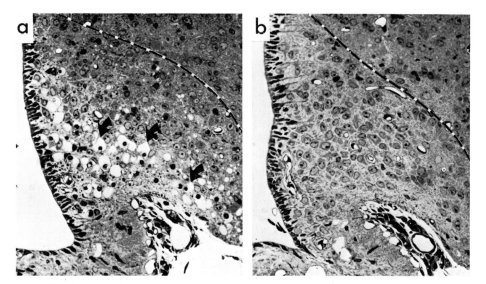

FIG. 3. Light micrographs of arcuate hypothalamic region (under dashed line) of brain from a control 25-day-old mouse (a) treated 4 hr previously with N-methyl aspartate (NMA) (50 mg/kg sc) and experimental (b) treated with the same dose of NMA plus D-APV (5 mg/kg sc). The presence of acutely necrotic neurons (arrows) in (a) but not (b) illustrates the powerful blocking activity that D-aminophosphonovalerate exerts against NMA toxicity. For quantitative evaluation of this blocking action compared to that of D-α-aminoadipate, see the table. From Olney et al.[11]

Glutamate. It was shown in early studies[29] that Glu, Asp, or cysteic acid are equipotent as systemic AH lesioning agents, but a higher incidence of toxic side effects associated with the latter two agents, especially cysteic acid, makes Glu the most useful of the three. Various schedules for Glu treatment of neonatal rodents have been used, the aim being to expose CVO neurons periodically during the first 10 days of life to transiently elevated extracellular concentrations of Glu. The schedule we have usually used is 2 mg/kg sc on postpartum days 1 and 3, then 4 mg/kg on days 5, 7, and 9. Mice or rats thus treated have negligible mortality and sustain a loss of 75–80% of the neurons in the AH region. The loss is most complete (>95%) in the middle portions of the AH and less complete at the rostral and caudal extremes of the nucleus.[28,30] Efforts to achieve a more complete AH lesion by using a combination of excitotoxins or varying the treatment schedule have been unfruitful.[28] Therefore, we currently

[29] B. Schainker and J. W. Olney, *J. Neural Transm.* **35,** 207 (1974).
[30] H. Khachaturian, S. J. Watson, J. W. Olney, M. J. Brownstein, and H. Akil, unpublished observation, 1982.

regard Glu, when administered according to the above schedule, as the best approach for inducing a lesion in CVO brain regions of the immature rodent.

Systemic Ablative Approach

The literature on the systemic ablative use of ETs has been extensively reviewed.[5,31,32] Treatment of neonatal animals with Glu (or Asp in a few cases) has been the approach employed; see under Glutamate, above, for the treatment schedule we recommend for neonatal rodents. Under *N*-methyl aspartate above, we have recommended a treatment schedule for partial removal (30%) of AH neurons from adult rodent brain. Since the neuroendocrine sequelae of adult NMA treatment have not been studied, this approach will not be further discussed. Adult animals treated neonatally with Glu, including mice, rats, and hamsters,[33] manifest a complex syndrome of endocrine disturbances, including normophagic obesity, skeletal stunting, impaired reproductive capacity, reduced mass of the anterior pituitary and gonads, and reduced pituitary content of growth hormone (GH), prolactin (Prl), and luteinizing hormone (LH). Pulse amplitude of serum GH is markedly reduced, and basal serum levels of GH and LH also tend to be depressed. Responsiveness of hypothalamic Prl release mechanisms are altered—Prl response to estrogenic stimulation is weakened, but to serotonergic stimuli it is exaggerated. In Glu-treated rats, serum triiodothyronine and free thyroxine indices are in the hypothyroid range; and in Glu-treated mice, serum corticosterone levels are strikingly elevated. The Glu-treated rodent provides an interesting animal model for studying the relationship between the endocrine hypothalamus, obesity, and diabetes, since adiposity develops despite normal or subnormal food intake and is accompanied by hyperglycemia and hyperinsulinemia.[5,34,35]

Analysis of the mediobasal hypothalamus[30,32,36–38] after Glu treatment suggests that the deleted AH neurons comprise several subpopulations

[31] J. W. Olney and M. T. Price, *in* "Kainic Acid as a Tool in Neurobiology" (E. McGeer, J. W. Olney, and P. McGeer, eds.), p. 239. Raven, New York, 1978.
[32] J. S. Kizer, C. B. Nemeroff, and W. W. Youngblood, *Pharmacol. Rev.* **29**, 301 (1978).
[33] For dose schedule in treating hamsters, see A. Lamperti and G. Blaha, *Biol. Reprod.* **14**, 362 (1976).
[34] D. P. Cameron, T. K. Poon, and G. C. Smith, *Diabetology* **12**, 621 (1976).
[35] M. Utsumi, Y. Hirose, K. Ishihara, H. Makimura, and S. Baba, *Biomed. Res* **1**, Suppl., 154 (1980).
[36] M. Romagnano, W. Pilcher, C. Bennet-Clarke, T. Chafel, and S. Joseph, *Brain Res.* **234**, 387 (1982).
[37] M. Romagnano, T. Chafel, W. Pilcher, and S. Joseph, *Brain Res.* **236**, 497 (1982).
[38] D. Krieger, A. Liotta, G. Nicholsen, and J. Kizer, *Nature (London)* **278**, 562 (1979).

that collectively contain a number of neurotransmitter or modulatory substances (e.g., dopamine, acetylcholine, enkephalin, various members of the opiomelanocortin series, and somatostatin). Evidence that all β-endorphin (β-End)-containing cell bodies in brain are located in the AH region prompted studies[30,38] revealing that the Glu lesion is associated with 75% loss of AH β-End immunoreactive cell bodies and a marked reduction in β-End content both in the hypothalamus and in extrahypothalamic regions, where β-End fibers are thought to terminate. An increase in delta opiate receptors in the thalamus of rats treated neonatally with Glu has been interpreted as a possible opiate denervation supersensitivity phenomenon.[39]

Systemic Provocative Approach

Researchers exploring this approach have administered a single subtoxic dose of Glu (1 g/kg sc) or NMA (15–25 mg/kg sc) to weanling or adult rats or NMA (15 mg/kg iv) to adult female rhesus monkeys. Serum hormonal responses have usually been monitored in the acute posttreatment period (minutes to hours). The gonadotropin axis has been studied more comprehensively than any other neuroendocrine axis by the systemic provocative approach. Administering a single subtoxic dose of NMA to weanling or adult male rats[40] or adult female rhesus monkeys[41] results in a rapid (5–15 min) elevation of serum luteinizing hormone (LH). A similar but less striking release of LH is elicited in adult male rats by a single subtoxic dose of Glu.[42] In a previously reviewed series of studies,[43] the LH-releasing action of NMA has been characterized as reversible, rapid in onset, brief in duration, and dependent on AH neurons. The latter point is supported by the observation that the LH-releasing action of NMA is not demonstrable *in vivo* in rats whose AH neurons have been deleted by Glu treatment in infancy. The LH-releasing action of NMA is blocked by simultaneous subcutaneous administration of the NMA antagonist αAA or the inhibitory transmitter GABA. Although αAA also blocks the neurotoxic action of NMA on AH neurons,[11] GABA does not. In pituitary incubation experiments, neither NMA, Glu, nor GABA influences LH release from either rat or monkey pituitary.[44] From these several observations we have tentatively drawn certain conclusions (sche-

[39] E. Young, J. Olney, and H. Akil, *J. Neurochem.* **40**, 1558 (1983).
[40] M. T. Price, J. W. Olney, and T. J. Cicero, *Neuroendocrinology* **26**, 352 (1978).
[41] R. C. Wilson and E. Knobil, *Brain Res.* **248**, 177 (1982).
[42] J. W. Olney, T. J. Cicero, E. R. Meyer, and T. deGubareff, *Brain Res.* **112**, 420 (1976).
[43] J. W. Olney and M. T. Price, in "Glutamate as a Neurotransmitter" (G. DiChiara and G. L. Gessa, eds.), p. 423. Raven, New York, 1981.
[44] J. Tal, M. T. Price, and J. W. Olney, *Neurosci. Abstr.* **8**, 62 (1982).

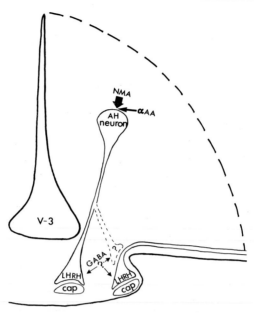

FIG. 4. Diagram of the arcuate hypothalamic (AH) region indicating the postulated loci of action of N-methyl aspartate (NMA), α-aminoadipate (αAA), and γ-aminobutyric acid (GABA) on the LH axis. We propose that the AH neuron has excitatory synaptic receptors on its dendritic and/or somal surfaces that are sensitive to NMA and through which it is physiologically innervated by fibers that use Glu or Asp as transmitter. By antagonist action at this receptor, αAA blocks all actions of NMA mediated through this receptor, i.e., its excitatory, toxic, and LH-releasing actions. GABA does not exert its inhibitory activity at the excitatory receptor, since it blocks one excitation-related action of NMA (LH release) but not the other (toxic action). We propose that GABA acts at a presynaptic locus on axons in the median eminence. By inhibiting at this locus, GABA prevents the excitatory stimulus from triggering secretion of luteinizing hormone-releasing hormone (LHRH) from these axons into the portal capillary system. Whether the LHRH-secreting axons originate from AH or non-AH neurons (or both) remains to be determined.

matically illustrated in Fig. 4) regarding the locus of action of these agents on the LH axis.

In addition to the above findings pertaining to the LH axis, it was reported that a systemic subtoxic dose of NMA induces a rapid elevation of serum growth hormone in adult male rats[45] and an elevation of Prl and follicle-stimulating hormone in adult female rhesus monkeys.[41] A single subtoxic dose of Glu induces an acute elevation of serum Prl but sup-

[45] C. Nemeroff, in "Proceedings of the Symposium on Excitotoxins" (K. Fuxe, P. J. Roberts, and R. Schwarcz, eds.). Pergamon, Oxford, in press.

presses pulsatile GH output in adult rats.[46] The basis for NMA and Glu having opposite actions on GH is unclear.

Direct Intracranial Injection of Excitotoxins

To date only limited use has been made of ETs as provocative or ablative tools for direct injection into neuroendocrine-related brain regions. Hansen et al.,[23] noting that sexual performance is altered by electrolytic lesioning of either the medial preoptic or lateral hypothalamic areas, ruled out a role for lateral hypothalamic neurons by injecting Ibo into both regions and reproducing the loss of sexual performance only in preoptic-lesioned rats. This serves as a useful example of the kind of information than can be obtained by direct intracranial application of an ET, although for reasons cited above, NMA may be the preferred agent for this type of application.

Simpson et al.[47] injected Glu into the paraventricular region of the hypothalamus to produce an axon-sparing lesion and described an increase in weight gain that they attributed to loss of neurons at the injection site. This must be interpreted with caution, since convincing histological evidence for neuronal loss was not presented and it is known that homeostatic mechanisms for uptake and inactivation of Glu are so efficient in adult brain that even an exceedingly large dose usually destroys only a very small number of neurons. Since NMA is intrinsically a more powerful ET than Glu and is less readily taken up and inactivated, it is preferable to Glu for inducing axon-sparing lesions by direct injection into adult brain. In our experience,[17] injecting NMA into the paraventricular hypothalamic nucleus destroys parvocellular neurons and spares magnocellular neurons. It would be of interest, therefore, to implant chronic cannulas bilaterally in the paraventricular nucleus and observe animals for acute changes in feeding behavior after injection of a subtoxic dose of NMA or for more long-term changes in food intake and weight gain after injection of a toxic dose. If only parvocellular neurons are sensitive to NMA, any changes induced could be ascribed to an effect mediated through these neurons. Preventing the NMA effect by preinjection of D-APV at the same site would further localize the effect to NMA-type receptors on the dendrosomal surfaces of these neurons.

Several groups have injected KA into the lateral hypothalamic area (LHA) to study the role of LHA neurons in feeding and drinking behaviors. Interpretational difficulties were encountered because of spread of

---

[46] L. C. Terry, J. Epelbaum, and J. B. Martin, *Brain Res.* **217**, 129 (1981).
[47] E. L. Simpson, R. M. Gold, L. J. Standish, and P. L. Pellett, *Science* **198**, 514 (1977).

the lesion into nontargeted brain areas. Grossman et al.[48] compared several regimens for injecting KA into LHA and reported that a low dose (2.35 nmol), small volume (0.1 µl), and long infusion period (12–15 min) resulted in optimal confinement of the lesion to LHA and maximal impact on feeding and drinking parameters. However, being concerned that "mechanical injection" of KA often causes unintended damage to nontargeted structures, they subsequently administered KA microelectrophoretically into LHA; this produced a partial LHA lesion (large neurons were spared) without extension of damage either to adjacent or remote structures.[49] However, Ruth[50] applied KA microelectrophoretically to the hippocampus and found that, even at the lowest delivery rate required for destroying local hippocampal neurons, distant lesions were evident in numerous brain regions (ventral subiculum, amygdala, hypothalamus, and certain neocortical and olfactory regions). In view of evidence[20,21] that this pattern of distant damage is seizure-mediated and that diazepam protects against such damage but not against local KA neurotoxicity at the injection site, the use of subcutaneous diazepam adjunctively with intracranial KA injection warrants exploration.

It is clear from the above discussion that the direct intracranial application of excitotoxins is an approach that, although theoretically appealing, is still in an embryonic stage of development. For investigators interested in furthering this development, the following admonition and tentative guidelines are offered.

Admonition. Careful histological monitoring of brain tissue (both local and distant) is a prerequisite to the intracranial use of ET for either provocative or ablative purposes.

Guidelines. Since NMA, KA, and QA putatively act at different receptor subtypes, any or all might theoretically be useful for lesioning or probing a given group of neurons—the agent(s) of choice for a given task can be determined only empirically. To prevent spread of ET effects to nearby nontargeted neurons, the agent should be delivered either microelectrophoretically or in small volumes (0.05–0.5 µl) over long infusion periods (10–15 min). For lesioning purposes the dose of KA should not exceed 2.5 nmol, and simultaneous subcutaneous administration of diazepam (10–20 mg/kg) warrants consideration. For QA or NMA lesioning, employ doses of 7.5–75 or 25–250 nmol, respectively. For provocative purposes KA might be tested at 2.5–250 pmol, QA at 5–500 pmol, and NMA at 7.5–750 pmol.

[48] S. P. Grossman, D. Dacey, A. Halaris, T. Collier, and A. Routtenberg, *Science* **202**, 557 (1978).
[49] S. P. Grossman and L. Grossman, *Physiol. Behav.* **29**, 553 (1982).
[50] R. E. Ruth, *Exp. Neurol.* **76**, 508 (1982).

Acknowledgments

This work was supported in part by NIMH Research Career Scientist Award MH-38894 (JWO) and USPHS Grants DA-00259 and NS-009156.

[25] Use of Excitatory Amino Acids to Make Axon-Sparing Lesions of Hypothalamus

By J. VICTOR NADLER and DEBRA A. EVENSON

Much of our information about hypothalamic control of hormonal secretion has been derived from studies in which specific hypothalamic nuclei were destroyed and the physiological effects of such lesions were then determined. The ablative approach continues to be extensively employed. However, the relatively nonselective techniques that have commonly been used to produce the desired lesion, such as knife cuts or electrocoagulation, destroy all neuronal elements, including cell bodies, afferent fibers, and fibers of passage traversing the target zone, and they inevitably damage the vasculature as well. In such circumstances, it is difficult to attribute any physiological change that is detected to ablation of any single neuronal element. Such ambiguities may be resolved or avoided altogether by destroying the nucleus in question with an "excitotoxic" amino acid.

Excitatory amino acids in general are toxic to neurons of the central nervous system.[1] Although the mechanism of amino acid-induced neuronal cell death has not been thoroughly elucidated, structure–activity studies have clearly suggested an essential relationship between neuroexcitatory and neurotoxic activities.[2-5] The major advantage of excitotoxin lesions lies in their specificity for structures that bear receptors for the amino acid. Nearly all cell bodies and dendrites are depolarized by at least some of the commercially available excitatory amino acids and thus can potentially be destroyed. However, fibers of passage and afferent fibers bear few or no excitatory amino acid receptors, and thus survive. Excit-

[1] J. W. Olney, in "Kainic Acid as a Tool in Neurobiology" (E. G. McGeer, J. W. Olney, and P. L. McGeer, eds.), p. 95. Raven, New York, 1978.
[2] J. W. Olney, O. L. Ho, and V. Rhee, *Exp. Brain Res.* **14,** 61 (1971).
[3] R. Schwarcz, D. Scholz, and J. T. Coyle, *Neuropharmacology* **17,** 145 (1978).
[4] J. V. Nadler, D. A. Evenson, and G. J. Cuthbertson, *Neuroscience* **6,** 2505 (1981).
[5] R. Zaczek and J. T. Coyle, *Neuropharmacology* **21,** 15 (1982).

atory amino acids do not kill glial cells and do not damage vascular elements (apart from incidental mechanical damage caused by the injection procedure itself). The relative specificity of excitotoxin lesions obviates many of the objections to the ablative approach. Neurons of the arcuate nucleus can be destroyed by parenteral injections of excitatory amino acid.[6] To destroy other hypothalamic neurons, however, one must apply the toxin directly to the target cells. Here we describe methods that have proved to be useful in our laboratory for injecting excitatory amino acids into discrete brain regions of rodents and for determining the extent of the resulting lesions.

Injection Procedure

Excitatory amino acids are delivered to hypothalamic nuclei by pressure injection, as described by Shipley.[7] The apparatus for injection includes a 1.0-μl syringe that can be filled through a side vent (Dynatech-Precision Sampling Corp., Baton Rouge, LA; No. 100011). The side vent is linked by flexible Teflon tubing (1.0 mm i.d.) to a 5-ml glass syringe. The needle of the microsyringe is connected by another length of flexible Teflon FEP tubing (0.794 mm i.d.) to a micropipette made from previously drawn "Kwik-Fil" microcapillary tubing (1.0 mm o.d.; WPI Inc., New Haven, CT) that has been bumped to a tip diameter of 15–20 μm. To fit snugly over the thin microsyringe needle, the tubing must first be warmed in a bunsen burner flame, quickly stretched, and then cut to a small terminal diameter. To insert the micropipette into the other end of the tubing, first heat the ends of both in a bunsen burner flame. Heating expands the end of the tubing and thus permits the undrawn end of the micropipette to be forced tightly into it. Apply household cement to the connection between micropipette and tubing to assure an airtight seal. Finally, clamp the syringes to support rods and fasten the micropipette to the electrode carrier of a stereotaxic instrument in a vertical position.

Just before commencing the injection procedure, connect the various components as follows. First, pull the plunger of the microsyringe nearly out in order to open the side vent. Then fill the 5-ml syringe and attached Teflon tubing with mineral oil and connect them to the side vent of the microsyringe. Fit the previously stretched end of the smaller-bore tubing over the microsyringe needle and glue it in place with rubber cement. By advancing the plunger of the 5-ml syringe, completely fill the tubing and micropipette with mineral oil. Finally, advance the microsyringe plunger

[6] J. W. Olney and M. T. Price, in "Kainic Acid as a Tool in Neurobiology" (E. G. McGeer, J. W. Olney, and P. L. McGeer, eds.), p. 239. Raven, New York, 1978.
[7] M. T. Shipley, Brain Res. Bull. **8**, 237 (1982).

to the 1.0-μl position, closing the side vent. If all parts have been properly connected, the system is entirely free of air bubbles.

A number of excitatory amino acids can be used to produce axon-sparing lesions in the rodent hypothalamus.[2] Thus far, however, we have experience only with kainic acid (KA) and N-methyl-DL-aspartic acid (NMA). Because of its high potency and solubility in aqueous media, KA may be dissolved directly in artificial cerebrospinal fluid.[8] On the other hand, NMA and most other excitatory amino acids of lesser potency and solubility should first be dissolved in one equivalent of NaOH before bringing the solution to volume with artificial cerebrospinal fluid. Amino acid solutions are equilibrated with 95% O_2 or compressed air–5% CO_2 immediately before use.

To fill the micropipette with amino acid solution, first advance the microsyringe plunger nearly to the end of its travel, to expel most of the mineral oil in the pipette. Then carefully place the pipette tip in the amino acid solution and advance the plunger as far as possible, expelling a small drop of oil into the solution. Fill the pipette with solution by withdrawing the plunger to the desired volume. After placing the pipette tip at the desired location in the brain under stereotactic guidance, amino acid solution is ejected by advancing the microsyringe plunger until the desired volume has been delivered. Because the micropipette tip is so fragile and difficult to clean, we prefer to use a different micropipette for each injection.

Comments

1. Because of its great potency, KA has most frequently been employed to make localized excitotoxin lesions in the brain. However, KA has two potential disadvantages: it preferentially destroys certain neurons over others, and, at sufficient dose, it initiates prolonged seizure activity that may be fatal to certain neurons of the limbic system.[9] Lesions distant from the injection site can be avoided by the discrete application of small doses of KA or by treating the animal with an anticonvulsant, such as diazepam.[10] However, it may be difficult or impossible to destroy whole hypothalamic nuclei with KA, since many hypothalamic neurons, notably those located most medially, survive exposure to even rather large doses of this excitant.[11] To avoid both of these problems, we advocate trying

[8] K. A. C. Elliott, in "Handbook of Neurochemistry" (A. Lajtha, ed.), Vol. 2, p. 103. Plenum, New York, 1969.
[9] J. V. Nadler, *Life Sci.* **29**, 2031 (1981).
[10] Y. Ben-Ari, E. Tremblay, O. P. Ottersen, and R. Naquet, *Brain Res.* **165**, 362 (1979).
[11] G. M. Peterson and R. Y. Moore, *Brain Res.* **202**, 165 (1980).

one or two of the more potent amino acid excitants that do not readily initiate seizures and that discriminate poorly, if at all, among the various neuronal types. In our experience, NMA[4,12] and ibotenic acid[4,13] are the best selections. Of the two, we prefer NMA, even though it is less potent than ibotenic acid, because of its greater chemical and biological stability and its much lower cost.

2. The extent of the lesion can be controlled by varying the concentration of the amino acid solution, the volume of solution injected, and the rate of injection. These parameters are best established empirically for the specific application. In general, we prefer to hold the volume delivered (0.1–0.2 μl) and delivery time (30 min) constant and to vary the amino acid concentration. The data of Nadler et al.[4] provide some guidance as to the useful concentration range for several amino acid excitants.

3. Restricted excitotoxin lesions could be made equally well by microiontophoresis or with commercially available micropressure devices. The major advantage of the device described here is its simplicity and low cost. These features should be especially attractive to researchers who need to make these lesions only occasionally or who would have no other use for the relatively expensive commercial equipment.

Histological Evaluation

In laminated regions of the brain, a Nissl procedure may be sufficient to reveal the location of degenerating neuronal cell bodies. When characterizing excitotoxin lesions in such poorly laminated regions as hypothalamus, however, we find a silver impregnation method to be indispensable. In addition, silver impregnation demonstrates the loci of terminal degeneration. Ideally, such a method vividly outlines the degenerating structures, while staining intact elements more lightly or not at all. The procedures of Fink and Heimer[14] and of de Olmos[15] have been successfully employed to characterize excitotoxin lesions, but they are quite time-consuming, often yield results of variable quality, and generally require the patience and skill of the full-time anatomist. We have found the shorter and simpler procedure described by Gallyas et al.[16] to yield impregnations of consistently superior quality. In contrast to the older methods, this procedure resulted from a systematic study of the experimental

[12] J. W. Olney, *Adv. Biochem. Psychopharmacol.* **27,** 375 (1981).
[13] R. Schwarcz, T. Hökfelt, K. Fuxe, and G. Jonsson, M. Goldstein, and L. Terenius, *Exp. Brain Res.* **37,** 199 (1979).
[14] R. P. Fink and L. Heimer, *Brain Res.* **4,** 369 (1967).
[15] J. de Olmos and W. R. Ingram, *Brain Res.* **33,** 523 (1971).
[16] F. Gallyas, J. R. Wolff, H. Böttcher, and L. Zaborszky, *Stain Technol.* **55,** 299 (1980).

conditions under which degenerating terminals, degenerating axons, intact axons, and lysosomes bind silver (become "argyrophilic") and under which the silver ions are subsequently reduced. It eliminates many of the lengthy, difficult, and arcane steps that discourage many investigators from employing silver impregnation as a tool for characterizing their lesions. We describe a modification of the procedure of Gallyas et al.[16] suitable for demonstrating both degenerating terminals and degenerating cell bodies.

Tissue Preparation

Perfuse the animal first with 0.9% (w/v) NaCl for 30 sec, then with 4% (w/v) paraformaldehyde in 0.1 M sodium phosphate buffer, pH 7.4, for 20 min (about 400–500 ml). About 1 hr after the end of the perfusion, carefully remove the brain and postfix it in cold, buffered paraformaldehyde solution for 3–8 days. Then cut the brain into serial frozen sections of 40-μm thickness and transfer the sections to individual compartments in staining racks that are immersed in the same paraformaldehyde solution. The sections may be stained immediately or refrigerated in the fixative for as long as 1 week.

Convenient staining racks can be constructed from 3 cm-deep hard plastic grids ("egg crates"), such as are used in some fluorescent ceiling lights. Cut these grids into rectangular racks of a size that just fits the staining dishes. Smoothe any rough spots by dissolving the excess plastic with chloroform. Cover the bottom of each rack with nylon mesh stocking. Stretch the nylon tightly and affix it to the rack by carefully dissolving the plastic with chloroform where it contacts the stocking.

Reagents

Stock solutions
 A: 9% (w/v) NaOH
 B: 16% (w/v) NH_4NO_3
 C: 50% (w/v) $AgNO_3$
 D: 1.2% (w/v) NH_4NO_3
 E: Dissolve 5 g of anhydrous Na_2CO_3 in 300 ml of 95% ethanol and 600 ml of H_2O. Dilute to 1 liter with H_2O.
 F: Dissolve 0.5 g of anhydrous citric acid in 15 ml of 37% formalin, 100 ml of 95% ethanol, and 700 ml of H_2O. Add solution A with stirring until the pH reaches 5.8–6.1 then dilute to 1 liter with H_2O.
 G: 0.5% (v/v) acetic acid

These solutions should be stored in tightly closed bottles at room temperature and protected from light. Under these conditions, they are stable for several weeks.

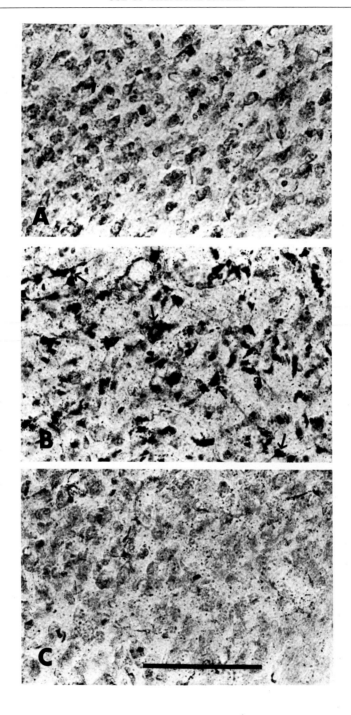

Working solutions (prepared no more than 1 hr before beginning the staining procedure)
Pretreating solution: Mix equal volumes of solutions A and D.
Impregnating solution: Add 1.5 volumes of solution A to each volume of solution B. Then add 0.5–0.6 ml of solution C for each 100 ml of total volume.
Washing solution: Mix 1 ml of solution D with each 100 ml of solution E.
Developing solution: Mix 1 ml of solution D with each 100 ml of solution F.

Staining Procedure

First wash the sections free of fixative by immersion in three changes of H_2O for 5 min each. Then transfer them sequentially into two changes of pretreating solution for 5 min each, impregnating solution for 10 min, three changes of washing solution for a *total* of 5 min, and finally developing solution for at least 1 min. Mount the sections from developing solution onto acid-cleaned glass slides that have been precoated with 0.05% (w/v) chromium potassium sulfate, 0.5% (w/v) gelatin. When the mounted sections have dried, immerse them in three changes of solution G for 10 min each, wash with H_2O, dehydrate with a graded series of ethanol solutions, clear with butanol and three changes of xylene, and enclose with coverslips.

The most uniform results are obtained by observing a few simple precautions. Working solutions and staining dishes must be covered at all times to prevent loss or absorption of NH_3. All glassware to be used in the staining procedure should be acid-washed (e.g., with 50% HNO_3) and stored free from dust. Racks are transferred from one solution to the next with the fingers or with plastic implements; use no metal at any point. Racks must be quickly drained of as much of the previous solution as possible before immersion in the next solution. Agitate the sections by hand for 1 min each time they are placed in a new solution, except when they are immersed in developing solution. All incubations through the developing step are best carried out at room temperature on a shaker set

---

FIG. 1. Cell body and terminal degeneration 1 day after a discrete injection of kainic acid (3.75 nmol in 0.1 μl injected over a 30-min period) into the ventromedial portion of the anterior hypothalamic nucleus. (A) Anterior hypothalamic nucleus contralateral to the injection. There are no degenerating somata. (B) Anterior hypothalamic nucleus just posterior to the injection site. Most, but not all, neuronal somata are degenerating. Some degenerating somata are indicated by arrows. (C) Terminal degeneration in the ipsilateral dorsal premammillary nucleus. Scale bar = 100 μm.

at a slow speed. A separate set of working solutions should be prepared for each rack of sections. Working solutions should not be reused. Do not allow the sections to fold over, get caught on the rack, or float on the surface of the solution. If, despite these precautions, results still prove to be unsatisfactory, consult the original reference for a list of the possible causes and corrective measures.

Comments

1. This procedure predominantly visualizes degenerating terminals and lysosomes. The rapid accumulation of lysosomes within degenerating somata presumably explains the ability of this procedure to visualize these structures as well (Fig. 1). Few degenerating axons stain under the conditions described here. Intact cell bodies and myelinated axons stain light yellow-brown and are easily distinguished from degenerating structures, which appear dark gray or black. One can manipulate the intensity of coloration of intact and degenerating structures by varying the proportion of solutions A, B, and C in the impregnating solution. We find that adding 0.5–0.6 ml of $AgNO_3$ solution to each 100 ml of a 1.5:1 mixture by volume of solutions A and B provides adequate contrast, while staining intact somata just intensely enough to obviate any need for counterstaining. Blood vessels and connective tissue accumulate some black silver deposit also, but they are unlikely to be confused with neuronal elements.

2. The procedure is very sensitive. In some cases, degenerating terminals can be visualized within a day after excitotoxin treatment and remain demonstrable for 2 months or more. On the other hand, degenerating cell bodies and dendrites stand out best if the animal is perfused between 8 hr and 2 days after treatment. Optimal survival times may vary somewhat, according to the brain region or pathway under study.

3. One serious disadvantage is that only tissue that has been fixed *in situ* can presently be stained with this procedure. Therefore one cannot ordinarily use this method to verify lesions in animals from which tissue is taken for biochemical analysis. Also, if survival times longer than a few days are required, degenerating somata will not be clearly demonstrable. Finally, the procedure has not yet been modified successfully for use with slide-mounted sections, although an on-the-slide modification of the Fink–Heimer procedure[17] has been employed in studies with excitotoxin lesions.[18]

---

[17] A. Hjorth-Simonsen, *Stain Technol.* **45**, 199 (1970).
[18] J. V. Nadler, B. W. Perry, C. Gentry, and C. W. Cotman, *J. Comp. Neurol.* **192**, 333 (1980).

[26] Use of Specific Ion Channel Activating and Inhibiting Drugs in Neuroendocrine Tissue

By. P. MICHAEL CONN

The availability of compounds that have direct actions on ion channels have made it possible to determine (1) the presence of such channels in a particular cell, (2) the relative numbers of channels that can be coupled to a biological response, and (3) the degree to which receptor-stimulated responses are mediated through ion channels. As such, these drugs are powerful tools in neuroendocrine research.

While many compounds are available that interact with ion channels, the present effort is directed toward only those that are among the best characterized and most specific. A summary is provided in the table. Local anesthetics—the actions of which are often through general pertubation of the plasma membrane, and compounds that have unclear or multiple actions on intracellular and plasma membranes (mersalyl acid, dantrolene)—have also been omitted. In some cases, only the best characterized of a group of compounds is described. Also omitted from discussion are the chemical ionophores that alter membranes to allow enhanced permeability. They do not activate endogenous channels so much as create new sites of ion entry.

Because of their profound actions on neural, endocrine, and muscle systems, drugs that activate and inhibit ion channels should be used with caution. They are frequently used as toxins in the plant and animal kingdoms. They are often present in the laboratory dissolved in dimethyl sulfoxide (DMSO). Such solutions are readily absorbed through the skin and therefore pose a special hazard.

Use

Characterization of Actions: General Considerations

Ion Channel Specificity. This is best demonstrated by manipulation of the ionic components of cell culture incubation medium. A compound believed to open the Na^+ channel (veratridine, for example) should have its action inhibited by depletion of extracellular Na^+. Medium containing 139 mM choline chloride, 10 mM HEPES, pH 7.4, 2 mM $CaCl_2$, 0.2 mM $MgCl_2$, and 0.3% dialyzed bovine serum albumin (BSA) is often suitable for this purpose. The dialysis of the albumin (against distilled water) is

DRUGS ACTING AT THE LEVEL OF ION CHANNELS

Drug	Chemical structure, molecular weight, source, and availability
Aconitine[a]	16-Ethyl-1,16,19-trimethoxy-4-(methoxymethyl)aconitane-3,8,10,11,18-pentol 8-acetate 10-benzoate, $C_{34}H_{47}NO_{11}$; molecular weight (M_r) 645.72. Obtained from *Aconitum napellus* L., Ranunculaceae (Sigma Chemical Company, St. Louis, MO).
Batrachotoxin[a-g]	Batrachotoxinin A, 20-(2,4-dimethyl-1H-pyrrole-3-carboxylate), $C_{31}H_{42}N_2O_6$; M_r 538.69. From the Colombian arrow poison frog, *Phyllobates*. Enhances Na$^+$ conductance.
Grayanotoxin	A mixture of dodecahydro-1,1-4β,8β-tetramethyl-7,9a-methano-9aαH-cyclopenta[b]heptalene-2,4,8,11,11aβ,12(1H)-hexol 12-acetate, dodecahydro-1,1,8β-trimethyl-4 methylene-7,9a-methano-9aαH-cyclopenta[b]heptalene-2,8,11,11aβ,12(1H)-pentol, and dodecahydro-1,1,4β,8β-tetramethyl-7,9a-methano-9aαH-cyclopenta[b]heptalene-2,4,8-11,11aβ,12(1H)-hexol; $C_{22}H_{36}O_7$, $C_{20}H_{32}O_5$, $C_{20}H_{34}O_6$, respectively. From *Rhododendron, Kalmia, Leucothoe,* Ericaceae leaves.
Methoxyl verapamil (D600)[h-o]	α-Isopropyl-a-[(N-methyl-N-homoveratryl)-γ-aminopropyl]-3,4,5-trimethoxyphenylacetonitrile-HCl, $C_{28}H_{41}ClN_2O_5$ (Knoll Pharmaceutical, Ludwigshafen-am-Rhein, Germany). Other antagonists including nifedipine, nimodipine, nisoldipine, nitrendipine, verapamil, YC-93, and others are available. Some have been described (see text footnote 5 and 6). They block Ca^{2+} channels in secretory, contractile, and neural tissues. Primary sources are Miles and Knoll Pharmaceutical Companies. (See also text footnotes 1–5, 13–15).
Pumiliotoxin C[p-s]	5-Methyl-2-propyl-*cis*-decahydroquinoline, $C_{13}H_{18}N$ (other synthetic derivatives are available). From poison frogs, Dendrobatidae. Interacts with acetylcholine receptor regulated ion channel.
Scorpion venom (tityustoxin)[t,u]	M_r 6995. Increases Na$^+$ permeability.
Tetrodotoxin[e,f,k]	Octahydro-12-(hydroxymethyl)-2-imino-5,9 : 7,10a-dimethano-10aH-[1,3]dioxocino[6,5-d]pyrimidine-4,7,10,11,12-pentol, $C_{11}H_{17}N_3O_8$; M_r 319.28. Obtained from ovaries and liver of puffer fishes, Tetraodontidae Sigma Chemical Company (St. Louis, MO). Inhibits Na$^+$ ion channels.
Veratridine[k,m,v]	4,9-Epoxycevane-3,4,12,14,16,17,20-heptol 3-(3,4-dimethoxybenzoate), $C_{36}H_{51}NO_{11}$. M_r 673.81 (although frequently supplied as the sulfate salts) M_r,

DRUGS ACTING AT THE LEVEL OF ION CHANNELS (continued)

Drug	Chemical structure, molecular weight, source, and availability
Veratrine	anhydrous, 771.9. Obtained from the seeds of *Schoenocanlon officinale* and rhizomes of *Veratrum album* L., Liliaceae (Sigma Chemical Company, St. Louis, MO). Erratically available and some batch variability; best to obtain a supply sufficient for planned studies. A mixture extracted from seeds of *Schoenocaulon officinale* or rhizomes of *Veratrum album* L., Liliaceae. In addition to veratridine, it contains, in undefined amounts, other alkaloids including cevadine, cevadilline, sabadine, and cevine (Sigma Chemical Company, St. Louis, MO).

[a] W. H. Herzog, *J. Gen. Physiol.* **47,** 719 (1964).
[b] E. X. Albuquerque and J. W. Daly, *in* "The Specificity and Action of Animal, Bacterial and Plant Toxins, Receptors and Recognition (P. Cuatrecasas, ed.), Ser. B, Vol. 1, p. 297. Chapman & Hall, London, 1977.
[c] E. X. Albuquerque, J. W. Daly, and B. Witrop, *Science* **172,** 995 (1971).
[d] E. X. Albuquerque, J. E. Warnick, and F. M. Sansone, *J. Pharmacol. Exp. Ther.* **176,** 511 (1971).
[e] E. X. Albuquerque and J. E. Warnick, *J. Pharmacol. Exp. Ther.* **180,** 683 (1972).
[f] J. R. Symthies, F. Benington, and R. D. Morin, *Nature (London)* **231,** 188 (1971).
[g] J. E. Warnick, E. X. Albuquerque, and F. M. Sansone, *J. Pharmacol. Exp. Ther.* **176,** 496 (1971).
[h] J. Cohen and Y. Gutman, *Br. J. Pharmacol.* **65,** 641 (1979).
[i] A. S. Fairhurst, M. L. Whittaker, and F. J. Ehlert, *Biochem. Pharmacol.* **29,** 155 (1980).
[j] C. J. Mayer, C. VanBreemen, and R. Casteels, *Pfluegers Arch.* **337,** 333 (1972).
[k] P. M. Conn and D. C. Rogers, *Endocrinology* **107,** 2133 (1980).
[l] P. M. Conn, *in* "Cellular Regulation of Secretion and Release" (P. M. Conn, ed.), p. 459. Academic Press, New York, 1982.
[m] D. E. Smith and P. M. Conn, *Life Sci.* **30,** 1495 (1982).
[n] J. Marian and P. M. Conn, *Mol. Pharmacol.* **16,** 196 (1979).
[o] D. J. Triggle, *Am. Phys. Soc.* **1,** 56 (1980).
[p] E. X. Albuquerque, J. E. Warnick, M. A. Maleque, F. C. Kauffman, R. Tamburini, Y. Minit, and J. W. Daly, *Mol. Pharmacol.* **19,** 411 (1981).
[q] L. E. Overman and P. J. Jessup, *J. Am. Chem. Soc.* **100,** 5179 (1978).
[r] Y. Nimitkitpaisan, J. W. Daly, P. J. Jessup, L. Overman, J. E. Warnick, and E. X. Albuquerque, *Trans. Am. Soc. Neurochem.* **11,** 233 (1980).
[s] J. E. Warnick, P. J. Jessup, L. E. Overman, M. E. Eldefrawi, Y. Nimit, J. W. Daly, and E. X. Albuquerque, *Mol. Pharmacol.* **22,** 565 (1982).
[t] J. E. Warnick, E. X. Albuquerque, and C. R. Diniz, *J. Pharmacol. Exp. Ther.* **198,** 155 (1976).
[u] W. A. Catterall and L. Beress, *J. Biol. Chem.* **253,** 7393 (1978).
[v] W. A. Catterall and M. Nirenberg, *Proc. Natl. Acad. Sci. U.S.A.* **70,** 3759 (1973).

needed, in most cases, to remove salts and salt chelators added during collection and purification of the BSA.

Blocking of Action. The action should be blocked by inhibitors of the ion channel under study. Tetrodotoxin, for example, should block the action of veratridine if the Na^+ channel is involved. D600 should not inhibit the action unless the Ca^{2+} channel is also involved (see the next paragraph).

Characterization of Indirect Actions. The gross perturbation of one ion channel can cause alterations of other channels. For example, veratridine stimulates release of pituitary gonadotropins,[1] an effect generally accepted to be mediated via the calcium ion channel.[2] Since the effect of veratridine can be blocked by tetrodotoxin or by D600, it appears that veratridine acts on the Na^+ channel, which results in Ca^{2+} entry through its own channel. This is similar to the response to depolarizing concentrations (50 mM) of KCl.

Measurement of Relative Numbers of Channels

Ion channel activating agents can be used to compare numbers of functional channels. Estrogens, for example, increase the sensitivity of LH release from cultured pituitary cells to gonadotropin-releasing hormone. The observation that veratridine stimulates LH release with the same efficacy and potency whether or not the cells have been estrogen-treated indicates that the estrogen-dependent change *precedes* the receptor-coupled ion channel.[3]

Implication of an Ion Channel as the Mediator of Receptor Actions

Pituitary gonadotropin release in response to gonadotropin releasing hormone (GnRH) can be blocked by D600 but not by tetrodotoxin.[1,4] These observations suggest that the Ca^{2+} channel, but not the Na^+ channel, mediates the receptor action. Similarly, depletion of extracellular Ca^{2+} inhibits responsiveness. Depletion of extracellular Na^+ does not result in inhibition. A needed control is the demonstration that tetrodotoxin actually inhibits the Na^+ channel in the pituitary gonadotrope cell. The observation that tetrodotoxin inhibits the release of gonadotropin in response to veratridine satisfies this requirement.

[1] P. M. Conn and D. C. Rogers, *Endocrinology* **107**, 2133 (1980).
[2] P. M. Conn, in "Cellular Regulation of Secretion and Release" (P. M. Conn, ed.), p. 459. Academic Press, New York, 1982.
[3] D. E. Smith and P. M. Conn, *Life Sci.* **30**, 1495 (1982).
[4] J. Marian and P. M. Conn, *Mol. Pharmacol.* **16**, 196 (1979).

Practical Aspects

It will be immediately obvious there are wide differences in responsiveness of secretory, neural, and muscle tissue to these drugs. This has been emphasized for calcium ion channel antagonists.[5] Accordingly, it is best to develop dose-response curves and determine inhibitory concentrations *de novo* for each tissue under study. Careful characterization of continued cell viability (reversibility of action, trypan blue exclusion, failure to "spill out" soluble enzymes and other contents) are necessary controls to exclude artifacts due to cell death.[6]

[5] D. J. Triggle, *Am. Phys. Soc.* **1,** 56 (1980).
[6] Conn, P. M., Rogers, D. C., and Seay, S. G., *Endocrinology* **113,** in press (1983).

Section V

Quantitation of Neuroendocrine Substances

[27] Development and Use of Ultrasensitive Enzyme Immunoassays

By GLENNWOOD E. TRIVERS, CURTIS C. HARRIS, CATHERINE ROUGEOT, and FERNAND DRAY

Enzyme immunoassays (EIAs) have been the most successful of the nonisotopic immunoassays that have been developed to supplement or replace radioimmunoassay (RIA). Because of their safety, low cost, simplicity, and comparable sensitivity, EIAs have undergone rapid and continuous growth in their varieties and their widespread applications for measuring small quantities of molecules of medical and scientific interest. Both laboratory and clinical studies have shown their usefulness in the fields of cancer research[1-6] and infectious disease.[7-9] Although EIAs of polypeptide hormones are still being developed and are not as yet available for routine clinical application, experience in other fields suggests that this will soon become a reality. In this chapter we describe the basic principles of EIAs and provide examples of hormones measured by EIAs. These assays employ reagents immobilized in the wells of Microtiter plates and are usually quantitated by measuring visible, fluorescent, or radioactive products of enzyme substrates. When highly sensitive EIAs were needed, ultrasensitive enzyme radioimmunoassays (USERIA) were developed.[9a] The sensitivity of these EIAs is in the femtomole range or less.

History and Development of the EIA

EIAs and RIAs are formulated on similar immunological principles. EIAs are primarily different in that enzyme activity is substituted for

[1] I. C. Hsu, M. C. Poirier, S. H. Yuspa, D. Grunberg, I. B. Weinstein, R. H. Yolken, and C. C. Harris, *Cancer Res.* **41,** 1091 (1981).
[2] I. C. Hsu, M. C. Poirier, S. H. Yuspa, R. H. Yolken, and C. C. Harris, *Carcinogenesis* **1,** 455 (1980).
[3] J. A. Roth and R. A. Wesley, *Cancer Res.* **42,** 3978 (1982).
[4] F. P. Perera, M. C. Poirier, S. H. Yuspa, J. Nakayama, A. Jaretzki, M. Curnen, D. M. Knowles, and I. B. Weinstein, *Carcinogenesis* **3,** 1405 (1982).
[5] C. C. Harris, R. H. Yolken, and I. C. Hsu, *Methods Cancer Res.* **20,** 213 (1982).
[6] A. Haugen, J. D. Groopman, I. C. Hsu, G. R. Goodrich, G. N. Wogan, and C. C. Harris, *Proc. Natl. Acad. Sci. U.S.A.* **78,** 4124 (1981).
[7] A. Voller, D. E. Bidwell, G. Huldt, and E. Engvall, *Bull. W. H. O.* **51,** 209 (1974).
[8] I. Ljungstrom, E. Engvall, and E. I. Ruitenberg, *Proc. Int. Congr. Parasitol. 3rd,* p. 1204 (1974).
[9] R. H. Yolken, *Yale J. Biol. Med.* **53,** 85 (1980).
[9a] C. C. Harris, R. H. Yolken, and I. C. Hsu, *Proc. Natl. Acad. Sci. U.S.A.* **76,** 5336 (1979).

radioactivity as marker for the presence and quantity of an immunoreactant. The enzyme is coupled to one of the reactants and amplifies the signal from the antigen–antibody reaction by rapidly converting enzymic substrates to product.

In 1969, Avrameas[10] reported the conjugation of several different enzymes to antibodies, using sodium periodate treatment for horseradish peroxidase and glutaraldehyde for alkaline phosphatase and several others. Avrameas used these new reagents to detect solubilized antigens[11] and antigens localized in tissue.[10] Engvall and Perlmann[12] combined the radioimmunosorbent technique (RIST) for insolubilizing antibody by coupling to cellulose and the Avrameas method for enzyme–antibody conjugation by glutaraldehyde treatment. The result was a fluid-phase (F-P) EIA for measuring alkaline phosphatase-labeled human IgG, and the introduction of the phrase "enzyme-linked immunosorbent assay" with the acronym ELISA. Soon after, they modified an RIA method used by Catt and Tregear[13] in which antibody was immobilized on the inner surfaces of polystyrene tubes. This "solid-phase" (S-P) modification[14] allowed the separation of "free" reactants by decantation and eliminated the necessity for time-consuming centrifugations required in their first study. Van Weeman and Schuurs[15] reported the first EIA for a hormone in 1971. However, it was not until 1975 that Dray et al.,[16] designed an EIA for a hormone that achieved the sensitivity of the RIA.

Often since its evolution, EIA has progressed in parallel with the RIA. However, a departure occurred in 1974 when, Voller et al.[7] designed an assay using polystyrene Microtiter plates coated overnight with solubilized malaria antigen for the purpose of monitoring malaria infections in the field. Microtiter plate assays have subsequently become one of the more popular and important S-P methods. In addition, methods have also been developed for coupling antigens and antibodies to other solid phases including agarose,[17] cellulose,[18] silicone rubber rods,[19] polyacrylamide and glass beads.[20] More recently, magnetic polyacrylamide agarose

[10] A. Avrameas, *Immunochemistry* **6**, 43 (1969).
[11] S. Avrameas, *Int. Rev. Cytol.* **27**, 349 (1970).
[12] E. Engvall and P. Perlmann, *Immunochemistry* **8**, 871 (1971).
[13] K. Catt and G. W. Tregear, *Science* **158**, 1570 (1967).
[14] E. Engvall, K. Jonsson, and P. Perlmann, *Biochem. Biophys. Acta* **251**, 427 (1971).
[15] B. K. Van Weeman and A. H. W. M. Schuurs, *FEBS Lett.* **15**, 232 (1971).
[16] F. Dray, J. M. Andrieu, and F. Renard, *Biochim. Biophys. Acta* **403**, 131 (1975).
[17] J. G. Streefkerk and A. M. Deelder, *J. Immunol. Methods* **7**, 225 (1975).
[18] B. Ferrua, R. Maiolini, and R. Masseyeff, *J. Immunol. Methods* **25**, 49 (1979).
[19] S. Ohtaki and Y. Endo, in "Enzyme Immunoassay" (E. Ishikawa, T. Kawai, and K. Miyia, eds.), p. 198. Igaka Shoin, Tokyo, 1981.
[20] F. Dray and C. Gros, in "Enzyme Immunoassay" (E. Ishikawa, T. Kawai, and K. Miyia, eds.), p. 146. Igaka Shoin, *Tokyo,* 1981.

beads[21] have been employed in order to eliminate the centrifugation normally required for washing and for separating "free" reagents.

One class of EIA can be performed in a single step. The "homogeneous" EIA (one antibody and one reaction mixture) developed by Rubenstein et al.[22] is based upon covalent coupling of a hapten to an enzyme in such a way that, either by steric hindrance or conformational changes, complexes of hapten and anti-hapten antibody inhibit the enzyme substrate interactions. Therefore, "free" haptens in a mixture of enzyme-labeled hapten and antihapten antibodies cause an increase in enzyme activity, giving evidence of the specific hapten in a standard preparation or an unknown sample.

Most current EIA methods are of the heterogeneous class (requiring one or more wash steps to separate "free" reactants and using one or more antibodies from different species). They are of several different types, including S-P assays in which the unlabeled target antigen or antibody is immobilized on insoluble surfaces, primarily plastic. As in RIAs, assays are designed for noncompetition (NC), competition (C), or competitive inhibition (immunometric[23]), with either the subject antigen or an antibody as the enzyme-labeled component. In competitive and noncompetitive assays the resulting measure may be direct or indirect with respect to the relationship of the enzyme marker to the molecule being assayed. In noncompetitive assays the antigen (Ag) or the primary antibody (Ab_1) is immobilized, and the Ag, Ab_1, or a heterologous antibody (Ab_2) is conjugated to the enzyme marker (see Table I). An example of a noncompetitive design is the "sandwich" EIA[24] (variation No. 6, Table I). Immobilized, unlabeled Ab_1 binds Ag from the fluid phase or unknown sample, and the addition of enzyme-labeled Ab_1 results in the direct quantitation of the number of Ag molecules bound in the S-P reaction complex. Performed in reverse, i.e., with labeled antigen, the assay works as well for the detection of antibody. Competitive assays have the basic design of noncompetitive assays except that the unbound specific reactant (either antibody or antigen, labeled or unlabeled) is mixed in F-P with known amounts of the bound reactant as standard or with unknown samples for identification. If neither F-P reactant is labeled, as in the USERIA, the measure is indirect and an additional step is required to introduce an enzyme-labeled heterologous reactant, usually antibody (variation No. 4, Table I).

---

[21] J. L. Guesdon and S. Avrameas, *Immunochemistry* **14**, 443 (1977).
[22] K. E. Rubenstein, R. S. Schneider and E. F. Ullman, *Biochem. Biophys. Res. Commun.* **47**, 846 (1972).
[23] R. Elkins, *Nature (London)* **284**, 14 (1980).
[24] R. Maiglini and F. R. Masseye, *J. Immunol. Methods* **8**, 223 (1975).

TABLE I
Variations in Solid-Phase EIA Procedures for Noncompetition (NC), Competition (C), and Competitive Inhibition (CI) Models Using Labeled (*) and Unlabeled Antigen (Ag) or Antibody (Ab) for Direct (D) and Indirect (ID) Measurements

(Assay type/ method)	Addition of reagents in sequential steps 1–5 after washing the immunosorbent				
	1	2	3	4	5
Immobilized antigen					
1. (NC/D)	*Ab_1	Substrate	Measure hydrolysis	—	—
2. (NC/ID)[a]	Ab_1	*Ab_2	Substrate	Measure hydrolysis	—
3. (CI/D)	*Ab_1 + Ag	Substrate	Measure hydrolysis	—	—
4. (CI/ID)[a]	Ab_1 + Ag	*Ab_2	Substrate	Measure hydrolysis	—
5. (C/D)	*Ab_1 + Ab_1	Substrate	Measure hydrolysis	—	—
Immobilized antibody					
6. (NC/D)[b]	Ag	*Ab_1	Substrate	Measure hydrolysis	—
7. (NC/ID)[c]	Ag	Ab_1	*Ab_2	Substrate	Measure hydrolysis
8. (CI/D)	*Ag + Ab_1	Substrate	Measure hydrolysis	—	—
9. (C/D)	*Ag + Ag	Substrate	Measure hydrolysis	—	—

[a] Indirect "sandwich"; the indirect "sandwich" (Nos. 2 and 4) is the variation used in the USERIA.
[b] Direct "sandwich."
[c] Amplified direct "sandwich."

The EIA literature is extensive and constantly expanding. Descriptions of many potentially applicable and new methods can be found in several excellent monographs, reviews, and proceedings of symposia[18,24–31] now available. Accessing the literature via computer searches

[25] E. Engvall, in "Biomedical Applications of Immobilized Enzymes and Proteins" (T. M. S. Chang, ed.), Vol. 2, p. 87. Plenum, New York, 1977.
[26] D. Watson, *Lancet* **2,** 570 (1976).
[27] A. Voller, D. E. Bridwell, and A. Bartlett, "The Enzyme Linked Immunosorbent Assay (ELISA)." Flowing Pub., Guernsey, England, 1977.
[28] E. T. Maggio, ed., "Enzyme Immunoassay." CRC Press, Boca Raton, Florida, 1980.

can be extremely difficult, yielding primarily RIA references, owing perhaps to the lack of differentiating nomenclature.

In the remaining discussion we will emphasize the development of the USERIAs.[9a]

General Considerations

Antibody Production

Methods for antibody production can be found in basic texts and in special publications on immunology techniques. This is an area that abounds with variations in method and success. Antibodies for heterologous EIAs are prepared by the procedures used for any other purpose. In our laboratory, as in most, rabbits are routinely used for polyclonal sera and mice for high-specificity monoclonal preparations.

Most proteins are immunogenic and can be given in their natural forms for antibody responses. Smaller molecules, such as polypeptides and other haptens, may require a carrier, an immunogenic larger protein, in the immunization mixture. Covalent and electrostatic coupling between molecules such as keyhold limpet hemocyanin (KLH) or bovine serum albumin (BSA) is frequently used with molecules of 5000 daltons or less. The value of an antiserum from conjugated immunogens is based upon the ratio of antihapten antibodies to those directed against the carrier–hapten "linker" or "bridge," which can exist as a part of the carrier molecule. It is also important that the enzyme-labeled antibody demonstrates good and stable affinity (avidity) for Ab_1, its target antigen.

Enzyme-Labeled Antibodies (Conjugates)

The most common procedures for coupling enzyme to antibody involve glutaraldehyde or sodium m-periodate; the former for a host of different enzymes, the latter for peroxidase conjugation.[1] Noncovalent coupling is also done using the avidin–biotin system.[31] Avidin has four sites for biotin to attach. If biotinylated Ab_1 is allowed to react with immobilized Ag, and avidin is subsequently added, the result is a potential amplification of the signal when biotinylated enzyme is introduced as the last reactant. This is a within-assay antibody-conjugation through avidin

---

[29] S. B. Pal, ed., "Enzyme Labelled Immunoassay of Hormones and Drugs." de Gruyter, Berlin, 1978.

[30] R. M. Nakamura, W. R. Dito, and E. S. Tucker III, eds., *Lab. Res. Methods Biol. Med.* **4**, (1980).

[31] A. Voller, A. Bartlett, and D. Bidwell (eds.), "Immunoassays for the 80's." University Park Press, Baltimore, Maryland, 1981.

in a biotin sandwich. Whatever the conjugation method is, the conjugate must have high titer and good specificity, both of which must be confirmed by experiment. The conjugate must also be stable for long-term storage and free of both unbound enzyme and antibody. This objective is apparently difficult to achieve, but the advantage in heterogeneous EIAs is that unbound enzyme in the conjugate is removed in the wash step and does not contribute to erroneous results. The most popular enzymes in use with EIAs are alkaline phosphatase (AP), horseradish peroxidase (HRP), glucose oxidase (GO), and β-galactosidase (BGS). These enzymes satisfy the much desired criteria of availability—low cost and easy conjugation—and, moreover, they have a chromogenic, fluorogenic, or radioactive substrate that is stable and easily measured. In addition, their substrates are cheap, safe, and soluble: p-nitrophenylphosphate (PNPP) for AP; o-phenylenediamine (OPD) + H_2O_2 for peroxidase; o-dianisidine for GO; β-D-galactoside for BGS. In a comparative study[32] of these four enzymes, AP and BGS were shown to be superior to GO and HRP in reproducibility and sensitivity, which were judged to be comparable to RIA. The difference seems largely due to the greater stability of the substrates used with AP and BGS. The substrates of GO and HRP are apparently particularly unstable in EIAs with extended incubation periods.

USERIA

The USERIA is a S-P EIA modified in the final step by substituting a radiolabeled substrate, i.e., tritiated adenosine 5-monophosphate ([³H]AMP), tritiated p-nitrophenylphosphate ([³H]PNPP), or tritiated o-nitrophenyl-β-D-galactoside ([³H]BGS) for nonradiolabeled enzyme substrates. This method (see Models 2 and 4, Table I) has been used to develop assays for several antigens including cholera toxin and rotavirus,[9a] carcinogen-DNA adducts,[1,6] and epidermal growth factor (EGF).[33] We describe the method using results that include our experiences with several new antigens.

Materials and Methods

Reagents

Antigens. Among the antigens used in the assay development studies are hormones or biochemical molecules with hormonal-type effects in *in*

---

[32] J. L. Guesdon, T. Ternynck, and S. Avrameas, *J. Histochem. Cytochem.* **27**, 1131 (1979).
[33] G. E. Trivers, C. C. Harris, M. Yamaguchi, I. C. Hsu, and R. Yolken, *J. Cell Biol.* **87**, 230 (1980).

vitro and *in vivo* human and animal systems. They include adrenocorticotropic hormone (ACTH), using the synthetic tetracosapeptide (1-24) (Bachem), human chorionic gonadotropin (β subunit) (HCG_b), prostaglandin (PG), thymosin α-1 (α-1) (from Dr. A. L. Goldstein, GWU, Washington, D.C.), and EGF (from Dr. H. Haigler, UCLA). The data obtained with ACTH will be used to describe the method.

Antisera. Two antisera were used in each EIA: one a rabbit antiserum or IgG specific for the antigen (Ab_1); the other a commercial preparation of alkaline phosphatase-conjugated goat IgG raised against rabbit IgG (Ab_2)(Miles Laboratories). Goat anti-rabbit IgG (URIA, Pasteur Institute) was used for immunoprecipitation in the competitive RIA, and sheep anti-rabbit IgG (URIA) in the insolubilized RIA.

Radioiodinated Labels

1. ^{125}I-labeled ACTH. Radiolabeling of ACTH was performed by the Iodogen method.[34] Two micrograms of Iodogen (Pierce), 1,3,4,6-tetrachloro-3α-6α-diphenylglycoluril, in 20 μl of dichloromethane was added to an Eppendorf tube, and the organic solvent was evaporated under N_2. Twenty microliters of 0.5 M phosphate buffer (pH 7.4) and 5 μl of 0.1 N acetic acid containing 1 μg or 0.4 nmol of ACTH were added successively and allowed to react for 5 min. Five microliters of $Na^{125}I$ ($\simeq$750 Ci; Radiochemical Centre, Amersham) was added; after 5 min the reaction mixture was chromatographed on CM-52 cellulose (Whatman). Pure monoiodinated ACTH (immunoreactively stable for at least 6 months) was eluted with 0.5 M ammonium acetate buffer (pH 5.2), collected in 0.05 M PBS-BSA, fractionated, and stored at 80° or after lyophilization. The specific activity was $\simeq$1500 Ci/mmol.

2. [^{125}I]Ab_2. Purified sheep IgG (Ab_2) was also labeled by the Iodogen method (10 μg + 1 mCi ^{125}I) and purified by gel filtration on AcA-34 (IBF). The specific activity was $\simeq$5000 Ci/mmol.

Substrates. PNPP was obtained from Sigma; [^{3}H]AMP was from Amersham and [^{3}H]PNPP from New England Nuclear (specific activity 15 and 30 Ci/mmol, respectively).

Equipment

Microtiter Plates. Solid-phase EIAs provide the much desired ease of handling, simplicity, and reductions in the quantities of valuable reagents. However, it should be noted that only relatively small quantities of a reagent attach to the plates, and denaturation and detachment of mole-

---

[34] S. Kochwa, M. Brownell, R. E. Rosenfield, and L. R. Wasserman, *J. Immunol.* **99**, 981 (1967).

cules may occur. Most important is the fact that the quantity of molecules absorbed can vary from lot to lot and, occasionally, from well to well of a Microtiter plate. In order to limit such occurrences, it is necessary to test a number of different lots or batches of the available products, then store sufficient stock to accommodate the assay for extended periods of time. For the assays described below we used several lots of Dynatec polyvinyl (PV) plates, each pretested for the efficiency of binding a particular antigen to be studied. Polystyrene (PS) plates were also tested but were in no instance chosen for an assay.

Instruments. The specialized equipment used included an automatic plate washer, and an EIA reader, but both processes are achievable manually.

Antigen Coating in Microtiter Plates

Antigens were coated in 100-μl volumes of carbonate buffer, pH 9.6, and 1× or 10× PBS, pH 7.4. Plates were (a) sealed with Parafilm, covered with a top and incubated at 4° at room temperature or at 37° for 16 hr or longer; (b) evaporated to dryness in uncovered plates at 37°. Coated plates were then stored at −20° until used.

Procedure: EIA/USERIA

1. Antigen-coated (central 60 wells), PV Microtiter plates are washed two or three times with phosphate-buffered saline, pH 7.4 (PBS). Then 200 μl of 1–2% horse serum (HS) in PBS containing 0.05% Tween 20 (PBS-Tween) is added to each well to coat residual binding sites and reduce nonspecific binding (NSB). The plates are covered with Parafilm and a plate top and are incubated for 1 hr at 37° in a humidified incubator or a moisture chamber in a convection incubator.

2. Plates are washed five times with PBS-Tween (5× wash) and 100 μl of Ab_1 diluted in 1% HS is added per well according to the protocol. HS is added to the outside wells. The plates are incubated for 1–2 hr as described above.

3. The 5× wash is repeated, 100 μl of Ab_2 diluted in 1% HS is added to the test wells, HS is added to the periphery, and the plates are incubated for 1 hr at 37°.

4. The 5× wash is repeated, then three washes are performed with 0.1 M diethanolamine (DEA) buffer, pH 9.8, to remove phosphates remaining from the PBS: substrate is added in 100 μl volumes containing 1 mg/ml PNPP in 1 M DEA buffer, pH 9.8 (EIA); or 30–100 pmol of [^{3}H]AMP (specific activity of 15–19 Ci/mmol) or [^{3}H]PNPP (29.9 Ci/mmol) in 0.02 M DEA buffer, pH 9.0 (USERIA).

5. (a) Hydrolyzed PNPP yields *p*-nitrophenol that is measured in the EIA reader at 405 nm. (b) Hydrolyzed [^{3}H]AMP releases [^{3}H]adenosine that is measured in 20-μl volumes from each well eluted with three washes of 1 ml each of 1:20 PBS from an 0.8-ml column of DEAE–Sephadex (A-25). Unhydrolyzed, charged [^{3}H]AMP is retained on the column. The eluates containing uncharged [^{3}H]adenosine are collected in 7-ml vials, 3 ml of scintillation fluid (with detergent) is added, and the radioactivity is counted. (c) Hydrolysis of [^{3}H]PNPP releases [^{3}H]PNP that is measured by adding 20 μl of sample volumes directly to scintillation vials containing 2 ml of PBS, pH 7.4. Three milliliters of a nondetergent scintillation fluid is added, and the resulting biphasic mixture separates the hydrolyzate by dissolution into the scintillation layer for counting. Unhydrolyzed [^{3}H]PNPP remains in the aqueous layer.

All tests are performed in duplicate or triplicate and each assay includes the following controls.

I. Noncompetitive EIA and USERIA
  Background: nonspecific binding of heterologous proteins and plastic by
  A. Ab_1: wells containing all reagents except Ag.
  B. Ab_2: wells containing all reagents except Ab_1 (zero control for calculating specific reactions).

II. Competitive EIA and USERIA
  1. Negative controls: same as I, B above.
  2. Positive controls—uninhibited, 100% controls: wells containing all reagents except inhibitor (pure antigen or sample).

III. Standard USERIA controls
  1. Spontaneous hydrolysis control: counts from an eluate of 20 μl of chromatographed stock [^{3}H]AMP or unchromatographed stock [^{3}H]PNPP collected at the time (zero) the substrate is added to the plate(s).
  2. Total count control: counts from 20 μl of unchromatographed stock [^{3}H]AMP or unincubated [^{3}H]PNPP.

Some controls are procedural checks that measure the stability of the immunosorbent and the antibody performances with respect to nonspecific activity, and are not used to calculate results.

Calculations

Arithmetic means and standard deviations are calculated for each duplicate or triplicate set. In the EIA, net values for noncompetitive test wells are obtained by the formula:

(Test mean) − (zero control mean) = net test value

Noncompetitive EIAs are done primarily to establish conditions for competitive assays.

The noncompetitive USERIA is also used primarily to design the competitive USERIA. However, some differences do exist. The background control in the Ag-coated wells of competitive USERIA usually doubles spontaneous hydrolysis. However, in the noncompetitive USERIA, which does not have Ag in the control wells, the control can be as high as four to five times the spontaneous hydrolysis controls, which provides an evaluation of substrate stability with time in storage.

In competitive assays (a) all test values are corrected for the controls without Ab_1, and (b) percentage of inhibition is computed by the following formula:

$$\left[1.0 - \frac{(\text{test mean}) - (\text{control mean})}{(100\% \text{ test mean}) - (\text{control mean})} \right] 100$$

Biological Extract Preparation

Tissue. Anterior pituitary lobes of decapitated adult male rats were dissected immediately (on ice) and homogenized in 5–10 ml of cold 2.5 N acetic acid. The homogenate was heated to 95° for 15 min, then centrifuged at 9800 g for 5 min. The supernatant was diluted with an equal volume of BSA (1 mg/ml of buffer) and lyophilized.

Peripheral Plasma. Blood samples (rat and human) were collected into chilled tubes containing anticoagulant and centrifuged immediately for 15 min at 3000 g and 4°. The plasma was separated and stored at $-20°$.

Chromatographic Purification. Plasma samples were chromatographed on Sep-Pak reversed-phase C_{18} cartridges (Waters Associates), selected because of the high percentage (>95%) of the hormone recovered from pretested biological samples (silicic acid (Mallinckrodt, 100 mesh) and gel filtration AcA-202 (IBF) were less efficient). With the top of the cartridge attached to a 3-ml disposable plastic syringe, the C_{18} reversed phase was activated and the sample (in 2 ml of methanol) was added, followed by 5 ml of distilled water. Two successive applications of the sample were made to the column. The silicic matrix was then washed with 5 ml of 1% trifluoroacetic acid (TFA) (Fluka), and the peptides (adsorbed on the surface of the matrix), were eluted with 5 ml of MeOH–H_2O–TFA (80:19:1, v/v/v). The eluate was evaporated to dryness under N_2, and the extract was dissolved in the appropriate buffer for the selected immunoassay and lyophilized.

Results

Microtiter Plates

The decision to immobilize the antigen eliminates the necessity for having to label it. This is of particular importance if, as in our laboratory, (a) most of the antigens of importance are in short supply and difficult to make or (b) the antigen exists as a part of a large host molecule, making specific labeling difficult.

The binding of proteins to polymers is dependent upon the nature of both the protein and the polymer.[34,35] Binding to polystyrene tubes is apparently independent of concentration up to approximately 1 mg (the upper limit of the "region of independence" or the maximum amount of protein that can bind in a single monolayer) and increases proportionately with temperature and time of incubation.[36] However, depending upon the polymer and the protein, a maximum of absorbed molecules can be reached, which then will decline with further incubation.[35] This may be due to dissociation in protein-to-protein attachments that occur beyond the "region of independence." These "bonds" are perhaps weaker than protein-to-plastic bonds. For that reason, it is important not to exceed the maximum amount of protein in the region of independent binding or 1000 ng if one is using polystyrene tubes (6.5 cm^3 of effective coating surface). The limits for plates and especially PV U-bottom plates are not known, but we have not generally found it necessary to exceed 1 μg in establishing acceptable levels of the immobilized proteins. Proteins may bind PS best at their isoelectric pH.[37] Molecular size appears to affect the efficiency of binding; i.e., it has been shown that the amount that binds is inversely proportional to the molecular weight.[36] Moreover, proteins absorbed at very low concentrations can react immunologically like denatured material.[34]

We tested both PV and PS plates with each new antigen–antibody system. In addition, we examined different "coating" conditions, i.e., antigen concentrations, Ab dilutions, pH, ionic strengths, temperature, and incubation time. The results of these tests depended most upon the characteristics of the individual antigens. However, purified polypeptides and proteins bound very poorly, if at all, to U-bottom PS plates (Fig. 1). Smaller molecules had lower binding efficiencies and greater molecular or immunogenic instability when bound. For example, EGF (M_r 6400) and prostaglandin E$_2$ (M_r 500) did not bind to PS in our hands and Alpha 1 (M_r

---

[35] R. G. Lee, C. Adamson, and S. W. Kim, *Thrombosis Res.* **4**, 485 (1974).
[36] L. A. Cantarero, J. E. Butler, and J. W. Osborne, *Anal. Biochem.* **105**, 375 (1980).
[37] I. Oreskes and J. M. Singer, *J. Immunol.* **86**, 338 (1961).

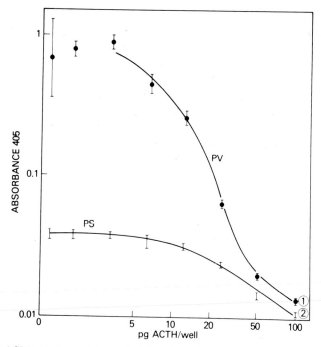

FIG. 1. ACTH coating efficiency in Microtiter plates of different plastics. EIA dose responses were performed in (1) polyvinyl plates from Dynatech or (2) polystyrene plates from Nunc.

3000) did so very poorly. Moreover, while DNA stores well for indefinite periods without loss of immunoreactivity when bound to PV plates, ACTH and Alpha 1 were used within a week after coating, and EGF was coated fresh before each assay.

Determining Optimal Assay Conditions

Our S-P Microtiter plate EIAs use Ag as the immobilized immunosorbent, rabbit antibodies (Ab_1) specific for the Ag, and commercial goat anti-rabbit IgG–alkaline phosphatase conjugate (Ab_2). Each component (the Ag and both Abs) must be optimized individually and in combination in order to obtain the quantities of each required for the most sensitive assay.

Titration of Ag and Ab_1. Optimization for the EIA was accomplished via a series of "checkerboard" titrations of the reagents, beginning with decreasing amounts of the Ag (e.g., 1000 to 1 ng/well in four 10-fold serial dilutions) versus one or more reasonably low dilutions (1 : 10^3, 1 : 10^4, etc.) of Ab_1, and a proven high concentration of the conjugate (1 : 10^3).

There were four major conditions used for coating the antigens (see Methods). Based upon prejudgments of available information, drying was used for the initial test of ACTH. ACTH (Fig. 2) was tested dry and gave acceptable readings from 0.2 ng to 1 ng using between 1:250 and 1:10,000 of Ab_1. This two-way titration (Ag vs Ab_1) established an acceptable low range of immobilized Ag, as immunosorbent, that was measurable by a reasonably high dilution of Ab_1. (Acceptable low Ag levels produce substrate conversions that are high enough for statistical accuracy and provide accurate and significant differences when decreased by the activity of an inhibiting reactant.) After refrigerated coating, 1 ng of HCG_B per well was measurable between 1:2000 and 1:10,000 dilution of Ab_1 (data not shown, DNS).

Ag and Ab_1 Titrations for Plate-Coating Conditions. Optimum conditions for binding the lower range of the Ag to Microtiter plates were tested. PS and PV plates were tested using various diluents (see Methods)

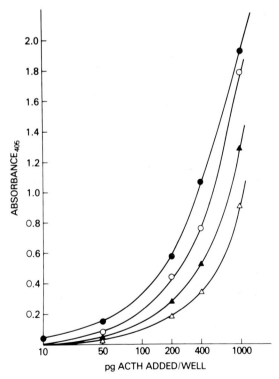

FIG. 2. Enzyme immunoassay "checkerboard" dose response of increasing amounts of ACTH-bound (37° to dryness) in polyvinyl plates versus decreasing concentrations of rabbit ACTH antibody: 1:250 (●——●), 1:2500 (○——○), 1:5000 (▲——▲), and 1:10,000 (△——△).

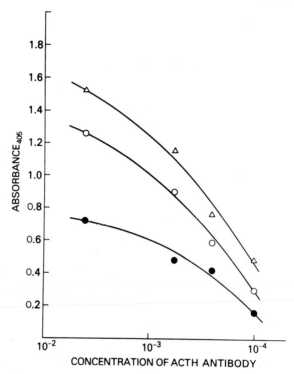

Fig. 3. Titration of anti-ACTH antibody in polyvinyl plates with 200 pg of ACTH added per well and incubated at 4° (●——●) or 37° (○——○) in Parafilm-covered plates at 37° until dry (△——△) for approximately 16 hr.

with different ionic strengths and pH values, and each combination at varying temperatures and for different periods of incubation.

Alpha 1 (3000 M_r) coated best when dried at 37° (DNS). In addition, as have all Ags tested to date, α-1 attached better when dissolved in PBS, pH 7.4, than it did in carbonate buffer, pH 9.8. PG coated when hydrated at room temperature dissolved in 0.001 ng of bovine IgG. The most resistant of any of the molecules we tested for PV coating, PG was not detectable when dissolved in 0.01% gelatin, 0.001% BSA, or the standard buffers and incubated in any manner described. Contrary to experience with PG, HCG_b (M_r 22,000) was the most efficient of the molecules tested for binding PV; i.e., it attached almost equally well under the major coating conditions in PBS, and drying was only slightly better than hydration (DNS).

ACTH attached to PV plates whether at 4°, at 37°, or dried. However, drying was the most effective (Fig. 3). EGF, however, is not detected

TABLE II
DETERMINATION OF OPTIMUM CONJUGATE CONCENTRATION FOR ACTH EIA

ACTH/well (pg)	Conjugate dilution factor			
	8000	4000	2000	1000[a]
	(1 : 200,000 anti-ACTH antiserum)			
NR[c]	0.006 ±	0.007 ±	0.014 ±	0.024 ±
0	0.006 ± 0.001	0.014 ± 0.001	0.024 ± 0.002	0.054 ± 0.006
10	0.005 ± 0.003	0.021 ± 0.000	0.030 ± 0.008	0.062 ± 0.011
30	0.012 ± 0.000	0.032 ± 0.000	0.059 ± 0.003	0.110 ± 0.001
100	0.061 ± 0.006	0.120 ± 0.010	0.193 ± 0.006	0.408 ± 0.024
1000	0.660 ± 0.050	1.232 ± 0.047	1.918 ± 0.013	2.000 ± 0.000
	(1 : 100,000 anti-ACTH antiserum)			
NR	0.000 ±	0.005 ±	0.011 ±	0.038 ±
0	0.009 ± 0.001	0.014 ± 0.005	0.020 ± 0.002	0.044 ± 0.001
10	0.010 ± 0.001	0.022 ± 0.010	0.029 ± 0.003	0.044 ± 0.004
30	0.014 ± 0.001	0.026 ± 0.007	0.036 ± 0.002	0.064 ± 0.004
100	0.035 ± 0.006	0.074 ± 0.016	0.108 ± 0.022	0.204 ± 0.016
1000	0.338 ± 0.059	0.628 ± 0.014	1.260 ± 0.149	1.930 ± 0.100

[a] Absorbance at 405 mm.
[b] Optimum conjugate dilution.
[c] NR, normal rabbit serum.

when dried on PV plates, and could not be stored after hydrated attachment. α-1 was immunologically stable for long periods when dried and stored at $-20°$. On the other hand, DNA, ACTH (M_r 37,000), and α-1 coated best when dried at $37°$.

Conjugate Titration. In the second titration we used the acceptable low antigen levels, in combination with one or more high dilutions of Ab_1, to titrate the conjugate. This titration examined the feasibility of higher dilutions of Ab_2 efficiently reacting with the lower range of acceptable amounts of Ag and Ab_1 (Table II). Decreases in substrate hydrolysis were directly related to the decreases in reactant concentrations. The uncorrected data (not adjusted for normal rabbit serum control) shows that within the dose response, differences between ACTH levels appeared best at the highest Ab_2 concentration (1 : 1000). However, decreases in the control values (normal rabbit serum) between conjugate amounts were directly proportional to decreases in conjugate concentrations. A plot of the data for 1 : 200,000 Ab_1 (Fig. 4) shows that although Ab_2 at 1 : 1000 was superior to the 1 : 2000 level in total activity, the latter was in the beginning of the linear response and produced acceptable substrate conversion. In addition, when the data were corrected for the control values (Table III), the result was higher test to control ratios at the lower Ab_2

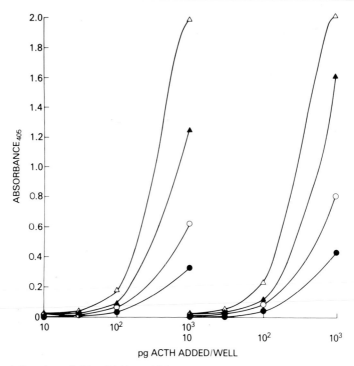

FIG. 4. Four-hour (left) and 24-hour (right) enzyme immunoassay "Checkerboard" titration of bound ACTH and 1:200,000 dilution of Ab_1 versus decreasing concentrations of the alkaline phosphatase-labeled goat anti-rabbit IgG antibody: 1:1000 (△——△); 1:2000 (▲——▲); 1:4000 (○——○); 1:8000 (●——●).

concentrations (2 and 4 × 10^{-3}) and lowering of the optimum reaction concentrations. Moreover, in 1:200,000 Ab_1, the optimum Ab_2 was placed at 1:2000 for a 4-hr incubation with the substrate. This combination also gave the best differentials between amounts of S-P ACTH. For the 24-hr assay, 1:4000 Ab_2 became the optimal for both dilutions of Ab_1. However, when the Ab_1 concentration was increased (10^{-5}), the optimum Ab_2 concentration decreased (1:4000). These results provided optimum levels of Ab_1 (i.e., 2 × 10^{-5}) for specific amounts of Ab_2 (1:2 × 10^{-3}) to be used with a specific and acceptable range of S-P Ag (200–500 pg/well). Similar conditions are required for sensitive immunometric assay of the Ag in fluid-phase competition for Ab_1.

Optimization of USERIA. Optimization did not end with achievement of specifications for an EIA. The USERIA could be performed with the conditions derived but would not be optimal. Rather, owing to the higher signal amplification generated by USERIA, the assay could, in some in-

TABLE III
Determining Optimum Conjugate Concentration for ACTH EIA Ratio of Test Control Mean A_{405}

ACTH/well (pg)	Conjugate dilution factor							
	8000		4000		2000		1000	
	4 Hr	24 Hr	4 Hr	24 Hr	4 Hr[a]	24 Hr	4 Hr	24 Hr
	1:200,000 anti-ACTH antiserum							
0	1.0	1.0	1.0	1.0[a]	1.0[a]	1.0	1.0	1.0
10	1.2	1.2	1.8	5.6[a]	2.5[a]	2.2	1.0	1.1
30	2.8	2.0	2.3	6.3[a]	3.6[a]	3.5	2.0	2.7
100	9.8	7.2	7.8	28.6[a]	15.6[a]	14.8	9.0	14.0
1000	110.0	83.0	78.0	267.0[a]	208.0[a]	197.0	95.0	122.0
	1:100,000 anti-ACTH antiserum							
0	1.0	1.0	1.0[a]	1.0[a]	1.0	1.0	1.0	1.0
10	.83	1.0	1.8[a]	1.8[a]	1.1	.7	1.0	1.0
30	2.0	1.8	3.1[a]	3.6[a]	3.0	2.6	1.7	2.2
100	10.2	8.2	12.6[a]	16.6[a]	11.4	11.0	11.6	12.2
1000	110.0	86.0	135.0[a]	176.0[a]	118.0	90.3	60.0	48.8

[a] Optimum conjugate and Ab_1 dilution for short and prolonged incubation periods.

stances, be overweighted with reagents and less sensitive than otherwise possible, with careful repetition of the procedure described, using [^{3}H]AMP as substrate. For example, in Fig. 5 the sensitivity of a competitive USERIA for HCG_B is shown to increase with the decrease in the amount of immobilized antigen. The 200 pg previously used in the HCG_B EIA was therefore reduced to 100 pg per well vs 1:50,000 Ab_1 for USERIA. Similarly, in determining conditions for a competitive USERIA for ACTH, we retested in USERIA the results achieved in the EIA. Figure 6 shows results of such a test of Ab_2 concentrations in the USERIA for ACTH, beginning below the EIA optimums and incubating with the [^{3}H]AMP for only 1 hr.

These tests for USERIA conditions resulted in specifics for competitive USERIA that were considerably different from those we established for the competitive EIA. An important influence on these differences derived directly from concerns for validation and reproducibility of the assay in practice. The conditions used are shown in Fig. 7, and include three major modifications of the standard protocol: (a) In the second step, 0.05% bovine γ-globulin (gG) or 0.1% HS was used instead of 1% FCS. Serum contains both ACTH and other molecules (i.e., BSA) that will bind ACTH and could interfere with the assay. (b) Also in step 2, the time and temperature of the immunological reaction was changed from 1 hr at 37°

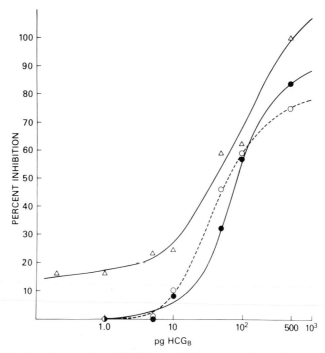

FIG. 5. Results of competitive USERIA for HCG_B showing increasing sensitivity when 200 pg (●——●), 100 pg (○---○), and 50 pg (△——△) of the solid-phase hormone was added to the plate. After the coupling of alkaline phosphatase-labeled Ab_2, the plate was incubated for 3 hr with [^{3}H]AMP.

to 16–24 hr at +4°. In the first trial of the new assay, conditions were altered to favor molecular integrity of the hormone and maximum sensitivity for detection of normally occurring very low concentrations of ACTH. (c) In step 4, incubation with the substrate was 24 hr instead of the 4 hr suggested by earlier preliminary tests. In order to improve reproducibility, we used the upper level of the acceptable low amounts of S-P Ag (200–500 pg). As a result, the optimum Ab_1 concentration was decreased to $1:2.4 \times 10^{-5}$, which gave very low kinetics in these conditions (+4°, 0.05% gG) and required longer incubation.

The initial test of the USERIA for ACTH was a comparative study (Fig. 8 and Table IV) including both EIA and RIA competitive measure of purified ACTH in standard curve determinations. The solubilized RIA (RIA_S) was performed with the double Ab method using ^{125}I-labeled ACTH in the F-P and propylene glycol to facilitate the precipitation of the labeled complex. In addition, an insolubilized RIA (RIA_I) was done with

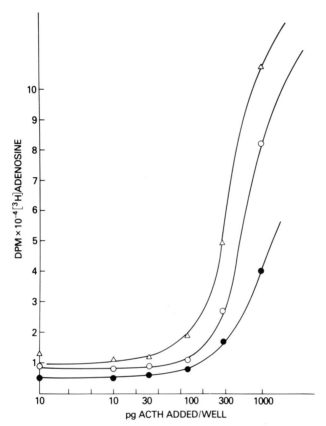

FIG. 6. A USERIA "checkerboard" titration of increasing solid-phase ACTH versus 1:50,000 Ab_1 and decreasing concentrations of alkaline phosphatase-labeled Ab_2: 1:1000 (△——△); 1:2000 (○——○), and 1:4000 (●——●). The plate was incubated with [^{3}H]AMP for 1 hr at 37°.

[^{125}I]Ab_2. The Ag, Ab_1, and Ab_2 used in each assay were from the same stock preparations; however, Ab_2 for the EIA and USERIA was AP-conjugated (Table IV).

Characteristics of the Standard Curves

The EIA optimized for the new conditions (+4° immunologic incubation in 0.05% gG) required 1000 pg in S-P, and 1:60,000 of Ab_1. The RIA_I had a 50% competitive inhibition (CI_{50}) at 194 pg, a detection limit (DL) at 50 pg, and was the least sensitive of the four assays. The RIA_S, EIA, and USERIA demonstrated successively higher sensitivity. With a CI_{50} of 5.6

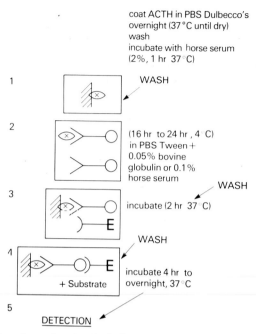

FIG. 7. Procedure for performing the indirect sandwich solid phase competitive enzyme immunoassay or ultrasensitive enzyme radioimmunoassay for ACTH: 1, Ag immobilized in polyvinyl plates; 2, addition of free antigen in solution with Ab_1; 3, addition of Ab_2–enzyme conjugate; 4, addition of nonradioactive or radioactive substrate; 5, quantitation of enzyme activity.

pg, and a DL of 1 pg, USERIA exhibited approximately 6-, 10-, and 50-fold greater sensitivity than did EIA, RIA_S, and RIA_I, respectively, while using lower amounts of reagents. Moreover, the data show (a) that, as a measure of ACTH, the radioactive substrate in USERIA provided encouraging improvement over classical EIA (colorigenic substrate), and (b) that the RIA_I used was inferior to the RIA_S.

Application to Biological Samples

Based upon the results described above, the RIA_I was considered to be too insensitive for inclusion in the comparative measure of ACTH in biological samples. As stated, each immunoassay for ACTH was optimized individually to compare the different capabilities for performing the standard curves. For this purpose, dilutions of Ab_1 and the amount of coated antigen were selected for each SP-EIA based upon reasonable reaction times required to develop acceptable specific amounts of the

TABLE IV
COMPARISON OF IMMUNOASSAYS FOR ACTH USING THE SAME STOCK REAGENTS

Assay	RIA Insolubilized (I) phase	RIA Soluble (S) phase	EIA Solid phase	USERIA Solid phase
Coating (pg)	2000	0 (10)[a]	1000	500
Rabbit anti-ACTH antibody	1:30,000	1:90,000	1:60,000	1:240,000
50% Competitive inhibition (pg)	194	76	24	5.6
Detection limit (pg)	50	10	6	1.0
			Alkaline phosphatase (Miles)	
Sheep or goat anti-rabbit IgG antibody (Ab$_2$)	1 µg/ml, ^{125}I-Ab$_2$	—	1 µg/ml	0.1 µg/ml
Substrate	—	—	PNPP	[^{3}H]AMP, [^{3}H]PPNP

[a] ^{125}I-ACTH.

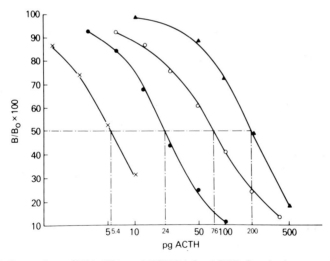

FIG. 8. Comparison of RIA, EIA, and USERIA for ACTH. Standard curves were generated using the methods described. RIA was done with insolubilized (I) ^{125}I-labeled Ab$_1$ (▲——▲) and solubilized (S) (○——○) Ab$_1$ using ^{125}I-labeled ACTH. The difference between the EIA (●——●) and the USERIA (×——×) is in the source of the signal only (colorimetric; radioactive). Based upon 50% inhibition, the sensitivity of the USERIA was greater than that of EIA, RIA$_I$, and RIA$_S$ by 4-, 14-, and 37-fold, respectively. See Table IV for details.

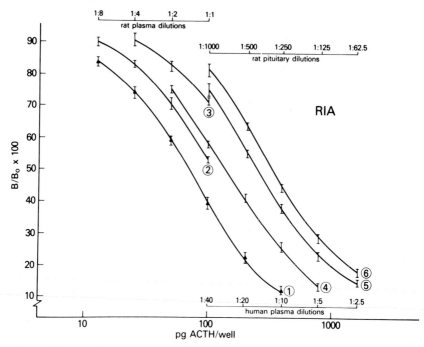

FIG. 9. Competitive RIAs for ACTH: (1) standard curve, (2) rat plasma after stress, (3) rat plasma in basal conditions, (4) human plasma after addition of 10 ng of synthetic ACTH (1–24), (5) human plasma after addition of 10 ng of synthetic ACTH and Sep-Pak purification, (6) acid extract of the anterior lobe rat pituitary. Sample dilutions are shown for each biological sample.

different signals. The effects of increasing amounts of purified ACTH and plasma on the immunologic reaction ($B/B_o \times 100$ vs log dose) in each method are shown in Figs. 9–11 for RIAs, EIA, and USERIA, respectively.

Immunoassays without Sample Purification

ACTH was measured directly in 2-fold, serial dilutions of peripheral plasma from resting and stressed male adult rats and after the addition of 10 ng of pure ACTH to 1 ml of plasma from adult men (Table V). Results in EIA were similar to competitive RIA but required a smaller volume of plasma (12.5 μl vs 100 μl for RIA) to obtain 50% inhibition. However, as shown in Fig. 10 (2), ELISA gave proportional values of ACTH only after the second dilution. On the other hand, due apparently to the interference of unknown constituents of plasma, CI was not achieved in the USERIA

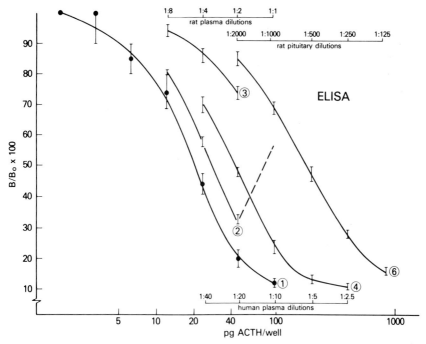

FIG. 10. ELISA for ACTH: (1) the standard curve, (2) rat plasma after stress, (3) rat plasma in basal conditions, (4) human plasma after addition of 10 ng of synthetic ACTH (1–24), and (6) acid extract of the anterior lobe of rat pituitary. Sample dilutions are shown for each biological preparation. Note that curves (1) and (2) are parallel only after the second dilution of the sample.

above a concentration of 0.4% of unpurified rat or human plasma (Fig. 11).

Immunoassays after Purification

Tissue. Extracts of rat anterior pituitary lobes were used as a source of ACTH. The values measured (Table V) were similar in each of the three assays. However, compared to the RIA, the amount of tissue required for measurement in the ELISA was reduced by a factor of 2, and in the USERIA by a factor of 8.

Plasma. The results of ACTH measurements with USERIA following plasma purification on Sep Pak columns are shown in Table V and Fig. 11. After the chromatographic step, USERIA gave results that were interpretable and similar to those from competitive RIA or ELISA. Again, however, the volume of plasma required for CI_{50} in the RIA was reduced

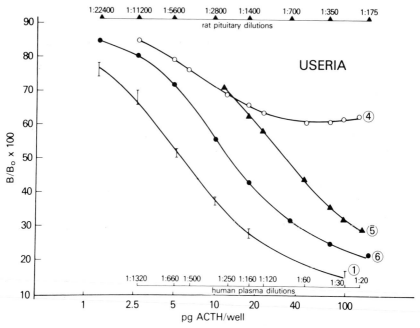

FIG. 11. USERIA for ACTH: (1) standard curve, (4) human plasma after addition of 10 ng of synthetic ACTH (1–24), (5) human plasma after addition of 10 ng of synthetic ACTH and Sep-Pak purification, (6) acid extract of the anterior lobe of rat pituitary. Sample dilutions are shown for each biological preparation. Note that there was no parallelism when USERIA was performed without plasma purification.

by a factor of 3 in the EIA and by a factor of 12 in the USERIA. These differences in volume are almost identical to the differences in sensitivity obtained in the standard curves.

Discussion

Microtiter plate EIAs require a good and reliable antibody, a good stock of the specific antigens, reagent (conjugate), and supplies (i.e., plates) vital to the ongoing ability to perform the assays. One of the important demonstrations in our experience is that PV U-bottom plates either bind antigens more efficiently (more in a shorter period) or, for geometric reasons, they retain more molecules than the flat-bottomed wells with right-angled walls. (In U-shaped wells gravity sedimentation could induce molecules to accumulate from the center or lowest point in the well. Binding may then occur in a "brick-laying" progression, terminating with securely bound molecules extending by vertical stacking

TABLE V
IMMUNOASSAYS FOR ACTH IN BIOLOGICAL SAMPLES

Sample	Solubilized RIA	ELISA	USERIA[a]
Without treatment			
Rat plasma			
Basal conditions	103.0 ± 11 pg/ml	88.0 ± 14 pg/ml	NI[b]
After stress	762.0 ± 46 pg/ml	801.0 ± 80 pg/ml	NI
Human plasma + 10 ng ACTH/ml	10.4 ± 0.6 pg/ml	10.9 ± 0.8 pg/ml	NI
After treatment			
Rat anterior pituitary lobes after acid extraction	319.0 ± 19 pg/ml	338.0 ± 38 ng/ml	324 ± 48.0 ng/ml
Human plasma + 10 ng ACTH/ml after Sep Pak column	11.9 ± 1.2 pg/ml	12.0 ± 1.7 ng/ml	10 ± 1.6 ng/ml

[a] USERIA required 8- and 4-fold less tissue and 12- and 4-fold less plasma than the ELISA and the RIA, respectively, to achieve 50% inhibition.
[b] NI, not interpretable.

along the walls of the wells.) Also important is the use of commercial conjugates that allows the investigator to concentrate on producing one good antibody for each system and more easily to standardize the assay systems.

EIAs are quite easy to perform. However, each antigen–antibody system must be approached as a new challenge, and each should be studied for optimum activity when allowed to react in experimental conditions. Only then can the maximum sensitivity of an EIA system be achieved.

As suggested by our ACTH data, the EIAs, especially USERIA, may potentially replace RIA for certain immunoassays of medical interest. In each case, however, the decision will require experimental evidence for the assay of choice. The sensitivity limits of the EIAs are restricted primarily by the quality of the antibody and the efficiency of the immunosorbent to immobilize on surfaces conducive to the performance of the reactions required.

As seen in these data, the success of any amplification procedure in enzyme immunoassays will depend on the last step, application to a given biological sample. In earlier studies, Dray *et al.*, failed in their effort to apply another ultrasensitive and novel immunoassay, i.e., viro-immunoassay, for measuring steroids in plasma directly or after solvent extraction.[38] Similar failures were experienced with USERIA for plasma ACTH when the assay was performed without purification.

Moreover, it was after correction for "blank" interference (the amount of substrate converted in the absence of the test antigen) following chromatographic purification of plasma that USERIA was not only operative but shown to be more sensitive than ELISA. In addition, plasma purification for ACTH measurement gave higher values of the hormone after purification than found before purification (Table V). The moderate increase observed could be explained by the release of ACTH bound to macromolecules such as serum albumin.[39]

For this reason, the requirement for an appropriate purification step, such as used to obtain USERIA measurements in plasma, becomes a general constraint for any immunoassay, whether directly measurable or not, because of the potential for the binding of peptides to macromolecules and the presence in plasma of related compounds (prohormones, metabolic products, etc.) that can recognize the hormone-specific antibody.

Establishing Microtiter plate S-P USERIA requires essentially the same preliminary efforts as for the EIA. However, USERIA requires additional optimization for the highly amplified radio signals that constitute the increased sensitivity of the method. The use of radioactive reagents may seem a reversal of the process of simplification and affordability we have stressed in earlier discussion of the EIA. However, this shift achieves increased sensitivity and a significant extension of the basic EIA principles described. Whereas Engvall et al. combined enzymes, antibodies, and plastics to achieve a safe and simple amplified marker of Ag-Ab reactions, Harris and co-workers have combined enzymes and radiolabeled substrates to achieve the magnification of the enzyme-amplified antigen–antibody reaction signal. Moreover, it is now generally accepted[23,40,41] that radiolabeled substrates provide valuable potential for effectively increasing the sensitivity of immunoassays. Developments of this kind are essential to our ability to investigate the more difficult problems in medical research, i.e., at the level of the gene and its interactions with nongenetic molecules. The greater sensitivity of USERIA justifies its additional requirements and will provide increased capabilities in many fields of investigation.

---

[38] F. Dray, E. Maron, S. A. Tillson, and M. Sela, *Anal. Biochem.* **50,** 399 (1972).
[39] J. E. Stouffer and I. C. Hsu, *Biochemistry* **5,** 1195 (1966).
[40] E. Engvall, *in* "Enzyme Immunoassay" (E. Ishikawa, T. Kawai, and K. Miyai, eds.), p. 1. Igaku Shoin, Tokyo, 1981.
[41] S. Avrameas *in* "Immunoassays for the 80's" (A. Voller, A. Bartlett, and D. Bidwell, eds.), p. 85. University Park Press, Baltimore, Maryland, 1981.

[28] Strategies for the Preparation of Haptens for Conjugation and Substrates for Iodination for Use in Radioimmunoassay of Small Oligopeptides

By WILLIAM W. YOUNGBLOOD and JOHN STEPHEN KIZER

The importance of small peptides to research in neuroendocrinology is rapidly increasing. Not unexpectedly, therefore, there has been a great increase in demand for assays for these substances. By far the most sensitive and simple technique for the quantification of small neuropeptides has been the radioimmunoassay, which permits reproducible measurements in the femtomole–picomole range. The major requirements for generating a radioimmunoassay for a small neuropeptide are 2-fold. First, one must be able to obtain a tracer of high specific activity; and second, one must be able to prepare a potent immunogen. Fortunately, most known oligopeptides possess moieties that both permit iodination to produce a suitable tracer and permit conjugation to a larger carrier protein to produce a potent immunogen. However, there are increasing numbers of small oligopeptides that possess neither of these two important features and yet whose physiological significance makes the development of radioimmunoassays for their measurement of major interest. In certain instances, although a residue suitable for iodination and conjugation may exist in a given oligopeptide, because of the residue's unique polarity or strategic location, conjugation at this site may greatly decrease the potency of the conjugated immunogen. Similarly, iodination at such a site may also decrease the affinity of the iodinated oligopeptide for its antisera. In still other instances, one may wish to raise specific C- or N-terminally directed antisera. To do so would require tethering of the oligopeptide, from either its C or N terminus, to a carrier protein, a procedure that is commonly not feasible in the native oligopeptide because of an N-terminal pyroglutamyl or C-terminal NH_2.

The purpose of this chapter, therefore, is to present strategies for, and specific examples of, the preparation of small oligopeptide haptens for iodination and conjugation both in circumstances where the native oligopeptide has no reactive amino acid residue and in circumstances where coupling or iodination through an available amino acid leads to an immunogen or tracer of poor quality. It is worth noting that most of our descriptions will apply to oligopeptides of fewer than 8–10 residues, because the greater the number of amino acids, the greater the probability that an appropriate residue for iodination or coupling can be found and the less is the likelihood that coupling through this residue will comprise the unique topography of the peptide.

Peptides Containing No Reactive Residues for Conjugation or Iodination

In this category of oligopeptides are included those having neither a tyrosine nor histidine nor free primary amine. Peptides that do contain one of these three can be readily iodinated or conjugated through a variety of techniques available in the literature (see the table).[1-3] Carboxyl groups are a special case. A peptide may be readily coupled to a carrier protein at those residues containing free COOH groups. Activated groups for introducing radiolabel at the C terminus could be activated aryl amines; however, such derivations may produce product mixtures because most aryl activating groups will compete with the amine for bonding with the peptide carbonyl. An example of a neuropeptide in this category is Pro-Leu-Gly-amide (MIF-1). In preparing a suitable molecule for both conjugation and iodination in this case, there are several considerations. First, Pro-Leu-Gly-NH_2 can be coupled directly to a carrier protein only with difficulty through its N terminus because of the relative lack of reactivity of the secondary amine of proline. Second, coupling through the N-terminal proline will undoubtedly result in tethering of the Pro-Leu-Gly close to the carrier molecule, and the angle of the proline peptide bond might leave little of the molecule exposed for immunological recognition. And, third, if coupling were to take place through the proline, one would still be left with the problem of developing an active tracer for use in an RIA. [Alternatively, one could produce an iodinated MIF tracer by allowing an iodinated hydroxyphenylacetyl active ester to react with Pro-Leu-Gly (iodination by the Hunter–Bolton reagent), but again left unsolved would be the problem of generating a hapten for conjugation to a carrier protein.] In solving this problem we elected to conjugate the N-terminal proline of MIF to a *p*-hydroxyphenylacetic acid moiety. This hapten may then be both conjugated to a carrier protein by the bisdiazotized benzidine method[1] and iodinated using standard techniques. It should be added that other molecules, such as γ-aminobutyric acid (GABA), could also be used to provide an N-terminal tether. These amino fatty acids can then be iodinated by reagents such as the Hunter–Bolton reagent. Such an approach, however, is more complex than is attachment of an activated aromatic amino acid (see below).

[1] R. M. Bassiri and R. D. Utiger, *Endocrinology* **90**, 722 (1971).
[2] S. Bauminger and M. Wilchek, this series, Vol. 70, p. 151.
[3] M. Reichlin, this series, Vol. 70, p. 159.

METHODS FOR CONJUGATION AND IODINATION

Method	Target residue	Reference[a]
Conjugation		
Bisdiazotized	Activated aromatic (tyrosine) imidazole (histidine)	1
Carbodiimide (water-soluble)	Free 1° or 2° amine carboxylic acids	2
Glutaraldehyde	Free 1° amine (primarily ε-amino of lysine)	3
Iodination		
Chloramine-T	Activated aryl (tyrosine) imidazole (histidine)	—
Lactoperoxidase	Activated aryl (tyrosine)	—
Hunter–Bolton	1° or 2° (low reactivity) amine	—

[a] Numbers refer to text footnotes.

The attachment of a prosthetic group to peptides is a technique that has been used for both small and large peptides. For example, [Tyr¹]somatostatin is an analog of somatostatin (SRIF) whose preparation and use for such a purpose is well known.[4] Thus, one approach is to attach an activated aromatic amino acid, such as tyrosine (or possibly histidine). As mentioned above, the approach favored by this laboratory is to acetylate the N-terminus of a protein with a small-chain fatty acid that bears an activated phenyl group. The appeal of this approach is 2-fold. First, it generates a molecule equally suitable for both iodination and conjugation; second, it obviates the need for working with N-blocked amino acids in the coupling step. Two such compounds suitable for N-peptide derivatization, which are inexpensive and may be obtained in high purity, are *p*-hydroxyphenylacetic acid (pHAA) and *p*-hydroxyphenylpropionic acid (pHPA). Also available are commercial preparations of the Hunter–Bolton reagent (an active ester or anhydride form of iondinated *p*-hydroxypropionic acid), which could be made to react with MIF or GABA-MIF to generate a suitable tracer.

In our development of a RIA for MIF, we prepared both *p*-hydroxyphenylacetyl-Pro-Leu-Gly-NH$_2$ (pHAA-MIF) and *p*-hydroxyphenylpropionyl-Pro-Leu-Gly-NH$_2$ (pHPA-MIF) using the coupling procedure detailed below. Immunogens were generated by bisdiazo coupling to bovine serum albumin (BSA) through the *p*-hydroxyphenyl moiety.[1] Both pPHAA-MIF and pPHPA-MIF were used in the preparation of iodinated tracer.

[4] A. Arimura, H. Sato, D. H. Coy, and A. V. Schally, *Proc. Soc. Exp. Biol. Med.* **148**, 784 (1975).

Coupling of p-Hydroxyphenylacetic Acid to Pro-Leu-Gly-NH$_2$ Using Ethyl Chloroformate (ECF)

pHAA (375 mg; 2.5 mmol) and triethylamine (TEA) (350 μl; 2.5 mmol) dissolved in 5 ml of anhydrous acetonitrile are cooled to $-10°$, and 200 μl of ECF (2.12 mmol) is added. After 30 min, MIF (284 mg; 1 mmol) is added, and the reaction is stirred at 4° for 24 hr.

pHAA-MIF is isolated by removing the solvent under vacuum, dissolving the residue in 25 ml of water, and adjusting the pH to between 2 and 3 using 5 M HCl. The aqueous phase is extracted twice with 5 ml of ether, which is discarded. The aqueous phase is freeze dried and chromatographed on silica gel using methanol–chloroform (20:80). Fractions are analyzed by silica gel thin-layer chromatography (TLC). Plates are developed in iodine vapor and/or by spraying with Pauli reagent, and chromatographically similar regions are pooled.

The simplicity of this method and the relative ease with which an activated aryl acyl may be coupled to a primary and secondary amine (such as the N-terminal secondary amine of proline) suggests that the activated aryl acyl be a useful general approach for preparing haptens for conjugation or inexpensive molecules for iodination. The general applicability of the method may be limited, however, to oligopeptides with only one R- or N-terminal amine because of the lack of selectivity of the activated aryl-acyl species. In these circumstances, more than one haptenic analog will result owing to indiscriminate coupling at all free amines. To circumvent these difficulties and extend the applicability of the method to peptides containing more than one free amine will require either sophistocated separative techniques carried out after the synthesis of the analog(s) or the introduction of R- or N-terminal amine blocking groups before the peptide is allowed to react with the activated aryl acyl.

For small peptides having no polar functional groups, such as MIF, another advantage of preparing the pHAA derivative is that it leads to a structure that can be analyzed by mass spectrometry. The structure and mass fragmentation pattern of the synthetic hapten *p*-hydroxyphenylacetylprolylleucylglycinamide are shown in Fig. 1. Figure 2 shows an RIA standard curve using antiserum generated against a conjugate of this hapten and using the ^{125}I-labeled hapten as tracer, demonstrating the utility of the methods that we have described for developing both immunogen and tracer for MIF.[5] Of interest is the observation that the pHPA-MIF was not suitable as a hapten or for iodination as a tracer in the development of the MIF-RIA, indicating that some trial and error is needed to obtain an appropriate peptide analog.

[5] P. J. Manberg, W. W. Youngblood, and J. S. Kizer, *Brain Res.* **241**, 279 (1982).

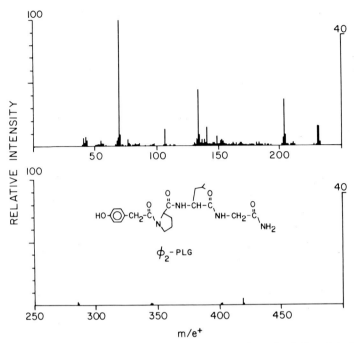

FIG. 1. Structure and mass spectrum of *p*-hydroxyphenylacetylprolylleucylglycinamide (ϕ_2-PLG). Mass fragmentation spectrum of solid probe sample was obtained using electron impact ionization in a Finnigan 3300 GC-MS. Reproduced, by permission, from Manberg et al.[5]

Preparation of Haptens That Do Not Compromise the Structural Uniqueness of a Given Oligopeptide

In many instances, an oligopeptide may possess only a single amino acid residue suitable for either iodination or conjugation to a larger protein carrier. Those amino acid residues most suitable for iodination or coupling are usually lysines with an ε-amino group, histidine with an imidazole ring, tyrosine with an activated aromatic ring, or glutamic and aspartic acids (coupling only) with exposed carboxyl groups. In small oligopeptides (which frequently contain only one of these moieties), the presence of these polar amino acids and their location within the oligopeptide chain are essential topographical features of the molecule that enhance its structural uniqueness and immunogenicity. Thus, in many circumstances, conjugation or iodination at these specific residues will result in either a tracer or an immunogen of poor potency.

A practical example of illustrating these considerations is the generation of antibodies to thyrotropin-releasing hormone, pyro-Glu-His-Pro-

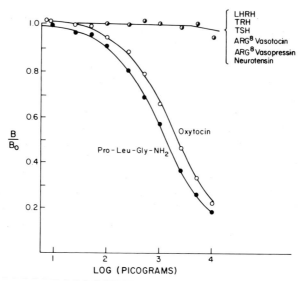

FIG. 2. Standard curve of radioimmunoassay for prolylleucylglycinamide using φ-PLG as tracer and hapten. Pro-Leu-Gly-NH$_2$ also represents the C-terminal tripeptide of oxytocin. Reproduced, by permission, from Manberg et al.[5]

NH$_2$. At present antisera for use in radioimmunoassay of TRH are raised against an immunogen produced by conjugating p-Glu-His-prolinamide and another larger protein through a diazotized benzidine bridge.[1] Immunization with this immunogen, however, has uniformly resulted in the raising of antisera of only moderately high affinity for TRH and that appear to be only weakly immunogenic, requiring between 4 and 8 months for the production of antisera with titers suitable for developing a radioimmunoassay.[6] Because the bisdiazotized benzidine procedure couples TRH to the bovine serum protein carrier through the imidazole ring of histidine, the poor immunogenicity of the coupled hapten may result either from the steric strain introduced into the coupled TRH molecule by attachment of the histidine ring and/or from the masking of the polar imidazole ring by the bulky BSA molecule. Thus, the most unique feature of the TRH molecule, its imidazole ring, may not participate to any great extent in the immunogenicity of this conjugate.

To circumvent these difficulties and to synthesize haptens that, when conjugated with a protein carrier, might result in a more immunogenic conjugate, one may synthesize analogs of TRH by attaching a GABA at

---

[6] W. W. Youngblood, L. J. Moray, W. H. Busby, and J. S. Kizer, *J. Neurosci. Methods* (in press).

either end of a TRH-like analog. This hapten may then be coupled to a protein carrier by a water-soluble carbodiimide, leaving the histidine and the adjacent C terminus or histidine and the preceding N terminus free for interaction with macrophages for antigen processing.

Our approach to preparing the N-terminus extended peptide is to couple a carbobenzyloxy N-blocked ω-amino fatty acid (Z-ω AFA) to the C-blocked protein, remove the N-blocking group by hydrogenation, and couple the AFA N-terminus to a larger protein. For protein amides, such as MIF or p-Glu-His-Pro-NH$_2$, there is no need to block the C terminus before coupling. Our preparation of the haptens pHAA-MIF, GABA-His-Pro-NH$_2$, and ACA-His-Pro-NH$_2$ are examples of how this C-terminal NH$_2$ may be used to an advantage. For peptides that have a free carboxy terminus or may contain glutamic or aspartic acids, it is necessary with most coupling methods (except the diazo coupling) to derivatize these groups as methyl esters (see below) before conjugating the peptide with the carrier. Failure to block the carboxylates in the hapten may result in self-condensation or in coupling of the carboxylate to lysine residues on the carrier. Removal of the methyl esters following conjugation with the carrier will not be necessary, since they will undergo spontaneous hydrolysis in the physiological environment. In our studies, we have prepared the hapten BSA-GABA-Gln-His-Pro-Gly-OMe, which was dialyzed after coupling and used without removing the methyl ester.

The selection of the AFA is also important. We have compared antibodies obtained using the conjugate BSA-NH-(CH$_2$)$_n$-CO-His-Pro-NH$_2$ where n values of 3 and 5 are used (see below). The antisera for TRH resulting from a chain length of 3 were much superior to that obtained from a carbon chain length of 5. Thus, it is important when developing haptens to synthesize several tethers of varying lengths and to select the one that gives the most potent immunogen after coupling. The attachment to N-terminal glutamine is another potential problem, since the tendency of glutamine to undergo internal cyclization rather than intermolecular coupling may be a major limitation in some instances. We have successfully prepared Z-GABA-Gln-His-Pro-Gly-OMe in moderate to high yields by hydrogenating Z-Gln-His-Pro-Gly-OMe and rapidly filtering the hydrogenation solution directly into the cold reactant vessel containing Z-GABA active ester.

Both *N*-carbobenzyloxy-ω-amino-*n*-caproic (Z-ωACA) and *N*-carbobenzyloxy-α-aminobutyric acids (Z-GABA) are commercially available; however, they will sometimes contain trace to minor amounts of benzyl alcohol, which may necessitate purification prior to use. We subject all batches of commercial preparations to thin layer, mass spectral, and amino acid analysis before using them. We have also used both Z-GABA and Z-ωACA prepared using the procedure outline below, which is a variation of the procedure of Carter et al.[7] This method should be suitable as a general preparation for Z-ωAFAs in the event that a suitable commercial source cannot be found.

Preparation of Carbobenzyloxy-γ-aminobutryic Acid (Z-GABA) and Carbobenzyloxy-ω-aminocaproic Acid (Z-ωACA)

Either GABA (1.03 g; 10 mmol) or ωACA (1.31 g; 10n mmol) is added to a vigorously stirred mixture containing 25 ml each of dioxane and water cooled to $-10°$. Two milliliters of 5 M NaOH (4 ml when using the amino fatty acid-HCl) is added to liberate the amine. A commercial preparation of carbobenzyloxy chloride (CBZ-Cl; Pierce Chemical Company, Rockford, IL) and 5.6 M NaOH are added to the vigorously stirred mixture maintained at $-10°$ as follows:

1. CBZ-Cl (800 μl; 956 mg; 5.6 mmol) is added, and the reaction is allowed to proceed for between 30 and 45 min. Then 1 ml of NaOH is added slowly to keep the temperature below $0°$. The pH is checked and adjusted if necessary to between 9 and 11.

2. The addition of CBZ-Cl followed in 30–45 min by the addition of NaOH is repeated using quantities of reagents given below:

Addition:	2	3	4	5	6
CBZ-Cl:	400 μl	200 μl	100 μl	100 μl	100 μl
NaOH:	500 μl	250 μl	125 μl	125 μl	125 μl

3. After the last addition, the pH is adjusted to between 9 and 11, and the mixture is stirred for an additional hour.

The reaction mixture is diluted to 200 ml with water, then extracted twice with 25 ml of diethyl ether (which is discarded). The pH of the aqueous phase is adjusted to between 2 and 3.5 using 1.0 M sulfuric acid, and the aqueous phase is extracted three times with 50 ml of dichloromethane. The combined organic phases are dried over sodium sulfate, after which the Z-GABA or Z-ωACA is recovered by removal of solvent under vacuum.

As with commercial preparations, the recovered product should be subjected to TLC analysis. The product is examined for the presence of benzyl alcohol and unreacted amino fatty acid (ninhydrin positive). Pure samples of Z-ωAFA are obtained by chromatography of the product on 24 g of activated silica gel using 100 ml each of 5, 10, and 20% methanol in chloroform.

[7] H. E. Carter, R. L. Frank, and H. W. Johnston, in "Organic Synthesis" (E. C. Hornig, ed.), Vol. III, p. 167. Wiley, New York, 1955.

Preparation of Carbobenzyloxy-γ-aminobutyrylhistidine Methyl Ester (Z-GABA-His-OCH$_3$)

Coupling of Z-GABA, or Z-ωACA, to the N terminus of histidine may be accomplished using two different coupling procedures. One procedure uses the classic dicyclohexylcarbodiimide (DCC) reagent used for preparation of peptide bonds. The second procedure uses an active ester coupling prepared by allowing the Z-GABA to react with ethyl chloroformate (ECD) in the presence of a tertiary amine. The DCC approach is advantageous when the peptide contains reactive amino acids, such as histidine, serine, or tyrosine. However, coupling may proceed more slowly than the active ester method. It is probably preferred when coupling to proline. The active ester method proceeds rapidly, and purification is sometimes easier than in the DCC procedure. Although the active ester method should be more sensitive to the presence of reactive amino acid R groups, we have attached both Z-GABA and Z-ωACA to histidine-containing peptides in very good yields using this method.

Coupling Using the Active Ester Prepared from Ethyl Chloroformate (ECF)

Z-GABA (1.185 g; 5 mmol) is dissolved in 25 ml of anhydrous N,N-dimethylformamide (DMF) containing 5 mmol of either anhydrous N-methylmorpholine (NMM) (550 μl) or triethylamine (700 μl) freshly distilled from ninhydrin. The mixture is cooled to −10°, and ECF (460 μl; 4.8 mmol) is added at the outer edge of the rapidly stirred solution. The reaction is stirred for 30–45 min, at which time 10 mmol of NMM (1.1 ml) or TEA (1.4 ml) is introduced. Histidine methyl ester dihydrochloride (1.162 mg; 3.8 mmol) is added to the reaction mixture, and the stirring is continued overnight at 4°.

Coupling Using the Dicyclohexylcarbodiimide (DCC)

Z-GABA (1.185 g; 5 mmol) is dissolved in 25 ml of anhydrous DMF containing 10 mmol of either anhydrous N-methylmorpholine (1100 μl) or triethylamine (1400 μl) freshly distilled from ninhydrin. Histidine methyl ester dihydrochloride (1.210 mg; 5.0 mmol) is added, and the reaction mixture is cooled to −10°. DCC (1.030 g dissolved in 2 ml of DMF) is introduced, and the mixture is stirred at 4° for 18–24 hr. The reaction mixture is then centrifuged or filtered to remove dicyclohexylurea, the precipitate is washed with DMF until 5 μl of filtrate no longer gives a positive Pauli test. The filtrate and washings are combined, and solvent is evaporated under vacuum.

Purification of Z-GABA-His-OCH₃ and Saponification

After evaporation of the solvent under reduced pressure, the residue is suspended in 20% methanol–2% TEA in chloroform; the suspension is applied to the top of a 50 × 1.5 cm column containing activated silica gel (24 g; slurried into the column using the eluate). Fractions are examined using silica gel TLC eluted with dry methanol–dichloromethane (40:60). Plates are developed with Pauli reagent, ninhydrin, and iodine. The R_f of Z-GABA-His-OMe is 0.85. Fractions are pooled, yielding between 50 and 75% recovery of highly pure Z-GABA-His-OCH₃ based upon histidine methyl ester. The weight of product should be determined and used for calculating the amount of proline amide to be used in the last coupling step. The methyl ester is rapidly converted to the free carboxylate by dissolving the Z-GABA-His-OCH₃ in water–methanol (90:10), adding one equivalent of NaOH, and stirring for between 10 and 15 min. The R_f of Z-GABA-His is 0.20 in MeOH–dichloromethane (40:60). The pH is adjusted to 6.0 using 1.0 M HCl, and the sample is used without further purification by freeze drying in the reaction vessel to be used for coupling to prolinamide.

Coupling of Z-GABA to Gln-His-Pro-Gly-OCH₃ Using Ethyl Chloroformate (ECF)

Z-GABA (0.140 g; 0.6 mmol) dissolved in 2 ml of anhydrous DMF containing 0.6 mmol of anhydrous N-methylmorpholine (65 µl) freshly distilled from ninhydrin is cooled to $-10°$, and ECF (50 µl; 0.5 mmol) is added to the rapidly stirred solution. The reaction is stirred for 45 min, during which time Z-Gln-His-Pro-Gly-OCH₃ (25 mg) dissolved in 1 ml of DMF is hydrogenated over 10% Pd/C at atmospheric pressure. The reaction mixture is force filtered through a disposable filter assembly into the reaction vessel containing the activated Z-GABA, and the mixture is stirred at 4° for 24 hr. The mixture is diluted to 25 ml with water, the pH is adjusted to 10, and the aqueous phase is extracted twice with 10 ml of diethyl ether (which is discarded). The pH is adjusted to between 2.5 and 3.0 using 1 M HCl, and the aqueous phase is extracted a second time with ether to remove the excess Z-GABA. The pH is adjusted to 7.0, and the sample is freeze dried. Hydrogenation of the sample at atmospheric conditions using 10% Pd/C liberates GABA-Gln-His-Pro-Gly, which is separated from contaminating p-Glu-His-Pro-Gly by chromatography. Structural identity of and purity of GABA-Gln-His-Pro-Gly and p-Glu-His-Pro-Gly are verified by color reagents sprayed on TLC plates (ninhydrin and Pauli reagents), and by amino acid analysis.

Reesterification of GABA-Gln-His-Pro-Gly and Attachment of BSA

GABA-Gln-His-Pro-Gly (5 mg) is dissolved in 10 ml of anhydrous methanol into which HCl gas has been dissolved to produce a concentration of between 1 and 2 M. Reesterification of the starting material is followed by TLC. The solvent is evaporated under dry nitrogen, and the GABA-Gln-His-Pro-Gly-OCH$_3$ is coupled to BSA using a water-soluble carbodiimide procedure. After coupling, the hapten is dialyzed and subjected to amino acid analysis. By comparing amino acid molar ratios of one of the known amino acids in BSA and the tether amino acid in hydrolyzates of the hapten, one can determine the number of peptide molecules that have been coupled to the BSA molecule. Thus, another advantage in preparing haptens by tethering with an amino fatty acid is that it also introduces a convenient tag for establishing the quality of the conjugate.

C-Terminus Extension

The coupling of AFAs to the C terminus of peptides necessitates the use of N-blocked peptides and C-blocked AFAs. The most common C-terminus blocking group for this purpose is clearly the methyl ester, whose appeal comes from its general ease of preparation and saponification. In selecting the amino acid to be used, it should be kept in mind that certain structures under the conditions outlined below for esterification will also undergo cyclization to form the lactam. Although the lactam's presence in the product will not interfere in the subsequent coupling of the methyl ester with the peptide, it will cause underestimation of the stoichiometric amounts of ester to be coupled with the peptide. We have observed that aminocaproic acid forms the lactam with such facility as to make it less desirable for this purpose. Thus, in our studies we have used GABA or shorter amino fatty acids, since they do not readily form cyclic lactams. Some small-chain amino fatty acids are commercially available as the methyl ester. As with the Z-ωAFA, the quality of the product should be checked before using. In preparing the methyl ester, the procedure cited below is an example of the classic esterification technique for an acid and alcohol using HCl.[8] For small quantities this method may be preferred to the thionyl chloride procedure, which may produce more product contaminants if the thionyl chloride is impure. Purity of product and ease of isolation are more important when preparing small quantities of the ester.

[8] T. H. Curtis, *J. Prakt. Chem.* **24**, 239 (1881).

Preparation of GABA-OCH$_3$

Anhydrous methanol is prepared by distillation from magnesium turnings. GABA (1.03 g; 10 mmol) is dissolved in dry methanol (100 ml), immersed in an ice bath, and saturated with dry HCl. The flask is fitted with a drying tube and allowed to stand overnight. The methyl-4-aminobutanoate hydrochloride is obtained by evaporation of the solvent and drying under reduced pressure. It is used without further purification.

Preparation of Pyro-Glu-His-Pro-GABA-OCH$_3$ and p-Glu-His-Pro-Gly-OCH$_3$

Z-Pro-Gly-OCH$_3$ and Z-Pro-GABA-OCH$_3$ were prepared by coupling Z-proline with either Gly-OCH$_3$ or GABA-OCH$_3$, using procedures described above for DCC and ECF preparation of Z-GABA-His-OCH$_3$. Chromatographic purifications were accomplished using silica gel and anhydrous methanol–chloroform (10:90). The N terminus was deblocked by hydrogenation using 10% Pd/C at atmospheric pressure. Both Pro-Gly-OCH$_3$ and Pro-GABA-OCH$_3$ were coupled to p-Glu-His using DCC in DMF, and the two tetrapeptides were separated from reacted dipeptides using silica gel chromatography. Methyl esters were saponified in water–methanol (90:10) at a pH of 10, and the saponification was followed by TLC. After saponification the solutions were neutralized and freeze dried. The two tetrapeptides were isolated by Sephadex G-25 chromatography.

When attempting to synthesize the same two tetrapeptides by coupling p-Glu-His-Pro to GABA-OCH$_3$ or Gly-OCH$_3$, yields were lower than those obtained by coupling the dipeptides p-Glu-His and pro-γABA-OCH$_3$; thus, failure to attach a tether to a peptide may necessitate a total synthesis of the desired analog.

After immunization with the two conjugates, BSA-GABA-His-prolinamide and p-Glu-His-Pro-Gly GABA, antisera were raised that recognized iodinated TRH. Antisera raised against the GABA-His-Pro-NH$_2$ hapten have an affinity for unlabeled TRH almost an order of magnitude greater than antisera raised against an immunogen prepared by the classical methods of conjugation through a bisdiazotized benzidine bridge. These antibodies have been used for the development of an improved and more sensitive radioimmunoassay for TRH.[6] The antisera raised against the p-Glu-His-Pro-Gly-GABA conjugate on the other hand, did not produce antiserum satisfactory for a radioimmunoassay, a finding that stresses the need for multiple approaches and testing of each antiserum for its value in a radioimmunoassay.

Comments

In certain circumstances, the site of iodination of the oligopeptide may differ from the site for conjugation. For example, coupling of a hapten that contains a tyrosine to BSA through an N-terminal NH_2 may raise an antiserum that will not recognize the oligopeptide when the tyrosine is iodinated. On the other hand, coupling of a hapten that contains lysine through a histidine or tyrosine may raise an antiserum that will not recognize the oligopeptide if iodination is carried out with the Hunter–Bolton reagent. Again, TRH is a good example of this particular technical predicament. Antisera raised against GABA-His-Prolinamide could have a very high affinity for native TRH, but recognize iodinated TRH somewhat more poorly. Under these circumstances, one may wish to derive an N-terminal analog that could be iodinated with the Hunter–Bolton reagent, leaving the histidine residue free to interact with the antiserum. On the other hand, a peptide antigen that has been coupled to an ε-amine of lysine or primary N-terminal amine may best be iodinated with a Hunter–Bolton reagent, thereby leaving the remaining amino acid residues in nearly the same structural configuration as in the immunogen. These possible difficulties can also be remedied by using the methods outlined above for the introduction of either a hydroxyphenylacetic or a GABA moiety.

In summary, we have tried to present a few strategies for use in the development of radioimmunoassays for very small oligopeptides. In doing so we have chosen the simplest techniques with the realization that few laboratories possess the necessary biochemical or chemical expertise to synthesize longer oligopeptides or to introduce amino acid substitutions within the peptide chain. We believe that these methods are satisfactory for application in a wide variety of circumstances and may form the most appropriate first cut at producing haptens for iodination and conjugation where the immunogencity of an oligopeptide will be compromised by, or its structure will not permit, iodination or conjugation of the native peptide.

Acknowledgments

This work was supported by Grants HD-14005 and HD-03110 from NICHD. J. S. Kizer is the recipient of Research Scientist Career Development Award MH-00114.

[29] Preparation and Use of Specific Antibodies for Immunohistochemistry of Neuropeptides[1]

By LOTHAR JENNES and WALTER E. STUMPF

Immunohistiochemical localization of various peptides in the central nervous system as well as in peripheral organs has gained substantial importance because more and more peptides are being discovered and these peptides are available in quantities sufficient for immunization of various animal species. Most neuropeptides show a large degree of interspecies homology in their amino acid composition, which allows one antibody to be used for the visualization of the peptide in many different animal species. This homology, and the relative small size of the peptides, however, are responsible for a low antigenicity. Although a few research groups have succeeded in inducing an immune response and production of specific antibodies against native, unconjugated peptides, most success was obtained when the peptide was coupled to a large immunogenic carrier protein. Usually, these carriers belong to the serum proteins, e.g., human serum albumin (HSA), bovine serum albumin (BSA), thyroglobulin (TG), or keyhole limpet hemocyanin (KLH).[1a] Conjugation of a peptide to a protein induces an increase in antigenicity and allows one to select a particular amino acid of the peptide hormone at which conjugation most likely will take place. The site of conjugation is of importance for the specificity of the antibody, since usually only those 4–6 amino acids of the antigen are recognized that are farthest away from the site of conjugation to the carrier protein.[2] This property of the immune system allows one to predict and to manipulate more or less precisely the antigenic determinant of the peptide against which antibodies are to be raised.

Conjugation of Peptide Hormone to Carrier Protein

The conjugation of a peptide to a carrier protein requires a bivalent coupling reagent that establishes covalent bonds between two amino acids of the two different molecules. Most commonly used are dialdehydes, such as glutaraldehyde or formaldehyde, carbodiimides, bisdiazotized

---

[1] This work was supported by NIH Grant NS17614.
[1a] B. F. Erlanger, this series, Vol. 70, p. 85.
[2] C. W. Parker, "Radioimmunoassay of Biologically Active Compounds." Prentice-Hall, Englewood Cliffs, New Jersey, 1976.

TABLE I
FUNCTIONAL GROUPS AND CORRESPONDING AMINO ACIDS
AT WHICH A PEPTIDE MAY BE CONJUGATED

Group	Amino acid
Amino, imino	Arginine, histidine, lysine, tryptophan, N-terminal
Amide	Asparagine, glutamine
Phenolic	Tyrosine
Carboxyl	C-terminal, glutamic acid, aspartic acid
Imidazo	Histidine
Phenyl	Phenylalanine, tryptophan, tyrosine
Guanidino	Arginine
Indolyl	Tryptophan
Thiol	Cysteine
Hydroxyl	Serine, threonine

benzidine, *p*-diazonium phenylacetic acid and toluene diisocyanate[3-5]; these are described in this chapter. Further information about other cross-linking reagents can be found in Butler and Beiser,[6] Kennedy *et al.*,[7] and Likhite and Sehon.[8] The methods presented here were developed for the conjugation of the neuropeptide GnRH (pyro-Glu-His-Trp-Ser-Tyr-Gly-Leu-Arg-Pro-Gly-NH$_2$) to keyhole limpet hemocyanin (KLH), but they are applicable for most neuropeptides that contain the particular amino acid at which the coupling takes place (Tables I and II).

Glutaraldehyde

Glutaraldehyde has found the widest application for conjugations of peptides and proteins, since it establishes convalent linkages under mild conditions. It is composed of five carbons, contains two aldehyde groups, and occurs in differently hydrated forms. Although the detailed chemistry of the reactions of glutaraldehyde with peptides or proteins is not fully understood, it is clear that most conjugations occur at the ε-amino group of lysine.[9] Three different reactions have been described: According to

---

[3] B. M. Jaffe and H. R. Behrman (eds.), "Methods in Hormone Radioimmunoassay," 2nd ed. Academic Press, New York, 1979.
[4] J. Barry, *Int. Rev. Cytol.* **60**, 179 (1979).
[5] A. H. W. M. Schuurs and B. K. Van Weemen, *Clin. Chim. Acta* **81**, 1 (1977).
[6] V. P. Butler and S. M. Beiser, *Adv. Immunol.* **17**, 255 (1973).
[7] J. H. Kennedy, L. J. Kricka, and P. Wilding, *Clin. Chim. Acta* **70**, 1 (1976).
[8] V. Likhite and A. Sehon, *in* "Methods in Immunology and Immunochemistry" (C. A. Williams and M. W. Chase, eds.), Vol. 1, p. 150. Academic Press, New York, 1967.
[9] M. Reichlin, this series, Vol. 70, p. 159.

TABLE II
DIVALENT COUPLING REAGENTS AND FUNCTIONAL GROUPS, AT WHICH
A COVALENT CROSS-LINK MAY BE ESTABLISHED

Reagent	Group
Glutaraldehyde	Amino, imidazo, phenol
Carbodiimides	Carboxyl, amino, phenolic, thiol, alcohols
Bisdiazotized benzidine	α-, ε-Amino, imidazo, phenol
Bisdiazotized phenylacetic acid	Imidazo, phenol, ε-amino
Toluene diisocyanate	ε-Amino, phenol, primary alcohols, sulfhydryl

Hardy et al.,[10] a dihydropyridine is formed that dimerizes to assume an anabasine structure; however, Richards and Knowles[11] proposed a reaction in which a Michael adduct is formed by the amine and the double bond of an unsaturated aldehyde. Previously assumed simple Schiff-base formation does not seem to occur, since the reaction product of glutaraldehyde and an amino group resists acid hydrolysis.[12] In addition to the above reactions with ε-amino groups of lysine, glutaraldehyde can react with α-amino groups, especially at the N-terminus of a peptide, with the phenol ring of tyrosine, or with the imidazole ring of histidine.[13] A large number of conjugation methods for binding a peptide to a protein via glutaraldehyde exist in the literature. These methods differ mostly in minor aspects, such as the molar ratio of peptide to protein, which generally ranges between 5 and 40:1.

Procedure. Four milligrams of GnRH are dissolved at room temperature (RT) in 0.5 ml of 0.1 M phosphate buffer, pH 7.2 (PB) and added to a solution of 20 mg of KLH in 0.5 ml of PB. Under constant stirring, 10 μl of a 50% aqueous glutaraldehyde (5 μM) is added; after 1 hr at RT, the reaction is stopped by the addition of 7.5 mg of sodium metabisulfite in 1 ml of PB. The solution is brought to a final volume of 10 ml with PB, then aliquoted into 0.5-ml portions containing 200 μg of peptide each, and stored until use at $-30°$.

[10] P. M. Hardy, A. C. Nicholls, and H. N. Rydon, *J. Chem. Soc. Perkins Trans.* **1**, 958 (1976).
[11] F. M. Richards and J. R. Knowles, *J. Mol. Biol.* **37**, 231 (1968).
[12] A. F. S. A. Habeeb and R. Hiramato, *Arch. Biochem. Biophys.* **126**, 16 (1968).
[13] J. D. Ford and A. J. Pesce, in "Enzyme Immunoassay" (E. Ishikawa, T. Kawai, and K. Miyai, eds.), p. 54. Igaku Shoin, Tokyo, 1981.

Carbodiimides

This family of bivalent coupling reagents, with the general formula R—N=C=N—R, encompasses many compounds, the most important of which are probably 1-ethyl-3-(3-dimethylaminopropyl)carbodiimide-HCl and 1-cyclohexyl-3-(2-morpholinyl-4-ethyl)carbodiimide metho-*p*-toluenesulfonate. Carbodiimides can couple under mild conditions several functional groups, such as carboxylic acids or ε-amino groups, whereas for the conjugation of thiols, phenols, or alcohols, more drastic conditions are required.[7,14] Although the mechanism is not fully elucidated, it has been proposed that several sets of reactions occur during the coupling procedure, which result in an amide bond between an amino and a carboxyl group of the protein and the peptide. In one set of reactions, carbodiimide binds to a carboxyl group of a protein to form an intermediate *O*-acyl urea. This intermediate reaction product can rearrange to a N-substituted urea at the carboxyl group or it can react with an amino group of another peptide or protein to establish a peptide bond between the conjugates. Since the acyl urea is formed preferentially at warmer temperatures, the reaction should be performed between 0 and 4° in order to favor peptide bond formation.[15] Most methods of conjugation of a peptide to a protein are based on the procedure of Goodfriend *et al.*[16] and contain only slight modifications.

Procedure. Four milligrams of GnRH and 20 mg of KLH are dissolved in 500 µl of distilled water, to which 100 mg of freshly dissolved 1-ethyl-3-(3-dimethylaminopropyl)carbodiimide-HCl in 250 µl of water is added. After 30 min under constant slight stirring, the reaction is stopped either by dialysis against water for 24 hr or by addition of 9.25 ml of PB with subsequent aliquoting into 20 portions of 500 µl, each containing 200 µg of GnRH. These aliquots are stored at −30°. During the conjugation procedure, a precipitate may form that is used together with the clear supernatant for the immunization.

Bisdiazotized Benzidine

Since the conjugations via glutaraldehyde or carbodiimides usually require the presence of a reactive amino or carboxyl group, these two reagents may not be applicable for the conjugation of all neuropeptides. An alternative coupling reagent is bisdiazotized benzidine, which is formed in a reaction of an aromatic amine with nitrous acid to form a

[14] K. L. Carraway and D. E. Koshland, Jr., *Biochem. Biophys. Acta* **160**, 272 (1968).
[15] S. Bauminger and M. Wilchek, this series, Vol. 70, p. 151.
[16] T. L. Goodfriend, L. Levine, and G. D. Fasman, *Science* **144**, 1344 (1964).

diazonium salt.[17-19] Bisdiazotized benzidine can react in an alkaline milieu with α- or ε-amino groups,[20] and with imidazole or phenol groups. Even a reaction with arginine and tryptophan has been reported.[21] Under equimolar conditions, azo linkages are established to result in monosubstituted histidine and tyrosine and in disubstituted lysine groups. If excessive diazonium salts are present, histidines and tyrosines will be disubstituted and additional reactions with tryptophan and arginine will occur.[22]

Procedure. The conjugations procedure is divided into two parts. In the first reaction, bisdiazotized benzidine is formed. Benzidine hydrochloride (230 mg) is dissolved in 45 ml of 0.2 N HCl, to which 175 mg of sodium nitrite in 5 ml of distilled water is added. The color of the mixture changes to orange, and the reaction is allowed to proceed at 4° for 60 min under constant slight stirring. Aliquots of 1.5 ml containing 7 mg of bisdiazotized benzidine are frozen in Dry Ice or liquid nitrogen and stored at $-30°$.

In the second reaction, a covalent bond is established between the peptide and the protein carrier. One aliquot (7 mg) of bisdiazotized benzidine is dissolved in 2.5 ml of aqueous 0.1 M borate and 0.13 M sodium chloride at a pH of 9.0. To this mixture are added 25 mg of KLH and 10 mg of GnRH, each dissolved in 2.5 ml of borate–sodium chloride. A brown precipitate appears immediately, and the conjugation is continued for 2 hr at 4° under constant stirring. The reaction is stopped by dialysis at 4° against water for 7 days with a final dialysis against 0.15 M sodium chloride for 1 day. Alternatively, the dialysis is omitted, and the reaction mixture is quickly frozen in 100-μl aliquots, each containing 200 μg of the peptide.

p-Diazonium Phenylacetic Acid Carbodiimide

This method for conjunction has been developed to improve the homogeneity of the antigen and the predictability of the amino acids at which the cross-link will be established. The reaction can be divided into two steps: first, a free carboxyl group is introduced into the peptide via attachment of *p*-diazonium phenylacetic acid; then this free carboxyl

---

[17] R. M. Bassiri and R. D. Utiger, *Endocrinology* **90**, 722 (1972).

[18] T. M. Nett, A. M. Akbar, G. D. Niswender, M. T. Hedlund, and W. F. White, *J. Clin. Endocrinol. Metab.* **36**, 880 (1973).

[19] G. F. Brice, *Immunochemistry* **11**, 507 (1974).

[20] M. Tabachnick and H. Sobotka, *J. Biol. Chem.* **235**, 1051 (1960).

[21] A. N. Howard and F. Wild, *Biochem. J.* **65**, 651 (1957).

[22] A. Nisonoff, in "Methods in Immunology and Immunochemistry" (C. A. Williams and M. W. Chase, eds.), Vol. 1. p. 120. Academic Press, New York, 1967.

group is conjugated via carbodiimide to an amino group of the carrier protein to establish a regular peptide bond. A higher degree in homogeneity of the amino acids involved in the conjugation is achieved with this method than with the bisdiazotized benzidine, since almost exclusively azohistidyl and azotyrosyl derivatives are formed. In the case of GnRH, a molar ratio of 7:3 of the derivatized azohistidyl and azotyrosyl has been reported. Furthermore, since azo derivatives with a free carboxyl group are formed only at the peptide, but not at the carrier protein, peptide polymers, peptide–protein polymers, or protein–protein polymers are not likely to occur. For the introduction of a free carboxyl group into histidine or tyrosine, the method of Koch et al.[23] may be followed.

Procedure. p-Aminophenylacetic acid (10 mM) in 25 μl of cold 2 N HCl is added to 10 μM sodium nitrite in another 25 μl of cold water. After 8 min at 4°, the pH is adjusted to pH 8.5 via 150 μl of cold 75 μM sodium bicarbonate, before addition of 10 μM GnRH in 90 μl of 60% aqueous N,N'-dimethylformamide containing 20 μM sodium bicarbonate. The solution changes its color to brown, and the reaction proceeds for 12 hr at 4°. The formation of the azohistidyl- or azotyrosyl-GnRH is stopped by adjusting the pH to pH 2 via 2 N HCl before the mixture is extracted in 2 volumes of ether. The ether is discarded, and the pH level of the aqueous phase is adjusted to pH 7.0 via 0.5 M sodium dicarbonate.

The now purified azo derivatives are conjugated to a carrier protein via the standard 1-ethyl-3-(3-dimethylaminopropyl)carbodiimide–HCl method. Ten micromolar azotized GnRH is added to a solution of 30 mg of KLH in 500 μl of distilled water. The coupling is performed at pH 5.0 by the addition of 250 mg of freshly dissolved 1-ethyl-3-(3-dimethylaminopropyl)carbodiimide-HCl in 250 μl of water. After 30 min of shaking at room temperature, the conjugation is stopped by dialysis against water for 24 hr. Aliquots containing 200 μg of conjugated hormone are frozen and stored at $-30°$.

Toluene 2,4-Diisocyanate

In an approach to achieve a high degree of homogeneity of a conjugate, the bivalent coupling reagent toluene 2,4-diisocyanate, which has two differentially active groups, has found wide application.[24] Toluene 2,4-diisocyanate is thought to react with free amino groups to establish a covalent, probably ureido, linkage.[25] Furthermore, the primary alcohol

[23] Y. Koch, M. Wilchek, M. Fridkin, P. Chobsieng, U. Zor, and H. R. Lindner, *Biochem. Biophys. Res. Commun.* **55**, 616 (1973).
[24] J. Barry, M. P. Dubois, and P. Poulain, *Z. Zellforsch. Mikrosk. Anat.* **146**, 351 (1973).
[25] S. J. Singer and A. F. Schick, *J. Biophys. Biochem. Cytol.* **9**, 519 (1961).

group of, for example, serine, and the phenol group of tyrosine have been suggested as possible sites of linkage. The advantage of toluene 2,4-diisocyanate lies in the different activities of its isocyanate groups. Since the isocyanate group in position 4 is approximately 8 times more reactive (in a reaction with *n*-butanol) than the one in position 2, a conjugation procedure can be conducted as a two-step reaction: first, the amino groups of the carrier react with the isocyanate in position 4 at 4° and, after removal of the excessive toluene 2,4-diisocyanate, the hormone is cross-linked to the carrier via the less reactive group in position 2. The second step is carried out at 37° to activate the less reactive isocyanate group.[26] The product of such a reaction is very homogeneous, and polymers are largely avoided.

Procedure. For the coupling of GnRH, 20 mg of KLH is dissolved in 5 ml of cold 0.1 M phosphate buffer, pH 7.4, to which 0.2 ml of toluene 2,4-diisocyanate is added. After 1 hr at 4°, the precipitate is discarded and the supernatant is brought together with 5 mg of GnRH in 0.5 ml of 0.1 M phosphate buffer with a pH of 7.4. The reaction is allowed to proceed for 1 hr at 37° before aliquots containing 200 μg of GnRH each are rapidly frozen and kept at −30° until use.

Unspecific Adsorption

Successful generation of antibodies against small, unconjugated neuropeptides, such as GnRH, has been achieved occasionally, although the rate of success is low and the amount of peptide required is usually high. In most cases, the hormone is protected from immediate degradation by an unspecific attachment to polyvinylpyrrolidone.[27-29] In this procedure, 5 mg of GnRH is dissolved in 1 ml of 0.9% sodium chloride and added to 2 ml of 50% aqueous polyvinylpyrrolidone (M_r 40,000). The mixture is stirred constantly for 2 hr at room temperature and used without freezing.

When we compare the properties of antibodies to GnRH that were raised against different conjugates, it becomes apparent that most antisera recognize the N terminus of the molecule. In addition, some antisera against GnRH conjugated via carbodiimide require an intact glycine in position 6, but usually modifications in position 1 and 2 are not recognized. Glutaraldehyde and toluene 2,4-diisocyanate conjugates result in

[26] A. F. Schick and S. J. Singer, *J. Biol. Chem.* **236**, 2477 (1961).
[27] R. B. Worobec, J. H. Wallace, and C. G. Huggins, *Immunochemistry* **9**, 229 (1972).
[28] A. Arimura, H. Sato, T. Kumasaka, R. B. Worobec, L. Debeljuk, J. Dunn, and A. V. Schally, *Endocrinology* **93**, 1092 (1973).
[29] A. Arimura, H. Sato, D. H. Coy, R. B. Worobec, A. V. Schally, N. Yanaihara, T. Hashimoto, C. Yanaihara, and N. Sakura, *Acta Endocrinol.* **78**, 222 (1975).

very similar antibodies; these require an intact N terminus, but not the first, second, or sixth amino acid. If GnRH was coupled via diazonium-phenylacetic acid carbodiimide, the antibody requires the first, second, sixth, and tenth amino acid to bind to the hormone. Immunization of several animals with the same antigen results, in our experience, in production of antibodies that often recognize different portions of the antigen, as can be derived from radioimmunoassay characterization. Thus, although a certain trend in the resulting antigenic determinante may be recognized, one should expect a substantial variety of antibodies directed against the same antigen.

Immunization

Before immunization of an animal is initiated, the antigen needs to be protected against rapid enzymic or cellular degradation, which starts immediately after the injection of the antigen. The most common procedure is to emulsify the antigen in Freund's adjuvant, that is a 8.5:1.5 mixture of mineral oil and an emulsifier, usually Aracel. To this mixture, heat-inactivated mycobacteria (complete adjuvant) are added to stimulate the general immune response.[30] A water-in-oil emulsion allows a steady slow release of the antigen, which is sufficient to stimulate the immune system to produce specific antibodies. The release of the antigen should be as slow as possible in order to allow only the highest-affinity antibodies to bind to it, and by this to limit the large-scale production of antibodies to the high-affinity clones.[31] The administration of too much antigen is a common cause for failure to induce specific, high-affinity antibodies.

We obtained very good emulsions with the following method. An aliquot of conjugated peptide (200 μg) in 1 ml of buffer is added to 1 ml of Freund's complete adjuvant, and the mixture is placed into a 5-ml syringe, which is connected via its Luer-lok and a three-way stopcock to a second syringe. If the plungers are pushed back and forth (about 20 times), the mixture changes its consistency into a stiff emulsion. The emulsion is collected in one of the two syringes and can be used for immunization without any further treatment.

For the initial inoculation, we prefer multiple intradermal injections. Simultaneously, we inject subcutaneously 0.5 ml of "triple antigen" containing inactivated diphtheria and tetanus toxins and pertussis vaccine to stimulate the general immune response. For later booster immunizations,

---

[30] J. S. Garvey, N. E. Cremer, and D. H. Sussdorf (eds.), "Methods in Immunology." Benjamin, Reading, Massachusetts, 1977.

[31] F. M. Burnett, "Immunology, Aging and Cancer." Freeman, San Francisco, California, 1976.

the antigen is emulsified in Freund's incomplete adjuvant and subcutaneously into 4–6 places. Triple antigen is omitted. A large number of immunization protocols exist in the literature; these may differ in the route of administration of the antigen, dosage, or timing of the injection-bleeding rhythm.[32,33]

Our typical immunization schedule is as follows: An initial injection of 200 μg of antigen in complete Freund's adjuvant is placed into multiple (approximately 50), intradermal sites; at the same time, 0.5 ml of triple antigen is given subcutaneously. After 4 weeks, a booster injection of 100–200 μg of antigen in incomplete Freund's adjuvant is applied subcutaneously at 4–6 places on the animal's back. Blood is drawn from the ear vein 7 days after the last injection and carefully tested for the presence of specific antibodies. The animal is boosted from now on every 2 weeks for the next 6–8 weeks with 100–200 μg of antigen. Usually, specific antibodies appear 8 weeks after the initial injection, i.e., after the second booster. The antibody titer increases substantially over the next 3–4 boosters. If an animal does not produce suitable antibodies within the first 3 months, the immunization of this animal is discontinued. Representative titer profiles of three antisera against GnRH, which were raised according to the above method, are shown in Fig. 1.

The specificity of an antiserum is critical for the validity of the results and should be carefully established, although inherent limitations related to basic mechanisms of antigen–antibody recognition and binding make it impossible to achieve absolute security.[34] A specific antibody recognizes an antigenic determinant and will bind to it regardless, whether this determinant is present exclusively in the assumed antigen or whether it occurs in an appropriate configuration as part of an unrelated molecule.[35] One way to minimize the probability of undetected cross-reactivity is to characterize the amino acids of the antigen, which are recognized by the antibody in a radioimmunoassay system. An example of such a characterization is given in Fig. 2. In this particular case, an antibody was tested that was raised against GnRH conjugated to KLH according to the method of Koch et al.[23] The specificity of the antiserum and the sites of the antigen that are recognized by the antibody are identified by competition of GnRH analogs with single amino acid substitution with radioactive GnRH. This antiserum requires an intact first, second, sixth, and tenth

---

[32] J. Vaitukaitis, J. B. Robbins, E. Nieschlag, and G. T. Ross, *J. Clin. Endocrinol. Metab.* **33**, 988 (1971).
[33] B. A. L. Hurn and S. M. Chantler, this series, Vol. 70, p. 104.
[34] E. A. Kabat, this series, Vol. 70, p. 3.
[35] W. M. Hunter, in "Handbook of Experimental Immunology" (D. M. Weir, ed.), Vol. 1, p. 14.1. Blackwell, Oxford, 1979.

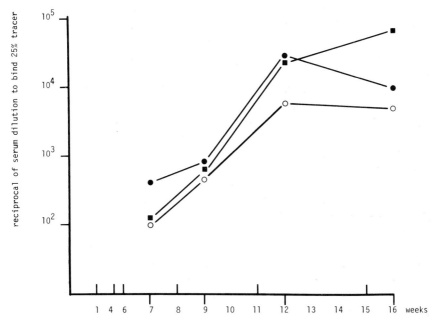

FIG. 1. Titer profiles of three GnRH antisera in relation to the time of immunization. The animals were boosted 4, 6, 8, 11, and 15 weeks after the initial immunization, and blood was collected after weeks 7, 9, 12, and 16.

amino acid, since analogs with modifications in these amino acids are not bound by the antibody.

If such a system is not available, the specificity of an antiserum may be tested in an immunohistochemical system via absorption of the antiserum with different amounts of antigen-related compounds as well as via addition of unrelated substances or tissue extracts.[36,37] If an antiserum is used optimally in an immunohistochemical staining procedure at a dilution of 1 : 1000, it is usually saturable with about 1–5 μM antigen per milliliter of the working dilution. In case of a GnRH antiserum, addition of 1 μg to 1 ml of working dilution should prohibit any staining, and addition of 10–100 μM unrelated peptides, such as somatostatin, enkephalin, or substance P, should not reduce the staining intensity. Characterization of the peptide residues that are recognized by the antibody, however, is very limited in an immunohistochemical system, since the tissue has to be stabilized or fixed, usually via aldehydes, before it can be exposed to the

[36] L.-I. Larsson, in "Modern Methods in Pharmacology" (S. Spector and N. Back, eds.), p. 1. Liss, New York, 1982.
[37] P. Petrusz, P. Ordronneau, and J. C. W. Finley, *Histochem. J.* **12**, 333 (1980).

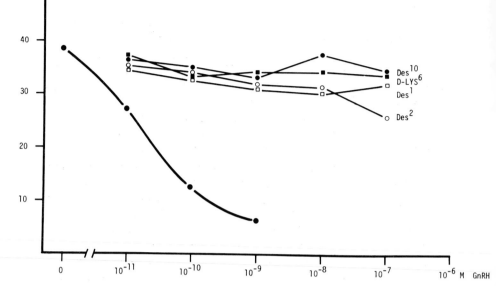

FIG. 2. Radioimmunoassay of a GnRH antiserum raised according to the method of Koch et al.[23] The ED_{50} for GnRH is 5×10^{-11} M, and the ED_{50} is greater than 10^{-7} for the GnRH analogs Des¹ (□——□), Des² (○——○), DLys⁶ (■——■), and Des¹⁰ (●——●).

antiserum. Such a chemical fixation modifies the tertiary and quarternary structure of a peptide in a relatively inconsistent and unpredictable way by cross-linking it, in part covalently, to other tissue components. Thus, it is problematic to relate changes in the intensity of the staining to a certain degree of absorption of the antibody by various amounts of different peptide fragments. Since both radioimmunoassay and immunohistochemistry have inherent limitations, the specificity of an antibody should be established optimally with both systems.

Staining

The description of methods for application of antisera in immunohistochemical staining procedures is subject of numerous publications[38,39] and will be limited here to the presentation of our routine procedure for im-

[38] G. R. Bullock and P. Petrusz (eds.), "Techniques in Immunocytochemistry," Vol. 1. Academic Press, New York, 1982.
[39] G. D. Johnson, E. J. Holborow, and J. Dorling, in "Handbook of Experimental Immunology" (D. M. Weir, ed.), Vol. 1, p. 15.1. Blackwell, Oxford, 1979.

munohistochemistry at the light microscopical level, which is based upon the Mason single-bridge technique.

If possible, we avoid embedding media, such as plastic or paraffin, since most of these media require alcohol dehydration of the tissue and peptides may be lost to varying degree during dehydration. For most purposes, we use 50–100 μm-thick vibratome sections of tissue that was perfusion fixed with paraformaldehyde. The sections are collected in scintillation vials and washed in 0.1 M phosphate buffer, pH 7.3 (PB), containing 0.2% Triton X-100 and 1% normal goat serum. After 30 min, this buffer mixture is exchanged with primary antiserum—for example, a rabbit anti-GnRH serum in a dilution of 1 : 1000 in PB—to which 0.2% Triton X-100, 0.1% sodium azide, and a small spatula tip of the carrier protein, to which GnRH was conjugated for the antibody production, are added. After 36–48 hr at room temperature, the sections are rinsed for 15 min in PB, incubated for 2 hr in a sheep anti-rabbit IgG serum in a dilution of 1 : 50, washed again in PB, and kept for another 2 hr in a rabbit antiperoxidase serum in a dilution of 1 : 100. The sections are washed again for 15 min and incubated for 2 hr in 0.5 mg of horseradish peroxidase per 100 ml of PB, washed again, and stained in a solution of 75 mg of 3,3'-diaminobenzidine tetrahydrochloride in 100 ml of Tris-saline buffer, 0.05 M, pH 7.6, and 7 μl of concentrated (30%) hydrogen peroxide. Staining is usually terminated after 10 min by washing the sections in PB. Finally, the sections are mounted on glass slides and coverslipped with Elvanol, a water-soluble medium of poly(vinyl alcohol), glycerol, and PB.

[30] Production of Monoclonal Antibodies Reacting with the Cytoplasm and Surface of Differentiated Cells

By RICHARD M. SCEARCE and GEORGE S. EISENBARTH

Hybridization of mouse myeloma cells and antibody-producing spleen cells was described by Kohler and Milstein in 1975.[1] The "immortality" of the parent myeloma coupled with the "specific" antibody production by the splenic lymphocytes enables investigators to produce large quantities of antibody reacting with "single" antigenic determinants.[2] While general methodology for production of murine monoclonals is relatively

---

[1] G. Köhler and C. Milstein, *Nature* (*London*) **256,** 495 (1975).
[2] G. S. Eisenbarth and R. A. Jackson, *Endocr. Rev.* 3:26–39 (1982).

simple, production and characterization of a specific monoclonal antibody usually requires 6–8 months of cell culture, assays, and biochemical analysis. Within a general framework, methods vary considerably among laboratories. In this chapter we describe our techniques that have been successful in creating monoclonal antibodies to neuronal,[3] T cell,[4] islet cell,[5] and thymic cell[6] antigens. The techniques we describe involve a minimum of tissue culture work, as we have found it uneccessary to feed our cultures (except for the addition of a single drop of medium at 7 days) until colonies are visible and ready for assay.

Materials

Media and Reagents

Dulbecco's modified Eagle's medium (DMEM) with high glucose (4.5 g/liter) is used as our base medium (medium described by Kennett and coworkers). A week prior to fusing, 5 liters of DMEM (GIBCO, Grand Island, NY, catalog No. 430-2100) is prepared with deionized, double-distilled water. This assures a fresh, standardized medium for the entire procedure.

Concentrated (100 ×) stocks of medium supplements should be prepared and frozen ($-20°$) in convenient aliquots.

HAT Solutions (Selective Reagents: Hypoxanthine, Aminopterin, Thymidine)

100X HT: 136 mg of hypoxanthine (Sigma, St. Louis, MO; catalog No. H9377) and 38.7 mg of thymidine (Sigma; catalog No. T9250) are prepared in 75 ml of distilled water and adjusted to pH 10.0 with NaOH before bringing the final volume to 100 ml with distilled water. The solution is sterilized by Millipore filtration and stored in 5-ml aliquots at $-20°$.

100X Aminopterin: 4.4 mg of aminopterin (Sigma; catalog No. A2255) is added to 75 ml of distilled H_2O and adjusted to pH 9.8 before bringing the final volume to 100 ml with distilled water. *Note:* When using the 63-Ag8 · 653 myeloma parent, dilute regular stock aminopterin by 1:25 when preparing Super HAT (see below). Final 100X molar concentration for P3 and NS · 1 lines is 10^{-4} M and for 653 is 4.0×10^{-5} M. Sterilize by Millipore filtration and pipette

---

[3] G. S. Eisenbarth, F. Walsh, and M. Nirenberg, *Proc. Natl. Acad. Sci. U.S.A.* **76**, 4913 (1979).

[4] G. S. Eisenbarth, B. Haynes, J. Schorer, and A. Fauci, *J. Immunol.* **124**, 1237 (1980).

[5] G. S. Eisenbarth, H. Oie, A. Gazder, W. Chick, J. A. Schultz, and R. M. Scearce, *Diabetes* **30**, 226 (1981).

[6] B. F. Haynes, R. Scearce, and L. L. Hensley, *Clin. Res.* **31**, 345A (1983).

into 5-ml aliquots. Aminopterin can be frozen at $-20°$ or stored in the dark at 4°. Care should be observed when handling aminopterin powder and concentrated aminopterin stock because of its toxicity.

100X Bovine insulin, 0.2 U/ml. Stock crystalline insulin 24 IU/mg (Sigma; catalog No. I5500); 0.83 mg should be solubilized in 5 ml of a 1 : 500 dilution of concentrated HCl in water before bringing to 100 ml distilled water final volume. Insulin should be stored at $-20°$ in 5-ml aliquots after sterile filtration.

The remaining supplements can be bought commercially as 100X stock solutions.

MEM sodium pyruvate solution: 100X 100 mM (GIBCO; catalog No. 320-1360)

MEM nonessential amino acids: 100X (GIBCO; catalog No. 320-1140)

L-Glutamine: 100X (GIBCO; catalog No. 320-5030)

NCTC, offered as 10× supplement (MA Bioproducts, Walkersville, MD; catalog No. 12-923A).

Selection Medium Formula (Super HAT)

Freshly prepared DMEM is used for preparing 3920 ml of Super Base. Super Base lacks HAT selective supplements and can be used to grow hybrid cells when selection has occurred or can be used for selection with the addition of HAT. Add 40 ml of 100X stock solutions of insulin, sodium pyruvate, nonessential amino acids, and L-glutamine and 400 ml of NCTC-109 to 3360 ml of DMEM to prepare Super Base. Super Base should be sterilized by Millipore filtration and stored in 500-ml sterile bottles. To prepare selective medium, Super HAT, from Super Base add 5 ml of HT and 5 ml of aminopterin 100X stock to 490 ml of Super Base. *Note:* When using 63-Ag8 · 653 myeloma parent, add only 200 μl of 100X aminopterin and 5 ml of 100X HT to 495 ml of Super Base. Twenty percent heat-inactivated fetal calf serum (HIFCS) (56° for 30 min) is used throughout the procedure for fusing, routine passaging, and cloning. Once a new hybridoma line has been cloned and safely frozen, the hybrid cells may be adapted to 10% HIFCS.

We have found that with proper sterile techniques and appropriate precautions that no antibiotics are necessary.

Equipment and Supplies

Disposable, individually wrapped plastics are used when possible to reduce the risk of contamination. Essential equipment includes a fully humidified 37° incubator with a controlled atmosphere of 10% CO_2 in air, a laminar-flow sterile hood equipped with pipette aid (Fisher Scientific; catalog No. 13-681-16), a centrifuge, and a liquid nitrogen tank for cell storage (we keep approximately 1 in. of distilled water with 1% sodium dodecyl sulfate in the bottom of the incubator).

Parental Myeloma Lines

We have had success with three of the available HAT-sensitive mouse myeloma lines, P3X63-Ag8 (P3), NS1-Ag4-1 (NS1),[7] and X63-Ag8 · 653 (653)[8]. P3 expresses immunoglobulin with chain composition of a γ heavy chain and κ light chain; NS1, a variant derived from P3, expresses only intracellular κ chains; and 653, a subclone of P3, does not express heavy- or light-chain immunoglobulin. It is recommended that several lines be used with fusing; regardless of the line used, it is essential that it be in its log phase of growth for a successful hybridization. It is best to obtain the cell line from an investigator who has successfully fused with the line, develop a large stock of frozen cells, and use cells for fusion after only a limited number of passages.

Immunization

Immunization methods and schedules vary tremendously, depending on the antigens and the investigator. A major advantage of the monoclonal antibody system is the ability to immunize with only partially purified antigens (e.g., whole cells!). An important requirement is that the fusion be performed 3–5 days after the last antigen injection.

We use female BALB/c mice 6–12 weeks old. Soluble antigens are normally injected first with complete, and subsequently with incomplete, Freund's adjuvant. When making monoclonal antibodies to cellular antigens, whole cells or dissociated cellular stroma can be used. Cells (2×10^7) are injected intravenously and intraperitoneally on day 0 and intraperitoneally only on days 7 and 14. The final injection should not employ adjuvant, and splenocytes are fused on day 17. We usually continue immunizations of other mice at 7-day intervals until the results of an initial fusion are clear. If the first fusion is not successful, another mouse is sacrificed and splenocytes are fused. Stroma for immunization is prepared by mincing a small piece of tissue (1 cm^2) with a scalpel and aspirating with 19-gauge needle until the stroma can easily be drawn into a 5-ml

[7] G. Kohler, S. C. Howe, and C. Milstein, *Eur. J. Immunol.* **6**, 511 (1976).
[8] J. F. Kearney, A. Radbruch, B. Leisegang, and K. Rajewski, *J. Immunol.* **123**, 1548 (1979).

syringe. The mouse is injected (intraperitoneally only) with a 19-gauge needle, using the same schedule as with cells.

After several immunizations the mouse can be test bled and the serum assayed for circulating antibody. Checking for serum antibody at this point serves two important purposes.

1. By performing serial serum dilutions the screening assay can be tested for sensitivity and last-minute technical problems can be worked out before fusing.
2. A mouse that has no serum antibody against an antigen is unlikely to produce specific hybrids. If no antibodies are produced, changing the strain of mice immunized may result in the generation of serum antibodies.

Fusion Protocol

Materials

Sterile pipettes 1, 5, 10, and 25 ml, individually wrapped (Costar, Cambridge, MA)
Sterile centrifuge tubes, 15 ml (Falcon; catalog No. 2099)
Sterile centrifuge tubes, 50 ml (Falcon; catalog No. 2070)
Microtiter plates, 96-well, flat-bottom (Linbro, Flow Laboratories) Hamden, CT; catalog No. 76-003-05)
Plates, 24-well (Costar; catalog No. 3524)
Flask, 25 cm^2 (Corning, Fisher Scientific, Pittsburg, PA; catalog No. 10-12-6-30)
Flask, 75 cm^2 (Corning, Fisher Scientific; catalog No. 10-126-31)
Petri dishes, 60 × 15 mm (Falcon, catalog No. 1007)
Sterile scissors and forceps, 2 sets
Dulbecco's PBS: Combine, per liter, 8 g of NaCl, 0.2 g of KCl, 1.15 g of Na_2HPO_4, 0.2 g of KH_2PO_4, 0.1 g of $MgCl_2 \cdot 6H_2O$, 0.1 g of $CaCl_2$, pH 7.4
Alcohol, 70%: 250 ml in a 500-ml beaker
DMEM, 20% HIFCS
Super HAT, 20% HIFCS
Polyethylene glycol 1000 (PEG) 50% (J. T. Baker Chemical Co., Pillipsburg, NJ; catalog No. 8-01218): Autoclave 5 g, allow to cool to 50° before adding 5 ml of DMEM.
Feeder spleen cells: Normal spleen cells are prepared as described in the following protocol and frozen as described below in the section on high-titered antibody in mice. Ten frozen vials of cells are pre-

pared from each spleen, yielding approximately 10^7 normal spleen cells per vial.

Procedure

1. Immunized mouse is sacrificed by cervical dislocation and dipped immediately in 70% ethanol for 1 min.
2. Spleen is aseptically removed using one set of instruments to incise the skin and another to incise the peritoneum and remove the spleen.
3. Spleen is placed in 5 ml of Dulbecco's PBS in a petri dish, and remnants of fat are carefully removed.
4. Intact spleen is then transferred to another petri dish with 5 ml of cold PBS and minced with fine scissors.
5. Triturate minced spleen 4 or 5 times with a 10-ml pipette and transfer to a 15-ml centrifuge tube.
6. While the spleen clumps are settling, 2×10^7 myeloma cells in 40 ml of DMEM 20% HIFCS are transferred to a 50-ml centrifuge tube.
7. Suspended spleen cells are carefully removed, not disturbing the gravity-sedimented tissue, and added directly to the myeloma cell centrifuge tubes with the myeloma cells.
8. Combined cells are pelleted (600 g) for 7 min at room temperature.
9. The supernatant is discarded, and the pellet is gently disrupted by tapping at the bottom of the tube.

The following steps are carried out at 37°

10. Slowly, 0.8 ml of 50% PEG is added with a 1-ml pipette over a 1-min period while swirling the suspension.
11. After an additional minute of swirling, 1 ml of DMEM is added over a 1-min period.
12. While continuing to swirl, 20 ml of DMEM is added over a 5-min period to dilute the PEG slowly.
13. The suspension is centrifuged at 600 g for 7 min, and the supernatant is discarded.
14. The pellet is disrupted by tapping the bottom of the tube and is resuspended in 2 ml of DMEM; 1 ml is transferred to another centrifuge tube, and 40 ml of Super HAT, 20% HIFCS, is added to both centrifuge tubes.
15. Normal spleen cells (10^7) from a frozen vial in 1 ml of DMEM are added to the fused cells.
16. Two drops from a 10-ml pipette (approximately 100 μl) are added to each well of ten 96-well plates.
17. Plates are taped with Scotch tape to prevent accidental opening when handled and placed in a humidified 10% CO_2 incubator at 37°.

18. After 7 days, plates are fed with an additional drop of Super HAT, 20% HIFCS.
19. Visible colonies should be apparent in approximately 14 days. When visible colonies (covering approximately one-fourth of the bottom of the well) are present, the supernatant is removed for assay.

Antibody Screening Assay

Selection of an appropriate assay to detect specific antibody in supernatants should be given careful consideration. The assay must be sensitive, yet flexible enough to screen hundreds of samples rapidly. Protocols of many successful assays are described in this series, Vol. 73 [1], by A. Galfré and C. Milstein. We will describe an additional standard assay that we have found to be particularly useful, namely, indirect immunofluorescence using acetone-fixed sections of tissue. This simple assay, which may lack sensitivity, is particularly useful in defining antibodies to specific cell subsets.

Materials

Glass slides, coplin jar: 75 × 25 mm, 1.0 mm (American Scientific Products, Charlotte, NC; catalog No. M6168-1)
Coverslips: 22 × 44 mm, No. 1 (American Scientific Products; catalog No. M60454)
Squeeze bottle
Moist chamber: 150 × 15 mm petri dish (Falcon No. 1058) with a moist tissue
Acetone
Dulbecco's phosphate-buffered saline (PBS) with 0.02% sodium azide
Glycerol, 30%: 30 ml of glycerol in 70 ml of PBS 0.02% azide.
FITC goat anti-mouse IgG (TAGO, Burlingame, CA; catalog No. 6250)
DMEM–2% bovine serum albumin (BSA): 2 g of BSA per 100 ml of DMEM

Procedure

1. Frozen sections are cut with a cyrostat from tissue that was quickly frozen in liquid nitrogen or on Dry Ice and placed two per slide.
2. Sections on slides are fixed in cold acetone ($-70°$) for 5 min.
3. Allow slides to air-dry. *Note:* At this point slides can be stored in a slide box at $-70°$ until needed.
4. In a moist chamber, incubate 25 μl of monoclonal antibody supernatant on the tissue sections for 30 min at room temperature.

5. Rinse antibody with a squeeze bottle of PBS and place slide into a rinse coplin jar containing PBS at 4° for 5 min. Transfer to a second similar container for 5 min and then to a third for an additional 5 min.
6. After final PBS rinse, gently dry slide surrounding the tissue.
7. Add to section goat anti-mouse IgG, FITC conjugated, diluted 1:100 in DMEM 2% BSA.
8. Incubate for 30 min at room temperature.
9. Repeat step 5.
10. After final PBS wash, rinse by briefly dipping slide into a coplin container with distilled water.
11. Add 2 drops of 30% glycerol to the slide and seal with a coverslip and nail polish.
12. Store slides at 4° in the dark until they are to be examined with a fluorescent microscope.

Cell Expansion, Cloning, and Freezing

When a desirable antibody is detected with the screening assay, the hybrid cells are transferred to 0.5-ml cultures in 24-well plates using a 0.1-ml pipette. When these wells become 50% confluent (4–5 days), a sample of supernatant is assayed. Positive colonies are then transferred to 25 cm^2 flasks containing 5 ml of medium, and the initial well is refed with 1 ml of medium. At this point, it is very important to clone positive colonies to ensure a homogeneous culture, thus preventing overgrowth by hybrid cells not producing the desired antibody. When confluent, cells from the 25-cm^2 flask should be frozen in dimethyl sulfoxide-containing medium to assure a permanent stock of antibody-producing cells. After the initial freeze, cells from the 25-cm^2 flasks should be expanded to 75-cm^2 flasks, and large densities of cells should be grown to harvest the supernatant and to develop frozen stocks of cells. If a cell line stops producing antibody, it can be retrieved from an early freeze and recloned. No lines should be discontinued until a frozen vial is thawed and grown up in culture and its supernatant is assayed.

Cloning by Limiting Dilution

Materials

Centrifuge tubes, 15 ml
Super HAT, 20% HIFCS medium containing 1×10^7 normal spleen cells per 50 ml
Pipettes, 1 and 10 ml
96-Well flat-bottom plates

Ten milliliter of Super HAT containing 20% HIFCS is added to a centrifuge tube, and 9 ml to an additional three tubes labeled 2, 3, and 4. Then 100 μl of cells from a confluent hybrid culture is added to centrifuge tube 1 and mixed thoroughly. One milliliter of mixed suspension is taken from tube 1 and added to tube 2; after several aspirations, 1 ml is taken from tube 2 and added to tube 3, and similarly to tube 4 from each dilution. Two drops are added to each of 24 wells of 96-well plates (2 drops per well) with a 10-ml pipette. Plates are taped and placed in a 37° incubator with 10% CO_2. In 1–2 weeks, colonies will be visible. Several colonies may appear in a single well in the lower diluted wells. Five to ten supernatant samples for assaying should be taken from the highest dilution wells with one visible colony per well. Several of the positive colonies should be expanded, frozen, and recloned. Recloning should continue until all resulting clones are positive.

Freezing and Thawing Hybridoma Cells

Materials

Freezing vials (Nunc, GIBCO; catalog No. 363401)
2 × freezing media: 85 ml of Super HAT–20% HIFCS + 15 ml of dimethylsulfoxide (DMSO; Mallinckrodt, Paris, Kent; catalog No. 4948), sterilized by Millipore filtration
DMEM, 20% FCS
Polystyrene box, 2-cm-thick walls
Flask, 25 cm^2
Pipettes, 1, 5, and 10 ml

Procedure. Approximately 10^6 cells in log phase are pelleted by centrifugation at 600 g for 5 min. The supernatant is removed, and the pellet is resuspended in 0.5 ml of DMEM, 20% FCS. Slowly add 0.5 ml of 2 × freezing media and transfer mixture to a Nunc vial. Place vial in a polystyrene box and keep at −70° overnight. The following day, transfer the vial directly into liquid nitrogen for storage.

Frozen vials should be thawed quickly by swirling in a 37° water bath. When thawed, centrifuge for 5 min at 600 g and sterilely aspirate the supernatant. Resuspend the pellet with 1 ml of Super HAT with 20% HIFCS a 1-ml pipette and transfer to an additional 4 ml in a 25-cm^2 flask. Cell viability usually ranges from 60 to 80%.

High-Titered Antibody in Mice

As an alternative to growing large volumes of monoclonal antibody by culturing cells, high-titered antibody in ascites form can be obtained by injecting viable cells into pristane (2,6,10,14-tetramethylpentadecane, Aldrich Chemical, Milwaukee, WI; catalog No. T2,280-2) primed BALB/c mice (0.25-ml of pristane injected 1–7 days prior to hybrid cells). Hybridoma cells *in vitro* produce from 10 to 50 µg of monoclonal antibody per milliliter, whereas ascites fluid often contains as much as 20 mg of antibody per milliliter. Occasionally even a contaminated hybrid culture can be rescued by growth as an ascites tumor in a BALB/c mouse.

Procedure. Inject 10^6 to 10^7 viable cells in 0.5 ml of PBS intraperitoneally into the lower abdomen of the pristane-treated mouse. Ascites fluid should begin to accumulate at 7 to 14 days after inoculation (the mouse develops a tense large abdomen). Ascites is removed by inserting a 19-gauge, 1-in. needle into the abdomen and allowing the fluid to drain into a centrifuge tube. Cells are separated from the fluid by centrifuging at 600 g for 5 min. Ascites fluid should be aliquoted and stored at $-20°$ to minimize repeated freezing and thawing. Ascites cells can be frozen using the same procedure as for cultured hybrid cells. An assay should be performed using various dilutions to determine the saturating titer of the ascites antibody. Some ascites containing monoclonal antibody can be as much as 1 : 100,000 without loss in binding activity, but when added undiluted no binding is detected. We recommend screening ascites at a 1 : 100 initial dilution.

Monoclonal Antibody Availability

As the use of monoclonal antibodies increases, more companies are marketing media, reagents, and devices for making monoclonal antibodies. Many hybridoma lines along with parent myeloma lines are available through the American Type Culture Collection (Rockville, MD) and the Human Genetic Mutant Cell Repository (Camden, NJ). Monoclonal antibody supernatants and ascites are also available from Accurate Chemical and Scientific Corporation (Westbury, NY), Pel-Freeze (Rogers, AR) and Becton-Dickenson (Sunnyvale, CA), Ortho Diagnostics (Raritan, NJ), and many other companies. There are many known cross-reactivities of existing monoclonal antibodies with unexpected tissues (e.g., anti-T cell antibody 4F2 reacting with pancreatic islet cells).[9] Thus, a monoclonal antibody may already exist that will serve the function (e.g., cell isolation) for which an investigator is considering monoclonal antibody production. Many investigators will share their monoclonal antibodies or even cell lines, and existing monoclonals can be useful to characterize screening assays.

[9] S. Srikanta, R. Dolinar, and G. S. Eisenbarth, *Am. Diabetes Assoc.* **31**, 19A (1982).

Techniques similar to these discussed allow the production of autoantibodies by fusing spleen cells from animals with autoimmune disease.[10] The production of human monoclonal autoantibodies is possible[11] but continues to be extremely difficult compared to producing murine monoclonals.

[10] J. Buse, S. Srikanta, B. Haynes, and G. S. Eisenbarth, *Clin. Res.* (in press).
[11] G. S. Eisenbarth, H. Linnenbach, R. Jackson, R. Scearce, and C. Croce, *Nature (London)* **300**, 264 (1982).

[31] Voltammetric and Radioisotopic Measurement of Catecholamines

By PAUL M. PLOTSKY

Tyrosine derived from dietary protein and phenylalanine synthesized by the liver is converted to catecholamines (CA) by specialized cells primarily located within the nervous system, adrenal gland, and certain tumor cells. The rate-limiting process of the synthetic pathway illustrated by Fig. 1 is at the initial hydroxylation catalyzed by tyrosine hydroxylase (tyrosine 3-monooxygenase, EC 1.14.16.2). The relevant end products, norepinephrine (NE), epinephrine (E), and dopamine (DA), subserve important physiological functions as neurotransmitters, hormones, and hypophysiotropic factors.

Advances in the biological sciences have resulted from the ability to resolve and quantitate these substances in biological tissues. Analysis of CA is difficult owing to their ease of oxidation upon exposure to light and atmospheric oxygen. However, they are relatively stable in acidic solutions. Many assay methods exist for determination of CA, including gas chromatography coupled with mass spectrometry,[1] thin-layer chromatography,[2] and fluorometric techniques.[3] Routine quantification is primarily accomplished by high-performance liquid chromatography coupled with electrochemical detection (HPLC-EC) or radioenzymic assay (REA). In addition, developing voltammetric microelectrode techniques allow direct *in situ* measurement of CA fluxes. These latter techniques are described in this chapter.

[1] H. G. Lovelady, *Biochem. Med.* **15**, 130 (1976).
[2] R. W. Stout, R. J. Michelot, I. Molnar, C. Horvath, and T. K. Coward, *Anal. Biochem.* **76**, 330 (1976).
[3] G. M. Anderson and J. G. Young, *Life Sci.* **28**, 507 (1981).

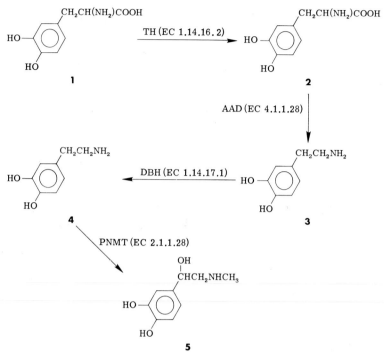

FIG. 1. Biosynthetic pathway of the catecholamines. (TH, tyrosine hydroxylase, tyrosine 3-monooxygenase; AAD, aromatic-L-amino acid decarboxylase; DBH, dopamine β-monooxygenase; PNMT, phenylalanine N-methyltransferase, noradrenaline N-methyltransferase; **(1)** l-tyrosine; **(2)** l-DOPA; **(3)** dopamine; **(4)** norepinephrine; **(5)** epinephrine.)

HPLC-EC

The use of liquid chromatography for analysis of CA is a recent phenomenon owing to the previous lack of adequately sensitive detection methods. This problem was solved with the development of thin-layer amperometric detection systems by Adams and co-workers.[4] These systems exploit the two-electron oxidation of CA to their corresponding quinones. HPLC-EC has found wide acceptance because of its modest cost, sensitivity, and versatility. Theoretical aspects of liquid chromatography[5] and electrochemistry[6] have been adequately reviewed elsewhere; however, a brief review will be helpful.

[4] C. J. Refshauge, P. T. Kissinger, R. Dreiling, L. Blank, R. Freeman, and R. N. Adams, *Life Sci.* **14**, 311 (1974).
[5] L. R. Snyder and J. J. Kirkland, "Introduction to Modern Liquid Chromatography." Wiley, New York, 1974.
[6] D. D. MacDonald, "Transient Techniques in Electrochemistry." Plenum, New York, 1977.

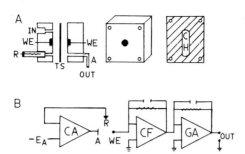

FIG. 2. (A) Thin-layer amperometric detector cell is illustrated in several views. The inflow (IN), outflow (OUT), glassy-carbon working electrodes (WE), silver/silver chloride reference (R), and stainless steel auxiliary (A) electrode positions are visible. A Teflon spacer (TS) fits between the detector blocks. The frontal view illustrates the interior face of the detector assembly and Teflon spacer with its flow channel (CH). (B) A simplified schematic of the potentiostat–amplifier circuit used for electrochemical measurements. The control amplifier (CA) senses the solution potential through an Ag/AgCl reference electrode (R). Current is supplied to the solution via the auxiliary electrode (A) in order to maintain the potential gradient dictated by the applied potential (E_A). A current follower (CF) allows measurement of oxidative current through the carbon WE. This is converted to a voltage and amplified via the gain amplifier (GA).

HPLC-EC may be divided into separative and quantitative aspects. Separation of CA is achieved using an analytical column packed with either cation-exchange or C_{18} reverse-phase material. Resolution of sample molecules occurs via differential interactions of these substances (solutes) with the carrier solvent (mobile phase) and the column-packing materials, such that distinct bands of solute molecules form during passage through the system. The degree of resolution is controlled by adjusting the pH, ionic strength, nature of the aqueous phase, and concentration of the organic component(s) of the mobile phase.

Use of a C_{18} reverse-phase column and suitable mobile phase permits analysis of CA precursors, CA, CA metabolites, serotonin and its primary metabolite.[7–10]

Quantitative detection is accomplished as the eluting solute bands flow through the low-volume electrochemical detector cell shown in Fig. 2A. This cell contains three electrodes connected to a potentiostat-amplifier, a diagram of which is presented in Fig. 2B.

The potential applied to the detector cell energetically favors oxidation of the CA. Each molecule oxidized loses two electrons. The electrons

[7] J. Wagner, P. Vitali, M. G. Palfreyman, M. Zraika, and S. Huof, *J. Neurochem.* **38**, 1241 (1982).
[8] I. N. Mefford and J. D. Barchas, *J. Chromatogr.* **181**, 187 (1980).
[9] C. L. Blank and R. Pike, *Life Sci.* **18**, 859 (1976).
[10] M. J. K. Rahmar, T. Nagatsu, and T. Kato, *Life Sci.* **28**, 485 (1981).

lost during oxidation transfer to the working electrode, which acts as an electron sink, inducing a measurable current. The relationship between the current and concentration of oxidized species is discussed in detail by Weber and Purdy[11] and can be expressed as

$$i_L = KC^b$$

where i_L is the maximal current, C^b is the solution concentration of reactive species, and K is the proportionality constant derived from the cell constant (electrode geometry and area), diffusion coefficient of the species of interest, Faraday constant, flow velocity, etc. Thus for a given and constant set of operating conditions, the oxidative current is directly proportional to the concentration of electroactive species present in solution.

Apparatus

The HPLC-EC system is composed of interchangeable, commercially available modules. The system consists of a Constametric III pump (Laboratory Data Control) and an injection valve with a 50-μl sample loop (Rheodyne No. 7125). This is followed by a 50 × 4.6-mm i.d. guard and a 250 × 4.6-mm i.d. stainless steel analytical column containing 5 μm Nucleosil SA cation-exchange resin (Alltech). Column effluent is routed through a Bioanalytical Systems TL-7A detector cell connected to an upgraded LC-2A potentiostat-amplifier from the same source. An applied potential of +0.72 V versus the reference electrode is used for CA detection. The detector output is electronically filtered through a Spectrum 921 filter and recorded on a conventional strip-chart recorder.

Mobile Phase

All solutions are prepared in deionized water redistilled from an alkaline permanganate solution. Chemicals are reagent grade and used without further purification.

Chromatographic separation is effected in the isocratic mode with a citrate–acetate (pH 5.2) mobile phase of the following composition: 0.08 M sodium acetate, 0.03 M citric acid, 0.05 M sodium hydroxide, and 10% methanol. This solution is filtered (0.45 μm) and vacuum degassed before use. It is pumped through the system at a constant flow rate of 1.0 ml min^{-1}.

Standards

The NE, E, and DA (Sigma) are prepared as stock solutions (1 × 10^{-3} M) in 100 ml of 0.1 M acetic acid that has been deoxygenated by

[11] S. G. Weber and W. G. Purdy, *Anal. Chim. Acta* **100**, 531 (1978).

bubbling with nitrogen or argon. The internal standard, α-methyldopamine, MDA (Merck, Sharp and Dohme) is prepared in a similar fashion. Stock solutions are stable at 4° for at least 3 weeks. Dilutions and diluted mixtures of these standards ($1 \times 10^{-7}\ M$) were prepared daily in cold, deoxygenated acid.

Sample Collection and Initial Preparation

Blood, incubation medium, or perfusion medium is collected into chilled tubes containing 10 μl ml^{-1} of a solution 0.2 M in EGTA and 0.2 M in glutathione. The samples must remain chilled until centrifugation (2000 g, 5 min, 4°). Plasma is transferred to new tubes and deproteinated by addition of 5 μl of 0.5 M perchloric acid per 100 μl of plasma, followed by brief mixing and centrifugation (2000 g, 5 min, 4°). Supernatants are transferred to new containers and may be stored at $-20°$ until assay.

Tissue is homogenized or ultrasonically disrupted in ice-cooled tubes containing 200 μl per 100 mg tissue net weight of a 0.2 M perchloric acid solution made to contain 2.7 mM EDTA. The homogenate is centrifuged (10,000 g, 10 min, 4°), and supernatant is transferred to new tubes for storage at $-20°$ until assay.

While it is possible to assay small volumes of these extracts by direct injection into the HPLC system, it is preferable to introduce a sample clean-up step utilizing activated aluminum oxide, prepared according to the procedure of Anton and Sayre.[12]

Alumina Extraction of Samples

Catecholamines are selectively separated from most other components of the sample matrix by adsorption onto acid-activated alumina. Supernatant (10–1000 μl) from tissue homogenate or plasma is added to centrifuge tubes containing 25 mg of activated alumina. In order, add 50 μl of 0.05 M reduced glutathione, 10–50 μl (1–5 pmol) of $1 \times 10^{-7}\ M$ internal standard (α-methyldopamine), and 1.0 ml of 1.0 M Tris buffer (pH 8.6) made to contain 0.2 M EDTA. Each tube is capped and mixed for 10 min. After brief centrifugation to pack the alumina, the supernatant is aspirated and discarded. The alumina is washed three times with 1.0-ml portions of distilled water–5% Tris buffer (1.0 M, pH 8.6) solution. Each wash is aspirated and discarded after centrifugation to pack the alumina. CA are desorbed by addition of 50–125 μl of 0.2 M perchloric acid followed by mixing for 10 min. The alumina is again packed by centrifugation, and the supernatant is transferred to new tubes for immediate assay

[12] A. H. Anton and D. F. Sayre, *J. Pharmacol. Exp. Ther.* **138**, 360 (1962).

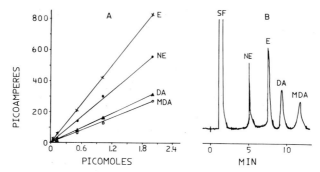

FIG. 3. (A) Calibration plot of authentic catecholamine (CA) standards illustrates linearity of detector response. The slopes of the lines in picoamperes picomole^{-1} are 306 (norepinephrine, NE), 425 (epinephrine, E), 159 (dopamine, DA), and 139 (α-methyldopamine, MDA). The intra- and interassay coefficients of variation are, respectively, 3.5 and 5.6% for plasma samples containing 0.3 pmol of each CA. (B) A typical chromatogram resulting from injection of 0.5 pmol of each compound. Retention times (min) are 5.0 (NE), 7.0 (E), 8.3 (DA), and 11.5 (MDA).

or frozen at $-20°$ until use. Overall recovery for the alumina extraction is 75–80%.

Calibration Procedure and Assay Calculations

Each new column and new or resurfaced detector electrode must be fully calibrated by injecting authentic solutions containing 0.05–10.0 pmol of CA. The first injection on a new or reconditioned electrode is usually spuriously high and is ignored. Calibration should be done with triplicate injections and plotted as mean oxidative current in picoamperes (or peak height) versus picomoles of CA injected. A typical calibration plot is shown in Fig. 3A. The limits of detection (signal-to-noise ratio >2) of the assay are 60 fmol of NE, 60 fmol of E, and 100 fmol of DA. A representative chromatogram is shown in Fig. 3B. It is good practice to inject calibration standards (0.1 pmol and 1.0 pmol) at the beginning, middle, and end of each working day.

When samples are prepared for the assay, standards should be added to sample medium containing no CA or known levels and run through the alumina extraction and assay. Duplicate sets of standards are added for every 15–20 unknowns. Neither plasma nor tissue extract appears to affect recovery from alumina or detector response to standards. Using the standard curve and the extracted standards, including internal standard, it is possible to correct for CA losses occurring during sample processing.

The CA content of the unknown samples may be calculated with the following formula:

$$\text{Picomoles ml}^{-1} \text{ CA} = \left[\frac{\frac{\text{picoamperes CA unknown}}{\text{picoamperes Internal Std.}}}{\frac{\text{picoamperes CA Std.}}{\text{picoamperes Internal Std.}}} \right] \left[\frac{\text{picomoles CA Std.}}{\text{ml plasma in sample}^{13}} \right]$$

It is absolutely essential to maintain a totally leak-free system. Even small, slow leaks of mobile phase can introduce disastrous noise into the system. The detector cell should be enclosed in a Faraday cage connected to a low-resistance ground. The chassis of the potentiostat-amplifier, filter, and recorder should be grounded only to the Faraday cage (not grounded through their power cords).

Radioenzymic Assay

The REA of NE and E was originally described by Engelman and Portnoy[14] and extended to include DA by Coyle and Henry.[15] The sensitivity was greatly enhanced by the work of Passon and Peuler,[16] and their assay is the basis of a commercially available kit[17] and the procedure to be described.

The assay entails conversion of the CAs to their O-[^{3}H]methyl derivatives by the enzyme catechol O-methyltransferase (COMT, EC 2.1.1.6). The methyl donor is S-adenosyl-L-[$methyl$-^{3}H]methionine (^{3}H-SAM). After enzymic reaction, the products are separated in a series of organic extractions and thin-layer chromatography (TLC). Spots corresponding to 3-[^{3}H]methoxytyramine (^{3}H-MT), [^{3}H]normetanephrine (^{3}H-NM), and [^{3}H]methanephrine (^{3}H-M), corresponding to DA, NE, and E are extracted, allowed to react to form [^{3}H]vanillin, and counted by liquid scintillation spectrometry. Numerous variants of this assay have been published[18,19] (see Cooper and Walker, this volume [32]). A flow chart of the steps in this assay is presented in Fig. 4.

[13] For picomole g^{-1} tissue, substitute grams of tissue per milliliter of acid-solution for homogenization.
[14] K. Engelman and B. Portnoy, *Circ. Res.* **26**, 53 (1970).
[15] J. T. Coyle and D. Henry, *J. Neurochem.* **21**, 61 (1973).
[16] P. G. Passon and J. D. Peuler, *Anal. Biochem.* **51**, 618 (1973).
[17] CAT-A-KIT, Upjohn Diagnostics, Kalamazoo, MI.
[18] N. Ben-Jonathan and J. C. Porter, *Endocrinology* (*Baltimore*) **98**, 1497 (1976).
[19] C. Gauchy, J. P. Tassin, J. Glowinski, and A. Cheramy, *J. Neurochem.* **26**, 471 (1976).

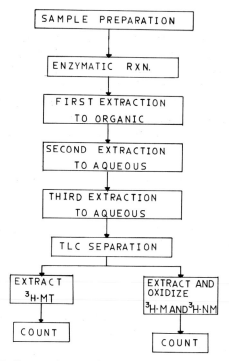

FIG. 4. Radioenzymic assay flow chart outlining assay procedure.

Preparation of COMT

Catechol O-methyltransferase (COMT, EC 2.1.1.6) is prepared using a modification of the Axelrod and Tomchick method.[20] Rat livers (100 g) are homogenized in 250 ml of cold 1.2% KCl. After centrifugation (2000 g, 4°, 30 min) the pellet is discarded and the supernatant is recentrifuged (90,000 g, 4°, 45 min). The supernatant volume is adjusted to 300 ml with 1.2% KCl and brought to pH 5.0 with addition of 1.0 M acetic acid. Ammonium sulfate (52 g) is added, and the solution is stirred for 10 min. After centrifugation (2000 g, 4°, 30 min), the pellet is resuspended in 50.0 ml of 1.0 mM phosphate buffer (pH 7.2) containing 1.0 mM dithiothreitol. The enzyme solution is dialyzed against 6.0 liters of the same buffer overnight (2.0 liters at a time). Any precipitate is removed by centrifugation. The solubilized enzyme is aliquoted into 0.5-ml fractions and stored at −70° for up to a year.

[20] J. Axelrod and R. Tomchick, *J. Biol. Chem.* **237**, 1657 (1962).

Preparation of Reagents

Catecholamine-free plasma is prepared from pooled rat plasma that is mixed with 20 μl ml^{-1} of 1.0 EGTA (pH 7.4). Plasma is centrifuged (2000 g, 4°, 15 min) and then dialyzed overnight at 4° against 100 volumes of 0.5 M phosphate-buffered saline (pH 7.4). The dialyzate is aliquoted and stored at −20°. Methoxylated derivative standards (NM, M, MT) are prepared as a combined stock solution, 250 mM in each compound in 0.001 M HCl. The CA standards of 200 mM are prepared in a similar manner. Aliquots of these solutions are stored at −20°. ^{3}H-SAM is obtained from New England Nuclear with a specific activity of 70 Ci mmol^{-1}. It is kept at −20° until dilution to a specific activity of 500 μCi ml^{-1}. A Tris-based buffer (pH 8.3) is prepared in distilled water to a final concentration of 1400 mM Tris, 160 mM EGTA, 410 mM Mg^{2+}. The solution used to stop the enzymic reaction is 1 M boric acid–0.1 M EDTA; it is adjusted to pH 11.0 with NaOH.

Assay Procedure

All assay tubes are run in duplicate and include blanks (50 μl of CA-free plasma), standards (50 μl of CA-free plasma), unknown samples (50 μl of plasma), and plasma pools (50 μl) containing a known amount of CA. The following is then added to each 13 × 100 mm glass tube, which is kept in an ice bath: 10 μl of 0.001 M HCl (standard tubes receive this volume containing 0.05–5 pmol of authentic CA standard), 10 μl of 10 mM glutathione solution, 1 μl of 22 mM pargyline solution, 12 μl of Tris–EGTA–Mg^{2+} buffer, 5 μl of ^{3}H-SAM solution, and 40 μl of COMT solution for a total volume of 150 μl per tube. Vortex each tube to mix. The tubes are incubated in a shaking water bath (37°) for 60 min, then are moved to an ice bath; 50 μl of the boric acid–EDTA solution is added to stop the reaction. Methoxylated derivative standard is added at this time (2 μl), and tubes are vortexed.

The first extraction is carried out by adding 2.0 ml of toluene–isopentyl alcohol (3:2, v/v) to each tube. After vortexing (15 sec), tubes are centrifuged (1300 g, 4°, 2 min). The aqueous phase is frozen in a Dry Ice–ethanol bath, and the organic phase is transferred to new tubes containing 100 μl of 0.1 M acetic acid. The aqueous phase is discarded.

The methoxylated amines are back-extracted into the acid aqueous phase by vortexing and centrifuging as above. After freezing the aqueous phase, the organic phase is removed and discarded. Samples may be safely stored at −20° at this stage.

The third extraction using 1.0 ml of toluene–isopentyl alcohol (3:2, v/v) further cleans up the aqueous phase. After mixing, centrifugation, and removal of the organic phase, 100 μl of absolute ethanol is added and each tube is briefly vortexed.

Extracts are spotted onto silica gel 60 F_{254} plates (20 × 20 cm × 0.2 mm), 10 extracts per plate. Plates are developed for 60 min in a tank containing 80 ml of chloroform, 15 ml of absolute ethanol, and 10 ml of 70% ethylamine. Plates are dried under a stream of nitrogen for 30 min; the spots are visualized under UV light and circled with pencil.

The ^{3}H-MT (top) spot is cut out and placed in a 16 × 125 mm glass tube. Spots corresponding to ^{3}H-M (middle) and ^{3}H-NM (bottom) are cut out and placed in separate scintillation vials. The gel is eluted by adding 1.0 ml of 0.05 M NH_4OH with occasional shaking for 20 min. The ^{3}H-M and ^{3}H-NM are oxidized to [^{3}H]vanillin with the addition of 50 μl of 4% sodium metaperiodate. The reaction is stopped after 5 min by addition of 50 μl of 10% glycerol. Acetic acid (1.0 ml, 0.1 M) is added to neutralize the solution. Add 10 ml of toluene–Liquifluor (1000:50, v/v) scintillation fluid to each vial; cap, vortex, and let stand for 30 min.

The tubes containing ^{3}H-MT receive 10 ml of toluene–isopentyl alcohol (3:2, v/v), are mixed and centrifuged (1300 g, 4°, 2 min). The lower phase is frozen, and the organic phase is transferred to scintillation vials. Liquifluor (300 μl) is added to each vial and mixed. All vials are counted for 5 min.

This assay has a limit of detection of approximately 0.01 pmol for each CA and is linear from 0.01 to at least 3 pmol of each amine. It is very sensitive to the volume of plasma (or tissue homogenate) used, as increasing volume inhibits COMT activity. Thus, all tubes must have the same amount of sample matrix added. For volumes greater than 100 μl, it is best to do an alumina extraction prior to REA.

Comparison of HPLC-EC and REA

Overall, the REA has a lower limit of detection than does the HPLC-EC technique. It requires a high degree of skill to run either of these assays at their limits. For most routine work, the assays are quite comparable. It is possible to analyze approximately 30 samples per working day using HPLC-EC and 40 samples every 2 days using REA. HPLC through-put is easily increased with the use of an autosampler whereas REA through-put is fixed. Reagents for HPLC are relatively inexpensive, but the REA requires purchase of ^{3}H-SAM and preparation or purchase of COMT. Figure 5 shows rat plasma CA levels determined by REA and

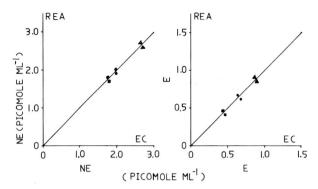

FIG. 5. Comparison of radioenzymic assay (REA) and high-performance liquid chromatography–electrochemical detection (HPLC-EC) of rat plasma catecholamines. Samples from three chronically catheterized male rats (■, ●, ▲) were split into duplicate aliquots (150 μl) and submitted to each assay. The correlation coefficient between the assays is 0.935. E, epinephrine; NE, norepinephrine.

HPLC-EC. Excellent agreement is demonstrated between these methods, as has been reported by many groups.[21,22]

In Situ Voltammetry

Many laboratories have demonstrated direct voltammetric recording of catecholamine flux in neural tissue.[23–25] The carbon-based microelectrodes used for *in vivo* voltammetry exhibit low residual current, wide anodic range, and sensitivity to compounds of interest. These useful characteristics derive from the interactions of CA and other molecules in solution with the carbon surface. The extent and specificity of these interactions are dependent upon the voltage waveform applied between the microelectrode and the solution in which it resides. As the magnitude of the applied potential is increased, CAs in a thin layer of solution adjacent to the electrode surface undergo oxidation (electron loss). The current generated is measured using a circuit similar to that described for HPLC-EC (Fig. 2B). The magnitude of the current generated is a function of the concentration of electroactive species in solution. This mathematical rela-

[21] P. Hjemdahl, M. Daleskog, and T. Kanan, *Life Sci.* **25,** 131 (1979).
[22] B.-M. Eriksson and B.-A. Persson, *J. Chromatogr.* **228,** 143 (1982).
[23] F. Gonon, M. Buda, R. Cespuglio, M. Jouvet, and J. F. Pujol, *Brain Res.* **223,** 69 (1981).
[24] A. G. Ewing, R. M. Wightman, and M. A. Dayton, *Brain Res.* **249,** 361 (1982).
[25] P. M. Plotsky, W. J. deGreef, and J. D. Neill, *Brain Res.* **250,** 251 (1982).

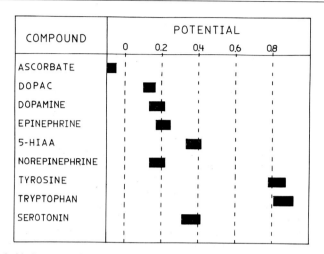

FIG. 6. Oxidation potentials of selected compounds at modified carbon fiber electrode. Ranges of potentials are shown, since potentials depend on the nature of the electrode surface and pH of the medium. Ranges given are for Thornell 300 carbon fiber electrodes in 0.5 M phosphate-buffered saline (pH 7.3) at a scan rate of 20 mV sec^{-1} in the differential pulse mode (50 mV pulse amplitude). 5-HIAA, 5-hydroxyindole acetic acid.

tionship has been discussed in detail.[26] The potential at which the current appears serves as a qualitative indication of the identity of the species, as illustrated in Fig. 6.

Electrochemical techniques are relatively insensitive to side-chain groups; thus it is not possible to differentiate among DA, E, and NE on the basis of electrochemical information alone. Judicious electrode placement and pharmacological intervention can be used to overcome this problem. Several types of carbon microelectrodes have been developed displaying differential sensitivity to compounds that may contribute to the oxidative current arising from CA (e.g., 3,4-dihydroxyphenylacetic acid, ascorbic acid).[27–30]

Equipment

A Princeton Applied Research Corporation Model 174A polarographic analyzer can be used in conjunction with a Model 175 Universal Programmer to generate potential waveforms and measure oxidative current. A

[26] A. G. Ewing, M. A. Dayton, and R. M. Wightman, *Anal. Chem.* **53**, 1842 (1981).
[27] F. G. Gonon, C. M. Fombariet, M. J. Buda, and J. F. Pujol, *Anal. Chem.* **53**, 1386 (1981).
[28] L. Falat and H.-Y. Cheng, *Anal. Chem.* **54**, 2108 (1982).
[29] G. Nagy, M. E. Rice, and R. N. Adams, *Life Sci.* **31**, 2611 (1982).
[30] P. M. Plotsky, *Brain Res.* **235**, 179 (1982).

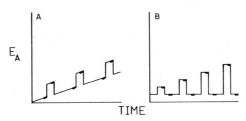

FIG. 7. (A) Potential waveforms used for *in vivo* voltammetry. Differential pulse voltammetry consists of square-wave potential pulses of fixed amplitude (50 mV) and duration (100 msec) superimposed on a linearly ascending ramp potential (−0.2 V to +0.6 V vs Ag/AgCl at 5–20 mV sec^{-1}). Current is sampled immediately prior to and during the final 33 msec of each pulse, the difference being displayed. (B) Backstep-corrected normal pulse chronoamperometry. A train of potential pulses (100 msec duration) of ascending amplitude is applied from an initial potential of −0.2 V vs Ag/AgCl. Current is sampled during the last 33 msec of each positively going pulse and from 77 to 100 msec of each subsequent negatively going pulse. These currents are digitally added to correct for charging and residual currents and displayed. E_A, applied potential; current sampled at times indicated by thick-line segments.

preamplifier is necessary for current levels less than 10 picoamperes. Output is displayed on a storage oscilloscope or an X-Y recorder. Alternatively, a laboratory constructed potentiostat-amplifier, similar in design to that described by Ewing *et al.*[31] is routinely used. Waveform formation, data storage, and analysis are performed on an Apple computer. Applied potential waveforms are explained in Fig. 7.

Electrode Construction

Voltammetric microelectrodes are constructed from Thornell 300 carbon fibers (Union Carbide) 6.9 µm in diameter. Individual fibers are inserted into glass capillary tubing (6 cm × 1.0 mm o.d.) filled with absolute ethanol. After aspiration of the ethanol through cotton, the capillary is mounted in a vertical pipette puller. The glass is drawn around the carbon fiber to a fine, long taper. The glass-to-carbon seal is formed by dipping each electrode tip into Spurr low-viscosity embedding epoxy (Polysciences) and allowing the tip to fill via capillary action. The epoxy is cured at 80° for 3 hr. Each electrode tip is cut under microscopic observation to expose a fresh surface, and the barrel is filled with mercury. A contact wire is glued into each barrel. Prior to use, each electrode tip is placed in 0.5 M citrate–phosphate buffer (pH 7.4) and scanned repetitively from 0 to +1.8 V vs silver/silver chloride cell (Ag/AgCl) at 100 mV sec^{-1} until the output is stable. An electrode is illustrated in Fig. 8A.

[31] A. G. Ewing, R. Withnell, and R. M. Wightman, *Rev. Sci. Instrum.* **52**, 454 (1981).

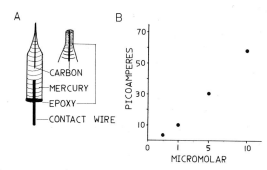

Fig. 8. (A) Carbon fiber voltammetric microelectrode. A 6.9 μm carbon fiber is epoxied into a glass capillary tube drawn to a fine tip diameter. Contact is made via a mercury pool. The fiber is cut flush with the capillary tip. (B) Calibration of modified carbon fiber microelectrode to dopamine in deoxygenated 0.5 M phosphate-buffered saline (pH 7.4). Differential pulse voltammetry was used at a scan rate of 20 mV sec^{-1} from an initial potential of -0.2 V to a final potential of $+0.6$ V vs Ag/AgCl and a modulation pulse amplitude of 50 mV.

The electrode surface may be modified to provide greater resolution of ascorbic acid and CA. Electrode tips are immersed in 70% chromic acid, and 250 μA of anodic current is passed for 30 sec. After a distilled-water rinse, tips are immersed in pH 7.4 phosphate-buffered saline and cycled from -0.2 to $+2.0$ V vs Ag/AgCl at a rate of 200 mV sec^{-1} for 2 min. Each electrode is then cycled from -0.2 to $+0.8$ V at 20 mV sec^{-1} until a stable background current response is achieved. Modified electrodes exhibit an ascorbic acid response at -0.150 V and a CA response at $+0.130$ V versus Ag/AgCl. They are 30-fold more sensitive to DA than to 3,4-dihydroxyphenylacetic acid, a metabolite.

Voltammetric microelectrodes should be calibrated immediately prior to and following *in vivo* use in phosphate-buffered saline (pH 7.4) containing 1–100 μM of each species alone and various mixtures of CA and ascorbate. A typical calibration curve is shown in Fig. 8B. Electrodes should be rinsed in distilled water after use and stored in a dry, closed container. Prior to reuse, they should be electrochemically conditioned (see above). Electrodes should be maintained at a resting potential of -0.2 V vs Ag/AgCl during implantation and while *in vivo*.

Silver/silver chloride reference electrodes are prepared by pulling glass capillary tubing to tip diameters of approximately 20 μM. Each electrode is filled with 1.5 M sodium chloride prepared in 10% gelatin. A 4.0 cm length of anodized silver wire is inserted into each electrode and epoxied in place. Electrodes are stored, shorted together, with tips immersed in 1.5 M sodium chloride solution. Auxiliary electrodes are constructed from a short length of silver wire or a stainless steel jeweler's screw in electrical contact with the preparation.

Carbon fiber voltammetric microelectrodes appear to remain stable *in situ* for at least 48 hr. Owing to their small size, they show minimal current contribution from consecutive electron transfer reactions following the initial oxidation of CA. Because of the time independence of the oxidative current at these electrodes, it is theoretically possible to sample continuously. However, practical aspects (electrode filming when maintained at positive potentials) do not permit this, although it is possible to sample at rates of 1 Hz. Thus the time resolution achieved with this technique greatly exceeds that obtained with other methods.

Overview

The analytical techniques reviewed allow accurate and sensitive determination of endogenous CA levels in biological tissue and fluids. They may be applied to the measurement of CA content, synthesis rate, secretion rate, and turnover rate in normal subjects and during behavioral and pharmacological manipulation. The information gained through the use of these probes provides valuable insights into the physiological function of peripheral and central catecholamines.

[32] Microradioenzymic Assays for the Measurement of Catecholamines and Serotonin

By RALPH L. COOPER and RICHARD F. WALKER

In vivo, dopamine (DA), norepinephrine (NE), and epinephrine (E) are O-methylated to form methoxytyramine, normetanephrine, and metanephrine, respectively, and serotonin (5-HT) is sequentially N-acetylated and O-methylated to form melatonin. Radioenzymic assays exploit, *in vitro*, these metabolic fates of the catecholamines and serotonin by incorporating tritium (^{3}H) into the structure of the O-methylated derivatives. This is accomplished by using the enzyme catechol O-methyltransferase (COMT), and the methyl donor S-adenosyl-L-[methyl-^{3}H]methionine (^{3}H-SAM). Similarly, the assay for serotonin is based on the ability of hydroxyindole O-methyltransferase (HIOMT) to transfer the ^{3}H-methyl group of ^{3}H-SAM to the hydroxyl group of N-acetylserotonin producing [^{3}H]melatonin ([^{3}H]methoxy-N-acetylserotonin). Prior to the O-methylation reaction, serotonin is acetylated enzymically by N-acetyltransferase and acetyl coenzyme A. After their formation, the radiolabeled products are extracted, further isolated by thin-layer chromatography, and quantified by liquid scintillation spectrometry.

Microenzymatic assays allow measurement of picogram levels of biogenic amines in animal tissue. When these assays are coupled with microdissection techniques, monoamine metabolism can be studied in sites as small as individual brain nuclei. Several reviews discuss the developmental, practical, and theoretical aspects of radioenzymic assays. The catecholamine assay described here is based primarily on the work of Coyle and Henry,[1] Beaven,[2] Ben-Jonathan and Porter,[3] Saller and Zigmond,[4] and Philips[5] as well as modifications introduced in our laboratories. The serotonin assay is based on the procedure described initially by Saavedra et al.[6] and discussed by Beavan[2] and Philips,[5] the major difference being that N-acetyltransferase obtained from Drosophila melanogaster[7] is used in lieu of a similar enzyme purified from rat liver. The advantages of Drosophila N-acetyltransferase are discussed in detail elsewhere.[8]

Radioenzymic Assay of Catecholamines

Sample Preparation

The catecholamine content of brain tissue or plasma can be determined by radioenzymic methods. All tissue samples should be processed immediately, as the transmitters are unstable at room temperature. Brain tissue is frozen immediately on Dry Ice. Catecholamines are extracted by homogenizing (or sonicating at 20 kilocycles, 35 W, 15 sec), on ice, in 1–200 volumes (w/v) of ice-cold 0.1 N perchloric acid. Prior to homogenizing, brains may be blocked or microdissected using the "punch" technique described by Palkovits[9] (see also this volume [23]). If microdissected, the tissue is suspended in 60–100 μl of ice-cold perchloric acid and sonicated. The homogenate is centrifuged for 10 min at 6500 g and 4°. Ten microliters of the homogenate is usually aliquoted prior to centrifugation for subsequent determination of protein content. The su-

[1] J. T. Coyle and D. Henry, *J. Neurochem.* **21**, 61 (1973).
[2] M. A. Beavan, in "Handbook of Psychopharmacology: Biochemical Principles and Techniques in Neuropharmacology" (L. L. Iversen, S. D. Iversen, and S. H. Snyder, eds.), p. 253. Plenum, New York, 1975.
[3] N. Ben-Jonathan and J. C. Porter, *Endocrinology* **98**, 1497 (1976).
[4] C. F. Saller and M. J. Zigmond, *Life Sci.* **23**, 1117 (1978).
[5] S. R. Philips, *Adv. Cell. Neurobiol.* **2**, 355 (1981).
[6] J. M. Saavedra, M. Brownstein, and J. Axelrod, *J. Pharmacol. Exp. Ther.* **186**, 508 (1973).
[7] M. McCaman, R. McCaman, and J. Stetsler, *Anal. Biochem.* **96**, 175 (1976).
[8] R. F. Walker, D. W. Friedman, and A. Jimenez, *Life Sci.* (in press).
[9] M. Palkovitz, *Brain Res.* **59**, 449 (1973).

pernatant can be stored at −50° for 1 month, but prolonged storage is not recommended.

Blood is collected in ice-cold heparinized tubes and centrifuged at 500 g for 10 min at 4°. Platelet-rich plasma is transferred to ice-cold tubes and 4 µl of freshly prepared ice-cold, 5.0 N perchloric acid (containing 0.25 M dithiothreitol) is added to each 96 µl of plasma. The deproteinized plasma may be assayed immediately or stored at −50°.

Preparation of Catechol O-Methyltransferase (COMT)

All steps of purification of COMT are performed at 0–4°. Freshly obtained rat livers are homogenized in 4 volumes (ml) (w/v) of 1.55 M KCl. The homogenate is filtered through glass wool to remove fat and centrifuged at 46,000 g for 30 min. After another filtration through glass wool and a subsequent centrifugation at 100,000 g for 60 min, the supernatant is again filtered,[9a] and stirred in an ice bath; the pH is adjusted to 5.3 with 1.0 M acetic acid. When the desired pH is obtained, the supernatant is allowed to equilibrate for 10 min, after which it is centrifuged at 15,000 g for 10 min at 4°. The supernatant is then adjusted to pH 6.8 with 0.5 M phosphate buffer (pH 7.0), and fractionated with ammonium sulfate (enzyme grade), as described by Nikodejevic et al.[10] This is accomplished by gently stirring the supernatant in an ice bath and slowly adding ammonium sulfate crystals to 30% saturation (i.e., 21 g of solid ammonium sulfate per 100 ml of supernatant). After equilibration for 30 min, the mixture is centrifuged at 14,000 g for 15 min at 4° and the precipitate is discarded. Ammonium sulfate is added to reach 45% saturation (i.e., 31.5 g per 100 ml of solution). After a 35-min equilibration period, the mixture is centrifuged at 14,000 g for 15 min. The supernatant is then discarded, and the final precipitate is dissolved by gently mixing in a minimal volume of 0.1 M phosphate buffer (pH 7.0), containing 0.1 M dithiothreitol, 0.1 mM sodium EDTA, and 0.01% sodium azide. This mixture is dialyzed overnight against 500 volumes of the same buffer. The dialyzate is divided into 0.05-ml aliquots and stored at −50°. Under these conditions, enzyme activity is retained for several months. Protein content and specific activity of the partially purified COMT is determined by the methods of Lowry et al.[11] and Coyle and Henry,[1] respectively. The partially purified enzyme is usually diluted (1:3, v/v) with distilled water for use in the assay,

[9a] At this point the supernatant can be stored for several months with minimum loss of activity.
[10] B. Nikodejevic, S. Senoh, J. W. Daly, and C. R. J. Creveling, *Pharmacol. Exp. Ther.* **174**, 83 (1970).
[11] O. H. Lowry, N. J. Rosebrough, A. L. Farr, and R. J. Randall, *J. Biol. Chem.* **193**, 265 (1951).

contains 15–20 mg of protein per milliliter, and has a specific activity of 100–150 units/ml.

Materials

Catechol *O*-methyltransferase (see above)
Perchloric acid (PCA), 0.1 N
Borate buffer, 1.0 N, pH 11.0
Tris-HCl buffer, pH 10.8
Magnesium chloride, 0.4 M
Dopamine: stock solution, 8 mg per 100 ml of 0.1 N PCA
Norepinephrine: 8 mg per 100 ml of 0.1 N PCA
Epinephrine: 8 mg per 100 ml of 0.1 N PCA
Dithiothreitol
Methoxytyramine (MTX)
Normetanephrine (NM)
Metanephrine (M)
[^{3}H]*S*-adenosylmethionine, specific activity 5–15 Ci/mmol
Chloroform
Ethyl alcohol, 100%
Methylamine, 40% in H_2O
Acetic acid, glacial
Ethyl acetate, chromotography grade
Toluene–isoamyl alcohol, 3:2, v/v

Perchloric acid, borate buffer, Tris-HCl, MgCl, and stock catecholamine solutions should be kept fresh and stored at 4°. MTX, NM, and M (with dithiothreitol) are mixed immediately prior to addition to the assay tubes. ^{3}H-SAM should be stored at −50° in convenient aliquots (i.e., 50–100 μl) for assay. COMT should also be stored at −50° in convenient aliquots. The incubation buffer (see below) should be prepared immediately prior to addition to the assay tubes.

Catecholamine Assay

All tubes are run in duplicate on ice. After extraction (or thawing, if supernatant is stored), the tissue supernatant is centrifuged at 11,000 g for 10 min at 4°. Aliquots (10 μl) of the supernatant are placed in 6 × 50-mm Dow-Corning tubes (catalog No. 9820, or Scientific Products, catalog No. T1310-1) containing 5 μl of 0.1 N PCA. In this assay, blanks contain 15 μl of 0.1 N PCA. In separate tubes, external standards are prepared by adding 125–1000 pg of DA, NE, and E (Sigma, St. Louis, MO) dissolved in a total of 15 μl of 0.1 N PCA. Internal standards are prepared by adding 10 μl of pooled tissue supernatant to tubes containing the CA standards dissolved in 5 μl of 0.1 N PCA. The O-methylation reaction is initiated by adding 25 μl of an incubation buffer enzyme substrate mixture bringing the final reaction volume to 40 μl. Sufficient incubation buffer for a 100-tube assay is obtained by mixing 100 μl of ^{3}H-SAM (100 μCi; 24 nM; 15 Ci/mmol, 300 μl of distilled H_2O, 134 μl of 0.4 M MgCl containing 6 mg of dithiothreitol, 874 μl of 1 M tris(hydroxymethyl)aminomethane buffer-HCl, pH 10.8, Trizma base), 800 μl of COMT (freshly thawed and appropriately diluted with distilled water). If the assay is for determination of plasma catecholamines, 2.72 mg of EGTA is added to the incubation buffer. EGTA is not necessary for brain catecholamine assays.

The tubes are vortexed, centrifuged briefly (<1 min at 1000 rpm) at 4° to assure that the mixture is at the bottom of the tube. The tubes are incubated at 37° for 45 min in a water bath. After incubation, the reaction is stopped by placing the tubes on ice and adding 30 μl of 1.0 M borate (pH 11.0). Then, 10 μl of a mixture containing MTX (2 mg), NM (2 mg), M (2 mg), and sodium metabisulfate (1 mg) per milliliter of distilled H_2O is added to facilitate recovery of the ^{3}H-labeled products. All tubes are vortexed twice.

The O-methylated products are extracted into 250 μl of toluene–isoamyl alcohol (3:2, v/v) by vortexing the mixture vigorously, twice, for 15 sec and centrifuging at 6500 g for 10 min. In a second extraction step, 200 μl of the organic phase is transferred to tubes containing 40 μl of 0.1 N PCA. The O-methylated derivatives are extracted into the aqueous phase by vortexing. The tubes are centrifuged again at 6500 g, and the organic layer is discarded (aspirated). Then 25 μl of the remaining acid solution is streaked on the loading zones of TLC plates (Whatman, LK6D) that were prespotted (30 min before) with 5 μl of the carrier mixture (described above) containing MTX, NM, M, and sodium metabisulfate. The TLC plates are placed in the dark and allowed at least 30 min to dry. After drying, the chromatograms are developed in a mixture containing 5 ml of 40% methylamine, 18 ml of ethanol (100%), and 40 ml of chloroform for approximately 1.5 hr. The plates are then removed and dried in a hood under light and a stream of air for 12 hr to facilitate autoxidation and visualization of the products. The spots corresponding to 3-methoxytyramine, metanephrine, and normetanephrine (approximate R_f values = 88.0, 74.0, and 65.0, respectively) are scraped into individual scintillation vials (5-ml capacity), and 0.5 ml of a mixture of acetic acid–ethyl acetate–H_2O (2:6:2, v/v/v) is added to elute the amines from the silica gel. The samples are then shaken gently for 5 min, after which 4.5 ml of scintillation cocktail is added. The vials are shaken again and placed in the dark for at least 1 hr before counting, as the samples are initially chemoluminescent.

The sensitivity (twice blank) of this microenzymic assay is approximately 8 pg for DA and NE and 15 pg for E. The assay is linear to at least 1.5 ng. Intraassay variations for determination of replicate samples containing added CA standards were 4.8%, 5.9%, and 6.4% for DA, NE, and E, respectively. Interassay variations for DA, NE, and E in five consecutive assays were 10.8%, 11.9%, and 12.2%, respectively.

The sensitivity of the assay allows determination of catecholamines after microdissection of most brain areas. However, the concentration of catecholamines in blood generally falls below the limit of sensitivity of the microassay as described above. With use of larger volumes of unknowns and appropriate adjustments of reagent concentrations, the assay will readily detect catecholamines in rat plasma. A macroradioenzymic assay for human plasma catecholamines has been published.[12] In addition, the sensitivity of the microassay for NE and E can be increased by adding a periodate oxidation step[4] as follows:

1. Silica corresponding to ^{3}H-labeled O-methylated NE and E is scraped into scintillation vials and 500 μl of 0.05 N ammonium hydroxide is added.
2. Twenty-five microliters of 4% sodium metaperiodate (w/v) is added, and 5 min later 25 μl of 10% glycerol (v/v) is added.
3. Acetic acid, 500 μl 0.1 N, and 4 ml of scintillation cocktail are added; samples are vortexed and counted.

Radioenzymic Assay for Serotonin

Sample Preparation

Based upon their structural similarities, catecholamines and serotonin are coextracted in acid. Thus, 0.1 N PCA or 0.1 N HCl extracts of tissue as described above will contain serotonin as well as catecholamines. Similar precautions to avoid oxidation or enzymic degradation of catecholamines apply to serotonin; therefore tissues are processed rapidly after collection, and extracts are stored at $-50°$ until they are analyzed.

Enzyme Preparation

In the original radioenzymic assay for serotonin,[6] N-acetyltransferase (NAT) was obtained from rat liver. Since rats are easily acquired, this

[12] V. K. Weise and I. J. Kopin, *Life Sci.* **19**, 1673 (1976).

tissue provides a readily available source of NAT. However, partially purified rat liver enzyme degrades rapidly, possibly due to contaminating proteases. Rat liver NAT also contains significant 5-hydroxytryptophan decarboxylase activity.[13] On the other hand, we[8] recently found that NAT partially purified from *Drosophila* is superior to rat liver NAT. When *Drosophila* NAT is substituted for rat liver NAT, many of the problems involving specificity, blank values, and sensitivity associated with the original 5-HT radioenzymic assay are reduced. However, at least 10–15 g of *Drosophila* are required for a sufficient yield of NAT. The relative difficulty in acquiring large numbers of *Drosophila* could make the use of this tissue impractical for many laboratories. If rat liver NAT is freshly prepared and properly stored with stabilizers, it can be used reliably in the 5-HT radioenzymic assay, albeit requiring more precautions than *Drosophila* NAT. Since both enzymes offer certain advantages, procedures for partial purification and use of each will be described.

Drosophila N-Acetyltransferase. N-Acetyltransferase is obtained from *Drosophila melanogaster*, wild type, according to the procedure described by McCaman et al.[7] *Drosophila* are collected live and frozen immediately at −50° until processing; they are then homogenized in 5 volumes (w/v) of ice-cold 0.04 M potassium phosphate buffer (pH 7.0), filtered through glass wool, and centrifuged at 20,000 g for 30 min at 4°. The supernatant is acidified (pH 5.0) by slowly adding 1.0 M acetic acid (while stirring on ice). After a 15-min equilibration period, the mixture is centrifuged at 20,000 g for 60 min. The supernatant is then neutralized (pH 7.0) by adding 1.0 M Tris base (pH 10.8). The active protein is isolated by ammonium sulfate fractionation. The fraction precipitated between 28 and 60% saturation is centrifuged at 20,000 g for 30 min. The supernatant is discarded, and the pellet is resuspended in a minimum amount of 0.02 M potassium phosphate buffer (pH 6.5) containing 1 mM 2-aminoethylisothiouronium bromide-HBr, 1 mM sodium EDTA, and 0.02% sodium azide and dialyzed overnight against 500 volumes of the buffer solution. The dialyzate is divided into 50-μl aliquots and stored at −50°.

Rat N-Acetyltransferase. Adult male rats (3–4 months old) are killed by decapitation; their livers are removed rapidly and minced, and NAT is prepared according to the methods of Saavedra et al.[6] Briefly, livers are homogenized in 2 volumes (w/v) of 0.1 M sodium phosphate buffer (pH 7.2) and centrifuged (78,000 g for 90 min). Ammonium sulfate is added to the supernatant, and the protein fraction precipitating between 40 and

[13] J. M. Saavedra, *Fed. Proc. Fed. Am. Soc. Exp. Biol.* **36**, 2134 (1977).

65% saturation is dialyzed, aliquoted, and stored under the conditions described above for *Drosophila* NAT.

Hydroxyindole O-Methyltransferase. Beef pineal hydroxyindole *O*-methyltransferase is prepared from 5–10 g of beef pineal (Pel-Freez Biological Inc., Rogers, AR) according to the procedure of Axelrod and Weissbach[14] and modified by Saavedra *et al.*[6] The pineals are homogenized in 5 volumes (w/v) of isotonic potassium chloride (1.15 g/100 ml). The homogenate is centrifuged at 16,000 g for 20 min. The supernatant is stirred slowly, and ammonium sulfate is added to 35% saturation. This mixture is allowed to equilibrate for 60 min, after which it is centrifuged at 16,000 g for 20 min. Sufficient ammonium sulfate is added to the supernatant while stirring slowly to bring the solution to 65% saturation. This mixture is equilibrated for 1 hr and centrifuged at 16,000 g for 20 min. The supernatant is discarded, and the precipitate is dissolved in a minimum amount of 50 mM sodium phosphate buffer (pH 7.9). The sodium phosphate buffer is added slowly until all the precipitate is dissolved. The dissolved precipitate is dialyzed overnight against the same sodium phosphate buffer (500 volumes). Then 1 mM 2-aminoethylisothiouronium bromide-HBr, 0.1 mM sodium EDTA, and 0.01% sodium azide are added as stabilizers, and 30-μl aliquots of the enzyme solution are stored at $-50°$. Enzyme activity is retained for several months.

Activity Determinations. Activities of *Drosophila* and rat liver NAT are determined by the method of Deguchi and Axelrod[15] utilizing tryptamine as a substrate. HIOMT activity is determined according to Axelrod *et al.*[16] using acetylserotonin as a substrate. The protein content of the enzyme preparations is determined according to Lowry *et al.*[11] Rat liver or *Drosophila* NAT are diluted with distilled water for use in the 5-HT radioenzymic assay. The amount of distilled water added to each enzyme is determined empirically based upon the dilution giving optimal acetylation and lowest blank values. Thus, appropriate dilution of the solutions generally occurs when *Drosophila* NAT and rat liver NAT dialyzates are diluted 1:5 and 1:1, respectively. At these dilutions, *Drosophila* NAT is 3–4 times more active than rat liver NAT. The *Drosophila* NAT preparation contains 5–8 mg of protein per milliliter with an activity of 3500–3600 units/ml as compared with 40–50 mg of protein per milliliter and 800–900 units/ml for rat liver NAT. HIOMT, diluted 1:3 with distilled water for use in the 5-HT assay, contains 10–12 mg of protein and 300–350 units/ml activity.

---

[14] J. Axelrod and J. Weissbach, *J. Biol. Chem* **236,** 211 (1961).
[15] T. Deguchi and J. Axelrod, *Anal. Biochem.* **50,** 174 (1972).
[16] J. Axelrod, R. J. Wurtman, and S. H. Snyder, *J. Biol. Chem.* **240,** 949 (1975).

Serotonin Assay

Materials

NaH$_2$PO$_4$ · H$_2$O, 0.2 N, pH 7.9
NaOH, 1.0 N
HCl, 0.1 N
Borate buffer, pH 11.0
N-Acetyltransferase
Bovine pineal HIOMT
[^{3}H]S-Adenosylmethionine
Serotonin stock solution
Melatonin
Toluene
Ethyl alcohol, 100%

Reagents and serotonin stock solution are prepared fresh and stored refrigerated until ready for use. N-Acetyltransferase, acetyl-CoA, HIOMT, and ^{3}H-SAM are stored in appropriate aliquots at −50° and prepared immediately before use.

Procedure. Samples are run in duplicate in 12 × 75 mm DisPo tubes on ice. A 10-µl aliquot of buffer solution, containing 1 ml of 0.2 N sodium phosphate buffer (pH 7.9) and 110 µl of 1 N NaOH, is added to each tube, including blanks. Blanks contain 0.1 N PCA, whereas external standards are prepared by adding 125–1000 pg of serotonin (serotonin creatine sulfate, Sigma) dissolved in 10 µl of 0.1 N PCA. Internal standards are prepared by adding authentic 5-HT to the pooled tissue supernatant and are then carried through the entire assay; 10.0 µl of the unknown, standard, or acid is added to the sample tubes containing 10 µl of incubation buffer. The first reaction is started by adding 5 µl of the first enzyme mixture containing equal parts of N-acetyltransferase and acetyl coenzyme A (2.5 mg per milliliter of distilled H$_2$O). The tubes are then shaken gently and incubated for 30 min at 37°. The tubes are placed back on ice, and a 5-µl aliquot of the second enzyme mixture containing 50 µl of bovine pineal 5-HIOMT, 50 µl of ^{3}H-SAM (2 nM, 15 Ci/mmol), and 175 µl of 0.2 N sodium phosphate buffer (pH 7.9) is added to each tube. The tube is shaken gently and incubated for an additional 30 min at 37°. The reaction is stopped by adding 25 µl of 0.5 N borate buffer (pH 11.0) and 5 µl of an alcoholic solution of melatonin (5 mg per milliliter of 50% ETOH).

The radioactive product is extracted by adding 1.3 ml of toluene to each tube and vortexing twice. The tubes are centrifuged at 11,000 g, and 1 ml of the organic phase is transferred to 5-ml minivials containing 1 ml of toluene. The vials are placed in a vacuum oven (28 psi, 80°) and allowed to dry overnight. After drying, 1.0 ml of ETOH (100%) is added to each vial and shaken gently for 1 min. Then 4.0 ml of scintillation cocktail is added to each vial, the contents are shaken, and the vials are allowed to sit for at least 1 hr before counting.

The sensitivity of this assay is 15 pg per tube. The assay is linear to at least 1.5 ng with an intraassay variance of 6.4% based on the determination of six replicate samples. The interassay variance for five consecutive assays was 12.7%.

General Discussion

McCaman et al.[7] first used *Drosophila* NAT in a microradioenzymic assay for dopamine. Under the conditions of their assay, the enzyme did not acetylate 5-HT. However, acetylation of 5-HT by *Drosophila* NAT readily occurs at higher pH and in different buffers.[8] In fact, substitution of *Drosophila* NAT for rat liver NAT in the 5-HT radioenzymic assay of Saavedra et al.[6] greatly improves the original assay. This improvement results from the greater activity, purity, and stability of *Drosophila* NAT when compared with rat liver NAT. For example, the fly enzyme is approximately four times more active, requiring addition of less protein to the reaction mixture. Thus, more product is formed and blank values are lower when *Drosophila* NAT is used. Since 5-HTP decarboxylase activity may also contribute to higher blanks,[13,17] the absence of this secondary enzyme activity from the partially purified *Drosophila* NAT may contribute to the improved sensitivity when this enzyme is used.

Although freshly prepared rat liver NAT is less active than *Drosophila* NAT, assays using either enzyme give comparable estimates of brain 5-HT content. However, rat liver NAT degrades rapidly, and storage for as little as 30 days leads to significant activity loss accompanied by increased blank values. On the other hand, *Drosophila* NAT remains stable for a year in storage, and we have used the enzyme as long as 18 months after preparation without significant changes in its activity. Thus, higher activity and greater purity and stability of *Drosophila* NAT undoubtedly increases sensitivity and reduces interassay variance in the 5-HT assay.

Interassay variance results in part from manual transfer of small volumes used in the microradioenzymic assay for 5-HT as well as the catecholamines. Since the total reaction volumes of the above monoamine assays are small, requiring the addition of 5-μl and 10-μl fractions, error is introduced by hand pipetting. We have been successful in reducing the intraassay variance to a minimum with the aid of a computerized diluter–pipettor (Hamilton Microlab P, Hamilton Co., Reno, NV). This allows all reagents for a single determination to be collected and mixed in one step.

[17] L. Hammel, Y. Naot, E. Ben-David, and H. Ginsberg, *Anal. Biochem.* **90**, 840 (1978).

With the extracts and reagents handled in this manner, intraassay variance is routinely less than 6%.

[³H]Melatonin extraction in the 5-HT assay is improved when 6 ml rather than 1.3 ml of toluene is used. Thus, for samples with low 5-HT content increased volumes of solvent will improve recovery and increase sensitivity of the assay. However, the assay is more economical when smaller volumes of toluene are used. The smaller volume provides adequate recovery of [³H]melatonin for accurate estimation of 5-HT in tissue such as whole hypothalamus or pineal. Thus, when performing large assays of 5-HT-rich tissue on a daily basis, the use of smaller solvent volumes is appropriate.

Acknowledgments

Portions of the work reported in this manuscript were supported by Research Grants 00566 (to R. L. C.) and 02867 (to R. F. W.) from the National Institute of Aging.

[33] Hypothalamic Catecholamine Biosynthesis and Neuropeptides

By DAVID K. SUNDBERG, BARBARA A. BENNETT, and MARIANA MORRIS

Abundant evidence suggests that the hypothalamic catecholamines norepinephrine and dopamine are intimately involved in the control of anterior pituitary function via their influence on the hypothalamic peptidergic neurons.[1,2] Several methods have been used to estimate central catecholaminergic activity and to correlate this with specific endocrine events.

The most commonly used technique has been to measure the disappearance of neuronal catecholamines after inhibition of their biosynthetic enzymes using various drugs. Thus, the "turnover" of hypothalamic norepinephrine and dopamine can be derived by plotting the log of the disappearance of tissue amines vs time after inhibition of tyrosine hy-

---

[1] S. M. McCann, L. Krulich, S. R. Ojeda, H. Negro-Vilar, and E. Vijayan, *in* "Central Regulation of the Endocrine System" (K. Fuxe, T. Hökfelt, and R. Lufts, eds.), p. 329. Plenum, New York, 1979.

[2] R. I. Weiner and W. F. Ganong, *Physiol. Rev.* **58,** 905 (1978).

droxylase (tyrosine 3-monooxygenase) by α-methyl-tyrosine (αmpt).[3,4] The time for 50% of the amine pool to be released ($t_{1/2}$) is determined by inspection of the plot, and the rate of release (K, hr^{-1}) is derived by dividing this value into the ln 2 ($K = 0.693/t_{1/2}$). Other investigators have modified this technique by measuring the disappearance of norepinephrine after inhibition of dopamine β-hydroxylase[5] (dopamine β-monooxygenase) or the appearance of L-DOPA after inhibition of L-amino acid decarboxylase.[6] These techniques are termed "non-steady state" methods since they result in a change in the intraneuronal concentrations of the transmitters and may affect feedback inhibition of biosynthetic enzymes and/or neuronal firing rates. While their application has resulted in invaluable information regarding catecholaminergic regulation of neuroendocrine function, these techniques require a minimum of two or three experimental animals per observation (i.e., 0, ½, and 1 hr after drug administration).

Other methods for studying catecholaminergic activity have been termed "steady state," since they do not alter the intraneuronal concentrations of transmitter. These techniques include measurement of the disappearance of administered labeled neurotransmitters that have been taken up in neurons or the appearance of newly synthesized transmitter after administration of a labeled precursor.[7-9] All of these techniques have been thoughtfully and critically evaluated.[10]

High-performance liquid chromatography with electrochemical detection (HPLC-EC) offers a reliable and relatively easy method for quantitating central catecholamine concentrations. The basic principles of this technique are covered in this volume.[11] Work in our laboratory has focused on adapting this technique (HPLC-EC) to measure steady-state concentrations and the biosynthesis of central norepinephrine (NE) and dopamine (DA) both *in vitro*[12] and *in vivo*.[13] Coupled with radioimmunoas-

[3] S. Spector, A. Sjoerdsma, and S. Udenfriend, *J. Pharmacol. Exp. Ther.* **147**, 86 (1965).
[4] B. B. Brodie, E. Costa, A. Dlabac, N. H. Neff, and H. H. Smookler, *J. Pharmacol. Exp. Ther.* **154**, 493 (1966).
[5] L. O. Farnebo, B. Hamberger, and J. Johnson, *J. Neurochem.* **18**, 2491 (1971).
[6] K. T. Demarest, R. H. Alper, and K. E. Moore, *J. Neurol. Trans.* **46**, 183 (1979).
[7] S. Udenfriend and N. P. Yaltzman, *Science* **142**, 394 (1963).
[8] G. C. Sedvall, V. K. Weise, and T. J. Kopin, *J. Pharmacol. Exp. Ther.* **159**, 274 (1967).
[9] N. H. Neff, S. H. Ngai, C. T. Wang, and E. Costa, *Mol. Pharmacol.* **5**, 90 (1969).
[10] N. Weiner, in "Neuropsychopharmacology of Monoamines and Regulatory Enzymes" (E. Usdin, ed.), p. 143. Raven, New York, 1974.
[11] P. M. Plotsky, this volume [31].
[12] D. K. Sundberg, B. A. Bennett, O. T. Wendel, and M. Morris, *Res. Commun. Chem. Pathol. Pharmacol.* **29**, 599 (1980).
[13] B. A. Bennett and D. K. Sundberg, *Life Sci.* **28**, 2811 (1981).

says for hypothalamic peptides, one can concurrently monitor catecholaminergic and peptidergic activity.[14,15]

I. Preparation and Incubation of Hypothalamic

Several laboratories have used an isolated hypothalamic incubation model to investigate peptide release *in vitro*. These models have included both long-term[16,17] and short-term[18-21] static incubation systems, as well as superfusion systems.[22,23] The superfusion system has the advantage in that there is little buildup in the media of significant amounts of peptidases and amine-catabolyzing enzymes. The static incubation system, however, has an advantage, since one can minimize the amount of tritiated tyrosine needed to study catecholamine biosynthesis.

Incubation Media

These studies employ a simple balanced salt solution such as Earle's or Hanks' previously gassed with water-saturated 95% O_2 and 5% CO_2 and containing 5 μCi of [2′,6′-^{3}H]tyrosine (30–40 Ci/mmol, NEN). The temperature and pH are maintained at 36–37° and 7.2, respectively, in a humidified incubator in order to most closely approximate the *in vivo* environment. Fortification of the media with an amino acid mixture has been avoided so as not to dilute the specific activity of the [^{3}H]tyrosine. Furthermore, some amino acids, such as glutamate and glycine, are candidates for central neurotransmitters.

Other investigators have used [3′,5′-^{3}H]tyrosine to measure its incorporation into catecholamines and H_2O.[24,25] Hydroxylation of the 3 position by tyrosine hydroxylase results in the formation of ^3H_2O, which can

[14] M. Morris, J. A. Wren, and D. K. Sundberg, *Peptides* **2**, 207 (1981).
[15] M. Morris and D. K. Sundberg, *Clin. Exp. Hypertens.* **3**, 1165 (1981).
[16] C. D. Sladek and K. M. Knigge, *Endocrinology (Baltimore)* **101**, 411 (1977).
[17] W. Shoemaker, W. Vale, and R. Peterfreund, this volume [21].
[18] M. W. B. Bradbury, J. Burden, E. W. Hillhouse, and M. T. Jones, *J. Physiol (London)* **239**, 269 (1974).
[19] W. H. Rotsztejn, J. L. Charli, E. Pattou, J. Epelbaum, and C. Cordon, *Endocrinology (Baltimore)* **99**, 1663 (1976).
[20] D. K. Sundberg and K. M. Knigge, *Brain Res.* **139**, 89 (1978).
[21] A. Negro-Vilar, S. R. Ojeda, and S. M. McCann, *Endocrinology (Baltimore)* **104**, 1749 (1977).
[22] L. W. Kao and J. Weiss, *Endocrinology (Baltimore)* **100**, 1723 (1977).
[23] E. Gallardo and V. D. Ramirez, *Proc. Soc. Exp. Biol. Med.* **155**, 79 (1977).
[24] M. Besson, A. Cheramy, and J. Glowinski, *J. Pharmacol. Exp. Ther.* **177**, 196 (1971).
[25] T. Westfall, M. Besson, F. Giorguieff, and J. Glowinski, *Naunyn-Schmiedeberg's Arch. Pharmacol.* **292**, 279 (1976).

be isolated easily by distillation. The use of the 2′,6′-³H label, however, results in a product with twice the specific activity. The labeled tyrosine should be obtained often (monthly) or repurified in order to avoid problems with breakdown. The purity of the labeled precursor can be routinely monitored by HPLC (Section III) and is found slowly to break down to ³H₂O. Labeled L-[³H]dihydroxyphenylalanine (L-DOPA) can also be used, although this would bypass the rate-limiting enzyme, tyrosine hydroxylase, and might have pharmacological effects by itself.

Tissue Dissection and Incubation

Sprague–Dawley rats are rapidly decapitated, the calvarium removed with rongeurs, and the entire brain retracted with curved forceps. The medial basal hypothalamus (MBH) is dissected with iris scissors using the landmarks of the posterior edge of the optic chiasm to the mammillary bodies; laterally it extends 1.5–2 mm from the midline and is approximately 1.5 mm deep, weighing 11.7 ± 2.3 mg. This tissue contains the median eminence, arcuate nucleus, and, depending on the depth of the cut, part of the ventral medial nucleus and anterior hypothalamic area. The specific dissection is discretionary and can be extended anteriorly to include the neurohypophysial nuclei. In some studies the neurohypophysis is also removed and incubated for the study of neurohypophysial dopamine synthesis and peptide release.[15] After dissection, the tissues are immediately placed in 0.5 to 1.0 ml of previously prepared incubation media in a 24-well (Falcon) disposable tissue incubation plate. The sacrifice-dissection time should be kept to a minimum, such that the tissue is placed in oxygenated media within 1–3 min. With a team of two investigators 20 hypothalamic samples can be procured within 30 min.

Time of Incubation

The temporal sequence of incubation is left to the discretion of the investigator. A 15- to 30-min preincubation followed by a 90-min incubation period has been found to circumvent the complication of rapid changes that occur immediately after dissection.

In order to determine the best sequence for tissue incubation, a time course should be performed. Figure 1 demonstrates a typical time course of incubation and its effect on tissue concentration (upper panel) and accumulation of newly synthesized (lower panel) NE and DA. From this information one can choose an appropriate length of incubation. For example, in studies of neurohypophysial DA biosynthesis, a 30-min preincubation in media with no precursor followed by a 90-min incubation period in media containing 5 μCi of [³H]tyrosine was chosen.[15]

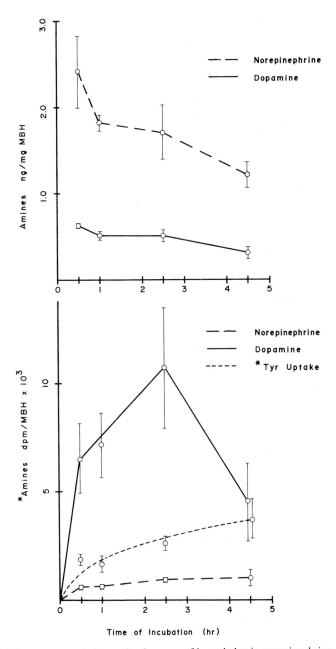

FIG. 1. The upper panel shows the decrease of hypothalamic norepinephrine (NE) and dopamine (DA) content (ng/mg) as a function of time of incubation. The lower panel shows the appearance of newly synthesized NE, DA, and tissue uptake of [^{3}H]tyrosine in the same experiment. D. K. Sundberg and M. Morris, unpublished data. MBH, median basal hypothalamus.

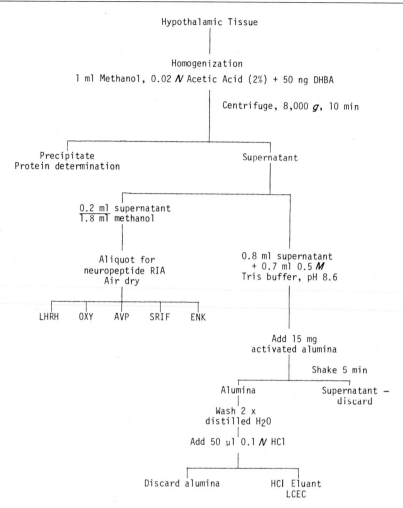

FIG. 2. A schematic diagram showing the extraction procedure used to monitor hypothalamic or neurohypophysial catecholamines and neuropeptides concurrently.

II. Extraction of Tissue and Media Neuropeptides and Biogenic Amines

Tissue Extractions

Owing to the high sensitivity of peptide radioimmunoassays and the large concentration of NE and DA in the hypothalamus, it is possible to quantitate all these parameters on a single hypothalamic sample.[14] The hydrophobic nature of most neuropeptides, however, excludes extraction with some dilute acids (perchlorate) that are commonly used for the isolation of tissue catecholamines. A simple modification of the isolation technique of Anton and Sayers[26] allows for the quantification of both hypothalamic neuropeptides and catecholamines. This technique has been useful for the concurrent measurement of the hypothalamic peptides, vasopressin (AVP), oxytocin (OXY), the enkephalins (ENK), luteinizing hormone releasing hormone (LHRH), and somatostatin (SRIF), as well as the biogenic amines, NE, epinephrine (EPI), and DA, from a single hypothalamic sample.[14] A schematic diagram of our extraction protocol is shown in Fig. 2.

After incubation, the tissues are placed in 1 ml of 98% methanol, 0.02 N acetic acid containing 50 ng of the internal amine standard dihydroxybenzylamine (DHBA). The tissue is essentially dehydrated, since methanol replaces the H_2O, stopping any continuing anabolic or catabolic events. The neural tissue, now quite "rubbery," is weighed and homogenized using a 1.0-ml Potter–Elvehjem ground-glass tissue grinder. The homogenates are centrifuged (8000 g), and the supernatant is divided. Twenty percent is further diluted in methanol, aliquoted appropriately into 12 × 75 glass tubes, and dried under a stream of air. This is sufficient to measure all the neuropeptides shown in Fig. 2 in duplicate or triplicate. The remaining methanolic supernatant (8/10) is adjusted to pH 8.6 with 0.7 ml of 0.5 M Tris-HCl buffer. Activated alumina (15 mg) is added to extract the biogenic amines. A small precalibrated scoop can be used to add the alumina, since the internal standard (DHBA) corrects for amine recovery. It is important to wash the alumina at least twice with distilled H_2O prior to eluting the catechols with 50 μl of 0.1 N HCl. This step removes most of the oxidizable noncatechol compounds that elute in the void volume. A comparison of this extraction vs the standard perchlorate procedure[26] for the catecholamines, NE and DA, and neuropeptides, LHRH and OXY, is shown in Table I. The methanolic extraction results in significantly better recovery of all these compounds.

Protein measurement[27] of the tissue pellet also demonstrates that a methanolic extraction precipitates 2–3 times more protein than several acidic extraction procedures (unpublished data).

Media Extraction for Peptides

The incubation media can also be deproteinated and aliquoted for the measurement of neuropeptide release *in vitro*. To remove peptidases,

---

[26] A. H. Anton and D. F. Sayers, *J. Pharmacol. Exp. Ther.* **138**, 360 (1962).
[27] O. H. Lowry, N. J. Rosebrough, A. L. Farr, and R. J. Randall, *J. Biol. Chem.* **193**, 265 (1951).

TABLE I
RECOVERY OF HYPOTHALAMIC CATECHOLAMINES AND NEUROPEPTIDES USING THE
STANDARD PERCHLORATE EXTRACTION VS METHANOLIC ACETIC ACID[a]

	Amine (ng/mg)		Peptide (ng/hypothalamus)	
	NE	DA	LHRH	OXY
Perchlorate, 0.1 M	1.55 ± 0.09	0.545 ± 0.032	7.90 ± 0.91	10.6 ± 2.1
METOH:HAc[b]	2.10 ± 0.05*	0.876 ± 0.169*	14.60 ± 0.75*	28.4 ± 5.2*

* $p < 0.05$.
[a] D. K. Sundberg and M. Morris, unpublished data.
[b] METOH:HAc = 98% methanol, 0.02 N acetic acid.

which rapidly degrade the neurohormones, several procedures are recommended. Acidification to 0.1 N HCl followed by boiling for 10 min or deproteination in at least five volumes of methanol is sufficient to eliminate peptidase activity. Nonetheless, the presence of these enzymes in the media during incubation results in an underestimation of neuropeptide release *in vitro*.[20]

III. HPLC-EC Measurement of Catecholamine Content and Biosynthesis

Reverse-phase high-performance liquid chromatography with electrochemical detection offers an excellent tool for the concurrent determination of neuronal catecholamine levels and biosynthesis.[12,13,15] This technique, originally developed by Adams and co-workers,[28] has the advantage of excellent sensitivity with on-line monitoring for neurotransmitters.[29] The following protocol was developed and found to be very useful for measurement of NE and DA content and biosynthesis both *in vitro*[12,15] and *in vivo*.[13]

Reagents and Equipment

Mobile-phase reservoir containing 0.1 M NaH$_2$PO$_4$ (7.9 g/liter); 1.0 mM EDTA (372 mg/liter); 2.5 mM heptane sulfonate (6.3 mg/liter); distilled, deionized water; vacuum filter for particles and dissolved gasses
High-pressure pump (Eldex Model A-30-S)

[28] P. T. Kissinger, C. J. Refshauge, R. Dreiling, and R. N. Adams, *Anal. Lett.* **6**, 465 (1973).
[29] P. T. Kissinger, C. S. Bruntlett, and R. E. Shoup, *Life Sci.* **20**, 455 (1981).

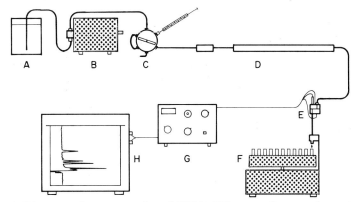

FIG. 3. Diagrammatic representation of HPLC–EC system for central catecholamine content and biosynthesis measurements. A, mobile-phase reservoir; B, high-pressure pump; C, sample injection valve; D, guard column and reverse-phase HPLC column; E, glassy carbon electrode (TL-8A) and reference (RE-3, BAS) electrode; F, fraction collector; G, electronic controller (LC-4, BAS); H, strip-chart recorder. D. K. Sundberg and M. Morris, unpublished data.

Sample injection valve (Valco, Model CV-6)
C_{18} μ-Bondapak reverse-phase column with guard column
Glassy Carbon, TL-8A Electrode with RE-3 reference electrode (Bioanalytical Systems, BAS)
Fraction collector
LC-4 electronic controller (Amperometric Detector, BAS)
Two-pen strip chart recorder

The medial basal hypothalamus contains noradrenergic, adrenergic, and dopaminergic systems. The chromatographic technique described here is easily capable of resolving all these neurotransmitters as well as an internal standard (DHBA) with 0% overlap and very low background noise. The basic set up is diagrammatically represented in Fig. 3.

Separation System

Many investigators have found ion-exchange HPLC columns to be useful for biogenic amine separation.[11] We find that reverse-phase chromatography coupled with an ion pairing agent (heptane sulfonate)[30] offers some specific advantages. The basic catecholamine structure has distinct hydrophobic as well as ionizable groups. Since reverse-phase chromatography preferentially retains lipophilic compounds, the elution becomes dependent on the number of charged groups and their pK_a. Therefore,

[30] T. P. Moyer and N. S. Jiang, *J. Chromatogr.* **153**, 365 (1978).

retention can be remarkably predicted on the basis of hydrophobicity, the number and the pK_a of ionizable groups, the pH of the eluent, and the concentration within the mobile phase of the ion pairing agent. The aliphatic sulfonates are ideal pairing agents for resolving catecholamines in that the negatively charged sulfonate ionically attracts the terminal amino groups of NE, DA, and DHBA ($pK_a = 8–10.3$) while the aliphatic chain is retained by the stationary hydrophobic phase of the column. This ensures the resolution of EPI from DHBA, and when the column loses efficiency, NE can still be separated from the void volume.[30]

A monobasic (0.1 M) sodium phosphate eluent is also quite convenient, since the resulting pH is around 4.0 and requires no adjustment. Changes of 1 pH unit can markedly effect the retention of these amines. It is very important to avoid the use of dibasic sodium phosphate (Na_2HPO_4 or buffers with a pH > 7) to prevent damage to the silica and bonded phases of the column.

The resolution of four biogenic amines and several of their metabolites is shown in Fig. 4. Since the detector signal output is negative, the amines elute from right to left. This chromatograph also demonstrates the predictability of this reverse-phase system. For example, 3-O-methylation of NA (A) to normetanephrine (C) increases its hydrophobicity and retention. This chromatograph also shows that the methylated metabolites of catecholamines and serotonin (I) are electrochemically reactive and can easily be measured using other isolation techniques.

Using this separation system 15–20 samples can easily be processed in 8 hr. The preliminary isolation technique also ensures that a single reverse-phase column can be used for 9–12 months.

Detection System

Traditional reference electrodes for electrochemical detection were designed so that the column eluent after passing the working (oxidizing) electrode was diluted in a larger volume of column mobile phase. In 1977, Freed and Asmus[31] described a low-volume reference electrode ideally suited for fraction collection of the column eluent. The TL-8A glassy carbon with an RE-3 reference electrode decreases the dead volume to less than 10 μl, so that the specific compound being measured by the detector can be collected and analyzed.

When medial basal hypothalami are incubated in the presence of [^{3}H]tyrosine as described in Sections I and II, newly synthesized catecholamines accumulate in the tissue. After extraction, 40 μl (4/5) of the alumina eluent is injected onto the column at a flow rate of 0.6 ml/min.

[31] C. R. Freed and P. A. Asmus, *Anal. Chem.* **49**, 2379 (1977).

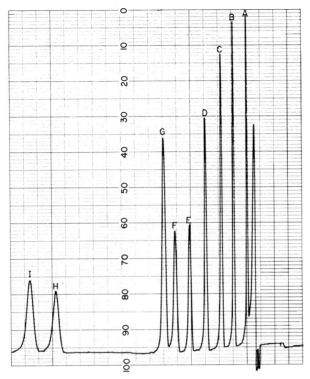

FIG. 4. This chromatograph shows the resolution of nine amines by reverse-phase chromatography. Elution is shown from right to left. A, norepinephrine; B, epinephrine; C, normetanephrine; D, dopamine; E, 3-methoxytyrosine; F, metanephrine; G, dihydroxyphenylacetic acid; H, 3-methoxytyramine; I, serotonin. D. K. Sundberg and M. Morris, unpublished data.

The catecholamines are quantitated by oxidation to their respective quinones at an applied voltage of 0.7 V. The electrochemical controller is set to detect changes in electron flow of 20 nA/V and monitored on a dual-pen recorder set at 1 and 0.1 V full scale. The fraction collector is set to collect 0.5-ml fractions directly into liquid scintillation "minivials" and is activated as soon as the void volume (negative deflection) appears on the strip-chart recorder (set at 10 cm/hr). After collection of the eluent a water-compatible liquid scintillation cocktail (6.5 ml) is then directly added to the vials, which are counted for presence of ^{3}H. Figure 5(I) shows the elution of newly synthesized NE and DA after a 2.5-hr incubation of four hypothalami. Also shown [Fig. 5(II)] is the resolution of an alumina-extracted standard of 10 ng to compare recoveries, which are: NE, 47%; L-DOPA, 11.4%; DHBA, 52%; EPI, 50%; and DA, 57%. In this

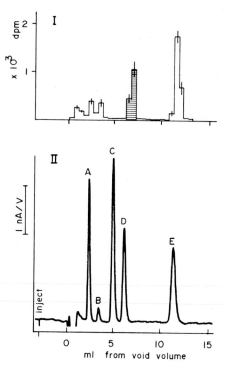

FIG. 5. (I) The elution profile of tritiated compounds (±SE) after a 2.5-hr incubation of rat hypothalami ($n = 4$) in media containing [³H]tyrosine. Each fraction consists of 0.5 ml of eluent. The shaded area represents [³H]tyrosine. (II) The elution profile of an alumina-extracted standard of equal amounts (10 ng) of NE (A), L-DOPA (B), EPI (C), DHBA (D), and DA (E), processed immediately after the four hypothalamic extracts above. From Sundberg et al.,[12] with permission of PJP Publications Ltd. (copyright 1980).

system a small percentage of the tritiated precursor carries through the extraction procedure (crosshatched peak); however, it conveniently elutes after the internal standard (DHBA). Elution and breakdown of [³H]tyrosine should be checked routinely by applying the precursor (spiked with standards) to the chromatographic system. Another good control to ensure that one is indeed measuring newly synthesized NE and DA is to incubate hypothalami in the presence of the tyrosine hydroxylase (monooxygenase) inhibitor αmpt.[12] In one instance, αmpt (10^{-4} M) decreased incorporation of precursor into NE from 392 ± 47 to 71.8 ± 8.9 dpm per hypothalamus and into DA from 973 ± 87 to 41.3 ± 12.8 dpm per hypothalamus.

Data Calculation

When measuring catecholamine content by HPLC-EC, peak height has been found to be directly proportional to amine quantity. In these experiments, content and biosynthesis are calculated using two essential parameters: (1) peak height or disintegrations per minute per peak, and (2) recovery of DHBA. Since a known amount of the internal standard (50 ng) is added, the effectiveness of extraction and application to the chromatography system are automatically corrected. Standards are run before and after analysis of each experiment, and peak height (mm/ng) is calculated for each standard. The quantity of measured amine in an unknown sample (Q) is then calculated by Eq. (1).

$$Q \text{ (ng/sample)} = \text{mm of unknown/mm per ng of standard} \quad (1)$$

Since recovery varies within an experiment and also between experiments (40–80%), this value is corrected for the measured amount of DHBA (R) [Eq. (2)].

$$R \text{ (\%)} = \text{mm DHBA in sample/mm 50 ng DHBA standard} \quad (2)$$

The actual concentration of tissue amine (C) can then be calculated, which automatically corrects not only for recovery but also for the volume of sample injected onto the column [Eq. (3)].

$$C \text{ (ng/tissue)} = Q/R \quad (3)$$

Likewise, the corrected quantity of newly synthesized amines (A, dpm/tissue) is corrected (A = dpm per sample/R). Values can then be expressed as nanograms or disintegrations per minute of amine/MBH with a coefficient of variation (C.V.) of 20–35%, as nanograms or disintegrations per minute of amine per milligram of tissue with a C.V. of 7–15%, or as a specific activity (S, dpm/ng) with a C.V. of usually less than 10% [Eq. (4)].

$$S \text{ (dpm/ng)} = A \text{ (dpm/tissue)}/C \text{ (ng/tissue)} \quad (4)$$

In this case tissue or protein weights cancel each other out, giving a very useful measurement of neuronal activity, but not necessarily of neuronal density or content. Reporting both specific activity (S) and corrected content (C) should allow readers to deduce both the activity and steady-state status of a neuronal system.

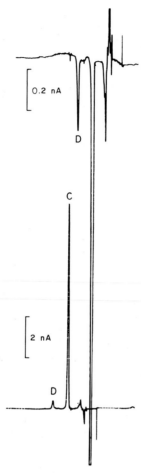

FIG. 6. Dual-pen, dual-sensitivity recording of a posterior pituitary extract by HPLC-EC. The top tracing was set at 0.1 V full scale (2 nA/V) for amplification of dopamine (D). The lower tracing was recorded at 1 V full scale (20 nA/V) for measurement of DHBA (C). From Morris and Sundberg,[15] reprinted with permission of Marcel Dekker Inc. (copyright 1981).

IV. Comparison with Non-Steady State Turnover

The most commonly used technique for measuring central and peripheral catecholaminergic activity has been to measure the disappearance of biogenic amines after inhibition of tyrosine hydroxylase with αmpt. It is crucial to compare these two techniques in order to determine their applicability to *in vivo* systems.

This can be done by administering the tritiated precursor directly into

TABLE II
STEADY STATE VS NON-STEADY STATE TURNOVER RATES[a,b]

Catecholamine and brain area[c]	Non-steady state			Steady state synthetic rate (fg/mg hr^{-1})
	$t_{1/2}$ (hr)	K (hr^{-1})	Synthetic rate (pg/mg hr^{-1})	
Norepinephrine				
AH	5.3	13.2	203	135
MBH	3.0	23.5	296	74
BS	3.0	23.1	37	68
C	>24	<0.03	—	17
Dopamine				
AH	1.5	46.8	262	178
MBH	2.1	33.0	185	260
BS	0.7	107.0	0.80	39
C	3.0	23.1	1,825	258

[a] From B. A. Bennett and D.K. Sundberg,[13] reprinted with permission of Pergamon Press Ltd. (copyright 1981).
[b] Values for half-life ($t_{1/2}$) and rate constants (K) were obtained from the disappearance curves between 1 and 3 hr. Non-steady state turnover is expressed as [CA] × K. Steady state turnover values were obtained from the amount of conversion of [^{3}H]tyrosine to [^{3}H]amine (pg/mg) at 10 min corrected to hours.
[c] AH, anterior hypothalamus; MBH, median basal hypothalamus; BS, brainstem; C, caudate area.

the cerebral ventricular system and studying amine biosynthesis in different brain areas.[13] The technique for *in vivo* cannulation of the ventricular system is described in this volume.[32]

When studying several brain areas simultaneously, a dual-pen recorder set at 1 and 0.1 V full scale is particularly convenient in order to circumvent the need for constantly switching the electrochemical detector's sensitivity range. Figure 6 illustrates that by measuring at two different potentials (in this case 2 and 10 nA/V), one can amplify the neurohypophysial DA peak (D, upper graph) while still keeping the DHBA peak (C) on scale. This is also important when studying caudate NE and DA, where there is a 100-fold difference in their concentrations (in favor of DA).

The steady state and non-steady state "turnover" rates of anterior (AH) and medial basal hypothalamic (MBH), brainstem (BS), and caudate (C) catecholamines are shown in Table II.[13] Non-steady state rates are

[32] D. K. Sundberg, M. Morris, and K. A. Gruber, this volume [35].

determined by injecting groups of rats with 250 mg/kg of αmpt intraperitoneally and sacrificing and quantitating the disappearance of biogenic amines at 0, 1, and 3 hr postinjection. Steady state synthesis rates are determined by administering 5 μCi of the tritiated precursor directly into the lateral ventricle. Groups of rats are sacrificed at 5, 10, and 15 min postinjection, and newly synthesized catecholamines in these same brain areas are determined by HPLC-EC.

The calculated synthetic rates derived from these two methods are shown in Table II. As is found *in vitro,* the medial basal hypothalamus synthesizes 3–4 times more dopamine than norepinephrine. The calculated synthetic rates using the αmpt method, however, would suggest that MBH NE synthesis is higher than that of DA. Likewise, in the caudate the non-steady state technique appears to overestimate DA synthetic rates. This might, in fact, be expected, since many brain catecholaminergic systems possess autoreceptors that presynaptically inhibit neuronal activity.[33] Nonetheless, this method has provided much valuable information, since both control and experimental animals receive the same pharmacological treatment.

V. Determination of Neuronal Tyrosine Concentration and Specific Activity

Theoretically, neuronal tyrosine concentration could be altered in different physiological states such as age, disease, drug administration, or differences in diet. Phenylalanine, methionine, tryptophan, L-dopa and arginine have all been shown to inhibit the brain uptake of [^{14}C]tyrosine in a competitive manner.[34] In addition, brain L-amino acid transport mechanisms might be altered in various states. For example, the central abnormalities accompanying genetic phenylketonuria might in part be due to an inhibition of essential amino acid uptake into the brain.

These considerations could lead to difficulties in interpreting catecholamine biosynthesis studies either *in vitro* or *in vivo.* The specific activity of the precursor available for catecholamine biosynthesis would obviously affect the absolute synthetic rate observed. Using the chromatographic system previously described in this chapter, tyrosine coelutes with EPI such that the specific activity of tyrosine and epinephrine biosynthesis could not be measured.

In the last few years significant advances have occurred resulting in increased efficiency of reverse phase high pressure columns. Column

---

[33] T. C. Westfall, *Physiol. Rev.* **57,** 659 (1977).
[34] M. Bradbury, "The Concept of the Blood-Brain Barrier." Wiley, New York, 1979.

efficiency is usually measured in "theoretical plates" and in our previous studies[12–15] we used a 10-μm C-18 packing having approximately 3000 plates/column.

Recently smaller diameter spherical packing materials have been developed which result in columns possessing 10,000 to 15,000 theoretical plates. This has made it possible to produce high efficiency short columns (3 μm, 10 cm) which speeds separation, decreases solvent consumption, and also increases (HPLC-EC) sensitivity since compounds are eluted and oxidized in smaller volumes.

These high efficiency columns now make it possible to resolve all of the catecholamines as well as their common precursor tyrosine as shown in Fig. 7. At an applied amperometric potential of 720 mV, the phenolic tyrosine oxidizes to its respective quinone and can be easily quantified. This separation is accomplished using a 25-cm 5-μm C-18 column (15,000 plates, Rainin) and increasing the concentration of the ion pairing agent (heptane sulfonate) to 16 mM (6 to 40 mg/liter). All the primary amines NE, EPI, DHBA, and DA are markedly retarded while the amino acid tyrosine is not affected. Acceleration of the entire separation is then accomplished by the addition of 1.5% methanol to the eluant.

Using this separation regime the biosynthesis and content of hypothalamic NE, DA, and now EPI can be monitored and corrected for differences of the specific activity of the precursor [^{3}H]tyrosine. Our modification of the alumina extraction results is partial extraction of tyrosine (4 ng/10 mg of hypothalamus). Thus, by collecting the tyrosine fractions for ^{3}H measurement and dividing by the measured concentration, precursor specific activity can be derived.

In application, it is important to rinse the incubated hypothalami in fresh cold media to remove surface tritium. After homogenization a 20-μl aliquot of the extract is removed to determine total [^{3}H]tyrosine content. We have found that the tissue uptake of the precursor (5 μCi or 27.2 ng/ml) averages 4.8% ± 0.05 after 1 hr. After HPLC-EC the specific activity of tissue tyrosine was found to be 110.3 ± 7.4 cpm/ng. By dividing the total tissue radioactivity (cpm/mg hypothalamus) by this specific activity (cpm/ng) the total tissue concentration of tyrosine is determined. In our experience the ratio of unlabeled/labeled tyrosine at the end of a 60-min incubation averaged 1107 ± 73.

Therefore, using this modified chromatographic procedure, variations in the content and specific activity of the catecholamine precursor can be determined. Assuming that the ratio of unlabeled/labeled tyrosine is evenly distributed in peptidergic and aminergic neurons, the absolute synthetic rates can be derived and would appear to be much higher than depicted in previous studies.

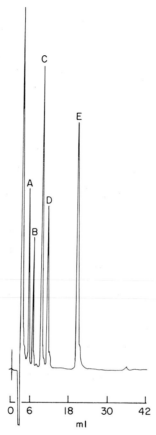

FIG. 7. Utilizing a high efficiency (15,000 plates) 5-μm spherical reverse phase column the primary catecholamines and their precursor tyrosine are resolved. This is accomplished utilizing an identical eluant which also contains 16 mM heptane sulfonate and 1.5% methanol. A, Norepinephrine; B, tyrosine; C, epinephrine; D, DHBA; E, dopamine.

VI. Summary

In conclusion, both steady state and non-steady state methods for investigating the turnover of central catecholamines can provide valuable information regarding the central control of neuroendocrine function. The system described here, utilizing HPLC with electrochemical detection, offers a relatively easy and reliable method for determining steady state catecholamine biosynthesis in the hypothalamus and other brain areas.

Acknowledgments

This work was supported in part by NIH Grants HD-10900 to D. K. S. and HL-22411 to M. M. and by a Grant-in-aid from the American Heart Association with funds contributed in part by the North Caroline affiliate. This work was done during the tenure of an established investigatorship from the American Heart Association (to M. M.) with funds contributed in part by the North Carolina affiliate. The author wishes to acknowledge the secretarial assistance of Ms. Stephanie Burgoyne.

[34] Methods for the Study of the Biosynthesis of Neuroendocrine Peptides *in Vivo* and *in Vitro*

By JEFFREY F. MCKELVY, JAMES E. KRAUSE, and JEFFREY D. WHITE

Study of the dynamics of neuroendocrine peptide metabolism, i.e., determination of rates of peptide biosynthesis and turnover, is a highly desirable approach with which to gain an understanding of the mechanisms by which "final common pathway" hypothalamic neuron systems achieve neuroendocrine integration. However, it has been extremely difficult to gain this type of information by the most direct experimental approach, i.e., by isotope incorporation studies using hypothalamic tissue. This has been for several reasons: (a) the extremely small size of pools of these peptides, challenging the sensitivity of chemical and even radiochemical detection of the synthesis of small numbers of molecules; (b) difficulties in purifying such small amounts of radioactivity sequentially, so as to be certain of radiochemical purity; and (c) lack of knowledge of the cytoarchitectonics of hypophysiotropic neuron systems.

In the past 5 years, progress has been made with respect to all three of these problem areas: more amino acids of high specific radioactivity are available; the use of high-performance liquid chromatography (HPLC) has revolutionized peptide purification, and the complete trajectories of several hypophysiotropic neuron systems projecting terminals to the median eminence are known.

In this chapter, we describe techniques for studying the *de novo* biosynthesis of the neurohypophysial hormones oxytocin and vasopressin and their neurophysins, both to their posterior pituitary and brainstem terminal fields, and of the hypophysiotropic hormone luteinizing hormone-releasing hormone (LHRH). These experimental protocols reflect the progress alluded to above. In addition, in their emphasis on *in vivo* studies, they reflect our 13 years of experience with labeling studies of neuroendocrine peptides. We have found that *in vivo* biosynthetic studies involving isotope administration to sites of cell bodies of origin of a peptide system of interest gives the highest levels of isotope incorporation and the most meaningful context in which to carry out studies on regulation. This approach has proved to be valuable to the study of other peptidergic systems in the central nervous system of the rat, such as the striatonigral substance P system,[1] and the hypothalamic proopiomelanocortin system.[2] In some cases, however, e.g., in order to observe the directionality of the effects of feedback signals, especially steroid hormones, on neuroendocrine peptide biosynthesis, it is convenient to carry out studies *in vitro*. To this end, we have included description of techniques for the study of LHRH biosynthesis in primary dispersed cell cultures of perinatal rat hypothalami.

General Aspects of *in Vivo* Biosynthetic Studies

In vivo peptide biosynthetic experiments can provide information on the dynamic nature of these peptide-secreting neuronal systems. Thus, a tissue pool of peptide can be radiolabeled, and its transit through a neuronal subsystem can be assessed. Several experimental strategies can be adopted to yield such information; however, the limiting factor is the amount of radioactive amino acid incorporation into the tissue pool of the peptide. The optimal use of (1) methods of isotopic amino acid delivery, (2) tissue sampling strategies, and (3) peptide purification techniques will allow sufficient incorporation of radioactive amino acids into the peptide, such that peptide synthesis, transport, and turnover rates may be assessed.

For *in vivo* peptide synthetic studies, it is of paramount importance that the cell body-containing region, the site of peptide precursor biosynthesis, is administered relatively large amounts of isotopic amino acids of the highest specific activity (SA) available (for ^{35}S-labeled amino acids, SA > 1000 Ci/mmol; for ^{3}H-labeled amino acids, SA > 100 Ci/mmol). Administration of the amino acid(s) is best performed by providing high amounts of activity in low volumes. Discrete cellular regions can be either continuously infused with radioactive amino acid(s) in order to equilibrate tissue pools with the radiolabeled peptide, or can be pulse-infused, followed by an appropriate chase time, in order to assess the dynamics of synthesis, processing, and intraaxonal transport. Since many hypophysiotropic peptidergic projections encompass distances of 2–8 mm, it is quite feasible to sample discrete regions of the peptidergic projection to correlate *de novo* peptide biosynthesis with the transit of the peptide

[1] J. E. Krause, J. P. Advis, and J. F. McKelvy, *Soc. Neurosci. Abstr.* **8**, 13 (1982).
[2] A. Liotta, J. P. Advis, J. E. Krause, D. Krieger, and J. F. McKelvy, submitted for publication.

along the course of the neuron. Thus, the cell body-containing region, the region of axons projecting to the median eminence, and the peptide terminal field (either the median eminence or neural lobe) can be discretely sampled. The tissue samples are then analyzed in order to ascertain the amount of radiolabeled peptide synthesized *de novo*. In this context, the synthetic rates, transport times, and turnover rates of neuronal peptides may be assessed.

Cannula Implantation

The implantation of cannulas is performed using standard stereotaxic techniques[3] while the rat is under ether anesthesia. The cannulas used are 26–30-gauge stainless steel (Small Parts, Inc., Miami, Florida) with slightly beveled edges. Each cannula has a short length of microbore tubing (inner diameter = 0.010 in.) attached to the end to facilitate administration of the radioactive amino acid solution. The implanted cannulas are affixed to the skull with ivory grip cement (L. D. Caulk Co., Milford, MA) and a machine screw (Small Parts, Inc.) attached to the calvarium a few millimeters caudal to the implanted cannulas. A piece of Tygon tubing is centered around the implanted cannulas for protection and is affixed with grip cement to the cannulas' cement support. The skin around the implanted cannulas is sutured, and the rat is allowed to recover, usually for 1–3 days prior to isotopic amino acid administration.

Radioactive Amino Acid Administration

For these experiments, the infusion of large amounts of radioactive amino acid(s) in small volumes has been quite useful. Thus, 250–500 μCi of [^{35}S]cysteine or [^{3}H]leucine and [^{3}H]proline is administered via each cannula. The rat is again lightly anesthetized with ether and placed in the stereotaxic device. Prior to the amino acid infusions, the inside of each cannula is carefully cleaned with a fine wire that has been premeasured to extend to the end of the cannula. For pulse-labeling studies, we resuspended aliquots of the lyophilized amino acid in 6 μl of artificial extracellular fluid.[4] Three microliters of this solution is drawn through a short piece of steel tubing connected to a long (25 cm) piece of P-10 tubing. This assembly is attached to an airtight 25–50 μl syringe (Hamilton). The steel connecting piece on the syringe assembly is attached to the P-10 tubing on the cannula, and the amino acid-containing solution is slowly injected into the cannula (approximately 5 min). For pulse-labeling studies, we rou-

[3] J. E. Skinner, "Neuroscience: A Laboratory Manual." Saunders, Philadelphia, 1971.
[4] K. A. C. Elliot, *in* "Handbook of Neurochemistry" (A. Lajtha, ed.), Vol. 2, p. 103. Plenum, New York, 1969.

tinely pulse with the amino acid solution for 2 hr using an osmotic minipump (Alzet No. 2001) delivery system. Thus, 1 µl of the original 3 µl is directly infused into the brain region, and the remaining 2 µl is administered at a rate of 1 µl/hr with the osmotic minipumps. The skin on the back of the animal is opened for subcutaneous implantation of the osmotic minipump. The osmotic minipump is connected with P-10 tubing that contains artificial extracellular fluid. This tubing is connected to the plastic tubing that is attached to the implanted cannulas. Prior to infusion, the osmotic minipumps are filled with a 0.2% Coomassie Brilliant Blue solution, to serve as a tracking dye, and are incubated in saline at 37° for at least 8 hr. For experiments involving continuous infusion of the radioactive amino acid solution, the radioactive amino acid solution (20–100 µCi/cannula hr^{-1}) can be administered over the time period desired, generally 8–16 hr total infusion time.

Tissue Harvest and Peptide Extraction

Tissue samples are obtained after the appropriate labeling time. For this purpose, the rat is lightly anesthetized with ether. The minipumps are removed from the back of the animal, the cannulas and cement support are removed, and the animal is killed by decapitation. The brain is removed as quickly as possible, and surface structures (i.e., median eminence) are microdissected from the fresh tissue. The remainder of the brain is frozen at $-80°$ so that discrete hypothalamic regions may be dissected from frozen coronal brain sections. The anterior and posterior-neurointermediate lobes of the pituitary are removed separately from the skull. Tissue samples are homogenized (in the presence of 10–50 µg of carrier peptide) in 100–200 µl of extraction buffer in small ground-glass microhomogenizers (MH microhomogenizers, Micro-metric Instrument Co., Cleveland, OH), and the homogenizing tubes are rinsed with the same buffer. The samples are acid-extracted twice over the course of 4–24 hr. The total extraction volume is 200 µl to 1 ml. All samples to be analyzed by HPLC are adsorbed to C_{18} Sep-Pak cartridges (Waters Associates) in 0.1% trifluoroacetic acid (TFA); generally 3–10 ml of 0.1% TFA is used to rinse the nonadsorbing material from the cartridge. Peptides and other cellular constituents adsorbed to the cartridge are eluted with 3 ml of 50% 1-propanol in 0.1% TFA into 15-ml conical-bottom polypropylene tubes or into flint-glass test tubes. The samples are then frozen and lyophilized prior to sequential purification by HPLC.

Peptide Purification

The peptides of interest are purified by HPLC using reverse-phase C_8 or C_{18} columns. The systems utilized include those involving organic elution and ion-pairing solvents in either isocratic or gradient chromatographic systems. In all cases elution of the carrier peptide from the column is monitored by absorbance at 210 nm. Peptide recoveries at each step are estimated on the basis of the peak area of the carrier peptide, as determined by digital integration, compared to known amounts of standard, or by radioactivity recovery measurements if it is known that the peptide is homogeneous. Specific chromatographic systems for various neuropeptides are presented in the sections on neurohypophysial peptide biosynthesis and LHRH biosynthesis.

In Vivo Neurohypophysial Peptide Biosynthesis

The nonapeptide hormones arginine vasopressin and oxytocin and their associated neurophysins are synthesized in the magnocellular neurons of the paraventricular nuclei (PVN) and supraoptic nuclei (SON) of the hypothalamus. The primary projection of these cell bodies is to the posterior lobe of the pituitary, although both rostral and caudal projections of these nuclei within the central nervous system have been described.[5] The methods described below have been developed for the neurohypophysial system but are generally applicable to these other projection systems.

For these studies we have concentrated on the paraventricular nucleus. Rats are bilaterally cannulated using a David Kopf stereotaxic instrument according to the coordinates A 0.5 mm, L 0.5 mm, V 7.3 mm, with respect to bregma. [^{35}S]Cysteine (>1000 Ci/mmol; 500 μCi total) is pulse-infused into the PVN for 2 hr using osmotic minipumps connected via polyethylene tubing to the implanted cannulas. After an appropriate chase time, the rat is decapitated. The collection of tissue samples is relatively easy for the oxytocin–vasopressin neurohypophysial system. The neural lobe of the pituitary is removed, and the PVN is punched from frozen coronal brain sections using the third ventricle and fornix as landmarks. Peptides and neurophysins are extracted from tissues homogenized in ice-cold 2 M acetic acid in ground-glass tissue homogenizers. The total extraction volume is 400 μl, and each sample is extracted twice for 4–8 hr. All samples for HPLC analysis receive 25 μg each of carrier oxytocin, vasopressin, and somatostatin-14.

After extraction, the samples are adsorbed to and eluted from C_{18} SEP-PAK cartridges as described above. In our hands, samples that cannot immediately be processed for HPLC show low degradation and loss of radiolabeled and carrier peptides if they remain in 2 M acetic acid at $-80°$ in flint-glass tubes.

[5] L. W. Swanson and H. G. M. Kuypers, *J. Comp. Neurol.* **194**, 555 (1980).

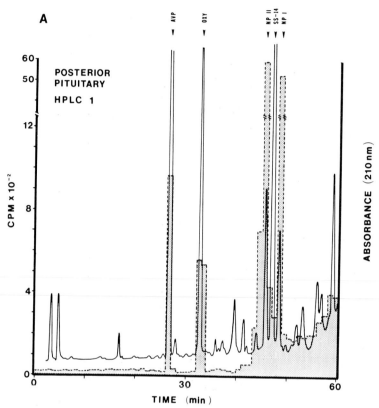

FIG. 1. (A) High-performance liquid chromatogram of the tissue extract from the posterior pituitary of a rat that had been administered 500 µCi of [^{35}S]cysteine over 2 hr followed by a 22-hr chase period. For HPLC conditions, refer to text and HPLC system No. 1. The elution positions of the carrier AVP, OXY, and SS-14, as well as endogenous neurophysins, are marked. ^{35}S Radioactivity is plotted on the left axis and is shown as the stippled bars. This radioactivity represents 10% of the total sample. Absorbance at 210 nm is plotted on the right-hand scale. Fractions 33 and 34 were pooled, lyophilized, and subjected to sequential HPLC purification for oxytocin. (B) Chromatogram of the performic acid-oxidized oxytocin taken from HPLC system No. 2 in the animal described for (A). The chromatographic system shown is HPLC system No. 3, described in the text. The elution of the ^{35}S radioactivity exactly corresponds with the elution of the carrier peptide. The radioactivity plotted represents the total radioactivity remaining in the sample. Axes are as in (A).

The HPLC purification steps use simple gradient systems that can be adapted to most gradient HPLC systems. HPLC system No. 1 is a linear gradient of 5 to 45% acetonitrile in 0.1 M monobasic sodium phosphate and 0.1 M phosphoric acid, pH 2.5 (buffer A), for 60 min at a flow rate of 1.0 ml/min. A representative example of such a chromatogram is shown

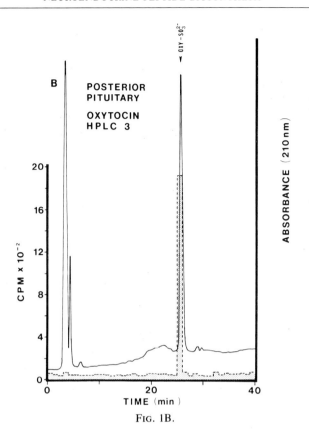

FIG. 1B.

in Fig. 1A. The peptides of interest—vasopressin, oxytocin, neurophysins I and II/III—are clearly separated from each other. Collection of 1-min fractions allows each peptide to be collected in a total volume of 2 ml. Aliquots of the chromatographic profile are taken for radioactivity determination and then the peptide-containing fractions are frozen and lyophilized in conical-bottom polypropylene tubes.

For the successive purification steps, oxytocin and vasopressin are treated separately; however, in each case the second purification is coupled with chemical modification (performic acid oxidation). The lyophilized fractions are resuspended in 50 μl of 0.1% TFA by vortexing and sonication and then cooled on ice; 50 μl of freshly prepared performic acid[6] is then added, and the reaction is allowed to proceed for 90 min on ice. At the end of the reaction, the entire 100 μl is injected directly onto the HPLC. The second purification system uses the buffer A–CH$_3$CN

[6] C. H. W. Hirs, this series, Vol. 11, p. 192.

system described above; however, the gradient is now an exponential gradient instead of the linear gradient, i.e., gradient 3 on a Waters Model 660 gradient programmer. For both oxytocin and vasopressin, the instrument remains at initial conditions (5% CH$_3$CN) for 5 min prior to starting the gradient to allow the performic acid to wash off the column. For oxytocin the gradient is 5 to 22% CH$_3$CN over 40 min at a flow rate of 1.0 ml/min, and for vasopressin the gradient is 5 to 18% CH$_3$CN. Under these conditions, on a C$_8$ reverse-phase column oxytocin elutes at 36 min, OXY-SO$_3^{2-}$ elutes at 25.6 min, vasopressin at 31 min, and AVP-SO$_3^{2-}$ at 28 min. Again, 1-min fractions are collected, and the fractions containing the carrier peak are frozen and lyophilized in conical-bottom polypropylene tubes.

This exponential gradient shape can be approximated as a series of linear gradients (see the tabulation).

% Time	% B
3.3	$0.33 (B_f - B_i) + B_i$
7.0	$0.50 (B_f - B_i) + B_i$
20.0	$0.67 (B_f - B_i) + B_i$
50.0	$0.84 (B_f - B_i) + B_i$
100.0	B_f

where B_i = %B at initial conditions and B_f = %B at final conditions.

The final HPLC purification step uses an ion-pairing buffer in the separation system. It is noteworthy that the retention behavior of the peptides in this system is distinctly different than that in previous systems; i.e., vasopressin elutes at a higher concentration of acetonitrile than does oxytocin.

The ion-pairing reagent used is triethylammonium phosphate (TEAP), which is prepared by the addition of 5.12 ml of triethylamine (Sigma) to 5.7 ml of phosphoric acid (Fisher). The solution is stirred for 15–30 min, after which time HPLC-grade water is added to make 500 ml. Of this stock solution, 20 ml is diluted with 980 ml of water to make the working buffer solution.

The oxytocin gradient system is a linear gradient of 5 to 30% CH$_3$CN over 40 min at a flow rate of 1.0 ml/min with an additional 5 min at final conditions. Under these conditions, OXY-SO$_3^{2-}$ elutes from a C$_8$ reverse-phase column at 24.6 min, and native oxytocin elutes at 44 min. The vasopressin system is a linear gradient of 5 to 40% CH$_3$CN over 40 min and flow rates are 1.0 ml/min. Vasopressin elutes from the column at 26.4 min, and AVP-SO$_3^{2-}$ elutes at 21.5 min.

The purification of the peptides through these steps has consistently yielded a single peak of radioactivity that coincides with the elution of the carrier peptide. Figure 1B shows a representative chromatogram for purified OXY-SO$_3^{2-}$.

The purification of neurophysins I, II, and III will not be discussed at length here. As shown in Fig. 1A, neurophysin I clearly separates from neurophysins II/III on the initial HPLC separation system. The peaks can be collected separately, then further purified by HPLC or immunoprecipitation and polyacrylamide gel electrophoresis. For additional information on purification and tryptic mapping, see Sherman and McKelvy.[7]

In Vivo LHRH Biosynthesis

The hypophysiotropic LHRH neurons projecting to the median eminence (ME) in the rat have the majority of their perikarya in the ventromedial preoptic area (POA). This hypothalamic region can be cannulated for biosynthesis studies of LHRH with little or no effects on reproductive cyclicity as assessed by serum luteinizing hormone levels[8] or by examination of vaginal cytology throughout the estrous cycle.

Female rats of the Sprague–Dawley strain (of defined reproductive status) are bilaterally implanted with 30-gauge cannulas in the POA (A, 1.1 mm; L, 0.5 mm; V, 8.0 mm—with bregma as reference) using a David Kopf stereotaxic instrument. A mix of [^{3}H]leucine and [^{3}H]proline (>100 Ci/mmol; 500 μCi total) is continuously infused into the POA for 8 hr using osmotic minipumps connected via polyethylene tubing to the implanted cannulas as described in detail above. After decapitation, the ME is microdissected, and the POA and medial basal hypothalamus are punched out from frozen coronal brain sections. The location of the indwelling bilateral POA cannulas is verified in the frozen POA-containing brain section by visual examination. Tissue samples are homogenized on ice in 0.1 M acetic acid containing 30 μg of carrier LHRH and are stored at −80° prior to analysis. All individual samples are analyzed for total tissue ^{3}H, 10% (w/v) trichloroacetic acid (TCA)-precipitable protein, and 10% TCA-soluble protein.

The TCA-soluble fraction contains the ^{3}H-labeled LHRH, and the decapeptide in this fraction is sequentially purified by HPLC. The ^{3}H-labeled LHRH in the acid extract is partially purified by adsorption to and elution from a C$_{18}$ SEP-PAK cartridge. The organic eluate from this step is lyophilized prior to HPLC. Systems we have used for ^{3}H-labeled LHRH purification are HPLC 1—chromatography on a C$_{18}$ column

---

[7] T. G. Sherman and J. F. McKelvy, *Ann. N. Y. Acad. Sci.* **394,** 82 (1982).
[8] J. E. Krause, J. P. Advis, and J. F. McKelvy, *Endocrinology* **3,** 344 (1982).

(Brownlee) using a linear 1 to 60% CH_3CN gradient in 0.1 M phosphoric acid and 0.1 M sodium phosphate, pH 2.5 (buffer A), for 50 min at 1.0 ml/min (V_e LHRH = 29 min); HPLC 2—chromatography on a C_8 column under isocratic conditions using 0.7 mM TEAP in 42% CH_3CN at 2.0 ml/min (V_e LHRH = 8.0 min); HPLC 3—chromatography on a C_{18} column using a 1 to 30% exponential CH_3CN gradient (program 3 on Waters 660 solvent programmer, or as described above) in buffer A for 50 min at a 1.0 ml/min flow rate (V_e LHRH = 32 min); HPLC 4—chromatography on a C_8 column using an exponential CH_3CN gradient (1 to 30%) in buffer A for 50 min at 1.0 ml/min (V_e LHRH = 26 min).

The table summarizes some aspects of the isolation and purification of ^{3}H-labeled LHRH from an *in vivo* biosynthetic experiment in which the hypothalamic POA was continuously infused with a total of 500 μCi [^{3}H]leucine and [^{3}H]proline over an 8-hr period. Incorporation of ^{3}H into TCA-precipitable tissue protein was greater than 80% of the total ^{3}H present in the tissue at the time of killing, with a protein specific radioactivity at the infusion site of 4×10^7 dpm of ^{3}H per milligram of TCA-

ISOLATION AND PURIFICATION OF ^{3}H-LABELED LUTEINIZING HORMONE RELEASING HORMONE[a]

	Hypothalamic region examined		
	Preoptic area	Medial basal hypothalamus	Median eminence
TCA-precipitable protein (μg)	310 ± 22 μg	269 ± 11 μg	33 ± 3 μg
Total tissue ^{3}H (dpm)	1.7 ± 0.6 × 10^7 dpm	1.1 ± 0.2 × 10^7 dpm	1.7 ± 0.4 × 10^5 dpm
TCA-precipitable protein (dpm)	1.3 ± 0.4 × 10^7 dpm	9.0 ± 2.0 × 10^6 dpm	1.4 ± 0.3 × 10^5 dpm
Protein specific activity	4.2 ± 1.4 × 10^7 dpm/mg	3.4 ± 0.6 × 10^7 dpm/mg	4.1 ± 0.8 × 10^6 dpm/mg
LHRH specific activity			
HPLC 3	2290 dpm/30 μg	1970 dpm/30 μg	270 dpm/30 μg
HPLC 4	1720 dpm/30 μg	1960 dpm/30 μg	270 dpm/30 μg

[a] From discrete hypothalamic regions of individual female rats after continuous infusion of [^{3}H]leucine and [^{3}H]proline into the ventromedial preoptic area. The values presented are the $\bar{X}$ ± SEM ($n = 7$). The protein specific activity is expressed as ^{3}H disintegrations per minute (dpm) per milligram of 10% TCA-precipitable protein. The HPLC systems used are as follows: HPLC 3: chromatography on a C_{18} column using a 1 to 30% exponential CH_3CN gradient (program No. 3 on Waters 660 solvent programmer) in 0.2 M sodium phosphate, pH 2.5, for 50 min at 1.0 ml/min; HPLC 4: chromatography on a C_8 column using an exponential CH_3CN gradient (1 to 30%) in 0.2 M sodium phosphate, pH 2.5, for 50 min at 1.0 ml/min (V_e LHRH = 26 min).

precipitable protein. ^{3}H-labeled LHRH was purified from the discrete hypothalamic regions of individual rats by sequential HPLC as described above. After three HPLC steps, the radiochemical purity of ^{3}H-labeled LHRH appears to be 100% for medial basal hypothalamic and ME samples, and greater than 75% for POA samples, when analyzed by HPLC in system 4.

Biosynthesis of Luteinizing Hormone Releasing Hormone in Primary Cell Cultures of Perinatal Rat Hypothalamus

Preparation and Growth of Cell Cultures

Sprague–Dawley rats of either sex and of ages from 1 to 15 postnatal days are used. Animals are decapitated adjacent to the cell culture room, the whole brain is removed (care being taken to leave the median eminence intact by releasing the dura covering the hypophysis), and placed in a sterile petri dish, which is taken to the operator who is seated at a laminar-flow hood in the cell culture room. All instruments, liquids, glass, and plasticware used are sterile. Whole hypothalami are dissected from the brains using sterile instruments (iridectomy scissors and a straight-edge razor blade fragment held in a mosquito hemostat) inside the laminar flow hood. The whole hypothalami are placed in sterile Spinner minimum essential medium (SMEM)–1% bovine serum albumin, pH 7.2 at 0° until 15 hypothalami have been collected. The medium covering the combined hypothalami is decanted, and the tissues are transferred with forceps to a 15 × 150-mm screw-cap polyethylene round-bottom tube in SMEM containing Difco trypsin, 0.1% (w/v) at 1 ml per hypothalamic equivalent.

Incubation is carried out at 37° for 20 min with rotation provided by an Ames aliquot mixer. After trypsin treatment, the tissue fragments are subjected to mechanical dispersion with Pasteur pipettes having decreasing bore diameters of 3, 2, and 1 mm i.d. Dispersion is carried out very gently, foaming being an indication of excessive force. Fragments that resist dispersion are collected by gravity and resuspended twice more to increase the harvest of cells. Residual debris are discarded. The resulting cell suspensions are centrifuged at 1500 rpm for 5 min at 22° in a Dynac II swinging-bucket rotor. The cell pellets are resuspended in Dulbecco's modified Eagle's medium (DMEM) containing 10% fetal calf serum (FCS), and 1% antibiotic–antimycotic solution (GIBCO), by gentle dispersion with a 3-mm and then a 2-mm i.d. Pasteur pipette. This procedure yields 1 to 2×10^8 cells per hypothalamus. The suspension is brought to a final dilution of two hypothalamic equivalents per milliliter, and 0.5 ml (one hypothalamic equivalent) is added to 2 ml of DMEM–10% FCS in a

60-mm Falcon plastic tissue culture dish. Before plating, the dishes are treated by adding 2 ml of poly(D-lysine), 10 μg/ml in sterile water, for 1 hr at 22°. The polylysine is aspirated away and the dishes are washed twice with water and once with culture medium before use.

The cultures are incubated at 37° in a humidified tissue culture incubator under 5% CO_2. After 24 hr of culture, cytosine 1-β-*arabino*-furanoside is added to a final concentration of 1×10^{-5} M (25 μl of a 10^{-3} M solution in DMEM per dish), and the cultures are exposed to the antimitotic agent for 48 hr to inhibit nonneuronal cell growth. After this time, the medium is removed and replaced with a defined serum-free medium,[9] modified by the addition of sodium selenite to 5×10^{-9} M. It should be noted that this defined medium contains progesterone, a steroid hormone involved in the regulation of LHRH expression. We have found that inclusion of this steroid is necessary for cell survival in defined medium. Thus, studies on steroid effects on LHRH synthesis in culture take place in the presence of constitutive levels of progesterone. However, the level of progesterone (1×10^{-9} M) is of the same order of magnitude as that achieved by attempting to rid fetal calf serum of progesterone by charcoal treatment. The cultures are grown in defined medium for an additional 72 hr (6-day cultures), after which they are used for LHRH biosynthesis studies. Cultures prepared in this way from 15-day-old females can be expected to contain from 5 to 15% of the normal hypothalamic content of LHRH with no LHRH detectable in the medium during the period in defined medium. Intact LHRH-containing neurons can be seen immunocytochemically at day 6, although they are few in number.

Exposure of Cultures to Radioactive Amino Acid Precursors

Tritiated L-proline and L-tyrosine of specific activities greater than 100 Ci/mmol (New England Nuclear or Amersham) are used. Enough isotope is lyophilized in a sterile conical-bottom polypropylene centrifuge tube to give a final concentration of each isotope of 75 μCi per milliliter of medium for 20 cultures, each covered by 1.5 ml of defined medium. The labeling of 20 hypothalamic equivalents, and as individual hypothalamic equivalents in 60-mm dishes, rather than in a single larger dish, are both necessary in order to observe adequate labeling of LHRH in primary cell culture. The labeling medium is equilibrated in the CO_2 incubator for 15 min prior to use. The cultures are removed from the incubator as an entire shelf, and the medium covering them is aspirated rapidly. Equilibrated labeling medium is then added, and the cultures are exposed to isotope for 8–20 hr. If experiments are to be conducted to test the effects of modula-

---

[9] J. E. Bottenstein and G. H. Sato, *Proc. Natl. Acad. Sci. U.S.A.* **76**, 514 (1979).

tory agents, such as sex steroids, these should be added to the cultures at the time of introduction of defined medium (i.e., after day 3). To protect against proteolytic breakdown of LHRH, bacitracin is added to the medium at a concentration of 0.1 mg/ml and pepstatin at 5 µg/ml.

Isolation of ^{3}H-Labeled LHRH: Treatment of Medium

The presence of ^{3}H-labeled LHRH can be expected in the culture medium upon treatment with estrogens, and possibly other regulatory agents. The media from 20 culture dishes (30 ml) are pooled in a 50-ml polypropylene centrifuge tube, cooled in ice, and acidified with 1 ml of 6 N HCl. Next, 30 µg of synthetic LHRH in 0.1 ml of water (freshly prepared) is added. After 1 hr at 0°, insoluble material is removed by centrifugation at 16,000 rpm for 20 min (SS-34 rotor in a Sorvall RC-2B). The supernatant is lyophilized, the residue is redissolved in 0.1% TFA, and the sample is subjected to sequential purification of LHRH by HPLC as described above for *in vivo* LHRH biosynthesis.

Isolation of ^{3}H-Labeled LHRH: Treatment of Cells

Immediately upon removal of the labeling medium, 750 µl of ice-cold 0.2 N HCl is added to each culture dish. The dish contents are rinsed into a 50-ml polypropylene centrifuge tube with a polypropylene "rubber policeman." Each dish is then rinsed with a further 250 µl of the acid; synthetic LHRH (30 µg in 0.1 ml of water) is added, and the suspension is then subjected to sonication, using a Branson sonicator at 22° for 20 min, and stored overnight at 4°. Insoluble material is removed by centrifugation at 16,000 rpm for 20 min, as above. The supernatant is lyophilized, the residue is redissolved in 0.1% TFA, and the sample is subjected to sequential purification of LHRH by HPLC, as described above for *in vivo* studies.

Acknowledgments

This work was supported by Grants BNS 7684506 from the National Science Foundation and a Research Career Development Award (AM 00751) from the National Institutes of Health.

[35] Methods for Investigating Peptide Precursors in the Hypothalamus

By DAVID K. SUNDBERG, MARIANA MORRIS, and KENNETH A. GRUBER

I. Predicting Peptide Synthetic Rates

Today, increasing evidence suggests that many of the hypothalamic and neurohypophysial peptides are synthesized ribosomally as much larger precursors or prohormones. The newly synthesized prohormones are packaged in vesicles within the neuronal bodies and transported axonally to the median eminence or posterior pituitary, where they are secreted. During this process these precursors are thought to be enzymically cleaved to yield the native neuropeptides as well as the precursor sequence. Several chapters in previous volumes of this series have dealt with methods for studying the isolation and synthetic rates of hormone precursors both in the endocrine pancreas[1] and parathyroid gland.[2]

Useful information regarding the rate and mechanism of peptide biosynthesis can be mathematically estimated by studying the dynamics of peptide content, distribution, and metabolism. For example, the static secretion rate of a neuropeptide or hormone can be derived using the equation

$$SR = [P] \cdot \ln 2 \cdot V_d / t_{1/2}$$

where SR = secretion rate (pg/min); $[P]$ = plasma hormone concentration (pg/ml); V_d = volume of distribution (ml); $t_{1/2}$ = plasma half-life (min); $\ln 2$ = 0.693. While this derivation does not take into account either pulsatile release kinetics or peptide secretion from more than one locus (i.e., somatostatin), it does provide an insight into the rate and mechanism of neuropeptide biosynthesis.

In a similar manner the tissue "turnover" of a neuropeptide (defined here as the time required to replace the entire neuronal pool) can be calculated using the formula:

$$\text{turnover} = TO = [T]/SR$$

where $[T]$ is the hypothalamic or neurohypophysial peptide concentration (ng/tissue).

---

[1] H. S. Tager, A. H. Rubenstein, and D. F. Steiner, this series, Vol. 37, p. 326.
[2] J. F. Habener and J. T. Potts, Jr., this series, Vol. 37, p. 345.

TABLE I
CALCULATED SECRETION RATES AND HYPOTHALAMIC
TURNOVER OF NEUROPEPTIDES

Peptide	[P] (pg/ml)	$t_{1/2}$ (min)	[T] (ng/tissue)	SR (ng/hr)	TO (hr)
AVP	2.2	2.9[a]	400[b]	1.6	250
OXY	9.3	3.4[a]	360[b]	5.7	63
LHRH	<5	2.5[c]	10.2[d]	4.2	2.4

[a] From D. Gazis.[3]
[b] Nanograms per neurohypophysis.
[c] From Heber and Odell.[4]
[d] Nanograms per medial basal hypothalamus.

The calculated values for secretion rates and tissue turnover for neurohypophysial vasopressin (AVP) and oxytocin (OXY) and hypothalamic luteinizing-hormone releasing hormone (LHRH) are depicted in Table I. In these calculations the volume of distribution (V_d) is assumed to be the extracellular fluid volume (20% of body weight), since the peptide receptors are located in extravascular spaces. The values for the $t_{1/2}$ were taken from the literature,[3,4] and other parameters are from our previous work[5] using the male Sprague–Dawley rat (250 g). However, the distribution volume of LHRH may be much higher, since peripheral membranes bind this peptide with a high affinity.[4] This would, in fact, underestimate the calculated synthetic rate.

By comparing these calculated turnover rates (Table I), one would predict that the synthesis and secretion of OXY is four times greater than that of AVP, and that neurohypophysial OXY would be completely replaced every 2.5 days. The small apparent storage pool of LHRH, however, would necessitate a total neuronal turnover every 2.4 hr. This might suggest that the neuropeptide prohormones may, in fact, represent a very important larger storage pool, which under appropriate physiological stimuli can be made available to meet the peripheral endocrine demands of an organism.

In the case of the neurohypophysial hormone precursors, a recent article shows some of the developments in this area and suggests some provocative mechanisms regarding how these events might occur.[5] In this

[3] D. Gazis, *Proc. Soc. Exp. Biol. Med.* **158**, 663 (1978).
[4] D. Heber and W. D. Odell, *Proc. Soc. Exp. Biol. Med.* **158**, 643 (1978).
[5] M. J. Brownstein, J. T. Russell, and H. Gainer, in "Frontiers in Neuroendocrinology" (W. F. Ganong and L. Martini, eds.), Vol. 7, p. 31. Raven, New York, 1982.

TABLE II
Hypothalamic LHRH Content before and after Incubation
as a Function of Sex and Age[a,b]

Sex	Age (days)	LHRH content (ng) Preincubation	LHRH content (ng) Postincubation	Δ (ng)
Male	19	1.27 ± 0.35 (4)	2.11 ± 0.42 (4)	0.84
	21	1.41 ± 0.04 (4)	2.06 ± 0.24* (4)	0.65
	58	6.28 ± 0.73 (5)	10.60 ± 1.09* (5)	4.32
Female	19	0.87 ± 0.06 (4)	1.53 ± 0.30* (4)	0.66
	27	2.92 ± 0.58 (4)	4.26 ± 0.64** (4)	1.31

[a] Sundberg and Knigge,[6] reprinted with permission of Elsevier Biomedical Press B.V. (copyright 1978).
[b] Values are means ± SEM. Numbers in parentheses show number of samples. Δ = difference between pre- and postincubation content. * $p < 0.05$ and ** $p < 0.01$: significance vs preincubation content.

chapter we will outline the methodology used in our laboratory to investigate the synthetic rates of hypothalamic peptides and prohormones.

II. Tissue Content of Neuropeptides after Incubation *in Vitro*

If appreciable concentrations of peptide precursors were present within the hypothalamus, one might expect under appropriate circumstances to observe changes in the absolute content of the neuropeptide under investigation. This assumes that the rate of processing of the precursor would be different from the rate of release of the peptide, such that "steady state" levels are altered.

Indeed, earlier evidence demonstrated that this is true for LHRH.[6] For these studies, the medial basal hypothalami are dissected and either incubated or immediately extracted as described in this volume.[7] The hypothalamic methanolic extracts are aliquoted and dried under a stream of air; the LHRH content is measured by radioimmunoassay. The postincubation content of this peptide in hypothalami from both male and prepubertal female rats of different ages increases by a mean of 60.3 ± 6.3% after 2 hr of incubation (Table II). Additional information is obtained by determining the time and temperature dependence of this increase in content. In the case of LHRH, the tissue accumulation continues for up to 3

[6] D. K. Sundberg and K. M. Knigge, *Brain Res.* **139**, 89 (1978).
[7] D. K. Sundberg, B. A. Bennett, and M. Morris, this volume [33].

TABLE III
CHANGES IN TISSUE CONTENT OF HYPOTHALAMIC PEPTIDES
DURING A 2-HR INCUBATION[a]

Peptide	Preincubation content (ng)	Postincubation content (ng)	Δ Change[b] (ng)	%[c]
SRIF	12.6 ± 0.4	31.0 ± 1.29	18.4	146
AVP	15.7 ± 2.7	26.1 ± 7.5	10.4	66
LRF	6.49 ± 1.25	11.82 ± 0.74	5.33	82
OXY	47.3 ± 7.2	102.4 ± 10	55.1	116

[a] D. K. Sundberg and Mariana Morris, unpublished data.
[b] Absolute change in hypothalamic content.
[c] Percentage increase in preincubation content.

hr, but does not occur when the hypothalami are incubated at 0° (ice).[6] To ensure that this increase is not due to *de novo* peptide biosynthesis, protein synthesis inhibitors such as cycloheximide or actinomycin D are added to the media. The increase in LHRH tissue content is still observed, suggesting a postribosomal step.[8]

With regard to the other hypothalamic neuropeptides, similar experiments show that the postincubation tissue content of hypothalamic vasopressin, oxytocin, and somatostatin (SRIF) all increase during acute incubation (Table III). These types of experiments offer presumptive evidence that the hypothalamic peptides are stored in a larger precursor pool, which continues to be processed during an *in vitro* incubation.

III. Central Delivery of Tracer Amino Acids

In order to investigate the incorporation of labeled amino acids into hypothalamic prohormones or native peptides, central delivery is, by far, the preferred route of administration. Peripheral injection leads to a massive dilution of the labeled amino acid resulting in low specific activity due to dilution in blood and uptake into organs. Perhaps more important, the "blood–brain barrier" greatly restricts the active entry of amino acids into the brain.

Choice and Preparation of Tracer Amino Acids

Several considerations should be kept in mind when choosing an amino acid for studying biosynthesis of brain peptides and prohormones.

[8] D. K. Sundberg, K. A. Gruber, and M. Morris, *Adv. Pharmacol. Ther. Proc. Int. Congr. Pharmacol. 7th 1978* (1979).

A number of amino acids such as glycine and glutamate are candidates for central neurotransmitters and might induce pharmacological effects. An ideal amino acid should possess high specific activity, have no pharmacological effects of its own, and be found in relatively low concentrations in the brain to prevent dilution. For example, glutamate, aspartate, and taurine are found to be 10–100 times greater in concentration in the brain than in plasma. On the other hand, proline, lysine, leucine, isoleucine, valine, tryptophan, and arginine are lower in concentration in the brain than in blood.[9]

If one wants to compare the synthetic rate of two or more peptides, precursor amino acids that are common and in the same proportion should be chosen. For example, LHRH, AVP, and OXY have one tyrosine and proline residue each, so a mixture of these would be useful for comparing the synthetic rates of these peptides.

In the case of tritiated amino acids, high specific activity L-[ring 2,3,4,5,6-^{3}H]phenylalanine (103 Ci/mmol), L-[3,4,5-^{3}H]leucine (158 Ci/mmol), and L-[2,3,4,5-^{3}H]proline (102 Ci/mmol) (New England Nuclear) are lyophilized and resuspended in sterile saline (0.9%). These precursors are diluted such that 40–120 μCi are contained in a 4-μl volume.

Sulfur-containing amino acids (methionine and cysteine) are also useful because of the high specific activity obtained by labeling with ^{35}S. In the early studies only L-[^{35}S]cystine was available, so it had to be previously reduced to L-cysteine.[10] Today [^{35}S]cysteine with a specific activity greater than 500 Ci/mmol is available for use.[11,12] This label is also lyophilized and taken up in a small volume (4 μl) of 0.9% NaCl containing 10 mM dithiothreitol. Dithiothreitol is necessary to prevent oxidation of the sulfur-containing amino acid. It should be kept in mind that [^{35}S]cysteine is quite unstable and breaks down at a rate of 5% per week (manufacturer). Coupled with a half-life of 87 days, it should be used and processed soon after arrival.

Ventricular Cannulation and Injection of Labeled Precursors

For the third ventricular injection of labeled amino acids, a 23-gauge stainless steel cannula is chronically implanted about 1 week prior to an experiment. For this procedure the rat is anesthetized with pentobarbital

---

[9] A. V. Paladin, Y. V. Belik, and N. M. Polyakova, in "Studies in Soviet Science, Protein Metabolism of the Brain" (A. Lajthe, ed.). Consultants Bureau, New York, 1977.
[10] H. Sachs, *J. Neurochem.* **10**, 299 (1963).
[11] J. T. Russell, M. J. Brownstein, and H. Gainer, *Brain Res.* **205**, 299 (1981).
[12] K. A. Gruber, D. K. Sundberg, and M. Morris, "Advances in Physiological Sciences" (E. Stark, G. B. Makara, Z. Acs, and E. Endroczi, eds.), Vol. 13, p. 111. Akademiai Kiado, Budapest, 1981.

(50 mg/kp, i.p.) and placed in a Kopf stereotaxic instrument. A small hole is drilled in the cranium, and the cannula is lowered 1.8 mm posterior to the bregma, 0.0 mm lateral and 9.1 mm ventral (deep) into the brain. The cannula is secured in place by dental cement and intracranial screws. To prevent infection and leakage of cerebrospinal fluid, a metal stylet the same length as the cannula is inserted.

When the labeled precursor is injected, a 10-μl (Glenco) gastight syringe is coupled to a 27-gauge injection stylet (the same length as the cannula) by polyethylene (PE-10) tubing. The syringe is rinsed and filled with saline, a small air bubble to prevent mixture (<1 μl), and then the previously prepared ^{3}H-labeled amino acid. The animals are restrained in a towel while the solution is injected over a 5-min period. After another 5 min the syringe is removed, the stylet is replaced, and the animal is returned to its home cage for the correct experimental period. This procedure ensures complete delivery and minimal or no back leakage of the labeled precursor.

IV. Microdissection and Extraction of Hypothalamic Nuclei

After a specified interval of time, the rats are sacrificed by decapitation, and the brains and posterior pituitaries are removed and frozen on Dry Ice. Using a dissection technique as described below, the specific hypothalamic nuclei are removed. Trunk blood is also collected for determination of radioactivity. This is useful in the evaluation of the efficacy of the intraventricular injection.

Since most high-performance liquid chromatography (HPLC) systems are analytical in nature and because the entire hypothalamus would contain a large amount of unincorporated labeled amino acid, a microdissection technique allows one to study discrete areas in which the neuropeptides are synthesized.

A more in-depth coverage of this technique, originally developed by Palkovits, is found in this volume.[13] After the brain has been frozen, serial 300-μm sections are cut in the frontal plane in a cryostat at $-15°$. Each section is removed and placed on a frozen glass slide. Under a dissecting scope, the specific hypothalamic nuclei can be visualized and are removed with a chilled 300-μm stainless steel punch. From each section the nuclei under investigation are dissected and placed in 500 μl of 0.2 N acetic acid. Thus, one can specifically examine the paraventricular, supraoptic, arcuate/median eminence, and posterior pituitary content and biosynthesis of hypothalamic hormones.

[13] M. Palkovits, this volume [23].

The pooled punches are homogenized in the 0.2 N acetic acid, and a small aliquot is removed (20 μl) for total tissue radioactivity. Although the intraventricular delivery of amino acid provides a very reproducible labeling of nuclei, their anatomical location and the cell body population dictates the amount of label present. For example, 6 hr after administration of label, 29,130 cpm of tritiated amino acid was found in the paraventricular nucleus, which is located close to the ventricular surface. On the other hand, in the homogenate of the supraoptic nucleus only 15,560 cpm are found even though they contain about the same amount of protein. In brain areas that are primarily axonal in nature, large amounts of label are usually found. For example, in the median eminence, which is also close to the ventricle, 882,260 cpm were detected.

After centrifugation at 8000 g for 5 min, the sample is subjected to an initial purification on Sep-Pak C_{18} cartridges (Waters Associates, Inc.). The small columns are first washed with 10 ml of distilled H_2O. The sample (0.5 ml) is then placed on the column, which is washed with 10 ml of distilled H_2O and 5 ml of 10% methanol. This removes most of the unincorporated amino acids. The newly synthesized peptides can then be eluted with 5 ml of 98% methanol that is lyophilized and resuspended for HPLC (Sections V and VI). Measurement of recovery from Sep-Pak columns was found to be almost 100% using ^{125}I-labeled hormones and between 70 and 80% for synthetic peptide standards.

V. Molecular Weight Size-Exclusion Chromatograph for Neuropeptide Precursors

Molecular size exclusion is an ideal first step in the chromatographic isolation of potential hormone precursor proteins since it requires little knowledge of the protein's chemical properties. Since size-exclusion columns can be eluted in a wide variety of polar or nonpolar solvents, it is possible to chromatograph plasma or tissue samples with a solvent system similar to the ones used to extract them. In addition, most proteins can be chromatographed under aqueous conditions that preserve their tertiary structure.

Equipment and Methodology

A general approach for the isolation of precursor proteins is as follows: a Synchropak GPC-100 column (25 × 0.46 cm, obtained from Synchrom, Inc., Linden, IN) is eluted with 0.1 M formic acid (pH 2.3) at a flow rate of 50–100 μl/min. This solvent system has been shown to be capable of separating low molecular weight peptides from proteins in chromatographic systems.[14] The solvent reservoir is a pressurized (10–25 psi N_2) borosilicate glass bottle (Biolab, Inc., Dover, NJ). The column pump can be as simple as a Milton Roy Simplex minipump (Laboratory Data Control, Riviera Beach, FL) or as sophisticated as a Waters M-45 solvent delivery system custom-modified to permit a 50-μl/min flow rate. The low flow rates used are to ensure resolution of a large spectrum of molecular weight compounds. Previous work has shown that at flow rates of 0.5 ml/min and changing fraction collector tubes every minute resulted in a larger number of molecular weight fractions in each test tube.[15] Therefore, flow rates are set at 50–100 μl/min. Samples are applied to the column with a six-part injection valve (Valco Instruments, Houston, TX).

Although UV detection is an acceptable technique for monitoring chromatographic column effluents, it does not compare to a preparative fluorescamine detection system as regards sensitivity for primary amines. The system used is modeled after that first described by Böhlen et al.[16] Every movement of a fraction collector initiates the injection of a small aliquot of column effluent into a fluorescamine detection system. In the system, the sample is first mixed with 0.3 M borate buffer (pH 9.3) and then fluorescamine (20 mg% in spectranalyzer grade acetone). The resulting solution is then passed through a flow cell fluorometer with fluorescence plotted on a strip-chart recorder.

Chromatography of Standards and Tissue Extracts

The elution time of a series of standards is plotted in Fig. 1. There is a linear relation between molecular weights (M_r) of 5000 to approximately 700,000. Using standards, several other aspects of the Synchropak GPC-100 column that are applicable to biological extracts can be investigated.

For example, an important aspect of HPLC columns is the sample volume that can be applied to them. The absoute concentration of substances in biological samples is usually not as great a problem as the large applied volumes of solvent. With this column, injection of more than 200 μl of sample will cause the peak width of standards to widen significantly. Thus, if the injection of a large volume (300–400 μl) of sample is necessary, then when the substance of interest is eluted from the column it can be lyophilized, dissolved in 150–200 μl of column eluent, and rechromatographed to ascertain its precise molecular weight.

Another advantage to the use of a dilute formic acid eluent is that

---

[14] B. Y. B. Frankland, M. D. Hollenberg, D. B. Hope, and B. A. Schachter, *Br. J. Pharmacol. Chemother.* **26**, 502 (1966).
[15] K. A. Gruber, J. M. Whitaker, and M. Morris, *Anal. Biochem.* **97**, 176 (1979).
[16] P. Böhlen, S. Stein, J. Stone, and S. Udenfriend, *Anal. Biochem.* **67**, 438 (1975).

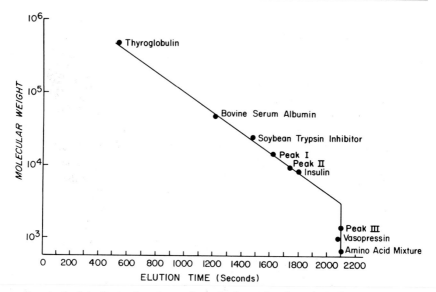

FIG. 1. A plot of retention times of standards on Synchropak GPC-100 against their molecular weight. The retention times of the radioactivity incorporating peaks I–III from the pituitary are included. However, a separate chromatographic run using a small sample aliquot (8%) and a slower flow rate (50 μl/min) was used to determine their exact molecular weight. From Gruber and Morris,[18] reproduced with permission of Marcel Dekker, Inc. (copyright 1980).

relatively large areas of the chromatogram can be collected and immediately chromatographed on a cation-exchange resin (Partisil SCX, Whatman, Inc., Clifton, NJ). This column can be used for the chromatography of proteins.[17] The initial solvent for the Partisil SCX column (5×10^{-3} M pyridine acetate, pH 3) is compatible with the formic acid eluent of the Synchropak GPC-100 column. This approach allows two HPLC separations to be run consecutively within a 2–3-hr period, permitting a large degree of purification to be accomplished in a short period of time.

A demonstration of the ability of this HPLC column and procedure to provide rapid information can be seen in the following series of experiments. The peptides and proteins of the hypothalamoneurohypophysial tract were labeled by infusion of 2 mCi of [^{35}S]cysteine (New England Nuclear, Boston, MA) into the third ventricle of beagle dogs as previously described.[18] After 6 hr a dog was killed by an injection of KCl, the calvaria

---

[17] A. N. Radhakrishnan, S. Stein, C. Licht, K. A. Gruber, and S. Udenfriend, *J. Chromatogr.* **132**, 552 (1977).
[18] K. A. Gruber and M. Morris, *Endocr. Res. Commun.* **7**, 45 (1980).

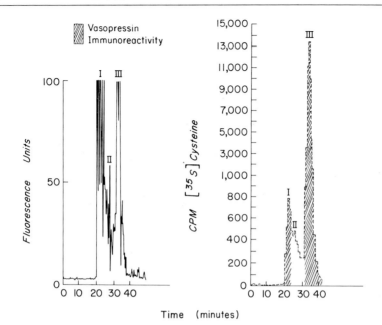

FIG. 2. Chromatogram of 20% of a dog hypothalamus on Synchropak GPC-100. There are high and low molecular weight areas of fluorescence (left) and radioactivity (right), similar to the pituitary, and there is also high molecular weight immunoreactivity. The lack of resolution in peak I is due to the inner-filter effect of fluorescence. From Gruber and Morris,[18] reproduced with permission of Marcel Dekker, Inc. (copyright 1980).

was opened, and the pituitary and hypothalamus were removed and frozen on dry ice.

The tissues were homogenized in 0.2 M acetic acid–0.2 N hydrochloric acid, centrifuged at 10,000 g; the supernatant was decanted, and its pH was adjusted to 2.3. Aliquots (approximately 200 μl) of each sample were applied to the Synchropak GPC-100 column. The chromatogram of a hypothalamic extract is depicted in Fig. 2. The left panel depicts the fluorescamine fluorescence, and the right panel depicts the radioactivity elution profile and the distribution of vasopressin immunoreactivity. Proteins appeared in an area of the chromatogram with less than 30 min retention time. Peaks I and II, respectively, refer to molecular weight areas of 20,000 and 10,000. These are the molecular weights of a previously reported neurophysin precursor[19] and neurophysins. Peak III represents compounds of less than M_r 5000. Incorporation of radioactivity appeared in both protein peaks, as well as the low molecular weight area.

[19] H. Gainer, Y. Sarne, and M. Brownstein, *Science* **195**, 1354 (1977).

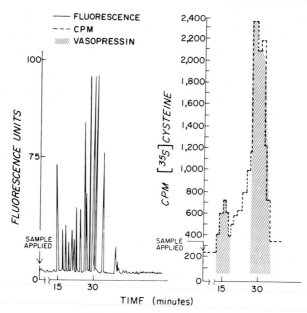

FIG. 3. Trypsin digest of the 20,000 M_r peak from a dog hypothalamus. The discontinuous appearance of the chromatogram is due to the preparative fluorescamine monitoring system. Note shift of fluorescamine-positive material and radioactivity to the low molecular weight area of the chromatogram. Vasopressin immunoreactivity also appears in the <5000 M_r peak. From Gruber and Morris,[18] reproduced with permission of Marcel Dekker, Inc. (copyright 1980).

Vasopressin immunoreactivity was detected in the M_r 20,000 and less than 5000 areas of the chromatogram.

This type of result had two possible explanations. The first was that low molecular weight compounds such as vasopressin are binding to proteins and are not being resolved during chromatography. Several lines of evidence speak against this. The formic acid eluent used has been shown to separate vasopressin and proteins in other molecular sieve chromatographic systems.[14] This solvent has been investigated in the chromatography of a mixture of standard vasopressin and neurophysin (unpublished data) as well as other peptide–protein mixtures,[15] and complete separation was achieved. A chemical extraction technique is useful in the demonstration that low molecular weight substances are not being "carried" on chromatographed proteins. In the case at hand, the chromatographed M_r 20,000 peak can be precipitated with 10% trichloroacetic acid. This technique has been shown to be an effective means of separating vasopressin from carrier proteins.[20] After the extraction, no immunoreactive vasopressin was detected in the supernatant. A final piece of evidence from our laboratory against vasopressin being carried on proteins was that no immunoreactivity was detected in the area of the chromatogram in which neurophysins, known carriers of vasopressin, elute. Independent confirmation of high molecular weight vasopressin immunoreactivity has been provided by Nicolas et al.[21]

A rapid means of enzymically digesting a precursor, especially if the disappearance of substrate and appearance of product can be monitored by radioimmunoassay (RIA), is to use immobilized trypsin on CM-cellulose beads (Miles Laboratories). This approach will allow for immediate rechromatography of the enzymic digest after the trypsin is removed by centrifugation. The use of immobilized enzymes also ensures that a limited digestion occurs, since the reaction can be easily ended without the addition of cold acid. Minimizing the extent of enzymic action is important, since the products released may also be sensitive to the enzyme that cleaved them from their precursor.

When the putative vasopressin precursor was digested and rechromatographed, the chromatogram revealed a shift of fluorescamine-positive material, radioactivity, and vasopressin immunoreactivity to the low molecular weight area of the chromatogram (Fig. 3). Another approach to monitoring the release of a hormone from its precursor would be by an appropriate bioassay. However, it is important to establish by a chromatographic method the molecular weight of the assayed substance.

VI. Reverse-Phase Isolation of Newly Synthesized Peptides

Isocratic reverse-phase liquid chromatography has been found to be a useful means for studying the biosynthesis of compounds with a molecular weight range from 200 to 2000.[7] In the case of hypothalamic peptides separation is dependent on their size and hydrophilic nature.

This chromatographic setup includes a high-pressure pump, injection valve, and a C_{18} μBondapak column (Waters Associates, Inc.). The column eluent is monitored by a UV spectrophotometer (Gilson) at 280 nM. Column fractions (1 ml) are collected for RIA of OXY[22] and AVP[23]

---

[20] R. Archer and P. Fromageot, "The Neurohypophysis" (H. Heller, ed.), p. 39. Butterworth, London, 1957.
[21] P. Nicolas, M. Camier, M. Lauber, M. F. O. Masse, J. Moltring, and P. Cohen, *Proc. Natl. Acad. Sci. U.S.A.* **77**, 2587 (1980).
[22] M. Morris, S. W. Stevens, and M. R. Adams, *Biol. Reprod.* **23**, 702 (1980).
[23] M. Morris, *Hypertension* **4**, 161 (1982).

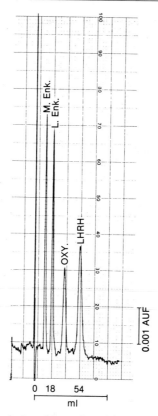

FIG. 4. Reverse-phase separation of 5 μg of synthetic methionine enkephalin (M. Enk.), leucine enkephalin (L. Enk.), oxytocin (OXY), and luteinizing hormone releasing hormone (LHRH). Peptides were monitored by UV absorbance at 280 nm and 0.01 AUF.

and for liquid scintillation counting of incorporated ^{3}H. The column is eluted with an isocratic 0.1 M monobasic sodium phosphate (NaH$_2$PO$_4$) : 40% methanol mobile phase. The chromatographic system should initially be characterized using synthetic peptides. Five to ten micrograms of most tyrosine- or tryptophan-containing peptides can easily be detected at 280 nm and 0.01 AUF. Figure 4 demonstrates the resolution of methionine and leucine enkephalin, OXY, and LHRH in this system. Vasopressin migrates slightly after methionine enkephalin. An isocratic system is convenient in that one need not reequilibrate or recharacterize the system between runs.

After characterization with synthetic peptides, it is advisable to wash the column several times if the eluent is to be monitored by RIA. This

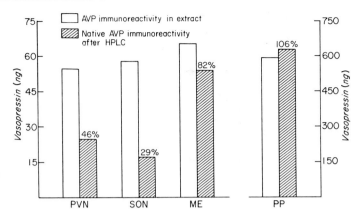

FIG. 5. Vasopressin immunoreactivity before and after HPLC separation. Tissue extracts of the rat paraventricular nucleus (PVN), supraoptic nucleus (SON), median eminence (ME), and posterior pituitary (PP) were prepared (pool of four animals). Vasopressin was measured by radioimmunoassay in the tissue extract and in the HPLC column effluent (vasopressin peak). % = (content post HPLC)/(content before separation). M. Morris and D. K. Sundberg, unpublished data.

chromatographic system is designed to resolve only the native peptides, whereas the larger precursors are highly retained by the column.

Some RIAs developed for the neuropeptide do detect the precursors as well.[12,21] For example, AVP immunological activity was determined in the paraventricular nucleus (PVN) and supraoptic nucleus (SON), median eminence (ME), and posterior pituitary (PP), microdissected as previously described. The remaining extract was chromatographed, and vasopressin immunoreactivity was measured in the eluent (Fig. 5). Whereas all the immunologically active AVP in the PP coeluted with vasopressin, immunoreactivity in the nuclei was retained by the column, indicating a larger molecular weight. In the ME, 82% of the immunoreactive AVP eluted with the natural peptide. This is consistent with the mechanism described by Brownstein and others that the precursor is slowly processed within the axon as it is transported to its site of release.[5]

When 120 μCi of a combination of ^{3}H-labeled proline, leucine, and phenylalanine (40 μCi each) is injected intraventricularly into male rats and the animals sacrificed 24 hrs later, significant amounts of radioactivity do coelute with vasopressin and oxytocin (Fig. 6). In this case each peptide contains one proline residue and one of the other amino acids (AVP–phenylalanine and OXY–leucine). In the other microdissected areas the labeled proteins were not found to coelute with the neurohypophysial

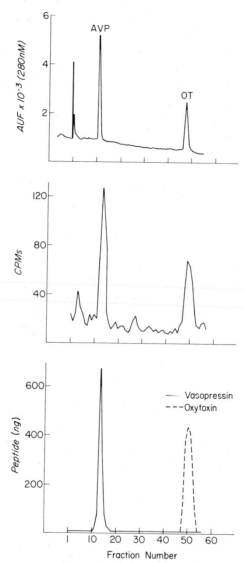

FIG. 6. The top panel shows the UV absorbance of 5 μg of synthetic vasopressin and oxytocin standards. The middle panel shows the radioactive peptides in the posterior pituitary after intraventricular injection of 120 μCi of ^{3}H-labeled proline, leucine, and phenylalanine 24 hr prior to the sacrifice. The lower panel shows the elution profile of immunoreactive vasopressin and oxytocin as measured by radioimmunoassay. These fractions are the same as those from the chromatograph shown in the middle panel. M. Morris and D. K. Sundberg, unpublished data.

peptides; again it is consistent with the hypothesis that these peptides are associated with a higher molecular weight precursor.

Of particular interest is the fact that when [^{3}H]proline (40 μCi) is administered, most incorporation is found in association with OXY (M. Morris and D. K. Sundberg, unpublished data). This ratio of incorporation is similar to that reported by Russell et al.[11] using [^{35}S]cysteine and demonstrates that peptide synthetic rates may be predicted by studying the dynamics of peripheral distribution and elimination (Section I).

VII. Summary

In conclusion, as in the aminergic nervous system,[7] the hypothalamic neuropeptides "turn over" at a rate sufficient to maintain a "steady state" concentration within the hypothalamus. However, unlike catecholamine-secreting neurons, the brain peptides are synthesized and secreted by a much more circuitous mechanism, involving ribosomal biosynthesis as larger prohormones, axonal transport to the site of secretion, and enzymic processing to the biologically active hormones. The methods described here, including *in vitro* incubation and high-performance size-exclusion and reverse-phase chromatography, have been found to be useful for examining and comparing the rate and mechanism of biosynthesis of several hypothalamic neuropeptides.

Acknowledgments

This work was supported in part by NIH Grants HD-10900 to D. K. S. and HL-22411 to M. M. as well as by a grant-in-aid from the American Heart Association with funds contributed in part by the North Carolina Affiliate, and NSF Grant BNS 77-25988 to K. A. G. This work was done during the tenure of an established investigatorship (M. M.) from the American Heart Association and with funds contributed in part by the North Carolina affiliate. The authors wish to acknowledge the scientific assistance of Ms. Suzanne Gallimore and the secretarial assistance of Ms. Stephanie Burgoyne.

[36] Measurement of the Degradation of Luteinizing Hormone Releasing Hormone by Hypothalamic Tissue

By JAMES E. KRAUSE and JEFFREY F. MCKELVY

This chapter describes the strategies and methodologies used to assess the enzymic degradation of luteinizing hormone releasing hormone (LHRH) by discrete regions of the rat hypothalamus. The presence in the central nervous system (CNS) of peptidase activities able to degrade various peptides of CNS origin has been known for some time.[1,2] At present, however, it is not certain whether these inactivation enzymic activities play a significant role in the determination of biological response or activity related to the storage, release, or action of the neuropeptide. At the hypothalamic level, for instance, intracellular peptidases may regulate the amount of peptide available for release from neuroendocrine nerve terminals in the hypothalamic median eminence (ME), for example, in the interaction between secretory granules and lysozomes during active neurosecretion. However, in order to assess the physiological relevance of such peptidase activity, it is important to examine enzymic activity in relation to the activity of LHRH secretory neurons. The present chapter describes our approach to the assessment of hypothalamic peptidase activity acting on LHRH and on its possible physiological relevance as related to the activity of the LHRH hypophysiotropic neuronal system that regulates gonadotropin secretion. The strategies described herein should be readily adaptable for the study of peptidase activities acting on a variety of neuroendocrine peptides in discrete hypothalamic regions, or, indeed, for the investigation of neuropeptidase activity at any site in the nervous system.

Physiological Modeling

The reasons for selecting the LHRH hypophysiotropic neuronal system are that the primary structure of the decapeptide is known as pGlu-His-Trp-Ser-Tyr-Gly-Leu-Arg-Pro-Gly-NH$_2$,[3,4] and that both the anatomy and physiology of this peptidergic projection is reasonably well understood. For example, immunohistochemical evidence indicates that the largest accumulation of LHRH neuronal perikarya projecting to the ME in the female rat is located in the ventromedial preoptic area (POA).[5] Also, it is known that LHRH is secreted from its terminal field into the portal vessels in the hypothalamic ME[6] and that the peptide elicits the release of both luteinizing hormone (LH) and follicle-stimulating hormone from the anterior pituitary.[7] Thus, LHRH peptidase activities may be assessed at different anatomical levels of this projection system: (1) at the

[1] E. C. Griffiths and J. A. Kelly, *Mol. Cell. Endocrinol.* **14**, 3 (1979).
[2] L. B. Hersh and J. F. McKelvy, *Brain Res.* **168**, 553 (1979).
[3] A. V. Schally, *Science* **202**, 18 (1978).
[4] R. Guilleman, *Science* **202**, 390 (1978).
[5] Y. Ibata, K. Watanabe, H. Kinoshita, S. Kubo, and Y. Samo, *Cell Tissue Res.* **198**, (1979).
[6] J. E. Levine and V. D. Ramirez, *Endocrinology* (*Baltimore*) **107**, 1782 (1980).
[7] A. V. Schally, T. W. Redding, H. Matsuo, and A. Arimura, *Endocrinology* (*Baltimore*) **90**, 1561 (1972).

level of LHRH neuronal cell bodies in the POA, (2) at the level of LHRH axons in the medial basal hypothalamus, (3) at the level of LHRH neuronal terminals in the hypothalamic ME, and (4) at the level of the main target organ of LHRH, the AP.

Adaptive changes in peptide hydrolase activities should correlate with physiological end points used to assess the biological effect(s) of a defined peptide. The closer the end point to be measured is to the site at which the peptide is being released, the better these correlations will be. If changes in peptidase activities correlate with functional changes elicited by the defined peptidergic system, they may represent enzymic activities of physiological relevance. More definitive conclusions should be obtained from *in vivo* studies in which functional changes in the peptidergic system are produced by experimental manipulation of the peptidase activity.

Peptidases degrading LHRH may provide a site of action at which feedback mechanisms could exert regulatory controls on LHRH and thus on LH secretion. Several models can and have been used to test the physiological relevance of possible components of the neuroendocrine system controlling gonadotropin secretion. These models relate to natural reproductive cycles,[8,9] to the tonic regulation of LH secretion[10] (negative steroid feedback), and to phasic regulation of LH secretion[11] (positive steroid feedback).

Hypothalamic Tissue Dissection and Preparation of Tissue Extracts

Discrete regions of the hypothalamus of the rat may be obtained by the use of a micropunch technique[12] whereby specific nuclei (discrete neuronal cell body-containing regions) are punched with a hollow needle of specified diameter from either frozen or fresh coronal brain sections. To obtain hypothalamic ME samples, the proximal stump of the stalk ME is held with a pair of fine forceps, while the tip of an iridectomy scissors is introduced into the third ventricle.[13] Two longitudinal cuts are then made in a rostral direction using as a reference point the lateral limits of the infundibular recess. A last frontal cut is made behind the optic chiasm.

To obtain the preoptic area (POA) sample, each brain is frozen on Dry

---

[8] S. R. Ojeda, J. E. Wheaton, H. E. Jameson, and S. M. McCann, *Endocrinology* (*Baltimore*) **98**, 630 (1976).

[9] L. G. Nequin, J. Alvarez, and N. B. Schwartz, *J. Steroid Biochem.* **6**, 1007 (1975).

[10] J. P. Advis, S. M. McCann, and A. Negro-Vilar, *Endocrinology* (*Baltimore*) **197**, 892 (1980).

[11] L. Caligaris, J. J. Astrada, and S. Taleisnik, *Endocrinology* (*Baltimore*) **89**, 331 (1971).

[12] M. Palkovits, *Brain Res.* **59**, 449 (1973).

[13] J. P. Advis, J. E. Krause, and J. F. McKelvy, *Endocrinology* (*Baltimore*) **110**, 1238 (1982).

Ice immediately after removal of the ME. Each frozen brain is then positioned with its dorsal aspect in contact with a flat surface and is allowed to thaw partially, and two coronal sections are made with a razor blade, one immediately rostral to the anterior border of the optic chiasm and the other at the level of the posterolateral aspect of the optic chiasm. The POA sample is punched out with a 200-μl capillary-type micropipette (Corning Glass Works, Corning, NY), using as a dorsolateral reference point at each side the anterior commissure converging toward the midline and as midventral reference point the dorsal aspect of the optic chiasm.

Samples for LHRH peptidase activity are homogenized in ice-cold 20 mM potassium phosphate, pH 7.4[13-15] Homogenization volumes are 30 μl for ME, 100 μl for POA, and 150 μl for AP. Homogenates are centrifuged for 5 min at 10,000 g in a desk top Brinkmann Eppendorf microcentrifuge (No. 5412, Brinkmann Instruments, Westbury, NY) at room temperature. The supernatant fraction is used as the source of peptidase activity and can be stored at $-80°$ until assay (see below). Thus, in these studies, a sample buffer without additions is used to display enzyme activity as influenced by the metabolic state of the tissue at the time of sampling, rather than by exogenous agents (thiol protectors, chelators, etc.).

Total LHRH Peptidase Assay

Total LHRH-degrading activity is assessed by incubating exogenous LHRH (270 μM) with the tissue fractions in capped polypropylene tubes (Eppendorf) in a final volume of 20 μl of 25 mM potassium phosphate, pH 7.2. Peptidase activity is generally determined at three enzyme concentrations (10, 5, and 2.5 μl of tissue extract). Under these conditions, total LHRH peptidase activity is linear with respect to tissue concentration when a 30–60-min incubation time is used. It is essential that comparisons of degradation activity between different physiological states be made on the basis of linearity of degradation with respect to tissue concentration. The reaction is initiated by the addition of LHRH. A preboiled (110° for 10 min) tissue sample is included for each tissue preparation. Reactions are halted by heating (100° for 6 min) in a Reacti-Therm heating apparatus (Pierce Chemical Co., Rockford, IL), followed by cooling on ice and brief centrifugation before storage at $-80°$ until HPLC analysis.

[14] J. P. Advis, J. E. Krause, and J. F. McKelvy, *Anal. Biochem.* **125**, 41 (1982).
[15] J. E. Krause, J. P. Advis, and J. F. McKelvy, *Biochem. Biophys. Res. Commun.* **108**, 1475 (1982).

High-Performance Liquid Chromatographic Fractionation of LHRH Peptidase Products

The HPLC fractionation of peptidase sample digests is performed using a Waters Associates liquid chromatograph equipped with a Model U6K injector, two Model 6000A pumps, and a Model 660 solvent programmer. Detection is carried out using a Waters Model 450 variable-wavelength absorbance detector at 210 nm, the isosbestic point of the peptide bond $\pi \rightarrow \pi^*$ transition. Peak areas are determined using a

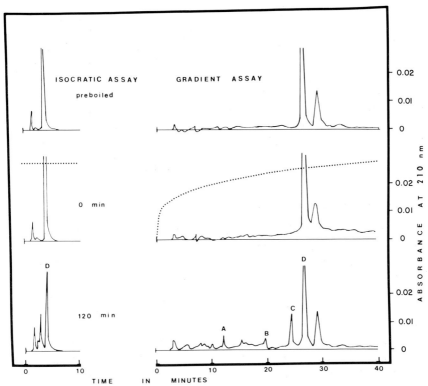

FIG. 1. Isocratic and gradient high-performance liquid chromatography (HPLC) assay of LHRH peptidase activity in rat median eminence. Tissue extracts of hypothalamic median eminence are incubated in the presence of LHRH, and the extent of total LHRH degradation is assessed by isocratic HPLC as described in the text. Gradient fractionation of the peptidase products can be performed to ascertain the specific site(s) of peptide bond cleavage. The following peaks were identified by amino acid analysis and by comigration with standard peptides: peak A, $LHRH_{6-10}$; peak B, $LHRH_{1-3}$; peak C, $LHRH_{1-5}$; peak D, LHRH; peak P, puromycin.

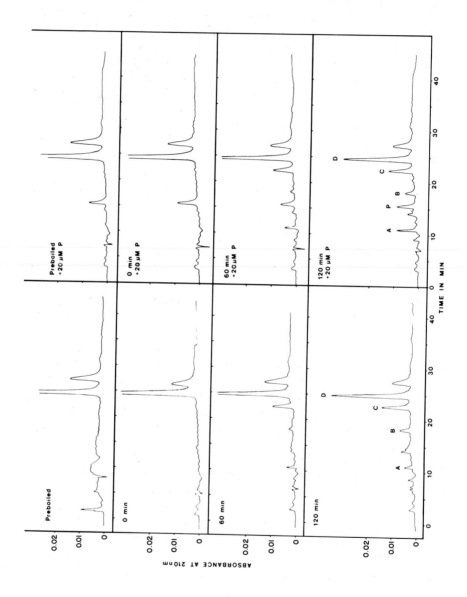

Hewlett-Packard 3380 integrator or by manual triangulation. Chromatography is performed with a 4 mm × 25 cm reverse-phase C_8 or C_{18} column with a 10-μm particle size (Brownlee Laboratories, Santa Clara, CA). Incubation samples for total LHRH peptidase activity are routinely processed in 0.7 mM triethylamine phosphate (TEAP) containing 42% acetonitrile using isocratic conditions at a flow rate of 3 ml/min.[13] A stock TEAP solution (25 mM) is added to distilled water and filtered *in vacuo* prior to use. LHRH is eluted under these conditions with a retention volume of approximately 15 ml. This relatively fast isocratic HPLC system provides an estimate of total LHRH-degrading activity, since LHRH is clearly separated from its peptidase products as shown in Fig. 1.[13,14] The products of the major peptidase activity in the total degradation assay are not clearly distinguished with the relatively rapid isocratic HPLC assay; however, they are clearly separated by exponential gradient HPLC elution analysis as described below.

Gradient elution HPLC is carried out on a C_8 column equilibrated in 0.1 M sodium phosphate–0.1 M phosphoric acid, pH 2.5, and eluted with a 1 to 30% exponential gradient of acetonitrile (Waters Solvent Programmer No. 3) at 1.0 ml/min for 50 min. Detection is again carried out at 210 nm. The elution times of LHRH and LHRH partial sequences are presented in the table, and they demonstrate the power of exponential gradient elution techniques for the separation of structurally related LHRH sequences.

Since the initial cleavage products of LHRH in hypothalamic tissue extracts are subject to further action by proteases, primarily by aminopeptidases, incubations of tissue extracts can be performed in the presence of aminopeptidase (and/or carboxypeptidase) inhibitors. In Fig. 2, we show a time course of LHRH degradation by hypothalamic ME tissue extracts in the presence and the absence of the aminopeptidase inhibitor puromycin[16] at a concentration of 20 μM. Note that the presence of puromycin in the incubation medium protects the LHRH peptidase product peak A, suggesting that this LHRH degradation product is susceptible to aminopeptidase activity present in tissue extracts.

Amino acid analysis of the LHRH peptidase products is used to establish unequivocally the site of peptidase cleavage of the decapeptide.[15] These results indicate that under these conditions of measurement an

---

[16] R. K. Barclay and M. A. Phillips, *Biochem. Biophys. Res. Commun.* **81,** 1119 (1978).

FIG. 2. Cleavage of LHRH by median eminence extract from prepubertal Sprague–Dawley rats in the presence and absence of 20 μM puromycin. LHRH digests and gradient elution HPLC was performed as described in the text.

EXPONENTIAL ACETONITRILE GRADIENT
HIGH-PERFORMANCE LIQUID CHROMATOGRAPHY
OF LHRH AND LHRH PARTIAL SEQUENCES

Peptide	Elution time (min)	$V_e(\text{rel})^a$
LHRH	24.3	—
LHRH_{1-6}	20.4	0.84
LHRH_{1-5}	22.1	0.91
LHRH_{1-3}	18.2	0.75
<Glu	7.1	0.29
LHRH_{2-10}	18.9	0.78
LHRH_{3-10}	22.3	0.92
LHRH_{4-10}	13.7	0.56
LHRH_{6-10}	10.7	0.44
LHRH_{4-5}	10.3	0.42

[a] The elution of each peptide is calculated relative to LHRH according to the following formula: $V_e(\text{rel}) = V_e(\text{peptide})/V_e(\text{LHRH})$.

endopeptidase cleaves the $\text{Tyr}^5\text{-Gly}^6$ bond of LHRH, and that subsequent cleavage of LHRH_{1-5} at either $\text{His}^2\text{-Trp}^3$ or $\text{Trp}^3\text{-Ser}^4$ occurs at a much lower rate. The experimental approach to neuropeptide degradation studies reported here has the desirable property of using the native peptide as a substrate, rather than synthetic peptide derivatives bearing chromophoric groups with unknown interactions with enzyme active sites. The strategy of selecting conditions for rapid isocratic assay of total degradation in concert with gradient analysis of fragmentation patterns is of general applicability. It allows for the assessment of the possible physiological regulation of overall degradation and the discerning of specific bond cleavages in such studies, and is also rapid and sensitive enough for use in enzyme purification.

Post Proline Cleaving Enzyme Assay

Post proline cleaving enzyme (PPCE) is a serine protease that catalyzes the hydrolysis of peptide bonds on the carboxyl side of L-prolyl residues in neuroendocrine peptides.[17] Although this enzymatic activity cleaves the $\text{Pro}^9\text{-Gly}^{10}$ bond of LHRH *in vitro,* its activity has been known to be invariant with respect to the secretory activity of the LHRH neu-

[17] H. Knisatschek and K. Bauer, *J. Biol. Chem.* **254,** 10936 (1979).

ron.[13-19] Thus, this enzymic activity may serve as a control for general metabolic changes accompanying steroid action on the hypothalamus. PPCE activity is determined by incubating the fluorogenic substrate Z-Gly-Pro-aminomethylcoumarin (Z-Gly-Pro-AMC) in polypropylene tubes (Eppendorf) in a final volume of 20 μl of 250 mM sodium–potassium phosphate, pH 7.8. This synthetic substrate can be synthesized from Gly-Pro-AMC (Peninsula) and carbobenzoxychloride using standard techniques. Tissue fractions for PPCE are generally assessed at three concentrations of tissue supernatant (10, 5, and 2.5 liters). The reaction is initiated by the addition of Z-Gly-Pro-AMC. A preboiled (110° for 10 min) tissue sample is included for each tissue preparation. Reactions are stopped by heating (110° for 6 min) in a Reacti-Therm heating apparatus, cooled on ice, and briefly centrifuged. Before fluorescence determinations, 1 ml of 250 mM sodium phosphate, pH 8.0, is added to each sample. Fluorescence measurements are made with an Aminco-Bowman spectrofluorometer. Wavelengths of 380 and 460 nm are used for excitation and emission, respectively.

Acknowledgments

Work performed in the authors' laboratory was supported by a grant from the National Science Foundation (BNS 7684506).

[18] J. E. Krause, J. P. Advis, and J. F. McKelvy, *Endocrinology* (*Baltimore*) **111**, 344 (1982).
[19] J. P. Advis, J. E. Krause, and J. F. McKelvy, *Endocrinology* (*Baltimore*) **112**, 1147 (1983).

[37] Measurement of β-Endorphin and Enkephalins in Biological Tissues and Fluids

By Jau-Shyong Hong, Kazuaki Yoshikawa, and R. Wayne Hendren

Three different protein precursors of various opioid peptides have been identified. Preproopiomelanocortin is the common precursor of corticotropin (ACTH), MSH, and β-lipotropin (β-LPH), from which β-endorphin is formed.[1,2] Preproenkephalin A contains four copies of

[1] S. Nakanishi, A. Inoue, T. Kita, M. Nakamura, A. C. Y. Chang, S. Cohen, and S. Numa, *Nature* (*London*) **278**, 423 (1979).
[2] A. C. Y. Chang, M. Cochet, and S. N. Cohen, *Proc. Natl. Acad. Sci. U.S.A.* **77**, 4890 (1980).

[Met^5]enkephalin and one copy each of [Leu^5]enkephalin, [Met^5]enkephalin-Arg^6-Phe^7 and [Met^5]enkephalin-Arg^6-Gly^7-Leu^8,[3-5] and preproenkephalin B is the common precursor for β-neoendorphin, dynorphin, and a copy of [Leu^5]enkephalin with a carboxyl extension.[6] Owing to its great sensitivity and specificity, radioimmunoassay (RIA) has been the major method used to determine the concentrations of various opioid peptides in biological tissues. However, because of the similarity in chemical structures among these three families of peptides (e.g., [Met^5]enkephalin is a fragment of β-endorphin, and [Leu^5]enkephalin is a fragment of dynorphin), antiserum raised against one peptide may cross-react with other peptides. Therefore, it is imperative to separate peptides before RIA is performed. Since enkephalins and β-endorphin are the most extensively studied peptides so far, these peptides will be used as prototypes to illustrate the general procedures for separation and RIA of opioid peptides.

For more detailed information concerning the technique of RIA, readers are referred to review articles.[7,8]

Production of Antibodies

Conjugation of Peptides to Carrier Protein

Reagents

Buffer: 25 mM sodium phosphate, pH 7.4
Enkephalin- or β-endorphin-related peptides (Peninsula Laboratories, San Carlos, CA)
Bovine thyroglobulin, type 1 (Sigma Chemical Co.)
1-Ethyl-3-(3-dimethylaminopropylcarbodimide · HCl) (ECDI; Sigma Chemical Co.)
^{125}I-labeled enkephalin- or β-endorphin-related peptide (see below).

Procedure

1. Dissolve peptide (5 mg) and thyroglobulin (20.5 mg) in 750 μl of buffer.

---

[3] M. Noda, Y. Furutani, H. Takahashi, M. Toyosato, T. Hirose, S. Inayama, S. Nakanishi, and S. Numa, *Nature* (*London*) **295**, 202 (1982).
[4] U. Gubler, P. Seeburg, B. J. Hoffman, L. P. Gage, and S. Udenfriend, *Nature* (*London*) **295**, 206 (1982).
[5] M. Comb, P. H. Seeburg, J. Adelman, L. Eiden, and E. Herbert, *Nature* (*London*) **295**, 663 (1982).
[6] H. Kakidani, Y. Furutani, H. Takahashi, M. Noda, Y. Morimoto, T. Hirose, M. Asai, S. Inayama, S. Nakanishi, and S. Numa, *Nature* (*London*) **298**, 245 (1982).
[7] B. A. L. Hurn, and S. M. Chantler, this series, Vol. 70 [5].
[8] J. G. Loeber and J. Verhoef, this series, Vol. 73 [18].

2. Add about 100,000 cpm of ^{125}I peptide to use as a measure of conjugation of peptide to thyroglobulin.
3. Add ECDI (1.2 mg) in 200 μl of buffer to the reaction and stir slowly at 4° for 20 hr. Prepare ECDI fresh, check pH, and neutralize with NaOH if necessary.
4. Remove unreacted peptide and ECDI by dialysis against water for 24 hr. Count aliquot of dialyzate to determine the percentage conjugation.
5. Lyophilize the dialyzate.

Immunization Procedures

Preparation of Freund's Emulsions. The peptide conjugate is dissolved in 0.15 M NaCl at a concentration of 0.2–0.5 mg/ml. An equal volume of complete Freund's adjuvant (Calbiochem, La Jolla, CA) is mixed with the conjugate by one of the following methods.

Method a: Mechanical mixing, e.g., by means of a mixer (such as Sorvall Omnimixer or ultrasonic probe).

Method b: Manual mixing. This is performed by using two syringes connected by a double-hub needle. To form the emulsion, begin by squirting the aqueous phase into the oil as vigorously as possible, then continue squirting the total contents from one syringe to the other until a stable "water-in-oil" emulsion is formed.

Dose and Route of Injection. To produce antisera with high avidity, a large amount of immunogen should not be injected at one time.[7] One or two milliliters (0.4–1.0 mg of peptide conjugate) of emulsion is injected intradermally into the back of a rabbit at 20–40 sites as described by Vaitukaitis et al.[9]

Timing of Injections. After the primary injection, booster injections (half-dose of primary injection) can be given every 4 weeks. At 7–10 days after the second booster injection, blood can be drawn from an ear vein to check the titer and avidity of antisera.

Collection and Storage of Antisera

Blood is drawn from an ear vein and allowed to clot at room temperature until the clot has retracted. Serum is separated by low speed centrifugation (5000 g, 10 min). Antiserum is stable for a long period of time in a $-20°$ freezer. It is recommended that antiserum be stored in small aliquots to prevent deterioration due to frequent freezing and thawing.

[9] J. Vaitukaitis, J. B. Robbins, E. Nieschlag, and G. T. Ross, *J. Clin. Endocrinol.* **33**, 988 (1971).

Determination of Antiserum Titer and Avidity

The titer of antiserum is checked by RIA as described below. A series of dilutions of antiserum, beginning with 1 : 500, can be used to determine the necessary dilution that will bind 30–40% of total labeled antigen. This dilution of antiserum can then be used to evaluate the avidity of the antiserum by performing a competition curve.

Preparation of RIA Tracers

Radioiodination of Peptides

Reagents

Phosphate-buffered saline (PBS): 50 mM sodium phosphate, pH 7.4, 0.9% (w/v) NaCl
Chloramine-T, 0.33 mg/ml in 0.5 M sodium phosphate, pH 7.4
Sodium metabisulfite, 1 mg/ml in PBS
Na^{125}I, 5 mCi in 50 µl of dilute NaOH, pH 10. Add 200 µl of PBS to the vial to give a final concentration of 5 mCi/250 µl.
Tracer buffer: 20 mM sodium phosphate, pH 7.5, 0.15 M NaCl, 0.01% (w/v) BSA, 0.01% (w/v) thimerosal, 0.1% (w/v) gelatin, 0.1% (v/v) Triton X-100, 2 mM dithiothreitol

Procedure

1. Add 2 µg of peptide dissolved in 10 µl of 1 M acetic acid to a 12 × 75 mm polypropylene or siliconized glass culture tube. Evaporate to dryness under a stream of nitrogen.
2. Dissolve peptide in 40 µl of PBS.
3. Add 1 mCi (50 µl) of Na^{125}I.
4. Add 60 µl of chloramine-T.
5. Mix. Wait 30 sec.
6. Add 200 µl of sodium metabisulfite.
7. Mix. Wait 60 sec.
8. Add reaction mixture to a 10-ml Sephadex G-50 column equilibrated in tracer buffer. Elute with same buffer at 20 ml/cm²/hr.

Analysis of ^{125}I-Labeled β-Endorphin. The specific activity of ^{125}I-labeled β-endorphin is determined prior to chromatographic purification by mixing an aliquot of the reaction mixture with rabbit serum and measuring the radioactivity precipitated by trichloroacetic acid (TCA).[10] For

---

[10] M. Chen, J. R. Berman, M. C. McCarroll, and I. W. Chen, *Clin. Chem.* **27**, 632 (1981).

use as a tracer in RIA, the ^{125}I-labeled β-endorphin should have an initial specific activity of approximately 700–800 Ci/mmol.

The quality of ^{125}I-labeled β-endorphin tracers for RIA may be rapidly assessed with the talc–resin–TCA test.[11] Although this test was developed for peptide hormones unrelated to the opioid peptides, we have successfully used it to predict the suitability of ^{125}I-labeled β-endorphin and β-lipotropin as RIA tracers. Radioiodinated peptides suitable for use as a tracer in RIA will exhibit greater than 90% adsorption to talc, less than 25% adsorption to a quaternary ammonium anion-exchange resin, greater than 90% precipitation by TCA, and agreement of the talc adsorption and TCA precipitation values within 3%.

Purification of ^{3}H-labeled [Met^5]Enkephalin

Commercially available tritium-labeled [Met^5]enkephalin may contain substantial quantities of radiolabeled impurities. If desired, ^{3}H-labeled [Met^5]enkephalin may be purified by HPLC as described below prior to being used as an RIA tracer.

Materials and Reagents

^{3}H-labeled [Met^5]enkephalin, 250 μCi/250 μl, 30–60 Ci/mmol, dissolved in 70% ethanol, 0.2% 2-mercaptoethanol, and 0.3% triethylamine (New England Nuclear, Boston, MA)

μBondapak C$_{18}$ HPLC column, 3.9 mm × 30 cm (Waters Associates, Milford, MA)

Sample diluent: 7 mM trifluoroacetic acid in H$_2$O

Initial solvent: 3.5% (v/v) acetonitrile in 7 mM aqueous trifluoroacetic acid

Final solvent: 49% (v/v) acetonitrile in 7 mM aqueous trifluoroacetic acid

Procedure

1. Dilute the ^{3}H-labeled [Met^5]enkephalin solution 1 : 30 with sample-diluent to lower the sample ethanol concentration below 3%.
2. Inject the diluted sample onto the HPLC column previously equilibrated with initial solvent. The sample volume is not critical, because [Met^5]enkephalin is quantitatively adsorbed to the column under these conditions.
3. Elute the [Met^5]enkephalin with a 20-min linear gradient from the initial solvent to the final solvent at 1 ml/min at room temperature.

---

[11] B. B. Tower, M. B. Sigel, R. E. Poland, W. P. Vanderlaan, and R. T. Rubin, this series, Vol. 70 [23].

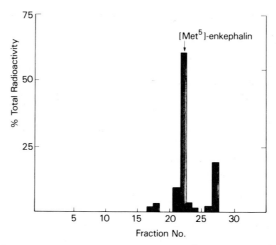

FIG. 1. Purification of a commercial preparation of ^{3}H-labeled [Met^5]enkephalin (250 μCi).

Collect 1-ml fractions and determine radioactivity by liquid scintillation counting. Under these conditions, the retention time of [Met^5]enkephalin is 18.5 min.

4. Evaporate fractions containing ^{3}H-labeled [Met^5]enkephalin in a centrifugal vacuum concentrator or under a stream of inert gas such as argon or nitrogen. Store the purified tracer dry under an inert gas atmosphere at 4°.

Note: This procedure has been used successfully to purify as much as 7.2 ml of diluted ^{3}H-labeled [Met^5]enkephalin containing 250 μCi in a single injection (Fig. 1).

Sample Preparation Procedures

Extraction of Enkephalins and β-Endorphin from Different Tissues

Brain and Pituitary Gland

[Met^5]enkephalin, [Leu^5]enkephalin, and β-endorphin are extracted from various regions of the brain or from the pituitary according to the following protocol.

1. Kill the animals by either decapitation or microwave irradiation.

2. Dissect the brain or the pituitary, freeze the tissues on Dry Ice, and weigh them quickly. The frozen tissues are kept at $-70°$ until processed.
3. Suspend the tissue in 2 N acetic acid (20 ml/g tissue) and homogenize in a polypropylene tube (12 × 75 mm) with a Polytron homogenizer; hot acetic acid (90–95°) is recommended for the homogenization of non-microwave-irradiated tissues. Use a sonic cell disruptor for the homogenization of small tissues such as pituitary lobes or punched cerebral nuclei.
4. Immerse the tube of homogenate in a boiling water bath for 5 min, cool in an ice bath, and centrifuge at 2.5 × 10^4 g for 20 min. Transfer an aliquot of the supernatant fluid to another polypropylene tube and lyophilize.
5. Reconstitute the residue with RIA buffer and centrifuge at 2.5 × 10^4 g for 20 min to remove the denatured protein.
6. Use an aliquot of the supernatant for the determination of [Met^5]enkephalin, [Leu^5]enkephalin, and β-endorphin levels by RIA as described below.

Plasma

Extraction of [Met^5]Enkephalin from Plasma with Octadecylsilylsilica (ODS)[12]

1. Per milliliter of plasma, add 150 μl of 1.6% (w/v) glycine in 1.0 M HCl and 13 μl of 90% (w/w) formic acid. [Plasma is obtained from blood collected in heparinized tubes containing sufficient Trasylol to produce a concentration of 1000 kallikrein inhibitor units (KIU) per milliliter.]
2. Add acidified plasma (1–2 ml) to a prewetted Sep-Pak C_{18} cartridge (Waters Associates, Inc., Milford, MA). (Sep-Pak C_{18} cartridges are prewetted with 2 ml of methanol followed by 5 ml of distilled water.)
3. Wash cartridge with 1 ml 0.9% (w/v) NaCl, 1% (w/w) formic acid, 0.024% (w/v) glycine, 15 mM HCl followed by 0.8 ml of 1% (w/w) formic acid.
4. Elute adsorbed [Met^5]enkephalin with 1.6 ml of 80% methanol containing 0.2% formic acid and collect eluent in a 10 × 75 mm polypropylene tube containing 100 μg Polypep (Sigma Chemical Co.,

[12] V. Clement-Jones, P. J. Lowry, L. H. Rees, and G. M. Besser, *Nature (London)* **283**, 295 (1980).

St. Louis, MO; catalog No. P-5115) and 10 mg of mannitol. (Precoat each polypropylene tube by adding 100 µl of an aqueous solution containing 1 mg of Polypep per milliliter and 100 mg/ml mannitol and drying in a centrifugal vacuum concentrator for approximately 1 hr.)

5. Evaporate eluted samples to dryness in a centrifugal vacuum concentrator (Savant Instruments, Inc., Hicksville, NY); this requires approximately 3 hr.
6. Reconstitute dried samples in 1 ml of 0.1 M sodium phosphate, pH 6.4, containing 0.1% (w/v) bovine serum albumin (boiled at least 5 min to destroy proteolytic activity) and 0.01% Merthiolate.
7. Quantify recovered [Met^5]enkephalin by RIA as described below.

Extraction of β-Endorphin from Plasma with Silica Gel[13]

Reagents

Buffer C: 50 mM sodium phosphate buffer, pH 7.4, 50 mM NaCl, 0.1% (w/v) BSA, 0.9% (v/v) Triton X-100

Procedure

1. Maintain temperature of 4° throughout procedure.
2. Prepare silica gel (100–200 mesh, Fisher Chemical) by washing several times with 1 M HCl, then with distilled H$_2$O until neutral pH is achieved. Dry for 18 hr at 100°.
3. Add 1 ml 1 M HCl to 5 ml of fresh or thawed plasma in polypropylene tubes to adjust pH to 4.0.
4. Add 300 mg silica gel and tumble end over end for 30 min.
5. Centrifuge at 8000 g for 10 min.
6. Wash pellet with 5 ml of distilled H$_2$O *twice* and then with 5 ml of 1 N HCl (centrifuge after each washing at 8000 g for 10 min).
7. Elute β-endorphin with 1 ml of 50% acetone and tumble end-over-end for 30 min.
8. Centrifuge at 8000 g for 10 min.
9. Transfer acetone–water phase to a siliconized tube using a polypropylene pipette.
10. Repeat elution process with 1 ml of 50% acetone.
11. Combine acetone–water supernatants and evaporate to dryness in a water bath at 45° under a stream of N$_2$ or in a centrifugal vacuum concentrator.

[13] V. E. Ghazarossian, R. R. Dent, K. Otsu, M. Ross, B. Cox, and A. Goldstein, *Anal. Biochem.* **102,** 80 (1980).

Extraction of β-Endorphin from Plasma with ODS Columns[14]

1. Attach a Sep-Pak C_{18} cartridge (Waters Associates, Milford, MA) to a 12-ml disposable syringe and prepare the column by washing with 5 ml of methanol. Rinse with 20 ml of distilled or deionized water.
2. Acidify 1 ml of EDTA plasma with 100 µl of 1 M HCl. Load the sample in the syringe and push slowly, over a period of 1 min, through the column.
3. Rinse the column with 20 ml of 4% acetic acid.
4. Elute the peptide with 3 ml of methanol, pushing slowly so that the methanol remains in contact with the ODS-silica for at least 3 min. Collect the eluate in 16 × 100 mm tubes.
5. Put through another 1 ml of methanol to ensure complete elution.
6. Evaporate the methanol eluate to dryness in a 37° water bath under a nitrogen stream or in a centrifugal vacuum concentrator.
7. Reconstitute the sample with 0.5 or 1 ml of 0.1 M H_3BO_3–NaOH, pH 8.4, containing 1% bovine serum albumin and 0.001% Merthiolate. Vortex and incubate at 37° for 10 min to ensure reconstitution of the dried sample. Vortex again.
8. Assay 200 µl in duplicate by RIA as described below.

Cerebrospinal Fluid: Extraction of Opioid Peptides

Although β-endorphin[15] and [Met^5]enkephalin[16] may be extracted from cerebrospinal fluid (CSF) by methods similar to those used for plasma samples, extraction prior to chromatographic separation is not required because the protein concentration of CSF is several hundredfold lower than that of plasma. Opioid peptides in CSF may be analyzed directly[17,18] or after lyophilization[19,20] by the procedures described above for tissue extracts.

[14] Immuno Nuclear Corporation, Stillwater, Minnesota, Catalog No. 16-L1 (1980).
[15] H. Przuntek, J. P. Stasch, M. Graf, K. W. Pflughaupt, N. Gropp, and M. Witteler, *J. Neurol.* **224**, 203 (1981).
[16] V. Clement-Jones, L. McLoughlin, P. J. Lowry, G. M. Besser, L. H. Rees, and H. L. Wen, *Lancet* **2**, 380 (1979).
[17] W. J. Jeffcoate, L. H. Rees, L. McLoughlin, S. J. Ratter, J. Hope, P. J. Lowry, and G. M. Besser, *Lancet* **2**, 119 (1978).
[18] M. M. Wilkes, R. D. Stewart, J. F. Bruni, M. E. Quigley, S. S. C. Yen, N. Ling, and M. Chretien, *J. Clin. Endocrinol. Metab.* **50**, 309 (1980).
[19] K. Nakao, S. Oki, I. Tanaka, K. Horii, Y. Nakai, T. Furui, M. Fukushima, A. Kuwayama, N. Kageyama, and H. Imura, *J. Clin. Invest.* **66**, 1383 (1980).
[20] K. Nakao, Y. Nakai, S. Oki, S. Matsubara, T. Konishi, H. Nishitani, and H. Imura, *J. Clin. Endocrinol. Metab.* **50**, 230 (1980).

Chromatographic Separation of Enkephalins and β-Endorphin prior to RIA

In most cases, antisera against one opioid peptide will cross-react with other opioid peptides to some extent; e.g., antiserum against [Met^5]enkephalin cross-reacts with β-endorphin, β-lipotropin, [Leu^5]enkephalin, or dynorphin. Samples prepared from various regions of the brain, the pituitary, or plasma contain multiple species of these opioid peptides. In addition, the precursor molecule and the metabolized intermediates also show some interference. The immunoreactive components in tissue or plasma extracts are thus separated by chromatography prior to measurement by RIA.

Gel Filtration

For the separation of β-endorphin and related peptides, gel filtration chromatography is commonly employed.

1. Prepare the tissue extract in 2 *N* acetic acid as described above.
2. Apply the sample to a column (polypropylene or silicone-coated glass, 1 × 60 cm) of BioGel P-30 (P-60) or Sephadex G-50 (G-75) equilibrated with 1 *N* acetic acid.
3. Elute the peptides with 1 *N* acetic acid or 0.1 *N* HCl at room temperature. Collect the fractions (1 ml) at a flow rate of 0.1 ml/min.
4. Lyophilize the fractions and reconstitute with RIA buffer.
5. Use an aliquot of each fraction for β-endorphin RIA.

Elution profiles of the pituitary extract reveal three major peaks of β-endorphin immunoreactivity. The first peak (I) appears in the void volume. The second (II) and the third peaks (III) coelute with β-lipotropin and synthetic β-endorphin standards, respectively. The profiles of β-endorphin immunoreactivity in the extracts from various tissues or biological fluids exhibit different patterns. In the case of the anterior pituitary, the extraction condition is reportedly critical in determining the elution pattern (i.e., the ratio between II and III).[21]

High-Performance Liquid Chromatography (HPLC)

Reverse-phase HPLC on octadecylsilylsilica (ODS) columns is a very powerful technique for separating complex mixtures of opioid peptides. Samples are adsorbed to the column in an acidic aqueous solvent and

[21] A. S. Liotta, T. Suda, and D. T. Krieger, *Proc. Natl. Acad. Sci. U.S.A.* **75**, 2950 (1978).

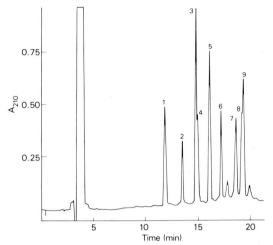

FIG. 2. High-performance liquid chromatography separation of an opioid peptide mixture by gradient elution. The injected sample contained approximately 5 μg of each of [Met0^5]enkephalin (1), β-LPH 77–91 (2), [Met^5]enkephalin (3), α-endorphin (4), [Leu^5]enkephalin (5), γ-endorphin (6), β_c-endorphin (7), β_h-endorphin (8), [Leu^8]-β_h-endorphin (9). The small peaks at 18 and 20 min are solvent impurities.

sequentially eluted with a gradient of acetonitrile,[22,23] methanol,[8] or n-propanol.[24] Provided that the acidic buffer components are volatile, the HPLC-purified fractions may be lyophilized for subsequent structural analysis or quantitation by RIA. Although maximum resolution of a wide variety of peptides is achieved by gradient elution, isocratic separation of closely related peptides is simpler and faster. The isocratic HPLC procedures described below have been optimized for purification of [Met^5]- and [Leu^5]enkephalin and for β-endorphin. As shown in Fig. 2, the gradient elution procedure is suitable for separating a variety of opioid peptides and is also much more rapid than previously reported gradient methods.

Isocratic HPLC Procedures

SEPARATION OF [Met^5]- AND [Leu^5]ENKEPHALIN

Materials and Reagents

ODS-5 S reverse-phase HPLC column, 4 mm × 150 cm (Bio-Rad Laboratories, Richmond, CA)

[22] J. R. McDermott, A. I. Smith, J. A. Biggins, M. Chyad Al-Noaemi, and J. A. Edwardson, *J. Chromatogr.* **222**, 371 (1981).
[23] J. L. Meek, *Proc. Natl. Acad. Sci. U.S.A.* **77**, 1632 (1980).
[24] R. V. Lewis, S. Stein, and S. Udenfriend, *Int. J. Pept. Protein Res.* **13**, 493 (1979).

Sample buffer: 0.01 M sodium acetate buffer, pH 4.15
Elution buffer: acetonitrile/0.01 M sodium acetate buffer, pH 4.5 (30:70, v/v)

Procedure

1. Extract enkephalins from tissue or plasma sample as described above, lyophilize, and redissolve in sample buffer.
2. Inject an aliquot of the sample onto the HPLC column previously equilibrated with elution buffer.
3. Elute the column with elution buffer at 2 ml/min at room temperature and collect the effluent in 1-ml fractions. Under these conditions, the retention times for [Met^5]- and [Leu^5]enkephalin are 6.5 and 12.5 min, respectively.
4. Lyophilize fractions containing [Met^5]- and [Leu^5]enkephalin and quantify by RIA as described below.

β_h-ENDORPHIN

Materials and Reagents

μBondapak C$_{18}$ HPLC column, 3.9 mm × 30 cm (Waters Associates, Milford MA)
Sample solvent: 1 M acetic acid
Elution solvent: acetonitrile–0.08% (v/v) trifluoroacetic acid in H$_2$O (35:65, v/v)

Procedure

1. Extract β-endorphin from tissue or plasma as described above, lyophilize, and redissolve in sample solvent.
2. Inject sample onto the HPLC column previously equilibrated with elution solvent. Elute at 1 ml/min at room temperature and collect 1-ml fractions. Under these conditions, the retention time of β_h-endorphin is 6.5 min.
3. Lyophilize fractions containing β-endorphin and quantify by RIA as described below.

Gradient Elution HPLC

Materials and Reagents

μBondapak C$_{18}$ HPLC column, 3.9 mm × 30 cm (Waters Associates, Milford, MA)
Sample solvent: 1 M acetic acid

Initial solvent: acetonitrile–0.08% (v/v) trifluoroacetic acid in H_2O (10:90, v/v)

Final Solvent: acetonitrile–0.08% (v/v) trifluoroacetic acid in H_2O (60:40, v/v)

Procedure

1. Extract opioid peptides from tissue or plasma as described above, lyophilize, and redissolve in sample solvent.
2. Inject sample onto the HPLC column previously equilibrated with initial solvent. Elute peptides with a 20-min linear gradient from the initial solvent to final solvent at 1 ml/min at room temperature and collect 1-ml fractions. Under these conditions, various opioid peptides are separated as shown in Fig. 2.
3. Lyophilize desired fractions and quantify by RIA as described below.

RIA Procedures for Enkephalins and β-Endorphins

Measurement of enkephalins and β-endorphin concentrations by RIA is based on the competition of unlabeled peptide in the sample with a radiolabeled peptide tracer for binding sites on specific antibodies. The general theory and practice of RIA are the subjects of several reviews.[25-27] The procedures described below have been used successfully in our laboratories.

Preparation of Standards and Tissue Blanks for RIA of Opioid Peptides

All immunoassay calculations are ultimately based on comparison of samples with a series of standards after subtraction of appropriate blanks.

Standards. With the exception of β_h-lipotropin, most of the opioid peptides of interest are available commercially. Several commercial suppliers offer opioid peptides on the basis of actual peptide weight as determined by quantitative amino acid analysis. Standard solutions should never be prepared by reweighing the solid peptide, because the total weight often contains as much as 20% water or salts. If confirmation of the peptide concentration is desired, standards may be dissolved initially

---

[25] T. Chard, *in* "Laboratory Techniques in Biochemistry and Molecular Biology" (T. S. Work and E. Work, eds.), Vol. 6, p. 293. North-Holland Publ., Amsterdam, 1978.
[26] E. A. Kabat, this series, Vol. 70 [1].
[27] H. Van Vunakis, this series, Vol. 70 [10].

in 1 M acetic acid and analyzed qualitatively by HPLC as described above and quantitatively by subjecting a hydrolyzed aliquot to amino acid analysis. For hydrolysis, an aliquot containing approximately 2–8 nmol of peptide in 1 M acetic acid is transferred to a 13 × 100 mm borosilicate glass tube containing 10 nmol of norleucine as an internal standard. After the solvent is evaporated in a centrifugal vacuum concentrator, the residue is dissolved in 0.5 ml of 6 N HCl, and the solution is degassed by two or three freeze–pump–thaw cycles with a high-vacuum pump. The tube is flame-sealed under vacuum while the degassed solution is frozen on Dry Ice. The sealed tube is then heated for 16–24 hr at 110° in an aluminum block. Note, however, that complete liberation of all isoleucine residues requires a 72-hr hydrolysis. After cooling, the tube is opened, the HCl is removed in a centrifugal vacuum concentrator, and the residue is dissolved in 60 μl of sodium citrate sample dilution buffer, pH 2.20 (Pierce Chemical Co., Rockford, IL). A 20-μl aliquot of the hydrolyzed sample is analyzed with a Durrum D-500 automatic amino acid analyzer (Durrum Instrument Corp., Palo Alto, CA), which is calibrated with a standard solution aliquot containing 10 nmol of each amino acid (Pierce Chemical Co., Rockford, IL). Peptide concentrations are determined from the amino acid analysis after correction for overall recovery, usually greater than 90%, based on the norleucine internal standard.

Standards of endorphins and cross-reacting materials have been prepared and analyzed by these methods and distributed to investigators under a contract from the National Institute on Drug Abuse.[28]

For long-term storage, peptide standards should be stored dry in sealed ampules under an inert gas atmosphere at 4° or below. For short-term storage, standards may be stored in 1 M acetic acid at $-70°$; alternatively, 0.1-ml aliquots of peptide dissolved in 0.5% (w/v) heat-treated (5 min at 95°) bovine serum albumin may be lyophilized and stored at $-20°$. Storage of peptide standards dissolved in plasma is not recommended.

Tissue Blanks. If peptides are to be measured in tissue extracts without prior purification, a tissue blank solution must be added to each standard to correct for interfering materials extracted from tissue. The following procedure is recommended for the measurement of opioid peptide levels in brain tissue. Tissue blanks are prepared from cerebellum tissue, where opioid peptide immunoreactivity is barely detectable. A rat cerebellum is frozen and thawed twice, incubated at 37° for 30 min, then homogenized in 1 ml of 2 M acetic acid. The homogenate is heated at 95°

---

[28] Project Officer: Dr. Richard L. Hawks, Acting Chief, Research Technology Branch, Rm. 10-A-19, Division of Preclinical Research, National Institute on Drug Abuse, 5600 Fishers Lane, Rockville, MD 20857.

for 5 min and centrifuged at 25,000 g for 20 min, and the supernatant is lyophilized. The lyophilized supernatant is dissolved in H_2O at a concentration of 1 ml/100 mg of original tissue and clarified by centrifugation at 25,000 g for 20 min. The clarified supernatant is then treated with 0.1 volume of 5% charcoal slurry in 0.3 M Tris-HCl, pH 7.4, containing 0.5% dextran and incubated at 4° for 30 min. The charcoal is removed by centrifugation at 4000 g for 10 min, and the supernatant is used as the tissue blank solution. A volume of tissue blank solution equivalent to the weight of sample tissue is added to each standard tube.

Enkephalin RIA Protocol

Reagents

Buffer (used to dilute samples, antiserum, and tracer): 0.1 M sodium phosphate, pH 7.5, 0.15 M NaCl, 0.1% (w/v) gelatin, 0.01% (w/v) bovine serum albumin, 0.01% (w/v) thimerosal
Human γ-globulin, 0.5% (w/v)
Polyethylene glycol 6000, 25% (w/v)
Charcoal suspension, 1.5% (w/v): 0.3 M Tris-HCl, pH 7.4, containing 0.15% (w/v) dextran

Procedure

1. Add 20 μl of ^{125}I- or ^3H-labeled [Met^5]enkephalin or ^{125}I- or ^3H-labeled [Leu^5]enkephalin to a 12 × 75 mm polypropylene tube (10,000 cpm per tube).
2. Add 20 μl of sample or standard.
3. Add 200 μl of antiserum, diluted to produce an initial binding of 30–50% of tracer.
4. Vortex and incubate for 24 hr at 4°.
   Do *either* 5a or 5b
5a. Add 100 μl of 0.5% (w/v) human γ-globulin and 1 ml of 25% (w/v) polyethylene glycol 6000 at 4°.
5b. Add 0.2 ml of 1.5% (w/v) charcoal suspension.
6. Vortex and incubate for 30 min at 4° for polyethylene glycol or 10 min at 4° for charcoal.
7. Centrifuge 15 min at 3000 g at 4°.
8. Decant and count supernatants when charcoal is used and pellets when polyethylene glycol is used. Supernatants are counted in a liquid scintillation counter after being mixed with 10 ml of 0.6% (w/v) Omnifluor (New England Nuclear, Boston, MA) in 33% (v/v) Triton X-100–toluene when ^3H-labeled tracers are used.

RIA for β-Endorphin

Precautions. β-Endorphin is a very basic peptide and thus tends to bind tightly to untreated glass surfaces. All tubes, pipette tips and other surfaces contacting the peptide solutions should be restricted to polypropylene or siliconized glass. In addition, priming the pipette tips by filling and emptying several times is recommended whenever β-endorphin solutions are pipetted. All buffers used for β-endorphin RIA must contain Triton X-100 and gelatin or bovine serum albumin to minimize losses due to surface adsorption.

Small quantities of glassware may be siliconized by allowing the glassware to stand in contact with 5% (v/v) dimethyldichlorosilane (Pierce Chemical Co., Rockford, IL) in toluene for 15 min. The glassware is then rinsed sequentially with toluene, methanol, and water and dried. Larger quantities of glassware are more conveniently siliconized with hexamethyldisilazane (HMDS) (Pierce Chemical Co., Rockford, IL) vapors in a vacuum oven.[29] The oven is filled with glassware, heated for 2 hr at 225°, then evacuated with a high-vacuum pump for 30 min. The oven vacuum is then partially released (to 20 in. of Hg) by allowing $CaSO_4$-dried air to enter after bubbling through a glass frit immersed in liquid HMDS, thus filling the oven with HMDS vapors. The glassware is then allowed to remain in contact with the HMDS vapors for 1–16 hr at 225°. The HMDS is pumped from the oven and condensed in a Dry Ice trap under high vacuum, and the siliconized glassware is ready for use.

β-Endorphin Radioimmunoassay Protocol Using Charcoal Separation[30]

Reagents

Buffer A: 0.02 M sodium phosphate, 0.15 M NaCl, 0.01% (w/v) bovine serum albumin, 0.1% (w/v) gelatin, 0.01% (w/v) thimerosal, pH 7.5

Buffer B: buffer A + 0.1% (v/v) Triton X-100

Charcoal: buffer A + 1% (w/v) charcoal + 0.5% (w/v) bovine serum albumin (RIA grade, Sigma Chemical Co.)

Procedure

1. Add 300 μl buffer A to a 12 × 75 mm polypropylene tube.
2. Add 50 μl of sample or β-endorphin standard in buffer B.

---

[29] D. C. Fenimore, C. M. Davis, J. H. Whitford, and C. A. Harrington, *Anal. Chem.* **48**, 2289 (1976).

[30] V. Hollt, R. Przewlocki, and A. Herz, *Naunyn-Schmiedeberg's Arch. Pharmacol.* **303**, 171 (1978).

3. Add 100 µl of antiserum in buffer A, diluted to bind approximately 40% of the added tracer in the absence of β-endorphin.
4. Add 50 µl of ^{125}I-labeled β-endorphin in buffer B (10,000 cpm).
5. Incubate for 12–72 hr at 4°.
6. Add 500 µl of charcoal suspension.
7. Vortex for 5 sec.
8. Let stand for 10 min.
9. Centrifuge for 5 min. at 3000 g at 4°.
10. Count 800 µl of the supernatant in a gamma counter.

β-Endorphin Radioimmunoassay Protocol Using Second Antibody Separation[31]

Reagents

Buffer A: 0.02 M sodium phosphate, 0.15 M NaCl, 0.01% (w/v) Thimerosal, pH 7.5
Buffer B: 0.1% (w/v) gelatin in buffer A
Buffer C: 0.01% (w/v) crystalline bovine serum albumin in buffer B
Buffer D: buffer C containing 0.1% (v/v) Triton X-100, pH adjusted to 7.5 with 5 N NaOH

Protocol

1. Add β-endorphin standards in 100 µl of buffer C or unknown samples in 10–100 µl of buffer C.
2. Dilute standards and unknowns to a total volume of 500 µl with buffer C (B_0 and NSB tubes receive 500 µl of buffer C).
3. Add 100 µl of ^{125}I-labeled β-endorphin (5000 cpm) in buffer D and mix.
4. Add 100 µl of β-endorphin antiserum in buffer B supplemented with rabbit serum to give a total concentration of 0.5% and mix. Antiserum should be diluted to give 30–40% maximum binding (B_0) of tracer.
5. Incubate for 4–24 hr.
6. Add 50 µl of goat anti-rabbit γ-globulin diluted with buffer B to give maximal precipitation and mix.
7. Incubate for 16–24 hr.
8. Add 2 ml of buffer A.
9. Centrifuge for 30 min at 3000 rpm at 4°.
10. Aspirate supernatants and count pellets in a gamma counter.

[31] R. Guillemin, N. Ling, and T. Vargo, *Biochem. Biophys. Res. Commun.* **77**, 361 (1977).

Calculations

The radioactivity bound to antibody may be linearly related to the standard peptide concentrations by the logit-log transformation.[32]

$$\ln[(B - \text{NSB})/(B_0 - B)] = m \ln P + k \tag{1}$$

in which B is the cpm bound to antibody in the presence of peptide concentration P, NSB is the nonspecifically bound cpm, B_0 is the cpm bound to antibody in the absence of added peptide, and $e^{-k/m}$ is the peptide concentration that displaces 50% of the bound radioligand. Since the logit-log transformation results in nonuniformity of variance with a large increase at the ends of the regression line, a weighted regression analysis should be used on such data.[33,34] Equation (1) may be converted to the four-parameter logistic formulation

$$B = [a/(1 + bP^c)] + d \tag{2}$$

by defining the parameters as follows: $a = B_0 - \text{NSB}$; $b = e^{-k}$; $c = -m$; $d = \text{NSB}$. Computerized methods for weighted least-squares analysis of the four-parameter model have been described.[33,35]

Acknowledgments

The authors wish to thank Dr. A. R. Jeffcoat for advice on HPLC analysis of peptides, Dr. C. E. Cook for advice and encouragement during preparation of the manuscript, Ms. A. F. Gilliam for technical assistance with radioimmunoassays and HPLC, and Ms. T. W. Erexson for expert typing assistance. Portions of this work were supported by contract No. 271-81-3801 from the National Institute on Drug Abuse.

---

[32] D. Rodbard, W. Bridson, and P. L. Rayford, *J. Lab. Clin. Med.* **74**, 770 (1969).
[33] D. Rodbard, *Clin. Chem.* **20**, 1255 (1974).
[34] D. Rodbard, R. H. Lenox, H. L. Wray, and D. Ramseth, *Clin. Chem.* **22**, 350 (1976).
[35] K. Ukraincik and W. Piknosh, this series, Vol. 74 [33].

[38] Assay of Corticotropin Releasing Factor

By Wylie Vale, Joan Vaughan, Gayle Yamamoto,
Thomas Bruhn, Carolyn Douglas, David Dalton,
Catherine Rivier, and Jean Rivier

Ovine corticotropin releasing factor (oCRF, amunine, corticoliberine) was isolated from sheep hypothalamic extracts and characterized as a 41-residue peptide.[1] CRF is a potent stimulator of ACTH, β-endorphin, and other proopiomelanocortin (POMC) products *in vitro* and *in vivo* in various species including human beings. Rat CRF has been characterized and found to differ from oCRF by 7 residues.[2] Both peptides are related to sauvagine (frog) and urotensin-I (fish); all exhibit ACTH releasing and hypotensive actions when administered peripherally.[3]

Immunoneutralization studies with anti-CRF serum and observations of CRF dynamics in portal blood strongly support the key involvement of CRF in the regulation of ACTH secretion under basal and stressful circumstances. Other weaker ACTH secretagogues, catecholamines, vasopressin, and perhaps angiotensin II, enhance the response to CRF *in vitro* and *in vivo* and appear to modulate the action of CRF during stress.[3]

Immunoactivity of CRF type is widely distributed throughout the brain, particularly in the limbic system and brainstem, where it may participate in the regulation of the autonomic nervous system. Intracranial administration of CRF stimulates sympathetic activities and inhibits parasympathetic outflow. These results, taken together with arousal-type behavioral effects observed following low doses of the peptide, suggest that CRF may be involved in the integration of endocrine, visceral, and behavioral responses to stress.[3]

The CRF can be assayed by a variety of biological and immunological techniques. Two methods will be described here: an *in vitro* assay employing primary anterior pituitary cell cultures and a radioimmunoassay (RIA) procedure.

Cell Culture Assay

Reagents

HEPES dissociation buffer (HDB): 137 mM NaCl, 5 mM KCl, 0.7 mM Na$_2$HPO$_4$, 25 mM HEPES, 10 mM glucose. The pH is adjusted to 7.3.

---

[1] W. W. Vale, J. Spiess, C. Rivier, and J. Rivier, *Science* **213**, 1394 (1981).
[2] J. Rivier, J. Spiess, and W. Vale, *Proc. Natl. Acad. Sci. U.S.A.* (in press).
[3] W. Vale, C. Rivier, M. R. Brown, J. Spiess, G. Koob, L. Swanson, L. Bilezikjian, F. Bloom, and J. Rivier, *Recent Prog. Horm. Res.* **39** (in press).

Collagenase–DNase: 0.4% collagenase (Worthington, type II), 80 µg of DNase II (Sigma) per milliliter, 0.4% bovine serum albumin (BSA), 0.2% glucose in HDB

Viokase, 0.25% (GIBCO) in HDB

Poly-D-lysine (Sigma), 20 µg per milliliter of distilled H_2O, filter sterilized

Human fibronectin, sterile, 10 µg (Collaborative Research per milliliter of β-PJ)

β-PJ medium[4]: powdered SFRE-199-2/Earle's salts (Kansas City Biologicals), dissolved in an appropriate amount of deionized water and containing $NaHCO_3$ (2.2 g/liter), HEPES (3.575 g/liter), 100X MEM vitamins (10 ml/liter), DL-lipoic acid (210 µg/liter), linoleic acid (84 µg/liter), penicillin (100,000 U/liter), streptomycin (100 mg/liter), trace salts (ref), as well as hormones and growth factors (ref). Trace salts include: aluminum chloride (1.2070 µg/liter), silver nitrate (0.1699 µg/liter), barium acetate (2.554 µg/liter), potassium bromide (0.119 µg/liter), cadmium chloride (2.284 µg/liter), cobalt chloride (2.379 µg/liter), chromic sulfate (0.6624 µg/liter), copper sulfate (2.497 µg/liter), sodium fluoride (4.199 µg/liter), iron sulfate (534 µg/liter), germanium oxide (0.529 µg/liter), potassium iodide (0.166 µg/liter), manganese sulfate (0.241 µg/liter), molybdate solution (0.120 µg/liter), nickel chloride (1.20 µg/liter), rubidium chloride (1.209 µg/liter), selenious acid (4.00 µg/liter), sodium silicate (0.2841 µg/liter), stannous chloride (0.110 µg/liter), titanium chloride (0.9486 µg/liter), vanadic acid (0.590 µg/liter), zinc sulfate (762.6 µg/liter), and zirconium oxychloride (3.22 µg/liter).

β-PJ incubation medium (IM): β-PJ medium + 0.1% crystalline BSA

β-PJ culture medium: β-PJ medium plus the following hormones and growth factors added immediately prior to use: insulin (5 mg/liter), transferrin (5 mg/liter); parathyroid hormone (0.5 µg/l), T_3 (30 pM), fibroblast growth factor (1 µg/liter). The culture medium usually includes fetal bovine serum.

Dissociation Procedure[5]

Pituitary glands are removed from rats sacrificed by rapid decapitation. Anterior lobes are separated from neurointermediate lobes. The

---

[4] W. Vale, J. Vaughan, M. Smith, G. Yamamoto, J. Rivier, and C. Rivier, *Endocrinology* (*Baltimore*) (in press).

[5] W. Vale, G. Grant, M. Amoss, R. Blackwell, and R. Guillemin, *Endocrinology* (*Baltimore*) **91**, 562 (1972).

intermediate lobe adheres to the neural (posterior) lobe when it is teased away from the anterior lobe. Both anterior and intermediate lobes contain CRF-responsive corticotropic cells that can be readily established in culture. The method for dissociating 50 anterior pituitary glands or a similar quantity of tissue is described below.

Halved anterior lobes are rinsed several times with sterile HDB and placed in a sterilized 100- or 50-ml Spinner suspension flask (Bellco Glass Co.) containing collagenase–DNase solution. Generally 50 ml of this enzyme solution is used for 50 anterior pituitary glands (from 200–250 g male rats).

The suspension flask is partially immersed in a water bath (37°), placed on an immersible magnetic impeller (TRI-R Instruments, Inc.), and the glands are stirred at 100–200 rpm. The tissue fragments are gently drawn in and out of a siliconized Pasteur pipette every 30–40 min throughout the entire procedure. At a point where most of the fragments have dissociated and the remainder are small and thready, cells in the collagenase–DNase solution are transferred to a sterile plastic tube and centrifuged at 475 g for 8 min. The pelleted cells are then resuspended in 30 ml of Viokase solution, returned to the suspension flask, and incubated and stirred until fragments and threads disappear, but for no longer than 10 min. The Viokase solution and cells are then transferred to a sterile plastic tube, centrifuged as before, and resuspended in incubation medium with 2% fetal bovine serum. Cells are washed by repeated centrifugation and resuspension in 40 ml of culture medium (3–5 times). During the washing procedure visible clumps should be permitted to settle at unit gravity and the supernatant be removed with a Pasteur pipette. After the final wash, an aliquot is counted by Coulter counter or hemacytometer, and the cells are suspended in an appropriate volume of culture medium + 2% fetal bovine serum to yield ca. 3.3×10^5 cells per milliliter. Of this suspension, 1 or 0.5 ml is distributed with periodic swirling to each well of 24-well Linbro or 48-well Costar (3548) tissue culture plates (76-033-05). The plates are placed in a water-jacketed incubator maintained at 37° and with an atmosphere of 93% air and 7% CO_2. Yields are ca. 2×10^6 cells per anterior pituitary gland; therefore ca. 6 wells can be plated with cells from one rat pituitary.

Substratum Modification. For some applications it may be desirable to establish cultures under serum-free conditions. To do so, the substratum of the culture surface must be modified. We have used the following procedure: To each well, add 0.4 ml of filtered poly-D-lysine solution. After setting at room temperature for 2 hr, the poly-D-lysine solution is aspirated and the wells are rinsed twice with incubation medium. Then, 1 ml of human fibronectin solution is added for 45 min before removal by aspiration. Dispersed cells in serum-free culture medium are immediately placed into the coated wells.

Neurointermediate Lobe Cultures. Neurointermediate lobe cells are dispersed in a manner similar to that for anterior lobes. For 50 neurointermediate lobes, use 25 ml of collagenase–DNase solution. Neurointermediate lobes are generally plated in culture medium + 2% fetal bovine serum onto wells coated with poly-D-lysine as described above. We have not found it to be of value to precoat wells with fibronectin for neurointermediate lobe cells. Usually each well is plated with cells derived from one neurointermediate lobe.

Experimental Incubation Procedures. Cells are generally used from 3–5 days after plating. At that time cells are firmly attached and can be washed by repeated medium change. Medium can be removed either by aspiration (if sterility must be maintained) or by rapid inversion of the plates. Cells are washed three times with 1.0 ml of β-PJ incubation medium (IM), returned to the incubator for a 1-hr preincubation, and then washed one more time with IM prior to addition of test substances. The CRF and other treatments are added in small volumes (≤ 50 μl) to triplicate wells, and the cells are incubated for 3–4 hr prior to the removal of medium by aspiration and its storage for future ACTH[6,7] or β-endorphin radioimmunoassay.[7] As shown in Fig. 1, doses of CRF ranging from 0.004 nM to 2.5 nM are routinely chosen to develop the linear portions of a standard curve. This bioassay can typically detect between 0.005 and 0.01 nM or 5–10 fmol of added CRF. The cell culture assay was the exclusive assay used for the isolation and subsequent characterization of ovine CRF[1] and has thereby been proved to be useful for the detection and quantification of CRF in crude or partially purified extracts. The method has been applied for a wide range of studies investigating the mode of action of CRF and its interactions with other physiological modulators, such as glucocorticoids, catecholamines, and neurohypophysial peptides.[4] This method is useful for the assay of both agonist and antagonist analogs of CRF and of highly purified CRF-like peptides of various mammalian (Fig. 1) and nonmammalian species.

The presence of toxic materials, ACTH, or non-CRF-related substances that can weakly release ACTH on their own and enhance the response to CRF often complicates the interpretation of results with impure natural samples. The meaningful bioassay of native CRF in biologi-

---

[6] D. Orth, in "Methods of Hormone Radioimmunoassay" (B. M. Jaffe and H. R. Behrman, eds.), 2nd ed., p. 245. Academic Press, New York, 1979.

[7] C. Rivier, M. Brownstein, J. Spiess, J. Rivier, and W. Vale, *Endocrinology* (*Baltimore*) **110**, 272 (1982).

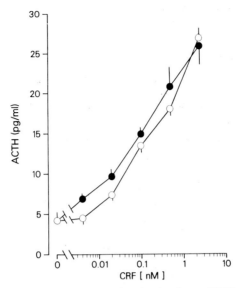

FIG. 1. Effect of synthetic rat corticotropin releasing factor (CRF) (●) and ovine CRF (○) on secretion of ACTH by cultured rat anterior pituitary cells.

cal tissues generally involves purification of the assay sample and always requires extensive validation.

CRF Radioimmunoassay

For the most routine applications, CRF levels will be measured by radioimmunoassay. The principal requirements are the availability of a suitable antiserum and radiolabeled tracer.

Development of CRF Antisera

CRF antibodies have been raised in rabbits in response to a variety of immunogens (see the table). The method followed in producing antiserum rC-68 will be described.

rCRF is coupled by bisdiazotization to human α-globulins. Human α-globulins (12.65 mg; US Biochemicals, fraction IV) are dissolved in 8 ml of borate buffer (0.13 M NaCl, 0.16 M boric acid, pH to 9.0 with NaOH) in a 20-ml scintillation vial kept in an ice-water bath and stirred by a magnetic spin bar. To this add 10.05 mg of rCRF in a volume of 1 ml of distilled H_2O (dH_2O), then add 1.34 ml of a cold 2.5 mM bisdiazotized benzidine solution dropwise with rapid stirring. Continue stirring for 3 hr

Corticotropin Releasing Factor (CRF) Immunogens

Rabbit/antiserum no.	Immunogen
oC-24	[Tyr^{22},Gly^{23}]-oCRF(1-23) conjugated to human (h) α-globulins with bisdiazotized benzidine
oC-28	oCRF conjugated to h α-globulins with glutaraldehyde
oC-29	Same as 28
oC-30	Same as 28
oC-31	Same as 28
oC-33	oCRF polymerized with 1-ethyl-3-(3-dimethylaminopropyl) carbodiimide HCl (EDCI)
oC-35	Same as 33
rC-68	rCRF conjugated to h α-globulins with bisdiazotized benzidine
rC-69	Same as 68
rC-70	Same as 68
rC-71	rCRF conjugated to h α-globulins with glutaraldehyde
rC-72	Same as 71
rC-73	Same as 71

in the ice bath. Transfer the mixture to dialysis tubing (Spectrapor No. 6, molecular weight cutoff 1000) and dialyze against 4 liters of 0.15 M NaCl with four changes. After dialysis the rCRF-BDB-h α-globulin conjugate is diluted with saline to give a final concentration of 1 mg of total protein per milliliter and is aliquoted and frozen.

The immunogens are prepared by emulsifying with a Polytron Freund's complete adjuvant–modified *M. butyricum* (Calbiochem) with an equal volume of saline containing either 1.0 mg of conjugate per milliliter (for initial injections) or 0.5 mg/ml (for boosters). For each immunization, each rabbit receives a total of 1 ml of emulsion in >30 intradermal sites. The initial 1 ml injection contains 0.5 mg of conjugate. Every 2 weeks thereafter rabbits are boosted with 1 ml of emulsion containing 0.25 mg of conjugate. Animals are bled through an ear vein 10–14 days after each booster. Blood is allowed to clot, and serum is separated. The antiserum from each bleeding is characterized with respect to titer and affinity, often versus several candidate tracers.

Preparation of CRF Tracers

Reagents

PO^4 buffer: 0.5 M Na$_2$HPO$_4$, pH adjusted to 8.0 with 0.5 M NaH$_2$PO$_4$
triethyl ammonium formate (TEAF): 11.5 ml of formic acid (88%

HCOOH) per liter; pH adjusted to 3.0 with triethylamine (MCB reagents)

Acetonitrile (HPLC grade)

Filter all HPLC solvents through 0.45 μm Millipore type HA filter; thoroughly degas prior to use.

The stock peptide used for the iodination is diluted to 1 mg/ml in 0.1 M HOAc, 50-μl aliquots of this stock are placed in glass tubes, lyophilized, and then stored frozen. Immediately prior to iodination, an aliquot is reconstituted to 1 mg of CRF per milliliter with 0.1 M HOAc. Chloramine-T and sodium metabisulfite are diluted in dH$_2$O at a concentration of 1 mg/ml immediately prior to the iodination. All additions are done using Hamilton microliter syringes: To a 1-mCi vial of Na-^{125}I (New England Nuclear) at pH 8–10 is added 8 μl of 0.5 M PO$_4$ buffer, pH 8.0, and 4 μl of the peptide (1 mg/ml). This mixture is vortexed, and 2 μl of chloramine-T (1 mg/ml) is added. The iodination is allowed to proceed for 30 sec, at which point it is immediately quenched with 5 μl of sodium metabisulfite (1 mg/ml) and again vortexed for 30 sec. Then 100 μl of 10% crystalline BSA (Pentex) is added to adsorb any excess free iodine from the reaction and vortexed for 40 sec. The mixture is then diluted to 1 ml with 0.5 N HOAc before being applied to a Sep-Pak C$_{18}$ cartridge (Waters Associates) in order to separate the iodinated peptide from free iodine and BSA. The Sep-Pak C$_{18}$ cartridge is prewetted with 5 ml of MeOH, then 5 ml of 0.5 N HOAc and subsequently the iodination mixture is applied. Five milliliters of dH$_2$O followed by 5 ml of 0.5 N HOAc are passed through the column. These two washes will remove most of the free iodine.

The iodinated peptide is then eluted from the column into a glass tube by the addition of 5 ml of 75% acetonitrile, 25% (0.5 N HOAc). The spent cartridge is discarded. The volume of the peptide fraction is then reduced to approximately 0.5 ml with a Speed-vac (Savant). This is necessary in order to remove most of the acetonitrile so that the peptide will adsorb to the HPLC column. TEAF (1.0 ml, pH 3.0; solvent A of the HPLC system) is then added to the tube containing the peptide, and the iodinated peptide is purified using an Altex Model 332 HPLC system with a Rheodyne injector equipped with a 2-ml sample loop. The column used is a reverse-phase μBondapak C$_{18}$ (Waters Associates). The solvent for pump A is TEAF, pH 3.0; pump B contains 60% acetonitrile, 40% TEAF, pH 3.0. The acetonitrile is distilled in glass and obtained from Burdick & Jackson; the triethylamine is from MCB Reagents. The HPLC system is equilibrated to 40% "B" and the peptide is then applied. After 2 min, a gradient is run from 40% B to 95% B in 30 min at a flow rate of 1.5 ml/min. Fractions are collected every 0.5 min, and small aliquots of these fractions are counted for radioactivity. The peak tubes of radioactivity, which normally have a retention time of approximately 22 min, are pooled and made 0.1% in crystalline BSA. In the absence of standard peptides synthesized with cold iodide, all major peaks should be screened with antisera. rCRF has two potential iodination sites, His^{13} and His^{32}, each of which could have up to two iodines. It would not be unexpected for radioactive peptide tracers with ^{125}I on His^{13} versus His^{32} to select different antibody populations within the same polyclonal antiserum. Thus the choice of peak may drastically influence the specificity, sensitivity, and other characteristics of the resultant RIA. By employing consistent iodination and purification conditions, similar peaks should be produced each time and therefore be available for RIAs. The iodinated tracer is stored at 4° and typically used in the radioimmunoassay for 2 months.

Procedure for Rat CRF RIA

Reagents

RIA buffer (SPEAB/0.1% Triton X-100): 0.1 M NaCl; 0.05 M Na$_2$HPO$_4$–NaH$_2$PO$_4$, pH 7.3; 0.025 M EDTA; 0.1% azide; 0.1% BSA (crystalline); 0.1% Triton X-100. Adjust to pH 7.3

Standard: Frozen 20-μl aliquots of 1 mg of rCRF per milliliter in 0.1 N HOAc–0.1% BSA. Just prior to assay, the standard is serially diluted with assay buffer.

CRF antibody: anti-rat CRF rC-68 (see above). The 5/2/83 bleeding is used at a final dilution of 1/1,000,000. Antibody is diluted 1/100 in RIA buffer, aliquoted, and frozen. Fresh aliquot is used for each assay.

CRF tracer: ^{125}I-labeled rCRF is prepared as described above and stored at 4°.

Pansorbin (Calbiochem) diluted 1/80 with assay buffer

Saline–EDTA: 0.15 M NaCl, 5 mM disodium ethylenediaminetetraacetic acid.

All steps of the assay are carried out with chilled reagents; tubes are partially immersed in ice water. On the first day, 300 μl of buffer with antibody (diluted 1/600,000) is added to borosilicate glass tubes containing standard or test samples or buffer-only blanks in a volume of 100 μl. All treatments are tested in duplicate. Standards ranging from 0.5 to 150 fmol are used; samples are generally given at 3–4 dose levels. To control for "block effects," standard curves are included at beginning and end of all assays. After these additions, tubes are shaken and placed in the cold room at 4° for 18–24 hr.

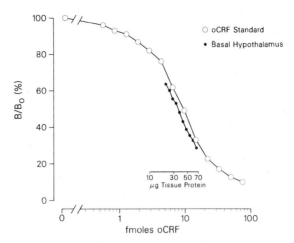

FIG. 2. Radioimmunoassay of acid extract of basal rat hypothalamus using antiserum oC-24 and ^{125}I-oCRF tracer and oCRF standard.

On day 2, ca. 10,000 cpm of ^{125}I-labeled rCRF tracer are added in a volume of 100 μl to all tubes, and the assay is shaken and returned to the cold room.

On day 3, free tracer is separated from tracer bound to antibody by the addition of fixed *Staphylococcus aureus* (Pansorbin). Add 100 μl of Pansorbin (1/80), shake racks, and incubate at room temperature for 20 min. Add 1 ml of saline–EDTA and centrifuge for 45 min at 2400 g at 4°. Decant supernatants, and count pellet in a gamma counter. Results are calculated using Model Two of the logit/log radioimmunoassay data processing program RIA PROG obtained from the Biomedical Computing Technology Information Center at Vanderbilt Medical Center.

With the availability of partial information on the sequence of ovine CRF, we prepared an N-terminal fragment, [Tyr^{21},Gly^{22}]oCRF(1-22), which could be (nonexclusively) coupled and iodinated at the Tyr^{22} residue. Immunization of rabbits with this analog coupled by bisdiazotization to human α-globulins led to antiserum oC-24, which has been our most useful antiserum for the detection of CRF-like peptides in various species including sheep, rat (Figs. 2 and 3A), dog, human being, monkey, frog, and fish. Radioimmunoassays with C-24 detect synthetic rat CRF, sauvagine, and urotensin-I although the latter two peptides give competition curves that are nonparallel with oCRF. The other antisera oC-28, oC-29, oC-30, C-33, and C-35 were subsequently developed against coupled or polymerized oCRF and, when used with ^{125}I-labeled oCRF tracer, the

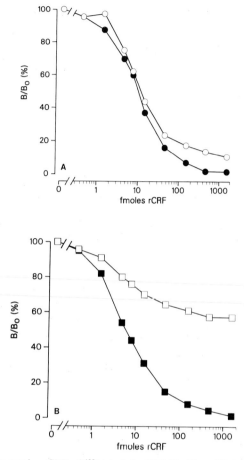

FIG. 3. Displacement by rCRF of ^{125}I-labeled rCRF (●, ■) or ^{125}I-labeled oCRF (○, □) binding to antiserum oC-24 (panel A) or antiserum oC-30 (panel B).

resultant RIAs are middle and C-terminally directed and detect nonovine CRFs very poorly. Following the characterization of rat CRF,[2] we have developed antisera to synthetic rat CRF. One of these, rC-68 (Fig. 4) is used in an RIA at a final dilution of 1/1,000,000 with ^{125}I-labeled rCRF as a tracer; the RIA exhibits an IC_{50} of 3 fmol and can be routinely used to detect ≤1 fmol of rCRF.

Although RIAs with antisera such as oC-30 detect rCRF very poorly with ^{125}I-labeled oCRF as tracer, when ^{125}I-labeled rCRF is used with oC-30 the resultant RIAs are quite sensitive to rCRF (Fig. 3B). Thus the two tracers select different populations of antibodies from this polyclonal anti-

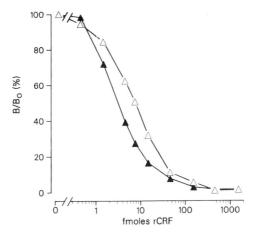

FIG. 4. Displacement by rCRF of ^{125}I-labeled rCRF binding to antiserum oC-24 (△) or antiserum rC-68 (▲).

serum. Our RIA results with oC-30 are now reconciled with immunohistochemical results, which showed good staining of rat brain with oC-30.[8]

In order to develop the most sensitive assay for a particular application, various combinations of antisera and tracers should be tried, particularly if a homologous system is not available. For example, we find an RIA using oC-30 with ^{125}I-labeled rCRF to give the highest readings of partially purified human CRF fractions.

Extraction of CRF

A variety of procedures have been employed for the extraction and/or concentration of CRF in tissues and biological fluids. An acid extraction method for tissue fragments and a Bond Elute method for aqueous samples are given.

Reagents

1.0 N HOAc + 0.1 N HCl
Triethyl ammonium formate (TEAF): 11.5 ml of HCOOH (88%) per liter of dH$_2$O; pH adjusted to 3.0 with triethylamine
Acetonitrile, HPLC grade
Bond Elute columns: 6 ml containing 500 mg of C$_{18}$ sorbent
TEAF–CH$_3$CN (4:6): 4 volumes TEAF mixed with 6 volumes of CH$_3$CN

[8] L. W. Swanson, P. E. Sawchenko, J. Rivier, and W. W. Vale, *Neuroendocrinology* **36**, 165 (1983).

Acid Extraction

Tissues are removed, dissected, and quickly frozen on Dry Ice. The frozen tissue is subsequently added to a glass tube containing a 20-fold excess (v/w) of a hot (>90°) mixture of HOAc (1.0 N) and HCl (0.1 N) and heated for 5 min in a boiling water bath. The tube is then cooled on ice, and the tissue is homogenized with a Beckman Polytron. The samples are centrifuged at 2000 g and the supernatants are removed and lyophilized.

Bond Elute Extraction

Plasma, serum, cerebral spinal fluid, serum containing cell culture medium, clear tissue extracts (containing ≤50 mg of protein per milliliter) can be extracted by adsorption to hydrophobic derivatized silica. Usually a ≤1 ml sample is treated with ≤50 μl of 2% Triton X-100 (to give 0.1%) and then added to 1 ml of TEAF and applied to a prewetted 6-ml Bond Elute column containing 500 mg of C_{18} sorbent. The column has been prewetted by passing through 6 ml of TEAF–CH_3CN (4:6) followed by 6 ml TEAF. The sample is applied at a rate of 2 ml per 5 min. An additional 6 ml TEAF is slowly applied to "wash" the column, and then the CRF-containing fraction is eluted very slowly with 4 ml of TEAF–propanol (1:1). The eluted fraction is collected in glass borosilicate tubes and evaporated in a Savant Speed Vac. Recoveries should be ascertained and can sometimes be improved by adding Triton X-100 or BSA to samples prior to evaporation. Triton X-100 should obviously not be added if the sample is to be bioassayed or in amounts that would interfere with an RIA.

When extracting, by hydrophobic adsorption, samples with high levels of lipid, a defatting step may be desirable. Plasma samples or tissue extracts can be defatted by adding 7 volumes of diethyl ether followed by vortexing, centrifuging, and aspirating and discarding the upper phases (this step can be repeated if necessary).

Levels in CRF-rich tissues, such as hypothalamus, can be measured by RIA of acidic extracts. In samples such as portal plasma, where the concentration of CRF per weight sample is much lower, purification must be accomplished prior to assay in order to avoid nonspecific interference with the RIA. With proper validation, using one or both of these extraction methods, CRF can be assayed in biological samples by radioimmunoassay and in limited situations by cell culture bioassay. Although most routine results would be obtained with the RIA, major findings should be confirmed with the bioassay. Under some circumstances, the CRF-like peptides should be highly purified and subjected to analytical procedures including HPLC analysis, peptide mapping, amino acid analysis, and ultimately sequence analysis.

Acknowledgments

This research was supported in part by NIH Grants AM26741, HD13527, AA03504, and The Rockefeller Foundation. The research was conducted in part by the Clayton Foundation for Research, California Division. Drs. W. Vale and C. Rivier are Clayton Foundation Investigators.

The authors wish to thank Dr. D. Orth and H. Friesen for the gifts of antisera to ACTH and β-endorphin. We further recognize the assistance of M. Tam, A. Corrigan, R. Kaiser, R. Galyean, and S. McCall.

[39] Direct Radioligand Measurement of Dopamine Receptors in the Anterior Pituitary Gland

By MARC G. CARON, BRIAN F. KILPATRICK, and DAVID R. SIBLEY

The physiological actions of catecholamines are mediated at the cellular level in target tissues by specific receptors. Originally, Ahlquist[1] classified the responses to the classical adrenergic catecholamines as α- and β-adrenergic responses. Much later came the notion that dopamine, a precursor of norepinephrine and epinephrine, elicited physiological effects that might be mediated via specific dopaminergic receptors,[2,3] especially in the central nervous system (CNS). More recently it has become evident that the multiple effects of dopamine in the CNS or in peripheral tissues are mediated by at least two different subclasses of receptors referred to as D_1 and D_2.[4] D_1-dopamine receptors are biochemically coupled to the stimulation of adenylate cyclase and are found typically in the brain caudate nucleus region and in the parathyroid gland, where dopamine stimulates the release of parathyroid hormone. D_2-dopamine receptors have been shown to be coupled to the inhibition of adenylate cyclase and are also found in brain caudate nucleus as well as in the pituitary gland, where dopamine inhibits the secretion of prolactin from the anterior lobe and α-MSH from the intermediate lobe. The purpose of

[1] R. P. Ahlquist, *Am. J. Physiol.* **153**, 586 (1948).
[2] J. L. McNay and L. I. Goldberg, *J. Pharmacol. Exp. Ther.* **151**, 23 (1966).
[3] R. B. Godwin-Austen, E. B. Tomlinson, C. C. Frears, and H. W. L. Kok, *Lancet* **2**, 165 (1969).
[4] J. W. Kebabian and D. B. Calne, *Nature (London)* **227**, 93 (1979).

this essay is to assemble methods and procedures that have been developed to study these D_2-dopamine receptors by direct ligand binding in the anterior pituitary gland.

General Considerations

The principle behind the technique of direct ligand binding has been reviewed previously.[5,6] Briefly, a radiolabeled ligand, which should possess the same pharmacological properties as the unlabeled ligand, is incubated with a particulate, a dispersed cell preparation, or a solubilized preparation derived from the tissue to be studied. The ligands that have been the most extensively used to study dopamine receptors in the anterior pituitary gland include [^{3}H]dihydroergocryptine,[7] [^{3}H]spiroperidol,[8] [^{3}H]domperidone,[9] [^{3}H]apomorphine,[10] and [^{3}H]N-propylapomorphine.[11] A large number of other ligands have been used to study dopamine receptors, mainly in brain tissue preparations. These studies have been reviewed by Seeman.[12] [^{3}H]Dihydroergocryptine behaves as a dopaminergic agonist in the anterior pituitary gland; i.e., it inhibits prolactin release from dispersed cells in primary culture.[7] However, in ligand binding studies [^{3}H]dihydroergocryptine does not behave as a classical agonist.[13] Thus, mainly for technical reasons, the agonist [^{3}H]N-propylapomorphine and the high-affinity antagonist [^{3}H]spiroperidol have been the more widely used ligands in this system.[14-16]

After incubation of the tissue preparations with the labeled ligand for a period sufficient to allow equilibrium binding, the bound ligand is usually separated from free ligand by one of two techniques. For studies of receptors on dispersed cells or membrane preparations, a vacuum filtration technique over glass-fiber filters represents the method of choice to quantify bound ligand. This technique provides a rapid way to separate bound

[5] L. T. Williams and R. J. Lefkowitz, "Receptor Binding Studies in Adrenergic Pharmacology." Raven, New York, 1978.
[6] E. Burgisser and R. J. Lefkowitz, "Methods in Neurobiology" (P. J. Marangos, ed.), Academic Press, New York, in press.
[7] M. G. Caron, M. Beaulieu, V. Raymond, B. Gagne, J. Drouin, R. J. Lefkowitz, and F. Labrie, *J. Biol. Chem.* **253**, 2244 (1978).
[8] I. Creese, R. Schneider, and S. H. Snyder, *Eur. J. Pharmacol.* **46**, 377 (1977).
[9] M.-P. Martres, M. Baudry, and S. C. Schwartz, *Life Sci.* **23**, 1781 (1978).
[10] I. Creese, T. Prosser, and S. H. Snyder, *Life Sci.* **23**, 495 (1978).
[11] M. Titeler and P. Seeman, *Eur. J. Pharmacol.* **56**, 291 (1979).
[12] P. Seeman, *Pharmacol. Rev.* **32**, 229 (1980).
[13] D. R. Sibley and I. Creese, *Mol. Pharmacol.* **23**, 585 (1983).
[14] A. De Lean, B. F. Kilpatrick, and M. G. Caron, *Mol. Pharmacol.* **22**, 290 (1982).
[15] B. F. Kilpatrick, A. De Lean, and M. G. Caron, *Mol. Pharmacol.* **22**, 298 (1982).
[16] D. R. Sibley, A. De Lean, and I. Creese, *J. Biol. Chem.* **257**, 6351 (1982).

from free ligand that minimizes the dissociation of bound ligand during the separation, allowing estimation of equilibrium binding conditions. For studies of receptors in solubilized preparations, the use of molecular exclusion chromatography with small Sephadex G-50 (fine) (0.6 × 12 cm) columns at 4° provides an effective means of quantifying bound ligand under equilibrium binding conditions. These techniques of separation are adequate when the rate of dissociation of the labeled ligand used is slow enough so that the time required for the separation of bound from free ligand is negligible relative to the half-time of dissociation. For the D_2-dopamine receptor of the pituitary gland, the ligands [^{3}H]N-propylapomorphine and [^{3}H]spiroperidol have relatively slow dissociation rates with dissociation constants (K_d values) in the range of 50–200 pM for the receptor. Moreover, these slow dissociation conditions can be favored by performing the separation at temperature of 0–4° as described below.

For both ^{3}H-labeled agonist and ^{3}H-labeled antagonist binding, determination of nonspecific (nonreceptor) binding can be obtained by carrying out parallel incubations in the presence of 1 μM (+)-butaclamol. At this concentration the antagonist (+)-butaclamol blocks the receptor binding of both classes of ligand so that only nonspecific binding is observed. The specific or receptor binding of the ligand is obtained by subtracting this nonspecific binding from the total ligand binding observed in the absence of (+)-butaclamol.

Dopamine Receptors in Anterior Pituitary Membranes

Membrane Preparation from Anterior Pituitary Glands

The procedure described below has been applied to both bovine and porcine anterior pituitary tissue. Fresh pituitary glands are obtained from a local slaughterhouse and kept in ice-cold buffered saline or 25 mM Tris-HCl–2 mM MgCl$_2$ (pH 7.4 adjusted at 25°), until preparation of the membranes. Alternatively, pituitary glands can be quickly frozen in liquid N$_2$ immediately after removal and stored for up to 1–2 weeks. The anterior lobes are dissected from the posterior and intermediate lobes, finely minced, and homogenized with ~12 strokes of a glass–Teflon homogenizer in 250 mM sucrose–2 mM MgCl$_2$–25 mM Tris-HCl (pH 7.4 at 4°) (buffer A) (20 ml/g wet weight of tissue). All steps are performed at 0–4°. The homogenate is filtered through two layers of cheesecloth and centrifuged at 300 g for 10 min. The supernatant is retained, and the pellet is rehomogenized and centrifuged again. The combined supernatants are layered over 50% sucrose–2 mM MgCl$_2$–50 mM Tris-HCl (pH 7.4 at 4°) (10 ml) and centrifuged at 30,000 g for 30 min in an SS-34 Sorvall rotor.

The sucrose–buffer interface is collected, diluted with buffer A, and centrifuged at 30,000 g for 20 min. The pellet is washed by resuspension in buffer A and centrifugation at 30,000 g for 20 min. The resulting pellet is suspended and homogenized in 50 mM Tris-HCl–6 mM MgCl$_2$–2 mM EDTA–100 mM NaCl–0.1% ascorbate–10 μM pargyline (pH 7.4 at 25°) (buffer B) at a concentration of 2–3 mg of protein/ml (2 ml of buffer per gram wet weight of tissue). The particulate preparations thus obtained can be frozen in liquid nitrogen and stored at −90° for 1–2 weeks without loss of binding activity. This particulate fraction, which contains elements of plasma membranes as well as other cellular organelles but is essentially devoid of secretory granules, contains a major proportion (75–80%) of the overall dopamine receptor binding activity of this tissue.[7,14]

Radioligand Binding Assays in Anterior Pituitary Membrane Preparations

One hundred microliters of the membrane preparation (0.2–0.3 mg of protein/ml) in 50 mM Tris-HCl–6 mM MgCl$_2$–1 mM EDTA–100 mM NaCl–0.1% ascorbate–10 μM pargyline (pH 7.4 at 25°) is added to 12 × 75 mm glass tubes containing the radioligand and any unlabeled drugs in a total volume of 1 ml of the same buffer. The incubation is initiated by the addition of membranes, carried out for 60 min at 25° and then terminated by the addition of 5 ml of ice-cold 25 mM Tris-HCl–2 mM MgCl$_2$ (pH 7.4 at 4°) buffer and rapid vacuum filtration on Whatman GF/C or GF/B filter disks with four additional 5-ml washes. The filtration and washing procedure usually takes less than 10 sec. Bound radioactivity trapped on the filters is then counted by liquid scintillation spectrophotometry.[14]

Characteristics of Dopaminergic Ligand Binding to Anterior Pituitary Membranes

The kinetics of ligand binding to anterior pituitary membranes are rapid. In the porcine pituitary system, the binding of [^{3}H]N-propylapomorphine and [^{3}H]spiroperidol over a wide range of concentrations (10–200 pM) is at equilibrium after 60 min at 25°. At 25°, the half-time of dissociation of both ligands is about 20–30 min. In the bovine system most studies have been performed at 37°.[16] Under these conditions the binding of both [^{3}H]N-propylapomorphine and [^{3}H]spiroperidol reaches equilibrium within 10–15 min and dissociates with a half-time of 5–10 min. Thus, under these conditions equilibrium binding can be assessed by the procedures described above.

The pharmacological characteristics of the binding of both labeled agonists and antagonists, delineated by performing agonist and antagonist

TABLE I
PROPORTIONS OF DOPAMINERGIC SITES LABELED IN ANTERIOR PITUITARY
BY AGONIST AND ANTAGONIST LIGANDS

	Number of sites labeled	
Ligand	Porcine anterior pituitary membranes (fmol/mg protein)	Bovine anterior pituitary membranes (pmol/g tissue)
[^{3}H]Spiroperidol	82.4 ± 1.4[a]	4.3 ± 0.2[b]
[^{3}H]N-Propylapomorphine	44.7 ± 3.1	2.3 ± 0.1

[a] Results taken from DeLean et al.[14]
[b] Results taken from Sibley et al.[16]

competition curves, indicate that the anterior pituitary contains a single class of dopaminergic receptors characterized as the D_2 subtype.[14,16] However, this receptor exists in two affinity states that are differentially labeled by agonists and antagonists. In both porcine and bovine anterior pituitary membranes,[14,16] [^{3}H]spiroperidol labels about twice as many sites as the agonist [^{3}H]N-propylapomorphine (Table I). This is explained by the fact that in saturation experiments [^{3}H]spiroperidol (10–2000 pM) labels the totality of the receptor sites with a K_d = 50–300 pM, whereas the agonist [^{3}H]N-propylapomorphine labels only the portion (~50%) of these sites that exhibits high affinity (K_d ~ 200 pM) for the agonist. The remaining portion possesses an affinity for agonists that is too low (K_d ~ 15–30 nM for N-propylapomorphine) to allow labeling by direct ^{3}H-labeled agonist binding under the experimental conditions employed.[14,16] These two affinity forms of the receptor can also be evidenced by agonist competition for [^{3}H]spiroperidol binding. Competition curves of agonists for [^{3}H]spiroperidol binding are complex and have low slope factors (n_H < 1); they are best explained by assuming that the agonists discriminate between two states of the receptor with high and low affinity. As shown in Table II, the K_d values for the high-affinity agonist state of the receptor correlate particularly well with the K_d values of these agents calculated from their ability to inhibit prolactin release from rat anterior pituitary cells in culture.[7] These data suggest that the high-affinity agonist state of the receptor may be the form of the receptor mediating the physiological effect of dopaminergic agents.

Data Analysis of Ligand Binding

Saturation and competition binding data can be analyzed by a nonlinear least-squares curve-fitting procedure based on a generalized model for

TABLE II
Comparison of the Dissociation Constants of Dopaminergic Agonists for the Dopamine Receptor in Various Preparations of Anterior Pituitary

	Agents				
Preparations[a]	Dopamine	Apomorphine	ADTN	N-Propyl-apomorphine	Pergolide
Porcine membranes[b]					
K_H	66	4.7	2.3	0.081	2.0
K_L	4300	160	190	18	35
Solubilized porcine membranes[c]					
K_D	5700	310	54	19	—
Bovine membranes[d]					
K_H	190	6.3	31	0.27	3.8
K_L	15000	350	1800	26	137
Solubilized bovine membranes[e]					
K_D	3200	115	120	13	—
Dispersed bovine cells[f]	12000	380	3200	43	—
Prolactin release from rat anterior pituitary cells					
K_D	35[g]	3[g]	1.2[h]	—	4[i]

[a] K_H and K_L values represent dissociation constants for the high- and low-affinity states of the receptor. Where a single value (K_D) is shown, it indicates that both ligand and competitor bound to a single affinity state of the receptor (see text for discussion). Calculated dissociation constants are reported in nM.
[b] From DeLean et al.[14]
[c] From Kilpatrick and Caron.[29]
[d] From Sibley et al.[16]
[e] From Kilpatrick and Caron.[31]
[f] From Sibley et al.[27]
[g] From Caron et al.[7]
[h] J. Rick, M. Szabo, P. Payne, N. Kovathana, J. G. Gannon, and L. A. Frohman, *Endocrinology* (Baltimore) **104**, 1234 (1979).
[i] G. Delitala, T. Yeo, A. Grossman, N. R. Hathaway, and G. M. Besser, *J. Endocrinol.* **87**, 95 (1980).

complex ligand–receptor systems according to the law of mass action.[17-19] For saturation binding data, a model for a radioligand binding to one or two forms of the receptor (R_H and R_L) with high-affinity (K_H) or low-

[17] R. S. Kent, A. De Lean, and R. J. Lefkowitz, *Mol. Pharmacol.* **17**, 14 (1980).
[18] A. De Lean, A. A. Hancock, and R. J. Lefkowitz, *Mol. Pharmacol.* **21**, 5 (1982).
[19] P. J. Munson and D. Rodbard, *Anal. Biochem.* **107**, 220 (1980).

affinity (K_L), respectively, for the ligand can be considered.[14] In the case of competitive binding data, the curves can be initially analyzed according to a four-parameter logistic equation in order to determine the IC_{50} and the steepness factor of the curve as described.[20] The competition curves are then analyzed according to a model for one or two forms of the receptor potentially discriminated by both the radioligand and the competitor. Statistical analysis comparing "goodness of fit" between one- and two-affinity state models can be provided and are used to determine the most appropriate model for the ligand being examined. For ligands whose interaction with the dopamine receptor were best described by two states of binding, K_H and K_L represent the high- and low-affinity dissociation constants whereas R_H and R_L represent the proportion of receptor corresponding to the high- and low-affinity states, respectively. The validation and the advantages as well as the limitations of the quantitative analysis have been examined and documented by De Lean et al.[18]

Regulation of Ligand Binding to Anterior Pituitary Membranes

By Guanine Nucleotides. In membranes from both bovine and porcine anterior pituitary glands, guanine nucleotides can interconvert the portion of the receptor having high affinity for agonists into a single-receptor population having low affinity for agonists.[14,16] This effect of nucleotides can be demonstrated by (1) a shift to the right and a steepening of agonist–^{3}H-labeled antagonist competition curves, or (2) a decrease in the number of receptor sites labeled by ^{3}H-labeled agonist binding, or (3) an accelerated dissociation of ^{3}H-labeled agonist ligands in the presence of nucleotides. In the porcine anterior pituitary,[14,21] but not in the bovine system,[16] the antagonist spiroperidol has been found to discriminate between these two forms of the receptor in a manner reciprocal to agonists that is also sensitive to nucleotides. However, the discrimination of these two states of the receptor by the antagonist is much more subtle than for agonists. The variance between the two systems may stem from small differences in the stability of the interaction of the receptor with the putative protein, which confers on the system its nucleotide responsiveness.

By Divalent and Monovalent Cations. Divalent cations have been shown to promote agonist–receptor interactions in many catecholamine receptor systems.[22,23] In the pituitary D_2-dopamine receptor system, the divalent cations Mg^{2+}, Ca^{2+}, and Mn^{2+} have been demonstrated to poten-

[20] A. De Lean, P. J. Munson, and D. Rodbard, *Am. J. Physiol.* **235**, E97 (1978).
[21] A. De Lean, B. F. Kilpatrick, and M. G. Caron, *Endocrinology (Baltimore)* **110**, 1064 (1982).
[22] L. T. Williams, D. Mullikin, and R. J. Lefkowitz, *J. Biol. Chem.* **253**, 2984 (1978).
[23] B. S. Tsai and R. J. Lefkowitz, *Mol. Pharmacol.* **14**, 540 (1978).

tiate the agonist-induced high-affinity receptor binding state.[24] In the complete absence of all divalent cations there is very little [^{3}H]N-propylapomorphine binding to the high-affinity receptor state comprising only about 25% of that which is maximally obtainable. The effect of divalent cations is to increase both the affinity (decrease in K_d) and binding capacity of N-[^{3}H]N-propylapomorphine. Mg^{2+} exhibits an EC_{50} of about 1 mM for this effect, the maximum potentiation occurring at about 10 mM, at which point the binding capacity of [^{3}H]N-propylapomorphine is 50–55% of that for [^{3}H]spiroperidol (see above). Similar effects of divalent cations are observed in agonist–[^{3}H]spiroperidol competition experiments where Mg^{2+} has been shown to increase the agonist affinity (decrease in K_H) for, and the proportion of, the high-affinity receptor state (% R_H).[24]

In contrast to the effects of divalent cations, monovalent cations, including Na^+, K^+, and Li^+, are without effect in potentiating agonist–receptor interactions. This suggests that the effects of divalent cations are not due simply to changes in ionic strength. Furthermore, monovalent cations do not affect the potentiation of agonist binding by divalent cations.[24] Neither divalent nor monovalent cations affect the binding of [^{3}H]spiroperidol or other antagonist ligands to the pituitary dopamine receptor. One report, however, has indicated that the receptor binding of substituted benzamide antagonists may be regulated by sodium.[25]

In summary, the major effect of divalent cations on pituitary D_2-dopamine receptor agonist binding is to increase both the proportion and affinity of the high-affinity agonist binding state as reflected in direct ^{3}H-labeled agonist saturation isotherms as well as in indirect agonist/^{3}H-labeled antagonist competition experiments. Any investigation of agonist interactions with pituitary dopamine receptors should thus be performed using a maximally effective divalent cation concentration.

By Ascorbate. A larger number of the studies that have examined dopamine receptor binding in anterior pituitary have included as a standard condition in their membrane preparations or assays the antioxidant ascorbate. The presence of ascorbate does not affect the binding of antagonists such as [^{3}H]spiroperidol, but as documented by Leff *et al.*[26] ascorbate is required to observe high-affinity agonist interactions with the receptor. The effect of ascorbate appears to be a dramatic reduction in the nonspecific binding of a ^{3}H-labeled ligand such as a N-propylapomorphine. Only in the presence of ascorbate is the binding of [^{3}H]N-propylapomorphine freely and completely reversible.[26] The same results are

[24] D. R. Sibley and I. Creese, *J. Biol. Chem.* **258**, 4957 (1983).
[25] E. Stefanini, Y. Clement-Cormier, F. Vernaleone, P. Deroto, A. M. Marchisio, and R. Collu, *Neuroendocrinology* **32**, 103 (1981).
[26] S. Leff, D. R. Sibley, M. Hamblin, and I. Creese, *Life. Sci.* **29**, 2081 (1981).

observed with the dissociation of N-[^{3}H]propylapomorphine in solubilized preparations of porcine pituitary membranes (unpublished observations). The effective concentration of ascorbate on the binding of [^{3}H]N-propylapomorphine to bovine anterior pituitary membranes appears to be between 1 and 10 mM or 0.02 and 0.2% ascorbate. This effect is also observed in other dopaminergic systems, such as corpus striatum preparations.[26] Thus, routinely ascorbate or another antioxidant should be included in assays of dopamine receptor binding.

Dopamine Receptors on Dispersed Anterior Pituitary Cells

One of the advantages of utilizing dispersed cell preparations for the characterization of pituitary dopamine receptors is that such investigations can be conducted under conditions closely approximating the normal physiological state. Thus, ligand binding studies of dopamine receptors in intact cells offer complementary information to that derived from membrane (see above) or solubilized (see below) preparations. In addition, the use of dispersed cell preparations allows the direct correlation between ligand binding properties and physiological events, such as prolactin release, in the same experimental system. Techniques of radioligand binding using dispersed cells are similar to those employing washed membrane preparations, the major difference being the method of tissue preparation.

Preparation of Dispersed Anterior Pituitary Cells

The following procedure for preparing dispersed bovine anterior pituitary cells has been used successfully by Sibley *et al.*[27] for the characterization of ligand binding to the dopamine receptor in intact cells. Fresh pituitary glands are obtained from a local slaughterhouse. The anterior lobes are removed and finely minced, using a scalpel, into pieces approximately 1 mm^3. The minced tissue is washed three times with an ice-cold buffer consisting of 25 mM HEPES, 137 mM NaCl, 5 mM KCl, 0.7 mM Na$_2$HPO$_4$, 0.2% glucose, 50 μg/ml gentamicin sulfate, pH 7.4, at 25° (buffer A). The tissue is then transferred to a Wheaton Celstir suspension culture flask and resuspended to a final concentration of 100 mg wet weight per milliliter at 37° in buffer A containing 0.4% collagenase (Worthington type II), 0.4% BSA (Sigma fraction V), and 10 μg/ml DNase II (Sigma type IV). The tissue suspension is then stirred constantly for 2 hr at 37° with periodic removal of the gelatinous material accumulating on the stir arm. After 2 hr, the suspension is vigorously triturated using a

[27] D. R. Sibley, L. C. Mahan, and I. Creese, *Mol. Pharmacol.* **23**, 295 (1983).

small-bore pipette followed by filtration through two layers of cheesecloth. The filtrate is subsequently centrifuged at 300 g for 10 min. The resulting cell pellet is resuspended in standard Dulbecco's modified Eagle's medium supplemented with 10% newborn calf serum, 50 μg/ml gentamicin sulfate, and 20 mM HEPES, pH 7.4, at 25° (buffer B). The cells are spun down as before and washed twice more with buffer B. Before the final centrifugation, the cells are suspended in buffer A and cell counts and viability are determined. Primarily single cells as well as occasional clumps of two–six cells are present. Routine yields vary between 2×10^7 and 7×10^7 cells per gram wet weight of anterior pituitary tissue. After the final centrifugation, the cells are resuspended in 37° assay buffer consisting of 50 mM Tris-HCl, pH 7.7, at 25°, 120 mM NaCl, 5 mM KCl, 2 mM CaCl$_2$, 1 mM MgCl$_2$ and 5.7 mM ascorbic acid (buffer C) for direct use in the binding assay. This assay buffer is approximately isotonic and maintains the viability of the cells throughout the radioligand binding assay as described by Sibley et al.[27]

Radioligand Binding Assays to Dispersed Anterior Pituitary Cells

Approximately 4×10^6 cells in buffer C are incubated in the presence of ligand and competing drugs in a final volume of 1 ml for 15 min at 37°. The incubation is terminated by diluting the incubation with 5 ml of ice-cold 50 mM Tris-HCl, pH 7.4, at 25° and filtering under vacuum on GF/B glass fiber filters followed by three more 5-ml washes of the filters. The total filtration and washing procedure is accomplished in less than 10–15 sec. Cell concentrations higher than 5×10^6 cells/ml have a tendency to slow the rate of filtration considerably and thus should be avoided.

Characteristics of Dopaminergic Ligand Binding to Dispersed Anterior Pituitary Cells

Using this dispersed cell preparation, dopamine receptors can be easily identified with the antagonist ligand [^{3}H]spiroperidol. [^{3}H]Spiroperidol binding on dispersed bovine anterior pituitary cells exhibits properties expected for binding to a D$_2$-dopamine receptor, including appropriate kinetics, high affinity, saturability, stereoselectivity, and pharmacological specificity of binding.[27] These binding properties are essentially identical to those found in washed membrane preparations, suggesting that the receptors identified in these two experimental paradigms are identical. In contrast to membrane preparations, however, there is no detectable high-affinity, stereospecific, or saturable binding of the radiolabeled agonist, [^{3}H]N-propylapomorphine, to dispersed pituitary cells. This lack of high-affinity ^{3}H-labeled agonist binding on intact cells has been shown not to be due to rapid destruction or uptake of ligand.[27] Furthermore, washed membrane preparations derived from these cells do exhibit high-affinity [^{3}H]N-propylapomorphine binding, suggesting that the receptors are not damaged during the cell dispersion procedure.

An explanation for this lack of [^{3}H]N-propylapomorphine binding on intact cells can be found by examining agonist–[^{3}H]spiroperidol competition curves using these cells. In contrast to the two-component agonist competition curves seen in membrane preparations (see above), in intact cells agonist competition curves are monophasic and exhibit affinities similar to the low-affinity agonist binding state seen in membranes (Table II).[27] Addition of guanine nucleotides does not alter the agonist–[^{3}H]spiroperidol curves in intact cells. These data suggest that in intact cells endogenous GTP regulates agonist binding in an identical fashion as exogenous guanine nucleotides in membrane preparations. Thus, any high-affinity agonist binding will be rapidly reversed by the intracellular GTP, directly eliminating ^{3}H-labeled agonist binding or the high-affinity component in agonist–^{3}H-labeled antagonist competition experiments.

Under these conditions, the binding capacity of [^{3}H]spiroperidol was about 9.0 fmol/10^6 cells in these dispersed cell preparations. Assuming that the population of mammotrophs in this preparation might be around 20% of the total cell population, and assuming that dopamine receptors are present only on these prolactin-secreting cells, the number of 2.7×10^4 sites per cells can be tentatively estimated.

Dopamine Receptor in Solubilized Preparations of Anterior Pituitary Membranes

Ligand binding in dispersed cell or membrane preparations allows the pharmacological identification and characterization of a membrane-bound receptor site. However, the biochemical characterization of such a receptor requires the solubilization of the sites with retention of their ability to interact specifically with ligands. The solubilization of the D_2-dopamine receptor and characterization of soluble receptor binding has been documented for both porcine and bovine anterior pituitary receptors.[28–31] Successful solubilization with the retention of the intrinsic binding properties of the receptor fulfills the necessary requirement for the further characterization of the receptor and its interactions with other components of the effector system.

[28] M. G. Caron, B. F. Kilpatrick, and A. De Lean, in "Dopamine Receptors" (J. W. Kebabian and C. Kaiser, eds.), pp. 73–92. ACS Symposium Series 224, 1983.
[29] B. F. Kilpatrick and M. G. Caron, submitted.
[30] B. F. Kilpatrick and M. G. Caron, Fed. Proc., Fed. Am. Soc. Exp. Biol. **42**, 1875 (1983).
[31] B. F. Kilpatrick and M. G. Caron, J. Biol. Chem., in press (1983).

Solubilization of Anterior Pituitary Membranes

Membrane preparations stored frozen are thawed and centrifuged at 30,000 g for 10 min, and the pellet is resuspended in one-third the original volume of the membrane suspension with ice-cold 100 mM NaCl–2 mM MgCl$_2$–25 mM Tris-HCl–1% digitonin (Gallard-Schlessinger) (w/v), pH 7.4 at 4°. The ratio of membrane protein to detergent is usually 9–10 mg of protein per 1 ml of 1% digitonin. The suspension is stirred slowly at 0–4° for 30 min and centrifuged at 40,000 g for 45 min. Centrifugation at 100,000 g for 1 hr gives similar results. The supernatant containing the soluble receptor is kept at 4° and used immediately. Nonspecific binding is determined in the presence of 1 μM (+)-butaclamol. Typically 20–25% of the available membrane-bound receptor can be solubilized as assessed by [^{3}H]spiroperidol binding to the solubilized preparations.

Radioligand Binding Assay in Solubilized Anterior Pituitary Membranes

Solubilized receptor preparations, 200 μl (0.3–0.8 mg of protein per 200 μl), are incubated with [^{3}H]spiroperidol in the presence or the absence of competing ligand in a total assay volume of 0.5 ml of 100 mM NaCl–2 mM MgCl$_2$–25 mM Tris-HCl–0.1% ascorbate (pH 7.4 at 4°) for 2 hr (bovine preparations) and 6–18 hr at 4° (porcine preparations). At the end of the incubation, bound ligand can be separated from free ligand by Sephadex G-50 chromatography.[32] Nonspecific binding is determined in the presence of 1 μM (+)-butaclamol.

Characteristics of Dopaminergic Ligand Binding to Solubilized Preparations

In solubilized preparations of anterior pituitary membranes [^{3}H]spiroperidol binds to a saturable site of high affinity ($K_d \sim 450$ pM). Nonspecific binding as measured in the presence of 1 μM (+)butaclamol is approximately 25% of total [^{3}H]spiroperidol binding. Binding of the antagonist ligand displays stereoselectivity and D$_2$-dopaminergic specificity. Thus, the K_d values of the (+) and (−) isomers of butaclamol for competing for [^{3}H]spiroperidol binding are 3 and 2500 nM, respectively, and agonists compete with a typical D$_2$ order of potency (Table II). Agonist competition curves for [^{3}H]spiroperidol binding indicate that agonist binds to a single-affinity state of the receptor with K_d values corresponding essentially to the low-affinity form of the receptor documented in

---

[32] M. G. Caron and R. J. Lefkowitz, *J. Biol. Chem.* **251,** 2374 (1976).

membrane binding for these agonists. As expected, ^{3}H-labeled agonist binding cannot be demonstrated in solubilized preparations and guanine nucleotides do not affect agonist–^{3}H-labeled antagonist competition curves[29–31]; this suggests that the interactions of the receptor with the component which confers nucleotide sensitivity to the system are lost in solubilized preparations. However, a stable [^{3}H]N-propylapomorphine–receptor complex can be demonstrated in solubilized preparations if the ^{3}H-labeled agonist–receptor complex is preformed in the membrane prior to solubilization.[29–31] The agonist–receptor complex solubilized in this way retains its nucleotide sensitivity in that nucleotides can accelerate the rate of dissociation of [^{3}H]N-propylapomorphine from the receptor–agonist complex. These same observations on soluble receptor binding have been documented for the D_2-dopamine receptor in the caudate nucleus[33] as well as the β-adrenergic[34] and α_2-adrenergic receptors.[35,36]

Conclusions

The D_2-dopamine receptor of the anterior pituitary represents a useful model system in which to study both the biochemical mechanisms and physiological regulation of this subtype of dopamine receptors. Direct ligand binding can be correlated with the ability of a dopaminergic agent to inhibit the release of prolactin from pituitary cells. Moreover, ligand binding data can now be correlated with dopaminergic-mediated inhibition of adenylate cyclase in this tissue.[37] Table II presents a comparison of the dissociation constants (K_d) for a series of dopaminergic agonists obtained by ligand binding studies to membranes and solubilized preparations of porcine and bovine anterior pituitary gland and to dispersed bovine pituitary cells. These data are further compared with the EC_{50} of these same agonists for their ability to inhibit prolactin release in dispersed anterior pituitary cells. A pharmacologically similar system (D_2-receptor) exists in the intermediate lobe of the rat pituitary gland, where dopamine can inhibit the β-adrenergically stimulated secretion of α-melanocyte-stimulating hormone.[38] However, this system presents some limitations for the eventual biochemical characterization of the receptor.

[33] S. E. Leff and I. Creese, *Biochem. Biophys. Res. Commun.* **108**, 1150 (1982).
[34] L. E. Limbird and R. J. Lefkowitz, *Proc. Natl. Acad. Sci. U.S.A.* **75**, 228 (1978).
[35] T. Michel, B. B. Hoffman, R. J. Lefkowitz, and M. G. Caron, *Biochem. Biophys. Res. Commun.* **100**, 1131 (1981).
[36] S. Smith and L. E. Limbird, *Proc. Natl. Acad. Sci. U.S.A.* **78**, 4026 (1981).
[37] G. Giannasthasio, M. E. DeFerrari, and A. Spada, *Life. Sci.* **28**, 1605 (1981).
[38] M. Munemura, R. L. Eskay, and J. W. Kebabian, *Endocrinology (Baltimore)* **106**, 1795 (1980).

The anterior pituitary represents a suitable source for the purification and characterization of the D_2-dopamine receptor, as it has been possible to solubilize the receptor with retention of its pharmacological functions.

On a more practical basis, this system represents an ideal screening system for new potential dopaminergic agonists and antagonists, as both pharmacological and biochemical activities can be evaluated in the same system.

[40] Ornithine Decarboxylase: Marker of Neuroendocrine and Neurotransmitter Actions

By THEODORE A. SLOTKIN and JORGE BARTOLOME

Ornithine decarboxylase (ODC; EC 4.1.1.17) catalyzes the formation of putrescine from ornithine, the initial step in the synthesis of the polyamines spermidine and spermine (Fig. 1).[1,2] The polyamines are ubiquitously distributed in tissues and appear to be associated intimately with tissue growth and with cellular replication and differentiation.[3,4] Stimulation of animal and bacterial polymerase *in vitro* by physiological concentrations of polyamines has been well documented, and polyamines play a regulatory role in protein synthesis.[5–9] The ODC molecule itself may participate in growth regulation, as it appears to be an initiation factor for RNA polymerase I.[10] Polyamine, nucleic acid, and protein synthesis have been shown to vary in a parallel fashion in virtually all growing systems, ranging from bacteria through mammalian tissues.[11–28] These and other

---

[1] A. E. Pegg and H. G. Williams-Ashman, *Biochem. J.* **108**, 533 (1968).
[2] D. H. Russell and S. H. Snyder, *Proc. Natl. Acad. Sci. U.S.A.* **60**, 1420 (1968).
[3] U. Bachrach, "Function of Naturally Occurring Polyamines." Academic Press, New York, 1973.
[4] D. H. Russell and B. G. M. Durie, "Polyamines as Biochemical Markers of Normal and Malignant Growth." Raven, New York, 1978.
[5] K. A. Abraham, *Eur. J. Biochem.* **5**, 143 (1968).
[6] P. L. Ballard and H. G. Williams-Ashman, *J. Biol. Chem.* **241**, 1602 (1966).
[7] A. Raina and J. Janne, *Fed. Proc. Fed. Am. Soc. Exp. Biol* **29**, 1568 (1970).
[8] A. Raina and J. Janne, *Med. Biol.* **53**, 121 (1975).
[9] H. Tabor and C. W. Tabor, *Pharmacol. Rev.* **16**, 245 (1964).
[10] C. A. Manen and D. H. Russell, *Science* **195**, 505 (1977).
[11] S. S. Cohen, N. Hoffner, M. Jansen, M. Moore, and A. Raina, *Proc. Natl. Acad. Sci. U.S.A.* **57**, 721 (1967).
[12] A. Raina and S. S. Cohen, *Proc. Natl. Acad. Sci. U.S.A.* **55**, 1587 (1966).
[13] A. Raina, M. Jansen, and S. S. Cohen, *J. Bacteriol.* **94**, 1684 (1967).

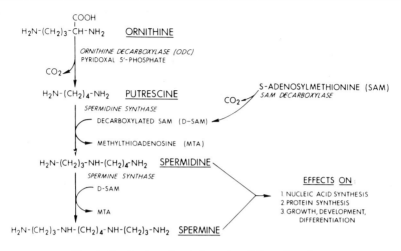

FIG. 1. Synthesis and putative roles of the polyamines.

studies suggest that ODC activity is stimulated in conditions in which growth is induced.

In addition to stimulation during general growth, increases in ODC activity have been demonstrated to occur in target organs exposed to specific hormones or neurotransmitters.[29-34] ODC is unusual in that it

---

[14] A. S. Dion and E. J. Herbst, *Proc. Natl. Acad. Sci. U.S.A.* **58**, 2367 (1967).
[15] A. S. Dion and E. J. Herbst, *Ann. N. Y. Acad. Sci.* **171**, 723 (1970).
[16] A. Raina, *Acta Physiol. Scand. Suppl.* **218** (1963).
[17] C. M. Caldarera, M. S. Moruzzi, C. Rossoni, and B. Barbiroli, *J. Neurochem.* **16**, 309 (1969).
[18] D. H. Russell and J. B. Lombardini, *Biochim. Biophys. Acta* **240**, 273 (1971).
[19] W. G. Dykstra and E. J. Herbst, *Science* **149**, 428 (1965).
[20] A. Raina, J. Janne, and M. Siimes, *Biochim. Biophys. Acta* **123**, 197 (1966).
[21] J. Janne and A. Raina, *Biochim. Biophys. Acta* **174**, 769 (1969).
[22] L. T. Kremzner, V. Iliev, and R. M. Starr, *Int. Congr. Ser.—Excerpta Med.* **129**, 1179 (1967).
[23] L. T. Kremzner, R. E. Barrett, and M. J. Terrano, *Ann. N. Y. Acad. Sci.* **171**, 735 (1970).
[24] L. A. Pearce and S. M. Schanberg, *Science* **166**, 1301 (1969).
[25] T. R. Anderson and S. M. Schanberg, *J. Neurochem.* **19**, 1471 (1972).
[26] T. A. Slotkin, G. Barnes, C. Lau, F. J. Seidler, P. Trepanier, S. Weigel, and W. L. Whitmore, *J. Pharmacol. Exp. Ther.* **221**, 686 (1982).
[27] T. A. Slotkin, F. J. Seidler, P. A. Trepanier, W. L. Whitmore, L. Lerea, G. A. Barnes, S. J. Weigel, and J. Bartolome, *J. Pharmacol. Exp. Ther.* **222**, 741 (1982).
[28] T. A. Slotkin, W. L. Whitmore, L. Lerea, R. J. Slepetis, S. J. Weigel, P. A. Trepanier, and F. J. Seidler, *Int. J. Dev. Neurosci.* **1**, 7 (1983).
[29] J. Bartolome, C. Lau, and T. A. Slotkin, *J. Pharmacol. Exp. Ther.* **202**, 510 (1977).
[30] C. Lau and T. A. Slotkin, *J. Pharmacol. Exp. Ther.* **208**, 485 (1979).

exhibits extremely rapid turnover ($t_{1/2}$ ca. 15–20 min in most tissues)[21,35,36]; consequently, interaction of a trophic agent with its target tissue usually results in a prompt and massive (severalfold) increase in ODC activity. Frequently, ODC stimulation is detectable less than 1 hr after administration of the agent, peaks within 3–6 hr and then declines to basal levels within 12–24 hr[4,34]; in some cases, elevation of ODC may require somewhat longer time periods.[37] In general, the rapid rise of ODC activity has enabled this enzyme to be useful as a marker for tissue stimulation. The widespread applicability of this approach is illustrated in Tables I and II, which are comprehensive, but not exhaustive, lists of cell types and tissues in which ODC stimulation by neurotransmitters and hormonal agents has been utilized.

Assay Procedures

The procedure originally devised by Russell and Snyder[2] has proved to be the most widely applicable method for assay of ODC; this technique uses carboxyl-labeled radioactive ornithine as a substrate and traps the evolved $^{14}CO_2$ for evaluation of ODC activity. With minor modifications,[37] the relative simplicity of the assay permits large numbers of samples to be analyzed in a relatively short time with minimal personnel and low cost. Typical requirements are described below.

Equipment and Supplies

Tissue disrupter
Centrifuge
Incubation bath set at 37°
Liquid scintillation counter
Glass vials, 4-dram, 24 × 62 mm; may be discarded or washed for reuse
Plastic-center wells with stem; not reusable (Kontes, Inc., Vineland, NJ, part No. 882320-0000)
Rubber serum stoppers, 19 mm stem length; these are used inverted with a permanent hole punched in the center plug to accommodate the stem of the center well; reusable after washing

---

[31] J. Janne, A. Raina, and M. Siimes, *Biochim. Biophys. Acta* **166**, 426 (1968).
[32] D. H. Russell, S. H. Snyder, and V. J. Medina, *Endocrinology (Baltimore)* **86**, 1414 (1970).
[33] D. V. Maudsley, *Biochem. Pharmacol.* **28**, 153 (1979).
[34] T. A. Slotkin, *Life Sci.* **24**, 1623 (1979).
[35] D. H. Russell and S. H. Snyder, *Mol. Pharmacol.* **5**, 253 (1969).
[36] H. Antrup and N. Seiler, *Neurochem. Res.* **5**, 123 (1980).
[37] C. Lau and T. A. Slotkin, *Mol. Pharmacol.* **16**, 504 (1979).

TABLE I
STIMULATION OF ORNITHINE DECARBOXYLASE BY NEUTRANSMITTERS AND HORMONAL PEPTIDES IN VARIOUS CELL TYPES in Vitro

Tissue	Trophic factors	References[a]
Brain cells	Insulin, growth hormone	1
Chondrocytes	Parathyroid hormone	2
Endothelial cells	Thrombin, serotonin	3
Fibroblasts	Insulin	4
Granulosa cells	FSH, LH, prostaglandins E_1 and E_2, epinephrine, epidermal growth factor, fibroblast growth factor	5–7
Hepatocytes	Glucagon, insulin, growth hormone, dexamethasone	8, 9
Hepatoma cells	Insulin, glucagon, growth hormone, dexamethasone	8, 10, 11
Kidney cells	Diethylstilbestrol, progesterone	12
Mammary gland	Insulin, prolactin, growth hormone	13, 14
Mammary tumors	Prolactin	15
Neuroblastoma	γ-Aminobutyric acid	16
Ovary cells	LH	17
Pelvic cartilage	Parathyroid hormone, prostaglandin E_1, triiodothyronine, insulin	18
Pheochromocytoma	Nerve growth factor, epidermal growth factor	19, 20

[a] Key to references
1. J. W. Yang, M. K. Raizada, and R. E. Fellows, *J. Neurochem.* **36**, 1050 (1981).
2. M. Takigawa, H. Ishida, T. Takano, and F. Suzuki, *Proc. Natl. Acad. Sci. U.S.A.* **77**, 1481 (1980).
3. P. A. Damore, H. B. Hechtman, and D. Shepro, *Thromb. Haemostasis* **39**, 496 (1978).
4. G. K. Haselbacher and R. E. Humbel, *J. Cell. Physiol.* **88**, 239 (1976).
5. J. Osterman and J. M. Hammond, *Endocrinology (Baltimore)* **101**, 1335 (1977).
6. J. Osterman and J. M. Hammond, *Horm. Metab. Res.* **11**, 485 (1979).
7. J. D. Veldhuis and S. Sweinberg, *J. Cell. Physiol.* **108**, 213 (1981).
8. L. Lumeng, *Biochim. Biophys. Acta* **587**, 556 (1979).
9. M. R. Klingensmith, A. G. Freifeld, A. E. Pegg, and L. S. Jefferson, *Endocrinology (Baltimore)* **106**, 125 (1980).
10. B. Hogan and A. Blackledge, *Biochem. J.* **130**, 78P (1972).
11. D. M. Moriarity, D. M. Disorbo, G. Litwack, and C. R. Savage, Jr., *Proc. Natl. Acad. Sci. U.S.A.* **78**, 2752 (1981).
12. Y. C. Lin, J. M. Loring, and C. A. Villee, *Biochem. Biophys. Res. Commun.* **95**, 1393 (1980).
13. R. P. Aisbitt and J. M. Barry, *Biochim. Biophys. Acta* **320**, 610 (1973).
14. J. A. Rillema, L. Y. Wing, and C. M. Cameron, *Horm. Res.* **15**, 133 (1981).
15. R. P. Frazier and M. E. Costlow, *Exp. Cell Res.* **138**, 39 (1982).
16. K. A. Diekema, P. P. McCann, and B. J. Lippert, *Am. J. Physiol.* **243**, C35 (1982).
17. M. Costa, M. Meloni, and M. K. Jones, *Biochim. Biophys. Acta* **608**, 398 (1980).
18. W. M. Burch and H. E. Lebovitz, *Am. J. Physiol.* **241**, E454 (1981).
19. H. Hatanaka, U. Otten, and H. Thoenen, *FEBS Lett.* **92**, 313 (1978).
20. G. Guroff, G. Dickens, and D. End, *J. Neurochem.* **37**, 342 (1981).

TABLE II
STIMULATION OF ORNITHINE DECARBOXYLASE BY NEUTRANSMITTERS AND HORMONAL PEPTIDES IN VARIOUS TISSUES *in Vivo*

Tissue	Trophic factors	References[a]
Adenohypophysis	Thyrotropin releasing hormone	1
Adrenal cortex	ACTH, growth hormone	4, 5
Adrenal gland	Prolactin, ACTH, growth hormone	2, 3
Adrenal medulla	Acetylcholine	6
Brain	Growth hormone, nerve growth factor, prolactin, placental lactogen, thyroxine, cortisol, vasopressin, angiotensin II, insulin, aldosterone, dexamethasone, histamine, norepinephrine	7–12
Fat	Growth hormone	13
Heart	Growth hormone, prolactin, isoproterenol, norepinephrine, triiodothyronine	3, 13–18
Kidney	Growth hormone, prolactin, ACTH, cortisol, parathyroid hormone, calcitonin, vasopressin, angiotensin, serotonin, β-endorphin, estradiol, isoproterenol, aldosterone, triiodothyronine, testosterone	3, 13, 19–25
Liver	Growth hormone, hydrocortisone, insulin, glucagon, thyroxine, fetal calf serum factor, prolactin, placental lactogen, vasopressin, angiotensin, serotonin, isoproterenol	3, 9, 25–29
Lung	Growth hormone	13
Ovary	LH, chorionic gonadotropin, prostaglandin E_2, FSH, estradiol, lutropin	30–34
Oviduct	Estradiol, diethylstilbestrol	35
Pubic symphysis	Relaxin	36
Salivary glands	Isoproterenol	37
Seminal plasma	Testosterone	38
Skin	Epidermal growth factor	39
Small intestine	Insulin	40
Spleen	Growth hormone, prolactin	3, 13
Superior cervical ganglion	Nerve growth factor	41, 42
Testis	Epidermal growth factor, FSH, LH, prostaglandin E_2, epinephrine, norepinephrine	43–45
Thymus	Growth hormone, prolactin	3, 13
Thyroid	Thyroid-stimulating hormone	46–48
Uterus	Estradiol, diethylstilbestrol, relaxin	36, 49

[a] Key to references
1. W. Combest, R. B. Chiasson, H. Klandorf, G. A. Hedge, and D. H. Russell, *Gen. Comp. Endocrinol.* **35,** 146 (1978).
2. J. H. Levine, W. E. Nicholson, G. W. Liddle, and D. N. Orth, *Endocrinology (Baltimore)* **92,** 1089 (1973).

TABLE II References (continued)

3. J. F. Richards, *Biochem. Biophys. Res. Commun.* **63,** 292 (1975).
4. R. Richman, C. Dobbins, S. Voina, L. Underwood, D. Mahaffee, H. J. Gitelman, J. Van Wyk, and R. L. Ney, *J. Clin. Invest.* **52,** 2007 (1973).
5. K. J. Catt and M. L. Dufau, *Adv. Exp. Med. Biol.* **36,** 379 (1973).
6. C. V. Byus and D. H. Russell, *Life Sci.* **15,** 1991 (1974); *Science* **187,** 650 (1975).
7. L. J. Roger, S. M. Schanberg, and R. E. Fellows, *Endocrinology (Baltimore)* **95,** 904 (1974).
8. T. R. Anderson and S. M. Schanberg, *Biochem. Pharmacol.* **24,** 495 (1975).
9. T. Ikeno and G. Guroff, *J. Neurochem.* **33,** 973 (1979).
10. L. J. Roger and R. E. Fellows, *Endocrinology (Baltimore)* **106,** 619 (1980).
11. M. A. Cousin, D. Lando, and M. Moguilewsky, *J. Neurochem.* **38,** 1296 (1982).
12. G. Morris, F. J. Seidler, and T. A. Slotkin, *Life Sci.* **32,** 1565 (1983).
13. R. K. Sogani, S. Matsushita, J. F. Mueller and M. S. Raben, *Biochim. Biophys. Acta* **279,** 377 (1972).
14. J. W. Warnica, P. Antony, K. Gibson, and P. Harris, *Cardiovasc. Res.* **9,** 793 (1975).
15. R. W. Fuller and S. K. Hemrick, *J. Mol. Cell. Cardiol.* **10,** 1031 (1978).
16. J. Bartolome, C. Lau, and T. A. Slotkin, *J. Pharmacol. Exp. Ther.* **202,** 510 (1977).
17. C. Lau and T. A. Slotkin, *J. Pharmacol. Exp. Ther.* **208,** 485 (1979).
18. J. Bartolome, J. Huguenard, and T. A. Slotkin, *Science* **210,** 793 (1980).
19. W. E. Nicholson, J. H. Levine, and D. N. Orth, *Endocrinology (Baltimore)* **98,** 123 (1976).
20. G. Scalabrino and M. E. Ferioli, *Endocrinology (Baltimore)* **99,** 1085 (1976).
21. M. K. Haddox and D. H. Russell, *Life Sci.* **25,** 615 (1979).
22. H. Nawata, R. S. Yamamoto, and L. A. Poirier, *Life Sci.* **26,** 689 (1980).
23. K. A. Pass, J. E. Bintz, J. J. Postulka, and H. L. Vallet, *Proc. Soc. Exp. Biol. Med.* **167,** 270 (1981).
24. J. E. Seely, H. Poso, and A. E. Pegg, *J. Biol. Chem.* **257,** 7549 (1982).
25. J. Bartolome, P. A. Trepanier, E. A. Chait, G. A. Barnes, L. Lerea, W. L. Whitmore, S. J. Weigel, and T. A. Slotkin, *J. Neurotoxicol. Neurobiol.* (in press).
26. D. H. Russell, S. H. Snyder, and V. J. Medina, *Endocrinology (Baltimore)* **86,** 1414 (1970).
27. W. B. Panko and F. T. Kenney, *Biochem. Biophys. Res. Commun.* **43,** 346 (1971).
28. C. G. D. Morley, *Biochim. Biophys. Acta* **362,** 480 (1974).
29. S. R. Butler, T. W. Hurley, S. M. Schanberg, and S. Handwerger, *Life Sci.* **22,** 2073 (1978).
30. Y. Kobayashi, J. Kupelian, and D. V. Maudsley, *Science* **172,** 379 (1971).
31. A. M. Kaye, I. Icekson, S. A. Lamprecht, R. Gruss, A. Tsafriri, and H. R. Lindner, *Biochemistry* **12,** 3072 (1973).
32. S. A. Lamprecht, U. Zor, A. Tsafriri, and H. R. Lindner, *J. Endocrinol.* **57,** 217 (1973).
33. C. S. Sheela Rani and N. R. Moudgal, *Endocrinology (Baltimore)* **104,** 1480 (1979).
34. C. S. Teng and C. T. Teng, *Biochem. J.* **188,** 313 (1980).
35. S. Cohen, B. W. O'Malley, and M. Stastny, *Science* **170,** 336 (1970).
36. S. A. Braddon, *Biochem. Biophys. Res. Commun.* **80,** 75 (1978).
37. H. Inoue, H. Tanioka, K. Shiba, A. Asada, Y. Kato, and Y. Takeda, *J. Biochem.* **75,** 679 (1974).

(*continued*)

TABLE II REFERENCES (continued)

38. P. G. Sayatilak, A. R. Sheth, A. N. Thakur, and D. S. Pardanani, *Ind. J. Biochem. Biophys.* **13,** 186 (1976).
39. M. Stastny and S. Cohen, *Biochim. Biophys. Acta* **204,** 578 (1970).
40. D. V. Maudsley, J. Leif, and Y. Kobayashi, *Am. J. Physiol.* **231,** 1557 (1976).
41. P. C. MacDonnell, K. Nagaiah, J. Lakshmanan, and G. Guroff, *Proc. Natl. Acad. Sci. U.S.A.* **74,** 4681 (1977).
42. I. A. Hendry and R. Bonyhady, *Brain Res.* **200,** 39 (1980).
43. M. Stastny and S. Cohen, *Biochim. Biophys. Acta* **261,** 177 (1972).
44. P. R. Reddy and C. A. Villee, *Biochem. Biophys. Res. Commun.* **65,** 1350 (1975).
45. R. Madhubala and P. R. Reddy, *Biochem. Biophys. Res. Commun.* **102,** 1096 (1981).
46. S. Matsuzaki and M. Suzuki, *Endocrinol. Jpn.* **21,** 529 (1974).
47. R. Richman, S. Park, M. Akbar, S. Yu, and G. Burke, *Endocrinology (Baltimore)* **96,** 1403 (1975).
48. D. R. Zusman and G. N. Burrow, *Endocrinology (Baltimore)* **97,** 1089 (1975).
49. A. M. Kaye, I. Icekson, and H. R. Lindner, *Biochim. Biophys. Acta* **252,** 150 (1971).

Glass syringes fitted with 19-gauge needles
Paper wicks, 13 × 20 mm, cut from Whatman No. 1 filter paper and inserted into center wells

Reagents

L-[*carboxyl*-^{14}C]Ornithine; 0.25 μCi, ca. 50 mCi/mmol

Pyridoxal 5'-phosphate stock solution; 24.7 mg per 100 ml of H_2O, made fresh; final concentration in assay = 50 μM

Dithiothreitol stock solution: 62 mg per 10 ml of H_2O, made fresh; final concentration in assay = 2 mM

Tris-HCl buffer, 10 mM, pH 7.2; store refrigerated up to several weeks

10% Trichloroacetic acid (or equivalent citric acid, perchloric acid, etc.)

Hyamine hydroxide, 1 M in methanol; store refrigerated and protected from light; keep tightly closed

Scintillation fluids: one for counting aqueous samples (assay mix) and one for nonaqueous samples ($^{14}CO_2$ adsorbed on Hyamine-saturated paper wick)

Tissues are homogenized in 20 volumes of Tris-HCl buffer and centrifuged for 10 min at 26,000 g. The supernatant solution is used for the assay; it typically contains 0.9 ml of tissue supernatant, to which 50 μl of a mixture containing 45 μl of dithiothreitol and 5 μl of [^{14}C]ornithine are added, with final addition of 50 μl of the pyridoxal 5'-phosphate solution; an aliquot of the dithiothreitol–[^{14}C]ornithine mixture is placed in a scin-

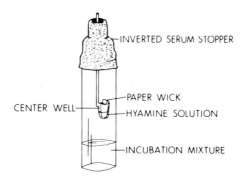

FIG. 2. Scintillation vial set up for incubations.

tillation vial with the aqueous scintillation counting cocktail to verify the amount of substrate added. The vial is then capped (Fig. 2) and incubated at 37° for 30–60 min; under these conditions the assay is linear with time and tissue concentration. The assay is terminated by chilling on ice and injecting (through the side of the serum stopper) 0.5 ml of trichloroacetic acid into the incubation mixture and 0.2 ml of Hyamine hydroxide into the center well containing the paper wick. A further incubation of 30 min is carried out at 37° to trap the radioactive CO_2 on the wick. At the end of the second incubation, the stopper is removed, the underside of the center well is wiped with an absorbent tissue (to remove any splashed material from the incubation mixture), the stem is cut, and the well is placed in the nonaqueous scintillation fluid.

The assay may be scaled up to accommodate larger volumes of tissue preparation or scaled down to assay small amounts; in either case, it is desirable to maintain the same tissue dilution as well as the same relative volumes and concentrations of tissue preparation, assay mixture, and pyridoxal 5'-phosphate solution to minimize nonspecific decarboxylation (see below). In cases where the tissue dilution must perforce be greater (small amounts of tissue available), nonspecific decarboxylation can be minimized by replacement of tissue supernatant with albumin to maintain protein concentrations equivalent to that in the assay as originally presented. In addition, when total incubation volumes are to be less than 0.25 ml, smaller incubation vials (and matching serum stoppers) should be substituted to avoid evaporation of sample during the incubation.

Assay Blanks. Two nonspecific processes contribute to evolution of $^{14}CO_2$ in the ODC assay: nonenzymic decarboxylation[38] and participation of other decarboxylase enzymes.[39] Since these may account for as much

[38] L. J. Roger, Ph.D. Dissertation, Duke University, Durham, North Carolina, 1976.
[39] B. J. Murphy and M. E. Brosnan, *Biochem. J.* **157**, 33 (1976).

as one-fourth of the measured activity under some conditions,[40] it is vital to minimize both and to utilize a blank that accurately assesses the proportion of $^{14}CO_2$ evolution unrelated to ODC activity. Conditions that promote the nonenzymic process include low protein concentration, high pyridoxal 5′-phosphate concentration, and high ornithine concentration; to minimize nonenzymic decarboxylation, one should maintain a constant ratio of assay protein to pyridoxal cofactor (if necessary, by adding albumin to replace tissue supernatant when very small tissue samples are to be assayed), use a subsaturating concentration of ornithine (see below), and add the pyridoxal 5′-phosphate reagent to the assay mixture *last* (cofactor concentrations would be transiently high if added to the mixture before the tissue supernatant). Obviously, the use of a "no-tissue" blank will give an artificially high blank value because nonenzymic decarboxylation will be excessive compared to that in the tissue-containing sample.

Because decarboxylase enzymes are selective but not totally specific, some $^{14}CO_2$ evolution from radiolabeled ornithine may involve enzymes other than ODC.[40] The use of an assay subsaturated with ornithine will minimize the participation of other enzymes with lower affinity for ornithine, but it is still critical to demonstrate that the measured activity represents the contribution only of ODC. Again, a "no-tissue" blank would be inappropriate, since in this case the estimated value would not correct for contributions of these additional enzymic processes. A blank utilizing boiled tissue could also yield an incorrect value, since all enzymic decarboxylation would be arrested, whether ODC-derived or not. Consequently, ODC activity measurements require, in some cases, unusual assay blank procedures that correct properly for both nonenzymic decarboxylation and nonspecific enzymic components. One such procedure[37,40] makes use of the uniquely rapid turnover of ODC ($t_{1/2}$ 15–20 min); after administering protein synthesis inhibitors such as puromycin or cycloheximide and waiting 1–2 hr, nearly all ODC activity disappears whereas there is little effect on total protein concentration or on other, longer-lived enzymes. The residual activity thus accurately represents nonenzymic decarboxylation plus enzymic decarboxylation not attributable to ODC per se. More recently, the development of α-difluoromethylornithine (DFMO), a specific irreversible ("suicide") inhibitor of ODC,[41–43] enables an additional pharmacological determination of an accurate assay blank. Within 2 hr of administration of a maximally effective

---

[40] C. Lau and T. A. Slotkin, *Mol. Pharmacol.* **18**, 247 (1980).
[41] R. R. Rando, *Science* **185**, 320 (1974).
[42] B. W. Metcalf, P. Bey, C. Danzin, M. J. Jung, P. Casara, and J. P. Vevert, *J. Am. Chem. Soc.* **100**, 2251 (1978).
[43] P. Bey, in "Enzyme Activated Irreversible Inhibitors" (N. Seiler, M. J. Jung, and J. Koch-Weser, eds.), p. 27. Elsevier/North-Holland, Amsterdam, 1978.

dose of DFMO, all the ODC has been inactivated,[27,28,44] and residual activity thus represents only the nonenzymic plus nonspecific components of ornithine decarboxylation.

In most instances then, the use of a protein synthesis inhibitor or a DFMO-generated assay blank for ODC is superior to the more usual no-tissue or boiled-tissue procedures. The latter, however, may prove to be more convenient when ODC activity is high (as in situations in which massive stimulation has occurred) and where a misestimate of blank values will have little impact on total measured activity.

Saturated vs Nonsaturated Assays. The choice of ornithine concentration to be used in the ODC assay is of some importance. As discussed above, high ornithine concentrations tend to enhance nonspecific decarboxylation, thus increasing the assay blank. In addition, the assay becomes more expensive as increasing amounts of [^{14}C]ornithine are used, or, if ornithine concentration is raised by addition of unlabeled ornithine, the assay becomes less sensitive owing to isotopic dilution resulting in decreased specific activity. These factors operate in favor of using an assay with subsaturating concentrations of ornithine (generally below 15 μM). What is perhaps a more important advantage of subsaturated assays is the ability to detect ODC stimulation resulting from posttranslational control of enzyme activity; one mechanism by which ODC is stimulated in some tissues involves a lowering of the K_m for ornithine,[37,40,44] which would not be detected if a saturated assay were used.

With these advantages in mind, it is necessary to discuss why assays using saturating concentrations of ornithine might be useful in some circumstances. First of all, tissue concentrations of ornithine, putrescine, and/or the polyamines may be high, especially in growing tissues or after stimulation with trophic hormones. Because tissue supernatant is used in the assay, the measured ODC activity could be artificially low, owing either to end-product competitive inhibition by endogenous putrescine or to isotopic dilution with endogenous unlabeled ornithine.[1] Neither factor would affect assays saturated with [^{14}C]ornithine, but these factors would influence measured activity in subsaturated assays. Consequently, stimulation of ODC seen with subsaturated assays may require additional verification that enzyme activity was actually affected, as opposed to increased apparent activity due to shifts in endogenous ornithine or putrescine. This can be done through experiments with enzyme preparations dialyzed to remove ornithine and putrescine (overnight dialysis is recommended against 10 mM Tris-HCl, pH 7.2, containing 1.5 mM dithiothreitol)[44,45]; alternatively, supernatants from stimulated and nonstimulated tissues can

---

[44] C. Lau and T. A. Slotkin, *Eur. J. Pharmacol.* **78**, 99 (1982).
[45] G. Morris, C. Lau, and T. A. Slotkin, *Eur. J. Pharmacol.* **88**, 177 (1983).

be mixed together in different proportions prior to assay.[44,45] If apparent ODC stimulation disappears upon dialysis or is disproportionately reduced by dilution with nonstimulated tissue preparations, alterations in tissue content of endogenous substrate or product (or other dialyzable factors, such as small molecule activators or inhibitors) may have contributed to the observed increase in activity.

Thus, there are advantages and disadvantages to the use of assays subsaturated vs saturated with ornithine, with higher sensitivity, lower cost, and greater latitude of stimulatory mechanism detectable with the subsaturated method, but also a greater potential chance of artifactual "stimulation." Ultimately, the most reliable approach to studying the actual mechanism by which ODC is stimulated by any trophic factor would utilize a range of ornithine concentrations spanning subsaturated and saturated conditions.[37,40,44,45]

Just as ornithine concentration is a variable that needs to be considered in utilizing ODC as a tissue stimulation marker, so too does the concentration of the pyridoxal 5'-phosphate cofactor. Multiple forms of enzyme with different K_m values for pyridoxal 5'-phosphate have been reported to occur in some instances.[46] Although the same factors tend to operate as for ornithine concentration (increased nonspecific decarboxylation at high concentration, problem of endogenous pyridoxal 5'-phosphate influencing overall concentration), most investigators use assays containing saturating concentrations of pyridoxal 5'-phosphate, since it is a cofactor and not the actual precursor substrate of the polyamines. Under typical conditions, it is important only that sufficient exogenous pyridoxal 5'-phosphate be added to obviate or overwhelm any differences in endogenous tissue levels, such that conditions are always met for optimal ODC activity. The concentration described in the assay here (50 μM final) is sufficient to do so without causing excessive nonspecific decarboxylation of ornithine.

Avoiding Chemiluminescence. Since the $^{14}CO_2$ evolved during the assay is adsorbed using an alkaline solution (Hyamine hydroxide in methanol) prior to liquid scintillation counting, chemiluminescence (which may be long-lived) is an occasional problem in sample determinations. In general, this can be avoided by preventing evaporation of the solvent (methanol) in which the Hyamine hydroxide adsorbent is dissolved. Thus, chemiluminescence is more likely to occur if the Hyamine is injected into the center well at the beginning of the assay as opposed to injection after the first incubation; if the second incubation (after addition of trichloroacetic acid) is prolonged far beyond the 30-min period indicated; if the

---

[46] J. L. A. Mitchell, S. N. Anderson, D. D. Carter, M. J. Sedory, J. F. Scott, and D. A. Varland, *Adv. Polyamine Res.* **1**, 39 (1978).

paper wick is too large; if the stock bottle of Hyamine hydroxide has been left open or is old. In any case, results should always be examined for potentially chemiluminescent (and hence inaccurate) samples by standard techniques (e.g., channels-ratio method). Overnight chilling of the scintillation samples in the dark usually eliminates such problems when they occur. While it is also possible to reduce the risk of chemiluminescence through use of a less volatile, aqueous adsorbent solution (e.g., KOH in water), the subsequent scintillation measurement would then require a water-compatible counting cocktail, as opposed to the water-incompatible one indicated here; in that circumstance, any of the radioactive incubation mixture splashed on the underside of the center well and not removed by wiping would also dissolve in the scintillation fluid, which would invalidate the sample. In contrast, the water-incompatible cocktail would less readily dissolve (and thus not count) the splash.

Mechanisms of ODC Stimulation

Many mechanisms have been proposed to explain regulation of ODC activity. These include transcriptional,[2,4,32,35,47,48] posttranscriptional,[49] translational,[50,51] and posttranslational[52,53] processes. The induction of ODC seen in many tissue systems is often due to increased transcription, probably involves cyclic AMP-dependent protein kinase, and is usually prevented by actinomycin D.[4,54] The possible additional participation of induction at the level of translation is somewhat more difficult to assess, as the rapid turnover of ODC leads to prompt and complete loss of enzyme activity after administration of protein synthesis inhibitors[37,40]; since the decline after inhibition of translation occurs more rapidly than does stimulation of ODC by trophic factors, standard procedures for assessing the role of translational events in ODC regulation are not readily utilizable.

Posttranslational control mechanisms for ODC involve either end-product feedback inhibition by the polyamines (particularly putrescine),[1,55] modulation by intracellular amino acids such as asparagine and

[47] D. H. Russell and R. L. Taylor, *Endocrinology (Baltimore)* **88**, 1397 (1971).
[48] C. V. Byus and D. H. Russell, *Science* **187**, 650 (1975).
[49] P. P. McCann, C. Tardif, M. C. Duchesne, and P. S. Mamont, *Biochem. Biophys. Res. Commun.* **76**, 893 (1977).
[50] J. L. Clark and J. L. Fuller, *Biochemistry* **14**, 4403 (1975).
[51] Y. Yamasaki and A. Ichihara, *J. Biochem. (Tokyo)* **80**, 557 (1976).
[52] M. J. Sedory and J. L. A. Mitchell, *Exp. Cell Res.* **107**, 105 (1977).
[53] M. F. Obenrader and W. F. Prouty, *J. Biol. Chem.* **252**, 2860 (1977).
[54] D. H. Russell, *Pharmacology* **20**, 117 (1980).
[55] J. Janne and H. G. Williams-Ashman, *J. Biol. Chem.* **246**, 1725 (1971).

glutamine,[56] macromolecular activators or inhibitors,[57–60] or appearance of new forms of the enzyme with different affinities for substrate or cofactor.[37,40,44,46] No one model has been satisfactory in elucidating control of ODC activity, probably because mechanisms may differ in different tissues or cell types. For instance, cAMP is a potent inducer of ODC in rat liver, a tissue subject to transcriptional control,[4] but is ineffective in the heart, where new RNA synthesis is not necessary for stimulation of ODC.[37] Even within the same tissue or cell system, several mechanisms modulating ODC activity may operate simultaneously.[45,61,62] As was true for assessing potential translational control of ODC activity, the extremely short half-life renders the use of standard protein synthesis inhibitors impracticable, since all enzyme activity is rapidly lost after inhibition of translation. Consequently, much of the evidence supporting post-translational models has been obtained in the presence of active protein translation. Nevertheless, an increasing amount of experimental data favors posttranslational modulation as a major mechanism of control of ODC activity.[63] Multiple forms of ODC, small molecular effectors, and macromolecules regulating ODC activity have been described and partially characterized in both mammalian and nonmammalian systems,[1,45,46,50,53,56–58,64–67] and results from recent studies[37,40,44,45] indicate that posttranslational mechanisms may operate in modulating ODC activity in the rat heart and brain, where increases in ODC occur in response to neuronal, hormonal, or ontogenetic stimulation of growth. It thus appears that multiple mechanisms contribute to ODC stimulation in any particular tissue responding to a trophic agent, and it is incorrect to assume that ODC stimulation always occurs by induction of new enzyme molecules.

Of great importance in utilizing ODC as a tissue stimulation marker is

[56] K. Y. Chen and E. S. Canellakis, *Proc. Natl. Acad. Sci. U.S.A.* **74**, 3791 (1977).
[57] D. A. Kyriakidis, J. S. Heller, and E. S. Canellakis, *Proc. Natl. Acad. Sci. U.S.A.* **75**, 4699 (1978).
[58] W. F. Fong, J. S. Heller, and E. S. Canellakis, *Biochim. Biophys. Acta* **428**, 456 (1976).
[59] J. S. Heller, W. F. Fong, and E. S. Canellakis, *Proc. Natl. Acad. Sci. U.S.A.* **73**, 1858 (1976).
[60] J. S. Heller, D. Kyriakidis, W. F. Fong, and E. S. Canellakis, *Eur. J. Biochem.* **81**, 545 (1977).
[61] A. Kallio, *Acta Chem. Scand.* **B32**, 759 (1978).
[62] P. P. McCann, C. Tardif, J. M. Hornsperger, and P. Bohlen, *J. Cell. Physiol.* **99**, 183 (1979).
[63] E. S. Canellakis, D. Viceps-Madore, D. A. Kyriakidis, and J. S. Heller, *Curr. Top. Cell. Regul.* **15**, 155 (1979).
[64] J. E. Kay and V. L. Lindsay, *Biochem. J.* **132**, 791 (1973).
[65] E. Holtta, J. Janne and J. Pispa, *Biochem. Biophys. Res. Commun.* **59**, 1104 (1974).
[66] C. G. D. Morley and H. Ho, *Biochim. Biophys. Acta* **433**, 551 (1976).
[67] J. L. A. Mitchell, *Adv. Polyamine Res.* **3**, 15 (1981).

to define carefully how the stimulation comes about. Unless care is taken to ensure that ODC stimulation is directly linked to the interaction of the trophic factor with its target tissue, misleading interpretations of the meaning of the ODC stimulation may result. For example, administration of insulin *in vivo* causes an increase in cardiac ODC activity, but the stimulation is not due to direct actions of insulin on the heart; rather, insulin-induced hypoglycemia results in sympathetic nerve activation, and it is this neuronal reflex that causes the elevation of cardiac ODC.[29,30,34,68] The neuronal origin of this ODC effect can be demonstrated clearly by the fact that sympathectomy or ganglionic or adrenergic receptor blockade can prevent nearly all of the insulin effect. In this case, then, cardiac ODC by insulin is independent of the response of cardiac cells to insulin per se, but is instead dependent upon sympathetic innervation of the heart. The general susceptibility of ODC to stimulation by neurotransmitters or hormonal agents is thus advantageous in that it provides a marker of stimulation that is of broad application, but also requires that the exact mechanism of stimulation be verified unequivocally, either by demonstration of direct stimulation of the target tissue by trophic agents *in vitro,* or through utilizing receptor agonists and antagonists to define the site of action. With judicious application and appropriate prior elucidation of the mechanism of stimulation, ODC activity will continue to expand its utility as a general marker of interactions of trophic substances with their target tissues.

[68] D. L. Bareis and T. A. Slotkin, *J. Pharmacol. Exp. Ther.* **205**, 164 (1978).

Section VI

Localization of Neuroendocrine Substances

[41] Secretion of Hypothalamic Dopamine into the Hypophysial Portal Vasculature: An Overview

By JOHN C. PORTER, MARIANNE J. REYMOND, JUN ARITA, and JANICE F. SISSOM

The neurohypophysis of the brain of mammals can be considered to consist of the infundibulum, the infundibular stem, and the infundibular process. The infundibular process, also called the pars neuralis, constitutes the portion of the pituitary gland that is derived from the central nervous system. In contrast to the neurohypophysis, the adenohypophysis is derived from Rathke's pouch, a dorsal outpocketing of the embryological stomodeum. The pars distalis, pars intermedia, and pars tuberalis are subdivisions of the adenohypophysis.

The ventromedial region of the infundibulum is known as the median eminence and is a region of the brain that is vascularized by a dense network of capillaries that are conjoint with veins lying, for the most part, on the surface of the infundibular stem, a structure comprising much of the pituitary stalk. These blood vessels, known as the hypophysial portal veins, are also confluent with the sinusoids of the pars distalis. Inasmuch as the direction of blood flow in this vasculature is from the median eminence to the pars distalis, it is evident that venous outflow of the median eminence is the blood supply of the pars distalis. Neurosecretory substances released in the median eminence by hypothalamic cells diffuse into blood within capillaries of the median eminence and are directly transported to the sinusoids of the pars distalis by way of the hypophysial portal veins. Several of these neurosecretory substances are known to affect the release of hormones by cells of the pars distalis.

Neurosecretory Substances in Hypophysial Portal Blood

Peptides. The presence in hypophysial portal blood of a biologically active substance—a substance possessing corticotropin releasing activity—was first demonstrated in 1956 by Porter and Jones.[1] The finding of these workers was subsequently confirmed and extended.[2-4] A peptide having corticotropin releasing activity, which is called the corticotropin

---

[1] J. C. Porter and J. C. Jones, *Endocrinology* (*Baltimore*) **58,** 62 (1956).
[2] J. C. Porter and H. W. Rumsfeld, Jr., *Endocrinology* (*Baltimore*) **58,** 359 (1956).
[3] H. W. Rumsfeld, Jr. and J. C. Porter, *Endocrinology* (*Baltimore*) **64,** 942 (1959).
[4] J. C. Porter and H. W. Rumsfeld, Jr., *Endocrinology* (*Baltimore*) **64,** 948 (1959).

releasing factor (CRF), was isolated from ovine hypothalami and characterized by Vale et al.[5] and Rivier et al.[6] Using a radioimmunoassay, Gibbs and Vale[7] demonstrated the presence of immunoreactive CRF in hypophysial portal blood of rats.

A hypothalamic peptide that stimulates the release of the gonadotropic hormones, luteinizing hormone and follicle stimulating hormone, is known as the gonadotropin releasing hormone or luteinizing hormone releasing hormone[8] and has been demonstrated in hypophysial portal blood of rats,[9-12] monkeys,[13,14] and rabbits.[15] Thyrotropin releasing hormone[9,16,17] as well as somatostatin,[18] an inhibitor of growth hormone release and possibly other hormones as well, has been demonstrated in portal blood of rats.[19-21] β-Endorphin is present in high concentrations in hypophysial portal blood of monkeys.[22,23] The relationship of β-endorphin of hypothalamic origin to the function of the pars distalis is unknown at present. Vasoactive intestinal peptide (VIP) has been shown to be present

[5] W. Vale, J. Spiess, C. Rivier, and J. Rivier, *Science* **213**, 1394 (1981).
[6] C. Rivier, M. Brownstein, J. Spiess, J. Rivier, and W. Vale, *Endocrinology (Baltimore)* **110**, 272 (1982).
[7] D. M. Gibbs and W. Vale, *Endocrinology (Baltimore)* **111**, 1418 (1982).
[8] H. Matsuo, Y. Baba, R. M. G. Nair, A. Arimura, and A. V. Schally, *Biochem. Biophys. Res. Commun.* **43**, 1334 (1971).
[9] R. L. Eskay, C. Oliver, N. Ben-Jonathan, and J. C. Porter, *in* "Hypothalamic Hormones: Chemistry, Physiology, Pharmacology and Clinical Uses" (M. Motta, P. G. Crosignani, and L. Martini, eds.), p. 125. Academic Press, New York, 1975.
[10] G. Fink and M. G. Jamieson, *J. Endocrinol.* **68**, 71 (1976).
[11] R. L. Eskay, R. S. Mical, and J. C. Porter, *Endocrinology (Baltimore)* **100**, 263 (1977).
[12] M. Ching, *Neuroendocrinology* **34**, 279 (1982).
[13] P. W. Carmel, S. Araki, and M. Ferin, *Endocrinology (Baltimore)* **99**, 243 (1976).
[14] J. D. Neill, J. M. Patton, R. A. Dailey, R. C. Tsou, and G. T. Tindall, *Endocrinology (Baltimore)* **101**, 430 (1977).
[15] R. C. Tsou, R. A. Dailey, C. S. McLanahan, A. D. Parent, G. T. Tindall, and J. D. Neill, *Endocrinology (Baltimore)* **101**, 534 (1977).
[16] W. J. de Greef and T. J. Visser, *J. Endocrinol.* **91**, 213 (1981).
[17] G. Fink, Y. Koch, and N. Ben Aroya, *Brain Res.* **243**, 186 (1982).
[18] P. Brazeau, W. Vale, R. Burgus, N. Ling, M. Butcher, J. Rivier, and R. Guillemin, *Science* **179**, 77 (1973).
[19] P. Gillioz, P. Giraud, B. Conte-Devolx, P. Jaquet, J. L. Codaccioni, and C. Oliver, *Endocrinology (Baltimore)* **104**, 1407 (1979).
[20] K. Chihara, A. Arimura, and A. V. Schally, *Endocrinology (Baltimore)* **104**, 1434 (1979).
[21] K. Chihara, A. Arimura, C. Kubli-Garfias, and A. V. Schally, *Endocrinology (Baltimore)* **105**, 1416 (1979).
[22] S. L. Wardlaw, W. B. Wehrenberg, M. Ferin, P. W. Carmel, and A. G. Frantz, *Endocrinology (Baltimore)* **106**, 1323 (1980).
[23] W. B. Wehrenberg, S. L. Wardlaw, A. G. Frantz, and M. Ferin, *Endocrinology (Baltimore)* **111**, 879 (1982).

in hypophysial portal blood of rats.[24,25] The release of VIP into portal blood is potentiated in the presence of serotonin.[26] The existence of VIP in portal blood in concentrations that are appreciably greater than those in arterial blood is suggestive of a role for this peptide in the stimulation of the release of prolactin.[27–29] Vasopressin has also been shown to be present in high concentrations in portal blood of monkeys[30] and rats.[31]

Catecholamines. The role of blood-borne catecholamines in the control of the secretion of hormones from the pars distalis has long been a subject of interest to investigators. However, it was little more than a decade ago that it was shown under *in vitro* as well as *in vivo* conditions that in the presence of dopamine the release of prolactin by pituitary cells was inhibited. The effect of norepinephrine or epinephrine on the release of prolactin is markedly less than is that of dopamine.

Monoaminergic, especially dopaminergic, neurons are present in the ventromedial hypothalamus, and nerve endings of such neurons are abundant in the median eminence.[32–36] In 1968, van Maanen and Smelik[37] hypothesized that dopamine released into hypophysial portal blood might act as an inhibitor of the release of prolactin. Subsequently, MacLeod *et al.*[38] and Birge *et al.*[39] demonstrated an inhibitory action for dopamine on prolactin release under *in vitro* conditions.

However, the crucial issue of whether dopamine is actually released into hypophysial portal blood and whether there is a relationship between

[24] S. I. Said and J. C. Porter, *Life Sci.* **24**, 227 (1979).
[25] A. Shimatsu, Y. Kato, N. Matsushita, H. Katakami, N. Yanaihara, and H. Imura, *Endocrinology* (*Baltimore*) **108**, 395 (1981).
[26] A. Shimatsu, Y. Kato, N. Matsushita, H. Katakami, N. Yanaihara, and H. Imura, *Endocrinology* (*Baltimore*) **111**, 338 (1982).
[27] Y. Kato, Y. Iwasaki, J. Iwasaki, H. Abe, N. Yanaihara, and H. Imura, *Endocrinology* (*Baltimore*) **103**, 554 (1978).
[28] M. Ruberg, W. H. Rotsztejn, S. Arancibia, J. Besson, and A. Enjalbert, *Eur. J. Pharmacol.* **51**, 319 (1978).
[29] C. J. Shaar, J. A. Clemens, and N. B. Dininger, *Life Sci.* **25**, 2071 (1979).
[30] E. A. Zimmerman, P. W. Carmel, M. K. Husain, M. Ferin, M. Tannenbaum, A. G. Frantz, and A. G. Robinson, *Science* **182**, 925 (1973).
[31] C. Oliver, R. S. Mical, and J. C. Porter, *Endocrinology* (*Baltimore*) **101**, 598 (1977).
[32] A. Carlsson, B. Falck, and N. A. Hillarp, *Acta Physiol. Scand. Suppl.* **196**, 1 (1962).
[33] K. Fuxe, *Z. Zellforsch. Mikrosk. Anat.* **61**, 710 (1964).
[34] K. Fuxe and T. Hökfelt, *Acta Physiol. Scand.* **66**, 245 (1966).
[35] W. Lichtensteiger and H. Langemann, *J. Pharmacol. Exp. Ther.* **151**, 400 (1966).
[36] A. Björklund, B. Falck, F. Hromek, C. Owman, and K. A. West, *Brain Res.* **17**, 1 (1970).
[37] J. H. van Maanen and P. G. Smelik, *Neuroendocrinology* **3**, 177 (1968).
[38] R. M. MacLeod, E. H. Fontham, and J. E. Lehmeyer, *Neuroendocrinology* **6**, 283 (1970).
[39] C. A. Birge, L. S. Jacobs, C. T. Hammer, and W. H. Daughaday, *Endocrinology* (*Baltimore*) **86**, 120 (1970).

the concentration of dopamine in portal blood and the release of prolactin was not resolved until an assay for dopamine in plasma was developed that had both specificity and great sensitivity. Such an assay was described in 1976 by Ben-Jonathan and Porter.[40] This assay, a radioenzymic assay, was an adaptation of the procedure of Coyle and Henry[41] for the quantification of catecholamines in brain tissue. In 1977, Ben-Jonathan et al.[42] not only demonstrated dopamine in plasma of hypophysial portal blood of rats, but also showed that the concentration of dopamine in portal plasma was appreciably greater than that in arterial plasma, leading to the conclusion that dopamine of hypothalamic origin was released into hypophysial portal blood. The validity of this conclusion has been repeatedly affirmed by us and others using rats[16,43–47] and monkeys.[48]

In contrast to dopamine, there is no evidence to support the view that norepinephrine or epinephrine is released into hypophysial portal blood by the hypothalamus. Even so, the possibility that these two catecholamines in the circulation are involved in hypothalamic–pituitary regulation ought not be totally dismissed. Reymond and Porter[49] observed that the concentrations of norepinephrine and epinephrine in plasma of blood collected from a single hypophysial portal vessel of rats are significantly less than those in plasma of arterial blood collected simultaneously from the same animals. This finding is suggestive that these two catecholamines in the general circulation are extracted by the brain as arterial blood traverses the median eminence. Thus, norepinephrine and epinephrine in the arterial circulation may in this manner affect the function of cells of the hypothalamus.

Differential Release of Dopamine into Various Portal Vessels

The results of many studies on the release of hypothalamic dopamine into hypophysial portal blood pertain to dopamine in blood collected from

[40] N. Ben-Jonathan and J. C. Porter, *Endocrinology (Baltimore)* **98**, 1497 (1976).
[41] J. T. Coyle and D. Henry, *J. Neurochem.* **21**, 61 (1973).
[42] N. Ben-Jonathan, C. Oliver, H. J. Weiner, R. S. Mical, and J. C. Porter, *Endocrinology (Baltimore)* **100**, 452 (1977).
[43] P. M. Plotsky, D. M. Gibbs, and J. D. Neill, *Endocrinology (Baltimore)* **102**, 1887 (1978).
[44] O. M. Cramer, C. R. Parker, Jr., and J. C. Porter, *Endocrinology (Baltimore)* **104**, 419 (1979).
[45] O. M. Cramer, C. R. Parker, Jr., and J. C. Porter, *Endocrinology (Baltimore)* **105**, 636 (1979).
[46] O. M. Cramer, C. R. Parker, Jr., and J. C. Porter, *Endocrinology (Baltimore)* **105**, 929 (1979).
[47] G. A. Gudelsky, D. D. Nansel, and J. C. Porter, *Brain Res.* **204**, 446 (1981).
[48] J. D. Neill, L. S. Frawley, P. M. Plotsky, and G. T. Tindall, *Endocrinology (Baltimore)* **108**, 489 (1981).
[49] M. J. Reymond and J. C. Porter, *Endocrinology (Baltimore)* **111**, 1051 (1982).

all of the portal vessels after transection of the pituitary stalk. Such results are reflective of the concentration of dopamine in the mixture of blood from all the vessels of the stalk. Since each portal vessel appears to receive blood from a discrete region of the median eminence, the question arises: How does the concentration of dopamine in plasma of a portal vessel located laterally on the pituitary stalk compare with the concentration of dopamine in plasma of a portal vessel located medially on the pituitary stalk?

We collected blood through a microcannula from a lateral portal vessel or a medial portal vessel of female rats. When the plasma of this blood was analyzed, it was found that the concentration of dopamine in plasma from a medial portal vessel was two times that in plasma from a lateral portal vessel.[50]

When the rate of release of dopamine into various portal vessels was calculated, it was ascertained that the rate of release of dopamine into a medial portal vessel was two and one-half times greater than that into a lateral portal vessel. Although the basis for this disparity is not fully understood, it would appear that dopaminergic neurons with terminals in the lateral median eminence release less dopamine into the capillary plexus joining a lateral portal vessel than do dopaminergic neurons with terminals in the medial median eminence.

It has long been recognized that blood of a given portal vessel is not distributed homogeneously throughout the pars distalis.[51,52] Although exceptions exist, it appears to be generally true that blood from a portal vessel located near the sides of the pituitary stalk perfuses the lateral regions of the pars distalis; whereas blood from a portal vessel located medially on the pituitary stalk perfuses the more medial regions of the pars distalis. Thus, it is interesting to speculate that some aberrant functions of the pars distalis may be consequences of a disparity in the concentrations of dopamine in the portal blood delivered to the sinusoids of the pars distalis. It is possible that in some instances the concentration of dopamine in portal blood perfusing a particular group of prolactin-secreting cells is sufficiently low that the lactotrophs are not suppressed. In view of the ability of dopaminergic agonists to cause a reduction in the size of some prolactin-secreting pituitary tumors,[53–58] it is interesting to

[50] M. J. Reymond, S. G. Speciale, and J. C. Porter, *Endocrinology (Baltimore)* **112**, 1958 (1983).
[51] J. H. Adams, P. M. Daniel, and M. M. L. Prichard, *Neuroendocrinology* **1**, 193 (1965/66).
[52] J. C. Porter, R. S. Mical, J. G. Ondo, and I. A. Kamberi, *Acta Endocrinol. (Copenhagen)* Suppl. **158**, 249 (1972).
[53] S. K. Quadri, K. H. Lu, and J. Meites, *Science* **176**, 417 (1972).
[54] R. M. MacLeod and J. E. Lehmeyer, *Cancer Res.* **33**, 849 (1973).
[55] C. Davies, J. Jacobi, H. M. Lloyd, and J. D. Meares, *J. Endocrinol.* **61**, 411 (1974).

speculate that inadequate suppression of the lactotrophs may result not only in hyperprolactinemia, but in hypertrophy of these cells as well.

Control of the Release of Hypothalamic Dopamine

Hormones, Neurotransmitters, and Neuromodulators. The rate of release of dopamine by hypothalamic neurons into hypophysial portal blood is subject to modification by a variety of agents and conditions. There is often a marked difference in the concentration of dopamine in portal blood of female and male rats, being greater in the female than in the male.[42,59] Yet, the concentration of prolactin in the circulation is generally higher in the female than in the male. This fact would appear to contradict the view that the concentration of dopamine in portal blood is inversely related to the rate of secretion of prolactin.

Since the flow of hypophysial portal blood in the female rat is not appreciably different from that in the male rat, one can conclude that the rate of release of dopamine into the portal blood of female rats is greater than that of male rats. Indeed, even among female rats, the concentration of dopamine in portal plasma of animals undergoing regular ovulatory cycles is greatest on the day of estrus and least on the day of proestrus,[42] suggestive of a role for ovarian hormones in the control of the release of dopamine. After male rats are treated for several days with estradiol, the concentration of dopamine in portal blood is often increased by a factor of three to four.[59] Similarly, treatment of female rats with estradiol results in an increase in the concentration of dopamine in their hypophysial portal blood.[60]

It is noteworthy that one action of estradiol is to interfere with the action of dopamine on the lactotrophs of the pars distalis.[60–63] Thus, in this sense it can be said that estradiol is an anti-dopaminergic agent. A conse-

---

[56] R. A. Prysor-Jones and J. S. Jenkins, *J. Endocrinol.* **88**, 463 (1981).
[57] M. O. Thorner, W. H. Martin, A. D. Rogol, J. L. Morris, R. L. Perryman, B. P. Conway, S. S. Howards, M. G. Wolfman, and R. M. MacLeod, *J. Clin. Endocrinol. Metab.* **51**, 438 (1980).
[58] L. G. Sobrinho, M. C. Nunes, C. Calhaz-Jorge, J. C. Mauricio, and M. A. Santos, *Acta Endocrinol. (Copenhagen)* **96**, 24 (1981).
[59] G. A. Gudelsky and J. C. Porter, *Endocrinology (Baltimore)* **109**, 1394 (1981).
[60] G. A. Gudelsky, D. D. Nansel, and J. C. Porter, *Endocrinology (Baltimore)* **108**, 440 (1981).
[61] V. Raymond, M. Beaulieu, F. Labrie, and J. Boissier, *Science* **200**, 1173 (1978).
[62] L. Ferland, F. Labrie, C. Euvrard, and J. P. Raynaud, *Mol. Cell. Endocrinol.* **14**, 199 (1979).
[63] D. D. Nansel, G. A. Gudelsky, M. J. Reymond, and J. C. Porter, *Endocrinology (Baltimore)* **108**, 903 (1981).

quence of this action of estradiol is the occurrence of hyperprolactinemia in the treated animals. It has not been ascertained, however, whether the increased rate of release of dopamine by tuberoinfundibular neurons in estradiol-treated animals is a consequence of a direct action of the estrogen on the tuberoinfundibular dopaminergic neurons, an indirect action mediated through prolactin, or both.

It has been shown by several investigators that treatment of rats with prolactin results in a turnover of dopamine in the median eminence that is greater than that in untreated rats.[64–68] Moreover, hyperprolactinemia induced by treatment of animals with an antagonist of dopamine results in an increase in the turnover of dopamine in the median eminence.[69,70] It is of interest that prolactin treatment appears to affect the dopaminergic neurons terminating in the median eminence, but not the noradrenergic neurons.[68]

When prolactin is injected intracerebroventricularly into rats, there is within 16 hr an increase of 4- to 5-fold in the concentration of dopamine in portal blood.[71] Hyperprolactinemia due to the presence of prolactin-secreting tumors or hyperprolactinemia induced by treatment with the dopamine antagonist haloperidol results in an increase in the release of hypothalamic dopamine into portal blood of rats.[45,71] When, prior to treatment with haloperidol, rats are passively immunized against prolactin to neutralize circulating prolactin, the effect of haloperidol on dopamine release is significantly attenuated, supporting the view that the effect of haloperidol on dopamine release is mediated by way of prolactin.[71]

Thus, there is support for the perception that prolactin and dopamine are involved in a feedback cycle that results in tonic suppression of prolactin release. In this cycle, prolactin stimulates the release of dopamine into portal blood, and dopamine in turn inhibits the release of prolactin. When the concentration of dopamine in hypophysial portal blood is too low to suppress prolactin release, the secretion of prolactin increases, resulting in an increase in the amount of prolactin impinging on brain cells (tuberoinfundibular dopaminergic neurons?) and in turn in an increase in the release of hypothalamic dopamine into portal blood. If the resulting

---

[64] T. Hökfelt and K. Fuxe, *Neuroendocrinology* **9**, 100 (1972).
[65] G. A. Gudelsky, J. Simpkins, G. P. Mueller, J. Meites, and K. E. Moore, *Neuroendocrinology* **22**, 206 (1976).
[66] F. A. Wiesel, K. Fuxe, T. Hökfelt, and L. F. Agnati, *Brain Res.* **148**, 399 (1978).
[67] L. Annunziato and K. E. Moore, *Life Sci.* **22**, 2037 (1978).
[68] M. Selmanoff, *Endocrinology (Baltimore)* **108**, 1716 (1981).
[69] G. A. Gudelsky and K. E. Moore, *J. Pharmacol. Exp. Ther.* **202**, 149 (1977).
[70] J. S. Kizer, J. Humm, G. Nicholson, G. Greeley, and W. Youngblood, *Brain Res.* **146**, 95 (1978).
[71] G. A. Gudelsky and J. C. Porter, *Endocrinology (Baltimore)* **106**, 526 (1980).

concentration of dopamine in portal blood is sufficient to suppress the release of prolactin, the hyperprolactinemia will be ameliorated.

The efficacy of this feedback cycle is known to be modified by several factors. These include agents that interfere with the inhibitory action of dopamine on the lactotrophs, e.g., estradiol or antagonists of dopamine, such as haloperidol. In addition, brain cells, as effectors of prolactin action, may in some conditions be unable to respond to the stimulation of prolactin. Thus, it is emphasized that this feedback cycle involving the interrelationship of the release of prolactin and dopamine is operative under conditions that may, and probably do, vary from time to time. Should the cycle be interrupted, hypersecretion of dopamine could coexist with hypersecretion of prolactin. The concentrations of prolactin that seem to be effective in stimulating the release of hypothalamic dopamine are high when compared to the concentrations of prolactin that are commonly seen in the general circulation. However, in studies on retrograde transport of pituitary hormones to the hypothalamus, Oliver et al.[31] found that prolactin can be transported in high concentrations to the hypothalamus by way of the pituitary stalk.

When rats are treated with morphine, administered systemically or intracerebroventricularly, there is a prompt reduction in the release of hypothalamic dopamine into the hypophysial portal vessels. There is, however, a marked difference in the efficacy of morphine administered by these two routes. Gudelsky and Porter[72] injected systemically a relatively large dose (10 mg/kg body weight) of morphine sulfate in order to effect a suppression of the release of hypothalamic dopamine. In contrast, Reymond et al.[73] were able to suppress almost completely the hypothalamic release of dopamine by the intracerebroventricular administration of only 60 ng of morphine sulfate. Between 0 and 60 ng of morphine sulfate, there was a dose-dependent inhibition of dopamine release.

In order to test the possibility that the inhibitory effect of morphine on the release of dopamine is a consequence of an action of morphine on hypothalamic dopaminergic neurons, Haskins et al.[74] instilled morphine sulfate by microiontophoresis into the anterior portion of the arcuate nuclei of rats and found that, during the 30-min period of instillation of morphine as well as the 30-min period following the instillation, the release of dopamine into hypophysial portal blood was markedly depressed. When the animals were pretreated with naloxone, an opiate antagonist, the inhibitory effect of morphine on dopamine release was completely abolished.

[72] G. A. Gudelsky and J. C. Porter, *Life Sci.* **25**, 1697 (1979).
[73] M. J. Reymond, C. Kaur, and J. C. Porter, *Brain Res.* **262**, 253 (1983).
[74] J. T. Haskins, G. A. Gudelsky, R. L. Moss, and J. C. Porter, *Endocrinology* (*Baltimore*) **108**, 767 (1981).

Inasmuch as the perikarya of the tuberoinfundibular dopaminergic neurons are concentrated in the anterior regions of the two arcuate nuclei,[34–36] it is reasonable to speculate that morphine inhibits the release of dopamine by these dopaminergic cells by interacting with the soma of these cells, possibly through opiate receptors. Although it seems unlikely that this action of morphine is revealing of a physiological requirement for opiates, these results may indicate that endogenous opiate-like peptides such as β-endorphin and the enkephalins are involved in the regulation of the secretion of dopamine. Indeed, after the intracerebroventricular injection of either β-endorphin or an analog of enkephalin, D-Ala^2-enkephalinamide, the release of dopamine into hypophysial portal blood is suppressed in a manner that is similar to that seen with morphine.[72] Since β-endorphin is present in portal blood,[22,23] presumably as a result of the release of the peptide by hypothalamic neurons, a role for endorphins in the control of the secretion of hypothalamic dopamine is a possibility.

The release of hypothalamic dopamine is inhibited by serotonin,[75] a neurosecretory substance of neurons having terminals in the ventromedial hypothalamus. In association with the inhibition of dopamine release following the administration of serotonin, there is an immediate increase in the release of prolactin. Thus, the hyperprolactinemia that occurs in response to treatment with serotonin is attributable in part to the suppression of the release of dopamine into hypophysial blood.

Relationship of Dopamine Release to Dopamine Synthesis. The rapidity with which the release of dopamine into portal blood can be suppressed seems at first paradoxical in view of the relatively large quantity of dopamine that is present in median eminence tissue. This fact in association with the observation that treatment of rats with α-methyltyrosine, an inhibitor of the activity of tyrosine hydroxylase (tyrosine 3-monooxygenase), results in a prompt and nearly complete cessation of the release of dopamine into portal blood led Gudelsky and Porter[76] in 1979 to propose that newly synthesized dopamine, not stored dopamine, is released into portal blood. Additional data in support of this view were subsequently provided by Reymond and Porter.[49] When the conversion of L-dihydroxyphenylalanine (DOPA) to dopamine is prevented by treatment of rats with 3-hydroxybenzylhydrazine (NSD 1015), an inhibitor of the activity of aromatic-L-amino-acid decarboxylase,[77,78] the release of dopamine into portal blood is immediately inhibited. These observations indicate that the rate of release of dopamine into portal blood is dependent

[75] N. S. Pilotte and J. C. Porter, *Endocrinology (Baltimore)* **108**, 2137 (1981).
[76] G. A. Gudelsky and J. C. Porter, *Endocrinology (Baltimore)* **104**, 583 (1979).
[77] A. Carlsson, J. N. Davis, W. Kehr, M. Lindqvist, and C. V. Atack, *Naunyn-Schmiedeberg's Arch. Pharmacol.* **275**, 153 (1972).
[78] A. Carlsson and M. Lindqvist, *J. Neural Transm.* **34**, 79 (1973).

on the rate of formation of dopamine from DOPA. Inasmuch as the rate of transformation of DOPA to dopamine is rapid, it is reasonable to assume that the availability of DOPA limits the rate of formation, and hence the rate of release, of dopamine.

Thus, the question arises: What controls the availability of DOPA? Inasmuch as tyrosine is the precursor of DOPA, it is interesting to consider the effect of the availability of this substrate on the rate of release of dopamine. When rats are treated systemically with tyrosine, the rate of release of dopamine into portal blood is unchanged compared to vehicle-treated rats.[79] On the other hand, when DOPA is administered to rats, the concentration of dopamine in portal blood is increased 50-fold compared to that in vehicle-treated animals, whereas the concentration of dopamine in arterial blood is hardly affected. These findings are supportive of the view that the rate of release of hypothalamic dopamine is increased as a consequence of a rapid conversion of DOPA to dopamine in the median eminence, presumably in the tuberoinfundibular dopaminergic neurons. Thus, it appears that the rate of release of dopamine, under physiological conditions, is not limited by the availability of tyrosine but is dependent on the rate of conversion of tyrosine to DOPA.

To test the hypothesis that the availability of the pterin cosubstrate, presumably tetrahydrobiopterin, limits the rate of formation of DOPA and, in turn, the rate of release of dopamine into hypophysial portal blood, rats were injected intracerebroventricularly with the pterin analog 6-methyl-5,6,7,8-tetrahydropterin. This analog of the pterin cosubstrate had no stimulatory effect on the rate of formation of DOPA or on the release of dopamine into portal blood. Thus, if our present perception of the reaction involving the transformation of tyrosine to DOPA is correct, it would appear that the reaction is regulated by the activity of tyrosine hydroxylase. It is interesting that Joh et al.[80] observed that the activity of tyrosine hydroxylase is dependent on the extent of its phosphorylation. In light of these considerations, it is interesting to speculate that the rate of phosphorylation of tyrosine hydroxylase in tuberoinfundibular dopaminergic neurons regulates the rate of release of dopamine into hypophysial portal blood by controlling the rate of synthesis of DOPA.

Influence of Aging on Dopamine Release

In rats of the Long–Evans strain, the capacity of the tuberoinfundibular dopaminergic neurons to secrete dopamine is appreciably modified by the aging process. In old male rats as well as old female rats, the concen-

[79] M. J. Reymond and J. C. Porter, *Brain Res. Bull.* **7,** 69 (1981).
[80] T. H. Joh, D. H. Park, and D. J. Reis, *Proc. Natl. Acad. Sci. U.S.A.* **75,** 4744 (1978).

tration of dopamine in hypophysial portal blood is much less than that in young animals.[47,79] In old rats of both sexes of this particular strain, there is also a marked hyperprolactinemia that appears to be a consequence of the reduced concentration of dopamine in the blood perfusing the pars distalis. If the concentration of dopamine in portal blood is increased by treatment of old rats with DOPA, the release of prolactin is suppressed and the hyperprolactinemia is abolished.[79]

The basis for the reduced release of dopamine into portal blood in old rats appears to be due to a reduced rate of formation of DOPA in tuberoinfundibular neurons as shown by Demarest et al.[81] and confirmed by us. The reduced formation of DOPA could be a consequence of one of several factors. Owing to neuronal degeneration, the number of dopaminergic neurons available to synthesize and release dopamine may be less in old rats than in young rats.[82] In addition, the capacity of existing neurons to synthesize and release dopamine may be suppressed in old rats. Such suppression could be the result of reduced activity of tyrosine hydroxylase, but it does not appear to be due necessarily to a lack of the cosubstrates tyrosine and tetrahydrobiopterin.

The reduced ability of old rats of the Long–Evans strain to synthesize and release dopamine is unexpected in view of the persistent hyperprolactinemia in these animals. As noted earlier, treatment of young rats with prolactin or with haloperidol to induce hyperprolactinemia results in an increase in the rate of release of dopamine into portal blood. If this action of prolactin is mediated by way of prolactin receptors in the plasma membranes of brain cells—a reasonable assumption in view of the mechanism by which prolactin acts on cells of mammary tissue[83,84] and in view of the presence of prolactin receptors in the hypothalamus of rabbits[85] as well as in the choroid plexus and median eminence of rats[86,87]—it is possible that prolactin, when circulating in high concentrations for a prolonged period of time, down-regulates the prolactin receptors as has been shown in liver and mammary gland.[88] It is also possible that altered lipid membrane fluidity due to aging results in a reduction in the number of exposed prolactin receptors. Studies by Heron et al. are supportive of such an interpretation.[89,90]

[81] K. T. Demarest, G. D. Riegle, and K. E. Moore, *Neuroendocrinology* **31,** 222 (1980).
[82] D. K. Sarkar, P. E. Gottschall, and J. Meites, *Science* **218,** 684 (1982).
[83] R. P. C. Shiu and H. G. Friesen, *Science* **192,** 259 (1976).
[84] R. P. C. Shiu and H. G. Friesen, *Biochem. J.* **157,** 619 (1976).
[85] R. Di Carlo and G. Muccioli, *Life Sci.* **28,** 2299 (1981).
[86] R. J. Walsh, B. I. Posner, B. M. Kopriwa, and J. R. Brawer, *Science* **201,** 1041 (1978).
[87] M. van Houten, B. I. Posner, and R. J. Walsh, *Exp. Brain Res.* **38,** 455 (1980).
[88] J. Djiane, H. Clauser, and P. A. Kelly, *Biochem. Biophys. Res. Commun.* **90,** 1371 (1979).
[89] D. S. Heron, M. Shinitzky, M. Hershkowitz, and D. Samuel, *Proc. Natl. Acad. Sci. U.S.A.* **77,** 7463 (1980).

Conclusions

1. Dopamine, but not norepinephrine or epinephrine, is released by the hypothalamus, presumably by tuberoinfundibular dopaminergic neurons, into hypophysial portal blood.
2. The rate of release of dopamine into the various hypophysial portal vessels is greatest in the vessels located on the medial aspect of the pituitary stalk and least in the vessels located laterally on the pituitary stalk. Consequently, the concentration of dopamine in portal blood of a medial vessel is greater than that in a lateral portal vessel.
3. The rate of release of dopamine into portal blood is stimulated by prolactin, by agents that cause hyperprolactinemia, such as dopamine antagonists, by estradiol, and by DOPA, the precursor of dopamine.
4. The rate of release of dopamine into portal blood is inhibited by morphine, by opiate-like peptides, and by serotonin.
5. The rate of release of dopamine into portal blood is markedly suppressed by agents that inhibit the biosynthesis of dopamine, e.g., α-methyltyrosine and NSD 1015.
6. In rats of the Long–Evans strain, aging is accompanied by a diminution in the rate of synthesis and release of hypothalamic dopamine.

Acknowledgments

The authors wish to thank Anita Crockett for editorial assistance. Funding to support the experiments reported in this chapter was provided in part by Grants AM01237 and AG00306 of the National Institutes of Health, Bethesda, Maryland.

---

[90] D. Heron, M. Israeli, M. Hershkowitz, D. Samuel, and M. Shinitzky, *Eur. J. Pharmacol.* **72**, 361 (1981).

[42] Neurotransmitter Histochemistry: Comparison of Fluorescence and Immunohistochemical Methods

By ROBERT Y. MOORE and J. PATRICK CARD

Our knowledge of the organization of the nervous system has emerged largely from two significant extensions of the neuron doctrine. First, the parcellation of neurons into systems with complex and highly organized connections has been accomplished by application of neuroanatomical methods that utilize general properties of the neuron and are not dependent on chemical specificity among neurons. Second, we have evolved a considerable body of information derived from analysis of the nervous system by methods that depend upon chemical specificity expressed as a feature of chemical transmission at the synapse. The purpose of this chapter is to review and compare methods that rely on neurotransmitter specificity and have been used to study the organization of catecholamine (CA) neuron systems. There are two principal CA neuron systems in the mammalian brain.[1] Dopamine neurons are most numerous and form a series of discrete, highly organized systems that are located primarily in rostral midbrain, diencephalon, and telencephalon. Norepinephrine neurons are present in the medulla and pons and project in two major systems over widespread areas of the neuraxis. These systems have been studied so extensively that we now have more information concerning their organization than for any neural system save some primary sensory and motor systems. The acquisition of this knowledge was dependent upon the development of two methods for investigation of CA neuron systems; fluorescent histochemical methods and immunocytochemical methods using antibodies raised against the major synthetic enzymes for CAs.

The fluorescence histochemical methods were the first developed for analysis of neural organization using chemical specificity of synaptic transmitter substances.[2] These methods employ the reaction of formaldehyde, or glyoxylic acid, with a primary or secondary amine to form a fluorophor, which is viewed in a fluorescence microscope. This reaction was first applied to nervous system histochemistry by Falck and Hillarp, and their associates,[3,4] using gaseous formaldehyde and dried tissue. Sub-

[1] R. Y. Moore, *Ann. Neurol.* **12,** 321 (1981).
[2] R. Y. Moore, *in* "Neuroanatomical Tract-Tracing Methods" (L. Heimer and M. J. Robarts, eds.), p. 441. Plenum, New York, 1981.
[3] B. Falck, N. A. Hillarp, G. Thieme, and A. Torp, *J. Histochem. Cytochem.* **10,** 304 (1962).
[4] B. Falck, *Acta Physiol. Scand.* **56,** *Suppl.* **197** (1962).

sequent work in Falck's laboratory led to the establishment of glyoxylic acid as a second compound that could be used for histochemistry.[5] The advantages of the fluorescence histochemical methods are their specificity and sensitivity. The available methods demonstrate only dopamine, norepinephrine, epinephrine, and serotonin in the nervous system. There are few epinephrine neurons in brain, and serotonin neurons are best analyzed by an immunocytochemical method. Consequently, fluorescent histochemical methods are best applied to study of dopamine and norepinephrine neuron systems. The formaldehyde methods generally are less sensitive than those using glyoxylic acid, either alone or in combination with formaldehyde, but all the newer methods are sufficiently sensitive to demonstrate the entire CA neuron, including cell body, dendrites, preterminal axon, and terminal axonal plexus. Several fluorescence histochemical methods are described below. Their major disadvantages are that they are somewhat cumbersome and generally require specialized equipment, particularly a fluorescence microscope. In contrast, immunohistochemical methods require less specialized equipment but do necessitate that a well characterized, specific, and sensitive antibody is available to the investigator. The application of these methods to the study of CA neuron systems is described below.

Fluorescence Histochemistry

The ALFA Method for Adult Brains.[6] The animal is perfused under general anesthesia via the ascending aorta using a preperfusion solution followed by the aluminum perfusion. The preperfusion is to rinse the vasculature of blood, which precipitates with the aluminum solution. The solution consists of either ice-cold Tyrode's buffer or a 2.0% glyoxylic acid solution at pH 7.0, which is introduced at a pressure of 0.4–0.8 bar/cm^2; 150–250 ml is suitable for a small animal such as the rat.

The aluminum perfusion is performed with one of two solutions. Aluminum alone: 10 g of $Al_2(SO_4)_3 \cdot 18H_2O$ per 100 ml is dissolved in Tyrode's buffer with the pH of the solution adjusted to about 3.8 by adding 0.38 g of $Na_2B_4O_7 \cdot 10H_2O$ per 100 ml. Aluminum and formaldehyde: 20 g of paraformaldehyde is dissolved in 500 ml of distilled water with 4–6 drops of 1 N NaOH added. The solution is placed in boiling water until all paraformaldehyde is dissolved. After cooling to room temperature, it is mixed with 500 ml of ice cold Tyrode's buffer; 10 g of $Al_2(SO_4)_3 \cdot 18H_2O$ per 100 ml is added, and the pH is adjusted with sodium borate as noted above. When

[5] O. Lindvall and A. Björklund, *Histochemistry* **39**, 97 (1974).
[6] O. Lindvall, A. Björklund, B. Falck, and I. Loren, *Histochem. J.* **13**, 583 (1981).

switching from preperfusion to perfusion solutions, the pressure is gradually raised to 2.0 bar/cm^2 and a volume of 300–450 ml of perfusion solution is used. The perfusion requires about 30 sec, and marked swelling of the head and neck indicates that the perfusion is bad owing to intravascular precipitation of the aluminum solution.

The brain is removed and dissected into pieces that are wrapped in gauze and frozen in a liquid propane–propylene mixture cooled to the temperature of liquid nitrogen. The samples are freeze-dried and processed for fluorescence microscopy by the Falck–Hillarp method (see Moore[1] for details).

The Glyoxylic Acid (GA) Cryostat Method.[7] The animal is sacrificed by decapitation under general anesthesia. Fresh tissue is removed and cut in 5–10 mm slabs, which are immediately placed on a precooled cryostat chuck. The chuck is quickly returned to the cryostat, which is maintained at $-30°$. Cryostat sections, 16–32 μm, are cut and picked up by pressing a clean glass slide against the knife holding the section.

The cut section is immediately dipped three times, about 1 dip per second, in a beaker containing the SPG solution at room temperature. Excess solution is wiped off the bottom and edges of the slide using absorbent paper. The solution is made up by adding distilled water to 10.2 g of sucrose, 4.8 g of monobasic KH_2PO_4, and 1.5 g of glyoxylic acid to make 100 ml. The solution is stirred until clear, and 1 N NaOH is added to reach pH 7.4. Distilled water is added to make 150 ml of solution. The solution must be made fresh each day. The sections are dried on the slides under a cool hair dryer until dry or for as long as 15–30 min.

After drying, 1–2 drops of light USP mineral oil are placed on the slide covering the tissue. The slides are placed in a prewarmed oven set at 95° for 2.5 min. Slides in the oven should lie on a copper plate to obtain even heating. The oven must be kept between 90 and 95° to obtain an optimal reaction. The slides are then removed, and excess mineral oil is blotted from the edges of the slide; 1–2 drops of fresh mineral oil are placed on the slides, and they are coverslipped. The sections are examined in a fluorescence microscope.

Alternate sections may be placed on albuminized slides and processed for staining with cresyl violet, hematoxylin and eosin, or another stain.

The Vibratome–Formaldehyde Method.[8] The Vibratome formaldehyde method affords the advantages of well fixed tissue with maximal preservation of histological integrity. The necessary special instrumentation is a Vibratome (Oxford Instruments, San Mateo, CA).

---

[7] J. C. de la Torre, *J. Neurosci. Methods* **3**, 1 (1980).
[8] T. Hökfelt and A. Lungdahl, *Histochemie* **33**, 231 (1972).

Two solutions are utilized in the Vibratome–formaldehyde method. The first is a buffered formaldehyde solution made by combining 83 ml of a solution of monobasic sodium phosphate (2.26 g/100 ml) with 17 ml of a sodium hydroxide solution (2.52 g/100 ml), adding 4 g of paraformaldehyde, and heating to 60° to dissolve the paraformaldehyde. The pH is adjusted to 7.2–7.4 with sodium hydroxide. The second solution is a calcium-free Tyrode solution (NaCl, 0.8 g; KCl, 0.02 g; $MgCl_2$, 0.01 g; NaH_2PO_4, 0.05 g; Na_2HPO_4, 0.1 g; dextrose, 0.1 g added to 100 ml of distilled H_2O; the pH is adjusted to 7.2–7.4).

The animal is perfused under general anesthesia via the ascending aorta with 100–200 ml of the buffered formaldehyde solution at 4°. The volume of perfusion solution depends on the size of the animal (rat) used. The brain is removed rapidly and stored in ice-cold, buffered formaldehyde solution for approximately 20 min. The brain is then dissected into pieces approximately 4 mm in diameter that are mounted for sectioning. Other pieces of tissue may be maintained for several hours in the fixative and give satisfactory results when sectioned subsequently. Sections are usually cut at 10–25 µm as measured by the Vibratome scale, generally with the vibration rate at 6–9 scale units and a feeding speed of 1–3 units. The trough of the Vibratome is filled with Tyrode solution kept at or below 5°. Immediately after being cut, the sections are placed on a clean glass slide, blotted around the edges to remove excess Tyrode solution, dried under the warm air stream of a hair dryer for 15 min, and placed in a desiccator overnight under vacuum and over phosphorus pentoxide. The following day the sections are treated for 1 hr at 80° in an oven with paraformaldehyde equilibrated to 70–80% relative humidity. The sections are mounted in liquid paraffin or immersion oil and examined in the fluorescence microscope. If kept refrigerated between use, the sections may be examined for 1–2 weeks without loss of quality. Sections prepared in this manner can be stained for histological study.

Fluorescence Microscopy. The fluorescence microscopy required for the methods described above can be performed with an ordinary light microscope equipped for fluorescence work. The microscope may use either transmitted or incident (epi) illumination. Transmitted light is usually more advantageous for work with low-power objectives, whereas incident light illumination works best for high-power objectives with a numerical aperture greater than 0.75. Current advances in incident illumination technology suggest that this may become preferred for fluorescence microscopy. The light source for the fluorescence microscope is usually a mercury vapor lamp (e.g., Ostram HBO 200), which requires a special housing and power supply. Light emitted from the lamp passes through a heat-absorption filter (Schott KGl) and through Schott BG 12 primary filters to produce light in wavelengths at the activation maximum for the monoamine fluorophores. The secondary, or barrier, filter should have an absorption around 475 nm to permit viewing of the fluorophore but exclude ultraviolet light. Photomicrography is carried out with a camera attachment using Kodak Tri-X or Technical Pan 2415 film for black and white and High Speed Ektachrome for color. Exposure times must be determined empirically.

Selection of Fluorescent Histochemical Methods. The methods presented above provide a wide selection among the variants now available for fluorescence histochemistry. For the investigator who has only limited or intermittent use for fluorescence histochemistry, the glyoxylic acid cryostat (SPG) method is probably optimal. A cryostat is usually available in an anatomical or histochemical laboratory, and the method is quick and simple and requires a minimum of other equipment. In addition, the method is reliable, and consistent results are usually obtained (Fig. 1). The major limitation of the SPG method is that it can be applied to only relatively small areas of the nervous system, and it is cumbersome, if not impossible, to process several brains at one time. For the investigator who plans extensive use of fluorescence histochemistry, it is ideal to be able to employ several methods. The ability to use the ALFA method, the Vibratome–formaldehyde method, and the GA cryostat method will provide maximum flexibility. These require the availability of a freeze-dryer, a Vibratome, and a cryostat as equipment, but allow the investigator to obtain the capacity for large and multiple samples afforded by the ALFA method, the tissue preservation of the Vibratome-formaldehyde method, and the quickness and ease of application of the GA cryostat method. All methods have equal specificity for catecholamines, and the ALFA and GA cryostat methods offer maximum sensitivity within the limitations of current methodology. It should be noted that monoamine fluorescence histochemistry can be combined with immunohistochemistry.

Immunohistochemistry

The immunohistochemical method is based upon the principle of utilizing specific antibodies as a means of localizing tissue antigens. The antigen–antibody complex is visualized by either an immunofluorescence[9] or immunoperoxidase[10] method, and each of these techniques has been

---

[9] A. H. Coons, *in* "General Cytochemical Methods" (J. F. Danielli, ed.), Vol. 1, p. 399. Academic Press, New York, 1958.

[10] L. A. Sternberger, "Immunohistochemistry." Prentice-Hall, Englewood Cliffs, New Jersey, 1979.

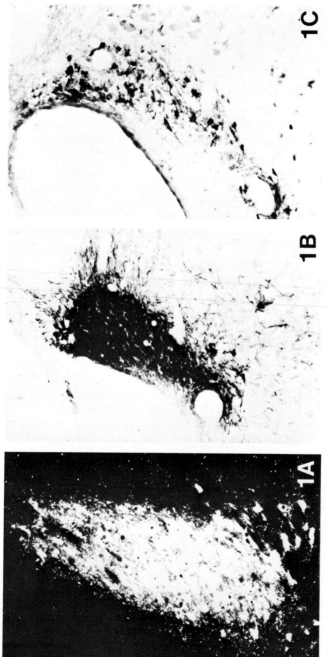

FIG. 1. The appearance of the locus coeruleus (LC) in tissue prepared for fluorescence histochemistry using the glyoxylic acid cryostat method of de la Torre[7] (A), and immunohistochemical staining with antisera generated against tyrosine hydroxylase (B; antiserum provided by Dr. Tong Joh), and avian pancreatic polypeptide (C; antiserum provided by Dr. J. R. Kimmel). The catecholaminergic nature of neurons of the LC is clearly established by both the fluorescence histochemical and immunohistochemical methods (A and B, respectively). Immunohistochemical staining for APP (C) demonstrates that this peptide is restricted to a subpopulation of LC neurons. Two studies[24,25] have demonstrated that it coexists with noradrenaline.

described thoroughly in reviews.[11,12] Our intent is to compare the usefulness of the unlabeled antibody enzyme method of Sternberger[10] with fluorescence histochemical techniques for the demonstration of catecholamine neuron systems and their relations to other peptide and neurotransmitter systems.

The unlabeled antibody enzyme method relies upon antigen-antibody interactions to attach peroxidase–antiperoxidase (PAP) complexes to antigenic sites. These sites are made visible by incubating tissue with hydrogen peroxide and 3,3-diaminobenzidine tetrahydrochloride (DAB) to produce an insoluble brown oxidation product of DAB. In the basic procedure, fixed tissue is incubated in high dilutions of the primary antiserum for 30–60 min at room temperature or for 24–48 hr at 4°. Antiserum to immunoglobulin G (IgG) then serves as a bridge linking the primary antibody and the PAP complex. The primary antiserum and the antiperoxidase moiety of the PAP complex are generally raised in the same species, with the secondary (IgG) antiserum produced in a second species and directed against the immunoglobulin fraction of the two homologous antisera. Erlandsen and collaborators[13] have demonstrated that heterologous antisera may be used in each step of the unlabeled antibody enzyme method as long as the coupling antisera show sufficient cross reactivity with the immunoglobulin fraction of the primary antisera and the PAP complex. This method has been used by Grzanna et al.[14] to demonstrate specific staining of central noradrenergic neurons with guinea pig anti-rat dopamine β-hydroxylase (DBH) and a heterologous secondary antibody (goat anti-rabbit IgG) and PAP (rabbit) complex. Therefore, although coupling of homologous antisera (primary antiserum and PAP complex) remains the method of choice in the unlabeled antibody enzyme method, it is clear that specific staining can be achieved with heterologous antisera provided the linking antiserum exhibits adequate cross-reactivity with the primary antiserum and PAP complex.

A fundamental concern in the immunohistochemical detection of tissue antigens is establishment of the specificity of the antigen–antibody interaction. Specificity of the primary antibody is generally demonstrated by an absence of immunoreactive staining when the antibody is preab-

---

[11] F. Vandesande, *J. Neurosci. Methods* **1**, 3 (1979).
[12] J. E. Vaughn, R. P. Barber, C. E. Ribak, and C. R. Houser, *in* "Current Trends in Morphological Techniques" (J. E. Johnson, Jr., ed.), Vol. III, p. 33. CRC Press, Boca Raton, Florida, 1981.
[13] S. L. Erlandsen, J. A. Parsons, J. P. Burke, J. A. Redick, D. E. Van Orden, and L. S. Van Orden, *J. Histochem. Cytochem.* **23**, 666 (1975).
[14] R. Grzanna, M. E. Molliver, and J. T. Coyle, *Proc. Natl. Acad. Sci. U.S.A.* **75**, 2502 (1978).

sorbed with the antigen it is directed against. It is also desirable to demonstrate that staining is not due to cross reactivity of the primary antiserum with other tissue antigens. This question can be resolved by demonstrating that preabsorbing the primary antiserum with other antigens does not alter the pattern of immunohistochemical staining normally achieved with the primary antiserum. Tests of antiserum cross-reactivity should be examined when the pattern of immunohistochemical staining obtained with the primary antiserum closely parallels that observed in the same area of study with another antiserum or when the antigen against which the primary antiserum is directed possesses structural similarity with other compounds known to exist in high concentration within the area of interest. Antibody specificity also can be demonstrated by immunoelectrophoresis, as has been shown with the catecholamine synthesizing enzymes dopamine β-hydroxylase and phenylethanolamine N-methyltransferase.[15]

Another major consideration in the immunohistochemical detection of antigens is fixation of tissue. Specificity and interpretation of immunohistochemical staining requires that the fixative stabilize the antigen without compromising antigenicity or redistributing the antigen within the tissue while preserving cellular morphology. Currently, the majority of fixatives employed in light microscopic immunohistochemical localizations include paraformaldehyde as a primary constituent. In 1974, McLean and Nakane[16] introduced a periodate–lysine–paraformaldehyde (PLP) fixative that has been widely utilized in both light and electron microscopic immunohistochemistry. In tissue perfused with this fixative, periodate oxidizes carbohydrate moieties of tissue to form aldehyde groups that are then cross-linked by lysine. The paraformaldehyde is added in low concentrations (2–4%) in order to stabilize proteins and lipids and because it does not undergo polymerization to form stable complexes with lysine. Other aldehydes have been used as immunohistochemical fixatives with varying degrees of success. Glutaraldehyde improves preservation of tissue morphology, but must be used in exceedingly low concentrations (on the order of 0.1%) owing to the adverse effect it exerts on tissue antigenicity. Studies[17,18] have demonstrated that fixation of tissue with the monoaldehyde acrolein results in excellent preservation of cellular morphol-

---

[15] L. S. Van Orden III, J. P. Burke, J. A. Redick, K. E. Rybarczyk, D. E. Van Orden, H. A. Baker, and B. K. Hartman, *Neuropharmacology* **16**, 129 (1977).
[16] I. W. McLean and P. K. Nakane, *J. Histochem. Cytochem.* **12**, 1077 (1974).
[17] P. F. Smith and D. A. Keefer, *J. Histochem. Cytochem.* **30**, 1307 (1982).
[18] J. C. King, R. M. Lechan, G. Kugel, and E. L. P. Anthony, *J. Histochem. Cytochem.* **31**, 62 (1983).

ogy at both the light and electron microscopic level and preserves tissue antigenicity to a variety of antisera.

The following procedure has resulted in optimal staining of central CA neurons with antiserum generated against tyrosine hydroxylase (raised in rabbit and generously provided by Dr. Tong Joh, Cornell Medical Center, New York, New York) and serves to illustrate the basic methods involved in processing tissue according to the unlabeled immunoperoxidase technique. Under general anesthesia, the animal is subjected to two-stage transcardiac perfusion at controlled perfusion pressure. In the initial stage of the perfusion, blood is washed from the vasculature of the head and upper extremities (the descending aorta is clamped) with 30–50 ml of a 6.0% solution of dextran in 0.1 M sodium phosphate buffer at pH 7.2. This is immediately followed by infusion of 300–500 ml of 4.0% paraformaldehyde in 0.1 M borate buffer at pH 11.0. The brain is then removed and postfixed in the primary fixative for 1 hr at 4°. Thereafter, the brain is washed in 0.1 M phosphate buffer (pH 7.2) for 1 hr at 4°, carried through sequential hourly changes of 10 and 20% phosphate-buffered sucrose and stored overnight at 4° in 30% phosphate-buffered sucrose.

Serial sections through the area of interest are cut at 30 μm with a freezing microtome and collected in 0.1 M phosphate buffer, where they are stored at 4° for up to 2 days prior to processing for immunohistochemistry. Tissue sections are then incubated in the primary antiserum diluted to 1 : 500 with 0.1 M sodium phosphate buffer containing 1.0% normal goat serum and 0.3% Triton X-100 (PBGT). The incubation is generally carried out at 4° for 24 hr, but can be conducted at room temperature for 30–60 min. Tissue is then thoroughly washed in several changes of phosphate buffer at room temperature over a period of 30 min, after which it is incubated in the goat anti-rabbit IgG diluted to 1 : 50 with PBGT for 30 min. Incubation in the secondary antibody is done at room temperature with continuous agitation and is followed by several washes in phosphate buffer over a period of 30 min. Tissue sections are then incubated for 30–45 min in rabbit PAP diluted to 1 : 100 with phosphate buffer and containing 0.3% Triton X-100. The PAP incubation is also done with continuous agitation and is followed by several phosphate buffer washes over a period of 30 min.

After the final wash, tissue is preincubated in a saturated solution of DAB in 0.1 M sodium phosphate buffer for 10 minutes at room temperature. In this instance, a saturated solution is defined as the amount of DAB that will go into solution in 5 min at room temperature with continuous stirring, and the solution is filtered prior to tissue incubation. After preincubation, 110 μl of hydrogen peroxide per 100 ml of DAB solution is added. The DAB reaction is monitored visually and is terminated by transferring tissue through several changes of fresh phosphate buffer. Tissue sections are then mounted on glycerin-coated slides, dehydrated in ethanol, cleared in xylene, and coverslipped with Permount. In most instances, the DAB reaction product is intensified and stabilized by immersing reacted tissue in a 0.1% solution of osmium tetroxide for 15–30 sec prior to dehydration of the tissue.

Although the above procedure has resulted in optimal staining of tyrosine hydroxylase-containing neurons in our laboratory, it should be emphasized that the quality of immunohistochemical staining varies with different antisera, and it is therefore necessary to screen a number of different parameters in order to determine the procedures that will result in the best staining with any given antiserum. Both specificity and intensity of immunohistochemical staining can be influenced by several different factors. As previously noted, fixation of tissue will often compromise the antigenicity. Consequently, poor staining can often be improved by altering the concentration of different aldehydes within the fixative, varying the pH, or changing buffers.

Another common problem with the immunohistochemical technique is either a total absence of staining or very faint staining of neuronal perikarya. This is often due to low concentrations of the antigen within the cell body and can usually be solved by intraventricular injections of colchicine (50 μg in 10 μl of sterile saline is generally adequate for an animal the size of a rat) 24–48 hr prior to sacrifice. Colchicine blocks axonal transport and thereby increases intracellular concentrations of the antigen. Faint immunohistochemical staining of both cells and processes can also be improved by applying methods for intensification of DAB-based reaction product. Adams[19,20] has reported two methods in which heavy metals can be utilized to intensify DAB-based HRP reaction product. We have found these methods extremely effective in improving the intensity of immunohistochemical staining, especially when combined with intraventricular administration of colchicine. Finally, the source and age of immunohistochemical reagents should be carefully monitored.

Summary and Conclusions

The above procedures clearly demonstrate the effectiveness of the fluorescence histochemical and immunohistochemical methods in analysis of central CA neuron systems. Each method is well characterized and possesses sufficient versatility to permit a variety of experimental applica-

[19] J. C. Adams, *Neuroscience* **2**, 141 (1977).
[20] J. C. Adams, *J. Histochem. Cytochem.* **29**, 225 (1981).

tions. The fluorescence histochemical technique is extremely reliable, produces good cell and fiber morphology, and has served as the fundamental procedure used to define the organization and distribution of CA-containing neurons throughout both the central and peripheral nervous system. However, the basic method makes no distinction between individual catecholamine neuron systems unless combined with mechanical or chemically induced lesions. The immunohistochemical technique relies upon the availability of antisera generated against the catecholamine-synthesizing enzymes tyrosine hydroxylase (TH), dopamine β-hydroxylase (DBH), and phenylethanolamine N-methyltransferase (PNMT). Each of these enzymes catalyzes different steps in catecholamine metabolism and therefore can be used in conjunction with one another to selectively delineate the organization and distribution of neuronal cells and processes containing dopamine (TH), noradrenaline (TH and DBH), and adrenaline (PNMT). In addition, the immunohistochemical method may be coupled with the fluorescence histochemical technique, dual-labeling immunohistochemical methods, and retrograde tract tracing methods to provide further information on the interrelations of individual CA systems with one another and other chemically distinct systems of neurons.

The usefulness of the immunohistochemical method in elucidating the organization of separate systems of CA-containing neurons is illustrated in a recent study by Swanson and collaborators.[21] Utilizing antisera raised against the previously mentioned CA synthesizing enzymes, they analyzed the organization of CA systems within the paraventricular and supraoptic nuclei of the hypothalamus. By combining this analysis with immunohistochemical staining with antisera generated against vasopressin and oxytocin, they were able to demonstrate differential distribution of adrenergic and noradrenergic fibers within each nucleus which could be correlated with the distribution of vasopressin-containing neurons. In addition, through the combined use of immunofluorescence and retrograde transport of the tracer dye true blue, they were able to show that a small percentage of TH-containing neurons within the paraventricular nucleus project to the region of the dorsal motor vagal complex and/or thoracic levels of the spinal cord. Although the later finding relied upon fluorescence microscopy, Bowker and colleagues[22] have recently described a dual-labeling technique that permits demonstration of retrogradely transported horseradish peroxidase in neurons stained with the unlabeled antibody enzyme method.

---

[21] L. W. Swanson, P. E. Sawchenko, A. Berod, B. K. Hartman, K. B. Helle, and D. E. Van Orden, *J. Comp. Neurol.* **196**, 271 (1981).
[22] R. M. Bowker, H. W. M. Steinbush, and J. D. Coulter, *Brain Res.* **211**, 412 (1981).

Immunohistochemical techniques have also proved to be invaluable in demonstrating the coexistence of peptides and classical neurotransmitters, such as noradrenaline.[23] For example, two investigations[24,25] have reported coexistence of noradrenaline and avian pancreatic polypeptide (APP) within brainstem catecholamine neurons. In these studies, catecholamine neurons containing APP and tyrosine hydroxylase immunoreactivity were demonstrated within a subpopulation of the neurons of the locus coeruleus and within neurons of the medullary lateral tegmental CA cell groups known to project to hypothalamus. No colocalization was demonstrated in the pontine cell groups (A5 and A7) of the lateral tegmental catecholamine system, groups that do not project significantly to hypothalamus.[26,27] Our own data[28] have demonstrated marked overlap of axons and terminals within the hypothalamus, which contain APP and noradrenaline, suggesting that APP and noradrenaline may also coexist within preterminal axons and terminal fields arising from brainstem catecholamine cell groups. Although this has not yet been demonstrated directly, the ability to couple immunohistochemical detection of peptides and neurotransmitters with the histological detection of retrogradely transported HRP[22] provides a means of addressing this question and at the same time illustrates the effectiveness and versatility of the immunohistochemical method in extending fluorescence histochemical observations on the organization and interrelations of catecholamine neuron systems.

---

[23] T. Hökfelt, O. Johansson, A. Ljungdahl, J. M. Lundberg, and M. Schultzberg, *Nature (London)* **284**, 515 (1980).

[24] J. M. Lundberg, T. Hökfelt, A. Anggard, J. Kimmel, M. Goldstein, and K. Markey, *Acta Physiol. Scand.* **110**, 107 (1980).

[25] S. P. Hunt, P. C. Emson, R. Gilbert, M. Goldstein, and J. R. Kimmel, *Neurosci. Lett.* **21**, 125 (1981).

[26] T. Sakumoto, M. Tohyama, K. Satoh, Y. Kimoto, T. Kinugasa, O. Tanizawa, K. Kurachi, and N. Schimizu, *Exp. Brain Res.* **31**, 81 (1978).

[27] M. Palkovits, L. Zaborszky, A. Femminger, E. Mezey, M. I. K. Fekete, J. P. Herman, B. Kanyicska, and D. Szabo, *Brain Res.* **191**, 161 (1980).

[28] J. P. Card, N. Brecha, and R. Y. Moore, *J. Comp. Neurol.* **217**, 123 (1983).

[43] Simultaneous Localization of Steroid Hormones and Neuropeptides in the Brain by Combined Autoradiography and Immunocytochemistry

By MADHABANANDA SAR and WALTER E. STUMPF

We have introduced a combined technique of autoradiography and immunocytochemistry for localization of radioactively labeled steroid hormones and antibodies to protein hormones, neuropeptides, or neurotransmitter synthesizing enzymes in the brain.[1-6] This method enables the simultaneous visualization of radioactively labeled cells and peptide or neurohormone producing cells in the same tissue preparation. With this procedure, anatomical relationships between sites of action and sites of production of two different substances have been demonstrated. Steroid hormones influence brain functions by genomic activation of target cells. These target cells and their topographical distribution have been identified by the use of autoradiography.[7-10] In certain areas of the brain, sex-steroid hormone-sensitive cells are coexistent with peptidergic and aminergic neurons. Also, morphological relationships between steroid hormone target cells and peptidergic or aminergic neurons appear to exist. This chapter describes in detail the combined technique of steroid hormone autoradiography and immunocytochemistry.

Description of Methods

In the combined technique of steroid hormone autoradiography and immunocytochemistry, autoradiograms are prepared by the thaw-mount or dry-mount technique[7,10] after injection of radioactively labeled hormone. Subsequently the autoradiograms are stained by the peroxidase–

[1] M. Sar and W. E. Stumpf, *J. Histochem. Cytochem.* **26**, 277 (1978).
[2] M. Sar and W. E. Stumpf, *Cell Tissue Res.* **203**, 1 (1979).
[3] M. Sar and W. E. Stumpf, *Neurosci. Lett.* **17**, 179 (1980).
[4] M. Sar and W. E. Stumpf, *Nature (London)* **289**, 500 (1981).
[5] M. Sar and W. E. Stumpf, *J. Histochem. Cytochem.* **29**, 1A, 161 (1981).
[6] M. Sar and W. E. Stumpf, *in* "Handbook of Chemical Neuroanatomy" (A. Björklund and T. Hökfelt, eds.), Vol. 1, p. 442. Elsevier/North-Holland Biomedical Press, Amsterdam.
[7] W. E. Stumpf and M. Sar, this series, Vol. 36, p. 135.
[8] W. E. Stumpf and M. Sar, *in* "Receptors and Mechanism of Action of Steroid Hormones, Modern Pharmacology–Toxicology" (J. Pasqualini, ed.), Vol. 8, p. 41. Dekker, New York, 1976.
[9] W. E. Stumpf and L. J. Roth, *J. Histochem. Cytochem.* **14**, 274, 1966.
[10] W. E. Stumpf, *Acta Endocrinol. (Copenhagen) Suppl.* **153**, 205 (1971).

antiperoxidase (PAP) immunocytochemical staining technique using unlabeled antibodies. This technique has been described previously.[2,5,6]

Autoradiography

Thaw-mount autoradiography is used generally. Some of the steps for the preparation of autoradiograms are identical in both thaw-mount and dry-mount techniques, which include freezing, storage, and sectioning of tissues as well as photographic development and fixation. Thaw-mount autoradiograms are easier to prepare, and the exposure time appears shorter when compared with dry-mount autoradiograms. The autoradiographic procedure includes four steps: (1) freezing of brain blocks, (2) sectioning of 4-μm frozen sections in a cryostat at $-35°$ to $-40°$, (3) picking up frozen sections on dried emulsion-coated slides (thaw-mount) or mounting of freeze-dried sections on dried emulsion-coated slides (dry-mount), and (4) photographic exposure of the slides at $-15°$ in a freezer.

Animal Preparation. The animals are gonadectomized or adrenalectomized, or both, for at least 48 hr, prior to the injection of labeled compound in order to reduce the endogenous level of steroid hormone(s). Twenty-four hours prior to the injection of labeled steroid hormone, the rats are injected intraventricularly with colchicine (25–50 μg per 100 g body weight of the animal). Colchicine treatment enhances perikaryal staining of neuropeptides.[11] Under normal conditions certain peptidergic neurons are difficult to detect by immunostaining. At 25–50 μg per 100 g body weight, colchicine does not appear to block nuclear uptake of radioactivity in neurons after injection of estradiol.

The radioactively labeled steroids are generally supplied with benzene–ethanol or toluene–ethanol. The solvent is evaporated, and the steroid is dissolved in 10% ethanol–isotonic saline and injected intravenously at a physiological dose under ether anesthesia. The animal is killed by decapitation at the specified time after injection.

Autoradiographic Procedure

Freezing of Brain Blocks. The brain is removed and dissected into two or three blocks on a petri dish cooled on ice. A tissue block is placed on a brass mount with minced liver as an adhesive. The mounted brain block is frozen in liquefied propane ($-180°$) and then stored in a liquid-nitrogen storage container until sectioning. Propane is liquefied by liquid nitrogen prior to the experiment and kept in a flask cooled by liquid

---

[11] M. Sar, W. E. Stumpf, R. J. Miller, K.-J. Chang, and P. Cuatrecasas, *J. Comp. Neurol.* **182**, 17 (1978).

nitrogen. Liquid propane is then poured into a beaker cooled with liquid nitrogen. When propane is not available, isopentane or Freon cooled with liquid nitrogen can be used.

Sectioning of Frozen Sections. Four-micrometer serial sections are cut in a cryostat maintained at $-32°$ to $-35°$ knife temperature. A dissecting microscope, equipped with fiber optics light mounted to the side of the cryostat, is swung over the knife in order to facilitate sectioning.

Picking up Frozen Sections on Dried Emulsion-Coated Slides. Two or three sections are picked up from the knife with a dried emulsion-coated slide under safelight. The slide with sections is then placed in a black desiccator box kept at room temperature. The box containing the section-mounted slides is sealed with black tape and stored in a freezer or refrigerator for autoradiographic exposure.

For dry-mount autoradiography, the frozen sections are freeze-dried and then mounted onto dried emulsion-coated slides. Frozen sections are transferred with the bristles of a fine brush from knife to vials which are kept near the knife. The vials are covered with a fine wire mesh with a hole in the center. The wire mesh helps to dislodge the sections from the brush and to prevent loss of sections during vacuum freeze drying and breaking of the vacuum. The vials with frozen sections are transferred to a vacuum flask within the cryostat for subsequent freeze-drying with a cryosorption pump.[12] Freeze-drying of the sections with the cryosorption pump may also be done outside of the cryostat. For this two Dewars are prepared, one filled with liquid nitrogen and the other with Dry-Ice alcohol slush. A vacuum between 10^{-4} and 10^{-5} Torr is maintained. Freeze-drying is done at a temperature below $-35°$. After 24-hr freeze-drying, the cryosorption pump is removed from the cryostat and the specimen chamber is kept at room temperature while the cryosorption chamber remains immersed in liquid nitrogen. The vacuum is broken with dry nitrogen gas, and the vials containing freeze-dried sections are placed in a desiccator. For mounting, freeze-dried sections are placed on a cleaned Teflon piece and under safelight are pressed against a dried emulsion-coated slide. Before mounting, freeze-dried sections may be checked under a dissecting microscope in order to prevent or remove folds. Folds may introduce pressure artifacts. The Teflon and the slide are pressed together between forefingers and thumb. After release of pressure, the Teflon falls and the tissue adheres to the emulsion-coated slide. The slide is stored in a light-proof desiccator box for exposure. The length of the exposure time depends upon several factors, such as the specific activity, the dose, and the specific tissue concentration of the labeled compound.

[12] W. E. Stumpf and L. J. Roth, *Stain Technol.* **39**, 219 (1964).

Preparation of Emulsion Coated Slides. Emulsion (Kodak NTB-3) is melted in a waterbath maintained at a temperature of 42°–45°. The cleaned slides are coated with liquid emulsion diluted (1 : 1 with distilled water) or undiluted in a darkroom under safelight. The slides are air-dried overnight and then kept in lightproof desiccator boxes for storage in a refrigerator.

Immunocytochemistry

Autoradiograms prepared by dry-mount or thaw-mount procedure may be used for immunocytochemical staining. The procedure consists of three steps: (1) fixation of exposed slides prior to autoradiographic development, (2) autoradiographic development, and (3) immunoperoxidase staining using PAP complex.

Fixation of Exposed Slides prior to Autoradiographic Development. At the end of exposure, the lightproof slide box is removed from the freezer and allowed to equilibrate to room temperature for 30–60 min under safelight. The slide box is opened, and the slides are removed and fixed for 30–60 sec in 2.5–4% paraformaldehyde solution in 0.01 M phosphate buffer–isotonic saline (PBS), pH 7.2–7.4. After fixation the slides are rinsed briefly with PBS. Fixative and PBS are kept in an ice bath. Several other fixatives, such as 10% buffered formalin or Bouin's fluid, have been tried. However, mild fixation with paraformaldehyde solution appears to be satisfactory. This mild fixation preserves tissue structures and retains antigenicity of the tissues. Also, it does not have deleterious effects on the undeveloped autoradiogram such as fading of latent image or fogging of the emulsion. The same fixation procedure is followed when freeze-dried sections are used except that the slides are breathed over the sections prior to the fixation in order to increase the adherence of sections to the emulsion and to prevent loss of sections during photographic processing.

Autoradiographic Development. After fixation, the slides are briefly rinsed in PBS and then developed in Kodak D-19 developer, diluted 1 : 1 with PBS, for 40–60 sec, rinsed in PBS, and followed by fixation for 5 min with Kodak fixer. The slides are washed in two changes of PBS, 5 min each, and subsequently processed for immunoperoxidase staining.

Immunocytochemical Staining. In our laboratory routinely the PAP procedure is used for immunostaining of autoradiograms. This technique is sensitive for the detection of antigen and yields little or no tissue background staining. In this procedure the unlabeled antibody technique[13] was

---

[13] T. E. Mason, R. F. Phifer, S. S. Spicer, R. A. Swallow, and R. B. Deskin, *J. Histochem. Cytochem.* **17**, 563 (1969).

modified using PAP complex,[14] instead of single-step incubation with anti-horseradish peroxidase serum and horseradish peroxidase serum. The immunocytochemical procedure consists of four steps. In step 1, the tissue antigen is localized by incubating with primary antiserum specific for the antigen, produced in a suitable species. In step 2, the tissue is incubated in excess of secondary antiserum specific for the immunoglobin of the primary antiserum. In step 3, the tissue is incubated with PAP complex produced in the species in which primary antiserum is produced. In step 4, PAP is allowed to react with diaminobenzidine tetrahydrochloride (DAB) to form a brown reaction product that can be visualized under the microscope.

Prior to incubation with the primary antiserum the autoradiograms are treated either with hydrogen peroxide–egg albumin[15] or phenylhydrazine[16] followed by egg albumin. Phenylhydrazine treatment reduces nonspecific tissue background staining and to some extent reduces endogenous peroxidase activity in frozen tissue. The slides are incubated with freshly prepared 0.05% phenylhydrazine solution in 0.01 M PBS adjusted to pH 7.1 for 1 hr at 37°. After incubation the slides are washed in three changes of PBS, 5 min each, followed by 5–10 min treatment with 5–10% egg albumin in PBS. The autoradiograms are then incubated with the primary antiserum. For the detection of peptides or enzymes in the perikarya by immunostaining, penetration of antibodies into tissue sections may be a limiting factor. For this reason, after washing with PBS the tissue sections are treated with 0.3% Triton X-100 in PBS for 10 min and washed in PBS three times, 5 min each, followed by 5–10 min of treatment with 5–10% egg albumin or 1% normal serum in PBS of the species in which secondary antibody is produced.

The PBS around the tissue section is wiped off. The slides are placed on plastic frame that can carry slides. The primary antibodies are used at optimal dilution. One or two drops depending on the size of the section, are placed over the section. The frame with the slides is transferred to a stainless steel rack that can carry three frames. At one time, one can incubate 18 slides containing serial section autoradiograms. The rack with frames is kept in a humidity chamber at 4° for 24–48 hr. The frame, rack, and humidity chamber are components of Gelman immunoelectrophoresis kit and commercially available. After incubation with the primary antiserum, the slides are washed with PBS and 1% normal serum (of the species in which secondary antibody is produced) in PBS for 3–5 min

---

[14] L. A. Sternberger, "Immunocytochemistry." Prentice-Hall, Englewood Cliffs, New Jersey, 1974.
[15] D. R. Zehr, *J. Histochem. Cytochem.* **26,** 415 (1978).
[16] W. Straus, *J. Histochem. Cytochem.* **27,** 1349 (1979).

each, followed by incubation for 10–15 min with secondary antibodies (anti-rabbit γ-globulin if the primary antibodies are raised in rabbit). Depending upon the source, secondary antibodies are used at a dilution of 1:50 or 1:100 in PBS. After a 3–5-min wash in PBS, the autoradiograms are incubated with PAP complex diluted in PBS 1:50 for 20–25 min and washed in PBS for 5 min. The autoradiograms are then incubated for 10 min in freshly prepared DAB under constant stirring. DAB solution is prepared by dissolving 150 mg of DAB in 200 ml of 0.05 M Tris-HCl buffer, pH 7.6, and then adding 15 μl of 30% hydrogen peroxide. After incubation with the substrate, the slides are washed with Tris-HCl buffer and PBS, 5 min each, followed by counterstaining with 1% methyl green in PBS for 1 or 2 min. Methyl green stains nuclei, and brown DAB reaction product is seen over cytoplasm. Counterstaining of immunoautoradiograms is not essential. After counterstaining, the sections are dehydrated through ascending grades of alcohol, cleaned in two changes of xylene for 3 min each, and mounted with Permount and coverslip.

Dilution of antisera is made in 0.01 M PBS (pH 7.6) containing 0.1 sodium azide and 1% normal serum of the species in which the secondary antibodies are produced. PAP is diluted in 0.01 M PBS containing 1% normal serum without sodium azide, since it destroys peroxidase activity. For each antiserum the optimal dilution is determined to obtain satisfactory results.

Different types of control are used to determine specificity of the immunocytochemical reaction. Tissue is processed for immunostaining without incubation with primary antiserum. Preimmune serum is used instead of specific antiserum. Antiserum preabsorbed with appropriate antigen is used for tissue incubation. Specificity of immunocytochemical staining has been discussed previously.[17]

Other immunocytochemical methods may be tried for immunostaining of autoradiograms. The ABC procedure for the localization of a variety of antigens has been developed by Hsu and co-workers.[18] This method is more sensitive than the PAP method. This technique employs primary antibody, biotinylated secondary antibody, and preformed avidin-biotinylated horseradish peroxidase. It is expected that the ABC procedure is applicable for immunostaining of autoradiograms.

Applications

Using the combined technique of autoradiography and immunocytochemistry, localization of [^{3}H]estradiol has been demonstrated in

[17] P. Petrusz, M. Sar, P. Ordronneau, and P. DiMeo, *J. Histochem. Cytochem.* **24**, 1110 (1976).

[18] S. M. Hsu, L. Raine, and H. A. Fanger, *Am. J. Clin. Pathol.* **75**, 734 (1981).

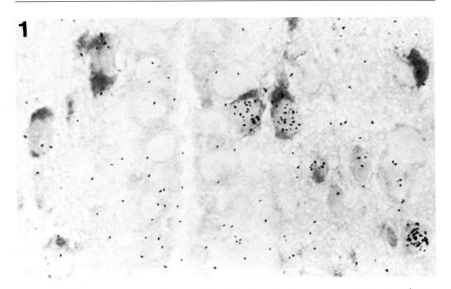

FIG. 1. Thaw-mount autoradiogram of anterior hypothalamus prepared 1 hr after [^{3}H]estradiol injection into ovariectomized colchicine-pretreated rat and stained by immunoperoxidase method with somatostatin antibodies. Note the nuclear concentration of [^{3}H]estradiol in somatostatin neurons in the nucleus periventricularis hypothalami. 4 μm, ×870.

neurophysin 1-producing and vasopressin-producing cells of mouse supraoptic and paraventricular nuclei[3] and neurophysin-producing cells of guinea pig preoptic periventricular and supraoptic nuclei.[5] With antibodies to dopamine β-hydroxylase (dopamine β-monooxygenase) (DBH), the enzyme that converts dopamine to noradrenaline, we showed that certain DBH-positive cells in the rat lower brainstem concentrate [^{3}H]estradiol.[4,19] Similarly, some of the tuberoinfundibular neurons that are positively immunostained with tyrosine hydroxylase (tyrosine 3-monooxygenase) antibodies show nuclear concentration of [^{3}H]estradiol.[19a] Sar et al.[20] reported that certain γ-aminobutyric acid-ergic neurons in the preoptic region concentrate [^{3}H]estradiol. With this technique, radioactively labeled estradiol and antibodies to cytosolic estradiol receptor are localized in rat anterior pituitary cells (M. Sar et al., unpublished observation). Further, the technique has been applied to characterize steroid hormone target cells in the rat and hamster anterior pituitary,[2,21] 1,25-(OH)$_2$-vitamin D$_3$ target cells in the rat anterior pituitary[22] and pancreas.[5] Figure 1 shows an example of simultaneous localization of [^{3}H]estradiol and

[19] M. Sar and W. E. Stumpf, *Exp. Brain Res., Suppl.* **3**, 29 (1981).
[19a] M. Sar, *Physiologist* **26**, A14 (1983).
[20] M. Sar, W. E. Stumpf, and M. L. Tappaz, *Fed. Proc.* **42**, 495 (1983).
[21] A. R. Munn, M. Sar, and W. E. Stumpf, *Brain Res.* (in press).
[22] M. Sar, W. E. Stumpf, and H. F. Deluca, *Cell Tissue Res.* **209**, 161 (1980).

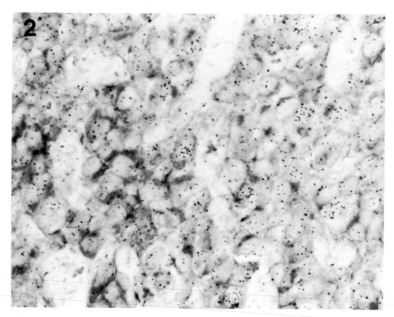

FIG. 2. Autoradiogram of rat anterior pituitary prepared after injection of [³H]estradiol and immunostained with antibodies to calf cytosolic estradiol receptor protein. Note the nuclear concentration of [³H]estradiol in anterior pituitary cells and the brown (dark) staining of cytoplasm with antibodies. Some cells are positively stained with antibodies but do not concentrate [³H]estradiol. 6 μm, ×860. Unpublished work of M. Sar, W. E. Stumpf, and E. E. Baulieu.

somatostatin antibodies in neurons of the rat nucleus periventricular hypothalami. Figure 2 shows localization of [³H]estradiol and cytosolic estradiol receptor antibodies in cells of the rat anterior pituitary.

Acknowledgments

This work was supported by U.S. Public Health Service Grants NS 17479 and NS 00914. We thank Robert P. Elde, Department of Anatomy, University of Minnesota, for his gift of antiserum to somatostatin.

[44] Immunocytochemistry of Steroid Hormone Receiving Cells in the Central Nervous System

By JOAN I. MORRELL and DONALD W. PFAFF

Introduction to Steroid Hormone Autoradiography

Identification of Steroid Hormone Receiving Cells

Steroid hormones exert powerful effects on the brain, influencing neuroendocrine aspects of reproduction and many aspects of behavior, particularly sexual behaviors. Steroid hormones are thought to exert their effects via a steroid-specific receptor protein. Once in the cell, the steroid hormone binds to the receptor in the cytoplasm. Subsequently, the steroid hormone–receptor complex moves into nucleus, binding at a specialized acceptor site to the DNA, where it exerts its effect on the genome of the cell.

Extensive biochemical experiments have identified specific areas of the brain that contain high-affinity, saturable, steroid-specific receptor proteins [1-3] that are virtually identical to those identified in such steroid target organs as the uterus, oviduct, epididymus.[4-7] The exact identification of the brain cells that contain the steroid receptor is possible with steroid autoradiographic methods,[8,9] which produce results completely consonant with the biochemical data.

Identification of steroid hormone concentrating cells has been attempted by other methods. A small number of studies, mostly restricted

---

[1] B. S. McEwen, *Science* **211**, 1303 (1981).

[2] T. C. Rainbow, B. Parsons, N. J. MacLusky, and B. S. McEwen, *J. Neurosci.* **2**, 1439 (1982).

[3] B. Parsons, T. C. Rainbow, N. J. MacLusky, and B. S. McEwen, *J. Neurosci.* **2**, 1446 (1982).

[4] K. L. Keiner, H. J. Kirchich, and E. J. Peck, Jr., *Endocrinology* **111**, 1986 (1982).

[5] E. Mulder, M. J. Peters, W. M. O. van Beurden, and H. J. van der Molen, *FEBS Lett.* **47**, 209 (1974).

[6] E. Mulder, M. J. Peters, J. de Vries, and H. J. van der Molen, *Mol. Cell. Endocrinol.* **2**, 171 (1975).

[7] V. Hansson, O. Djoseland, A. Attramadal, O. Trygstad, F. S. French, W. E. Stumpf, M. Sar, W. S. McLean, A. A. Smith, S. C. Weddington, A. L. Steiner, P. Petrusz, S. N. Nayfeh, E. M. Ritzen, and L. Hagenas, *Acta Pathol. Microbiol. Scand. Sect. A., Suppl.* **248**, 75 (1974).

[8] J. I. Morrell and D. W. Pfaff, *in* "Neuroendocrinology of Reproduction" (N. T. Adler, ed.), p. 519. Plenum, New York, 1981.

[9] W. E. Stumpf and L. D. Grant, "Anatomical Endocrinology." Karger, Basel, 1975.

to the examination of tumors, have been done using immunocytochemical (ICC) methods with an antibody to estradiol and a fluorescent marker or a peroxidase–antiperoxidase (PAP) visualization procedure.[10,11] The usefulness and validity of these methods remain unproved.[12]

Using a fundamentally different approach, experiments have been aimed at producing antibodies to purified steroid receptor protein.[13–16] Attempts to use these antibodies to identify steroid hormone receptor-containing neurons with standard ICC methods have yet to yield convincing results.[15–18] However, this basic approach could be very useful at both the light and electron microscopic level and should yield data as better receptor purification and monoclonal methods for antibody production are applied.

Thus, presently, there is only one well established method for cell-by-cell identification of steroid hormone receiving cells—steroid autoradiography. This method has been described[8,19,20] and used in many experiments.[21,22] What follows are examples of the data uncovered with this method.

CNS Distribution of Gonadal Steroid Hormone Receiving Cells in Vertebrates

Steroid hormone autoradiography has made possible the examination of the entire CNS of many vertebrates for hormone receiving cells. Mem-

---

[10] L. P. Pertschuk, E. H. Tobin, P. Tanapat, E. Gaetjens, A. C. Carter, N. D. Bloom, R. J. Macchia, and K. B. Eisenberg, *J. Histochem. Cytochem.* **28**, 799 (1980).

[11] I. Nenci, M. D. Beccati, A. Piffanelli, and G. Lanza, *J. Steroid Biochem.* **7**, 505 (1976).

[12] G. C. Chamness, W. D. Mercer, and W. L. McGuire, *J. Histochem. Cytochem.* **28**, 792 (1980).

[13] G. L. Greene, F. W. Fitch, and E. V. Jensen, *Proc. Natl. Acad. Sci. U.S.A.* **77**, 157 (1980).

[14] G. L. Greene, C. Nolan, J. P. Engler, and E. V. Jensen, *Proc. Natl. Acad. Sci. U.S.A.* **77**, 5115 (1980).

[15] S. Raam, E. Nemeth, H. Tamura, D. S. O'Brian, and J. L. Cohen, *Eur. J. Cancer Clin. Oncol.* **18**, 1 (1982).

[16] G. Morel, P. Dubois, C. Benassayag, E. Nunez, C. Radanyi, G. Redeuilh, H. Richard-Foy, and E.-E. Baulieu, *Exp. Cell Res.* **132**, 249 (1981).

[17] S. Reichlin, J. Connolly, G. L. Greene, E. V. Jensen, and R. J. Robbins, *Endocrine Soc., Abstr.*, Abstr. #121 (1980).

[18] J. I. Morrell and D. W. Pfaff, unpublished observation.

[19] D. Pfaff and M. Keiner, *J. Comp. Neurol.* **151**, 121 (1973).

[20] J. I. Morrell, C. H. Rhodes, and D. W. Pfaff, in "Neuroactive Drugs in Endocrinology" (E. E. Muller, ed.), p. 3. Elsevier/North-Holland Biomedical Press, Amsterdam, 1980.

[21] J. I. Morrell and D. W. Pfaff, *Am. Zool.* **18**, 447 (1978).

[22] J. I. Morrell and D. W. Pfaff, *Science* **217**, 1273 (1982).

bers of most of the major classes of vertebrates have been examined, including two teleosts, two amphibians, three birds, two reptiles, several rodents, a carnivore, and a primate.[21,23]

Based on these studies, the following generalizations have been drawn. (1) Gonadal steroid hormone concentrating neurons have been found in every species thus far examined. Cellular steroid hormone concentration has even been identified in the unicellular eukaryotic organism, the yeast.[24] (2) The vast majority of gonadal hormone concentrating cells are found in the medial preoptic area and medial hypothalamus, in particular limbic structures such as the septum and amygdala, and in the midbrain, deep to the tectum. (3) The majority of these regions have been demonstrated, with other experimental methods, to be involved in the control of neuroendocrine aspects of reproduction or sexual behaviors. This correlation has been established in many of the particular species that have been examined. (4) In some species, a small number of hormone-concentrating neurons have been uncovered in locations unique to that species or class. Often these neurons appear to be positioned to regulate a specific sexual behavior of that species. (5) In general, genetic sex is not an important factor in determining the number or distribution of hormone-concentrating neurons. There are two interesting exceptions to this generally true conclusion.[25,26] (6) Regions and, in some cases, specific cell groups have been uncovered that contain cells that concentrate estradiol but do not contain cells that concentrate androgen. The reverse has also been uncovered; that is, cell groups containing cells that concentrate androgen, but not estrogens.

The species that we have examined most recently[27] is the garter snake (genus *Thamnophis*), and examples of the generalities listed above are clearly present in the data. Gonadal steroid hormone concentrating cells were found in the brains of male and female garter snakes after either [^{3}H]estradiol or [^{3}H]testosterone injection. The majority of the hormone concentrating cells were found in the ventral amygdaloid nucleus, medial nucleus sphericus (amygdala), septum, paleostriatum, preoptic area, the retrobulbar pallium, in particular hypothalamic cell groups, the anterior, periventricular, ventromedial, and arcuate nuclei, and in the central gray and tegmentum of the midbrain (Fig. 1). Many of these regions have been demonstrated in snakes and other reptiles to be involved in control

---

[23] D. W. Pfaff, "Estrogens and Brain Function." Springer-Verlag, Berlin and New York, 1980.
[24] D. Feldman, Y. D. A. Burshell, P. Stathis, and D. S. Loose, *Science* **218**, 297 (1982).
[25] A. P. Arnold, *J. Comp. Neurol.* **189**, 421 (1980).
[26] S. M. Breedlove and A. P. Arnold, *Science* **210**, 564 (1980).
[27] M. Halpern, J. I. Morrell, and D. W. Pfaff, *Gen. Comp. Endocrinol.* **46**, 211 (1982).

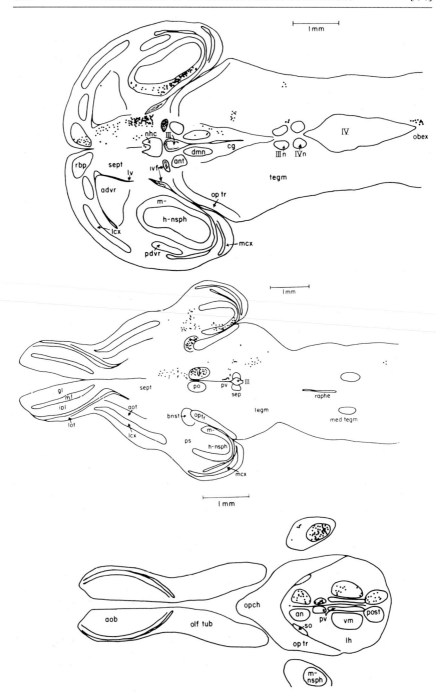

of reproductive endocrinology or sexual behaviors.[27,28] No striking differences in distribution of hormone concentrating cells or hormone concentrated per cell were found when males and females were compared.

The vomeronasal system is critical to sexual displays in the garter snake, particularly via rapid tongue flicking and subsequent stimulation of the vomeronasal organ by substances picked up on the tongue. Hormone-concentrating cells were found in every major CNS component of the vomeronasal system, except the accessory olfactory bulb. In addition, hormone-concentrating cells were found in the caudal medulla, near the obex. The location of these cells suggests that they are related to gustatory or motor activity of the tongue. Perhaps these collections of hormone receiving cells play a role in the modulation of sexually relevant sensory information received by the vomeronasal system, via the tongue.

Quantitative Studies: Number and Capacity of Steroid Hormone Receiving Cells

In addition to revealing the CNS distribution of hormone concentrating cells, steroid autoradiography can yield cell-specific information on their exact number, binding capacity, and specificity. For these stud-

---

[28] J. I. Morrell, D. Crews, A. Ballin, A. Morgentaler, and D. W. Pfaff, *J. Comp. Neurol.* **188**, 201 (1979).

FIG. 1. Representative projection drawings of horizontal sections from the brain of the garter snake (genus *Thamnophis*), injected with [³H]estradiol. All steroid hormone-retaining cells on one side of the brain are represented with black dots. Each black dot represents one labeled cell. Abbreviations: ac, anterior commissure; advr, anterior dorsal ventricular ridge; an, anterior hypothalamus; an pit, anterior pituitary; ant, anterior nucleus of thalamus; aob, accessory olfactory bulb; aot, accessory olfactory tract; arc, arcuate nucleus; bnst, bed nucleus of stria terminalis; cerbl, cerebellum; cg, central gray; dmn, dorsomedial nucleus of thalamus; gl, glomerular layer of accessory olfactory bulb; gr, granular layer of accessory olfactory bulb; h-nsph, hilar layer of nucleus sphericus; ipl, inner plexiform layer of accessory olfactory bulb; ivf, interventricular foramen; l, lateral hypothalamic nucleus; lcx, lateral cortex; lh, lateral hypothalamus; lot, lateral olfactory tract; lv, lateral ventricle; mcx, medial cortex; med tegm, medullary tegmentum; mhab, medial habenular nucleus; m-nsph, mural layer of nucleus sphericus; mt, mitral cell layer of accessory olfactory bulb; nhc, nucleus of hippocampal commissure; olf tub, olfactory tubercle; op ch, optic chiasm; opn, optic nerve; op tr, optic tract; pdvr, posterior dorsal ventricular ridge; po, preoptic area; post, posterior periventricular nucleus; ps, paleostriatum; pv, periventricular nucleus; ramy, rostral amygdala; rbp, retrobulbar pallium; sco, subcommissural organ; sep, secretory ependyma; sept, septum; so, supraoptic nucleus; tegm, tegmentum of midbrain; vamy, ventral amygdaloid nucleus; vm, ventromedial hypothalamic nucleus; III, third ventricle; IIIn, third cranial nerve nucleus; IV, fourth ventricle; IVn, fourth cranial nerve nucleus. From Halpern *et al.*[27]

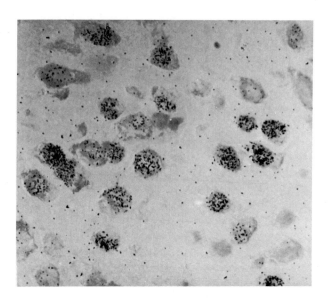

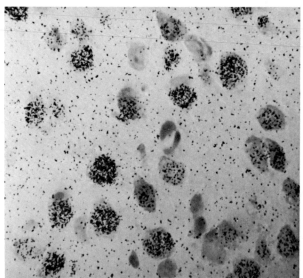

Fig. 2. Representative photomicrographs of autoradiograms from the VL-VM. These autoradiograms are typical of those made and analyzed in the quantitative experiments (Morrell et al.[30]). The autoradiograms were prepared from ovariectomized, adrenalectomized rats intravenously infused for 2 hr with [^{3}H]estradiol (E_2) (specific activity 80–102 Ci/mmol), 0.8 µg per 250 g body weight. This resulted in blood levels of 49–222 pg of E_2 per milliliter at sacrifice. The autoradiograms were prepared by standard methods (Morrell and Pfaff[8]) from 2-µm sections, and exposed for 6 months. A cresyl violet counterstain was used. The reduced silver grains, small black dots, are prominently clustered under many of the cells' perikarya.

TABLE I
Animals Administered [³H]Estradiol [a]

Region	Number of cells	% Estrogen concentrating cells	% Unlabeled cells
A. Animals administered [³H]estradiol alone			
Preoptic area, $n = 6$	22,353	24	76
VL-ventromedial nucleus, $n = 7$	12,752	29	71
Arcuate nucleus, $n = 7$	9,788	21	79
Amygdala-medial nucleus, $n = 7$	16,049	40	60
B. Animals administered [³H]estradiol plus 10X cold estradiol			
Preoptic area, $n = 2$	3,580	10	90
VL-ventromedial nucleus, $n = 2$	2,281	3.5	96.5
Arcuate nucleus, $n = 2$	1,108	.1	99.9
Amygdala-medial nucleus, $n = 2$	6,135	.3	99.7
C. Animals administered [³H]estradiol plus 100X cold estradiol			
Preoptic area, $n = 2$	644	0	100
VL-ventromedial nucleus, $n = 2$	679	0	100
Arcuate nucleus, $n = 2$	310	0	100
Amygdala-medial nucleus, $n = 2$	1,812	0	100

[a] Ovariectomized, adrenalectomized rats were intravenously infused for 2 hr with [³H]estradiol (E_2), 0.8 μg/250 g body weight. This resulted in blood levels of 49–222 pg of E_2 per milliliter at sacrifice. At least 10 different sections were analyzed per rat per anatomical region. Cells were considered labeled by the usual criterion: grains over cell bodies 5 times the number found over an adjacent cell-sized area of neuropil. VL, ventrolateral subdivision of ventromedial nucleus.

ies[29,30] the following modifications were made in the method. The highly radioactive estradiol was administered by intravenous infusion to adrenalectomized, ovariectomized rats that were sacrificed only after blood levels of [³H]estradiol had been kept at or above proestrous levels for more than 2 hr. The autoradiograms were prepared from 2-μm sections, thus eliminating the possibility that a cell which concentrated hormone would not be visualized as such because the β particle from the tritium travels only about 2 μm in the type of materials used.

The photomicrographs in Fig. 2 are representative of the autoradiograms produced and analyzed in this experiment. The first question was, In those areas of the hypothalamus and limbic system known to contain the largest numbers of hormone concentrating cells, did all cells concentrate hormone? The data in Table I show that in conditions that maxi-

[29] J. I. Morrell, T. D. Wolinsky, M. S. Krieger, and D. W. Pfaff, *Exp. Brain Res.* **45**, 144 (1982).

[30] J. I. Morrell, M. S. Krieger, and D. W. Pfaff, *Endocrinology* (*Baltimore*) (submitted for publication).

mized binding and the visualization of binding, between 21 and 40% of the cells retained hormone; a large number of cells in these regions did not. This means that the cells of the brain are heterogeneous with regard to their capacity to concentrate steroid hormone. That is, the brain has a limited number of cells that mediate its genomic regulation by steroid hormones. Table I also illustrates that the binding of the [^{3}H]estradiol can be effectively competed with nonradioactive estradiol added to the [^{3}H]estradiol infusion in either a dose of 10 times or 100 times the radioactive dose. This demonstrates that the receptors are saturable.

The amount of hormone retained per cell was examined by counting the number of grains per cell in a large sample of cells in the preoptic area, ventromedial nucleus, arcuate nucleus, and amygdala. An example of these results is illustrated in Fig. 3. The results from the other areas are virtually identical. The amount of hormone bound per cell varied over a wide range, and no modes were found. The distribution curve of the actual data varied considerably from the calculated Poisson. The fact that the data curve differs from the calculated Poisson curve strongly indicates that the cellular processes of hormone accumulation is not a random passive process such as diffusion, but an active, selective process in a particular subpopulation of the cells.

The lumbar and sacral regions of the spinal cord from these animals were also examined for the distribution, number, capacity, and specificity of the hormone receiving cells.[29] Estradiol concentrating cells were primarily found in the dorsal cord, and they were very similar to those in the hypothalamic and limbic regions. However, as illustrated in Fig. 4, one outstanding difference is clear. Even in these preparations that maximized the opportunity for binding and visualization of binding, the number of estradiol receiving cells in the spinal cord is very low compared to the number found in specific forebrain regions. This means that direct, receptor-mediated estrogen influence on spinal cord must be minor compared to that on the forebrain.

Introduction to the Combined Method: Steroid Autoradiography–Immunocytochemistry

Further characterization of hormone receiving cells is necessary in order to reveal the mechanisms through which steroids affect brain function. Cells of the brain contain a large variety of peptides that function as hormones, and in some cases these peptides are hypothesized to function as neurotransmitters or neuromodulators. Since many examples of steroid hormones affecting the cellular economy or release of these neuropeptides could be found, we thought that it was important to identify the

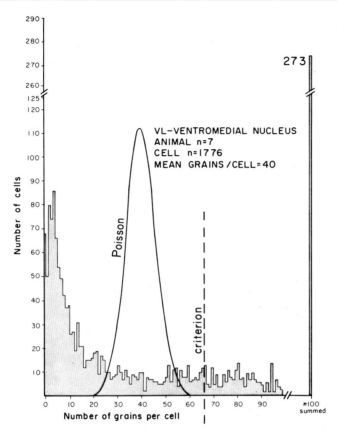

FIG. 3. Distribution of the number of grains per cell versus the number of cells found in the VL-VM, after experimental and autoradiographic conditions as described in the caption for Fig. 2. The grains under at least 200 cells per brain were counted. The photomicrographs in Fig. 2 are typical of autoradiograms analyzed. The criterion line marks the place in the distribution where a cell is considered to be labeled. A cell body is defined as labeled when the number of grains accumulated under it exceed a fixed quantitative criterion. This criterion is a ratio of grains under the cell body to grains under an adjacent cell-sized area of neuropil. When the value of the ratio for a given cell is five or greater, that cell is defined as labeled. Because this criterion requires that the number of reduced grains associated with a labeled cell far exceed the distribution of grains in the "background" (i.e., usually ≥6 standard deviations above mean background), false-positive designations of cells as labeled are highly improbable. It is also possible that cells with fewer grains under them accumulate hormone to a physiologically relevant extent; therefore, the criterion line must be viewed as a statistical point to provide objectivity in the analysis, not as a physiological statement. The Poisson line represents the hypothetical distribution curve, calculated on the basis of the actual mean number of grains per cell found in the data, which would hold if the distribution was representative of a random process. The curve based on the actual data differs significantly from the calculated Poisson. From Morrell et al.[30]

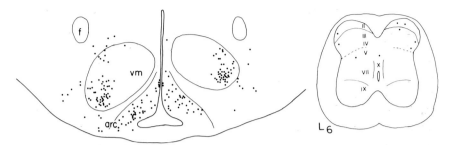

FIG. 4. *Left:* Diagram of a representative transverse section through the medial-basal hypothalamus; each black dot represents one estradiol-concentrating cell; all such cells ($n = 191$) present in the section are represented. Exposure time of autoradiograms is 6 months. ARC, arcuate nucleus; f, fornix; VM, ventromedial nucleus. *Right:* Diagram of a representative transverse section through L_6 of the spinal cord. Rexed lamina, II, III, IV, V, VII, IX, and X are represented. Exposure time of these autoradiograms is 6 months. Both autoradiograms are from preparations represented in Figs. 2 and 3. From Morrell et al.[29]

peptide content of steroid hormone receiving cells. We are interested in uncovering whether steroid hormones act directly, at the genomic level, on the cells in the brain that make, and release, neuroendocrinologically important peptides, or whether the steroids act indirectly, that is, via the genome of other brain cells that then exert their influence via synapse.

A key feature of steroid autoradiography is that its spatial resolution allows this steroid-specific endocrine characterization of the brain on a cell-by-cell basis. This is a critical feature of any method designed to characterize the brain, because neurons are very heterogeneous, even within small subregions. Therefore any peptide characterization of brain cells had to maintain this cell-by-cell level of resolution. Immunocytochemistry does have this level of cellular resolution, and it adds the dimension of identifying the peptide content of the cell. The methodological description that follows is complete for combined steroid autoradiography–immunocytochemistry. The changes, in principle, from the basic steroid autoradiography method[8,19] are few; however, there are a number of key additions.

How to Do Steroid Autoradiography Combined with Immunocytochemistry

Animals

Young adult animals of either sex of the species to be studied are usually gonadectomized and adrenalectomized 2–30 days before administration of ^{3}H-labeled hormone to reduce the endogenous circulating steroid hormones. The key is to reduce the blood and receptor-bound levels of the endogenous steroids so that the ^{3}H-labeled steroid will have an opportunity to bind to the receptors. However, removal of these glands may well have a profound effect on the amount or distribution of the cellular protein (the antigen) of interest. For example, Shivers et al.[31] found a profound reduction in the amount of luteinizing hormone-releasing hormone (LHRH) in the hypothalamic neurons of castrated as compared to intact male rats. As a result, far fewer LHRH positive cells (130 ± 58) were found in the castrated as compared to the intact (507 ± 52) males. The best solution, should these be opposing experimental demands, is to remove the glands sufficiently before isotope administration to reduce steroid levels in blood and receptors, but as short a time before isotope is given as possible so that the cellular economy of the antigen is not profoundly affected; 2–3 days may do.

It is often useful to administer 1–2 μl of colchicine (25–50 mg/ml PBS or dw) into the lateral cerebral ventricle, and to allow the animals to survive 24–48 hr before administering the isotope. Colchicine prevents the axonal transport of substances produced in the soma, without interfering with their production. Therefore greater amounts of somal antigen will be present after colchicine administration. We conclude that colchicine administered under this regime does not interfere with steroid hormone binding to the receptor, because there is no important alteration in the number or distribution of hormone receiving cells after its administration. Colchicine does make animals sick, particularly those already weakened by adrenalectomy, so some consideration should be given to this from both the point of view of total number of animals prepared and need for extra care in maintenance.

^{3}H-labeled Steroid Hormone

The relevant steroid hormone is determined from the physiological questions posed by experiment and the species to be studied; so far, the gonadal steroids, estradiol, testosterone, and dihydrotestosterone and a synthetic adrenal steroid, dexamethasone, have been used successfully. These hormones are available (New England Nuclear, e.g.) labeled with ^{3}H in two, four, six, or eight places. In general, the higher the specific activity the better; they range between 60 and 210 Ci/mmol. The isotope is usually purchased in absolute ethanol, and the amount of ethanol vehicle can be reduced by rapid evaporation in a vacuum-centrifuge evaporation apparatus. Dilution with equal amounts of physiological saline

---

[31] B. D. Shivers, R. E. Harlan, J. I. Morrell, and D. W. Pfaff, *Neuroendocrinology* **36**, 1 (1983).

reduces the ethanol concentration, an important consideration for intravenous administration. Generally, intravenous or intraperitoneal routes of administration are used. Although long (2 hr) intravenous infusions of a constant steroid dose may be necessary if maximal binding is a critical feature of the experiment,[29,30] it is usually sufficient to inject twice (each time one-half of the total dose) 30 min apart.

Gonadal and adrenal steroids administered in the range of 0.3 to 1 μg per 100 g body weight (this is between 25 and 200 μCi/100 g depending on the specific activity of the isotope) have yielded consistent, physiologically interpretable results. The dose of the steroid hormone should be chosen to result in blood levels within the physiological range (high end) for that steroid. It is also helpful if biochemical data for species and steroid describe a nuclear saturating dose for CNS receptors.

The steroid receptor (type I binding site) has the characteristics of high-affinity (K_d 10^{-11} to 10^{-8} M), saturable (low capacity), steroid-specific binding.[32] Although not found to any appreciable extent in the brain, much lower-affinity (K_d 10^{-7} to 10^{-5}), higher-capacity, less-specific binding sites, referred to as type II or type III binding sites, have been recognized in certain organs, for example, the uterus.[12] Supraphysiological doses of ^{3}H-labeled steroid should be avoided if the experiment is aimed at revealing the high-affinity, saturable, sites, because what may result is the additional filling of these lower-affinity sites, yielding results that are difficult to interpret.[12]

Sacrifice

The time between administration of isotope and sacrifice is usually 1–2 hr. The time is chosen when nuclear binding is still high, but when blood levels, and nonspecific binding, are reduced.

Unfixed, quickly extracted and frozen tissue obtained after sacrifice with a guillotine was originally the standard preparation for steroid autoradiography, and some tissue proteins survive this treatment in sufficient quantities to be visualized by immunocytochemistry. However, most CNS proteins of interest in these studies are better visualized after fixation. Therefore the animal is anesthetized (Nembutal), perfused with heparinized physiological saline, with or without phosphate buffer (PB), until the blood is rinsed out (usually 2–5 min), and perfused with the fixative of choice for 10–30 min. The brain is then extracted, blocked, and either postfixed (2–24 hr) or immediately put into a cryoprotectant solution (15–30% sucrose PB, pH 7.3–7.6) for 6–48 hr, typically at 4°. We have found that estradiol, testosterone, and dexamethasone binding in the CNS sur-

[32] J. H. Clark, J. W. Hardin, S. Upchurch, and H. Eriksson, *J. Biol. Chem.* **253**, 7630 (1978).

vives this regime when fixed with 10% PB formalin; 4% PB paraformaldehyde, pH 7.6; Zamboni's fixative and periodate–lysine–paraformaldehyde (PLP).[33] Typically a short postfixation period (2 hr) and overnight cryoprotectant soak is best to preserve the antigen. Steroid binding does not appear to survive fixation with Bouin's.

Freezing

The brain can be mounted onto a frozen cryostat chuck (prepared by freezing a platform of water and water-soaked gelfoam with Dry Ice) with a small drop of water, and then quickly frozen by covering in powdered Dry Ice. If done quickly, large ice crystal formation is minimal and morphology and cellular localization of the steroid and peptide are adequate.

The brain-cryostat chuck unit can also be frozen in rapidly stirred Freon 22 ($-41°$), cooled by liquid nitrogen (N_2) ($-196°$). This combination achieves a cooling rate of $-3200°$ per second,[34] and produces such small ice crystals that minimum disruption of internal structure of the cells takes place. The results with this method are good morphology and localizations. Gross cracks (common from directly freezing in nitrogen) in the tissue are minimized by slowly lowering the tissue into the Freon.

After freezing, the tissue can be transferred to liquid N_2 for storage until cutting. We have stored tissue from 2 days to 6 months in liquid N_2 without loss of morphology or steroid localization. Storing of the tissue in a standard freezer ($-20°$) is not recommended, because the tissue transforms into a dried powder in a few weeks. If the tissue is stored in nitrogen ($-196°$) it must be allowed to equilibrate slowly to cutting temperature ($-30°$) in order to prevent cracking of the tissue and to prevent the tissue coming off the chuck. This is accomplished by allowing 100 ml of N_2, in which the tissue is submerged, to evaporate over 2–4 hr in the fully equilibrated $-30°$ cryostat.

Preparation of the Autoradiogram

Microscope slides are washed (1 hr in 7X detergent in hot water; 3 hr in hot running water; three distilled-water rinses), dried, and dipped, under darkroom safelight conditions, into melted, undiluted (43°) nuclear tract emulsion. We most commonly use NTB-3 emulsion from Kodak, but grain size and sensitivity can be selected with different products. The emulsion-coated slides are air-dried in the darkroom at room temperature and stored until use in small black, lighttight boxes containing desiccant.

[33] I. W. McLean and P. K. Nakane, *J. Histochem. Cytochem.* **22,** 1077 (1974).
[34] L. Terracio and K. G. Schwabe, *J. Histochem. Cytochem.* **29,** 1021 (1981).

The tissue is cut on a cryostat (we have used both Model CTD Harris International Equipment Co., and a Bright by Hacker) under darkroom safelight conditions. Sectioning temperature depends on tissue preparation: about $-20°$ for fresh frozen; about $-25°$ after most fixatives; $-30°$ to $-35°$ for most fixed, sucrose solution-soaked brains. Tissue is typically sectioned at 2, 4, 6, 12, or 24 μm. The sections are picked up from the knife onto the dry, emulsion-coated slides.

The tissue section–emulsion-coated slide combination, the autoradiogram, is then stored in lighttight boxes containing desiccant at either $4°$, $-20°$, or $-70°$ for exposure times of 2–12 months. The storage temperature is selected to best support antigen survival over time in the stored sections. The longer the exposure, the greater the ^{3}H-labeled steroid signal; however, in some cases the immunocytochemistry signal decreases with storage, perhaps owing to denaturation of the antigen.

After exposure, the autoradiograms, still sealed in boxes containing desiccant, are allowed to equilibrate to room temperature. They are then photodeveloped under darkroom safelight conditions. As an aid to the survival of the antigen, the autoradiograms can be dipped briefly (30 sec) into fixatives: 3 or 4% PB paraformaldehyde, or Zamboni's pH ~7.6 at $4°$, $16°$, or $25°$ have worked. This fixation step *before* photodeveloping does not interfere with the ^{3}H-labeled steroid autoradiographic signal recognition; however, it is necessary only if the immunocytochemistry signal benefits. The autoradiogram is then rinsed in two changes of distilled water or PBS (phosphate-buffered saline) at $16°$ and photodeveloped in Kodak D-19 (pH >10) for 2 min at $16°$ or Kodak D-170 (pH 5.0) for 6–7 min at $16°$, stopped in Kodak Stop Bath (pH ~5) or distilled water for 1 min at $18–21°$, and photofixed in Kodak fixer (pH ~5.0), two changes for 10–18 min total at $18–21°$, followed by two rinses in distilled water for 5 min at $16–21°$. The lower pH of the D-170 may help the survival of the antigen; the extremely high pH of the D-19 may harm the antigen.

Immunocytochemical Procedures

Immediately after photofixing, the tissues are rinsed three times in PBS (10–25 mM, pH 7.6), for 30 min total, and then the primary antiserum (working dilution in PBS with 0.2–0.3% Triton) is applied to the sections *without delay*. From this point, the sections become increasingly fragile and are likely to come off the slides if they are handled roughly. It may be necessary gently to drip all the subsequent PBS rinses over the individual slides. The PBS should be gently drained off the slides, individually, and the slide around the tissue dried with Kimwipe. The primary antiserum is then added dropwise to form a bubble of the solution over the tissue. The slides are placed flat in a chamber that is sealed, then kept moist at 4° while the tissue incubates with the primary antibody. The working dilution depends on the antisera, but typically dilutions of 1 : 100 to 1 : 2000 have been used, for neurophysin, LHRH, and β-endorphin antibodies.

The tissue must *not* dry out during incubation with the primary antibody. The gelatin in the emulsion and the Triton in the antisera both tend to spread the antibody solution, preventing a bubble of it from forming over the section and making drying out more likely. A slide-sized piece of Parafilm placed gently over the tissue after the antibody is applied helps keep the solution in place. Alternatively, a line of nail polish can be quickly painted on the slide as a barrier around the tissue, and the antibody then applied. When no longer needed, the nail polish can easily be removed by peeling from the slide with forceps. It is good to check antibody levels halfway through the incubation.

After incubation of 24–48 hr at 4° in a moist chamber, the tissues are gently rinsed in PBS (three times) and the working dilution of the secondary antibody (1 : 100 to 1 : 200 in PBS, usually no Triton) is applied for 30 min (room temperature) to 24 hr (moist chamber; 4°). Then the tissues are rinsed in PBS (three times), and the working dilution of the peroxidase–antiperoxidase (PAP) (1 : 50 to 1 : 100 in PBS, no Triton) is applied for 30 to 90 min at room temperature. After rinsing in PBS (twice) and Tris (25–50 mM; pH 7.6) buffer (twice), the PAP is visualized using diaminobenzidine (DAB) (50 mg/100 ml in 25–50 mM Tris buffer; filtered, final pH 7.6) with 0.03% H_2O_2 for 5–10 min (with agitation), or less time if the tissue browns too quickly. The tissue is then rinsed (twice in Tris; twice in water), dehydrated through standard graded alcohols, cleared in xylene, and coverslipped. A light cresyl violet stain may be applied.

The avidin–biotin complex (ABC) visualization procedure is compatible with steroid autoradiography, and it can be substituted for Sternberger's[35] PAP method of visualizing the antibody–antigen recognition site described above. The ABC method[36–38] can be substituted after the primary antibody incubation step; the visualization with DAB is the same as for the PAP method. The ABC kits available from Vector (Burlingame, CA) are self-explanatory. Some investigators find that ABC visualizes lower concentrations of the antigen, with lower background staining.

---

[35] L. A. Sternberger, "Immunocytochemistry." Prentice-Hall, Englewood Cliffs, New Jersey, 1974.
[36] S. M. Hsu, L. Raine, and H. Fanger, *Am. J. Clin. Pathol.* **75**, 734 (1981).
[37] S. M. Hsu, L. Raine, and H. Fanger, *J. Histochem. Cytochem.* **29**, 577 (1981).
[38] S. M. Hsu and L. Raine, *J. Histochem. Cytochem.* **29**, 1349 (1981).

Analysis

The autoradiogram–immunocytochemistry (ICC) preparation is then scanned systematically with a light microscope (400×) for the presence of labeled cells. The ICC-labeled cell is one that contains the brown DAB reaction product evenly distributed in its cytoplasm and processes; the nucleus is devoid of brown stain. Our usual criterion for a steroid hormone labeled cell is one that has reduced silver grains over its nucleus and soma that are at least five times the number found in adjacent cell-sized areas of neuropil (background).[8]

Although we are aware that fewer grains per cell may be biologically meaningful, for example, see Fig. 3, we have chosen the five times criterion as one that is statistically unlikely to give a random false-positive result. The criterion[39,40] based on a Poisson analysis of grains over neuropil or over cells in brain regions in which cellular hormone concentration does not take place, is also a useful way to establish background. By using quantitative criteria such as these, it is possible to minimize the chance of falsely designating a cell as hormone concentrating when it is not, that is, a false-positive designation.

The autoradiogram–ICC preparation will reveal the exact cell-by-cell anatomical distribution of cells that are hormone concentrating and immunocytochemically positive, cells that are either independently, and cells that are neither. Quantitative analysis of the cells in each category is a natural extension of the distribution determination. In order to avoid a simple error due to double counting of parts of the same cell found in adjacent sections, brain areas are either sampled with sufficient (at least equal to the diameter of the largest cell in area of interest) micrometers of tissue skipped, or, if every section is taken and counted, the Abercrombie correction factor can be applied.[41] Consideration of the pitfalls described below may be helpful in the quantitative analysis of these preparations.

Since the β particle from tritium travels on average 2 μm, the only way to be sure that all β particles emitted have a chance to register in the emulsion is to use 2-μm sections. However, these can be technically very demanding, and our usual section thickness is 6 μm. This allows identification of a large number of steroid receiving cells, although it must be understood that only the 2 μm adjacent to the emulsion will register all its β particles. False-negative counting of a steroid concentrating nucleus is less likely the larger the cell, and more likely the thicker the section.

The false-positive identification of a particular protein, hence the false

[39] A. P. Arnold, *J. Histochem. Cytochem.* **29,** 207 (1981).
[40] M. L. Fine, D. A. Keefer, and G. R. Leichnetz, *Science* **215,** 1265 (1982).
[41] M. Abercrombie, *Anat. Rec.* **94,** 239 (1946).

characterization of a neuron, could be caused by lack of antibody specificity. This could be due to some gross contaminant or to identification of a very closely related compound, for example, Leu- versus Met-enkephalin. With complete characterization of an antibody such false-positive identifications can be rendered very unlikely. The use of an appropriate working dilution of the antisera and, if necessary, the usual procedures to limit nonspecific background staining,[35] should ensure that the ICC-positive neurons will be clearly visible compared to neuropil or the other non-antigen-containing cells.

The ICC signal is stronger in a thicker section. The thinner the section of the cell, the less likely that a sufficient amount of antigen will be present, and hence the more likely the false-negative designation of the cell's protein content. Therefore, for a particular ^{3}H-labeled steroid–antigen combination, one optimizes the results by choosing a section thickness that avoids the errors of the extremes, too thin or too thick.

Characterization of Steroid Hormone Receiving Neurons by Their Peptide Content

Do Oxytocin-, Vasopressin-, and Neurophysin-Producing Neurons Concentrate Estradiol?

We knew from previous experiments that some of the magnocellular neurons of the paraventricular nucleus concentrated estradiol.[19,30] These cells were found only in particular subdivisions of the nucleus. Although the hormone-concentrating magnocellular neurons were a small percentage of total population, the hormone binding was sufficient to be of physiological importance. We knew that the antigen content of these cells was robust, and we[42,43] had an antibody that recognized both the neurophysin associated with oxytocin and the neurophysin associated with vasopressin. Experiments demonstrating a rise in plasma neurophysin, oxytocin, and vasopressin levels as a response to estrogen treatment[44–47] indicated a possible physiological importance for these interactions.

We therefore applied the combined steroid autoradiographic–immunocytochemical method to unfixed brain sections from ovariecto-

---

[42] C. H. Rhodes, J. I. Morrell, and D. W. Pfaff, *Neuroendocrinology* **33**, 18 (1981).
[43] C. H. Rhodes, J. I. Morrell, and D. W. Pfaff, *J. Neurosci.* **2**, 1718 (1982).
[44] A. G. Robinson, *J. Clin. Invest.* **54**, 209 (1974).
[45] A. G. Robinson, C. Haluszczak, J. A. Wilkins, A. B. Huellmantel, and C. G. Watson, *J. Clin. Endocrinol. Metab.* **44**, 330 (1977).
[46] K. Yamaguchi, T. Akaishi, and H. Negoro, *Endocrinol. Jpn.* **26**, 197 (1979).
[47] W. R. Skowsky, L. Swan, and P. Smith, *Endocrinology (Baltimore)* **104**, 105 (1979).

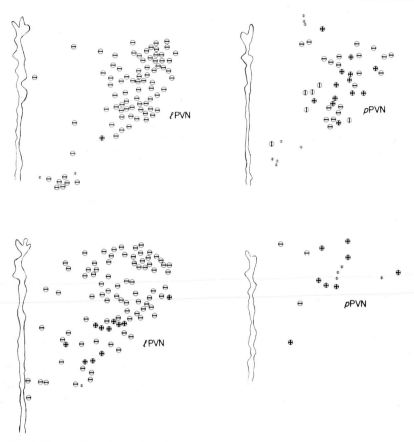

Fig. 5. Charts of a typical series of coronal sections of a normal rat through ACN and the PVN showing estrogen-concentrating and neurophysin-containing cells. Bar = 100 μm. ⊙, parvocellular estrogen-concentrating cell; ⊖, magnocellular neurophysin-containing cell; ⊕, magnocellular estrogen-concentrating cell; ⊕, magnocellular neurophysin-containing, estrogen-concentrating cell; ACN, anterior commissural nucleus; mPVN, medial subnucleus of PVN; lPVN, lateral subnucleus of PVN; pPVN, posterior subnucleus of PVN. From Rhodes et al.[42]

mized, normal, adult female rats, to explore the cellular nature of estradiol influence on these peptides.[42] We found magnocellular neurons in the PVN that concentrated estradiol in their nuclei and contained neurophysin in their cytoplasm. These neurons were principally found ventromedial to the lateral subdivision of the PVN, and in the posterior subdivision (Fig. 5). Estradiol concentration was rare in the neurophysin-containing neurons of the medial and lateral subdivisions of the PVN, the anterior

TABLE II
Distribution of Estrogen Binding and Neurophysin Production in Magnocellular Neurons (MNs) in Six Brattleboro Rats

Region[a]	% of MNs[a] that are neurophysin-containing	% of MNs that are estrogen concentrating	% of neurophysin containing MNs that are estrogen-concentrating	% of neurophysin-lacking MNs that are estrogen-concentrating	% of estrogen-concentrating MNs that are neurophysin-containing
ACN	100	0.8	0.7	—[b]	—
mPVN	89	2.3	0.3	16	—
lPVN	29	2.7	1.8	3.0	—
Region medial and ventral to lPVN	68	29	17	53	39
pPVN	61	67	64	74	56

[a] ACN, anterior comissural nucleus; mPVN, lPVN, pPVN, medial, lateral, and posterior subnucleus of periventricular nucleus.
[b] Dash indicates that denominator was less than 9 cells counted.

commissural nucleus, also called the rostral PVN, and the supraoptic nucleus.

To determine whether these estradiol-concentrating magnocellular neurons produced oxytocin or vasopressin, while continuing to use the same robust neurophysin antibody, we did the next set of experiments in the Brattleboro rat. The homozygous Brattleboro rat is unable to make vasopressin or its associated neurophysin, owing to a genetic deficiency. However, the magnocellular neurons that would, in the normal rat, produce vasopressin, are present in normal numbers and anatomical distribution in the Brattleboro; they simply do not contain vasopressin and its neurophysin. Therefore, the total number and distribution of magnocellular neurons is the same in normal and Brattleboro rats.

We applied the combined steroid autoradiographic–immunocytochemical method to unfixed brain sections from ovariectomized, homozygous Brattleboro adult female rats.[43] The photomicrograph in Fig. 6 shows estradiol-concentrating magnocellular neurons that contain neurophysin. Since these are in the Brattleboro rat, they are in fact oxytocin-producing neurons. Estradiol-concentrating magnocellular neurons without neurophysin in their cytoplasm were also found; these would produce vasopressin in the normal animal. Many non-estradiol-concentrating magnocellular neurons with and without neurophysin in their cytoplasm were also found. Table II shows the percentage of each type of cell

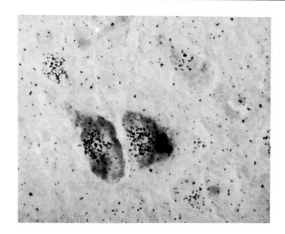

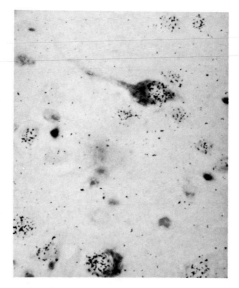

Fig. 6. Photomicrographs of combined autoradiographic–immunocytochemical preparation illustrating Brattleboro rat cells in the posterior subnucleus of the PVN, which both concentrate estradiol in the nucleus (as demonstrated by autoradiography) and contain neurophysin in the cytoplasm (as demonstrated by immunohistology). The reduced silver grains clustered in the nuclei of four neurons, two in each photomicrograph, are surrounded by dark neurophysin-positive cytoplasm. In addition, silver grain clusters are found over other neurons that are not magnocellular and without neurophysin in the cytoplasm. From Rhodes et al.[43]

found in the PVN and ACN. These results fit well with those from normal animals.

These data strongly indicate that a particular subpopulation of both oxytocin- and vasopressin-producing magnocellular neurons have their genomes directly regulated by estradiol. These neurons are located in PVN subdivisions in which the majority of neurons have been proved, using retrograde neuroanatomical tracing methods, to project to the caudal, medial medulla, and to the spinal cord.[48,49] An important implication of these facts is that estradiol may be genomically regulating oxytocin and vasopressin in neurons that use these peptides as neurotransmitters or neuromodulators within the CNS. The vast majority of the PVN neurons that project to the posterior pituitary are found in subdivisions that contain virtually no estradiol-retaining magnocellular neurons. Therefore, estradiol is probably not directly regulating neurons that make and release these peptides into the classical hypothalamoneurohypophysial system, and hence into the blood.

Does Estradiol Directly Regulate the Genome of Neurons That Produce Luteinizing Hormone-Releasing Hormone?

Hypothalamic control of pituitary secretion by releasing hormones is well documented. Luteinizing hormone releasing hormone (LHRH) is made in neurons primarily found in the diagonal bands of Broca, and in the medial preoptic area, particularly surrounding the area of the lamina terminalis.[31] This hormone is then axonally transported to the median eminence, where it is released into the portal blood, thus reaching the anterior pituitary and stimulating LH release.

Estradiol is known to have both positive and negative feedback effects on this system.[50,51] Estradiol can exert these effects on the hypothalamus and can also affect the sensitivity of the pituitary to LHRH.[52-54] Estrogen effects on the release of LHRH can partly be blocked by inhibitors of DNA-dependent RNA synthesis, or protein synthesis.[55,56]

---

[48] L. W. Swanson and G. J. Mogenson, *Brain Res. Rev.* **3,** 1 (1981).
[49] M. S. Fukuda, J. I. Morrell, and D. W. Pfaff, *Soc. Neurosci.* Abstr. No. 240.2 (1982).
[50] S. M. McCann, *Handb. Physiol. Sect. 7: Endocrinol.,* Vol. 4, Part 2 p. 489 (1974).
[51] G. Fink, *Annu. Rev. Physiol.* **41,** 571 (1979).
[52] S. M. McCann, L. Krulich, M. Quijada, J. Wheaton, and R. L. Moss, *in* "Anatomical Neuroendocrinology" (W. E. Stumpf and L. D. Grant, eds.), p. 192. Karger, Basel, 1975.
[53] G. H. Greeley, M. B. Allen, and V. B. Mahesh, *Neuroendocrinology* **18,** 233 (1975).
[54] G. H. Greeley Jr., T. G. Muldoon, and V. B. Mahesh, *Biol. Reprod.* **13,** 515 (1975).
[55] G. L. Jackson, *Endocrinology (Baltimore)* **91,** 1284 (1972).
[56] G. L. Jackson, *Endocrinology (Baltimore)* **93,** 887 (1973).

In view of these data, and because we also knew that the anatomical distribution of the LHRH-producing neurons overlapped significantly with one portion of the anatomical distribution of estradiol-concentrating neurons, we applied the combined steroid autoradiographic–immunocytochemistry to Zamboni's fixed brain sections from ovariectomized colchicine-treated female rats.[57,58] Because LHRH-containing perikarya can be difficult to demonstrate, we used a number of the best antibodies available for the ICC,[57,58] including the Benoit LR1. Between 600 and 1000 LHRH-containing cell bodies could be uncovered per brain,[31] indicating a high and useful level of sensitivity in the ICC procedures. The autoradiography showed the number and distribution of estradiol-retaining cells to be virtually identical to that uncovered in the maximized preparation (see the subsection on quantitative studies, above).

After extensive quantitative analysis, we found many LHRH-containing neurons and many estradiol-concentrating cells, often side-by-side, and yet virtually no (0.2%) LHRH-containing neurons that retained estradiol. Figure 7 illustrates these results.

The disjunction of these cellular functions strongly implies that estradiol does not act directly on the genome of LHRH-producing neurons. Rather, the estradiol could act on the genome of other neurons, possibly via estradiol-concentrating cells immediately adjacent to LHRH-containing cells, which subsequently act on the LHRH neurons.

Do β-Endorphin-Producing Neurons Concentrate Estradiol?

Among the compounds in the catalog of endogenous opiates that are now known, β-endorphin was one of the first uncovered. It consists of 31 amino acids that are cleaved from the 132 amino acid parent molecule, proopiomelanocortin.

β-Endorphin levels in the portal blood of intact female monkeys vary with the ovarian cycle, being high in the follicular and luteal phases and low during menstruation, and low after ovariectomy.[59] In rat, plasma and anterior pituitary levels of β-endorphin are reduced 3–5 weeks after gonadectomy. This effect is reversed with testosterone and partially reversed with estrogen.[60] Levels in the medial basal hypothalamus (MBH) and median eminence are increased in the intact female rat after estrogen

---

[57] B. D. Shivers, R. E. Harlan, J. I. Morrell, and D. W. Pfaff, *J. Histochem. Cytochem.* **29**, 901 (1981).

[58] B. D. Shivers, R. E. Harlan, J. I. Morrell, and D. W. Pfaff, *Nature (London)* **304**, 345 (1983).

[59] W. B. Wehrenberg, S. L. Wardlaw, A. G. Frantz, and M. Ferin, *Endocrinology* **111**, 879 (1982).

[60] F. Petraglia, A. Penalua, V. Locatelli, D. Cocchi, A. E. Panerai, A. R. Genazzani, and E. E. Muller, *Endocrinology (Baltimore)* **111**, 1224 (1982).

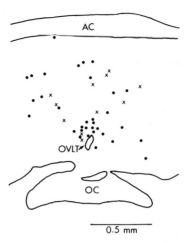

FIG. 7. Camera lucida drawing of autoradiographically localized estradiol-concentrating cells (●) and immunocytochemically localized luteinizing hormone releasing hormone (LHRH) cells (×) in a 6-μm transverse section from the brain of an ovariectomized, colchicine-treated rat. Note that no cell has both properties. Exposure time was 161 days. These cells are at rostral levels of the medial preoptic area. Dorsal is toward the top. AC, anterior commissure; OC, optic chiasm; OVLT, organum vasculosum lamina terminalis. From Shivers et al.[58]

administration,[60] but no change was noted after gonadectomy. Other workers have reported that hypothalamic levels of β-endorphin fall after gonadectomy, an effect only partially blocked with progesterone.[61] Opiates inhibit pulsatile LH secretion,[62] an effect that can be blocked by gonadectomy.[63] Thus, there is evidence that estrogen influences β-endorphin levels in neuroendocrinologically important tissues, and that this opiate may have influence over the pituitary's regulation of the ovarian cycle.

β-Endorphin is made and released by neurons located in the MBH and by the cells of the anterior pituitary.[64,65] The distribution of these cells in the MBH has a striking overlap with the distribution of gonadal steroid hormone-concentrating cells in this region. Because of this striking distribution overlap, and influenced by the data presented above, we[66] used the combined method to explore whether β-endorphin-containing neurons

[61] L. Sharon, S. L. Wardlaw, L. Thoron, and A. G. Frantz, *Brain Res.* **245**, 327 (1982).
[62] M. Ferin, W. B. Wehrenberg, N. Y. Lam, E. J. Alston, and R. L. vande Wiele, *Endocrinology (Baltimore)* **111**, 1652 (1982).
[63] R. Bhanat and W. Wilkinson, *Endocrinology (Baltimore)* **112**, 399 (1983).
[64] F. E. Bloom, *Sci. Am.* **245**, 148 (1981).
[65] S. Hisano, H. Kawano, T. Nishiyama, and S. Daikoku, *Cell Tissue Res.* **224**, 303 (1982).
[66] J. I. Morrell, J. McGinty, and D. W. Pfaff, *Soc. Neurosci. Abstr.* (in press) (1983).

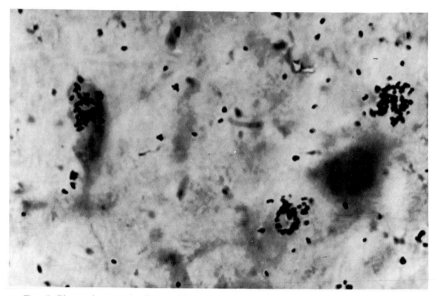

FIG. 8. Photomicrograph of a combined autoradiographic immunocytochemical preparation revealing (left) an estradiol-concentrating β-endorphin-containing neuron, as well as (right) two estradiol-concentrating neurons with no β-endorphin in the cytoplasm and one β-endorphin-containing cell without estradiol. The cells are in the medial-basal hypothalamus. This autoradiogram was exposed for 3 months. From Morrell, McGinty, and Pfaff, in preparation (1983).

retain estradiol. This was done in ovariectomized female rats, some treated with colchicine, using an antibody against β-endorphin (β-endo-2, made by R. Benoit).

Figure 8 illustrates that some β-endorphin containing neurons concentrate estradiol, while others do not. Not surprisingly, many estradiol concentrating cells, within the limits of the MBH distribution of the β-endorphin neurons, did not contain this peptide.

Therefore, we conclude that a subset of β-endorphin-producing MBH neurons accumulate estradiol in a manner consistent with a genomic influence of this hormone, but another subset does not. Discoveries of these chemical differences among neighboring neurons demand the high spatial and chemical resolution afforded by the combined steroid autoradiographic-immunocytochemical method.

Acknowledgments

We wish to thank Dr. Joan King for Arimura 743 and Jackson 29 LHRH antisera; Dr. Ann-Judith Silverman for Silverman-Nilaver Rabbit III LHRH antiserum; Dr. E. A. Zimmerman for the neurophysin antiserum; and Dr. Robert Benoit for LR 1 LHRH antiserum, and the β-endo-2 β-endorphin antiserum.

[45] Techniques for Tracing Peptide-Specific Pathways

By LARRY W. SWANSON

Immunohistochemical methods have provided invaluable information about the distribution of peptide-containing systems in the nervous system. It is now clear, for example, that each neuroactive peptide is synthesized by a unique set of cell bodies and is contained within a unique set of pathways, although these biochemically specified systems often overlap within individual nuclei or cortical areas, and individual neurons may contain more than one peptide. This very complexity has, however, limited the usefulness of immunohistochemistry as an anatomical tool when used by itself. As shown in Fig. 1, it is often not possible to determine the origin of immunostained terminal fields, because pathways arising in several different regions may converge and become indistinguishable in normal material. One approach to this problem involves the production of a lesion in the system, and the subsequent immunohistochemical mapping of pathological changes such as Wallerian (anterograde) degeneration, and retrograde changes in the cells of origin (Fig. 1). This strategy is an extension of the classical neuroanatomical approach to circuit analysis and suffers from two major limitations. First, the results are based on the interpretation of pathological changes that may be quite complex[1]; and second, the results are often confounded by the interruption of fibers-of-passage (preterminal axons) by the lesion. This general approach has largely been replaced in the last decade by pathway tracing methods that are based on the anterograde and retrograde axonal transport of markers in physiologically intact neural systems.[2] Several retrograde transport methods have been used in conjunction with immunohistochemistry to investigate experimentally the origin of specific peptidergic (and aminergic) terminal fields.[3-5] In this chapter one such method,[6] which is based on the use of fluorescent markers,[7] will be described in detail. The major advantages of this method are that (1) it is highly sensitive, (2) immunostained and retrogradely labeled cells are readily distinguishable,

---

[1] L. W. Swanson, *J. Histochem. Cytochem.* **29**, 117 (1981).
[2] E. G. Jones and B. K. Hartman, *Annu. Rev. Neurosci.* **1**, 215 (1978).
[3] T. Hökfelt, L. Terenius, H. G. J. M. Kuypers, and O. Dann, *Neurosci. Lett.* **14**, 55 (1979).
[4] R. M. Bowker, K. N. Westland, and J. D. Coulter, *Neurosci. Lett.* **24**, 221 (1981).
[5] D. van der Kooy and H. W. U. Steinbusch, *J. Neurosci. Res.* **5**, 479 (1980).
[6] P. E. Sawchenko and L. W. Swanson, *Brain Res.* **210**, 31 (1981).
[7] H. G. J. M. Kuypers, M. Bentivoglio, C. E. Catsman-Berrevoets, and A. T. Bharos, *Exp. Brain Res.* **40**, 383 (1980).

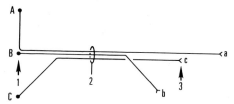

Fig. 1. The diagram illustrates some of the problems associated with determining the organization of a hypothetical, biochemically specific neural system in the brain. Three different cell groups (A–C), all of which contain the same peptide, project through the same pathway (2), such as the medial forebrain bundle, to three different terminal fields (a–c). In a series of normal immunostained sections, the origins of the terminal fields probably could not be determined because fibers from different sources become intermixed within the common pathway. If the organization of the system is examined experimentally by placing a lesion in one of the cell groups (as in B, arrow 1), the results may be confounded because fibers-of-passage from another cell group (A) are inadvertently cut as well. Alternatively, as described in this chapter, retrograde tracers may be injected into each terminal field, although with certain tracers (see the text) it is necessary to consider whether the terminal field also contains fibers-of-passage (arrow 3).

(3) the fiber-of-passage problem can be avoided with an appropriate retrograde tracer, (4) immunostained cells that send collateral projections to different terminal fields can be identified, and (5) it is relatively simple technically. The method will be described first, and then major problems associated with the interpretation of results will be discussed.

Choice of Retrograde Tracer

The purpose of this method is to identify the cell bodies that give rise to a biochemically specific terminal field in the central or peripheral nervous system. In principle, one simply places an injection of a suitable retrograde tracer within the terminal field of interest and prepares a series of sections through those parts of the nervous system that contain neurons synthesizing the antigen of interest. In practice, it is usually necessary first to generate (or to have access to) a detailed map of the distribution of immunostained cell bodies, pathways, and terminal fields. This is used to determine what parts of the nervous system to examine and to determine whether the injection may also label fibers-of-passage (see below). Depending on the organization of the system to be studied, one or more of the following fluorescent retrograde tracers may be used to advantage.

True Blue. This compound, which was introduced by Kuypers and his colleagues,[7] appears to be the most sensitive retrograde tracer available today[6] for use in the rat (and smaller animals). It produces bright blue labeling in the cytoplasm and nucleolus of retrogradely labeled cells when excited by ultraviolet light; is not degraded over time, so that very long survival times (months) can be used if necessary, as in developmental studies. Normally, however, survival times of 3–21 days are sufficient in the rat and are determined empirically, based on the length of the pathway involved, the pattern of terminal arborizations, and the degree of brightness desired in labeled cells. True blue is a lipophilic compound that probably diffuses across membranes; consequently its major limitation is that it is readily taken up and transported by intact (as well as by damaged) fibers-of-passage, although this feature can sometimes be used to advantage (by, for example, making large injections at upper cervical levels to label all descending projections to the spinal cord, or by applying it to peripheral nerve trunks). Once taken up, however, true blue appears to be bound within the cytoplasm, at least until histological processing.[8] True blue is normally injected by pressure into tissue sites as a 1–5% aqueous suspension, although crystals of the dye can be applied or implanted. If small injections (20–30 nl) of true blue are made over 10–15 min, relatively small injection sites, less than 500 μm in diameter, can be obtained.

Fast Blue. This tracer is quite similar to true blue, although it differs in the following respects. First, it is said to be transported effectively over longer distances, so that it is quite useful in larger animals, such as the cat and monkey.[7] Second, it is more soluble in water than true blue, and consequently it tends to produce larger injection sites. And third, it yields a blueish-white fluorescence in the cytoplasm that is less intense than true blue.

SITS.[9] This is a lipophobic molecule that contains a reactive isothiocyano group, and consequently forms covalent bonds with membrane (and other) proteins under mild aqueous conditions. SITS has three major advantages. First, it is taken up by axon terminals (presumably by endocytosis) and is transported to cell bodies, where it produces bluish-white granular fluorescence (ultraviolet excitation) in the cytoplasm. However, it does not appear to be taken up in detectable amounts by preterminal axons (fibers-of-passage). Second, because it is covalently bound within the cell, the labeling survives a wide range of histological procedures virtually intact. And third, its fluorescence fades slowly relative to other commonly used markers, such as true blue and fluorescein. On the other hand, its major disadvantages are that it is somewhat less sensitive, and its fluorescence less brilliant, than true blue. SITS (which is supplied as a brown powder) is injected as a 1–10% aqueous solution, and in general

[8] A. Björklund and G. Skagerberg, *J. Neurosci. Methods* **1**, 261 (1979).
[9] L. C. Schmued and L. W. Swanson, *Brain Res.* **249**, 137 (1982).

the injection site produced by a given volume is somewhat smaller than that for a comparable injection of true blue.

Propidium Iodide.[10] The major advantages of this tracer are that it produces orange-to-reddish cytoplasmic labeling when excited by green light and that it can be combined with immunofluorescence methods.[11] Therefore, it is readily distinguishable from the retrograde tracers discussed thus far and can be used in combination with them to determine whether individual immunostained cells give rise to axon collaterals that innervate two different terminal fields, although it has not yet been widely used for this purpose. The major disadvantages of propidium iodide are that it leaks out of retrogradely labeled cells at longer than optimal survival times, and may thus produce artifactual labeling of adjacent neurons, and that it is less sensitive than true blue in some systems. Propidium iodide is usually injected as a 1–5% aqueous solution, which is heated to 60° when prepared.[11] In general, the survival time for propidium iodide should be less than 1 week, since the brightness of retrogradely labeled cells tends to decrease at longer times.

Perfusion of the Animal

After an appropriate survival time, which must be determined empirically for each pathway and each retrograde tracer, the animal is deeply anesthetized and perfused with a fixative that is optimal for the immunohistochemical localization of the peptide (or other antigen) of interest, and at the same time binds the retrograde tracer within the tissue. The following protocol[6,12] has been found to demonstrate optimally most peptides (with the exception of thyrotropin-releasing hormone), as well as each of the tracers discussed above. In brief, the animal is cooled in an ice-bath for 5 min before the thorax is opened quickly and a cannula is passed through the left ventricle of the heart into the ascending aorta, where it is clamped in place to avoid backflow of fixative through the heart. The descending aorta is then clamped (when staining is to be limited to the brain), the right atrium is cut, and blood is rinsed from the vascular system with 50 ml of isotonic saline at room temperature, using a perfusion pump set to deliver 20 ml/min. The animal is then perfused with 100 ml of ice-cold (4°) 4% paraformaldehyde at pH 6.5 (in a 0.1 M acetate

---

[10] H. G. J. M. Kuypers, M. Bentivoglio, D. van der Kooy, and C. E. Catsman-Berrevoets, *Neurosci. Lett.* **12**, *1* (1979).

[11] D. van der Kooy and P. E. Sawchenko, *J. Histochem. Cytochem.* **30**, 794 (1982).

[12] L. W. Swanson, P. E. Sawchenko, J. Rivier, and W. W. Vale, *Neuroendocrinology* **36**, 165 (1983).

buffer), followed by 150 ml of ice-cold 4% paraformaldehyde–0.05% glutaraldehyde at pH 9.5 (in 0.1 M borate buffer). The brain is quickly removed and placed in an ice-cold solution of the second fixative, with 10% sucrose, until it is processed the next day. Overnight postfixation is necessary to bind true blue, fast blue, and propidium iodide within the retrogradely labeled cells.

It should also be mentioned that optimal perikaryal staining of some peptides requires treatment with colchicine 1–3 days before perfusion. The most effective dose, route of administration, and survival time may vary for different peptides and for different pathways containing the same peptide.

Immunofluorescence Staining

Frozen sections 20–30 μm thick are cut on a sliding microtome and collected in ice-cold 0.1 M potassium phosphate-buffered saline, pH 7.4 (KPBS). They are rinsed in collecting solution for at least 10 min before transfer to the wells of porcelain spot dishes that contain KPBS with 2% normal goat serum (to reduce nonspecific staining; NGS), 0.1–2% Triton X-100 (to delipidize and increase the penetration of antibodies), and an appropriate dilution of the primary antiserum (usually serial dilutions between 1/100 and 1/64,000 are tested initially). The dishes are placed in enclosed containers to reduce evaporation, and the sections are incubated for 1 hr at room temperature (for immediate preliminary results) or for 24–72 hr at 4° (for maximal staining). The sections are then rinsed in two changes (10 min each) of the incubation medium (without primary antiserum), and are transferred for 45 min to spot dishes containing a 1/100–1/300 dilution of affinity-purified goat anti-rabbit IgG that has been conjugated to fluorescein isothiocyanate (FITC-ARG; 6–8 mol of FITC per mole of protein), prepared according to Hartman[13] or purchased from Tago (Tago International, Burlingame, CA). The sections are then washed in two changes of KPBS (15 min each), mounted from distilled water onto gelatin-coated slides, and coverslipped with a nonfluorescent buffered glycerol mountant (pH 8.5).

It is essential to demonstrate insofar as possible the specificity of antigen–antibody reactions in immunohistochemical studies, and excellent discussions of this topic can be found in Jones and Hartman,[2] Hartman,[13] and Sterberger.[14]

[13] B. K. Hartman, *J. Histochem. Cytochem.* **21**, 312 (1973).
[14] L. A. Sternberger, "Immunocytochemistry," 2nd ed. Wiley, New York, 1979.

Localization of Fluorescent Markers

One factor contributing to the sensitivity of fluorescence methods is that light-emitting structures are viewed against a dark background, yielding a high signal-to-noise ratio. The dark background, however, is a serious limitation in neuroanatomical studies because it is essential to determine the precise location of labeled cell bodies and fibers. There are three simple ways of overcoming this problem. First, the sections can be counterstained with a fluorescent Nissl stain.[15] Ethidium bromide is particularly useful because it produces a red Nissl stain when excited by green wavelengths. Because true blue, fast blue, and SITS are excited by ultraviolet wavelengths, and FITC by blue wavelengths, the red counterstain does not mask retrograde tracer or immunofluorescence. Before mounting, immunostained sections are simply dipped in a 0.00005% aqueous solution of ethidium bromide for 15–120 sec and rinsed in KPBS for 30–90 sec. Second, sections are alternatively viewed with fluorescence and darkfield illumination.[12] When viewed properly with darkfield illumination, fiber tracts in frozen sections are revealed with the clarity of a silver stain, and major cytoarchitectonic features are visible because cell bodies appear as very dark spots against the somewhat brighter neuropil. And third, an adjacent series of sections is always cut and stained in the usual way with a conventional Nissl stain such as Thionin. This series is used to make projection drawings upon which the distribution of fluorescent cells and processes is plotted, either by eye, or with the aid of an X-Y plotter system.

Section Analysis

As outlined above, retrograde tracers, immunofluorescence, and a fluorescent Nissl stain are each viewed with a different filter system on the same section. Because of this feature, the fluorescence produced by one marker does not mask that from another marker (providing appropriately narrow-band filter systems are used). Therefore, singly, doubly, and triply-labeled neurons can be identified readily by switching filter systems. The appearance of such cells (unfortunately in black and white) is illustrated in Fig. 2.

It should also be pointed out that oil-immersion objectives yield significantly brighter images than do dry objectives, and that epi-illumination (or Ploem optics) is considerably more efficient than is illumination by transmitted light.

---

[15] L. C. Schmued, L. W. Swanson, and P. E. Sawchenko, *J. Histochem. Cytochem.* **30**, 123 (1982).

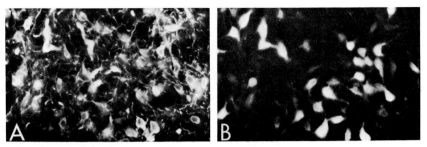

FIG. 2. A pair of photomicrographs to show the appearance of dopaminergic neurons in the ventral tegmental area that project to the nucleus accumbens in the rat. (A) Tyrosine hydroxylase (tyrosine monooxygenase)-stained neurons in the ventral tegmental area, detected with a fluorescein-conjugated secondary antiserum and viewed with filter system I_2. (B) True blue-labeled neurons in the same section viewed with filter system A. One week before the animal was sacrificed, true blue was injected into the ipsilateral nucleus accumbens. Both micrographs ×400.

Interpretation of Results

Two major limitations of this approach should be kept in mind. First, failure to observe doubly-labeled neurons (negative evidence) should be viewed with extreme caution because the sensitivity of the immunohistochemical method and of the retrograde tracer may not be sufficient to demonstrate all neurons that synthesize a specific peptide and all neurons that project to a specific terminal field. And second, when using most of the fluorescent retrograde tracers, it is necessary to consider the possibility that fibers-of-passage within the injection site have taken up the tracer (Fig. 1). Despite these caveats, however, it is clear that the development of combined retrograde transport–immunohistochemical methods allows a new level of analysis of neurotransmitter-specific pathways in the nervous system,[e.g., 16] and it seems likely that equally useful combined anterograde transport–immunohistochemical methods will be developed in the near future.

Acknowledgments

The original work described here was supported by Grants NS-16686 from the NIH and DA-00259 from ADAMHA; Dr. Swanson is a Clayton Foundation Investigator. I especially thank Dr. T. Joh (Cornell University) for the antiserum used to prepare the material shown in Fig. 2.

[16] L. W. Swanson and P. E. Sawchenko, *Annu. Rev. Neurosci.* **6**, 269 (1983).

[46] Immunocytochemistry of Endorphins and Enkephalins

By FLOYD E. BLOOM and ELENA L. F. BATTENBERG

Immunocytochemical methods have become increasingly important to the analysis of neuroendocrine peptide actions. These methods permit the localization of specific peptides, or more properly, immunoreactivities detected by antisera raised to specific synthetic oligopeptides, at both the light and electron microscope levels of resolution. Such structural information provides direct clues as to possible biological actions of the peptides by revealing the cell systems that can make, and presumably secrete, the peptide under examination. Knowledge of the sites of synthesis and release can often pinpoint likely target cells on which to evaluate in more detail the action of the peptide. The methods are further capable of following the end products of synthetic processing within cells, given antisera with selectivity against precursor and product peptides.

These applications apply to all neuropeptides. However, the endorphin family of endogenous opioid peptides[1-3] presents special problems of selectivity and specificity, since there are now at least three major groups of peptides that share partly homologous peptide sequences. Present evidence suggests that the peptide superfamily endorphin is composed of three main branches each of which has its own individual prohormone. β-Endorphin, derived from proopiomelanocortin,[4] has at its N terminus the pentapeptide Y-G-G-F-M, which was isolated and sequenced separately as [Met^5]enkephalin[5]; the latter peptide is not—as once believed—derived from β-endorphin, but rather from a large proenkephalin,[6] which contains six copies of [Met^5]enkephalin, one copy of [Leu^5]enkephalin, as well as a C-terminal heptapeptide, [Met^5,Arg^6,Phe^7]enkephalin.[1,2] [Leu^5]enkephalin, the other of the original enkephalin pentapeptides,[5] is also contained twice in a separate prohormone, termed prodynorphin,[7] where it occurs at the N terminus of both

---

[1] F. E. Bloom, *Annu. Rev. Pharmacol. Toxicol.* **23**, 344 (1983).
[2] J. Rossier, *Nature (London)* **298**, 221 (1982).
[3] A. Beaumont and J. Hughes, *Annu. Rev. Pharmacol. Toxicol.* **19**, 189 (1979).
[4] S. Nakanishi, A. Inoue, T. Kita, M. Nakamura, A. C. Y. Chang, S. N. Cohen, and S. Numa, *Nature (London)* **278**, 423 (1979).
[5] J. Hughes, T. W. Smith, H. W. Kosterlitz, L. A. Fothergill, B. A. Morgan, and H. R. Morris, *Nature (London)* **258**, 577 (1975).
[6] U. Gubler, P. Seeburg, B. J. Hoffman, L. P. Gage, and S. Udenfriend, *Nature (London)* **295**, 206 (1982).
[7] H. Kakidani, Y. Furutani, H. Takahashi, M. Noda, Y. Morimoto, T. Hirose, M. Asai, S. Inayama, S. Nakanishi, and S. Numa, *Nature (London)* **298**, 245 (1982).

dynorphin[8] and α-neoendorphin.[9] Thus, all endorphin peptides share a common N-terminal tetrapeptide, and the objectives of immunocytochemistry are to devise methods for the sensitive discrimination of these peptides at the cellular and subcellular levels.[10,11] (Figs. 1–3).

This chapter describes the specialized methods and considerations that have been demonstrated to be useful in the immunocytochemical analysis of the endorphin peptides. Two general approaches are advocated: immunofluorescence methods for rapid light microscopic localizations and immunoperoxidase methods for more detailed comparisons at the light and electron microscope levels.

Overview. There are several phases involved in successful immunocytochemistry. Since the raising of successful polyclonal and monoclonal antibodies to specific neuroendocrine peptides is covered elsewhere in this volume,[11a] our considerations here will assume that such specific antisera are available, already partially characterized for specificity by conventional radioimmunoassay, and ready for application to a problem requiring observations by immunocytochemistry. The process of immunocytochemistry will then require passage through these steps: tissue fixation, tissue sectioning, immune reactions for antigen detection, photomicrography or electron microscopy, and finally interpretation of the immunoreactivity observed. Two major interpretative problems will be faced most often: (1) specificity, the documentation that an observed immunoreactivity is selective for the antigen against which the antiserum was raised[12]; and (2) quantitation, the comparison between tissue samples (e.g., control versus experimental tissues, or between two regions of a given tissue, such as brain) for alterations in the amounts of immunoreactivity.[13]

Fixation

All forms of histochemistry require acceptance of a compromise between preservation of tissue structure and preservation of the cytochemi-

[8] A. Goldstein, W. Fischli, L. I. Lowney, M. Hunkapiller, and L. Hood, *Proc. Natl. Acad. Sci. U.S.A.* **78**, 7219 (1981).
[9] K. Kanagawa, N. Minamino, N. Chino, S. Sakakibara, and H. Matsuo, *Biochem. Biophys. Res. Commun.* **99**, 871 (1981).
[10] L. Sternberger, "Immunocytochemistry" 2nd ed. Wiley, New York, 1979.
[11] A. Kawamura, "Fluorescent Antibody Techniques and Their Applications." Univ. of Tokyo Press, Tokyo, 1969.
[11a] See this volume [Section V].
[12] D. F. Swaab, C. W. Pool, and F. W. van Leeuwen, *J. Histochem. Cytochem.* **25**, 388 (1977).
[13] L. F. Agnati, K. Fuxe, V. Locatelli, F. Benfenati, I. Zini, A. E. Panerai, M. F. E. Etreby, and T. Hökfelt, *J. Neurosci. Methods* **5**, 203 (1982).

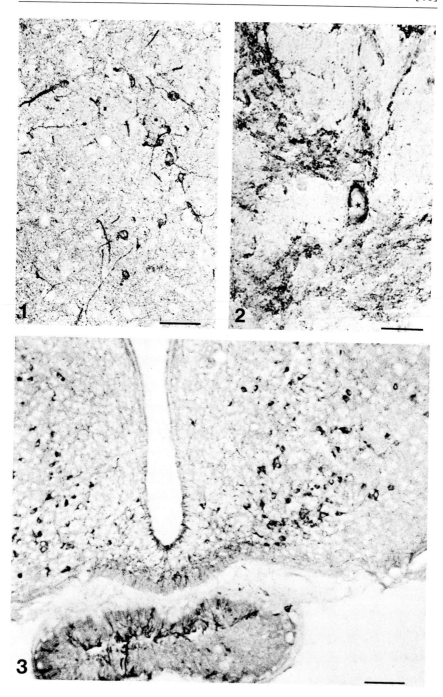

cal component (an antigen or enzymic activity) to be localized in its native location. Depending on the nature of the selected target, the choice of the fixative may be critical; fixation that gives excellent tissue preservation may render the antigen impermeant to or undetectable by the antibodies, whereas underfixation may retain immunoreactivity and sacrifice cytological detail. Furthermore, since the detection of immunoreactivity and cytological detail requires the completion of all of the phases of immunocytochemistry, the initial choice of the fixative may be only a temporary one designed to be adjusted once the optimal means for other steps have been determined. When initiating work with a new endorphin antiserum, we employ the routine methods of tissue fixation that have been successfully employed in prior localization studies, namely perfusion fixation with 4% paraformaldehyde.

Preparation of the Fixative. Make fresh fixative, allowing approximately 300 ml for each 150 g of rat to be fixed.

1. Determine the volume of fixative to be made; take half that volume of double strength (0.3 M; 2× Phosphate buffer, pH 7.4); for example, to perfuse 4 rats (300 ml per rat or 1200 ml total) take 600 ml of 2× buffer.

2. Heat the buffer to 70° in a wide-mouth Erlenmeyer flask on a hot plate, in a fume hood, with continuous magnetic stirring. The top of the flask should be sealed with foil or Parafilm to avoid evaporative loss.

3. Carefully weigh out paraformaldehyde powder as needed; use a top-loading balance in a well ventilated area (e.g., total volume in example will be 1200 ml; at 4 g/100 ml, weight out 48 g).

4. When the 2× buffer has been heated, turn off the heat, leave the flask on the warm hot plate, and, with continuous stirring, slowly add the formaldehyde powder to the buffer (take 3–5 min). The advantage of the

---

FIG. 1. Immunoperoxidase staining of endorphinergic neurons in rat central nervous system (see also Figs. 2 and 3). Several enkephalin immunoreactive cells in the paraventricular nucleus of the rat; from the cell bodies, thick dendritic and finer axonal processes can be seen. Rats were treated with intracisternally injected colchicine, 50 µg for 48 hr before fixation. Calibration bar = 50 µm.

FIG. 2. An enkephalin-immunoreactive neuron (*) in the border between caudate nucleus and globus pallidus is surrounded by immunoreactive processes; the thicker processes (e.g., projecting toward upper right corner) are presumably dendrites of the neuronal cell bodies that are not within the plane of this section. The finer varicose fibers at the upper and lower left edges represent typical terminal arborizations. Rat was treated with intracisternally injected colchicine, 50 µg, for 48 hr before fixation. Calibration bar = 25 µm.

FIG. 3. Low-power micrograph of β-endorphin immunoreactive cells and fibers in an untreated rat. Neurons are present within the arcuate nucleus bilaterally; their processes extend to the external palisade zone of the median eminence and are also visualized within the infundibular stalk below. Calibration bar = 100 µm.

wide-mouth Erlenmeyer is that the powder can be added directly into the buffer without littering the walls of the flask.

5. Once the paraformaldehyde has been depolymerized (gone into solution) the solution should be completely clear. If it has taken too long to add the powder, it may be necessary to reheat briefly. However, do not let the buffer come to a boil, as evaporative water loss will be difficult to estimate accurately.

6. Once the formaldehyde is in solution, add an equal volume of cool distilled water, (bringing the total volume to that desired, and lowering the temperature) and immerse the fixative in ice water; chill to 4°.

Perfusion. The following procedure applies to rodents. However, virtually the same procedure will be applicable to other animals, allowing for size and perfusion pressure. In our view, when a perfusion method has been practiced to the point of expertise it is unnecessary to "wash out" the vasculature with a physiological saline before perfusing the fixative; however, with less expertise, a preperfusion with saline will often yield a blood-free brain. The reason the latter approach is suboptimal is that some peptides of neuroendocrine importance will require prompt perfusion to be retained in their physiological locations and amounts. However, it is difficult to evaluate the efficacy of a perfusion unless erythrocytes can be used as a marker of inadequate fixation.

Rodents should be anesthetized; we prefer 350 mg of chloral hydrate per kilogram, i.p., as it has a high therapeutic index, yields a deep anesthetization, and maintains good spontaneous respirations. The animal is placed supine on a perfusion table, preferably in a hood or down-draft table to protect the investigator from formaldehyde fumes. The chilled fixative should be placed in a chilled reservoir and connected by plastic tubing to a large-bore needle (10–16 gauge, depending on rodent size).

When the animal is deeply anesthetized, as judged by lack of responsiveness to intense pressure stimuli, the heart is exposed by a midline cutaneous incision, followed rapidly by midaxillary transection of the rib cage bilaterally. Larger animals may benefit from ongoing artificial respiration through a tracheal intubation. The sternum is reflected upward, and the manubrium is fractured by placing a heavy hemostat horizontally across the sternum and the underlying internal mammary arteries, and quickly lifting the xiphoid end up. That hemostat is left in place. A second small hemostat is placed on the descending thoracic aorta. A third hemostat is placed on the right atrium. In a single continuous motion, the perfusion needle is inserted, into the left ventricle, the tubing is opened, and the right atrium is torn from the heart, allowing free outflow. The procedure should be practiced until no more than 45 sec is required between opening the thoracic cavity (which effectively ends spontaneous respiratory effort) and beginning the flow of fixative. For ultrastructural work, artificial respiration, using a small animal respirator and an endotracheal tube, may be useful.[14]

The perfusion is continued at a rate sufficient to pass 300–500 ml of fixative through the upper vasculature in approximately 7–10 min. We then remove the brain, dissect it into frontal or sagittal slabs of 3–5 mm thickness, and continue the fixation by immersion in the same fixative for another 3 hr at 4°. At the end of the immersion period, the tissues are washed in either phosphate buffer (when they are to be carried into electron microscopy) or in 12, 16, and 18% sucrose solutions (made in 0.15 M phosphate-buffer mixed with equal volumes of saline; this is called phosphate-buffered saline, or PBS). Each washing procedure requires changing solutions at 15-min intervals for 2–4 hr; this is accomplished by placing the tissues in a small vial (we use scintillation vials), decanting the fixative or wash solution, and quickly adding the fresh wash solution without permitting the tissues to dry. For electron microscopy, tissues should be processed within hours to retain optimal cellular fine structure.[15] However, for light microscopy, once tissues have been placed in 18% sucrose–PBS and stored at 4°, it is our experience that almost all immunoreactivity for neuroendocrine peptides is retained for at least 1 month without detectable decrement.

Sectioning

Frozen Sections

Frozen sections are essentially a means to obtain specimens hard enough to cut without the necessity to embed the material in a conventional sectioning medium, such as paraffin or epoxy. There are three sorts of devices available for obtaining frozen sections: cryostats, sliding microtomes, and pivoting (or surgical pathology) microtomes.

The cryostat is a rotary microtome built into a freezer cabinet, and it allows a wide range of cutting temperatures and section thicknesses. The size of the cabinet should be large enough to permit equilibration of several frozen specimens at once. This will allow placing sections from different tissue blocks onto the same slide (e.g., for comparing tissues fixed according to different protocols, with the same antibody, or for comparing tissues fixed in the same way, with the same antibody, but comparing blocks from experimental and control animals).

[14] K. R. Brizzee and S. K. Jirge, *Curr. Trends Morphol. Res.* **3**, 2 (1981).
[15] L. Y. Koda, E. L. F. Battenberg, and F. E. Bloom, *Current Trends Morphol. Res.* **3**, 2 (1981).

The cryostat can be used to prepare either slide-mounted or free-floating sections. The sliding microtome and the pivoting microtome obtain their specimen freezing capability from the specimen chuck, and the knife is generally at room temperature, as is the rest of the apparatus. These two microtomes are useful for cutting relatively thick specimens (>40 μm) to be handled as free-floating sections. They are much easier to use than the cryostat, but have far more limited temperature control.

Unfrozen Sectioning Methods

If frozen sections are not possible, two more choices exist for obtaining sections: (1) embed the material in paraffin-type embedments for more conventional microtomy; (2) use a vibrating sectioning device, which requires neither freezing or embedding.

The use of standard cytological embedments will generally render tissues far less immunoreactive than tissues not put through this process. However, excellent localizations on paraffin-embedded material have been published,[16] and the thinner sections made possible often yield spectacular contrast.

The vibrating sectioning devices[17] find their major application for sections of moderate thickness (50–75 μm) to be carried through the protocols for ultrastructural localization. The omission of the freezing–thawing, and the elimination of the cryoprotection washes, allow one to proceed more rapidly from fixed block to final reaction, thus preserving fine structure at many points.

Sectioning Pointers

Section thickness and quality depend on three factors: (1) quality of fixation and permeation by concentrated sucrose for cryoprotection (with embedded sections, the adequacy of dehydration and the embedment consistency are also critical); (2) speed of specimen movement past the knife edge; and (3) cutting temperature. The well fixed specimen will cut best at a cooler temperature than unfixed or poorly fixed brains, and this difference is enhanced by the cryoprotectants. This appears to be the result of the sucrose solution effectively lowering the freezing point of the specimen, such that blocks of the same hardness can be obtained at different temperatures depending on their treatment history. Good cutting quality is difficult to describe in words, but approaches the idea that the knife moves through the specimen smoothly, without resistance, and without

---

[16] F. Vandesande, *J. Neurosci. Methods* **1**, 3 (1979).
[17] R. E. Smith, *J. Histochem. Cytochem.* **18**, 590 (1970).

deformity of block or section. A venetian blind appearance with striations parallel to the knife edge, known as chatter, implies that the temperature is too cold (or the embedment too hard), the knife is dull, or the cutting speed is too fast. Vertical striations, perpendicular to the knife edge, imply debris or damage to the knife edge, which cause scratches; these may fracture the section, which can be disastrous with free-floating sections, and seems invariably to occur in a section just in the field you most want to photograph. When scratches occur, shift to a new spot on the blade, or remove the debris at once. When sections appear to "dissolve" on being touched to the water bath (for paraffin sections) or the collecting dish (for free-floating sections) this implies bad fixation, excessive ice crystals, inadequate cryoprotection, and a useless specimen. The same effect may be obtained with slide-mounted sections, which appear to be there when frozen, but disappear into a mushy globule when thawed. Time and antisera should not be wasted on badly fixed specimens without some other compelling reason.

Immune Reactions, Protocols

Standard Reagent Formulas

Stock Solutions
  Phosphate buffer, pH 7.4, double strength, 0.3 M (2× buffer): NaOH (dissolve first), 30.8 g; $NaH_2PO_4 \cdot H_2O$, 134.64 g; distilled H_2O to make 4.0 liters
  Phosphate buffer isotonic, 0.15 M (1× buffer): 2× phosphate buffer, 500 ml; distilled H_2O, 500 ml
  Phosphate-buffered saline (PBS): NaCl, 8.5 g; 2× phosphate buffer, 1 liter; distilled H_2O to make total volume 2 liters
  Sucrose solutions, 12%, 16%, and 18%. Store sucrose solutions at 4°. Discard solutions if bacterial growth is obvious.
  12% Sucrose: sucrose, 12 g; PBS, 100 ml
  16% Sucrose: sucrose, 16 g; PBS, 100 ml
  18% Sucrose: sucrose, 18 g; PBS, 100 ml
  Phosphate-buffered saline and Triton X-100: PBS, 100 ml; Triton X-100, 300 μl. Stir until the Triton is completely dissolved.
  Ethanolic-gelatin solution (0.3% gelatin in 40% ethanol): gelatin, 3 g; distilled H_2O, 500 ml; 80% ethanol, 500 ml

Procedure

1. Heat 250 ml of distilled H_2O to 50°.
2. Gradually add the gelatin to the H_2O with constant stirring and heat.

3. Stir until the gelatin is completely dissolved. The solution will appear opalescent.

4. Remove the solution from heat and add the remaining 250 ml of H_2O and stir in the 500 ml of 80% ethanol.

5. Mix well and store in a stoppered reagent bottle at room temperature.

Preparation of Gelatinized Slides

1. Clean glass microscope slides in 1% HCl–70% ethanol by immersion, and agitate for a few minutes.
2. Dip slides in 100% acetone and air-dry.
3. Fill a large glass staining dish with the ethanol–gelatin solution.
4. Fill a standard glass or metal slide rack with cleaned microscope slides. Dip the slides in the gelatin solution and agitate for a few minutes. Remove slides from solution and drain on paper towels.
5. Dry slides in oven at 60° for 2–24 hr.
6. Place dried, coated slides in a covered slide box until use.
7. Coated slides can be kept indefinitely.

Reagents Requiring Special Precautions

Reagents

H_2O_2–DAB solution
   H_2O_2 (final concentration is 1% H_2O_2): 3% H_2O_2, 75 μl; distilled H_2O, 150 μl. Mix the H_2O_2 and the H_2O together to yield a final solution of 1% H_2O_2.

   3,3′-Diaminobenzidine · HCl_4 solution (DAB) (final concentration is 0.05% DAB): DAB, 12.5 mg; 1× phosphate buffer, 6.25 ml; distilled H_2O added to a total volume of 25 ml.

Procedure

1. Mix the 1× buffer and H_2O to a total volume of 25 ml.
2. Add the DAB and stir continuously until dissolved. The solution will be slightly brown, and there should not be any visible precipitate; do not use free base.
3. Add 75 μl of 1% H_2O_2 stirring continuously.
4. The final solution, containing 0.05% DAB and 0.003% H_2O_2, should be used immediately.
5. Both the DAB solution and the 3% H_2O_2 stock solution should be stored at 4° in the dark. Both should be brought to room temperature before opening and use. The 0.05% DAB solution is stable for 2–3 hr, although it is recommended that the solution be prepared immediately

before use. Once the 1% H_2O_2 has been added to the DAB solution, the reaction media should be used immediately and any extra discarded. DAB is a known carcinogen and should be handled with extreme care. All work using DAB should be conducted in a well ventilated hood, and it is recommended that the researcher wear a face mask and gloves. At the end of the experiment, the mask, gloves, and any residual powdered DAB should be sealed in a plastic bag and disposed of by approved procedures.

Protocols

The protocols for immunohistochemistry, routinely used in our laboratory, are designed for free-floating sections. General techniques for handling free-floating cryostat sections are applicable to both immunofluorescence and immunoperoxidase reactions. Cryostat sections (40–60 μm) are cut; 1–3 sections are allowed to accumulate on the knife edge, flattened by an anti-roll plate. With a small camel or sable hair brush kept at room temperature, the frozen section is lifted from the cryostat knife and floated and thawed onto PBS in the incubation vessel. It is important that the sections float freely and are not curled over on themselves. The sections can also be collected in a variety of vessels: test tubes, centrifuge tubes, petri dishes, culture dishes, etc. In general, the chamber selected should minimize section handling during all of the incubation and rinse procedures. We use small open-ended plexiglass cylinders with nylon mesh bottoms, which rest within cylindrical cutouts in a solid plexiglass block. Once the sections have been collected in the mesh-bottom chambers, they can be moved through the reaction steps by lifting the screen carrier and transferring it with all its sections to a new depression dish containing the next incubation media. In this way, the sections are not physically handled until the end of the reaction, when they are slide-mounted. Once the cryostat sections have been collected in PBS they can be stored at 4° for several hours prior to incubation in the primary antisera.

Incubation Procedures

All primary and secondary antisera should be aliquoted and stored at −20°. Aliquots to be used should be thawed before opening and not refrozen, thus avoiding degradation from repetitive freezing and thawing. Unused antisera from a thawed aliquot should be stored at 4°.

The choice of the incubation apparatus determines the general method of incubation. Dilutions of antisera must be determined for each new antiserum, and the concentration of the antisera used will also affect the time and temperature of the incubations (see Bloom et al.[18] and Hökfelt et al.[19]).

Immunofluorescence
1. Collect free-floating cryostat sections in PBS as described.
2. Incubate in primary antiserum diluted with PBS containing TX100 (0.3%) and bovine serum albumin (BSA), 1 mg/ml. Incubation is carried out for 12–18 hr at 4° with continuous vibration. A tissue agitator or simple vibrating pad can be used for adequate agitation.
3. Wash in three 10 min changes of PBS at room temperature with vibration.
4. Incubate in goat-anti-rabbit IgG conjugated to FITC or rhodamine diluted 1 : 100 with PBS × TX100 (0.3%) for 1 hr at room temperature with vibration.
5. Rinse in PBS as above.
6. Float sections on an alcohol–gelatin solution containing 0.3% gelatin in 40% ethanol. Mount sections with a sable brush onto gelatinized glass slides, dry, and coverslip with paraffin oil.
7. View sections in a fluorescence microscope equipped with excitation and emission filters for fluorescein and rhodamine.

—

—

—

Immunoperoxidase
1. Collect free-floating cryostat sections in PBS as described.
2. Incubate in primary antiserum diluted with PBS containing TX100 (0.3%) and BSA at 1 mg/ml, for 12–18 hr at 4° with continuous vibration.
3. Wash in three 10-min changes of PBS at room temperature with vibration.
4. Incubate in GARHRP (goat antirabbit IgG conjugated to horseradish peroxidase) diluted 1 : 1000 with PBS + TX100 (0.3%) for 2 hr at room temperature with vibration.
5. Rinse in PBS as above.
6. Incubate in 0.05% DAB in 0.035 M 1× buffer containing 0.003% H_2O_2 for 8–15 min. at room temperature with constant vibration.
7. Rinse in PBS as above.
8. Incubate in 0.1% OsO_4 in 2× buffer for 30 sec followed by several PBS rinses (optional step to enhance reaction)[20,21]
9. Float sections on alcohol–gelatin solution as described under immunofluorescence (6), mount on gelatinized glass slides, dry, and coverslip with paraffin oil.
10. Sections may be dehydrated, counterstained, and coverslipped in Permount.

We also employ a four-step immunoperoxidase reaction, in which all antisera are unconjugated. After the primary rabbit antiserum step, un-

[18] F. Bloom, E. L. F. Battenberg, J. Rossier, N. Ling, and R. Guillemin, *Proc. Natl. Acad. Sci. U.S.A.* **75**, 1591 (1978).
[19] T. Hökfelt, A. Ljungdahl, L. Terenius, R. Elde, and G. Nilssen, *Proc. Natl. Acad. Sci. U.S.A.* **74**, 3081 (1977).
[20] F. Gallyas, T. Gorcs, and I. Merchenthaler. *J. Histochem. Cytochem.* **30**, 183 (1982).
[21] J. Adams, *J. Histochem. Cytochem.* **29**, 775 (1981).

conjugated goat anti-rabbit IgG is applied as above; the sections are then washed and allowed to react with an IgG fraction of a rabbit anti-HRP; last, the sections are washed, exposed to a dilute solutions of HRP, and allowed to react with DAB–peroxide to localize the antibody complex that binds the HRP. Since all antisera are unconjugated, immunoreactivity is optimal. Furthermore, all reagents are of minimum size, facilitating diffusion (see Vaughan et al.[22]) Many electron microscope studies have opted for the use of a peroxidase–antiperoxidase complex (PAP) to replace steps 3 and 4 of this indirect procedure. The demonstrated gain is that more molecules of HRP are bound per antigen–antibody complex. The disadvantage is that the very large size of the peroxidase anti-peroxidase complex has in our hands a detrimental effect on tissue penetrance.[23,24]

Comments

In general, fixed tissue sections are relatively impermeable to antibodies. The relatively large size of the immunoglobulin molecule, enhanced by conjugation with a marker protein like horseradish peroxidase further limits diffusion.[23,24] The relative impermeability of the plasma and intercellular membranes and the semiimpermeable matrix of cytoplasmic proteins formed by the reaction with the fixatives also contributes to limited antiserum penetrance. Although some antisera will yield good reactions without taking steps to improve tissue permeability, these instances are infrequent. In general, tissue immunoreactivity is enhanced by use of lipid extractions (chloroform–methanol[25]) or by addition of detergents (Triton X-100, 0.3%[26]) in all immune reagents.

Tissue permeation is also affected by tissue sectioning and incubation methods. In Figs. 4–7 we indicate the results obtained on the same region of rat CNS, from the same brain, and with the same anti-β-endorphin serum, when the tissues are exposed to the antiserum in different ways. In general, free-floating sections will exhibit more apparent immunoreactivity than sections incubated while slide mounted; the advantage of the free-floating sections presumably results from the fact that immunoreagents can penetrate the sections from both surfaces, and this advantage is further enhanced by the fact that slide mounting, and the drying needed to secure the sections to a slide, may further restrict diffusion

[22] J. E. Vaughan, R. P. Barber, C. E. Ryback, and C. R. Houser, *Curr. Trends Morphol. Res.* **3**, 33 (1981).
[23] J. Roth, M. Bendayan, and L. Orci, *J. Histochem. Cytochem.* **26**, 1074 (1978).
[24] J.-L. Guesdon, T. Ternynck, and S. Avrameas, *J. Histochem. Cytochem.* **27**, 1131 (1979).
[25] B. K. Hartman, D. Zide, and S. Udenfriend, *Proc. Natl. Acad. Sci. U.S.A.* **69**, 2722 (1972).
[26] L. W. Swanson and B. K. Hartman, *J. Comp. Neurol.* **163**, 467 (1975).

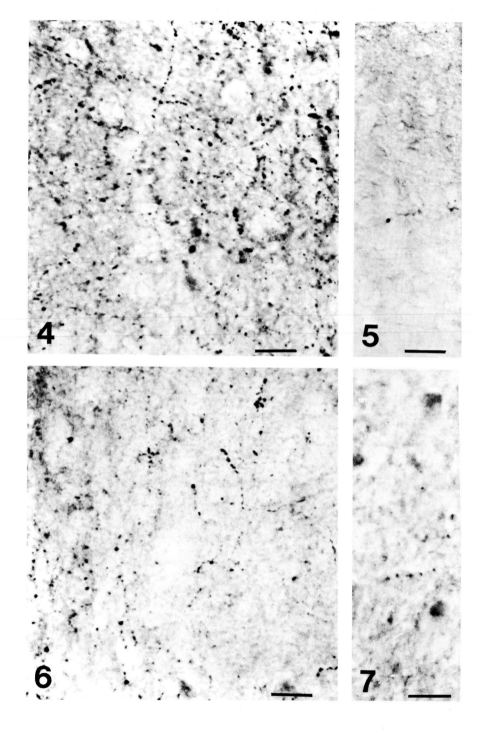

from the exposed surface. The improvement in immunoreactivity detected with detergents is often dramatic. However, because this process often renders disrupted, unacceptable fine structure, alternatives to detergents have been sought. One possibly successful method, which is less effective as well, however, is rapid freezing and thawing[27]; this may improve permeability by creating microfractures in membranes, and presumably accounts for the relatively greater immunoreactivity of frozen sections, over those cut on a Vibratome.

Photomicrography

The objective of immunocytochemistry is to determine the location of the immunoreactive structures revealed by the antisera being used. In order to prepare the results in a manner that can be used to document them for others, it is necessary to make photographic records through the microscope. The selection of the mode of optical detection can also influence the apparent degree of immunoreactivity and should be taken into consideration when comparing results among different laboratories. In Figs. 8–10, we present the same field of a single section, photographed with the tissue viewed in conventional bright-field microscopy and darkfield microscopy and by phase contrast. Aside from the inability to perceive the color of the horseradish peroxidase label, the degree of reactivity revealed by the black and white photographs varies considerably. When FITC- or rhodamine-labeled antisera are used, the final reaction may appear to be greater than that observed with immunoperoxidase methods. This is generally because the smaller immunofluorescent reagents penetrate better into the tissue section and because the final prod-

[27] P. Somogyi and H. Takagi, *Neuroscience* **7**, 1779 (1982).

FIGS. 4–7. Effects of tissue sectioning and exposure on apparent immunoreactivity. For each figure, Vibratome sections were cut from the same brain and exposed to the same antibody, at the same dilution, for the same period of time. Note that staining is improved both by freeze-thawing and by addition of the detergent. Calibration bars = 25 μm.

FIG. 4. Free-floating section, frozen and thawed to simulate cryostat sectioning and incubated with addition of Triton X-100 in primary antiserum and all subsequent immunoreagents.

FIG. 5. Section cut and incubated exactly as in Fig. 4, but Triton X-100 was omitted from all sera and washes.

FIG. 6. Section was cut and incubated exactly as in Fig. 4, but freeze–thaw step was omitted. Triton X-100 was included in all immunoreagents.

FIG. 7. Section was cut and incubated as in Fig. 4, but both freeze–thaw and addition of Triton X-100 was omitted.

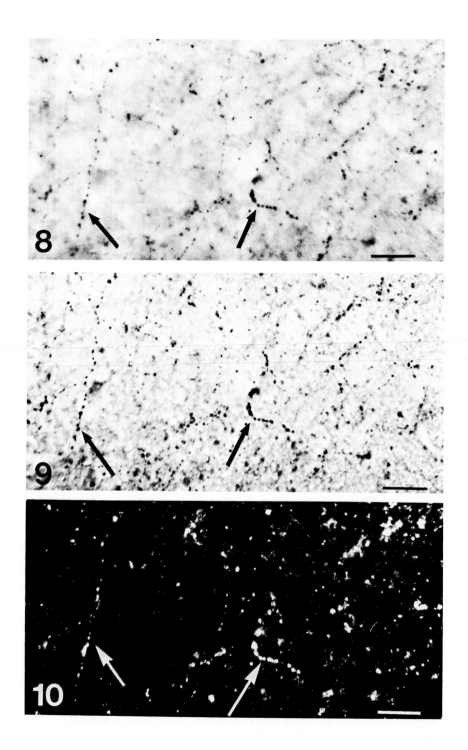

uct is viewed in dark-field, which gives a greater degree of contrast by a bright colored marker that will appear slightly larger than its bright-field image because of the tangential illumination of the dark-field condenser (see Figs. 8–10).

Immunohistochemistry can also be applied to epoxy-embedded thin sections or to materials embedded for thin sectioning in embedments that are more water soluble. For reactions to work well on epoxy-embedded material, the tissue antigen must be present in high concentrations (like the secretory granules of pancreatic endocrine cells),[28] and the fine structure must be preserved such that the embedment can be partially removed (called etching). This often weakens the ability of the specimen to resist the electron beam. The gold- or ferritin-conjugated antibody methods,[29,30] which are analogous to the fluorescent or enzyme-coupled indirect methods of light microscopy, have been reported to be successful when applied to partially etched thin sections, again localizing antigens that are present in high concentration within the specimen.

Interpretation

Verifying Immunoreactivity

The standard approach to verifying immunocytochemical specificity is the "blocking" procedure. In this protocol, a dose-response curve of immunoreactivity is determined; gradually increasing amounts of purified antigen are added to the maximum working dilution of the antiserum to be used for the staining. Several sets of sections are prepared from a well fixed tissue region known to have the desired immunoreactivity over a sufficiently large number of sections so that four sets of serial sections can be obtained.

[28] J. Roth, M. Bendayan, and L. Orci. *J. Histochem. Cytochem.* **27**, 55 (1980).
[29] C. L. Templeton, R. J. Douglas, and W. J. Vail, *FEBS Lett.* **85**, 95 (1978).
[30] J. Roth and M. Binder, *J. Histochem. Cytochem.* **26**, 163 (1978).

FIGS. 8–10. Effects of photomicrographic technique on apparent degree of immunoreactivity. The same section was photographed by different optical methods of illumination of the specimen: bright-field microscopy (Fig. 8); phase-contrast microscopy (Fig. 9); dark-field microscopy (Fig. 10). Two prominent immunoreactive axons are indicated by arrows in each micrograph for orientation. Note that comparison of dark-field (Fig. 10) and bright-field (Fig. 8) versions appear to show greater signal-to-noise of immunoreactivity and that phase contrast tends to obscure many fine immunoreactive processes. Tissue sections taken from rat hypothalamus were stained with β-endorphin antiserum. Calibration bars = 25 μm.

Set 1 is used to verify that the tissue is well fixed and that all reagents are working. Sets 2, 3, and 4 are then run the next day, using blocked and control sera: from radioimmunoassay (RIA) work, it should be known what the antigen binding capacity of the serum is at the high dilution end; the amount of immunogen that will be required to saturate all the binding in the working dilution of the antiserum can then be calculated. That amount of immunogen is added to the serum to be used for set 2, and 10 times more is added to the serum for set 3. A equal volume of unblocked serum is also raised to the working dilution, and all three sera are incubated overnight, sealed in tubes, in the cold room or refrigerator. These primary sera are then used to start the staining process on the three respective collections of tissue sections.

In an ideal blocking experiment, sets 1 and 4 will have full and intense staining, set two will be very light to minimal, and set 3 should have no staining at all. The value of the incubated, diluted serum is to have a sample that has not been tampered with kept under identical conditions, but allegedly not blocked. Unless this set works, when the presumably "blocked" sera (sets 2 and 3) fail to stain, it is difficult to assume that immunoabsorption has specifically blocked the staining.

If the blocking reaction does not block the staining, then one may conclude the staining to be nonspecific and unexplainable. However, if the blocking reaction does block the staining, that does not necessarily mean that all labeled structures contain the desired immunogen. At the very least, it means that those structures may contain some analogous molecule, as do all endorphins. To be safe, this is generally referred to as endorphin-like immunoreactivity. For further verification of specificity, (1) the regional distribution of the endorphin-like, enkephalin-like, or dynorphin-like staining should correlate closely with that found in RIAs of microdissected brain regions; (2) the same staining pattern, maybe more or less intensely, should also be obtained with three (or more) other antisera raised against the same immunogen. This is perhaps the most convincing of the easily applied specificity checks. Even when this holds, and the positive sera all work at 1:2500 or higher dilutions, additional proofs of specificity can be sought by demonstrating that the antisera being used do not recognize other homologous peptides added in comparable amounts in blocking protocols.[31,32] This is most useful in demonstrating structural separations between endorphin peptides of different precursors (e.g., β-endorphin versus enkephalins, or enkephalin versus

---

[31] J. F. McGinty and F. E. Bloom, *Brain Res.* (in press).

[32] J. F. McGinty, S. J. Henriksen, A. Goldstein, L. Terenius, and F. E. Bloom, *Proc. Natl. Acad. Sci. U.S.A.* **80**, 589 (1983).

dynorphin), where antisera can detect nonhomologous amino acid sequences unique to one branch of the peptide family. Demonstration of dual intracellular labeling for precursor products or possible dual end products (e.g., β-endorphin and corticotropin immunoreaction) requires much more extensive protocols beyond the scope of this survey.[32,33]

[33] S. J. Watson, H. Akil, C. W. Richards III, and J. D. Barchas, *Nature (London)* **275**, 226 (1978).

VII. Summary

[47] Previously Published Articles from Methods in Enzymology Related to Neuroendocrine Peptides

Some information on topics relevant to neuroendocrine peptides can be found in previous volumes of "Methods in Enzymology." Listed below are chapters, sections, and entire volumes by relevant topics.

Radioimmunoassay and Radioligand Assay

Vol. XXXVI [1]. Theory of Protein–Ligand Interaction. D. Rodbard and H. A. Feldman.
Vol. XXXVI [2]. Use of Specific Antibodies for Quantification of Steroid Hormones. G. D. Niswender, A. M. Akbar, and T. M. Nett.
Vol. XXXVII [1]. Statistical Analysis of Radioligand Assay Data. D. Rodbard and G. R. Frazier.
Vol. XXXVII [2]. General Considerations for Radioimmunoassay of Peptide Hormones. D. N. Orth.
Vol. XXXVII [16]. Methods for Assessing Immunologic and Biologic Properties of Iodinated Peptide Hormones. J. Roth.
Vol. 70. Immunochemical Techniques (Part A).
Vol. 73. Immunochemical Techniques (Part B).
Vol. 74. Immunochemical Techniques (Part C).
Vol. 84. Immunochemical Techniques (Part D: Selected Immunoassays).
Vol. 92. Immunochemical Techniques (Part E: Monoclonal Antibodies and General Immunoassay Methods).

Cell and Tissue Handling

Vol. XXXVII [5]. *In Vitro* Pituitary Hormone Secretion Assay for Hypophysiotropic Substances. W. Vale and G. Grant.
Vol. XXXIX [13]. Isolation, Cloning, and Hybridization of Endocrine Cell Lines. P. O. Kohler.
Vol. XXXIX [14]. Initiation of Clone Cultures of Rat Anterior Pituitary Cells Which Secrete a Single Gonadotropic Hormone. A. Steinberger.
Vol. XXXIX [17]. Methods for Studying Pituitary–Hypothalamic Axis *in Situ*. J. C. Porter.
Vol. XXXIX [27]. The Perfused Adrenal Gland. S. D. Jaanus and R. P. Rubin.
Vol. LVIII. Cell Cultures.

Microscopy

Vol. XXXVII [9]. Localization of Hormones with the Peroxidase-Labeled Antibody Method. P. K. Nakane.
Vol. XXXVII [10]. Autoradiographic Techniques for Localizing Protein Hormones in Target Tissue. H. J. Rajaniemi and A. R. Midgley, Jr.

Specific Assays

Vol. XXXVII [6]. Measurement of Somatomedin by Cartilage *in Vitro*. W. H. Daughaday, L. S. Phillips, and A. C. Herington.

Synthesis

Vol. XXXVII [35]. Solid Phase Synthesis of Luteinizing Hormone-Releasing Hormone and Its Analogues. D. H. Coy, E. J. Coy, and A. V. Schally.
Vol. XLVI. Affinity Labeling.

Electrophysiology

Vol. XXXIX [36]. Electrophysiological Techniques for the Study of Hormone Action in the Central Nervous System. B. J. Hoffer and G. R. Siggins.

Subcellular Fractionation

Vol. XXXI [1]–[4]. Section I. Multiple Fractions from a Single Tissue.
Vol. XXXI [5]–[16]. Section II. Isolation of Purified Subcellular Fractions and Derived Membranes (from Mammalian Tissue Excluding Nerve). A. Plasma Membrane.
Vol. XXXI [17]. Section II. Isolation of Purified Subcellular Fractions and Derived Membranes (from Mammalian Tissue Excluding Nerve). B. Golgi Complex.
Vol. XXXI [18]–[22]. Section II. Isolation of Purified Subcellular Fractions and Derived Membranes (from Mammalian Tissue Excluding Nerve). C. Rough and Smooth Microsomes.
Vol. XXXI [23]–[26]. Section II. Isolation of Purified Subcellular Fractions and Derived Membranes (from Mammalian Tissue Excluding Nerve). D. Nuclei.
Vol. XXXI [27]–[30]. Section II. Isolation of Purified Subcellular Fractions and Derived Membranes (from Mammalian Tissue Excluding Nerve). E. Mitochondria.
Vol. XXXI [31]–[35]. Section II. Isolation of Purified Subcellular Fractions and Derived Membranes (from Mammalian Tissue Excluding Nerve). F. Lysosomes.
Vol. XXXI [36]–[37]. Section II. Isolation of Purified Subcellular Fractions and Derived Membranes (from Mammalian Tissue Excluding Nerve). G. Peroxisomes.
Vol. XXXI [38]–[43]. Section II. Isolation of Purified Subcellular Fractions and Derived Membranes (from Mammalian Tissue Excluding Nerve). H. Secretory Granules.
Vol. XXXI [44]–[45]. Section II. Isolation of Purified Subcellular Fractions and Derived Membranes (from Mammalian Tissue Excluding Nerve). I. Specialized Cell Fractions.
Vol. XXXI [46]–[50]. Section III. Subcellular Fractions Derived from Nerve Tissue.
Vol. XXXII [51]–[60]. Section V. Isolation and Culture of Cells. A. Basic Methodology for Cell Culture.
Vol. XXXII [61]–[72]. Section V. Isolation and Culture of Cells. B. Isolation of Specific Cell Types.
Vol. XXXII [73]–[79]. Section V. Isolation and Culture of Cells. C. Isolation and Culture of Specific Cell Types.

Author Index

Numbers in parentheses are footnote reference numbers and indicate that an author's work is referred to although the name is not cited in the text.

A

Abe, H., 609
Abercrombie, M., 654
Abraham, K. A., 590
Abraham, M., 285, 287
Adachi, T., 321, 328
Adams, J., 680
Adams, J. H., 611
Adams, M. R., 535
Adams, P. R., 104, 107(39), 128
Adams, R. N., 470, 480, 500
Adams, T. E., 35(18), 37, 39, 42, 43
Adamson, C., 419
Adelman, J., 548
Adey, W. R., 376
Advis, J. P., 512, 519, 541, 542, 543(13, 14), 545(15), 547
Agin, D. P., 123
Agnati, L. F., 613, 671
Ahlquist, R. P., 577
Aisbitt, R. P., 593
Akaike, N., 102, 129
Akaishi, T., 655
Akbar, A. M., 36, 38(10), 44(10), 452, 691
Akert, K., 376
Akil, H., 198, 387, 388(30), 389, 687
Albuquerque, E. X., 402(b, c, d, e, g, p, r, s, t), 403
Aldrich, R. W., 165
Ali, M. A., 367
Allen, M. B., 659
Alonso, G., 200
Alper, R. H., 494
Alston, E. J., 661
Alvarez, J., 541
Amenomari, Y., 249
Amoss, M., 178, 249, 259, 296, 298(11), 566
Amsterdam, A., 68, 70(33), 258, 282(0), 283
Anctil, M., 367
Anderson, G. M., 469

Anderson, N. C., 254
Anderson, S. N., 600, 602(46)
Anderson, T. R., 590(25), 591, 594(8), 595
Andrieu, J. M., 410
Angeletti, P., 343
Angevine, Jr., J. B., 375
Anggard, A., 624(24), 630
Annunziato, L., 613
Antakly, T., 325(j), 326
Anthony, E. L. P., 626
Anton, A. H., 473, 499
Antony, P., 594(14), 595
Antrup, H., 592
Aquilera, G., 279, 281, 283, 285
Araki, S., 608
Arancibia, S., 609
Archer, R., 535
Argiro, V., 335, 340(1), 346, 347(1)
Arimura, A., 33, 35, 39, 176, 178, 437, 454, 540, 608
Armstrong, W. E., 133, 316(h), 317
Arnauld, E., 98
Arnold, A. P., 641, 654
Arnold, M. A., 179, 180(17), 181, 182(17), 185(17), 186(17)
Asada, A., 594(37), 595
Asai, M., 548, 670
Ash, J. F., 220
Asmus, P. A., 502
Assenmacher, I., 200
Astrada, J. J., 541
Atack, C. V., 615
Atkinson, D., 278(m, n), 279, 284(d, e), 285
Attramadal, A., 639
Aubert, M. L., 36, 40, 42(22), 43(22)
Auerbach, A., 112, 172
Avrameas, A., 410, 411, 415(10), 681
Avrameas, S., 410, 414, 434
Axelrod, J., 362, 476, 484, 488(6), 489(6), 490, 492(6)

B

Baba, S., 388
Baba, Y., 608
Bachrach, U., 590
Badger, T. M., 305
Baertschi, A. J., 319(i), 320, 328
Bailey, A. S., 61
Bailey, E., 341
Baird, A., 85
Baird, K. L., 49
Baker, B. L., 258
Baker, H. A., 626
Ballard, P. L., 590
Barany, G., 5, 7
Barber, R. P., 625, 681
Barbiroli, B., 590(17), 591
Barchas, J. D., 471, 687
Barclay, R. K., 543, 547(16)
Bareis, D. L., 603
Barkakati, N., 168
Barker, J. L., 103, 104, 106, 107, 108, 112, 114, 120, 121(48), 124(31), 125, 128(48), 129(48), 138(6), 139(6), 141(6), 143(6), 145(10)
Barnes, G. A., 590(27), 591, 594(25), 595, 599(27)
Barnes, G. D., 278(m, n), 279, 284(d, e), 285
Barnett, C. A., 294
Baron, M. D., 66
Barret, A., 330, 331
Barrett, R. E., 590(23), 591
Barry, J., 449, 453
Bartlett, A., 412, 413
Bartolome, J., 590(27), 591, 594(16, 18, 25), 595, 599(27), 603(29)
Bassini, R. M., 436, 437(1), 440(1)
Battenberg, E. L. F., 675, 680
Baudry, M., 578
Bauer, K., 546, 547(17)
Baulieu, E. E., 638, 640
Bauminger, S., 436, 437(2), 451
Bayley, H., 58, 63
Bayon, A., 198
Beaulieu, M., 251, 578, 580(7), 581(7), 582(7), 612
Beaumont, A., 670
Beavan, M. A., 484
Beccati, M. D., 639(11), 640
Becker, S. R., 39

Behrman, H. R., 449
Beiser, S. M., 449
Belik, Y. V., 528
Ben-Ari, Y., 371, 384, 392(21), 395
Ben Aroya, N., 608
Benassayag, C., 640
Bencosme, S. A., 258, 280(b), 281
Benda, P., 323(b), 326, 331, 332(20), 333
Ben-David, E., 492
Bendayan, M., 681, 685
Benefenati, F., 671
Benington, F., 402(f), 403
Benitez, H. H., 319(a), 320
Ben-Jonathan, N., 475, 484, 608, 610, 612(42)
Bennet-Clarke, C., 388
Bennett, B. A., 494, 500(12, 13), 504(12), 507, 509(12, 13), 526, 535(7), 539(7)
Bennett, H. P. J., 72, 77, 85(8)
Benoist, L., 278(o, p), 279, 280(g), 281, 282(f, g), 283, 286
Benoit, R., 82, 85(13), 87, 662
Bentivoglio, M., 663, 664(7), 665(7), 666
Beny, J. L., 319(i), 320, 328
Beraud, M. T., 280(g), 281, 282(f), 283, 286
Beress, L., 402(u), 403
Bergey, G. K., 114
Berman, J. R., 550
Berod, A., 629
Bertagna, X. Y., 334
Berthold, C.-H., 144
Berwald-Netter, Y., 113, 138(17), 143(17)
Besser, G. M., 553, 555, 582
Besson, J., 278(o, p), 279, 280(g), 281, 282(f, g), 283, 286, 609
Besson, M., 495
Bethea, C. L., 288
Bevington, P. R., 171
Bey, P., 598
Bhanat, R., 661
Bharos, A. T., 663, 664(7), 665(7)
Bialecki, H., 289
Biales, B., 104
Bidwell, D., 412(31), 413
Biggins, J. A., 557
Bilezikjian, L., 565
Binder, M., 685
Bintz, J. E., 594(23), 595
Bioulac, B., 332
Birdwell, C. R., 288

Birge, C. A., 609
Bishop, G. A., 127
Björklund, A., 609, 615(36), 620, 665
Blackledge, A., 593
Blackwell, R., 178, 249, 259, 296, 298(11), 566
Blaha, G., 388
Blanchard, S. G., 225, 226(15)
Blank, C. L., 471
Bleier, R., 319(b), 320
Bloom, F. E., 124(61), 125, 187, 357, 359(13), 362, 661, 670, 675, 686, 687(32)
Bloom, N. D., 639(10), 640
Bloom, S. R., 333
Bluet-Pajot, M. T., 278(o), 279, 280(g), 281, 282(f), 283, 286
Blum, J. J., 56, 57, 58(12)
Bocchini, V., 343
Boehme, R., 305, 306(2), 310(2)
Böhlen, P., 75, 78, 79, 82, 83, 85, 87, 89, 176, 531
Boissier, J., 612
Boltz, R. C., Jr., 277
Bonyhady, R., 594(42), 596
Borges, J. L., 258, 283, 287
Bornstein, M. B., 315, 319(j, k), 321, 341
Böttcher, H., 396, 397(16)
Bottenstein, J. E., 115, 312, 314, 330, 343, 350, 357, 522
Bourgoin, S., 333
Bourne, G. H., 376
Bourre, J. M., 330
Bower, J., 82
Bowker, R. M., 629, 663
Bradbury, M. W. B., 495
Braddon, S. A., 594(36), 595
Brandenburg, D., 63, 64(16), 65, 66, 69
Brawer, J. R., 617
Bray, D., 337
Bray, G. M., 306, 311(4)
Brazeau, P., 33, 79, 82, 83, 85(13), 87, 89, 176, 608
Brecha, N., 366, 630
Breedlove, S. M., 641
Brenneman, D. E., 113, 114(26), 115(22), 144(26)
Bressler, B. H., 323(a), 326
Bribkoff, V. K., 101
Brice, G. F., 452
Bridson, W., 564

Bridwell, D. E., 412
Brightman, M. W., 189, 193(6), 194(5), 196(5), 200(5), 205(4, 6), 206, 208(4), 209(4, 5, 6), 212(4, 6), 213(4, 6), 215(4, 6)
Brinck-Johnsen, T., 333
Brink, L., 77
Brizzee, K. R., 675
Broadwell, R. D., 188, 189, 192, 193(6, 60), 194, 196(5), 200(5, 7), 203, 205, 206, 208(4, 7), 209(4, 5, 6, 7, 8, 9, 60), 212, 213(4, 6), 215(4, 6, 58, 59), 218(58, 59, 60, 71)
Brodie, B. B., 494
Bronstein, M. J., 387, 388(30), 389(30)
Brosnan, M. E., 597
Brown, A. M., 102, 129
Brown, B. L., 278(m, n), 279, 284(d, e), 285
Brown, D. A., 104, 107(39), 128
Brown, K. D., 253
Brown, K. T., 122, 123, 137
Brown, L. L., 315, 319(k), 321
Brown, M., 178
Brown, M. R., 181, 565
Brown, T. H., 129
Browne, C. A., 72, 77, 85(8)
Brownell, M., 419
Brownfield, M. S., 195
Brownstein, M., 200, 484, 488(6), 489(6), 490(6), 492(6), 533, 568, 608
Brownstein, M. J., 362, 363, 369, 370(3), 371(3), 373(3), 374(3), 375, 525, 528, 537(5), 539(11)
Brugi, J. F., 555
Bruntlett, C. S., 500
Buchsbaum, S., 383
Buckley, D. I., 66
Buda, M. J., 480
Buijs, R. M., 199
Buisson, N., 333
Bullock, G. R., 458
Bullock, P. N., 112, 114, 115, 119(11), 124(30), 129, 139(11)
Bundgaard, M., 192
Bunge, M. B., 335, 342, 343, 345(29)
Bunge, R. P., 335, 340, 343, 345(29), 346, 347
Burch, W. M., 593
Burden, J., 495
Burgisser, E., 578

Burgus, R., 33, 608
Burke, G., 594(47), 596
Burke, J. P., 625, 626
Burlingame, P. L., 336
Burnett, F. M., 455
Burns, B. D., 97
Burrow, G. N., 594(48), 596
Burshell, Y. D. A., 641
Burt, D. R., 34
Burton, H., 335, 347
Busby, W. H., 440, 446(6)
Buse, J., 469
Butcher, M., 33, 176, 608
Butler, J. E., 419
Butler, S. R., 594(29), 595
Butler, V. P., 449
Byus, C. V., 594(6), 595, 601

C

Caldarera, C. M., 590(17), 591
Calhaz-Jorge, C., 611(58), 612
Caligaris, L., 541
Calne, D. B., 577
Calvet, J., 119
Calvet, M. C., 119
Cam, A. L., 69
Camardo, J. S., 150
Cambier, J. C., 233, 234
Cameron, C. M., 593
Cameron, D. P., 388
Camier, M., 331, 332(20), 535, 537(21)
Campbell, G. T., 319(j), 321
Campenot, R. B., 115, 335, 346(12)
Canellakis, E. S., 602
Canick, J. A., 323(e), 326
Cantarero, L. A., 419
Card, J. P., 630
Carey, D. J., 343, 345(29)
Carlisle, W., 333
Carlquist, M., 83
Carlsson, A., 609, 615
Carmel, P. W., 608, 609, 615(22)
Caron, M. G., 578, 580(7, 14), 581(7, 14), 582, 583, 587, 588, 589
Carpenter, J. L., 69
Carraway, K. L., 451
Carson, K. A., 306, 309(5), 310(5), 311(5), 312(5)

Carter, A. C., 639(10), 640
Carter, D. D., 600, 602(46)
Carter, H. E., 442
Casara, P., 598
Case, J. R., 61
Cassiman, J., 278(k), 279, 295
Casteels, R., 402(j), 403
Castillo, F., 75
Cataldo, A. M., 188, 189, 192(7, 9), 200(7), 203, 205(58, 59), 206(7, 58, 59), 208(7), 209(7, 9), 212(9), 215(58, 59), 218(58, 59)
Cate, C. C., 333
Catsman-Berrevoets, C. E., 663, 664(7), 665(7), 666
Catt, K., 38, 44, 258, 279, 281, 282(o), 283, 410, 594(5), 595
Catterall, W. A., 402(u, v), 403
Cespuglio, R., 479
Cesselin, F., 333
Chafel, T., 388
Chaikoff, I. L., 296
Chait, E. A., 594(25), 595
Chamness, G. C., 639(12), 640, 650(12)
Chandra, P., 281
Chang, A. C. Y., 547, 670
Chang, K. J., 49, 219, 220, 221, 222, 224, 225, 226(15), 312, 632
Chantler, S. M., 456, 548, 549(7)
Chapple, P. J., 327, 329
Chard, T., 559
Charli, J. L., 495
Chayvialle, J. A., 83
Cheek, W. P., 249
Chen, I. W., 550
Chen, K. Y., 602
Chen, M., 550
Cheng, H. Y., 480
Chermay, A., 475, 495
Chiasson, R. B., 594
Chick, W., 460
Chihara, K., 178, 608
Chikamori, M., 321, 328
Childs, G. V., 258, 282(o), 283
Ching, M., 608
Chino, N., 671
Chobsieng, P., 453, 456(23)
Choi, D. W., 114, 124(32), 125(32), 126
Chowdhry, V., 58
Chrambach, A., 35(14), 36
Chretien, M., 555

Christian, C. N., 112, 113, 119(11), 139(11)
Chu, L., 196, 206(39)
Chun, L. L. Y., 335
Chused, T. M., 239
Chyad-Al-Noaemi, M., 557
Cicero, T. J., 389
Citron, M. C., 123
Clark, B. R., 357
Clark, J. H., 650
Clark, J. L., 601, 602(50)
Clark, R. B., 150
Clauser, H., 617
Clayton, R. N., 38, 39, 40(21), 42, 43(21), 44, 46, 258, 282(o), 283
Clemens, J. A., 609
Clement-Cormier, Y., 584
Clement-Jones, V., 553, 555
Cocchi, D., 660, 661(60)
Cochet, M., 547
Cochran, M., 340
Codaccioni, J. L., 178, 608
Cohen, D., 288
Cohen, J., 402(h), 403
Cohen, J. L., 640
Cohen, L. B., 113
Cohen, P., 331, 332(20), 535, 537(21)
Cohen, S., 293, 547, 594(35, 39, 43), 595, 596
Cohen, S. N., 547, 670
Cohen, S. S., 590
Colburn, T. R., 103, 120, 121(48), 128(48), 129(48)
Cole, J. J., 334
Collier, T., 392
Collins, J. F., 381, 385(11), 386(11), 387(11), 389(11)
Collins, T. S., 258
Collu, R., 584
Colquhuon, D., 163, 165, 166(9), 168(9)
Comb, M., 548
Combest, I. W., 594
Cone, C. D., 241
Conn, P. M., 42, 43, 49, 53, 54, 55, 56, 57, 58(12), 242, 251, 363, 402(k, l, m, n), 403, 404
Connally, J. L., 354
Conne, B. S., 40, 42, 43(22)
Connor, J. A., 108
Conradi, S., 144
Constanti, A., 104, 107(39), 128
Conte-Devolx, B., 178, 608

Conway, B. P., 611(57), 612
Cooke, I. M., 327, 329
Coons, A. H., 139, 623
Cordon, C., 495
Corley, R. B., 234
Cornbrooks, C. J., 343, 345(29)
Costa, E., 494
Costa, M., 593
Costlow, M. E., 593
Cote, J., 195
Cotman, C. W., 101, 400
Cotran, R. S., 192
Coullard, P., 367
Coulter, J. D., 629, 663
Cousin, M. A., 594(11), 595
Coward, T. K., 469
Cox, B., 554
Coy, D. H., 437, 454
Coyle, J. T., 393, 475, 484, 485, 610, 625
Craig, L. C., 22
Cramer, O. M., 610, 613(45)
Creese, I., 578, 580(16), 582(16, 27), 583(16), 584, 585, 586(27), 587(27), 589
Cremer, N. E., 73, 455
Creveling, C. R. J., 485
Crissman, H. A., 234
Croce, C., 469
Cronin, M. J., 239, 258, 283, 287, 295
Cronkite, E. P., 281
Crowell, J. M., 232
Crowley, W. F., Jr., 305
Cuatrecasas, P., 49, 60, 63(8), 66, 219, 220, 221, 222, 224, 225, 226(15), 632
Cuello, A. C., 198
Culver, B., 350
Cummings, R., 143
Curnen, M., 409
Curry, K., 381, 385(11), 386(11), 387(11), 389(11)
Curtis, D. R., 379, 380(1)
Curtis, T. H., 445
Cuthbertson, G. J., 393, 396(4)
Czech, M. P., 66
Czernichow, P., 331, 332(20)

D

Dabney, L. G., 239, 258, 283, 287
Dacey, D., 392

Daikoku, S., 321, 325(h), 326, 328, 661
Dailey, R. A., 608
Daleskog, M., 479
Daly, J. W., 402(b, c, p, r, s), 403, 485
Damore, P. A., 593
Daniel, K. A., 296
Daniel, P. M., 611
Dann, O., 663
Danzin, C., 598
Darzynkiewicz, Z., 230, 234(4)
Daughaday, W. H., 609
Davies, C., 611
Davies, J., 381, 382(10), 384(10)
Davis, C. M., 562
Davis, J., 251, 258, 259, 264, 265(4), 267, 268, 278(h, i), 279, 280(c, d), 281, 285, 298
Davis, R., 376
Dayton, M. A., 479, 480
Dean, D. M., 306, 311(4)
Dean, P. N., 232
Debeljuk, L., 35(17), 39, 454
Deelder, A. M., 410
DeFerrari, M. E., 589
DeFesi, C. R., 258
DeGreef, W. J., 479, 608, 610(16)
de Gubareff, T., 383, 384(17), 389, 391(17)
Deguchi, T., 490
de la Torre, J. C., 621, 624
DeLean, A., 578, 580(14, 16), 581, 582, 583, 587
D'Eletto, R., 317
Delfs, J., 354
Delitala, G., 582
Deluca, H. F., 637
de Lucchi, M. R., 376
Demarest, K. T., 494, 617
DeMay, J., 198
de Monti, M., 279, 283
deMontigny, C., 383
Denef, C., 252, 258, 266, 278(k, l), 279, 280(h, i), 281, 282(h, i, j, k, l, m), 283, 284(f, g), 285, 295
Denizeau, F., 325(j), 326
Dennis, B. J., 376
Dent, R. R., 554
de Olmos, J., 396
Depasquale, J., 281
de Pasquier, P., 104
de Ray, C., 195

Deroto, P., 584
Deskin, R. B., 634
Desmazes, J. P., 103
Desnoyers, P., 195
Desy, L., 195
De Valois, R. L., 96
de Vitry, F., 323(b), 326, 331, 332, 333
de Vries, J., 639
Dewals, R., 252, 282(j, k), 283
DeWolf, A., 252, 258, 266, 280(h), 281, 22(i, k), 283, 284(f), 285
Diaconescu, C., 63, 64(16), 65(16)
Di Carlo, R., 617
Dichter, M. A., 104, 112, 138(5), 139(5), 141(5), 143(5), 144(5), 325(g), 326
Dickens, G., 593
Diekema, K. A., 593
Dierickx, K., 198
DiMeo, P., 636
Dingledine, R., 135
Dininger, N. B., 609
Diniz, C. R., 402(t), 403
Dion, A. S., 590(14, 15), 591
Dionne, V. E., 150, 165
DiPace, M., 325(f), 326, 329
Disorbo, D. M., 593
Dito, W. R., 412(30), 413
Djiane, J., 617
Djoseland, O., 639
Dlabac, A., 494
Dobbins, C., 594(4), 595
Docherty, J. J., 277
Dodd, J., 135
Dolinar, R., 468
Doran, A. D., 135
Dorling, J., 458
Douglas, C., 178
Douglas, R. J., 685
Douglas, W. H. J., 327, 329
Douglas, W. W., 93, 104, 110(36), 208
Dray, F., 410, 433, 434
Dreifuss, J. J., 132, 208, 319(h), 320, 350
Dreiling, R., 470, 500, 580
Dreyer, F., 125
Drouin, J., 578, 580(7), 581(7), 582(7)
Drujan, B. D., 367
Dua-Sharma, S., 376
Dube, D., 325(j), 326
Dubois, M. P., 453
Duchesne, M. C., 601

Dudek, F. E., 101
Dufau, M. L., 594(5), 595
Dufy, B., 95, 96, 99, 103, 104, 106, 107, 110(31, 32), 332
Dufy-Barbe, L., 95, 96, 99
duGubareff, T., 381
Dumont, O., 330
Duncan, J. A., 39, 40(21), 42(21, 24), 43, 46(21)
Dunlap, K., 128
Dunn, J., 35(17), 39, 454
du Pasquier, P., 104, 110(31, 32)
Dupont, A., 176
Durie, B. G. M., 590, 592(4), 601(4), 602(4)
Duval, J., 278(o, p), 279, 280(g), 281, 282(f, g), 283, 286
Dykstra, W. G., 590(19), 591

E

Eckel, R. W., 225
Edwards, C., 275, 281
Edwardson, J. A., 557
Eggena, P., 316(k), 317
Ehinger, B., 366
Ehlert, F. J., 402(i), 403
Ehrlich, P. H., 139
Eiden, L., 548
Eisenbarth, G. S., 459, 460, 468, 469
Eisenberg, K. B., 639(10), 640
Ekins, R. P., 278(n), 279, 284(e), 285
Elde, R., 680
Eldefrawi, M. E., 402(s), 403
Elkins, R., 411, 434(23)
Elliott, K. A. C., 395
Ellison, D. G., 258
Emmers, R., 376
Emson, P. C., 624(25), 630
End, D., 593
Endo, Y., 410
Engelhardt, D. L., 139
Engelman, K., 475
Engler, J. P., 640
Engvall, E., 409, 410, 412, 434
Enjalbert, A., 279, 283, 609
Epelbaum, J., 82, 391, 495
Eriksson, B.-M., 479
Eriksson, H., 650
Erlandsen, S. L., 625

Erlanger, B. F., 448
Esch, F., 79, 83, 85, 87, 176
Eskay, R. L., 589, 608
Essner, E., 194
Etreby, M. F. E., 671
Euvrard, C., 612
Evans, R. H., 381, 382(10), 384(10)
Evans, W. H., 251, 258, 259, 264, 265(4), 278(n), 279, 280(c), 281, 285, 298
Evans, W. S., 301, 304
Evarts, E. V., 96
Evenson, D. A., 393, 396(4)
Everett, J. W., 254, 295
Ewing, A. G., 479, 480, 481
Ezrin, C., 295, 296(8)

F

Fahimi, H. D., 192, 193, 195, 197
Fairhurst, A. S., 402(i), 403
Faivre-Bauman, A., 323(c), 326, 330, 331
Fajdiga, P. B., 305, 306(3)
Falat, L., 480
Falck, B., 609, 615(36), 619, 620, 621
Fanger, H., 196, 653
Farber, I., 113
Farnebo, L. O., 494
Farquhar, M. G., 259, 263, 294, 295, 298(3)
Farr, A. L., 485, 490(11), 499
Fasman, G. D., 451
Fathman, C. G., 232, 233(7)
Fauci, A., 460
Faulkner, C. S., 333
Faure-Bauman, A., 350
Fawcett, C. P., 249, 255(4), 279, 280(j), 281, 282(n), 283, 285
Fehlmann, M., 69
Feigin, R. D., 380, 381(8)
Fein, H., 121, 122
Fekete, M. I. K., 375, 630
Feldman, D., 641
Feldman, H. A., 691
Feldman, S. C., 315, 319(j, k), 321
Fellows, R. E., 593, 594(7, 10), 595
Femminger, A., 630
Fenimore, D. C., 562
Fenwick, E., 150, 154(2), 163(2), 171(2)
Fenwick, E. M., 114, 129(29), 305, 306(3)
Ferdinand, P., 317

Ferguson, W. J., 239
Ferin, M., 608, 609, 615(22, 23), 660, 661
Ferioli, M. E., 594(20), 595
Ferland, L., 612
Fermisco, J. R., 220
Fernandez, T., 234
Ferrua, B., 410, 412(18)
Field, P. M., 205
Fine, M. L., 654
Fink, G., 608, 659
Fink, R. P., 396
Finley, J. C. W., 457
Fischbach, G. D., 112, 114, 115, 124(32), 125(32), 126, 128, 138(7), 139(7), 141(7), 143(7)
Fischli, W., 198, 360, 671
Fishman, M. C., 113, 144(25)
Fitch, F. W., 639(13), 640
Fitzgerald, S., 113, 114(26), 144(26)
Flaming, D. G., 122, 123, 137
Flerko, B., 177
Fletcher, W. H., 254, 295
Fleury, H., 104, 110(31, 32), 332
Flier, J. S., 49
Flouret, G., 35(19), 39
Fombariet, C. M., 480
Fong, W. F., 602
Fontham, E. H., 609
Ford, J. D., 450
Foreman, J., 49
Fothergill, L. A., 670
Fox, T. O., 325(f), 326, 329
Francis, A. A., 381, 382(10), 384(10)
Frank, R. L., 442
Frankland, B. Y. B., 531, 534
Frantz, A. G., 608, 609, 615(22, 23), 660, 661
Fraser, H. M., 36
Frawley, L. S., 610
Frazier, G. R., 691
Frazier, R. P., 593
Frears, C. C., 577
Freed, C. R., 502
Freeman, R., 470
Freifeld, A. G., 593
French, F. S., 639
Frenk, S., 195
Freychet, P., 69
Fridkin, M., 453, 456(23)
Friedman, D. W., 484, 489(8), 492(8)
Friedman, S., 285

Friesen, H. G., 617
Frohman, L. A., 582
Fromageot, P., 535
Fujii, D. K., 288
Fukuda, M. S., 659
Fukushima, M., 555
Fuller, J. L., 601, 602(50)
Fuller, R. W., 594(15), 595
Fuller, T. A., 384, 385, 392(20)
Furshpan, E. J., 115, 116, 342
Furui, T., 555
Furutani, Y., 548, 670
Fuxe, K., 384, 385(22), 396, 609, 613, 615(34), 671

G

Gaddum, J. H., 178
Gadzer, A., 460
Gähwiler, B. H., 319(h), 320
Gaetjens, E., 639(10), 640
Gage, L. P., 548, 670
Gagne, B., 578, 580(7), 581(7), 582(7)
Gainer, H., 200, 525, 528, 533, 537(5), 539(11)
Gaines, R. A., 277
Gallardo, E., 495
Gallyas, F., 396, 397, 680
Gamse, G., 325(f), 326, 329, 356
Gamse, R., 325(f), 326, 329, 356
Gangloff, H., 375(17), 376
Gannon, J. G., 582
Ganong, W. F., 493
Gard, T. G., 278(m, n), 279, 284(d, e), 285
Garvey, J. S., 73, 455
Gauchy, C., 475
Gautvik, K. M., 104, 110(35)
Gazis, D., 525
Geddes, L. A., 123
Geller, H. M., 319(c, d), 320
Genazzani, A. R., 660, 661(60)
Gentry, C., 400
George, J. M., 135
Georgescauld, G., 103
Gerber, L., 77
Gerlach, J. L., 319(g), 320, 328
Gerstein, G. L., 138
Ghazarossian, V. E., 554
Giannasthasio, G., 589

Gibbs, D. M., 608, 610
Gibson, C. J., 323(a), 326
Gibson, K., 594(14), 595
Gilbert, R., 624(25), 630
Gilioz, P., 178
Gillies, G., 296, 305(14)
Gillioz, P., 608
Giloh, H., 219
Ginsberg, H., 492
Giorguieff, F., 495
Giranda, V., 277
Giraud, P., 178, 608
Girod, C., 295
Gitelman, H. J., 594(4), 595
Glaser, T., 333
Glazer, A. N., 239
Gliemann, J., 65
Globus, A., 124(60), 125
Glover, J. S., 34
Glowinski, J., 475, 495
Godwin-Austen, R. B., 577
Gold, R. M., 391
Goldberg, L. I., 577
Goldner, M. M., 128
Goldstein, A., 198, 360, 554, 671, 686, 687(32)
Goldstein, G., 232, 233(11)
Goldstein, M., 384, 385(22), 391(23), 396, 624(24, 25), 630
Gollapudi, G. M., 281
Goltzman, D., 72, 77(2)
Gonatas, N. K., 208, 212
Gonon, F. G., 480
Good, C. A., 232
Goodfriend, T. L., 451
Goodman, R., 314
Goodrich, G. R., 409, 414(6)
Gorcs, T., 680
Gosbee, J. L., 258
Gospodarowicz, D., 288, 289, 349
Gottlieb, D., 112
Gottschall, P. E., 617
Gourdji, D., 103, 104, 110(31, 32)
Graf, M., 555
Graham, R. C., 191, 192
Grant, E., 242
Grant, G., 48, 178, 249, 287, 296, 298(11), 566, 691
Grant, L. D., 639
Grant, M., 259

Greeley, G. H., 659
Greenberger, L. M., 191
Greenburg, G., 289, 295
Greene, G. L., 639(13), 640
Greenleaf, P. W., 317
Greenwood, F. C., 34, 62, 142, 146
Greuselle, D., 350
Griesbach, W. E., 296
Griesser, G.-H., 306, 311(7)
Griffiths, E. C., 540
Grimm-Jorgensen, Y., 333
Grindeland, R. E., 275, 281
Grinvald, A., 113
Griswold, E., 251, 258, 259, 264, 265(4), 278(h), 279, 280(c), 281, 285, 298
Groopman, J. D., 409, 414(6)
Gropp, N., 555
Gros, C., 410
Gross, G. W., 113
Grossman, A., 582
Grossman, L., 392
Grossman, S. P., 392
Grouselle, D., 330
Gruber, K. A., 507, 527, 528, 531, 532, 533, 534, 537(12)
Grumbach, M. M., 36
Grunberg, D., 409, 413(1), 414(1)
Gruol, D. L., 112, 125, 138(6), 139(6), 141(6), 143(6)
Gruss, R., 594(31), 595
Grzanna, R., 625
Gubler, U., 548, 670
Gudelsky, G. A., 610, 612, 613, 614, 615, 617(47)
Guesdon, J. L., 411, 414, 681
Guillemin, R., 33, 75, 79, 82, 83, 85, 87, 89, 176, 178, 198, 249, 259, 296, 298(11), 540, 563, 566, 608, 680
Gunn, A., 36
Guroff, G., 593, 594(9, 41), 595, 596
Guthrie, P. B., 113, 115(22)
Gutman, Y., 402(h), 403
Gyevaï, A., 327, 329

H

Haas, Y., 42(29), 43, 47(29)
Habeeb, A. F. S. A., 450
Habener, J. F., 524

Hackett, J. T., 252
Haddox, M. K., 594(21), 595
Hagenas, L., 639
Hagiwara, S., 102
Hagman, J., 62, 63(15), 65(15)
Halaris, A., 392
Hall, C., 327
Hall, M., 277, 281
Halpern, M., 641
Haluszczak, C., 655
Hamberger, B., 494
Hamblin, M., 584, 585(26)
Hamby, M., 242
Hamill, O. P., 102, 114, 129(28), 150
Hammel, L., 492
Hammer, C. T., 609
Hammond, J. M., 593
Hamon, M., 333
Hamprecht, B., 333
Hancock, A. A., 582, 583(18)
Handwerger, S., 594(29), 595
Hanker, J. S., 143, 191
Hanna, N., 285, 287
Hansen, J. J., 385
Hansen, P., 232, 233(11)
Hansen, S., 384, 391
Hansson, V., 639
Hardin, J. W., 650
Hardy, P. M., 450
Hardy, R. R., 227, 238(3)
Harlan, R. E., 649, 659(31), 660, 661(58)
Harrington, C. A., 562
Harris, C. C., 409, 413(1), 414, 434
Harris, G. W., 32
Harris, M. C., 98, 208
Harris, P., 594(14), 595
Hartman, B. K., 179, 626, 629, 663, 667, 681
Hartzell, H. C., 107
Haselbacher, G. K., 593
Hashimoto, T., 454
Haskill, J. S., 258, 280(b), 281
Haskins, J. T., 614
Hatanaka H., 593
Hatfield, J. M., 257, 258, 277, 286
Hathaway, N. R., 582
Hatton, G. T., 101
Haugen, A., 409, 414(6)
Haugland, R., 239
Hautekeete, E., 252, 258, 266, 280(h), 281, 282(h, i, j, k), 283, 284(f), 285

Havran, W. L., 234
Hawg, E., 104, 110(35)
Hawkes, A. G., 165
Hawrot, E., 335, 346
Haynes, B., 460, 469
Haynes, L. W., 321
Hayward, J. N., 96, 132, 133, 137, 138, 142, 143, 145, 146
Hazum, E., 47, 49, 59, 60, 61, 63, 64, 66, 67(17), 68, 69(23), 70, 71, 219, 220, 222, 224, 225(9)
Hearn, W. R., 249
Heber, D., 525
Hebert, N. C., 123
Hechtman, H. B., 593
Hedge, G. A., 181, 594
Hedlund, M. T., 36, 38(10), 44(10), 452
Heimer, L., 396
Helle, K. B., 629
Heller, J. S., 602
Hellmann, P., 258, 283, 287
Hendelman, W. J., 116
Hendry, I. A., 335, 594(42), 596
Henkart, M., 112, 115, 129
Henrick, S. K., 594(15), 595
Henriksen, S. J., 686, 687(32)
Henry, D., 475, 484, 485, 610
Hensley, L. L., 460
Herbert, E., 548
Herbst, E. J., 590(14, 15, 19), 591
Herman, J. P., 375, 630
Hernaez, L., 49
Heron, D. S., 617
Hersh, L. B., 540
Hershkowitz, M., 617, 618
Herz, A., 562
Herz, R., 305
Herzenberg, L. A., 227, 232, 233, 238(3), 239
Herzog, W. H., 402(a), 403
Higgins, D. H., 343, 345(29)
Hild, W., 313
Hill, C. E., 335
Hillarp, N. A., 609, 619, 621
Hillhouse, E. W., 495
Hiramato, R., 450
Hirooka, Y., 317
Hirose, T., 548, 670
Hirose, Y., 388
Hirs, C. H. W., 517

Hisano, S., 661
Hjemdahl, P., 479
Hjorth-Simonsen, A., 400
Hlivyak, L. E., 183
Ho, H., 602
Ho, O. L., 380, 381(6), 384, 385(6), 393, 395(2)
Hodos, W., 375
Hökfelt, T., 384, 385(22), 396, 609, 613, 615(34), 621, 624(24), 630, 663, 671, 680
Hoffer, B. J., 121, 125(50)
Hoffman, B. B., 589
Hoffman, B. J., 548, 670
Hoffman, G., 323(d), 326, 329
Hoffman, R. A., 232, 233(11)
Hoffner, N., 590
Hogan, B., 593
Holborow, E. J., 458
Holland, D. T., 36
Hollander, C. S., 317
Hollenberg, M. D., 531, 534
Holley, R. W., 253
Hollt, V., 562
Holtta, E., 602
Holtzman, E., 212
Honchar, M. P., 179
Honore, T., 385
Hood, L., 360, 671
Hope, D. B., 531, 534
Hope, J., 555
Hopkins, C. R., 255, 259, 294, 295, 298(3)
Horii, K., 555
Horn, R., 165, 174
Hornsperger, J. M., 602
Horvath, C., 469
Horvath, E., 295, 296(8)
Hostetter, G., 196, 206(39)
Houser, C. R., 625, 681
Housholder, D. E., 249
Hou-Yu, A., 139
Hou-Yu, E. A., 133
Howard, A. N., 452
Howards, S. S., 611(57), 612
Howe, N. B. S., 305, 306(3)
Howe, S. C., 462
Howell, S. L., 277, 281
Hromek, F., 609, 615(36)
Hsu, I. C., 409, 413(1), 414, 434
Hsu, S. M., 196, 636, 653

Huang, L. Y. M., 112, 125, 138(6), 139(6), 141(6), 143(6)
Hübner, K., 333
Huellmantel, A. B., 655
Huffman, R. D., 376
Hugenard, J., 594(18), 595
Huggins, C. G., 454
Hughes, J., 670
Huldt, G., 409, 410(7)
Humbel, R. E., 593
Humm, J., 613
Hunkapillar, M., 360
Hunkapiller, M., 671
Hunt, S. P., 624(25), 630
Hunter, W. M., 34, 62, 456
Huof, S., 471
Hurley, T. W., 594(29), 595
Hurn, B. A. L., 456, 548, 549(7)
Husain, M. K., 609
Hyde, C. L., 258, 279, 281, 283, 285
Hymer, W. C., 251, 257, 258, 259, 263, 264, 265, 267, 268, 269, 272, 275, 277, 278(h, i, j), 279, 280(b, c, d, e), 281, 285, 286, 298

I

Iacovitti, L., 335, 343, 345(29), 346, 347(14)
Ibata, Y., 540
Ichihara, A., 601
Ickeson, I., 594(31, 49), 595, 596
Ikeno, T., 594(9), 595
Iliev, V., 590(22), 591
Imura, H., 555, 609
Inayama, S., 548, 670
Ingram, W. R., 396
Inman, D. M., 323(a), 326
Inoue, A., 547, 670
Inoue, H., 594(37), 595
Ishida, H., 593
Ishihara, K., 388
Ishikawa, S. E., 316(i, j), 317
Israeli, M., 618
Iversen, L. L., 178
Iversen, S. D., 178
Iwasaki, J., 609
Iwasaki, Y., 609
Izumi, S., 196, 199(38)

J

Joanus, S. D., 691
Jackson, G. L., 659
Jackson, R., 469
Jackson, R. A., 459
Jackson, M. B., 113, 114(26), 144(26)
Jacobi, J., 611
Jacobs, H. L., 376
Jacobs, L. S., 609
Jacobs, S., 49, 60, 63(8), 66
Jacobson, S., 199, 206(54)
Jacoby, W. B., 112
Jaffe, B. M., 449
Jaffe, R. B., 293
Jameson, H. E., 541
Jamieson, M. G., 608
Janne, J., 590, 591, 592, 601, 602
Jansen, M., 590
Jaquet, P., 178, 608
Jaretzki, A., 409
Jarrett, D. B., 49
Jeffcoate, S. L., 36
Jeffcoate, W. J., 555
Jefferson, L. S., 593
Jenkins, J. S., 611(56), 612
Jensen, E. V., 639(13), 640
Jessell, T. M., 113, 114(21), 115(21)
Jessup, P. J., 402(q, r, s), 403
Jiang, N. S., 501, 502(30)
Jimenez, A., 484, 489(8), 492(8)
Jirge, S. K., 675
Jirikowski, G., 325(k, l), 327
Jörnvall, H., 83
Joh, T. H., 335, 347(14), 616, 624, 669
Johansson, O., 630
Johnson, A. B., 319(j), 321
Johnson, G. D., 458
Johnson, J., 494
Johnson, L. K., 295
Johnson, M., 335, 346
Johnson, M. I., 335, 340, 342, 343, 345(29), 346, 347
Johnston, G. A. R., 379
Johnston, H. W., 442
Jones, A. W., 381, 382(10), 384(10)
Jones, E. G., 663, 667
Jones, J. C., 607
Jones, M. K., 593
Jones, M. T., 495
Jones, R. H., 66
Jonsson, G., 384, 385(22), 396
Joseph, S., 316(l), 317, 388
Joshikawa, T., 375, 376(11)
Jouvet, M., 479
Jovin, T. M., 226
Joynt, R. J., 316(f, g), 317
Juan, S. I., 317
Judd, A. M., 255
Julius, M. H., 232, 233(7)
Jung, M. J., 598
Jutisz, M., 296

K

Kabat, E. A., 456, 559
Kaczorowski, G. J., 104, 110(37)
Kagawa, T., 249
Kageyama, N., 555
Kahn, C. R., 49
Kaiser, D. L., 304
Kakidani, H., 548, 670
Kallio, A., 602
Kallstrom, Y., 127
Kamberi, I. A., 611
Kamentsky, J. A., 243
Kanagawa, K., 671
Kanan, T., 479
Kandel, E. R., 98
Kaneko, A., 367
Kanyicska, B., 375, 630
Kao, L. W., 495
Kaplan, S. L., 36
Karavolas, H. J., 282(e), 283
Karnovsky, M. J., 191, 192
Karten, H. J., 375
Kastin, A. J., 33, 39(2)
Kasting, N. W., 179, 180, 182, 185(17), 186(17)
Katakami, H., 609
Kato, T., 471
Kato, Y., 594(37), 595, 609
Katz, G. M., 104, 110(37)
Kauffman, F. C., 402(p), 403
Kaur, C., 614
Kawamura, A., 671
Kawano, H., 325(h), 326, 661
Kay, J. E., 602

AUTHOR INDEX

Kaye, A. M., 594(31), 595
Kayser, B. F. J., 132
Kearney, J. F., 462
Kebabian, J. W., 577, 589
Keefer, D. A., 239, 258, 283, 287, 626, 654
Kehr, W., 615
Keinan, D., 66, 68, 69(23), 70(33), 71
Keiner, K. L., 639
Keiner, M., 640, 648(19), 655(19)
Kellerth, J.-O., 144
Kelly, J. A., 540
Kelly, J. E. T., 281
Kelly, J. S., 124(62), 125
Kelly, M. J., 101
Kelly, P. A., 42(23), 43, 617
Kenez-Keri, M., 66
Kennedy, J. H., 449, 451(7)
Kenney, F. T., 594(27), 595
Kent, R. S., 582
Kerdelhue, B., 196, 206(39), 296
Keri, G., 66
Keutmann, H. T., 181
Khachaturian, H., 387, 388(30), 389(30)
Kidokoro, Y., 104
Kiefer, H., 59
Kilian, A., 222
Kilpatrick, B. F., 578, 580(14), 581(14), 582, 583, 587, 589(29, 30, 31)
Kilpatrick, D. L., 306, 309(5), 310(5), 311(5), 312(5)
Kim, S. W., 419
Kimmel, J. R., 624(25), 630
Kimoto, Y., 630
Kimura, N., 104, 110(33)
Kimura, S., 77
King, A. C., 49
King, J., 127, 662
King, T. P., 22
Kinoshita, H., 540
Kinugasa, T., 630
Kirchich, H. J., 639
Kirkland, J. J., 470
Kirshner, A. G., 306, 309(5), 310(5), 311(5), 312(5)
Kirshner, N., 306, 309, 310(5), 311(5, 10), 312(5, 10)
Kissinger, P. T., 470, 500
Kita, T., 547, 670
Kizer, J. S., 388, 438, 439(5), 440, 446(6), 447, 613

Klandorf, H., 594
Klee, M. R., 128
Klein, M. J., 198
Kleinschmidt, D. C., 343, 345(29)
Kling, H., 286
Klingensmith, M. R., 593
Klippel, R. A., 375
Knigge, K. M., 133, 199, 316(e), 317, 323(d), 326, 329, 495, 500(20), 526
Knisatschek, H., 546, 547(17)
Knobil, E., 95, 389, 390(41)
Knowles, D. M., 409
Knowles, J. R., 58, 63, 450
Ko, C.-P., 335
Kobayashi, R. M., 375
Kobayashi, T., 249
Kobayashi, Y., 594(30, 40), 595, 596
Koch, Y., 42, 43, 47, 68, 70(33), 453, 456, 608
Kochwa, S., 419
Koda, L. Y., 675
Koestner, A., 135
Kohler, C., 384, 391(23)
Kohler, G., 462
Kohler, P. O., 691
Kok, H. W. L., 577
Koketsu, K., 103
Komatsu, N., 196, 199(38)
Konig, J. F., 375
Konig, W., 36, 42, 43(13)
Konishi, T., 555
Koob, G., 565
Kopin, I. J., 488
Kopin, T. J., 494
Kopriwa, B. M., 617
Kordon, C., 278(o), 279, 280(g), 281, 282(f), 283, 286
Koshland, D. E., Jr., 451
Kosterlitz, H. W., 670
Kostyuk, P. G., 129
Kovacs, K., 295, 296(8)
Kovathana, N., 582
Kozlowski, G. P., 195, 196, 206(39)
Kraicer, J., 251, 258, 259, 264, 265(4), 278(h), 279, 280(b, c, e), 281, 285, 298
Krause, J. E., 512, 519, 541, 542, 543(13, 14), 545(15), 547
Kreike, A. J., 208
Kremzner, L. T., 591
Krey, L. C., 199

Kricka, L. J., 449, 451(7)
Krieger, D., 325(i), 326, 329, 388, 389(38), 512
Krieger, M. S., 644(30), 645, 646(29), 647(30), 648(29), 650(29, 30), 655(30)
Krishtal, O. A., 129
Krnjevic, K., 124(59), 125
Krogsgaard-Larsen, P., 385
Krulich, L., 493, 659
Kuba, K., 103
Kubli-Garfias, C., 608
Kubo, S., 540
Kuffler, S. W., 107
Kugel, G., 626
Kuhnt, U., 101
Kumar, S., 350
Kumasaka, T., 35(17), 39, 454
Kung, P. C., 232, 233(11)
Kunze, M. E., 277
Kupelian, J., 594(30), 595
Kurachi, K., 630
Kuwayama, A., 555
Kuypers, H. G. J. M., 663, 664, 665, 666
Kyriakidis, D. A., 602

L

Labrie, F., 42(23), 43, 195, 251, 325(j), 326, 578, 580(7), 581(7), 612
Labruyere, J., 381, 385(11), 386(11), 387(11), 389(11)
Laemmli, U. K., 67
Lagace, L., 251
Lagowska, J., 384, 392(21)
Lai, M. F., 294
Lakshmanan, J., 594(41), 596
Lam, D. M. K., 367
Lam, N. Y., 661
Lamperti, A., 388
Lamprecht, S. A., 594(31, 32), 595
Landis, S. C., 115
Lando, D., 594(11), 595
Lange, K., 165
Langemann, H., 609, 615(35)
Lanza, G., 639(11), 640
Larsson, L.-I., 457
Lau, C., 590(26), 591, 592, 594(16, 17), 595, 598, 599, 600(37, 40, 44, 45), 601(37, 40), 602(37, 40, 44, 45), 603(29, 30)

Lauber, M., 535, 537(21)
Lavallee, M., 123
Leber, W., 128
Lebovitz, H. E., 593
Lecar, H., 113, 114(26), 125, 144(26)
Lechan, R. M., 199, 206(54), 626
Le Dafnit, M., 278(p), 279, 282(g), 283
Ledbetter, F. H., 306, 309(5), 310(5), 311(5), 312(5)
Lee, J. C., 376
Lee, K. S., 102, 129
Lee, R. G., 419
Lee, R. W. H., 306, 311(6)
Lee, Y. C., 333
Leeman, S. E., 323(e), 325(f, g), 326, 329
Lefebvre F.-A., 42(23), 43
Lefevres, G., 195
Leff, S., 584, 585(26)
Lefkowitz, R. J., 578, 581(7), 582, 583, 588, 589
LeGal LaSalle, G., 384, 392(21)
Legendre, P., 327, 329
Lehmeyer, J. E., 609, 611
Leibowitz, M. D., 150, 165
Leichnetz, G. R., 654
Leif, J., 594(40), 596
Leifer, A. M., 258, 282(o), 283
Leighton, J., 221
Leisegang, B., 462
Lemay, A., 325(j), 326
Lenox, R. H., 564
Lerea, L., 590(27, 28), 591, 594(25), 595, 599(27, 28)
Leuschen, M. P., 258, 284(b, c), 285
Levine, J. E., 186, 540
Levine, J. H., 594(19), 595
Levine, L., 333, 451
Levintow, L., 366
Levy, D., 60
Lewis, D., 239, 258, 283, 287
Lewis, R. L., 560
Lewis, R. V., 77, 557
Libber, M. T., 133, 142, 143, 147
Licht, C., 532
Lichtensteiger, W., 609, 615(35)
Liddle, G. W., 594
Lieberman, M., 112
Lightbody, J., 333
Likhite, V., 449
Lim, L., 327

Lim, R. K. S., 376
Limbird, L. E., 589
Lin, Y. C., 593
Lincoln, D. W., 99
Lindner, H. R., 68, 70(33), 453, 456(23), 594(31, 32, 49), 595, 596
Lindqvist, M., 615
Lindsay, V. L., 602
Lindstrom, S., 127
Lindvall, O., 620
Ling, N., 33, 75, 79, 82, 83, 85, 87, 89, 176, 555, 563, 608, 680
Linnenbach, H., 469
Linser, P., 366
Linthicum, G. L., 176
Liotta, A. S., 325(i), 326, 329, 556
Lipkowski, A., 35(19), 39
Lippert, B. J., 593
Lipsett, M. B., 35(14), 36
Litwack, G., 593
Liu, C., 376
Livett, B. G., 305, 306, 311(4)
Ljungdahl, A., 630, 680
Ljungstrom, I., 409
Lloyd, H. M., 611
Lloyd, R. V., 258, 280(d), 281, 282(c, d, e), 283
Locatelli, V., 660, 661(60), 671
Loeb, B. E., 113, 138(17), 143(17)
Loeber, J. G., 548, 557(8)
Loffeholz, K., 306, 311(7)
Loken, M. R., 227, 232, 233(10), 238(2), 239
Lombardini, J. B., 590(18), 591
Long, J. A., 336
Loose, D. S., 641
Loren, I., 620
Lorenzen, J. R., 258
Loring, J. M., 593
Lostra, F., 198
Loudes, C., 325(i), 326, 329, 330, 331
Loughlin, J. S., 305
Lovelady, H. G., 469
Lowney, L. I., 671
Lowry, O. H., 383, 485, 490, 499
Lowry, P. J., 252, 296, 298, 300, 305(14), 553, 555
Lu, K. H., 611
Lumeng, L., 593
Lund, J. P., 383
Lundberg, J. M., 624, 630

Lundgren, P., 275, 281
Lungdahl, A., 621
Luparello, T. J., 375(16), 376
Lux, H. D., 150
Lynch, G. S., 101

M

McBurney, R. N., 125
McCaman, M., 484, 489, 492
McCaman, R., 484, 489(7), 492(7)
McCann, F. V., 334
McCann, P. P., 593, 601, 602
McCann, S. M., 249, 255(4), 279, 280(j), 281, 282(n), 283, 285, 493, 495, 541, 659
McCarroll, M. C., 550
Macchia, R. J., 639(10), 640
MacDermott, A. B., 108
McDermott, J. R., 557
MacDonald, D. D., 470
MacDonald, J. F., 112, 125, 138(6), 139(6), 141(6), 143(6)
Macdonald, R. L., 112, 113, 125, 128(19), 138(19), 139(12, 19), 141(12), 143(12), 144(12)
MacDonnell, P. C., 594(41), 596
McEwen, B. S., 319(g), 320, 328, 639
McGarrigle, R., 120
McGeer, E., 381
McGeer, P., 381
McGinty, J. F., 686, 687(32)
McGuire, W. L., 639(12), 640, 650(12)
McIlwain, H., 134, 135(13)
McKeel, D. W., Jr., 258, 279, 280(k), 281, 283, 285
McKelvy, J. F., 316(l), 317, 325(i), 326, 329, 333, 512, 519, 540, 541, 542, 543(13, 14), 545(15), 547
MacKenna, D. G., 125
McLanahan, C. S., 608
McLean, I. W., 626, 651
McLean, W. S., 639
MacLeish, P. R., 115, 342
McLennan, L. S., 335
MacLoed, R. M., 252, 609, 611, 612
McLoughlin, L., 555
MacLusky, N. J., 639
McMartin, C., 298, 300(16)
McMillian, M., 242

McNay, J. L., 577
McNeil, R., 53, 54(10), 55(10), 56(10)
MacNichol, E. F., 97
McShan, W. H., 258, 280(d), 281, 282(c, d), 283
MacVicar, B. A., 101
Madhubala, R., 594(45), 596
Maggio, E. T., 412
Mahaffee, D., 594(4), 595
Mahan, L. C., 582(27), 585, 586(27), 587(27)
Mahesh, V. B., 659
Maiglini, R., 412
Maiolini, R., 410, 412(18)
Makara, G. B., 98
Maki, Y., 321, 328
Makimura, H., 388
Malamed, S., 316(c, d), 317
Maleque, M. A., 402(p), 403
Malmgren, L., 192
Mamont, P. S., 601
Manberg, P. J., 438, 439, 440
Manen, C. A., 590
Manocha, S. L., 376
Maranto, A. R., 127, 143
Marchak, D., 366
Marchalonis, J. J., 62
Marchisio, A. M., 584
Marian, J., 42, 43, 49, 242, 402(n), 403, 404
Mark, R. F., 113
Markey, K., 624(24), 630
Marks, B. H., 135
Marks, G. M., 277
Maron, E., 434
Marshall, J. C., 35, 36, 37, 39, 40(21), 42(21, 24), 43, 46(21)
Marshall, K. C., 113, 116
Marson, A. M., 321
Martes, M.-P., 578
Martin, J. B., 82, 176, 179, 180(17), 181, 182(17), 185(17), 186(17), 391
Martin, J. E., 251, 258, 279, 280(k), 281, 283, 285, 362
Martin, R., 286
Martin, W. H., 611(57), 612
Marty, A., 102, 114, 129(28, 29), 150, 154(2), 163(2)
Mason, T. E., 634
Mason, W. T., 315, 321, 328(5)
Massague, J., 66
Masse, M. F. O., 535, 537(21)

Masseye, F. R., 412
Masseyeff, R., 410, 412(18)
Massicotte, J., 251
Mastro, A., 251, 258, 259, 264, 265(4), 278(h), 279, 280(c), 281, 285, 298
Masurovsky, E. B., 319(a), 320
Mather, J. P., 293
Mathers, D. A., 107
Matsubara, S., 555
Matsumura, H., 325(h), 326
Matsuo, H., 540, 608, 671
Matsushita, N., 609
Matsushita, S., 594(13), 595
Matsuzaki, S., 594(46), 596
Maudsley, D. V., 591(33), 592, 594(30, 40), 595, 596
Maurer, L. H., 333
Mauricio, J. C., 611(58), 612
Mayer, C. J., 402(j), 403
Meares, J. D., 611
Mecklenburg, R., 35(14), 36
Medina, V. J., 591(32), 592, 594(26), 595, 601(32)
Meek, J. L., 19, 557
Mefford, I. N., 471
Mehrishi, J. N., 277
Meidan, R., 42, 43, 47, 68, 70(33)
Meiser, A., 350
Meissner, H. P., 241
Meites, J., 611, 613, 617
Melamed, M. R., 230, 234(4)
Meloni, M., 593
Mercer, W. D., 639(12), 640, 650(12)
Merchenthaler, I., 680
Merrifield, R. B., 4, 5, 7, 8, 10
Mesher, H. L., 288
Messer, A., 325(g), 326
Mesulam, M.-M., 191, 192
Metcalf, B. W., 598
Metz, C. B., 191
Meyer, E. R., 389
Meyers, M., 335, 347
Meyrs, R. D., 179
Mezey, E., 630
Mical, R. S., 608, 609, 610, 611, 612(42)
Michel, T., 589
Michelot, R. J., 469
Midgley, Jr., A. R., 691
Milito, R. P., 277
Miller, R. J., 198, 221, 225, 357, 632

Milner, R. J., 362
Milstein, C., 459, 462
Minamino, N., 671
Minit, Y., 402(p), 403
Mishell, B. B., 292
Mitchell, J. L. A., 600, 601, 602
Mithen, F. A., 340
Miyachi, Y., 35, 36
Miyazaki, S., 104, 110(34)
Mizuno, M., 249
Moffitt, R., 376
Moguilewsky, M., 594(11), 595
Molitch, M. E., 183
Moller, M., 192
Molliver, M. E., 625
Molnar, I., 469
Moltring, J., 535, 537(21)
Monnet, F., 290
Monnier, M., 375(17), 376
Monzain, R., 319(b), 320
Moody, A. J., 65
Moore, G. P., 138
Moore, K. E., 494, 613, 617
Moore, M., 590
Moore, R. Y., 395, 619, 621, 630
Moray, L. J., 440, 446(6)
Morel, G., 640
Morgan, B. A., 670
Morgenson, G. J., 659
Moriarity, D. M., 593
Moriarty, C. M., 258, 284(b, c), 285
Morimoto, Y., 548, 670
Morin, R. D., 402(f), 403
Morita, K., 103
Morley, C. G. D., 594(28), 595, 602
Morrell, J. I., 191, 639, 640, 641, 644, 645, 646(29), 647, 648, 649, 650(29, 30), 654(8), 655, 656(42), 657(43), 658(43), 659, 660, 661, 662
Morris, G., 594(12), 595, 599, 600(45), 602(45)
Morris, J. F., 208
Morris, J. L., 611(57), 612
Morris, M., 494, 495, 496(15), 497, 498(14), 499(14), 500, 501, 503, 504(12), 506, 507, 509(12, 14, 15), 526, 527, 528, 531, 532, 533, 534, 535, 537, 538, 539(7)
Mortel, R., 277
Moruzzi, M. S., 590(17), 591
Moschera, J., 79

Moscona, A. A., 366
Moss, R. L., 614, 659
Moudgal, N. R., 594(33), 595
Moule, M. L., 60, 62, 63(9, 14), 66, 67(14)
Mount, C. D., 334
Moya, F., 343, 345(29)
Moyer, T. P., 501, 502(30)
Muccioli, G., 617
Mudge, A. W., 114
Mueller, G. P., 613
Mueller, J. F., 594(13), 595
Muhlethaler, M., 132
Mulder, E., 639
Muldoon, T. G., 659
Mullaney, P. F., 232
Muller, E. E., 660, 661(60)
Muller, T., 310
Mullikin, D., 583
Munemura, M., 589
Munn, A. R., 637
Munson, P. J., 39, 40(21), 42(21), 43(21), 46(21), 582, 583
Murad, F., 252, 296
Muramoto, K., 60, 62, 63(15), 65(15), 66
Murphy, B. J., 597
Murray, M. R., 319(a), 320
Mutt, V., 83
Myers, R. D., 179

N

Nadler, J. V., 393, 395, 396, 400
Nagaiah, K., 594(41), 596
Nagatsu, T., 471
Nagy, G., 480
Nagy, K., 150
Nair, R. M. G., 608
Nairn, R. C., 238
Nakagawa, S. H., 35(19), 39
Nakai, Y., 555
Nakamura, M., 547, 670
Nakamura, R. M., 412(30), 413
Nakane, P. K., 194, 260, 296, 626, 651, 691
Nakanishi, S., 547, 548, 670
Nakano, H., 249, 255
Nakao, K., 555
Nakayama, J., 409
Nansel, D. D., 610, 612, 617(47)

Naor, Z., 258, 279, 280(j), 281, 282(n, o), 283, 285
Naot, Y., 492
Naquet, R., 395
Nastuk, W. L., 112, 146(4)
Natale, P. J., 243
Nawata, H., 594(22), 595
Nayfeh, S. N., 639
Neale, E. A., 113, 114, 128(19), 129, 138(19) 139(19)
Neff, N. H., 494
Negoro, H., 655
Negro-Vilar, A., 495, 541
Neher, E., 102, 114, 129(28, 29), 150, 154(2), 163(2), 171(2)
Neidel, J., 56
Neil, J., 168
Neill, J. D., 479, 608, 610
Nelson, P. G., 112, 113, 114, 115, 116, 119(11), 124(30), 125(12), 128(19), 129, 138(7, 19), 139(7, 11, 12, 19), 141(7, 12), 143(7, 12), 144(12, 25, 26)
Nemeroff, C., 390
Nemeskeri, A., 323(c), 326
Nemeth, E., 640
Nenci, I., 639(11), 640
Nequin, L. G., 541
Nestler, J. L., 199, 206(54)
Nett, T. M., 35(18), 36, 37, 38(10), 39, 44, 452, 691
Ney, R. L., 594(4), 595
Ngai, S. H., 494
Nicholls, A. C., 450
Nichols, C. W., 296
Nicholsen, G., 388, 389(38)
Nicholson, D. M., 315, 321, 328(5)
Nicholson, W. E., 334, 594(19), 595
Nicolas, P., 535, 537(21)
Niemer, W. T., 376
Nieschlag, E., 456, 549
Nikodejevic, B., 485
Nilaver, G., 198
Nilssen, G., 680
Nimitkitpaisan, Y., 402(r), 403
Nimrod, A., 63, 66(18)
Nioaver, G., 139
Nirenberg, M., 402(v), 403, 460
Nishanishi, S., 548
Nishi, N., 176
Nishitani, H., 555

Nishiyama, T., 661
Nisonoff, A., 452
Niswender, G. D., 36, 38(10), 44(10), 452, 691
Noda, M., 548, 670
Nolan, C., 640
Nomura, J., 343
Nordmann, J. J., 208
Norenberg, M. D., 366
Noritake, D., 305, 306(2), 310(2)
Norris, J. C., 129
Notter, M. F. D., 277
Nouvelot, A., 330
Novikoff, A. B., 189
Nowak, L. M., 125
Numa, S., 547, 548, 670
Numann, R., 150
Nunes, M. C., 611(58), 612
Nunez, E., 640

O

Obenrader, M. F., 601
O'Brian, D. S., 640
O'Brien, R. J., 115
Odell, W. D., 35, 36, 37, 525
Ogden, T. E., 123
Ohmori, H., 102
Ohtaki, S., 410
Oi, V. T., 239
Oie, H., 460
Ojeda, S. R., 42, 43, 493, 495, 541
Oki, S., 555
Okon, E., 68, 70(33)
Okun, L. M., 113, 115(23), 138(17), 143(17)
O'Lague, P. H., 115, 116, 342
Olds, J., 97, 98(10)
Oliver, C., 178, 188(8, 71), 189, 192(8), 193(6), 194(5), 196(5), 200(5), 205(6), 206(6), 209(5, 6), 212, 213(6), 215(6), 218(71), 608, 609, 610, 612(42)
Olney, J. W., 379, 380, 381, 382, 383, 384, 385, 386, 387, 388, 389, 391(17), 392(20), 393, 394, 395(2), 396
Olson, M. I., 335
Olsson, Y., 192
O'Malley, B. W., 594(35), 595
Ondo, J. G., 611
Ono, J. K., 125

Orci, L., 69, 681, 685
Ordronneau, P., 636
Orei, L., 208

Oreskes, I., 419
Orth, D. N., 594, 595
Osamura, R. Y., 196, 199(38)
Osborn, M., 67
Osborne, J. W., 419
Osinchak, J., 314
Osterman, J., 593
Oswaldo-Cruz, E., 375(15), 376
Otsu, K., 554
Otten, U., 593
Ottersen, O. P., 395
Overman, L. E., 402(q, s), 403
Owman, C., 609, 615(36)
Ozawa, S., 104, 110(33, 34)

P

Pace, M. D., 356
Page, R. B., 177
Pal, S. B., 412(29), 413
Palade, G., 189
Paladin, A. V., 528
Palfreyman, M. G., 471
Pakovits, M., 368, 369, 370(3), 371(3), 373(3), 374(3), 375, 529, 541, 630
Panerai, A. E., 660, 661(60), 671
Panko, W. B., 594(27), 595
Papermaster, D. S., 362
Pardanani, D. S., 594(38), 595
Parent, A., 325(j), 326
Park, D. H., 335, 616
Park, M. R., 128
Park, S., 594(47), 596
Parker, C. R., Jr., 610, 613(45)
Parks, D. R., 227, 232, 238(3), 239
Parsons, B., 639
Parsons, J. A., 625
Partouche, C., 99
Pass, K. A., 594(23), 595
Passon, P. G., 475
Pastan, I., 35, 219
Pataki, G., 24
Patel, Y. C., 327
Patlak, J., 174
Patterson, P. H., 114, 335, 346

Patton, J. M., 608
Pattou, E., 495
Pavasuthipaisit, K., 145
Paxinos, G., 375
Payne, P., 582
Peacock, J. H., 129
Pearce, L. A., 590(24), 591
Pearson, D., 316(c, d), 317
Pease, P. L., 96
Peck, E. J., Jr., 639
Pegg, A. E., 590, 593, 594(24), 595, 599(1), 601(1), 602(1)
Pelletier, G., 195, 325(j), 326
Pellett, P. L., 391
Penalua, A., 660, 661(60)
Penna, A., 242
Penttengill, O. S., 333, 334
Peper, K., 125
Perera, F. P., 409
Perez-Lopez, F. R., 145
Perez-Polo, J., 343
Perkel, D. H., 129, 138
Perlmann, P., 410
Perlow, M. J., 181
Perrie, S., 316(l), 317
Perrin, M. H., 42, 43, 46, 47, 48
Perry, B. W., 400
Perryman, R. L., 611(57), 612
Persson, B.-A., 479
Pertschuk, L. P., 639(10), 640
Pesce, A. J., 450
Peterfreund, R. A., 327, 329, 348, 349(1), 355
Peters, J., 298, 300(16)
Peters, K., 58
Peters, M. J., 639
Peterson, G. M., 395
Petraglia, F., 660, 661(60)
Petrusz, P., 457, 458, 636, 639
Peuler, J. D., 475
Pfaff, D. W., 191, 639, 640, 641, 644, 645, 646(29), 647(30), 648(8, 29), 649, 650(29, 30), 654, 655, 656(42), 657(43), 658(43), 659, 660, 661, 662(60)
Pfeiffer, S. E., 333
Pflughaupt, K. W., 555
Phelps, C. H., 316(l), 317
Phifer, R. F., 634
Philips, S. R., 484
Phillips, J. H., 309, 310(9)

Phillips, M. A., 543, 547(16)
Picart, R., 196, 199(37), 323(b), 326, 330, 331
Pidoplichko, V. I., 129
Pierantoni, R., 123
Pierce, Jr., G. B., 194
Piffanelli, A., 639(11), 640
Pike, R., 471
Piknosh, W., 564
Pilch, P. F., 66
Pilcher, W., 388
Pilgrim, C., 325(k, l), 327
Pilotte, N. S., 615
Pine, J., 113
Pispa, J., 602
Pitman, R. M., 103
Pittman, Q. J., 362
Pitts, A., 192
Plank, L. D., 277
Plotsky, P. M., 479, 480, 494, 501(11), 610
Poirier, L. A., 594(22), 595
Poirier, M. C., 409, 413(1), 414(1)
Poland, R. E., 551
Polson, A. X., 316(k), 317
Polyakova, N. M., 528
Pool, C. W., 671
Poon, T. K., 388
Portanova, R., 251
Porte, A., 198
Porter, J. C., 178, 475, 484, 607, 608, 609, 610, 611, 612, 613, 614, 615, 616, 617(47, 79), 691
Portnoy, B., 475
Posner, B. I., 617
Poso, H., 594(24), 595
Postulka, J. J., 594(23), 595
Potter, D. D., 115, 116, 342
Potts, Jr., J. T., 524
Peulain, D. A., 132, 133(3)
Poulain, P., 453
Pradayrol, L., 83
Prasad, J. A., 317
Pretlow, T. G., 192, 262, 272
Price, J. L., 206, 384, 392(20)
Price, M. T., 383, 388, 389, 394
Price, W., 35
Prichard, M. M. L., 611
Privat, A., 321
Prosser, T., 578
Prouty, W. F., 601
Prysor-Jones, R. A., 611(56), 612

Przewlocki, R., 562
Przuntek, H., 555
Pujol, J. F., 479, 480
Pun, R. Y. K., 113, 116
Purdy, W. G., 472
Purves, R. D., 112, 121, 122, 123(3), 125(3)
Puymirat, J., 330, 331, 350

Q

Quadri, S. K., 611
Quesenberry, P. J., 239
Quigley, M. E., 555
Quijada, M., 659

R

Raam, S., 640
Raben, M. S., 594(13), 595
Radanyi, C., 640
Radbruch, A., 462
Radhakrishnan, A. N., 532
Rahmar, M. J. K., 471
Raina, A., 590, 591, 592
Rainbow, T. C., 639
Raine, L., 196, 636, 653
Raisman, G., 205
Raizada, M. K., 593
Rajaniemi, H. J., 691
Rajewski, K., 462
Ramachandran, J., 60, 62, 63(15), 65(15), 66
Ramanovicz, D. K., 143
Ramirez, V. D., 186, 495, 540
Ramjattansingh, F., 281
Ramsdell, J. S., 290
Ramseth, D., 564
Ramsom, B. R., 112
Randall, R. J., 485, 490(11), 499
Rando, R. R., 598
Ransom, B. R., 114, 115, 119(11), 124(30, 31), 129, 139(11)
Ratter, S. J., 555
Rayford, P. L., 564
Raymond, U., 251
Raymond, V., 578, 580(7), 581(7), 582(7), 612
Raynaud, J. P., 612

Reaves, T. A., Jr., 132, 133, 142, 143
Redding, T. W., 176, 540
Reddy, P. R., 594(44, 45), 596
Redeuilh, G., 640
Redick, J. A., 625, 626
Rees, L. H., 553, 555
Rees, R. P., 335
Reeves, J. J., 39, 42, 43
Refshauge, C. J., 470, 500
Reichlin, M., 436, 437(3), 449
Reichlin, S., 313, 354, 355, 640
Reis, D. J., 616
Reisert, I., 325(k, l), 327
Renard, F., 410
Reppert, S. M., 181
Resene, D. L., 191
Reuben, J. P., 104, 110(37)
Reubi, J. C., 192
Reymond, M. J., 610, 611, 612, 614, 615, 616, 617(79)
Rhee, V., 380, 381(6), 384, 385(6), 393, 395(2)
Rhodes, C. H., 640, 655, 656, 657(43), 658
Ribak, C. E., 625
Ribet, A., 83
Rice, M. E., 480
Richard-Foy, H., 640
Richards, C. W., III, 687
Richards, F. M., 58, 450
Richards, J. F., 594(3), 595
Richardson, L. S., 277
Richardson, S. B., 317
Richman, R., 594(4, 47), 595, 596
Rick, J., 582
Riegle, G. D., 617
Riepe, R. E., 366
Rillema, J. A., 593
Ritzen, E. M., 639
Rivier, C., 33, 565, 566, 568, 577, 608
Rivier, J., 33, 77, 176, 565, 566, 568, 574(2), 575, 608, 666, 668(12)
Robbins, J. B., 456, 549
Robbins, R. J., 355, 640
Robinson, A. G., 139, 609, 655
Rocha-Miranda, C. E., 375(15), 376
Rodbard, D., 39, 40(21), 42(21), 43(21), 46(21), 564, 582, 583, 691
Rodbell, M. J., 64, 65
Roeder, P. E., 59

Rogawski, M. A., 108
Roger, L. J., 594(7, 10), 595, 597
Rogers, D. C., 42(30), 43, 53, 54(10), 55(10), 56, 402(k), 403, 404
Rogol, A. D., 611(57), 612
Role, L. W., 115
Romagnano, M., 388
Rorstad, O. P., 181, 362
Rosebrough, N. J., 485, 490(11), 499
Rosenbaum, E., 330, 350
Rosenberg, R. N., 333
Rosene, D. L., 192
Rosenfield, R. E., 419
Rosenheck, K., 305
Ross, D., 335, 340
Ross, G. T., 456, 549
Ross, M., 554
Ross, W. N., 113
Rosselin, G., 278(o), 279, 280(g), 281, 282(f), 283, 286
Rossier, J., 198, 670, 680
Rossoni, C., 590(17), 591
Roth, J., 35, 681, 685, 691
Roth, L. J., 631, 633
Rotsztejn, W. H., 278(o, p), 279, 280(g), 281, 282(f, g), 283, 286, 495, 609
Roufa, D. G., 342
Routenberg, A., 392
Rubenstein, A. H., 524
Rubenstein, K. E., 411
Rubenstein, M., 77
Rubenstein, S., 77
Ruberg, M., 609
Rubin, H., 253
Rubin, L., 282(h), 283
Rubin, R. P., 691
Rubin, R. T., 551
Ruitenberg, E. I., 409
Rumsfeld, Jr., H. W., 607
Ruoho, A. E., 59
Russell, D. H., 590, 591, 592, 594(6, 21, 26), 595, 601, 602(4)
Russell, J. A. G., 334
Russell, J. T., 525, 528, 537(5), 539
Rustioni, A., 191
Ruth, R. E., 392
Ryan, K. J., 323(e), 326
Ryback, C. E., 681
Rybarczyk, K. E., 626
Rydon, H. N., 450

S

Saavedra, J. M., 375, 484, 488(6), 489, 490, 492
Sachs, F., 112, 120, 125, 161, 166, 168, 172
Sachs, H., 314, 316(c, d), 317, 528
Sagar, S. M., 181
Said, S. I., 333, 609
Saito, T., 316(i, j), 317
Sakai, K. K., 135
Sakai, M., 127
Sakakibara, S., 671
Sakmann, B., 102, 114, 129(28), 150
Sakura, N., 454
Salcman, M., 192
Saller, C. F., 484, 488(4)
Salm, A. K., 135
Salzberg, B. M., 113
Salzman, G. C., 232
Samli, M. H., 294
Samo, Y., 540
Samuel, D., 617, 618
Sand, O., 104, 110(35)
Sandow, J., 36, 42, 43(13)
Sandoz, P., 319(h), 320
Sansone, F. M., 402(d, g), 403
Santha, R. T., 376
Santos, M. A., 611(58), 612
Sar, M., 631, 632, 636, 637, 638, 639
Sarkar, D. K., 617
Sarne, Y., 200, 533
Sarthy, P. V., 367
Sato, G. H., 115, 293, 312, 330, 343, 350, 357, 522
Sato, H., 35(17), 39, 437, 454
Satoh, K., 630
Sattler, C., 251, 258, 279, 280(k), 281, 283, 285
Satumoto, T., 630
Saunders, D., 63, 64(16), 65(16), 69
Savage, C. R., Jr., 593
Savion, S., 285, 287
Savoy-Moore, R., 42, 43
Sawchenko, P. E., 575, 663, 664(6), 666, 668, 669
Sawyer, W. H., 139
Sayatilak, P. G., 594(38), 595
Sayers, D. F., 499
Sayers, G., 251
Sayre, D. F., 473

Scalabrino, G., 594(20), 595
Scallet, A. C., 385, 386, 387(28)
Scearce, R., 460, 469
Schachter, B. A., 531, 534
Schaeffer, J. M., 362, 363
Schainker, B., 387
Schally, A. V., 33, 35(17), 39, 176, 178, 437, 454, 540, 608
Schanberg, S. M., 590(24, 25), 591, 594(7, 8, 29), 595
Schanne, O. F., 123
Scharrer, B., 187
Schechter, Y., 49, 60, 63(8), 66(8), 219
Schettini, G., 255
Schick, A. F., 453, 454
Schimizu, N., 630
Schinko, I., 199
Schlesinger, D. H., 176
Schlessinger, J., 49, 219, 226
Schlumpf, M., 357, 359(13)
Schmeckel, D. E., 363
Schmelz, H., 241
Schmued, L. C., 665, 668
Schneider, A. S., 305
Schneider, R. S., 411
Scholz, D., 393
Schorer, J., 460
Schrell, U., 199
Schrier, B. K., 114
Schubert, D., 333
Schulster, D., 277, 281
Schultz, J. A., 460
Schultzberg, M., 630
Schuurs, A. H. W. M., 410, 449
Schwabe, K. G., 651
Schwarcz, R., 384, 385(22), 393, 396
Schwartz, N. B., 42(24), 43, 258, 541
Schwartz, S. C., 578
Schwob, J. E., 384, 392(20)
Scott, B. S., 339, 346
Scott, D. E., 323(d), 326, 329
Scott, J. F., 600, 602(46)
Sedat, J. W., 219
Sedory, M. J., 600, 601, 602(46)
Sedvall, G. C., 494
Seeburg, P., 548, 670
Seely, J. E., 594(24), 595
Seeman, P., 578
Segal, D. M., 239
Seguin, C., 42(23), 43

Sehon, A., 449
Seidler, F. J., 590(26, 27, 28), 591, 594(12), 595, 599(27, 28)
Seiler, N., 592
Sela, M., 434
Selmanoff, M., 613
Senoh, S., 485
Shaar, C. J., 609
Shainberg, A., 316(c, d), 317
Shakespear, R. A., 39, 40(21), 42(21), 43(21), 46(21)
Shapiro, H. M., 227, 238(1), 240(1)
Sharma, S., 376
Sharon, L., 661
Sharpe, L. G., 179, 380, 381
Sharpless, T., 230, 234
Sharrow, S. O., 239
Shechter, Y., 220
Sheela, Rani, C. S., 594(33), 595
Sheffield, T., 56
Sheppard, M. S., 281
Shepro, D., 593
Sheridan, M., 316(l), 317
Sheriff, Jr., W. H., 113
Sherlock, D. A., 205
Sherman, T. G., 519
Sheth, A. R., 594(38), 595
Shiba, K., 594(37), 595
Shibasaki, T., 85
Shimatsu, A., 609
Shinitzky, M., 617, 618
Shipley, M. T., 394
Shiu, R. P. C., 617
Shivers, B. D., 649, 659(31), 660, 661
Shoemaker, W. J., 357, 359(13), 362
Shooter, E., 343
Shortman, K., 272, 274, 275
Shoup, R. E., 500
Sibley, D. R., 578, 580(16), 581(16), 582, 583(16), 584, 585, 586, 587(27)
Sidman, R. L., 375
Siegelbaum, S. A., 150
Siegel, M. B., 551
Signorella, A., 269
Sigworth, F. J., 102, 114, 129(28), 150, 163, 166(9), 168(9), 171(9)
Siigi, S. M., 292
Siimes, M., 590(20), 591, 592
Silverman, A. J., 195, 199, 662
Simpkins, J., 613

Simpson, E. L., 391
Singer, J. M., 419
Singer, S. J., 59, 453, 454
Sinha, Y. N., 281
Siperstein, E., 296
Sizonenki, P. C., 40, 42(22), 43(22)
Sjoerdsma, A., 494
Skagerberg, G., 665
Skaper, S. D., 115, 350
Skinner, J. E., 513
Skowsky, W. R., 655
Skutelsky, E. H., 294
Sladek, C. D., 133, 147, 316(e, f, g, h), 317, 495
Sladek, J. R., Jr., 323(d), 326, 329
Slayton, V. W., 333
Slepetis, R., 306, 309(5), 310(5), 311(5), 312(5), 590(28), 591, 599(28)
Slotkin, T. A., 590(26, 27, 28), 592, 594(12, 16, 17, 18, 25), 595, 598, 599, 600(37, 40, 44, 45), 601(37, 40), 602(37, 40, 44, 45), 603
Smelik, P. G., 609
Smith, A. A., 639
Smith, A. I., 557
Smith, B. M., 103, 120, 121, 125(50), 128(48), 129
Smith, D. E., 402(m), 403, 404
Smith, D. G., 321
Smith, G. C., 388
Smith, K. R., 178
Smith, M., 566, 568(4)
Smith, M. A., 300
Smith, P., 655
Smith, P. F., 626
Smith, R. E., 676
Smith, R. G., 42(30), 43
Smith, S., 589
Smith, T. G., Jr., 112, 120, 121(48), 125, 128(48), 129, 138(6), 139(6), 141(6), 143(6)
Smith, T. W., 670
Smith, W. A., 251
Smookler, H. H., 494
Snider, R. S., 376
Snodderly, Jr., D. M., 97
Snyder, G., 258, 272, 275(6), 279, 280(j), 281, 282(n), 283, 285
Snyder, J., 258, 267, 268, 278(i, j), 279, 280(d), 281

Snyder, L. R., 470
Snyder, S. H., 578, 590, 592, 594(26), 595, 601(2, 32, 35)
Sobkowicz, H. M., 319(b), 320
Sobotka, H., 452
Sobrinho, L. G., 611(58), 612
Sofroniew, M. V., 145, 199
Sogani, R. K., 594(13), 595
Sokol, H. W., 139
Soloman, S., 77, 85(8)
Somogyi, P., 683
Sonksen, P. H., 66
Sorenson, G. D., 333, 334
Spada, A., 589
Specht, P., 161
Speciale, S. G., 611
Spector, S., 494
Speiss, J., 176, 565, 574(2), 608
Spence, J. W., 281
Spencer, H. J., 101
Spicer, S. S., 634
Spies, H. G., 42, 43
Spiess, J., 33, 568, 608
Spitzer, N., 113
Springer, P. A., 113, 138(17), 143(17)
Spyer, K. M., 98
Srikanta, S., 468, 469
Stall, A. M., 227, 232(2), 238(2)
Stan, M. A., 65
Standish, L. J., 391
Staros, J. V., 58
Starr, R. M., 590(22), 591
Stasch, J. P., 555
Stastny, M., 594(35, 39, 43), 595, 596
Stathis, P., 641
Stean, J. P., 97
Stefanini, E., 584
Stein, S., 77, 78, 79, 531, 532, 557
Steinberger, A., 691
Steinbush, H. V. M., 384, 391(23), 629
Steiner, A. L., 639
Steiner, D. F., 524
Steinkamp, J. A., 234
Stell, W., 366
Stern, J., 242
Sternberger, L. A., 139, 198, 623, 625, 635, 653, 655(35), 667
Sternfeld, M. J., 239
Sterz, R., 125
Stetsler, J., 484, 489(7), 492(7)

Stevens, C. F., 108
Stevens, S. W., 535
Stewart, J. M., 5, 7(3, 4), 8(3, 4), 9(3, 4), 10(4), 11, 12(3, 4), 18(3, 4), 22(3, 4), 23(3, 4), 24(3, 4), 25(3, 4), 27(3, 4), 56
Stewart, R. D., 555
Stewart, W. W., 127, 137
Stickgold, R., 107
Stocker, M. P. G., 253
Stoeckart, R., 208
Stoeckel, M. E., 198
Stone, J., 78, 79(9), 531
Stouffer, J. E., 434
Stout, R. W., 469
Straus, W., 191, 192, 212, 635
Streetkerk, J. G., 410
Streit, P., 192
Strumwasser, F., 97
Stryer, L., 239
Stumpf, W. E., 631, 632, 633, 637, 638, 639
Suda, T., 556
Suffdorf, D. H., 73
Summerlee, A. J., S., 98
Sundberg, D. K., 494, 495, 496(15), 497, 498(14), 499(14), 500, 501, 503, 504, 506, 507, 509(12, 13, 14, 15), 526, 527, 528, 535(7), 537, 538, 539(7)
Surks, M. I., 258
Sussdorf, D. H., 455
Sutton, R. E., 355
Suzuki, F., 593
Suzuki, M., 594(46), 596
Suzuki, S., 317
Svaetichin, G., 367
Swaab, D. F., 195, 671
Swaiman, K. F., 114
Swallow, R. A., 634
Swan, L., 655
Swanson, L. W., 132, 206, 515, 575, 629, 659, 663, 664(6), 665, 666, 668, 669, 681
Swanson, N., 258, 267, 268, 280(d), 281
Swartz, B. E., 127
Sweet, R. G., 232, 233(7)
Sweet, W., 333
Sweinberg, S., 593
Swennen, L., 252, 278(l), 279
Sykes, J. A., 341
Symthies, J. R., 402(f), 403
Synder, S. H., 490

Szabo, D., 630
Szabo, M., 582

T

Tabachnick, M., 452
Taber Pierce, E., 375
Tabor, C. W., 590
Tabor, H., 590
Tager, H. S., 524
Tait, J. F., 278(m), 279, 284(d), 285
Takagi, H., 683
Takahashi, H., 548, 670
Takano, T., 593
Takeda, Y., 594(37), 595
Takigawa, M., 593
Tal, E., 285, 287
Tal, J., 389
Taleisnik, S., 541
Tamburini, R., 402(p), 403
Tamura, H., 640
Tanaka, I., 555
Tanapat, P., 639(10), 640
Tanioka, H., 594(37), 595
Tanizawa, O., 630
Tannenbaum, G. S., 176, 183
Tannenbaum, M., 609
Tappaz, M. L., 637
Taraskevich, P. S., 104, 110(36)
Tardif, C., 601, 602
Tarnavsky, G. K., 39
Tassin, J. P., 475
Tauber, J. P., 349
Taylor, R. L., 34, 601
Templeton, C. L., 685
Teng, C. S., 594(34), 595
Teng, C. T., 594(34), 595
Terada, S., 35, 39
Terenius, L., 384, 385(22), 396, 663, 680, 686, 687(32)
Ternynck, T., 414, 681
Terracio, L., 651
Terrano, M. J., 590(23), 591
Terry, L. C., 391
Thakur, A. N., 594(38), 595
Thamm, P., 63, 64(16), 65(16), 66, 69
Thaw, C., 317
Theodosis, D. T., 208
Thieme, G., 619

Thoenen, H., 593
Thomas, C. A., Jr., 113, 138(17), 143(17)
Thompson, C. J., 277
Thompson, S. H., 108
Thorner, M. O., 239, 252, 258, 283, 287, 296, 304, 611(57), 612
Thoron, L., 661
Tillson, S. A., 434
Tindall, G. T., 608, 610
Tischler, A. S., 333
Titeler, M., 578
Titus, J. H., 239
Tixier-Vidal, A., 104, 110(31, 32), 196, 199(37), 263, 296, 323(b, c), 326, 330, 331, 332, 333, 350
Tobin, E. H., 639(10), 640
Tobin, R. B., 258, 284(c), 285
Todd, P., 277
Tohyama, M., 630
Tomchick, R., 476
Tomlinson, E. B., 577
Toran-Allerand, C. D., 319(e, f, g), 320
Torp, A., 619
Tougard, C., 196, 199(37), 323(c), 326
Tower, B. B., 551
Toyosato, M., 548
Traganos, F., 230, 234(4)
Tregear, G. W., 410
Tremblay, E., 384, 392(21), 395
Trepanier, P. A., 590(27, 28), 591, 594(25), 595, 599(27, 28)
Trifaro, J. M., 306, 311(6)
Triggle, D. J., 402(o), 403, 405
Trivers, G. E., 414
Trojanowski, J. Q., 212
Trygstad, O., 639
Tsafriri, A., 594(31, 32), 595
Tsai, B. S., 583
Tsang, D., 82
Tsou, R. C., 608
Tucker, E. S., III, 412(30), 413
Tweedle, C. D., 135

U

Udenfriend, S., 77, 78, 79(9), 494, 531, 532, 548, 557, 670, 681
Ukraincik, K., 564
Ullman, E. F., 411

Underwood, L., 594(4), 595
Unsicker, K., 306, 310, 311(7)
Upchurch, S., 650
Utiger, R. D., 436, 437(1), 440(1), 452
Utsumi, M., 388

V

Vaccaro, D. E., 323(e), 325(f, g), 326, 329
Vaitukaitis, J., 456, 549
Vail, W. J., 685
Vale, W., 33, 42(29), 43, 46(31), 47(31), 48(31), 176, 178, 249, 259, 287, 296, 298(11), 300, 327, 329, 348, 349, 495, 565, 566, 568, 574(2), 575, 577, 608, 666, 668, 691
Valiquette, G., 139
Vallet, H. L., 594(23), 595
van Beurden, W. M. O., 639
Van Breemen, C., 402(j), 403
Vanderhaeghen, J. J., 198
van der Kooy, D., 663, 666
Vanderlaan, W. P., 551
van der Molen, H. J., 639
Vander Schueren, B., 252, 258, 266, 278(k), 279, 280(h), 281, 282(i), 283, 284(f), 285, 295
Vandesande, F., 198, 625, 676
vande Wiele, R. L., 661
Vandlen, R. L., 104, 110(37)
van Houten, M., 617
Van Leeuwen, F. W., 195, 671
van Maanen, J. H., 609
Van Orden, D. E., 625, 626, 629
Van Orden, L. S., 625
Van Vunakis, H., 559
Van Weeman, B. K., 410, 449
van Wimersma Greidanus, T. B., 198
Van Wyk, J., 594(4), 595
Vargo, T., 563
Varland, D. A., 600, 602(46)
Varon, S., 343, 350
Vaughan, J., 566, 568(4)
Vaughn, J. E., 625, 681
Vawdry, H., 195
Veale, W. L., 179
Veldhuis, J. D., 593
Verhaart, W. J. C., 376
Verhoef, J., 548, 557(8)

Vernadakis, A., 350
Vernaleone, F., 584
Vevert, J. P., 598
Viceps-Madore, D., 602
Vigny, A., 330, 331
Vijayan, E., 493
Villee, C. A., 593, 594(44), 596
Vincent, J. D., 95, 96, 98, 99, 104, 110(31, 32), 327, 329, 332
Visser, T. J., 608, 610(16)
Vitali, P., 471
Viveros, O. H., 306, 309(8), 310(8), 311(8), 312
Vivier, J., 33
Vlodavsky, I., 295
Voina, S., 594(4), 595
Voller, A., 409, 410, 412, 413

W

Wachtel, R. E., 128
Waggoner, A. S., 243
Wagner, H. G., 97
Wagner, J., 471
Wagner, T. O. F., 35, 39
Wahl, L. M., 258, 279, 283
Wakerley, J. B., 99, 132, 133(3)
Wakshull, E., 335, 347
Walicke, P. A., 335, 346(12)
Walker, R. F., 484, 489(8), 492(8)
Wallace, J. H., 454
Wallis, M., 277, 281
Walsh, F., 460
Walsh, R. J., 617
Wang, C. T., 494
Wang, K., 220
Wardell, J., 305, 306(2), 310(2)
Wardlaw, S. L., 608, 615(22, 23), 660, 661
Warnica, J. W., 594(14), 595
Warnick, J. E., 402(d, e, g, p, r, s, t), 403
Wasserman, L. R., 419
Watanabe, K., 196, 199(38), 540
Watkins, J. C., 379, 380(1), 381, 382(10), 384(10)
Watkins, W. B., 325(m), 327
Watson, C., 375
Watson, C. G., 655
Watson, D., 412
Watson, S. J., 198, 387, 388(30), 389(30), 687

Waymire, J. C., 305, 306(2), 310(2)
Waymire, K. G., 305, 306(2), 310(2)
Webb, A. C., 97
Weber, K., 67
Weber, S. G., 472
Weddington, S. C., 639
Wehrenberg, W. B., 87, 89, 176, 608, 615(22, 23), 660, 661
Weigel, S. J., 590(27, 28), 591, 594(25), 595, 599(27, 28)
Weight, F. F., 108
Weindl, A., 199
Weiner, H. J., 610, 612(42)
Weiner, N., 494
Weiner, R. I., 288, 290, 293, 493
Weinstein, I. B., 409, 413(1), 414(1)
Weinstein, J., 239
Weir, E. E., 192, 262, 272(3)
Weise, V. K., 488, 494
Weisel, T. N., 367
Weiss, J., 495
Weissbach, J., 490
Wen, H. L., 555
Wendel, O. T., 494, 500(12), 504(12), 509(12)
Wesley, R. A., 409
West, K. A., 609, 615(36)
Westfall, T. C., 508
Westheimer, F. H., 58
Westland, K. N., 663
Wetzstein, R., 199
Whatley, S. A., 327
Wheaton, J. E., 541
Whitaker, J. M., 531, 534
Whitaker, J. R., 24
White, W. F., 36, 38(10), 44(10), 452
White Smith, S., 42, 43
Whitford, J. H., 562
Whitmore, W. L., 590(26, 27, 28), 594(25), 595, 599(27, 28)
Whittaker, M. L., 402(i), 403
Wiegland, S. J., 206
Wiesel, F. A., 613
Wightman, R. M., 479, 480, 481
Wilchek, M., 451, 453, 456(23)
Wilchek, S. M., 436, 437(2)
Wild, F., 452
Wilding, P., 449, 451(7)
Wilkins, J. A., 655
Wilfinger, W., 258, 267, 268, 278(j), 279, 280(d), 281

Wilkes, M. M., 555
Wilkinson, M., 323(a), 326
Wilkinson, S., 220
Wilkinson, W., 661
Williams, E. E., 192
Williams, L. T., 578, 583
Williams-Ashman, H. G., 590, 599(1), 601, 602(1)
Willingham, M. C., 219
Wilson, C. B., 293
Wilson, R. C., 389, 390(41)
Wilson, S. P., 306, 309, 310(8), 311(8, 10), 312
Wilson, W. A., 128
Wing, L. Y., 593
Wisher, M. H., 66
Withnell, R., 481
Witrop, B., 402(c), 403
Witteler, M., 555
Wogan, G. N., 409, 414(6)
Wolbarsht, M. L., 97
Wolf, U., 306, 311(7)
Wolff, J. R., 396, 397(16)
Wolfman, M. G., 611(57), 612
Wolinsky, T. D., 645, 646(29), 648(29), 650(29)
Wong, R. K. S., 150
Wood, P., 343, 345(29)
Woody, C. D., 127
Worobec, R. S., 35(17), 39
Wray, H. L., 564
Wren, J. A., 495, 498(14), 499(14), 509(14)
Wurtman, R. J., 490
Wuttke, W., 101

Y

Yaksh, T. L., 179
Yaltzman, N. P., 494
Yamadu, T., 366
Yamaguchi, K., 655
Yamaguchi, M., 414
Yamamoto, C., 101, 134, 135
Yamamoto, G., 566, 568(4)
Yamamoto, R. S., 594(22), 595
Yamamura, H. I., 179
Yamasaki, Y., 601
Yanaihara, C., 454
Yanaihara, N., 454, 609

Yang, J. W., 593
Yarden, Y., 226
Yates, P. E., 191
Yeh, Y. C., 253
Yellen, G., 165
Yen, S. S. C., 555
Yeo, T., 582
Yeung, C. W. T., 60, 62, 63(9, 14), 66, 67(14)
Yip, C. C., 60, 62, 63(9, 14), 66, 67(14)
Yolken, R. H., 409, 413(1, 9), 414(1, 9)
Yoshida, S., 316(i, j), 317
Yoshikami, D., 107
Yoshimura, S., 196, 199(38)
Young, A. B., 125
Young, D. C., 35(19), 39
Young, E., 389
Young, J. D., 5, 7(3, 4), 8(3, 4), 9(3, 4), 10(4), 11, 12(3, 4), 18(3, 4), 19(3, 4), 22(3, 4), 23(3, 4), 24(3, 4), 25(3, 4), 27(3, 4)
Youngblood, W. W., 388, 438, 439(5), 440, 446(6)
Yu, S., 594(47), 596
Yuspa, S. H., 409, 413(1), 414(1)

Z

Zaborszky, L., 396, 397(16), 630
Zaczek, R., 393
Zakarian, S., 321
Zehr, D. R., 635
Zettergren, J. G., 262, 272(3)
Zide, D., 681
Zidovetzki, 226
Zigmond, M. J., 484, 488
Zigmond, R. E., 371
Zimmerman, E. A., 133, 139, 195, 609, 662
Zingg, H. H., 327
Zini, I., 671
Zor, U., 453, 456(23), 594(32), 595
Zraika, M., 471
Zusman, D. R., 594(48), 596
Zweig, G., 24
Zyzek, E., 103

Subject Index

A

D-α-AA, see D-α-Aminoadipate
AAP, see Avian pancreatic polypeptide
ABC method, see Avidin–biotinylated horseradish peroxidase complex method
Acetic acid, see also Methanolic acetic acid extraction
    use to prevent adsorption of peptide to glassware, 30, 32
N-Acetyltransferase
    activity determination, 491
    preparation, 488–489
    use in REA of serotonin, 491, 492
Aconitine, 402
Acrolein, as fixative, 626–627
ACTH, see Corticotropin
Adenohypophysis, 607
Adenoma, prolatin-secreting, culture medium for, 293
S-Adenosyl-L-[methyl-^{3}H]methionine, as methyl donor in REA, 475, 477, 483
Adipocytes, lipogenesis studies, 64–65
Adrenal medulla, see also Chromaffin cell
    separation of cell types, 307–309
    tissue disassociation, 306–307
Adrenocortical cell, steroidogenesis study, 65
Adrenocorticotropic hormone, see Corticotropin
Affinity chromatography, for peptide purification, 73, 81
Aging, effect on dopamine release, 616–617
Agonist, mechanism of action, 54–58
AH, see Hypothalamus, arcuate nucleus
Alanosine, structure, 380
ALFA method, 620–621
Alpha 1, binding to plastic, 419, 422–423
Alumina extraction, 473–474, 499
Aluminum perfusion, 620–621
Amacrine cell, isolation, 366–367

Amino acid
    Boc derivative
        attachment to resins, 10–11
        purity tests, 9
    excitotoxic, see Excitotoxin
    radiolabeled
        choice of, for tracer studies, 527–528, 539
        infusion technique, 513–514, 528–529, 532
        preparation, 528
Amino acid side chains, blocking groups for, 7
D-α-Aminoadipate, 381, 385, 386, 387
    site of action, 390
γ-Aminobutyric acid
    coupling to carbobenzyloxy chloride, 442
    effect on secretion of somatostatin, 355
    as reactive prosthetic group on small peptides, 436, 437, 440, 441
ω-Amino-n-caproic acid
    coupling to carbobenzyloxy chloride, 442
    as reactive prosthetic group on small peptides, 441
ω-Amino fatty acid, as reactive prosthetic group on small peptides, 441–446
α-Amino group, protection, 6
p-Aminophenylacetic acid, in coupling reaction, 453
D-Aminophosphonovalerate, 381, 385, 386, 387
Aminopterin, stock solution, 460–461
Angiotensin I, purification, 20
Angiotensin II, purification, 20
Antibody
    enzyme-coupling, 413–414
    production methods, 413, see also Antiserum
    in purification of peptides, 73
    radiolabeling procedure, 415

Antibody screening assay, for monoclonal antibody detection, 465–466
Antigen
 coating procedure in EIA, 416, 418–420
 distribution analysis by flow cytometry, 239–242
 preparation, 455
Antiserum
 against opioid peptides, 549–550
 collection and storage, 549
 determination of titer and avidity, 550
 specificity determination, 456–458
 use in Mason single-bridge technique, 458–459
Apomorphine
 dissociation constants in competitive binding studies, 581, 582
 tritiated, binding characteristics, 578
D-APV, see D-Aminophosphonovalerate
Arachidonic acid, in serum-free cell culture medium, 330, 331
Arginine vasopressin
 assay, using HPLC, 515–519
 coexistence with estradiol, 637, 655–659
Arginine vasopressin
 localization, 139, 141, 198, 199
 methanolic extraction procedure, 498–500
 presence in hypophysial portal blood, 609
 production by cloned cells, 332
 radioimmunoassay, 145
 release studies, 145–146
 tissue content after *in vivo* incubation, 526–527, 537–538
 turnover rates, prediction of, 524–526
Ascites cells, for monoclonal antibody production, 468
Ascorbic acid
 in chromaffin cell culture medium, 311, 312
 dopamine receptor-ligand binding and, 584–585
 oxidation potential, 480
L-Aspartate, see also Excitotoxin
 excitotoxic properties, 379, 387
 structure, 380
 use as lesion-sparing agent, 381–383

Autoradiography
 animal preparation, 632, 648–649, 650–651
 combined with immunocytochemistry to localize steroid hormones, 631–634, 646–655
 development of fixed slides, 634, 651–652
 dry-mount technique, 633
 interpretation of results, 654–655
 thaw-mount method, 632–633, 651–652
Avian pancreatic polypeptide, 630
Avidin-biotinylated anti-Ia antibody, 240, 241
Avidin-biotinylated horseradish peroxidase complex method, 196, 198, 413–414
 for immunostaining autoradiograms, 653
 sensitivity, 636
AVP, see Arginine vasopressin
Axial light-loss pulse width, measurement of, 233–238
Axonal transport, HRP studies of, 205–218
(4-Azidobenzoyl)-*N*-hydroxysuccinimide, 59, 60
N-5-Azido-2-nitrobenzoyloxysuccinimide, 51
p-Azidophenylacyl bromide, 51
p-Azidophenyl glyoxal, 51
N-(4-Azidophenylthio)phthalimide, 51

B

Batrachotoxin, 402
Benzidine, bisdiazotized
 coupling procedure, 451–452, 569–570
 preparation, 452
Benzidine dihydrochloride, 191
Bicuculine, 355
BioGel, for purification of peptides, 19–20, 22
BioGel P-2 polyacrylamide gel, as support in continuous perifusion cell culture technique, 299, 300, 302
Bis[2-(succinimidooxycarbonyloxy)ethyl] sulfone, 50
Boc, see *tert*-Butyloxycarbonyl group
Bovine serum albumin
 as carrier for small peptides, 437, 440, 445

Bovine serum albumin
  as gradient material for density gradient centrifugation, 272–273, 274–275, 307
  for unit gravity sedimentation, 265, 267, 269
Bradykinin, purification, 20
Brain block, freezing and sectioning, 632–633, 651, 667, 675–677
Brain explant
  culture medium, 357
  neuropeptide assays on, 358–361
  preparation, 357, 358
Brainstem, catecholamine turnover in, 506–508
Brain tissue, opioid peptide extraction from, 552–553
Bromobenzene, 277
Buserelin, 63–64, 67, 68, 69, 71
(+)-Butaclamol, 579, 588
tert-Butyloxycarbonyl group, 6–7

C

CA, see Catecholamine
Cannula, implantation technique, 513, 528–529
Carbobenzyloxy chloride, in preparation of immunogens, 442–443
Carbodiimide, coupling procedure, 451, 548–549
Carbonylbis(L-methionine-p-nitrophenyl ester), 50
Carboxytetramethylrhodamine, as intracellular stain, 127
Catecholamines, see also specific compounds
  assay by HPLC-EC, 470–475, 478–479
    by REA, 475–479, 483–488
    by voltammetric microelectrode techniques, 479–483
  biosynthetic pathway, 470
  conversion to o-[^{3}H]methyl derivatives, 475, 477
  in cultured chromaffin cells, 311–312
  presence in hypophysial portal blood, 609–610
Catecholamine neuron system, histochemical techniques for
  fluorescence, 620–623
  immunochemical, 623–630
Catecholamine turnover in hypothalamus
  determination, using HPLC-EC, 500–506
    comparison of steady state and non-steady state turnover, 506–508
    data analysis, 505–506
    detection system, 502–504
    reagents and equipment, 500–501
    separation system, 501–502
  determination, using RIA, 495–500
    culture media, 495–496
    dissection of hypothalamus, 496
    principles, 494–495
    time of incubation, 496–497
    tissue extraction procedure, 498–500
Catechol O-methyltransferase
  preparation, 476, 485–486
  use in REA of catecholamines, 475, 477, 483, 487
Cations, dopamine receptor-ligand binding and, 583–584
Caudate nucleus, catecholamine turnover in, 506–508
CCD, see Countercurrent distribution
CCK, see Cholecystokinin
Cell culture, see also Pituitary cell
  continuous perifusion technique, 297–305
  dispersion technique, choice of, 251, 260, 337–338, 349
  optimum conditions for pituitary cells, 249–257
  tissue source, choice of, 259, 289–292, 349
Cell culture, dissociated, 111–132
  advantages, 111–112
  culture conditions, selection of, 114–116
  intracellular recording, 118–132
  problems studies, 112–114
  tissue dispersion technique, 249–251, 259, 260, 289–292, 298, 338, 339, 349
  tissue source, 349
Cell culture, monolayer
  advantages of, 296–297
  for extracellular recording, 94, 102–111

Cell culture, primary dispersed
  advantages and disadvantages, 348–349
  attachment substratum, 349
Cell culture chamber, 340–341
Cell culture dish, 340–341
  attachment substratum, choice of, 288, 295, 341, 349
  collagen coating, 342
  production of extracellular matrix in, 288
Cell cycle, determination of phase of, 236–238
Cell density, effects of, 253–255, 339–340, 354
Cell diameter, determination of, 233–238
Cerebral cortex, collection from fetal rat, 352
Cerebrospinal fluid, extraction of opioid peptides from, 555
Chloramine T, 35, 36, 40–42, 550
Chlorine reagent, 24
Chloroform, alcohol-free, 13–14
Chloromethyl resin, 4, 10
Cholecystokinin, 33
  localization in neurons, 198
Chromaffin cell
  biochemistry in culture, 311–312
  isolation, 306–310
  morphology in culture, 311
  plating and maintenance in culture, 310–311
Cleavage vessel, 8
Cloning by limiting dilution, 466–467
Cobalt trifluoride, 17
Colchicine, use before immunohistochemical staining, 628, 632, 649, 667
Collagen
  application to culture dishes, 342
  preparation, 341–342
Collagenase
  for cell dispersion, 251, 259, 290, 307, 310, 339, 349, 353, 363, 566, 567, 585
  for removal of connective tissue, 150
COMT, see Catechol O-methyltransferase
Conjugation, to carrier, methods for, 437, 449–454
Corneal endothelial cells, dispersal and culture, 288–289
Corticosterone, in culture medium, 330, 331

Corticotroph, studies of separation of, 286
Corticotropin
  assay by EIA, 420–424, 427–434
    by RIA, 427–434
    by USERIA, 424–434
  binding to plastic, 419–423
  localization in neurons, 198, 199
  photoaffinity derivative
    applications, 63, 65
    preparation and characterization, 60–61
  radiolabeling procedure, 415
Corticotropin-producing cell, 258
Corticotropin-releasing factor, 33
  acid extraction, 576
  antiserum production, 569–570
  assay in pituitary cell culture, 565–569
    using RIA, 569–575
  Bond Elute column extraction, 576–577
  characteristics, 565
  coupling to α-globulin, 569–570
  effect on secretion of somatostatin, 345
  presence in hypophysial portal blood, 607–608
  radioiodination, 570–572
Countercurrent distribution, for purification of peptides, 19, 20–22
Counterstreaming centrifugation, 271–272
CRF, see Corticotropin-releasing factor
Cross-linkage agent, bifunctional, 49–53
Cryoprotectant solution, 650
Cryostat, 370
CT, see Chloramine T
Culture medium
  for brain explants, 357
  for catecholamine turnover studies, 495–496
  for hybridoma production, 460–461
  for long-term incubation, 251, 343
  for peptidergic neurons, 350, 351
  for sympathetic neurons, 343–346
  serum-free
    for brain explants, 357–358
    for hypothalamic cells, 329–331
    for pituitary cells, 566
    for prolactin-secreting adenoma cells, 293
    for sympathetic neurons, 343–344
Current-clamp technique, 104–107
CVO neurons, see Neurons, circumventricular organ

SUBJECT INDEX

Cyclic GMP, in retinal cells, 363, 364
Cysteic acid
  excitotoxic properties, 387
  structure, 380
Cysteinesulfinic acid, structure, 380
Cysteine-S-sulfonic acid, structure, 380
Cytocentrifuge, 265
Cytodex beads, as cell support matrix, 300
Cytofluorograf system 50H, 228–231
  applications, 231–245
Cytosine 1-β-arabino-furanoside, in culture medium, 522

D

D600, see Methoxyl verapamil
DCC, see Dicyclohexylcarbodiimide
DCM, see Dichloromethane
Density gradient centrifugation
  continuous gradient techniques, 274, 275–276
  discontinuous gradient techniques, 275
  for separation of adrenal medullary cells, 307–309
  of retinal cells, 363
Deoxyribonuclease, in digestion mixture, 250, 259, 290, 300, 349, 351, 566, 567, 585
DFMO, see α-Difluoromethylornithine
DHBA, see Dihydroxybenzylamine
3,3'-Diaminobenzidine tetrahydrochloride, 143, 191, 459, 625, 627, 636, 653
  stock solution, preparation and storage, 678–679
o-Dianisidine, 191
p-Diazonium phenylacetic acid carbodiimide, coupling procedure, 452–453
Dichloromethane, 7, 14
  purity test, 9
Dichloromethylsilane, 9
Dicyclohexylcarbodiimide, see also Carbodiimide
  coupling reaction, 10, 13–14, 15–16, 443
1,5-Difluoro-2,4-dinitrobenzene, 52
4,4'-Difluoro-3,3'-dinitrodiphenyl sulfone, 52
α-Difluoromethylornithine, inhibitor, of ODC, 598–599
Digitonin, for membrane dissociation, 588

Dihydroergocryptine, tritiated, binding characteristics, 578
Dihydroxybenzylamine, as internal standard, 499
L-Dihydroxyphenylalanine
  dopamine release and, 615–618
  effect of aging on formation of, 617
  synthetic pathway, 470
4,4'-Diisothiocyano-2,2'-disulfonic acid stilbene, 52
Dimethyl adipimidate · 2 HCl, 50
4-Dimethylaminopyridine, 14
Dimethyl-3,3'-dithiobis(propionimidate) · 2 HCl, 50
N,N-Dimethylformamide
  in preparation of immunogen, 443, 444, 453
  purity test, 9
Dimethyl pimelimidate · 2 HCl, 50
Dimethyl suberimidate · 2 HCl, 50
Dimethyl sulfoxide, in freezing solution for hybridoma cells, 467
3,3'-Dipentyloxacarbocyanine iodide, 243, 244
Direct ligand binding
  assay procedure, 580, 586
  data analysis, 581–583, 586–587
  membrane preparation, 579–580
  principles, 578–579
Disuccinimidyl (N,N'-diacetylhomocystine), 50
Disuccinimidyl suberate, 50
Disuccinimidyl tartrate, 50
4,4'-Dithiobisphenylazide, 52
Dithiobis(succinimidyl propionate), 50
Divalent ligands, synthesis, 49–53
DMAP, see 4-Dimethylaminopyridine
DMC, see Dichloromethane
DMF, see N,N-Dimethylformamide
Docosohexaenoic acid, in serum-free cell culture medium, 330, 331
Domperidone, tritiated, binding characteristics, 578
DOPA, see L-Dihydroxyphenylalanine
DOPAC, oxidation potential, 480
Dopamine
  assay, using HPLC-EC, 500–506
  cultured cell response to, 253–256
  dissociation constants in competitive binding studies, 581, 582
  effect of aging, 616–617

levels in different portal vessels, 610–612
oxidation potential, 480
presence in hypophysial portal blood, 609–618
regulation of release, 612–616
relationship between release and synthesis, 615–616
stock solution for HPLC standard, 472–473
synthetic pathway, 470
Dopamine β-hydroxylase, 637
Dopamine receptor
  direct radioligand measurement of, 577–590
    in anterior pituitary membranes, 579–585
    in dispersed cells, 585–587
    in solubilized membranes, 587–589
    regulation of ligand binding to, 583–585
Double-label technique, 139–141
Dye-marking, see Staining
Dynorphin, localization in neurons, 198

E

ECF, see Ethyl chloroformate
ECM, see Extracellular matrix
EGF, see Epidermal growth factor
EGS, see Ethylene glycol bis(succinimidyl succinate)
EIA, see Enzyme immunoassay
Electrode, see Microelectrode
Electron microscopy
  of HRP-labeled cells, 199–205, 207–218
  of triple-labeled neurons, 142–144
Elutriator, 271–272
Emulsion, radiographic, slide coating technique, 634
Endocytosis, HRP studies of, 207–218
β-Endorphin, see also Opioid peptides
  assay, using RIA, 562–564
  chromatographic purification methods, 556–557
  extraction from plasma, 554–555
  localization in neurons, 199
  presence in hypophysial portal blood, 608
  radioiodination, 550
  radiolabeled, specific activity determination, 550–551
  regulator, of dopamine release, 615
  secretion by brain explants, 359–360
β-Endorphin-producing neurons, estradiol concentration by, 660–662
Enkephalin, see also Leucine-enkephalin; Methionine-enkephalin; Opioid peptides
  assay, using RIA, 559–561
  localization in neurons, 198
  methanolic extraction of, 498–500
  production by chromaffin cells, 311–312
    by tumor cell line, 333
Enkephalin analog, effect on dopamine release, 615
Enkephalin receptor, visualization of, 224–225
Enzyme immunoassay
  for ACTH, 420–424, 427–434
  antigen coating, 416
  antigens for, 414–415
  antisera, 415
  biological samples
    assay results, 430–433
    preparation, 418
  data analysis, 417–418
  equipment, 415–416
  history and development, 409–413
  labeled substrates, 415
  optimal assay conditions, 420–424, 433–434
  plasma samples, preparation, 418
  procedure, 416–417
  standard curve, 427–429
  tissue extract for, preparation, 418
  types, 411–412
Epidermal growth factor, 293
  binding to plastic, 419
Epinephrine, see also Catecholamines
  assay, using HPLC-EC, 500–506
Epinephrine
  oxidation potential, 480
  stock solution for HPLC standard, 472–473
  synthetic pathway, 470
Erythritol biscarbonate, 50
Esterification
  of conjugated peptide, 445, 446
  of GABA, 445, 446
Estradiol
  biological activities, 659–662

SUBJECT INDEX

coexistence with neuropeptides, 636–638, 655–659
in serum-free culture medium, 330
receptor localization, 641–648
regulator, of dopamine release, 612–613, 614
Ethanolic-gelatin solution, 677–678
Ethidium bromide, staining procedure, 668
Ethyl 4-azido-1,4-dithiobutyrlmidate · HCl, 50
Ethyl chloroformate, in preparation of immunogens, 438, 443, 444
Ethylene glycol bis(succinimidyl succinate), 50, 53
Excitotoxin, *see also* L-Aspartate; L-Glutamate; Lesion, axon-sparing
  applications
    direct intracranial injection, 391
    systemic ablative approach, 388–389
    systemic provocative approach, 389–391
    treatment schedule, 387
  choice of, 383–388
  effective dose, factors affecting, 383
  injection procedure, 394–396
  injection route, choice of, 381–383
  lesion characteristics, 393–394
  provocative approach, 383
  toxicity mechanism, 379–381
  use as axon-sparing lesioning agents, 381–383
Extracellular matrix, production of, 288
Extracellular recording
  methods of, 93–94
  microelectrodes for, 96–98
  neuron identification *in vivo*, 98–101
  preparation of animal, 95–96

F

Fahimi medium, 193
Fast blue, 665
FGF, *see* Fibroblast growth factor
Fibroblast growth factor, 289, 566
Fibroblast growth inhibitor, 291, 345
Fibronectin, 566
  plate coating procedure, 567–568
Ficoll, as gradient material in unit gravity sedimentation, 265
Ficoll-Hypaque sedimentation, for recovery of pituitary cells, 292

Fixation
  for HRP cytochemistry, 193
  for immunohistochemical methods, 626, 650–651
  with paraformaldehyde, 138, 397, 626, 634, 651, 667, 673–675
  for silver impregnation, 397
  of tissue injected with Lucifer Yellow, 138
Flow cytometry, 227–245
  cell diameter measurements, 233–238
  immunofluorescence analysis, 238–241
  instrumentation, 228–231
  light scatter analyses of cell populations, 230, 232–233
  relative membrane potential, determination of, 241–245
Fluorescamine-stream sampling system, 78–79
Fluorescein, 239, 240, 667
Fluorescence, red, analysis of, in flow cytometer, 236–238
Fluorescence histochemistry
  advantages, 620, 629
  development of, 619–620
  procedures, 620–622
  selection of procedure for CA neuron system, 623
Fluorescence microscopy
  of immunochemically stained sections, 622–623, 683–685
  of radiolabeled cells, 138–141
Fluorescence microscopy, image-intensified, 219–227
  controls, 226–227
  fluorophore, choice of, 219–220
  instrumentation, 222–223
  interpretation of results, 225–226
  reagents, preparation of, 220–222
5-Fluorodeoxyuridine, 311, 328, 345–346
Fluoroinert, 277
4-Fluoro-3-nitrophenyl azide, 52, 60
Flurophore, choice of, 219–220, 239–240
Follicle-stimulating hormone-producing cell, 258
Formaldehyde perfusion, 622
Formalin fixation, 651
Freund's adjuvant
  in antigen preparation, 455
  emulsification with peptide conjugate, 549

G

GABA, see γ-Aminobutyric acid
Ganglion cells, retinal, isolation, 366
Garter snake, steroid hormone receptor studies in, 641–643
Gelatin coating
　of plastic culture vessels, 329
　of slides, 678
Gel filtration chromatography, for peptide purification, 73–74, 81, 87
GH, see Growth hormone
GHRIH, see Growth hormone release-inhibiting hormone
Glassware
　peptide adsorption to, 30
　silanization, 8–9, 351
α-Globulin, as carrier protein, 569–570
Glucose oxidase, in iodination of peptides, 35, 37, 42
L-Glutamate, see also Excitotoxin
　analogs, 380, 381
　excitotoxic properties, 379–383, 387–388, 389–391
　in HAT medium, 461
　structure, 380
　treatment schedule, 387
　use as lesion-sparing agent, 381–383
L-Glutamine, in culture medium, 332, 343, 345, 351
Glutamine synthetase activity, in Muller cells, 366, 367
Glutaraldehyde
　coupling procedure, 449–450
　as fixative, 193, 626
Glyoxylic acid cryostat method, 621
GnRH, see Gonadotropin-releasing hormone
Gonadal steroid hormone receiving cells, distribution of, in vertebrate CNS, 640–643
Gonadotropes, receptor localization in, 68–69
Gonadotroph, studies of separation of, 282–283
Gonadotropin-releasing hormone
　conjugation to KLH, 449–454
　cross-linking with EGS, 53–54
　iodination, 34–37
　photoaffinity derivative, 59, 61–62
　　applications, 63–64, 67–71

　preparation of antisera against, 455–458
　radioiodinated
　　biological activity, 35, 38–39
　　purification, 36–38
　　storage, 45–46
　　tritiated, 48
Gonadotropin-releasing hormone agonist analog
　iodination and purification, 39–43
　radioiodinated
　　maximum bindability, 45, 46
　　specific activity, 43–45
　　storage, 45–46
Gonadotropin-releasing hormone antagonist
　conversion to agonist, 56–58
　dimer–antibody conjugate, preparation, 55
　iodination, 46–47
Gonadotropin-releasing hormone receptor
　ion channel mediator of, 404
　localization, 68–69
　photoaffinity inactivation, 63
　regulation during estrus, 69–71
Gradient fractionator, using monocular microscope, 276, 277
Gradient material
　for density gradient centrifugation, 272–274, 307–309
　for unit gravity sedimentation, 265
Grayanotoxin, 402
GRF, see Growth hormone-releasing factor
Growth hormone, regulation of release, 176, 182–186
Growth hormone-producing cell, 258, 266, 267
Growth hormone-releasing factor, 33
　experimental use, 182–186
　isolation, 87–89
　properties, 176
Growth hormone release-inhibiting hormone, 33
Guanine nucleotides, dopamine receptor–ligand binding and, 583

H

Haloperidol, regulator, of dopamine release, 613, 614
HAT medium, preparation, 460–461
HBT, see 1-Hydroxybenzotriazole

Hepatocyte, insulin receptor in, 69
Hepes dissociation buffer, 350, 565, 585
Hepes saline, 259
Heptafluorobutyric acid, in HPLC mobile
    phase, 76, 77
HF, see Hydrofluoric acid
High-performance liquid chromatography
    for fractionation of LHRH peptidase
        products, 543–546
        of neuropeptide precursors, 530–535
        of opioid peptides, 556–559
    for purification of peptides, 19, 23, 75–
        79, 514–515
        of radioiodinated CRF, 571–572
        of tritiated met-enkephalin, 551–552
High-performance liquid chromatography,
    electrochemical detection
    column calibration, 474–475
    for detection of catecholamines, 500–
        510
        of tyrosine, 508–510
    for determination of catecholamine
        turnover, 500–506
    instrumentation, 471, 472
    mobile phase, 472
    principles, 470–472
    sample analysis, 474–475
    sample preparation, 473–474
    standards, preparation of, 472–473
High-performance liquid chromatography,
    reverse phase
    detection systems, 77–78
    mobile phases, 75–77
    for peptide purification, 75–79, 514–515
    in purification of growth hormone-
        releasing factors, 89
        of γ_1-melanotropin-like peptide, 85–
            86
        of somatostatin-28, 82–85
    recovery determination, 79
    stationary phases, 77
HIOMT, see Hydroxyindole O-methyl-
    transferase
Histidine methyl ester dihydrochloride, in
    preparation of immunogen, 443
Histogram fitting, 171–172
HNC, see Hypothalamoneurohypophysial
    complex
Homocysteic acid
    excitotoxic properties, 386
    structure, 380

Horseradish peroxidase, see also Perox-
    idase–antiperoxidase method
    as cytochemical tracer, 127–128, 187–
        218
    as intracellular stain, 127–128
    intracellular transport of, 205–218
    for staining neuropeptides, 459
    for staining retinal ganglion cells, 365,
        366
    visualization technique, 191–193
HPLC, see High-performance liquid chro-
    matography
HPLC-EC, see High-performance liquid
    chromatography, electrochemical
    detection
HRP, see Horseradish peroxidase
5-HT, see Serotonin
Hyaluronidase, for cell dispersion, 251,
    259, 290
Hyamine hydroxide, in assay of ODC, 596,
    597
Hybridoma, see also Monoclonal antibody
        production
    as ascites cells, 467–468
    cloning by limiting dilution, 466–467
    freezing and thawing, 467
    production of, 460–465
Hydrofluoric acid, 7, 8, 17–18
Hydrogen bromide-trifluoroacetic acid, 8,
    18
Hydrogen peroxide, 35, 37
    in iodination of peptides, 35, 37
    stock solution, preparation and storage,
        678–679
Hydrolysis, of peptide-resins, aminoacyl
    resins, and peptides, 24–25
3-Hydroxybenzylhydrazine, effect on
    dopamine release, 615
5-Hydroxyindole acetic acid, oxidation
    potential, 480
Hydroxyindole O-methyltransferase
    activity determination, 490
    preparation, 490
    use in REA of serotonin, 483, 491
1-Hydroxybenzotriazole, 13, 14
Hydroxymethylpolystyrene resin, attach-
    ment of Boc-amino acid to, 10–11
p-Hydroxyphenylacetic acid, as reactive
    prosthetic group, 436, 438
p-Hydroxyphenylpropionic acid, as reac-
    tive prosthetic group, 437

N-Hydroxysuccinimidyl 4-azidobenzoate, 51
N-Hydroxysuccinimidyl 4-azidosalicyclic acid, 51
Hypophysial portal blood
  catecholamines in, 609–618
  experimental problems, 177, 186
  neuropeptides in, 607–609
Hypothalamic cells
  clonal cell lines, preparation by *in vitro* viral transformation, 331–334
  culture techniques for, 313–334
  in vivo recording from, 95–101
  long-term culture studies, 322–329
  serum-free culture medium for, 329–331
  serum-supplemented media, 322, 324, 326, 328
Hypothalamic nuclei, microdissection by punch technique, 529, 541–542
Hypothalamoneurohypophysial complex
  explant preparation, 133–134
  in vitro analysis, 133–147
Hypothalamus
  collection from rat, 352
  crude extract preparation, 79–81
  dissection, 352, 541
  dopamine release from, 607–618
  explant cultures, long-term, 315, 318–321, 328
  organ culture, 314, 316–317, 495–496
  preparation of dispersed cells from, 521–522
  slice preparations, 101–102
Hypothalamus, anterior, catecholamine turnover in, 506–508
Hypothalamus, arcuate nucleus, lesion studies, 379, 381, 388–390
Hypothalamus, medial basal
  catecholamine turnover in, 506–508
  dissection of, from rat, 496, 514
Hypoxanthine, stock solution, 460

I

Ia antigen, analysis of expression of, 240–242
Ibotenic acid
  excitotoxic properties, 384–385, 396
  structure, 380
2-Iminothiolane · HCl, 52
Immunization procedure
  with peptide conjugates, 455–456, 549
  for production of activated splenocytes, 462–463
Immunocytochemical techniques, 139–144, 623–630, *see also* Avidin-biotinylated horseradish peroxidase complex method; Immunofluorescence staining; Immunoperoxidase staining; Peroxidase-antiperoxidase method
  advantages, 629–630
  applications, 629–630
    for identifying pituitary cell types, 266, 268
    for visualizing CA neurons, 625–630
  antibody specificity, 625–626, 671
  combined with autoradiography to localize steroid hormones, 631–634, 646–655
  double-label technique, 139–141
  interpretation of results, 681, 683, 685–687
  reagent preparation, 678–679
  staining procedure, 679–681
  tissue fixation, 626–627, 666–667, 671–675
  triple-label technique, 142–144
  using horseradish peroxidase-labeled antibodies, *see* Horseradish peroxidase
Immunofluorescence staining
  indirect
    assay for detecting monoclonal antibodies, 465–466
    for labeling neurons, 139
  of peptide-specific pathways, 667
  procedure, 679–681
Immunoperoxidase staining, *see also* Peroxidase-antiperoxidase method
  stain procedure, 679–681
Impalement procedure, 124
Indirect sandwich solid-phase competitive enzyme immunoassay, *see* Radioimmunoassay, ultrasensitive enzyme
Infundibular process, 607
Infundibular stem, 607

Infundibulum, 607
Insulin
  in culture medium, 293, 312, 315, 326, 330, 345, 358
  in HAT medium, 461
  photoaffinity derivatives
    applications, 63, 64–65, 67, 69
    synthesis and characterization, 60–61
Insulin receptor, localization, 69
Intracellular recording
  cell culture techniques, 114–116, 117
  cell impalement, 124
  control of vibration, 117–118
  data analysis, 136–138
  electronics for, 119–121
  explant preparation, 133–134
  grounding and interference problems, 121–122
  intracellular staining, 127–128
  iontophoresis, 124–125
  micromanipulators for, 118–119
  microperfusion, 125–127
  recording solutions, 116–117
  stimulation, 137
  support equipment, 116, 135
Iodination, 34–43, 46–47, 62, 415, 437, 550–552
Iodogen, 415
Ion channel
  action, characterization of, 404
  agents active on, 402–403
  determination of number of, 404
  specificity, determination of, 401, 404
  studies on, 401–405
Ion-exchange chromatography, for purification of peptides, 20
Iontophoresis, 124–125

K

Kainic acid
  excitotoxic properties, 384, 391–392, 395
  injection solution, 395
  structure, 380
Kainic acid receptor, 381
Kaiser test, 16–17
Ketamine hydrochloride, 95

Keyhole limpet hemocyanin, conjugation to GnRH, procedures, 449–454
KLH, see Keyhole limpet hemocyanin
Krebs' bicarbonate buffer, 363
Krebs-Ringer bicarbonate buffer, 249–250
Krebs-Ringer bicarbonate solution, Hepes-buffered, 351

L

Lactoperoxidase, in iodination of peptides, 35, 37, 42
Lateral hypothalamic area, excitotoxin lesion studies, 391–392
Lesion, axon-sparing, see also Excitotoxin
  with excitotoxin, 381–383
  histological evaluation, 396–400
Leucine-enkephalin, see also Enkephalin
  assay, using RIA, 559–561
  chromatographic purification methods, 556–558
  purification, 29
  secretion by brain explants, 359–361
  separation, by HPLC, 536–537
Leucine-enkephalin analog, fluorescent derivative of
  preparation, 220–221
  properties, 221–222, 224–225
LHRH, see Luteinizing hormone releasing hormone
Ligand dimerization, 49–58
Light microscopy
  of HRP-labeled cells, 198–199, 205–207
  of triple-labeled cells, 143–144
Light scatter analyses of cell populations, 230, 232–233
β-Lipotropin
  chromatographic purification methods, 556
  localization in neurons, 198
Locke's solution, 306
Lucifer Yellow, as intracellular stain, 127, 138, 139
Luteinizing hormone, effect of culture conditions on, 253–256
Luteinizing hormone-producing cell, 258
Luteinizing hormone releasing hormone
  cultured cell response to, 253–256
  enzymatic degradation products, isolation of, 543–546

isolation and purification
   from dispersed pituitary cells, 521–523
   from median eminence and preoptic area, 519–521
   using HPLC, 543–546
localization in neurons, 199
methanolic extraction of, 498–500
presence in hypophysial portal blood, 608
purification, 20, 27
regulation of secretion, 389–390, 659–660
synthesis, 27
synthetic rates, prediction of, 524–526
tissue content after *in vivo* incubation, 526–527
Luteinizing hormone-releasing hormone antagonist
   purification, 20, 28
   synthesis, 28
Luteinizing hormone-releasing hormone peptidase, assay, 542
Luteinizing hormone-releasing hormone superagonist
   purification, 28
   synthesis, 27–28
Lymphocyte, live–dead discrimination, 232, 233

M

Magnocellular neuroendocrine cell, immunocytochemical studies of, 132–147
m-Maleimidobenzoyl *N*-hydroxysuccinimide ester, 51
Mammotroph, studies of separation of, 278–279
MBHA, see *p*-Methylbenzhydrylamine resin
MDA, see α-Methyldopamine
Median eminence, 607
γ_1-Melanotropin-like peptide, isolation, 85–87
Membrane, terminal, studies of, 203–205
Membrane potential, relative, determination of, 241–245
Merrifield system, see Solid-phase peptide synthesis

Metanephrine, assay, using HPLC-EC, 500–506
Methanolic acetic acid extraction method, 498–500, 530
Methionine-enkephalin
   assay, using RIA, 559–561
   extraction from tissues, 552–553
   production by cloned cells, 333
   purification, 20, 29, 556–558
   separation, by HPLC, 536–537
   synthesis, 29
   tritiated, purification, using HPLC, 551–552
Methionine sulfoxide, analysis for, 26
Methocel, in culture medium, 357
Methoxyl verapamil, 402
3-Methoxytyramine, assay, using HPLC-EC, 500–506
3-Methoxytyrosine, assay, using HPLC-EC, 500–506
N-Methyl aspartate
   excitotoxic properties, 385, 390–392, 396
   injection solution, 395
   structure, 380
N-Methyl aspartate receptor, 381, 385
Methyl-4-azidobenzimidate · HCl, 51
p-Methylbenzhydrylamine resin, 6, 11
α-Methyldopamine, as HPLC internal standard, 473
Methyl green, staining procedure, 636
N-Methylmorpholine, in preparation of immunogen, 443, 444
6-Methyl-5,6,7,8-tetrahydropterin, effect on dopamine release, 616
α-Methyltyrosine, effect on dopamine release, 615
Metrizamide, as gradient material in density gradient centrifugation, 363
Microelectrode
   coating procedure, 151–152
   electronic problems with, 119–120
   for extracellular recording, 96–98, 129, 131
   fire-polishing, 152
   for intracellular recording, 122
   patch-clamp, 129, 131, 151–153
   resistance of, determination of, 153
   voltammetric, 481–483
Micromanipulators, 118–119
Microperfusion, 125–127

SUBJECT INDEX

Micropunch needle, 370
Micropunch sampling technique, 368–376
 advantages, 368–369, 375
 applications, 515, 519, 529–530, 541–542
 instrumentation, 369–370
 microdissection procedure, 370–371
 micropunch procedure, 371–375
Microscopy, see specific types
Microtiter plates, for EIA, 415–416, 418–420, 432
MIF-1, attachment of reactive group to, 436, 437, 438
Molecular sizing, for purification of peptides, 19–20
Molecular weight size-exclusion chromatography, for separation of neuropeptides, 530–535
 instrumentation, 530–531
 interpretation of results, 534–535
 procedure, 531–534
Monoclonal antibody
 availability, 468–469
 use in immunocytochemistry, 139
Monoclonal antibody production, 459–469
 antibody screening assay, 465–466
 ascites method, 467–468
 equipment and supplies, 461–462
 fusion protocol, 463–465
 hybridoma cloning, 466–467
 hybridoma storage, 467
 media, 460–461
 parental lines, preparation of, 462–463
Morphine, effect on dopamine release, 614–615
γ_1-MSH, see γ_1-Melanotropin-like peptide
Muller cells, isolation, 366
Myeloma lines, for hybridoma production, 462

N

Naloxone, effect on dopamine release, 614
NAT, see N-Acetyltransferase
Nerve growth factor, 114, 343, 345
Neurohypophysis, 607
Neuron
 degenerating, staining procedure for, 397–400
 identification in vivo
  functional, 99–101
  topographical, 98–99
 secretory process, 188–191
Neuron, circumventricular organ, effect of excitotoxins on, 379, 381–382, 383
Neuron, peptidergic, see also Brain explant
 culture conditions for, 347–362
 culture medium for, 350, 351
 isolation, 352–353
 morphology in culture, 353–354
Neuron, supraoptic, activity patterns, 136–137
Neuron, sympathetic, 334–347
 culture conditions for, 340–342
 isolation, 336–339
 morphology in culture, 344, 346–347
 plating density, 339–340
Neuropeptide, see also Opioid peptides; specific peptides
 acid extraction, 72–73, 498–500, 514, 530, 533
 adsorption to glass and plastic, 30
 analysis, 23–26
 attachment of prosthetic group to, 436–447
 conjugation to carrier protein, 448–454, 548–549
 C-terminus extension, 445–446
 coexistence with neurotransmitters, 630
 coexistence with steroid hormones, 636–638
 functional groups for conjugation, 448–449
 limitations in use of, 29–31
 microisolation, 72–89
 newly synthesized, isolation, 535–539
 N-terminus extension, 436–445
 photoaffinity derivatives, properties, 60–62
 purification, 19–23
 radiolabeled
  biological activity, 34
  preparation, 34–43
  storage, 45–46
 solubilization, 30
 stimulators of ODC, 593–596
 storage, 31–32
 synthesis, 4–18, 26–29
 synthetic rates, prediction of, 524–526

trifluoroacetic acid extraction, 72–73, 514
turnover studies
  in vitro, 521–523
  in vivo, 511–521
Neuropeptide gene, expression of, 333
Neuropeptide precursors, separation by molecular weight size-exclusion chromatography, 530–535
Neuropeptide-specific pathways, tracing of, 663–669
Neurophysin
  localization, 139, 141
  production by cloned cells, 332
Neurophysin I
  assay, using HPLC, 515–519
  coexistence with estradiol, 637, 655–659
  localization in neurons, 198, 200–205
Neurophysin II
  assay, using HPLC, 515–519
  localization in neurons. 198, 199, 200–205
Neurotensin, 33
  production by tumor cell line, 333
Neurotransmitters, see also Catecholamines
  coexistence with neuropeptides, 630
NGF, see Nerve growth factor
2-Nitro-5-azidophenylsulfenyl chloride, 60
p-Nitrophenyl 2-diazo-3,3,3-trifluoropropionate, 51
p-Nitrophenylphosphate, tritiated, 415
NMA, see N-Methyl aspartate
Noradrenaline, 630
Norepinephrine
  assay, using HPLC-EC, 500–506
  oxidation potential, 480
  stock solution for HPLC standard, 472–473
  synthetic pathway, 470
Normetanephrine, assay, using HPLC-EC, 500–506
NP, see Neurophysin
NSD 1015, see 3-Hydroxybenzylhydrazine
NT, see Neurotensin

O

Octadecyl silica chromatography, for peptide purification, 74–75, 85

ODAP, see β-N-Oxalyl-L-α,β-diaminopropionic acid
ODC, see Ornithine decarboxylase
Opioid peptides, see also specific peptides
  antiserum production, 549–550
  precursors, 547–548, 670–671
  radioiodination, 550–552
Opiate receptors, visualization of, 224–225
Ornithine
  in polyamine synthetic pathway, 591
  radiolabeled, in assay of ODC, 596
Ornithine decarboxylase
  assay, 592, 596–601
    blanks, 597–599
    chemiluminescence, avoidance of, 600–601
    equipment, supplies, and reagents, 592, 596
    sample preparation, 596
    substrate concentration, choice of, 599–600
  characteristics, 590–592
  for isolation of opioid peptides, 553–555, 557–558
  neuropeptide stimulators, 592–594
  regulation of, 601–603
β-N-Oxalyl-L-α,β-diaminopropionic acid, structure, 380
Oxytocin
  assay, using HPLC, 515–519
  coexistence with estradiol, 655–659
  localization in neurons, 198, 199
  methanolic extraction of, 498–500
  synthetic rates, prediction of, 524–526
  tissue content after in vivo incubation, 526–527, 537–538

P

Pancreas, crude extract preparation, 87
Papain, for removal of connective tissue, 150
PAP method, see Peroxidase-antiperoxidase method
Paraformaldehyde, 634
  fixation solution, preparation, 673–674
  perfusion of animal with, 397, 626–627, 666–667, 674–675
Parathyroid hormone, in culture medium, 566

Paraventricular nucleus, neuropeptide synthesis in, 515–519
Partition chromatography, column, for purification of peptides, 19, 22–23
Patch-clamp amplifier, 156–163
  frequency response, 160–161
  noise sources, 156–160
  patch potential changes, 161–162
  tracking operation, 162–163
Patch-clamp technique, 147–176
  advantages of, 149
  cell preparation, 150–151
  data acquisition techniques, 166
  data analysis, 163–175
    data characteristics, 154–155
    estimating event amplitudes, 172–174
    estimation of event durations, 169–171
    estimating number of channels, 174–175
    fitting histograms, 171–172
    low-pass filtering and event detection, 166–169
  determination of pipette resistance, 153
  estimating number of channels in, 174–175
  instrumentation, 156–163
  for intracellular recording, 129–131
  microelectrode construction, 151–153
  principle of, 149
  seal formation, 153–154
Pepstatin A, 72, 87
Peptidase, inactivation with hot acid, 359
Percoll, as gradient material
  for density gradient centrifugation, 273–274, 275–275, 307
  in unit gravity sedimentation, 265
Perfusion, with paraformaldehyde, 626–627, 666–667, 674–675
Pergolide, dissociation constants in competitive binding studies, 581, 582
Perifusion, continuous
  media and equipment, 299, 300–302, 303–304
  problems, 303–305
Periodate-lysine-paraformaldehyde fixative, 626, 651

Peroxidase–antiperoxidase method
  for immunostaining autoradiograms, 634–636, 652–653
  postembedding staining, 195–196
  preembedding staining, 196–197
  principle, 625
  procedure, 627–628, 652–653
  for visualizing CA neurons, 625, 627–630
PHAA, see p-Hydroxyphenylacetic acid
Phase-contrast microscopy, with intracellular recording, 116
Phencyclidine, 95
Phenylacetaamidomethyl polystyrene resin, 4–5
p-Phenylenediamine dihydrochloride-pyrocatechol, 191
Phenylhydrazine treatment, 635
Phenylmethylsulfonyl fluoride, 72, 81, 87
Photoaffinity labeling, 58–71
  applications, 63–65, 67–71
  general controls, 66
  photolysis procedure, 62–63
  synthesis and characterization of photoreactive hormones, 59–62
Photolysis, 62–63
Photomicrography, of immunochemically stained sections, 683–685
Photoreceptor cell
  isolation, 363
  morphology in culture, 364
PHPA, see p-Hydroxyphenylpropionic acid
Phycobiliproteins, 239
Picrotoxinin, 355
Pituitary cell
  culture on extracellular matrix, 287–293
  culture media, choice of, 251
  dispersion methods, 249–251, 259, 260, 289–292, 298, 338, 339, 349
  effect of culture duration, 255–256
    of initial plating concentration, 252–255
  optimum culture conditions, 249–257
  separation by affinity method, 287
    by density gradient centrifugation, 272–275
    by electrophoresis, 277, 287
    by Ficoll-Hypaque sedimentation, 292

Single-channel current, see Patch clamp technique, data analysis
SITS, 665–666
Slice technique, 94, 101–102
Sodium pentobarbital, 95
Solid-phase peptide synthesis, 4–29
  AE synthesis, 15–17
  apparatus, 7–9
  cleavage of peptide from resin, 17–18
  DCC synthesis, 12–14
  hydrolysis of peptides, 24–25
  starting materials, 9–11
  stepwise procedures, 11–18
  test for complete coupling, 16–17
  theory, 4–7
Solvent systems, countercurrent distribution, 20–22
Somatostatin
  in amacrine cells, 366–367
  attachment of reactive group to, 437
  coexistence with estradiol, 638
  effect on excitability of GH3/6 cells, 110–111
  experimental use, 182–186
  methanolic extraction procedure, 498–500
  presence in hypophysial portal blood, 608
  production by cloned cells, 333
  properties, 176
  secretion by cerebral cortex cells, 355–356
  tissue content after *in vivo* incubation, 526–527
Somatostatin-14, assay, using HPLC, 515–519
Somatostatin-28, isolation, 79–85
Somatotroph, separation from pituitary cells, 274–275, 277, 280–281
Soybean trypsin inhibitor, 290
Spermidine, biosynthetic pathway, 591
Spiroperidol, tritiated, binding characteristics, 578–579, 580–581, 583–584, 586–587, 588–589
Splenocytes
  as feeder cells, 463–464
  preparation and use in hybridoma production, 462–465
SPPS, see Solid-phase peptide synthesis
SRIF, see Somatostatin

SS, see Somatostatin
Steroid, removal from serum, 251
Steroid hormone
  coexistence with neuropeptides, 636–638
  localization with autoradiography and immunocytochemistry, 631–638, 646–655
  radiolabeled, injection procedure, 632, 649–650
Steroid hormone receiving cells
  characterization by peptide content, 655–659
  identification of, 636–638, 639–643, 654–655
  number and capacity of, 643–646
Staining
  with cyanine dye, 243–244
  with fluorophore, 139, 240
  intracellular, 127–128, 138
  with silver nitrate, 397–400
  with specific antiserum and HRP, 459
Substance P
  localization in neurons, 198
  purification, 20, 29
  synthesis, 28–29
N-Succinimidyl 6-(4′-azido-2′-nitrophenylamino) hexanoate, 51
N-Succinimidyl (4-azidophenyl)-1,3′-dithiopropionate, 51
Succinimidyl 4-(p-maleimidophenyl) butyrate, 52
Succinimidyl 4-(N-malemidomethyl) cyclohexane-1-carboxylate, 51
N-Succinimidyl 3-(2-pyridyldithio) propionate, 51
Superior cervical ganglion, culture of neurons of, 334–347
Supraopticoneurohypophysial system, cytochemical studies of, 190, 198–205
SV40, in viral transformation of hypothalamus cells, 332
Sylgard 184, for coating electrodes, 151–152

T

TEA, see Triethylamine
TEAF, see Triethylammonium formate
TEAP, see Triethylammonium phosphate

Tetramethylbenzidine, 191
2,6,10,14-Tetramethylpentadecane, 468
Tetramethylrhodamine, 139, 222, 239
Tetrodotoxin, 117, 402
Texas Red, 220–222, 239
TFA, see Trifluoroacetic acid
Thin-layer chromatography, for isolation of methylated catecholamines, 478, 487
Thymidine, in HAT medium, 460
Thymocyte nuclei, determination of diameter of, 234–238
Thyroglobulin, as carrier protein, 548–549
Thyrotroph, studies of separation of, 284–285
Thyrotropin releasing hormone
  attachment of reactive prosthetic group to, 439–442
  effect on excitability of GH3/6 cells, 110–111
  presence in hypophysial portal blood, 608
  production by tumor cells, 333
  purification, 20, 27
  synthesis, 26–27
Tityustoxin, 402
Toluene 2,4-diisocyanate, coupling procedure, 453–454
Toluene-isoamyl alcohol solvent system, in REA, 487
Toluene-isopentyl alcohol solvent system, in REA, 477–478
Trace salts, in culture medium, 566
Transferrin, human, in culture medium, 293, 312, 330, 345, 358
TRH, see Thyrotropin-releasing hormone
Triethylamine
  in neutralization reagent, 13
  in preparation of immunogen, 443, 444
Triethylammonium formate, 570, 571
Triethylammonium phosphate
  in HPLC mobile phase, 76, 77
  as ion-pairing reagent, 518, 520, 543
Trifluoroacetic acid
  in deprotection reagent, 7, 13
  in HPLC mobile phase, 76, 77
  neuropeptide extraction with, 72–73, 514
Triiodothyronine, in culture medium, 330, 331
Triple-label technique, 143–144

True-blue, 664–665
Trypsin
  for cell dispersion, 250, 251, 259, 298, 299, 338
  for removal of connective tissue, 150
  for tissue dissection, 521
Tryptophan, oxidation potential, 480
TSH-producing cell, 258
Tyrode solution, calcium-free, 622
Tyrosine
  assay, by HPLC-EC, 508–510
  dopamine release and, 616
  oxidation potential, 480
  triatiated, in catecholamine turnover studies, 495–496
Tyrosine hydroxylase
  dopamine release and, 615, 616
  immunohistochemical staining of, 627–628, 637

U

Unit gravity sedimentation, 260–271
  Celsep device, 266
  chamber, 262, 266
  gradient material, 265
  procedure, 262–265
  Staput device, 260–266
  streaming, 261–262
  theory, 260–262
USERIA, see Radioimmunoassay, ultrasensitive enzyme

V

[^{3}H]Vanillin, in REA of catecholamines, 475, 478
Vasoactive intestinal peptide
  presence in hypophysial portal blood, 608–609
  production by tumor cell line, 333
Vasopressin, see Arginine vasopressin
Velocity sedimentation, see also Unit gravity separation
  for pituitary cell separation, 260–272
Ventricular cannulation, 528–529
Veratridine, 402–403
Veratrine, 403
Vibration, control of, 117–118
Vibratome-formaldehyde method, 621–622

Viokase, 259, 290, 291, 566, 567
VIP, *see* Vasoactive intestinal peptide
Viral transformation, of hypothalamic cells, 331–332
Voltage-clamp technique
　for extracellular recording, 107–110
　for intracellular recording, 128–129
Voltammetry, *in situ*
　instrumentation, 480–483
　measurement procedure, 482
　principles, 479–480
VP, *see* Arginine vasopressin

W

Wheat germ agglutinin, conjugated to HRP, 208

X

XRITC, 239

Z

Zephiran chloride, 306